City of Westminster College

**Eileen Ramsden** BSc PhD DPhil

*Formerly of Wolfreton School, Hull*

# Chemistry

**Stanley Thornes (Publishers) Limited**

First published in 1994
This edition published in 1997 by:
Stanley Thornes (Publishers) Ltd,
Ellenborough House,
Wellington Street,
CHELTENHAM,
GL50 1YW

A catalogue record for this book is available from the British Library.

ISBN 0 7487 3009 5

98 99 00 01 / 10 9 8 7 6 5 4 3 2

Related titles
*Key Science: Biology* (0–7487–3007–9)
*Key Science: Physics* (0–7487–3008–7)
*Key Science: Chemistry Extension File* (0–7487–3006–0)

Designed by Maggie Jones
Artwork by Barking Dog Art, Peters & Zabransky, Harry Venning and Cauldron Design Studio
Typeset by Tech Set Ltd, Gateshead, Tyne & Wear
Printed and bound in Spain by Mateu Cromo

## Acknowledgements

I would like to express my thanks to Mr Gavin Cameron, Mr Steve Hewitt and Mrs Sheila Rogers for their constructive comments on the Earth Science topic, to Mr David Haslam for his advice on some of the chemical topics, and to Mr Roger Frost for his material on information technology. I thank all these people for their contributions to the text.

I would like to thank the following organisations and people who have supplied photographs.
Action Plus Photographic: p55; Ariel Plastics: 1.6D; British Rail: 20.5A; British Steel: 4.4A; Casio Electronics Co. Ltd.: 21.2A; Environmental Picture Library: 16.10A (David Townend), 17.1A (Alan Greig); Fisons plc: 10.4A; Gavin Cameron: 19.2C; Geological Museum: 19.5B; Geoscience Features Picture Library: 20.8F; ICI Chemicals and Polymers Ltd: 1.2A, 1.5D, 3.2A, 3.8B, 7.6C, 23.7B; Image Bank: xi (top – Bill Varie) (bottom – Jaime Villaseca); International Centre for Conservation Education: 16.4A, 17.4A (Andy Purcell), 17.6D, 28.1A; Mark Boulton: 29A, 22.1E; Martyn Chillmaid: 15.13A, 22.1G; Mary Evans Picture Library: p53, p54; Metropolitan Police: 30.1D; Mountain Camera: 1.5G; NASA: 13.4A, 22.3B, p124; National Portrait Gallery: portrait of Michael Faraday p103 reproduced by courtesy; Northern Irish Tourist Board: 19.1E; Oxford Scientific Films: 3.6A (G.I. Bernard), 30.1G (David Hall); Panasonic (UK) Ltd: 21.2B, 21.2C; Panos Pictures: 17.3A (J Hartley); Peter Newark's Western Americana: 4.2A; Pilkington Glass plc: 22.2B; Popperfoto: 2.1C, 17.6C; Science Photo Library: 2.1B (Prof. Stewart Lowther), 2.4B (Ben Johnson), 2.4C (Dr Mitsuo Ohtsuki), 2.4D (Dr M B Hursthouse), 4.1A (Ferranti Electronics/A Sternberg), 4.1C (Dr G M Rackham), 10.4D (CNRI), 13.1A (Susan Leavines) 13.5B (Alex Bartel), 17.6B (Simon Fraser), 19.2A, p208, 22.1D, p316 (Martin Bond), 19.3D (Sinclair Stammers), 20.1A (John Ross), 3.3B, 17.11B, 30.6B, 30.6C, p270; Sotherbys: 4.2B; Still Pictures: 20.11A (Martin Edwards); Sue Boulton: 29B, 30.1A; Sygma: 18.4A (Les Stone); Tony Stone Images: 4.3A (Rohan), 17.6D (Alan Levenson), 17.6E (Ben Osborne), 17.6A (David Woodfall), 24.18; Woodmansterne Ltd: 19.1D, 19.2F;

Every effort has been made to contact copyright holders to clear permission for reproduction of copyright material. The publishers should like to apologise if any such material has not been fully acknowledged and shall endeavour to rectify the situation at the earliest opportunity.

I thank the following examining groups for permission to reproduce questions from examination papers:
University of London Examinations and Assessment Council, Midland Examining Group, Northern Examinations and Assessment Board, Northern Ireland Council for the Curriculum Examination and Assessment Council. Southern Examining Group, Welsh Joint Education Committee.

The Examining Groups bear no responsibility for the answers to questions taken from their question papers contained in this publication.

The production of a science text book from manuscript involves considerable effort, energy and expertise from many people, and I wish to acknowledge the work of all the members of the publishing team. I am particularly grateful to Adrian Wheaton (Science Publisher), Helen Whitehead and Malcolm Tomlin (Senior Editor) for their advice and support.

Finally I thank my family for the tolerance which they have shown and the encouragement which they have given during the preparation of this book.

*Eileen Ramsden.*

# Contents

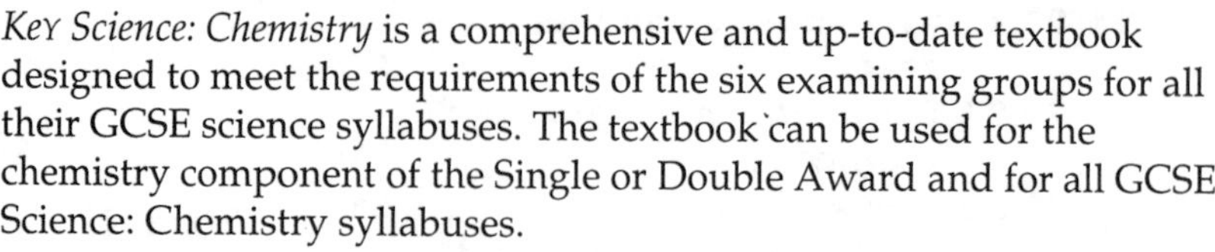

# Preface

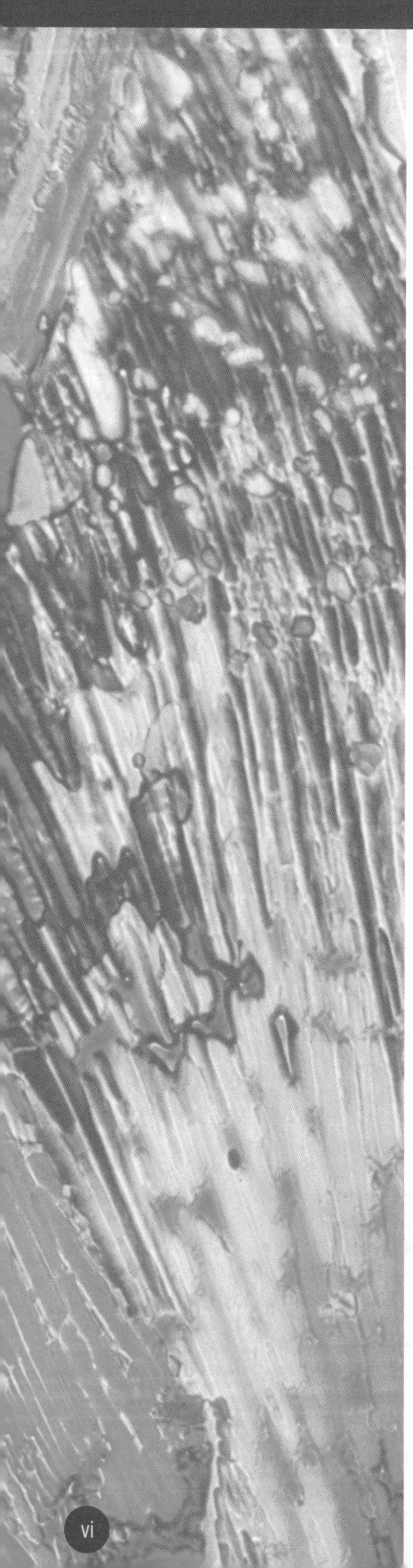

*Key Science: Chemistry* is a comprehensive and up-to-date textbook designed to meet the requirements of the six examining groups for all their GCSE science syllabuses. The textbook can be used for the chemistry component of the Single or Double Award and for all GCSE Science: Chemistry syllabuses.

Topics are differentiated into core material for Single or Double Science and extension material for Science: Chemistry (blue/grey margin). The examining groups have chosen different extension topics in addition to the Programme of Study for Key Stage 4; therefore teachers and students need to be familiar with the specific requirements of their own syllabus. *Key Science: Chemistry* contains the extension topics for all boards.

*Key Science: Chemistry* is organised in 7 Themes covering 30 major topics. Each topic is sub-divided into numbered sections covering all aspects of the topic. Each section includes special features, usually located in the margin, such as:

- **First Thoughts,** setting the scene and reminding students of the background knowledge to the section,
- **It's a Fact,** sharpening interest and enhancing text,
- **Key Scientist,** providing a historical perspective on how scientific ideas have developed,
- **Commentary and Summaries,** within each section of a topic and at the end of each section, helping students to review their work and highlighting the key points they should understand and learn,
- **Checkpoints,** testing knowledge and understanding through sets of short questions at regular intervals at the end of each section,
- **Theme Questions,** in the style of examination questions.

**Answers** to all numerical questions and Checkpoints and a comprehensive **Index** are at the end of the book.

*Key Science: Chemistry* is extended and supported by an Extension File. This contains a Teacher's Guide, a bank of photocopiable material and all Key Diagrams from the text in colour on CD-Rom. The extra resources include over 80 photocopiable Activities for Sc1 and Assignments for Sc 3 and Science: Chemistry topics. There are 60 Exam File questions for homework and revision. Mark Schemes are provided for all questions.

Guidance on Information Technology and using computers in science is provided in the Teacher's Guide where detailed notes include references to the World Wide Web, appropriate software and hardware for computer-assisted learning.

## Student's Note

I hope that this book will provide all the information you need and enough practice to enable you to do well in the tests and examinations in your GCSE course. However, although examinations are an important part of your course, they are not the only part. I hope also that you will develop a real interest in science, which will continue after your course has finished. Finally, I hope that you enjoy using the book. If you do, all the effort will have been worth while.

*Eileen Ramsden*

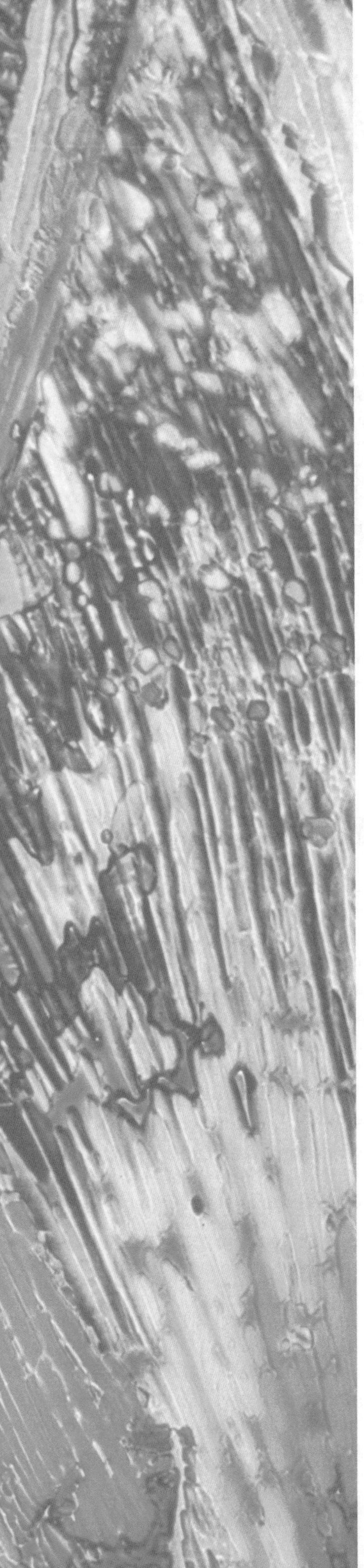



*Scientific research consists of seeing what everyone else has seen, but thinking what no one else has thought.*

(Szent-Gyögyi, 1893–1986)

*The compound whose structure is known but which has not yet been synthesised is to the chemist what the unclimbed mountain, the uncharted sea, the untilled field and the unreachable planet are to others.*

(Robert Woodward, 1917–79)

*Neither a gorgeous home nor security of occupation, nor fame, nor health appeals to me; for me rather my chemicals, amid the smoke, soot and flame of coals blown by bellows.*

(Johann Becher (1635–82)

# Introduction The world of science

## Adventures in science

### The Mirror of Galadriel

In the book *Lord of the Rings*, amazing things happen to Gandalf and the hobbits. They risk their lives on a perilous journey to save their world. They journey through secret tunnels and unknown lands and they witness astonishing events. In the early parts of their travels, they are invited to look into a magic mirror – a silver bowl filled with water. It shows 'things that were, and things that are, the things that yet may be.' But the viewer can't tell which it is showing!

Imagine you have travelled in time from two centuries ago to the present day. What would you make of television – a silver bowl showing events from the past and the present? Television is a product of the scientific age in which we live.

Science has many more amazing things to reveal. Imagine food 'grown' in a factory or round-the-world flight in a few hours or 'intelligent' computers that need no programming. These and many more projects are the subject of intense scientific research now. Fact is stranger than fiction.

Figure A What can you see?

### Can you believe what you see?

Look at the picture in Figure 1A. *What can you see?* Some people see an old lady; others see a young woman. Different people looking at the same picture see different images.

Witnesses of an event often report totally different versions of the event. Invent a harmless incident and try it out in front of unsuspecting witnesses without warning them. Then ask them to say what happened. Each person will probably have a different story to tell.

In science, there are countless events to observe. The story or 'account' of an event in science should not differ from one observer to another. Otherwise, who could you believe?

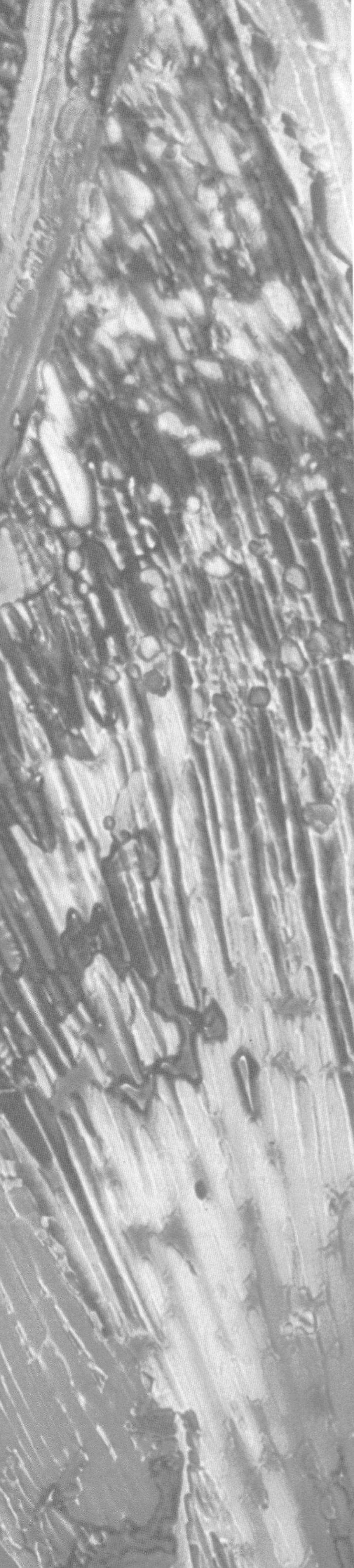

## How science works

Science involves finding out how and why things happen. Natural curiosity sometimes provides the starting point for a scientific investigation. Another investigation might set out to solve a certain problem and therefore have a definite aim. Even then, unexpected results may emerge to stimulate our natural curiosity.

The history of science has many examples where an important discovery or invention came about unexpectedly. The electric motor was invented when an electric generator was wired up wrongly at the Vienna Exhibition of 1873. X-rays, radioactivity and penicillin are further examples of unexpected scientific discoveries.

Discoveries in science are unpredictable. No-one can predict which research projects will lead to exciting new developments because no-one knows what new areas of science lie waiting to be discovered beyond the frontiers of knowledge.

## Chemistry as a key branch of science

Thank chemistry for a comfortable life. You can see the contribution which chemistry makes to our well-being in the materials that we use to make the clothes we wear, the houses we live in, the cars, boats and planes which transport us and the fuels which they burn. Fertilisers manufactured by the chemical industry help farmers to grow enough crops. Pesticides ensure that those crops survive to be harvested, and chemical preservatives ensure that the foods we eat are safe and apertising. Even the water we drink would not be safe without chemical treatment.

Thank chemistry for health. Chemotherapy (the use of chemical in the treatment of diseases) allows us to live longer, healthier lives, more free from pain than in any previous century. Most of us have occasion to use pain-killers such as aspirin and antiseptics such as TCP. For more serious complaints we have antibiotics such as penicillin, and surgery is made safe by anaesthetics such as fluothane. Thanks to the development of new materials, surgeons can now do 'spare part surgery'. Someone who needs a replacement part can be fitted with a teflon earbone, a dacron polyester artery, a hip joint of polymer and alloy, a duralium artificial leg, a polythene heart valve or a complete artificial heart of polyurethane plastic.

Thank chemistry for leisure activities. Chemists have developed new, cheaper materials to replace natural materials such as metals, wood and ox-gut and allow many people to acquire equipment that otherwise would have been beyond their means. When you surf on your polysterene and fibre glass surfboard, when you use your tennis raquet of polymer fibre strings on a frame of epoxy resin with graphite fibres, when you coast downhill on your skis of acrylic foam and polyester laminates on a polyurethane base, and when you simply go for a jog in your polyurethane trainers, think of the chemists who make things happen.

## A scientific race

On 18 March 1987, more than two thousand scientists crammed into a conference room in New York to listen to progress reports on research into superconductivity. The conference had to be relayed to hundreds of scientists in adjacent rooms.

Superconducting materials have no electrical resistance. Electric cables and electric motors made from superconductors would be far more efficient than those made from ordinary conductors. Superconducting computers and magnets would be much more powerful. However, before 1987, most scientists thought that superconductors would work only at very low temperatures. Materials that are superconducting at room temperature would be a technological breakthrough.

The first superconductors were discovered in 1911 when certain metals were cooled to below −270°C. In 1986, two European scientists, Georg Bednorz and Alex Müller, reported they had made a superconductor at 30°C higher (−240°C). Scientists round the world turned their attention to superconductors. Teams of scientists in different countries raced each other to find superconductors which would work at higher temperatures. By March 1987, the temperature limit had been raised to −179°C. By then the political leaders of the major industrial nations realised how important superconductors could become. The race continues because the benefits would be immense. The importance of Bednorz and Müller's work starting the race was recognised when they received the 1987 Nobel Prize for physics.

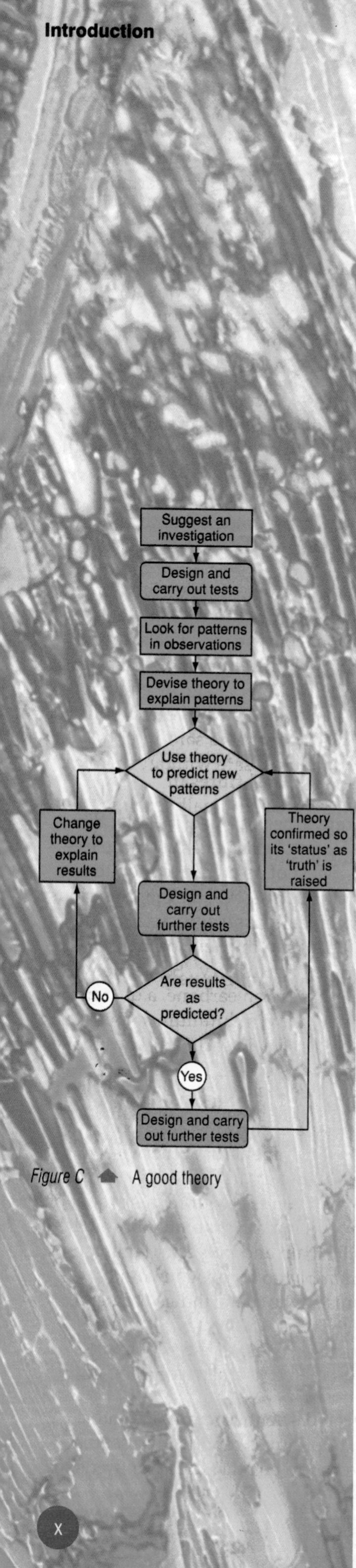


*Figure C* A good theory

## Models and theories

In 1987, scientists in laboratories throughout the world worked round the clock to try to make new superconductors. Other scientists were busy with computers trying to work out why the new materials were superconducting. They made models of how the atoms of the new materials might be fixed together. They tested and changed the models using their computers until they found a model that explained what was observed in the laboratories. Understanding how and why new superconductors work is important since it can give vital clues to better superconductors.

A good theory or model is one which gives predictions that turn out to be correct. The predictions are tested in the laboratory. If they do not pass the test, the theory or model must be altered or even thrown out. The more tests a model or theory passes, the more 'faith' scientists have in it. However, complete faith is never possible. Someone, somewhere might one day make a discovery which can't be explained by the theory.

## Beyond the frontiers

Who knows where a new discovery can lead? When Michael Faraday discovered how to generate electricity, he was asked 'What use is electricity, Mr Faraday?' He is reported to have replied 'What use is a new baby?' No one knows what might become of a new invention or a new discovery. More than 150 years later, life without electricity for most of us is unthinkable. Michael Faraday would have been truly astonished.

# Careers in science

Whatever career you choose to follow, you will almost certainly need to use your GCSE science skills and knowledge. Science is a feature of everyday life, at work or at home. Industry, hospitals, transport and agriculture are examples of major sectors of the economy that depend heavily on science. At home, almost all the things you do make use of science in some way.

Your GCSE science course is designed to give you a good scientific background so that you can, if you wish, carry on with science studies after GCSE.

To work in science requires specific personal qualities in addition to academic qualifications. Scientists are very creative and imaginative people and the work of an individual scientist can bring huge benefits to everyone. For example, Alexander Fleming's discovery of penicillin has saved countless lives. But do not be misled into thinking that the life of a scientist is one of continual discoveries. Scientists have to be very patient and methodical to discover anything; they have to be good at working together and at communicating their ideas to each other and to other people. The qualifications needed to become a scientist are outlined on p. xii; the qualities needed to become a scientist are just the same as you need in your GCSE course – enthusiasm, hard work, imagination, awareness and concern.

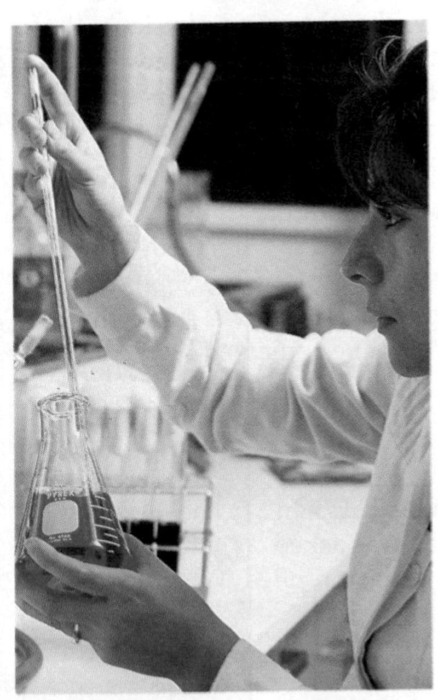

*What jobs are done by scientists?* In industry, scientists design, develop and test new products. For example, scientists in the glass industry are developing amazingly clear glass for use as **optical fibres** in communication links. In medicine, scientists are continually finding applications for scientific discoveries; for example, medical scientists have developed a high-power ultrasonic transmitter for destroying kidney stones, thus avoiding a surgical operation. These are just two examples of the work of scientists. You will find scientists at work in research laboratories, industrial laboratories, forensic laboratories, hospitals, schools, on field trips, expeditions, radio, TV and lots of other places. Scientists have to be very versatile as science is a very wide and varied field.

The skills and knowledge you gain through studying science will enable you to gain the benefits of new technologies. Ask your parents what aspects of life have **not** changed since they were children – they may be stuck for an answer! New technologies force the pace of change and if you do not learn to use them, you will not share their benefits. Studying science encourages you to develop an open mind and to seek new approaches. That is why a wide range of careers involve further studies in science.

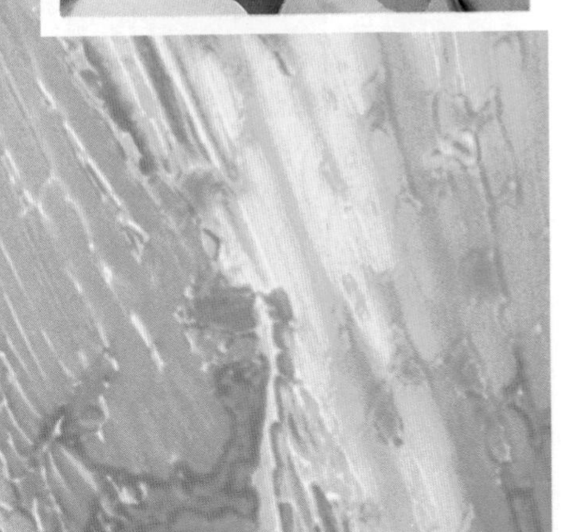

For many careers further studies in science are essential, for example, medicine, dentistry, pharmacy, engineering and computing. For other careers such as law, business studies, administration, the armed forces and retail management, studying science after GCSE will be helpful. Thus by continuing your science studies after GCSE, you keep many career options open, which is important when you are considering your choice of career.

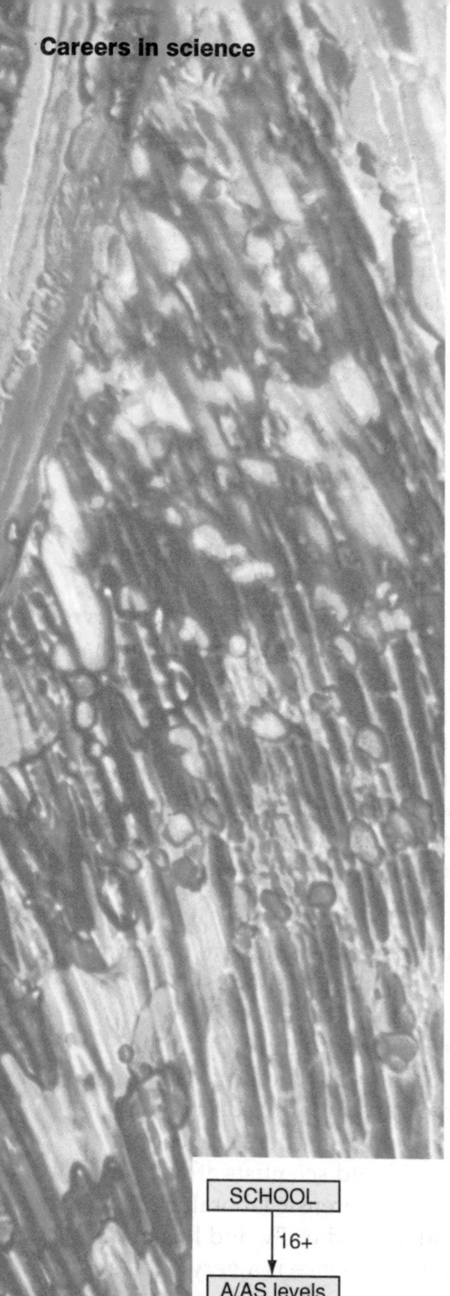

## The next steps after GCSE

Read this section carefully, bearing in mind that your working life will probably be about forty years. If you want your life ahead to be interesting, if you want to make your own decisions about what you do, if you want to make the most of your talents, then you should continue your studies. After GCSE, you can continue in full-time education at school or at college, or you can train in a job through part-time study. If you take a job without training, you will soon find that your friends who stayed on have much better prospects.

Most students aiming for a career in science or technology continue full-time study for two years, taking either GCE A- and AS-levels or a GNVQ course in Science. Successful completion of a suitable combination of these courses can then lead to a degree course at a university or a college of higher education or, alternatively, straight into employment.

**The A-level route** to higher education requires successful completion of a two-year full-time course, usually consisting of thee or more A-level subjects or an equivalent combination of AS and A levels. Students taking A-level Chemistry frequently choose their other subjects from Biology, Physics, Mathematics, Geography, Geology and Home Economics. Students with an interest in taking further courses in chemistry, biochemistry or chemical engineering are well advised to choose Mathematics. Students who plan further studies in chemistry or chemical engineering should choose Physics.

**The GNVQ science route** is also a two-year full-time course at advanced level, leading to a qualification equivalent to double award A-level science. All students study some biology, chemistry and physics and can then specialise. The course is assessed continuously through tests and laboratory work. Students who do not have grade CC in science to enter advanced level GNVQ can take the one-year GNVQ science intermediate course as a preparation for the advanced course.

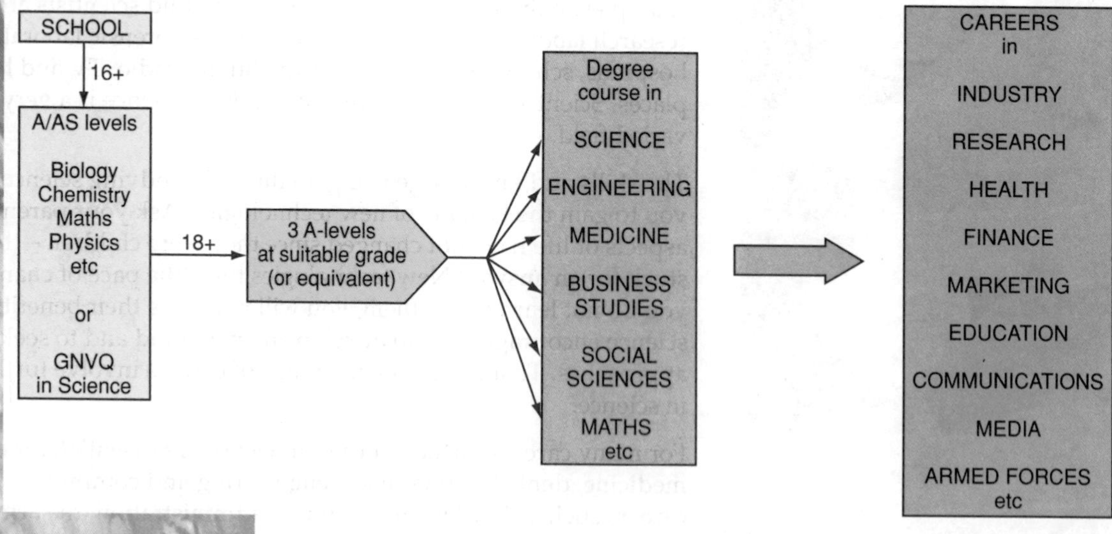


*Figure D* Career routes from GCSE

# Theme A

# Matter

Earth and all the other planets in the universe are made of matter. How many different kinds of matter are there? They are countless: far too many for one book to mention, but in Theme A we start to answer the question. What does matter do: how does it behave? The study of how matter behaves is called chemistry. Can one kind of matter change into a different kind of matter? The answer to the question is 'yes, it can': provided that it receives energy.

# Topic 1 Matter

## 1.1 The states of matter

FIRST THOUGHTS

What is a state of matter? How do states of matter differ? These are questions for you to explore in this topic.

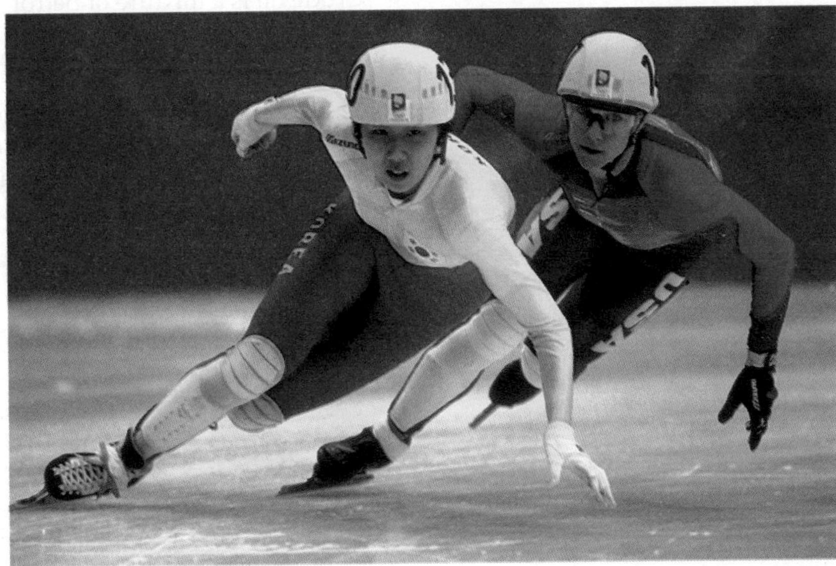

Figure 1.1A ▲ Ice skaters

It's lucky these skaters know how to change one state of matter into another! How do they do it?

Everything you see around you is made of matter. The skaters, the skates, the ice and all the other things in Figure 1.1A are different kinds of matter.

The skaters are overcoming friction as they glide across the frozen pond. *How do they manage this?* They change one kind of matter, ice, into another kind of matter, water. A thin layer of water forms between a skate and the ice, and this reduces friction and enables the skater to glide across the pond. The water beneath the skate refreezes as the skater moves on. *How are skaters able to melt ice? Why does it refreeze behind them?* Read on to find out.

The different kinds of matter are **solid** and **liquid** and **gaseous** matter. These are called the **states of matter**. Table 1.1 summarises the differences between the three chief states of matter. The symbols (s) for solid, (l) for liquid and (g) for gas are called **state symbols**. Later you will use another state symbol, (aq), which means 'in aqueous (water) solution'.

Table 1.1 ▼ States of Matter

| State | Description |
|---|---|
| Solid (s) | Has a fixed volume and a definite shape. The shape is usually difficult to change. |
| Liquid (l) | Has a fixed volume. Flows easily; changes its shape to fit the shape of its container. |
| Gas (g) | Has neither a fixed volume nor a fixed shape; changes its volume and shape to fit the size and shape of its container. Flows easily; liquids and gases are called **fluids**. Gases are much less dense than solids and liquids. |

SUMMARY

The three chief states of matter are:

- solid (s): fixed volume and shape,
- liquid (l): fixed volume; shape changes,
- gas (g): neither volume nor shape is fixed.

Liquids and gases are fluids.

## 1.2 Pure substances

Most of the solids, liquids and gases which you see around you are mixtures of substances.

- Rock salt, the impure salt which is spread on roads in winter, is a mixture of salt and sand and other substances.
- Crude oil is a mixture of petrol, paraffin, diesel fuel, lubricating oil and other liquids.
- Air is a mixture of gases.

Note the differences between a pure substance and a mixture.

Some substances, however, consist of one substance only. Such substances are **pure substances**. For example, from the mixture of substances in rock salt chemists can obtain pure salt which is 100% salt.

*Figure 1.2A* ▲ Rock salt and pure salt

SUMMARY

A pure substance is a single substance.

## 1.3 Density

Table 1.1 tells you that gases are much less **dense** than solids and liquids. What does **dense** mean? What is **density**? The two lengths of car bumper shown in Figure 1.3A have the same volume. You can see that they do not have the same mass. The steel bumper is heavier than the plastic bumper. This is because steel is a more **dense** material than the plastic; steel has a higher **density** than the plastic has.

$$\text{Density} = \frac{\text{Mass}}{\text{Volume}}$$

The unit of density is kg/m$^3$ or g/cm$^3$. The density values of some common substances are shown in Table 1.2.

*Table 1.2* ▼ Density

| Substance | Density (g/cm$^3$) |
|---|---|
| Air | $1.2 \times 10^{-3}$ |
| Aluminium | 2.7 |
| Copper | 8.92 |
| Ethanol | 0.789 |
| Gold | 19.3 |
| Hydrogen | $8.33 \times 10^{-5}$ |
| Iron | 7.86 |
| Lead | 11.3 |
| Methane | $6.67 \times 10^{-4}$ |
| Oxygen | $1.33 \times 10^{-3}$ |
| Silver | 10.5 |
| Water | 1.00 |

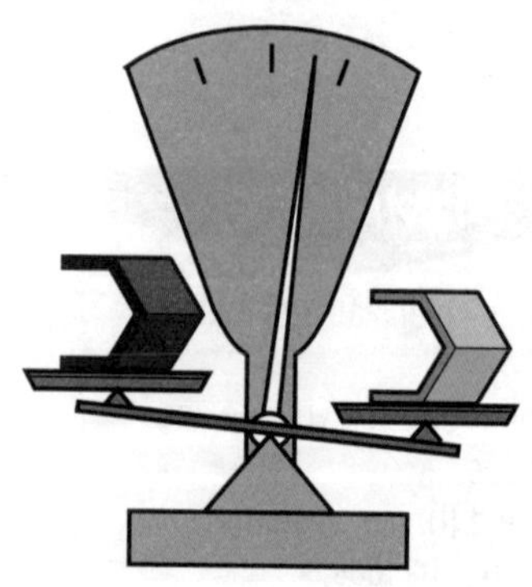

*Figure 1.3A* ▲ Two objects with the same volume

Archie says that aluminium is a 'light' material. Can you explain to him why this is incorrect? An aeroplane made of aluminium is not a light object! What term should Archie use in describing aluminium?

You can see that the gases are much less dense than any of the solid or liquid substances.

## SUMMARY

$$\text{Density} = \frac{\text{Mass}}{\text{Volume}}$$

Mass = Volume × Density

$$\text{Volume} = \frac{\text{Mass}}{\text{Density}}$$

The density triangle:

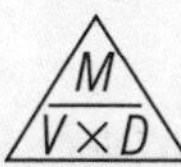


To find the quantity you want, cover up that letter.
The other letters in the triangle show you the formula.

## CHECKPOINT

1. A worker in an aluminium plant taps off 300 $cm^3$ of the molten metal. It weighs 810 g. What is the density of aluminium?
2. An object has a volume of 2500 $cm^3$ and a density of 3.00 $g/cm^3$. What is its mass?
3. Mercury is a liquid metal with a density of 13.6 $g/cm^3$. What is the mass of 200 $cm^3$ of mercury?
4. 50.0 $cm^3$ of metal A weigh 43.0 g
   52.0 $cm^3$ of metal B weigh 225 g
   Calculate the density of each metal. Say whether they will float or sink in water.

# 1.4 Change of state

Matter can change from one state into another. Some changes of state are summarised in Figure 1.4A.

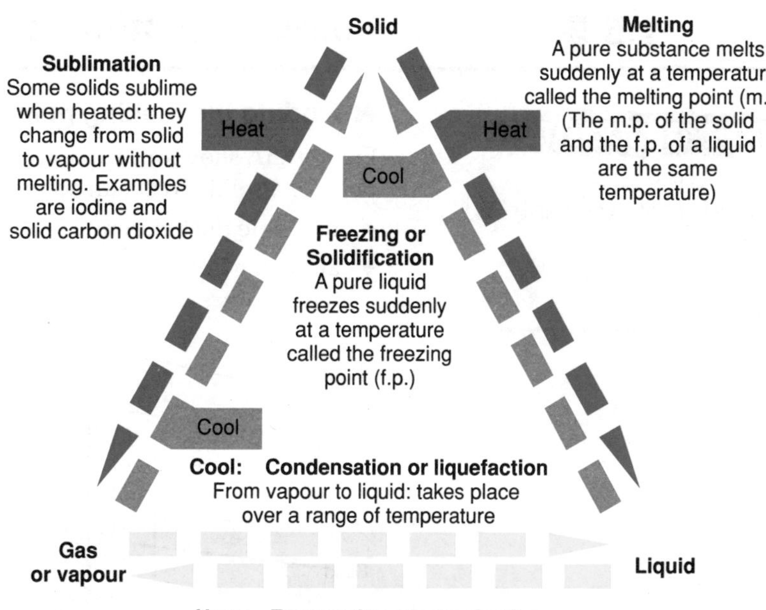

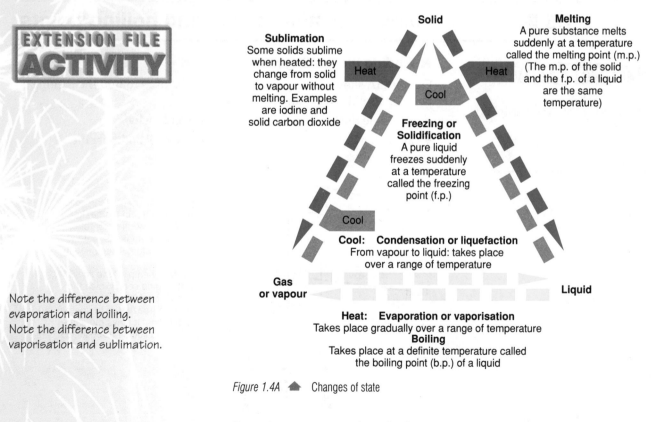

*Figure 1.4A* ▲ Changes of state

*Note the difference between evaporation and boiling.*
*Note the difference between vaporisation and sublimation.*

Sometimes gases are described as **vapours**. A liquid evaporates to form a vapour. A gas is called a vapour when it is cool enough to be liquefied.

$$\text{Vapour} \xrightarrow{\text{Either cool or compress without cooling}} \text{Liquid}$$

SUMMARY

Matter can change from one state into another.
The changes of state are:

- melting
- freezing or solidification
- evaporation or vaporisation
- condensation or liquefaction
- sublimation

The hotter a liquid is, the faster it evaporates. At a certain temperature, it becomes hot enough for vapour to form in the body of the liquid and not just at the surface. Bubbles of vapour appear inside the liquid. When this happens, the liquid is boiling, and the temperature is the boiling point of the liquid.

CHECKPOINT

1 Copy the diagram below. Fill in the names of the changes of state. (Some of them have two names.)

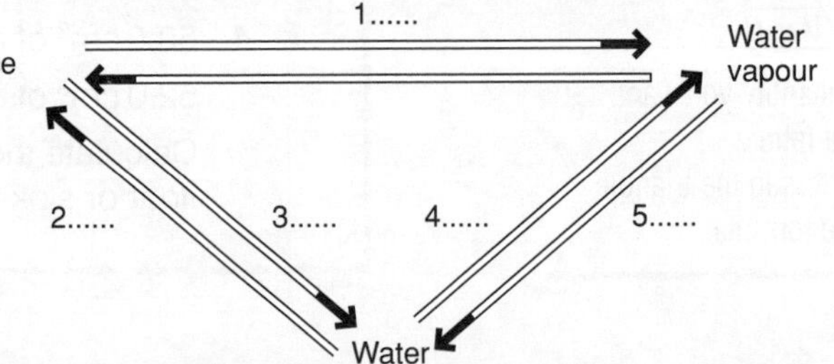


2 Give the scientific name for each of these changes.

(a) A puddle of water gradually disappears.
(b) A mist appears on your glasses.
(c) A mothball gradually disappears from your wardrobe.
(d) The change that happens when margarine is heated.

## 1.5 Finding melting points and boiling points

FIRST THOUGHTS

You can identify substances by finding their melting points and boiling points. Accuracy is essential, as you will see in this section

### Finding the melting point of a solid

Figure 1.5A shows an apparatus which can be used to find the melting point of a solid. To get an accurate result:

- first note the temperature at which the solid melts,
- allow the liquid which has been formed to cool and note the temperature at which it freezes.

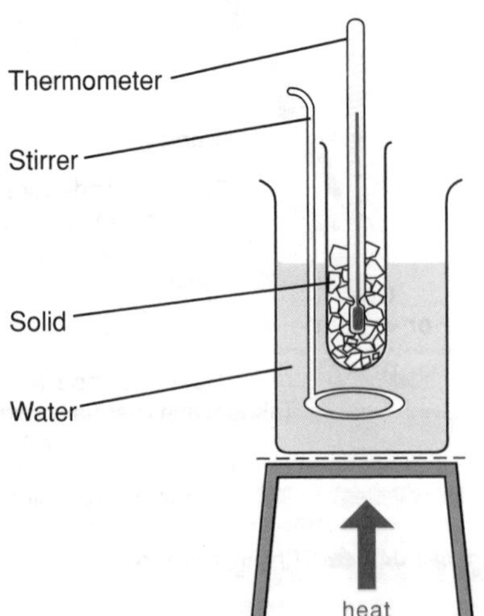


First, find the m.p. of the solid. Heat the water in the beaker. Stir. Watch the thermometer. When the solid melts, note the temperature. Stop heating

Now find the f.p. of the liquid. Let the liquid cool. Watch the thermometer. When the liquid begins to freeze, the temperature stops falling. It stays the same until all the liquid has solidified

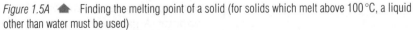

*Figure 1.5A* ▲ Finding the melting point of a solid (for solids which melt above 100 °C, a liquid other than water must be used)

The apparatus shown in Figure 1.5A will work between 20 °C and 100 °C. For solids with melting points above 100 °C, a liquid with a higher

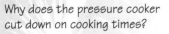

Why does the pressure cooker cut down on cooking times?

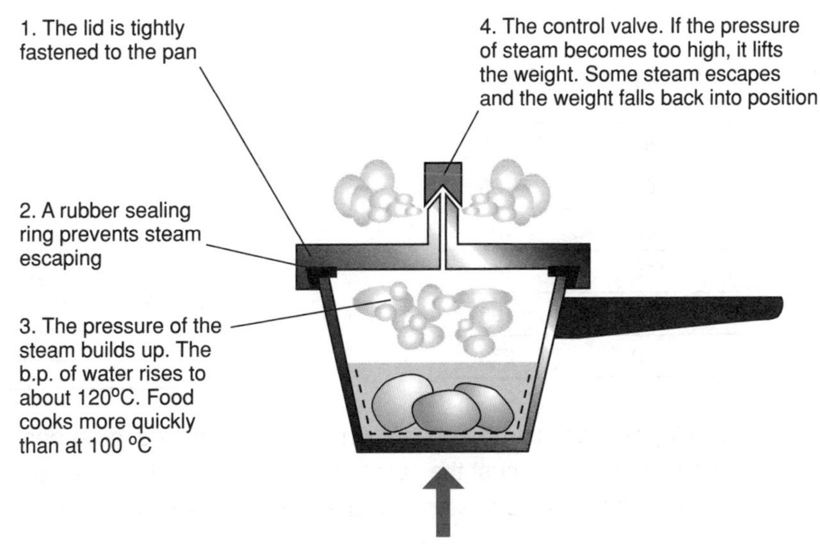

1. The lid is tightly fastened to the pan
4. The control valve. If the pressure of steam becomes too high, it lifts the weight. Some steam escapes and the weight falls back into position
2. A rubber sealing ring prevents steam escaping
3. The pressure of the steam builds up. The b.p. of water rises to about 120°C. Food cooks more quickly than at 100 °C
heat


*Figure 1.5H* ▲ How a pressure cooker works

Boiling points are stated at standard pressure (atmospheric pressure at sea level). Boiling points can be used to identify pure liquids. You take the boiling point of the liquid you want to identify; then you look through a list of boiling points of known liquids and find one which matches. *Which of the liquids could be the unknown substance X?*

| *Solid* | *Melting point (°C)* |
|---|---|
| Substance X | 111 |
| Benzene | 80 |
| Methylbenzene | 111 |
| Naphthalene | 218 |

No two substances have both the same boiling point and the same melting point.

## SUMMARY

The boiling point of a liquid is stated at standard pressure.

- At lower pressures, the boiling point is lower.
- At higher pressure, the boiling point is higher.

## CHECKPOINT

▶ **1** A pupil heated a beaker full of ice (and a little cold water) with a Bunsen burner. She recorded the temperature of the ice at intervals until the contents of the beaker had turned into boiling water. The table shows the results which the pupil recorded.

| *Time (minutes)* | 0 | 2 | 4 | 6 | 8 | 10 | 12 | 14 | 16 |
|---|---|---|---|---|---|---|---|---|---|
| *Temperature (°C)* | 0 | 0 | 0 | 26 | 51 | 76 | 100 | 100 | 100 |

(a) On graph paper, plot the temperature (on the vertical axis) against time (on the horizontal axis).
(b) On your graph, mark the m.p. of ice and the b.p of water.
(c) What happens to the temperature while the ice is melting?
(d) What happens to the temperature while the water is boiling?
(e) The Bunsen burner gives out heat at a steady rate. Explain what happens to the heat energy (i) when the beaker contains a mixture of ice and water at 0 °C, (ii) when the beaker contains water at 100 °C and (iii) when the beaker contains water at 50 °C.

▶ **2** Bacteria are killed by a temperature of 120 °C. One way of sterilising medical instruments is to heat them in an autoclave (a sort of pressure cooker: see the figure overleaf). The table shows the effect of pressure on the boiling point of water.

| Boiling point of water (°C) | Pressure (kPa) |
|---|---|
| 80 | 47 |
| 90 | 68 |
| 100 | 101 |
| 110 | 140 |
| 120 | 195 |
| 130 | 273 |

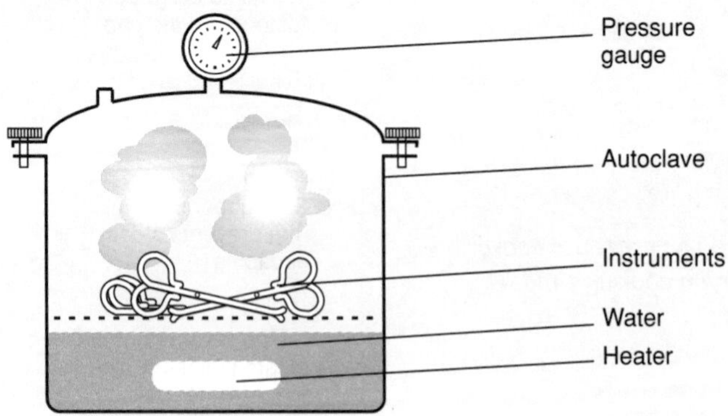


(a) Why are the instruments not simply boiled in a covered pan?
(b) What pressure must the autoclave reach to sterilise the instruments?
(c) What is the value of standard pressure in kPa (kilopascals)?

# 1.6 Properties and uses of materials

**FIRST THOUGHTS**

Properties of materials
- Hardness
- Toughness
- Strength
- Flexibility
- Elasticity
- Solubility
- Density (Topic 1.3)
- Melting point (Topic 1.5)
- Boiling point (Topic 1.5)
- Conduction of heat (KS: Physics, Topics 1.5 and 1.6)
- Conduction of electricity (Topic 7.1)

Types of matter from which things are made are called **materials**. Different materials are used for different jobs. The reason is that their different **properties** (characteristics) make them useful for different purposes. You will see a list of properties in the margin. Some of them have been mentioned earlier in this topic. Conduction of heat and electricity will be covered in later topics. Now let us look at the rest.

## Hardness

It is difficult to change the shape of a hard material. A hard material will dent or scratch a softer material. A hard material will withstand impact without changing. Table 1.3 shows the relative hardness of some materials on a 1 – 10 scale.

*Table 1.3* Relative hardness of materials

| Material | Relative hardness | Uses |
|---|---|---|
| Diamond | 10.0 | Jewellery, cutting tools |
| Silicon carbide | 9.7 | Abrasives |
| Tungsten carbide | 8.5 | Drills |
| Steel | 7–5 | Machinery, vehicles, buildings |
| Sand | 7.0 | Abrasives, e.g. sandpaper |
| Glass | 5.5 | Cut glass can be made by cutting glass with harder materials. |
| Nickel | 5.5 | Used in coins; hard-wearing |
| Concrete | 5–4 | Building material |
| Wood | 3–1 | Construction, furniture |
| Tin | 1.5 | Plating steel food cans |

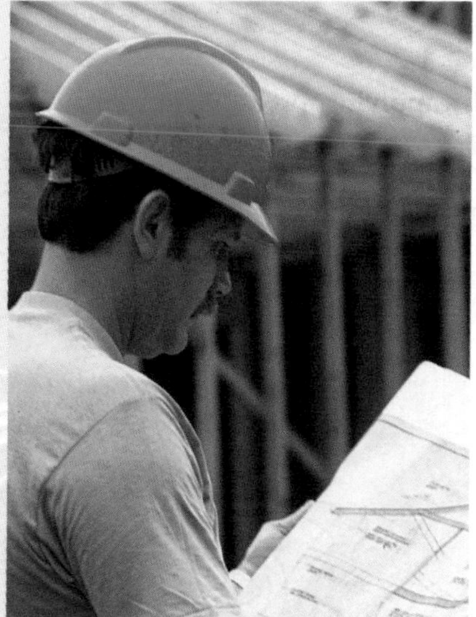

*Figure 1.6A* ▲ Hardness

## Toughness and brittleness

Construction workers on building sites wear 'hard hats' to protect themselves from falling objects. A hard hat is designed to absorb the energy of an impact. The hat material is **tough**, that is, it is difficult to break, although it may be dented by the impact. In comparison, a brick is difficult to dent and will shatter if dropped onto a concrete floor. The brick is **brittle**. Glass is another brittle material. These materials cannot absorb the energy of a large force without cracking. If a still larger force is applied the cracks get bigger and the materials shatter.

### *Composite materials*

To make brittle materials tougher you have to try to stop them cracking. Mixing a brittle material with a material made of fibres, e.g. glass fibre or paper, will often do this. The fibres are able to absorb the energy of a force and the brittle material does not crack. Plaster is a brittle material. Plasterboard is much tougher. It is made by coating a sheet of plaster with paper fibres. It is a **composite material**. Glass-reinforced plastic (GRP) is a mixture of glass fibre and a plastic resin.

Composite materials are widely used because they combine the good points of their components.

*Figure 1.6C* ▲ A GRP canoe

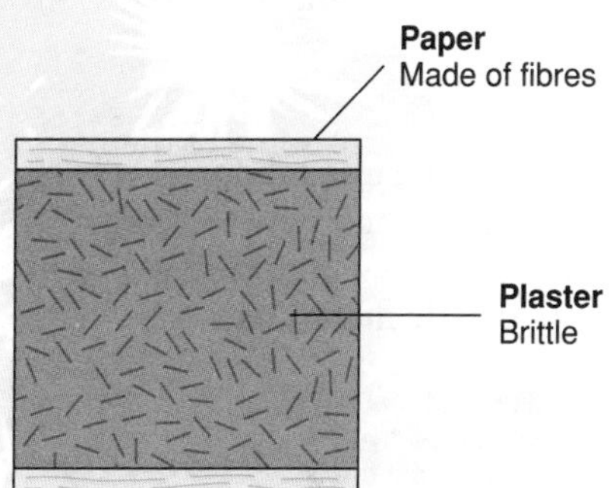


*Figure 1.6B* ▲ Plasterboard

*Figure 1.6D* ▲ Corrugated plastic roof

Concrete has great compressive strength, but it can crack if it is stretched. For construction purposes, **reinforced concrete** is used. Running through this composite material are steel rods which act like the glass fibres in GRP.

The shape of a piece of material alters its strength. Corrugated cardboard is used for packaging. Corrugated iron sheet and corrugated plastic sheet are used for roofing.

**SUMMARY**

Materials may be:

- hard – resistant to impact, difficult to scratch,
- tough – difficult to break, will 'give' before breaking,
- brittle – will break without 'giving',
- composite – made of more than one substance.

What do you mean exactly when you say a material is 'strong'? Can you stretch it? Can you squeeze it? Is it difficult to crush? Is it easy to bend without breaking? Note the difference between a plastic material and an elastic material.

## Strength

A strong material is difficult to break by applying force. The force may be a stretching force (e.g. a pull on a rope), or a squeeze (e.g. a vice tightening round a piece of wood), or a blow (e.g. a hammer blow on a lump of stone). A material which is hard to break by stretching has good **tensile strength**; a material which is hard to break by crushing has good **compressive strength**. The tensile strength of a material depends on its cross-sectional area.

## Flexibility

While a material is pulled it is being stretched: it is under **tension**. While a material is squashed it is being compressed: it is under **compression**. When a material is bent, one side of the material is being stretched while the opposite side is being compressed. A material which is easy to bend without breaking has both tensile strength and compressive strength. It is **flexible**.

*Figure 1.6E* ▲ It's flexible

## Elasticity

You can change the shape of a material by applying enough force. When you stop applying the force, some materials retain their new shapes; these are **plastic** materials. Other materials return to their old shape when you stop applying the force; these are **elastic materials**.

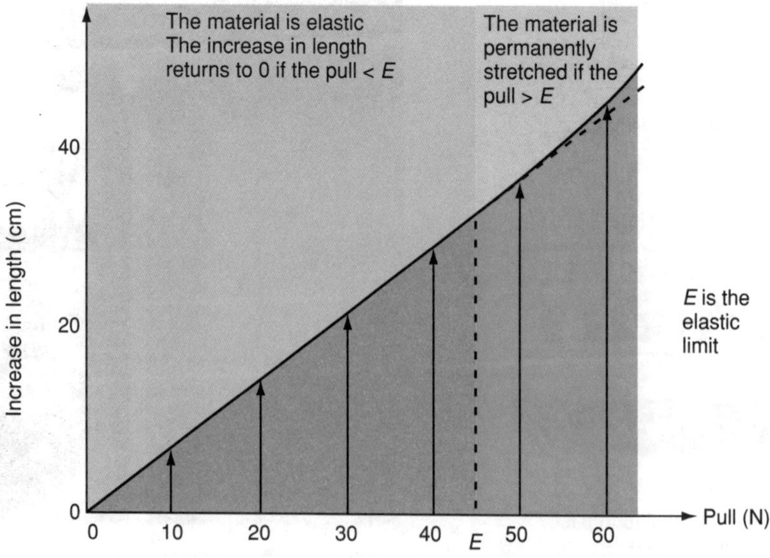


*Figure 1.6F* ▲ Increasing length

When you pull an elastic material, it stretches – increases in length. At first, when you double the pull, you double the increase in length. As the pull increases, however, you reach a point where the material no longer returns to its original shape. This pull is the **elastic limit** of the material. Increasing the pull still more eventually makes the material break (see Figure 1.6F).

### Science at work

Materials are tested to destruction to find their tensile strength, brittleness, etc. The results are fed into a computer to form a data base which can be used in the computer-aided design of articles from nuts and bolts to battleships.

### SUMMARY

Materials may have tensile strength (resistance to stretching), compressive strength (resistance to pressure), flexibility (both tensile and compressive strength) or elasticity (the ability to return to their original shape after being stretched).

# Topic 2 Particles

## 2.1 The atomic theory

FIRST THOUGHTS

What exactly is an atom, and why did the atomic theory take nearly 2000 years to catch on? Combine your imagination and your experimental skills in this section to find out why.

The idea that matter consists of tiny particles is very, very old. It was first put forward by the Greek thinker Democritus in 500 BC. For centuries the theory met with little success. People were not prepared to believe in particles which they could not see. The theory was revived by a British chemist called John Dalton in 1808. Dalton called the particles **atoms** from the Greek word for 'cannot be split'. According to Dalton's **atomic theory**, all forms of matter consist of atoms.

The atomic theory explained many observations which had puzzled scientists. Why are some substances solid, some liquid and others gaseous? When you heat them, why do solids melt and liquids change into gases? Why are gases so easy to compress? How can gases diffuse so easily? In this topic, you will see how the atomic theory provides answers to these questions and many others.

## 2.2 Elements and compounds

All the models in the photograph show molecules of elements. Each contains only one type of atom.

There are two kinds of pure substances: **elements** and **compounds**. An element is a simple substance which cannot be split up into simpler substances. Iron is an element. Whatever you do with iron, you cannot split it up and obtain simpler substances from it. All you can do is to build up more complex substances from it. You can make it combine with other elements. You can make iron combine with the element sulphur to form iron sulphide. Iron sulphide is made of two elements chemically combined: it is a compound.

The smallest particle of an element is an **atom**. In some elements, atoms do not exist on their own: they join up to form groups of atoms called **molecules**. Figure 2.2A shows models of the molecules of some elements.

*Figure 2.2A* ▲ Models of molecules (helium, He; oxygen, $O_2$; phosphorus, $P_4$; sulphur, $S_8$)

SUMMARY

Matter is made up of particles. Pure substances can be classified as elements and compounds. Elements are substances which cannot be split up into simpler substances. The smallest particle of an element is an atom. In many elements, groups of atoms join to form molecules.

A compound is a pure substance that contains two or more elements. The elements are not just mixed together: they are chemically combined. Many compounds consist of molecules, groups of atoms which are joined together by **chemical bonds**. All gases, whether they are elements or compounds, consist of molecules. Figure 2.2B shows models of molecules of some compounds.

The models in the photograph show molecules of compounds. You can see that each molecule contains more than one type of atom.

*Figure 2.2B* ▲ Models of molecules of some compounds (carbon dioxide, $CO_2$; methane, $CH_4$; ammonia, $NH_3$; water, $H_2O$; hydrogen chloride, HCl)

Some compounds do not consist of molecules; they are made up of electrically charged particles called **ions**. The word particle can be used for an atom, a molecule and an ion. Gaseous compounds and compounds which are liquid at room temperature consist of molecules.

**SUMMARY**

Compounds are pure substances that contain two or more elements chemically combined. Many compounds are made up of molecules; others are made up of ions. All gases consist of molecules.

## 2.3 ▶ How big are atoms?

Hydrogen atoms are the smallest. One million hydrogen atoms in a row would stretch across one grain of sand. Five million million hydrogen atoms would fit on a pinhead. A hydrogen atom weighs $1.7 \times 10^{-24}$ g; the heaviest atoms weigh $5 \times 10^{-22}$ g.

### How big is a molecule?

You can get an idea of the size of a molecule by trying the experiment shown in Figure 2.3A. This experiment uses olive oil, but a drop of detergent will also work. You can try this experiment at home.

Please try this experiment. You will be fascinated to find that you yourself can measure a molecule.

1. Fill a clean tea tray with clean water
2. Sprinkle fine talcum powder on the surface
3. Dip a fine piece of wire into the olive oil. Lift out a tiny drop of oil. Aim to get a drop about 0.5mm in diameter
4. Dip the wire into the water. The droplet of oil spreads out and pushes back the talcum powder. Measure as well as you can the area of the patch of olive oil

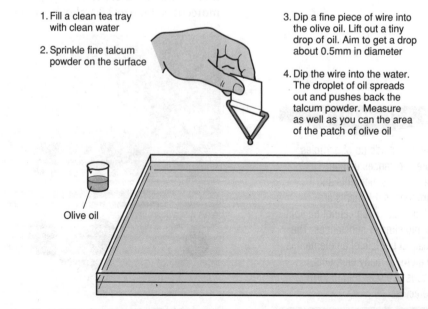

*Figure 2.3A* ▲ Estimating the size of a molecule

**Sample results**

Diameter of drop = 0.5 mm
Volume of drop = $(0.5\,\text{mm})^3 = 0.125\,\text{mm}^3$
Area of patch = $(25\,\text{cm})^2 = (250\,\text{mm})^2 = 6.25 \times 10^4\,\text{mm}^2$
Volume of patch = area × depth ($d$)
$0.125\,\text{mm}^3 = 6.25 \times 10^4\,\text{mm}^2 \times d$

$$d = \frac{0.125\,\text{mm}^3}{6.25 \times 10^4\,\text{mm}^2}$$

$$d = 2 \times 10^{-6}\,\text{mm}$$

The layer is only $2 \times 10^{-6}$ mm deep (two millonths of a millimetre). We assume that it is one molecule thick.

**SUMMARY**

Atoms are tiny! A pinhead would hold $5 \times 10^{12}$ hydrogen atoms. Olive oil molecules are $2 \times 10^{-6}$ mm in diameter.

# 2.4 The kinetic theory of matter

**FIRST THOUGHTS**

Particles in motion: what does this idea explain? The difference between solids, liquids and gases for a start, and the beauty of crystalline solids.

The **kinetic theory of matter** states that matter is made up of small particles which are constantly in motion. (Kinetic comes from the Greek word for 'moving'.) The higher the temperature, the faster they move. In a solid, the particles are close together and attract one another strongly. In a liquid the particles are further apart and the forces of attraction are weaker than in a solid. Most of a gas is space, and the particles shoot through the space at high speed. There are almost no forces of attraction between the particles in a gas.

Scientists have been able to explain many things with the aid of the kinetic theory.

## Solid, liquid and gaseous states

The differences between the solid, liquid and gaseous states of matter can be explained on the basis of the kinetic theory.

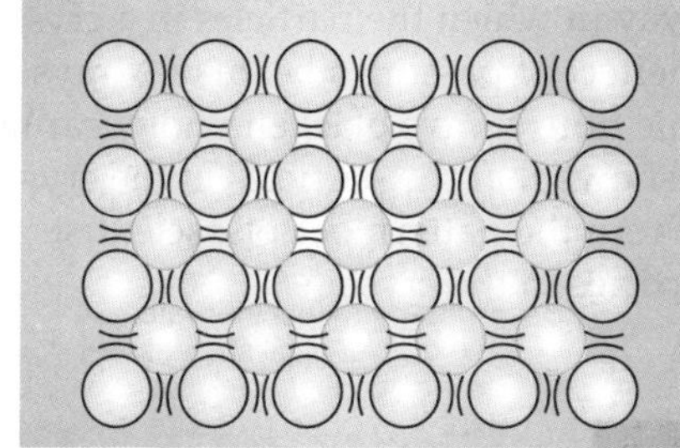

A solid is made up of particles arranged in a regular 3-dimensional structure. There are strong forces of attraction between the particles. Although the particles can vibrate, they cannot move out of their positions in the structure.

When a solid is heated, the particles gain energy and vibrate more and more vigorously. Eventually they may break away from the solid structure and become free to move around. When this happens, the solid has turned into liquid: it has melted.

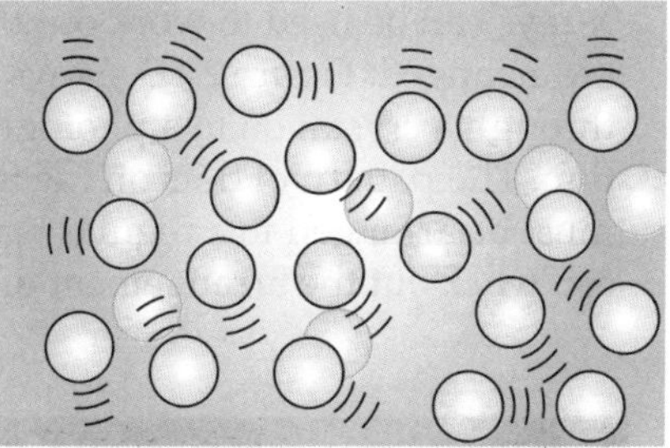

In a liquid the particles are free to move around. A liquid therefore flows easily and has no fixed shape. There are still forces of attraction between the particles.

When a liquid is heated, some of the particles gain enough energy to break away from the other particles. The particles which escape from the body of the liquid become a gas.

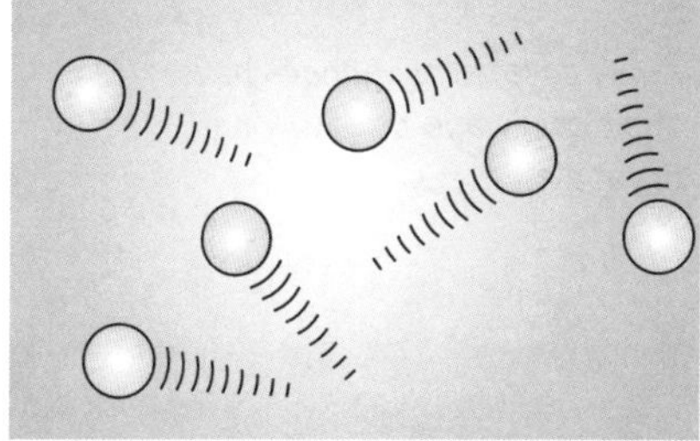

In a gas, the particles are far apart. There are almost no forces of attraction between them. The particles move about at high speed. Because the particles are so far apart, a gas occupies a very much larger volume than the same mass of liquid.

The molecules collide with the container. These collisions are responsible for the pressure which a gas exerts on its container.

*Figure 2.4A* ▲ The arrangement of particles in a solid, a liquid and a gas

## Crystals

Crystals are a very beautiful form of solid matter. A crystal is a piece of solid which has a regular shape and smooth faces (surfaces) which reflect light (see Figure 2.4B). Different salts have differently shaped crystals. *Why are many solids crystalline?* Viewing a crystal with an electron microscope, scientists can actually see individual particles arranged in a regular pattern (see Figure 2.4C). It is this regular pattern of particles which gives the crystal a regular shape.

You will enjoy watching crystals form under the microscope. The Extension File gives details.

*Figure 2.4B* ▲ Crystals of copper(II) sulphate

X-ray crystallographers have their own ways of studying crystals

X-rays can be used to work out the way in which the particles in a crystal are arranged. Figure 2.4D shows the effect of passing a beam of X-rays through a crystal on to a photographic film. X-rays blacken photographic film. The pattern of dots on the film shows that the particles in the crystal must be arranged in a regular way. From the pattern of dots, scientists can work out the arrangement of particles in the crystals.

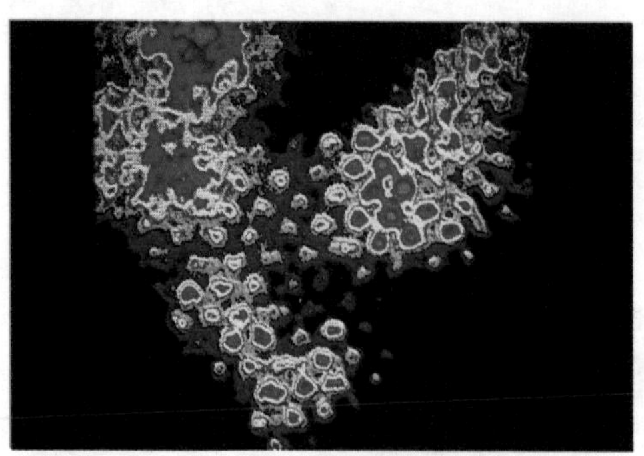

*Figure 2.4C* ▲ An electron microscope picture of uranyl ethanoate: each spot represents a single uranium atom

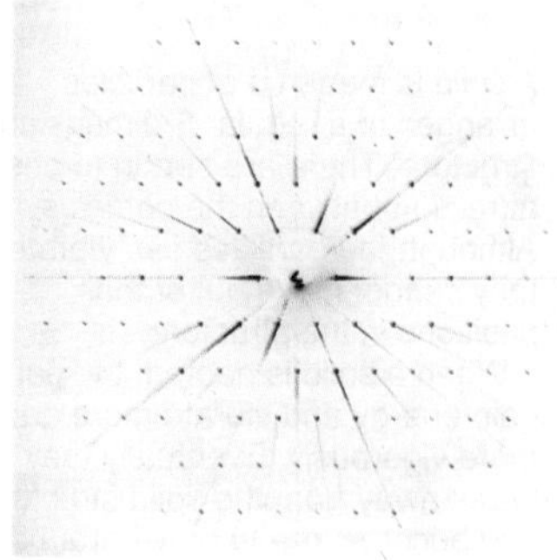

*Figure 2.4D* ▲ X-ray pattern from crystals of the metal palladium

SUMMARY

The kinetic theory of matter can explain the differences between the solid, liquid and gaseous states, and also how matter can change state. X-ray photographs show that crystals consist of a regular arrangement of particles.

FIRST THOUGHTS

How do solids dissolve, how do liquids vaporise and how do gases diffuse? Imagine particles in motion, and you will be able to explain all these changes.

## Dissolving

Crystals of many substances dissolve in water. You can explain how this happens if you imagine particles splitting off from the crystal and spreading out through the water. See Figure 2.4E.

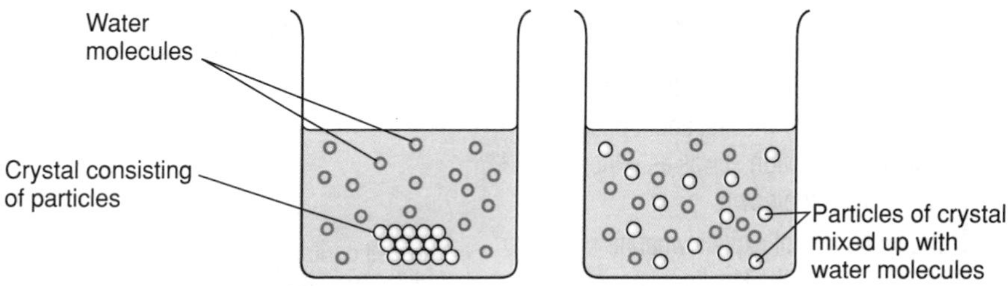


*Figure 2.4E* ▲ A coloured crystal dissolving

## Diffusion

*What evidence have we that the particles of a gas are moving?* The diffusion of gases can be explained. **Diffusion** is the way gases spread out to occupy all the space available to them. Figure 2.4F shows what happens when a jar of the dense green gas, chlorine, is put underneath a jar of air.

On the theory that gases consist of fast-moving particles, it is easy to explain how diffusion happens. Moving molecules of air and chlorine spread themselves between the two gas jars.

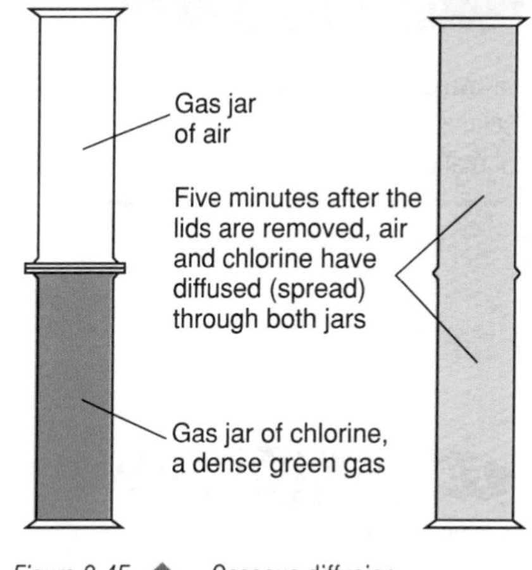


*Figure 2.4F* ▲ Gaseous diffusion

SUMMARY

Gases diffuse: they spread out to occupy all the space available to them. The kinetic theory of matter explains gaseous diffusion.

FIRST THOUGHTS

This section describes Brownian motion. You will see how neatly the kinetic theory can explain it.

## Brownian motion

Figure 2.4G shows a smoke cell and the erratic path followed by a particle of smoke.

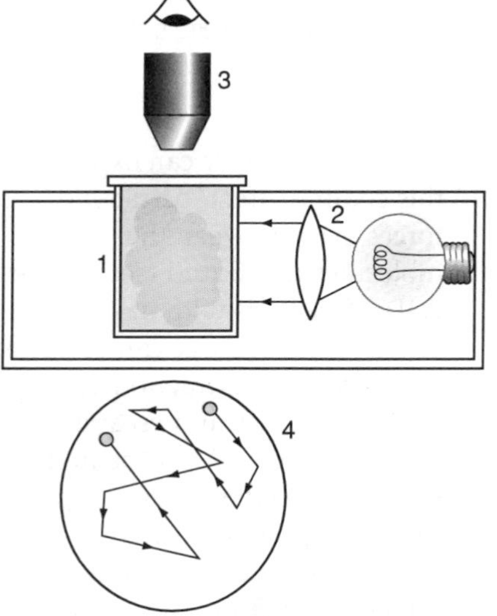


1 A small glass cell is filled with smoke

2 Light is shone through the cell

3 The smoke is viewed through a microscope

4 You see the smoke particles constantly moving and changing direction. The path taken by one smoke particle will look something like this

*Figure 2.4G* ▲ A smoke cell

## KEY SCIENTIST

**Brownian motion in a liquid**

In 1785, Robert Brown was using a microscope to observe pollen grains floating on water. He was amazed to see that the pollen grains were constantly moving about and changing direction. It was as if they had a life of their own.

Brown could not explain what he saw. You have the kinetic theory of matter to help you. Can you explain, with the aid of a diagram, what was making the pollen grains move?

## SUMMARY

Brownian motion puzzled scientists until the kinetic theory of matter offered an explanation.

We call this kind of motion **Brownian motion** after the botanist, Robert Brown, who first observed it. Figure 2.4H shows the explanation of Brownian motion.

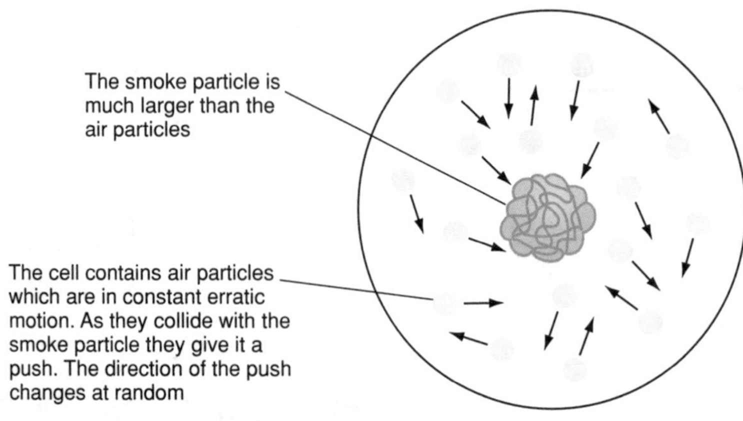

The smoke particle is much larger than the air particles
The cell contains air particles which are in constant erratic motion. As they collide with the smoke particle they give it a push. The direction of the push changes at random

*Figure 2.4H* ▲ Brownian motion

## Evaporation

When a liquid evaporates, it becomes cooler (see Figure 2.4I).

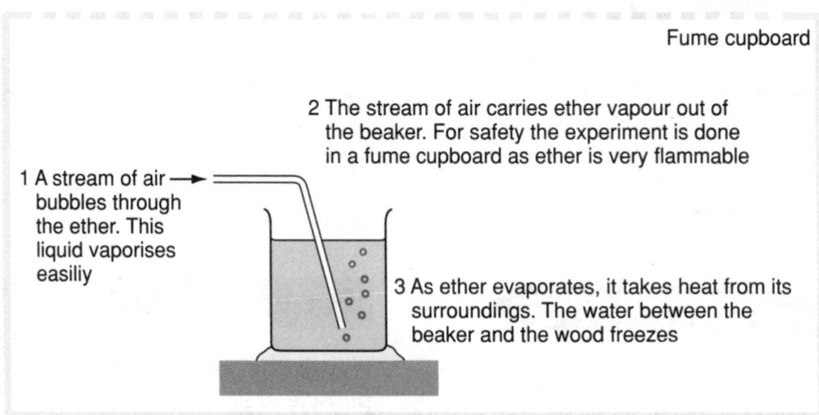

Fume cupboard
2 The stream of air carries ether vapour out of the beaker. For safety the experiment is done in a fume cupboard as ether is very flammable
1 A stream of air bubbles through the ether. This liquid vaporises easiliy
3 As ether evaporates, it takes heat from its surroundings. The water between the beaker and the wood freezes

*Figure 2.4I* ▲ The cooling effect produced when a liquid evaporates

## FIRST THOUGHTS

Why can you cool a hot cup of tea by blowing on it? Read this section to see if your idea is correct.

The kinetic theory can explain this cooling effect. Attractive forces exist between the molecules in a liquid (see Figure 2.4J). Molecules with more energy than average can break away from the attraction of other molecules and escape from the liquid. After the high energy molecules have escaped, the average energy of the molecules which remain is lower than before: the liquid has become cooler.

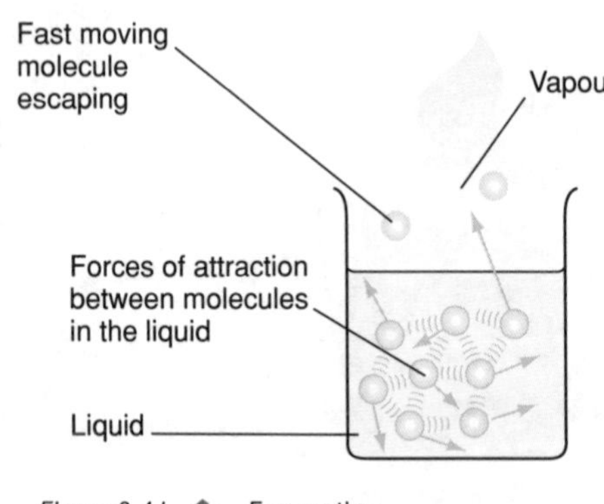


*Figure 2.4J* ▲ Evaporation

## SUMMARY

When a liquid evaporates, it takes heat from its surroundings. You can speed up evaporation by heating the liquid and by blowing air over it.

*What happens when you raise the temperature?* More molecules have enough energy to break away from the other molecules in the liquid. The rate of evaporation increases.

*What happens if you pass a stream of dry air across the surface of the liquid?* The dry air carries vapour away. The particles in the vapour are prevented from re-entering the liquid, that is, condensing. The liquid therefore evaporates more quickly.

## CHECKPOINT

▶ **1** Explain the following statements in terms of the kinetic theory.
(a) Water freezes when it is cooled sufficiently.
(b) Heating a liquid makes it evaporate more quickly.
(c) Heating a solid makes it melt.

▶ **2** The solid X does not melt until it is heated to a very high temperature. What can you deduce about the forces which exist between particles of X?

▶ **3** Of the five substances listed in the table below which is/are (a) solid (b) liquid (c) gaseous (d) unlikely to exist?

| *Substance* | *Distance between particles* | *Arrangement of particles* | *Movement of particles* |
|---|---|---|---|
| A | Close together | Regular | Move in straight lines |
| B | Far apart | Regular | Random |
| C | Close together | Random | Random |
| D | Far apart | Random | Move in straight lines |
| E | Close together | Regular | Vibrate a little |

▶ **4** Supply words to fill in the blanks in this passage.
A solid has a fixed ____ and a fixed ____. A liquid has a fixed ____, but a liquid can change its ____ to fit its container. A gas has neither a fixed ____ nor a fixed ____. Liquids and gases flow easily; they are called ____. There are forces of attraction between particles. In a solid, these forces are ____, in a liquid they are ____ and in a gas they are ____.

▶ **5** Imagine that you are one of the millions of particles in a crystal. Describe from your point of view as a particle what happens when your crystal is heated until it melts.

▶ **6** Which of the two beakers in the figure opposite represents (a) evaporation, (b) boiling? Explain your answers.

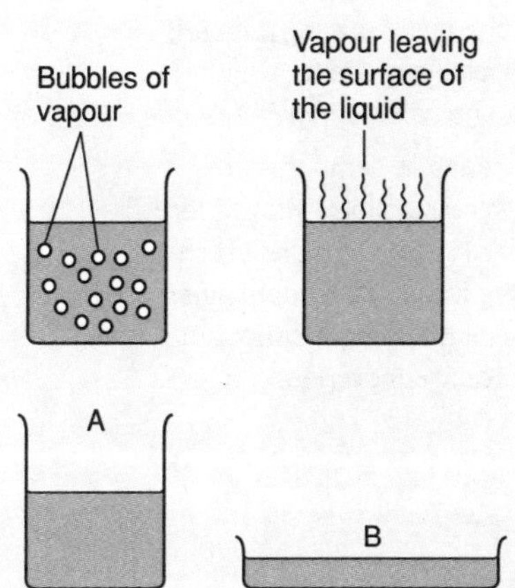


▶ **7** When a stink bomb is let off in one corner of a room, it can soon be smelt everywhere. Why?

▶ **8** Beaker A and dish B contain the same volume of the same liquid. Will the liquid evaporate faster in A or B? Explain your answer.

▶ **9** Why can you cool a cup of hot tea by blowing on it?

▶ **10** A doctor dabs some ethanol (alcohol) on your arm before giving you an injection. The ethanol makes your arm feel cold. How does it do this?

## 2.5 Gases

Gases have much lower densities than solids and liquids. The reason is that most of a gas is space. The particles are far apart, and move through the space at high speed. From time to time they collide with the walls of the container and with other particles. Gases are much more compressible than solids and liquids:

- Increase the pressure → the volume decreases.
- Decrease the pressure → the volume increases.

The kinetic theory explains this by saying that there is so much space between the particles that it is easy for the particles to move closer together when the gas is compressed.

A gas is mostly space! The particles of gas are far apart, with a lot of space between them. The particles move through the space at high speed. As a result, gases can be compressed.

Gases exert pressure because their particles are colliding with the walls of the container. If the volume of gas is decreased, the particles will hit the walls more often and the pressure will increase. The pressure of a given mass of gas changes with temperature:

- Increase the temperature, while keeping the volume constant → the pressure increases.
- Increase the temperature while keeping the pressure constant → the gas expands.

The kinetic theory explains these observations because, as the temperature rises, the particles have more energy, move faster and collide more frequently with the walls of the container.

One cylinder of a four-stroke car engine. During the compression stroke, the piston moves up the cylinder and decreases the volume of the mixture of petrol vapor and air; therefore the pressure increases

(a) The volume decreases therefore the pressure increases

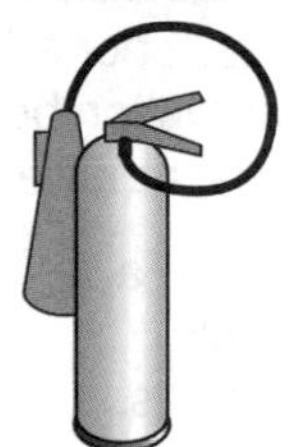

A cylinder of gas under pressure. When the valve is opened, the pressure decreases. The gas expands out of the cylinder

(b) The gas expands as the pressure decreases

An increase in pressure makes the particles move closer together, and the volume of the gas decreases.

When the pressure is reduced, the particles move further apart; the volume increases.

An increase in temperature makes the particles move more rapidly and collide more often with the walls of the container. If the volume is held constant the pressure increases.

When a balloon full of gas is heated, the pressure of gas increases until the balloon bursts

(c) The pressure increases as the temperature rises

A lump of bread dough contains air and carbon dioxide. When the dough is heated in an oven, the volume of gas increases and the dough 'rises'

(d) The gas expands as the temperature rises

*Figure 2.5A* Factors which determine the volume of a gas

The relationship between the volume, pressure and temperature of a gas is expressed by the equation:

$$\frac{\text{Pressure} \times \text{Volume}}{\text{Absolute temperature}} = \text{constant}$$

$$\frac{pV}{T} = \text{constant}$$

(for a fixed mass of gas).

The pressure, volume and temperature of a gas are related by the equation $pV/T$ = constant

### *Stating gas volume*

It would not be very informative to state the volume of a gas without stating the temperature and pressure at which the volume is measured. It is the custom to quote gas volumes either at standard temperature and pressure (s.t.p., 0 °C and 1 atm) or at room temperature and pressure (r.t.p., 20 °C and 1 atm).

### *Diffusion of gases*

A gas which has large molecules has a higher density than a gas which has small molecules. For example, chlorine molecules are larger than hydrogen molecules, and chlorine is denser than hydrogen. Under the same conditions of temperature and pressure, chlorine is 35.5 times as dense as hydrogen. Large, heavy molecules move more slowly than small, light molecules, and dense gases therefore diffuse more slowly than gases of low density. Under the same conditions, the rate of diffusion of chlorine is one-sixth that of hydrogen.

## CHECKPOINT

1. Explain why people are advised to let some of the air out of their car tyres when driving from the UK to a very hot country.
2. A weather balloon is released into the atmosphere, where it will rise to a height of 10 km. It is only partially inflated at ground level. Why should it not be filled completely?
3. A gas syringe holds 100 $cm^3$ of nitrogen at a pressure of 1 atm. The gas is allowed to expand into a 1 l flask. What can you say about the pressure of gas in the 1 l flask?

# Topic 3 Methods of separation

## 3.1 Raw materials from the Earth's crust

FIRST THOUGHTS

The Earth provides all the raw materials we use. The problem is to separate the substances we want from the mixture of substances which makes up the Earth's crust.

The Earth's crust and atmosphere provide us with all the raw materials that we use: metals, oil, salt, sand, limestone, coal and many other resources. We use these substances as the raw materials for the manufacture of the houses, clothing, tools, machines, means of transport, medicines and all the other goods which we need. Few useful raw materials are found in a pure form in the Earth's crust. It is usual to find raw materials that we want mixed up with other materials. Chemists have worked out methods of separating substances from mixtures. Table 3.1 summarises some of them.

The table summarises methods of separating the components of different types of mixture.

*Table 3.1* Methods of separating substances from mixtures

| Mixture | Type | Method |
|---|---|---|
| Solid + Solid | | Make use of a difference in properties, e.g. solubility or magnetic properties |
| Solid + Liquid | Mixture | Filter |
| Solid + Liquid | Solution | Crystallise to obtain the solid<br>Distil to obtain the liquid |
| Liquid + Liquid | Miscible (form one layer) | Use fractional distillation |
| Liquid + Liquid | Immiscible (form two layers) | Use a separating funnel |
| Solid + Solid | In solution | Use chromatography |

## 3.2 Pure substances from a mixture of solids

### Dissolving one of the substances

There are vast deposits of **rocksalt** in Cheshire. A salt mine is shown in Figure 3.2A. One method of mining salt is to insert charges of explosives in the rock face and then detonate them.

Rocksalt is crushed and used for spreading on the roads in winter to melt the ice. For many uses, pure salt (sodium chloride) is needed. It can be obtained by using water to dissolve the salt in rocksalt, leaving the rock and other impurities behind.

SUMMARY

A soluble substance can be separated from a mixture by dissolving it to leave insoluble substances behind.

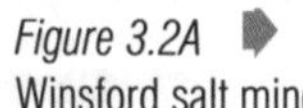
*Figure 3.2A* Winsford salt mine

## 3.3 ▶ Solute from solution by crystallisation

EXTENSION FILE ACTIVITY

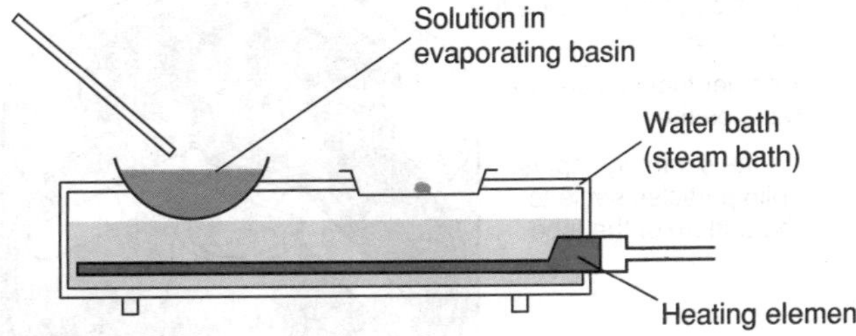

While the solvent is evaporating, dip a cold glass rod into the solution from time to time. When small crystals form on the rod, take the solution off the water bath and leave it to cool.
Solution in evaporating basin
Water bath (steam bath)
Heating element

*Figure 3.3A* ▲ Evaporating a solution to obtain crystals of solute

A laboratory method of evaporating a solution until it crystallises is shown in Figure 3.3A. The salt industry uses large scale evaporators which run non-stop. Figure 3.3B shows another method.

*Figure 3.3B* ▶ Salt pans in the Canary Islands. Sea water flows into the 'pans' and much of the water evaporates in the hot sun. When the brine has became a saturated solution, salt crystallises. The sea water is pumped through sluice gates where the salt crystals are filtered out.

SUMMARY

A solute crystallises out of a saturated solution. If a solution is unsaturated, some of the solvent must be evaporated before crystals form.

## 3.4 ▶ Filtration: separating a solid from a liquid

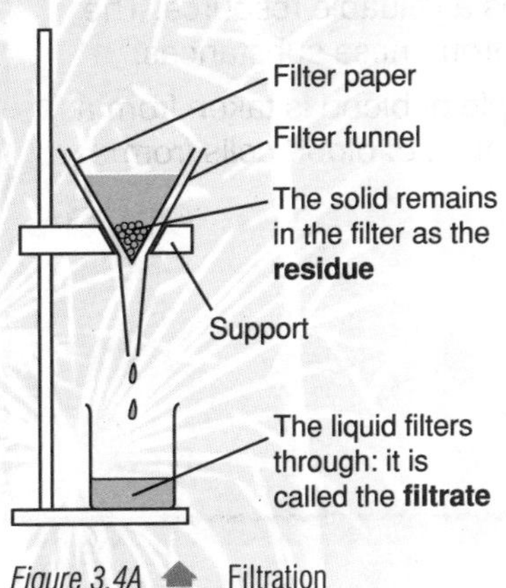


The solid remains in the filter as the **residue**

The liquid filters through: it is called the **filtrate**

*Figure 3.4A* ▲ Filtration

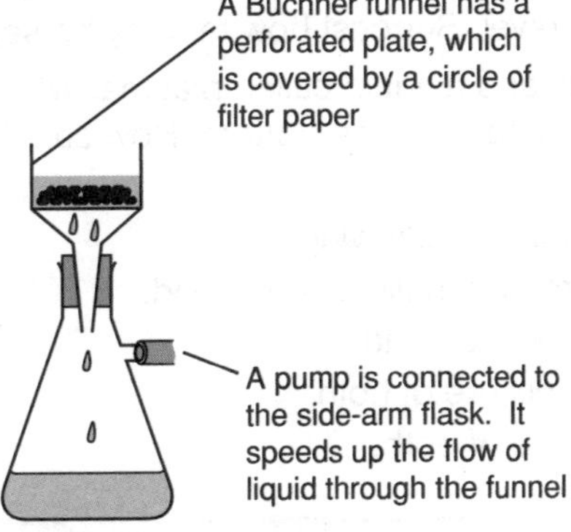


A Buchner funnel has a perforated plate, which is covered by a circle of filter paper
A pump is connected to the side-arm flask. It speeds up the flow of liquid through the funnel

*Figure 3.4B* ▲ Filtration under reduced pressure

Filtration can be used to separate a solid and a liquid. The filter must hold back solid particles but be fine enough to allow liquid to pass through. Figure 3.4A shows a laboratory apparatus for filtering through filter paper. Figure 3.4B shows a faster method: filtering under reduced pressure. In Topic 15.3, you will see how important filtration is in the purification of our drinking water.

## 3.5 ▶ Centrifuging

Biochemists have developed methods of growing bacteria as a source of **high-protein food**. When bacteria are allowed to grow in a warm solution of nutrients, they multiply so fast that they can double in mass every 20 minutes. From time to time, the harvest of bacteria is separated from the nutrient solution. Bacteria are too small to be separated by filtration. They are so small that they are **suspended** (spread out) in the liquid. They do not settle to the bottom as heavier particles would do, and they do not dissolve to form a solution. They can be separated by the method of **centrifuging** (centrifugation). The suspension of bacteria and liquid is placed inside a centrifuge and spun at high speed. The motion causes bacteria to separate from the suspension and sink to the bottom of the centrifuge tubes. Figure 3.5A shows a small laboratory centrifuge.

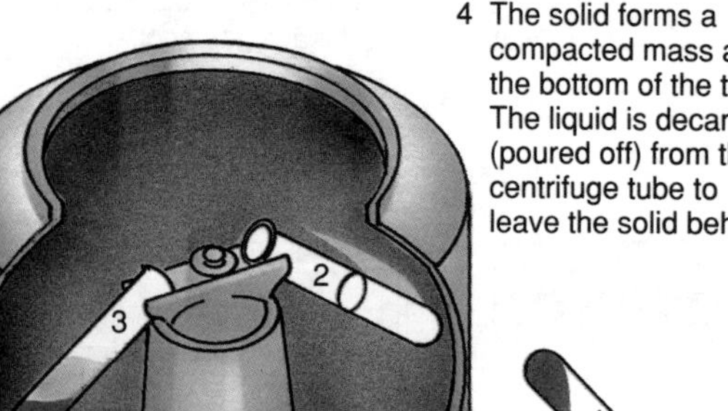

1 The suspension is poured into a glass tube inside the centrifuge
2 Another tube is used to balance the first
3 As the centrifuge spins, solid particles settle to the bottom of the tube
4 The solid forms a compacted mass at the bottom of the tube. The liquid is decanted (poured off) from the centrifuge tube to leave the solid behind
2
3


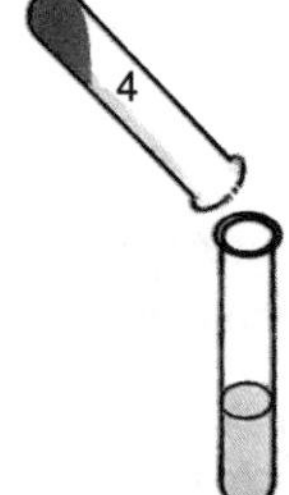


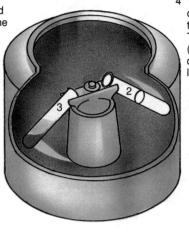

*Figure 3.5A* ▲ Centrifuging a suspension

## SUMMARY

- Filtration will separate a solid from a liquid.
- Centrifuging will separate a suspended solid from solution.

## CHECKPOINT

1. Describe how you would obtain both the substances in the following mixtures of solids.
   (a) A and B: Both A and B are soluble in hot water, but only A is soluble in cold water.
   (b) C and D: Neither C nor D is soluble in water. C is soluble in ethanol, but D is not.
2. In Trinidad and some other countries, tar trickles out of the ground. It is a valuable resource. The tar is mixed with sand and gravel. Suggest how tar may be separated from these substances.
3. Blood consists of blood cells and a liquid called plasma. When a sample of blood is taken from a person, the blood cells slowly settle to the bottom. How can the separation of blood cells from plasma be speeded up?
4. Suggest how you could obtain the following:
   (a) iron filings from a mixture of iron filings and sand,
   (b) wax from a mixture of wax and sand,
   (c) sand and gravel from a mixture of both,
   (d) rice and salt from a mixture of both.

## 3.6 Distillation

Sometimes you need to separate a solvent from a solution. In some parts of the world, drinking water is obtained from sea water. The method of **distillation** is employed. Figure 3.6A shows a laboratory scale distillation apparatus. The processes that take place are:

- in the distillation flask, **vaporisation**: liquid → vapour
- in the condenser, **condensation**: vapour → liquid.

Vaporisation followed by condensation is called **distillation**.

SUMMARY

Distillation is used to separate a solvent from a solution.

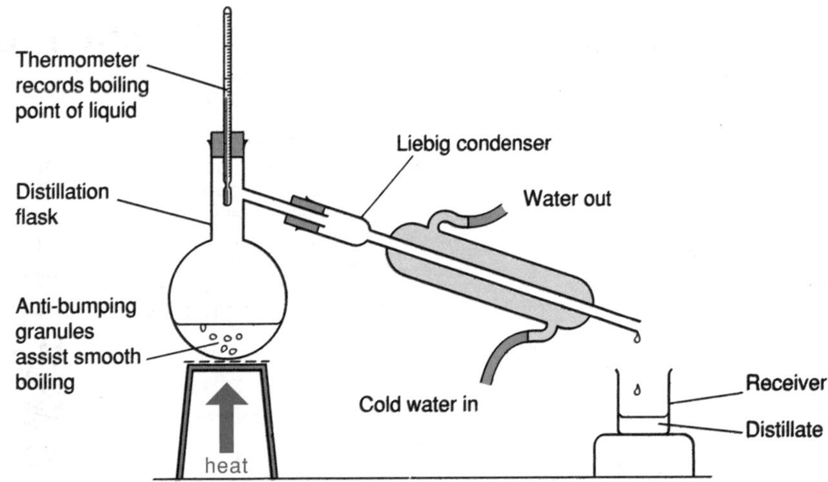


*Figure 3.6A* ▲ A laboratory distillation apparatus

CHECKPOINT

1. Why do some countries in the Persian Gulf obtain drinking water by distilling sea water? You will need to consider (a) the other sources of water and (b) the cost of fuel for heating the still.
2. Imagine that you are cast away on a desert island. You have two ways of obtaining drinking water. One is by separating pure water from sea water. The other is by collecting dew.
   (a) Think of a way of obtaining pure water from sea water. You do not have a proper distillation apparatus, but you have some matches and an empty petrol tin which were in the lifeboat. You find wood, bamboo canes, palm trees and coconuts on the island. Make a sketch of your design. With other members of your class, make a display of your sketches.
   (b) Every night on the island there is a heavy dew. How can you collect some of this dew? You have a sheet of plastic from the lifeboat. Again, sketch your idea. Then you can make another class display.
   (c) Write a letter to a 10-year-old, telling how you survived the shipwreck and how you obtained drinking water until you were rescued.

# 3.7 ▶ Separating liquids

## Using a separating funnel to separate immiscible liquids

When some liquids are added to each other, they do not mix: they form separate layers. They are said to be **immiscible**. They can be separated by using a separating funnel.

1 The mixture of immiscible liquids is poured in. It settles into two layers (or more) as the liquids do not mix.

2 The tap is opened to let the bottom layer run into a receiver.

3 The tap is closed and the receiver is changed. The tap is opened to let the top layer run out.

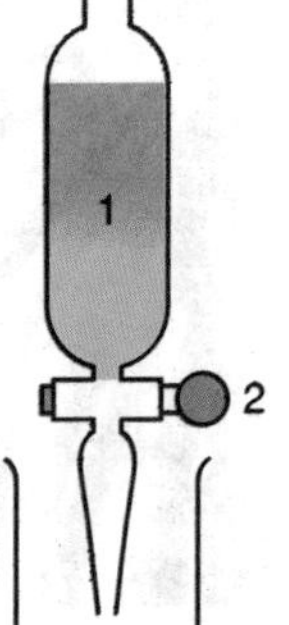


*Figure 3.7A* ▶ A separating funnel

SUMMARY

Immiscible liquids can be separated by means of a separating funnel.

# 3.8 Fractional distillation

The figure shows a solution of liquids being separated by fractional distillation by means of their different boiling points.

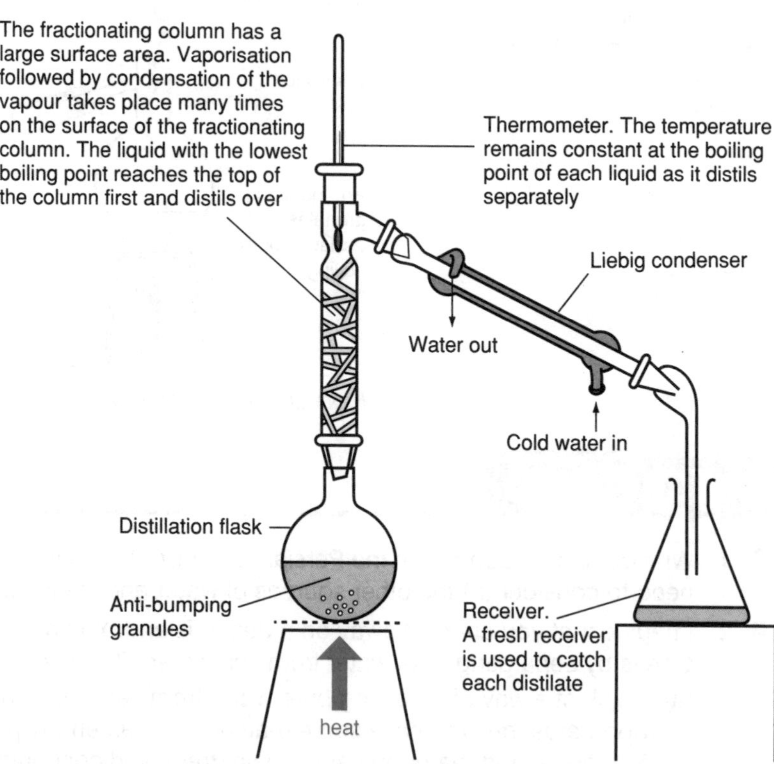
The fractionating column has a large surface area. Vaporisation followed by condensation of the vapour takes place many times on the surface of the fractionating column. The liquid with the lowest boiling point reaches the top of the column first and distils over
Thermometer. The temperature remains constant at the boiling point of each liquid as it distils separately
Liebig condenser
Water out
Cold water in
Distillation flask
Anti-bumping granules
Receiver. A fresh receiver is used to catch each distilate
heat

*Figure 3.8A* ▲ Fractional distillation

When liquids dissolve in one another to form a solution, instead of forming separate layers, they are described as **miscible**. Distillation can be used to separate a mixture of miscible liquids. Whisky manufacturers want to separate ethanol (alcohol) from water in a solution of these two liquids. They are able to do this because the two liquids have different boiling points: ethanol boils at 78 °C, and water boils at 100 °C. Figure 3.8A shows a laboratory apparatus which will separate a mixture of ethanol and water into its parts or **fractions**. The process is called **fractional distillation**. The temperature rises to 78 °C and then stays constant while all the ethanol distils over. When the temperature starts to rise again, the receiver is changed. At 100 °C, water starts to distil, and a fresh receiver is put into position to collect it.

*Figure 3.8B* ▲ An industrial distillation plant

## Continuous fractional distillation

Crude petroleum oil is not a very useful substance. By fractional distillation, it can be separated into a number of very useful products (see Figure 3.8C). In the petroleum industry, fractional distillation is made to run continuously (non-stop). Crude oil is fed in continuously, and the separated fractions are run off from the still continuously. These fractions are not pure substances. Each is a mixture of substances with similar boiling points. The fractions with low boiling points are collected from the top of the fractionating column. Fractions with high boiling points are collected from the bottom of the column.

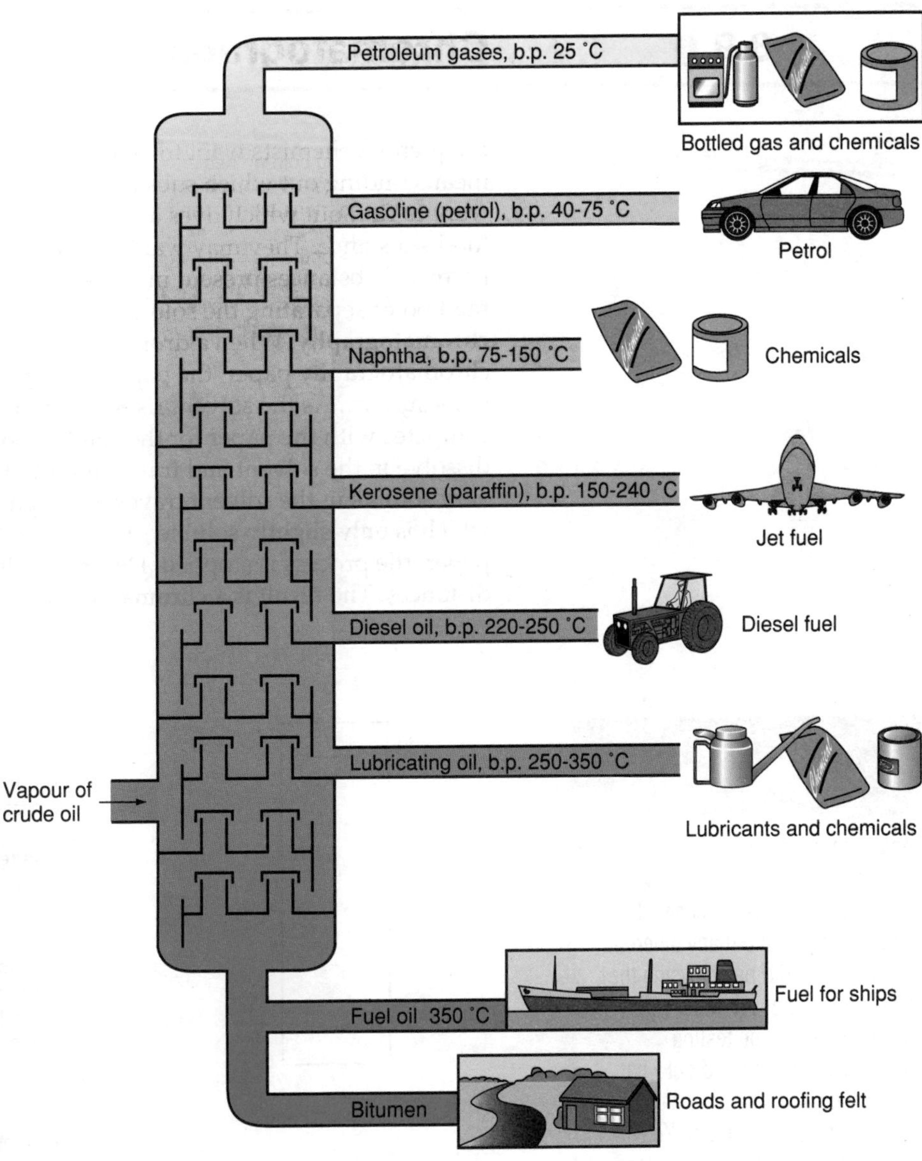

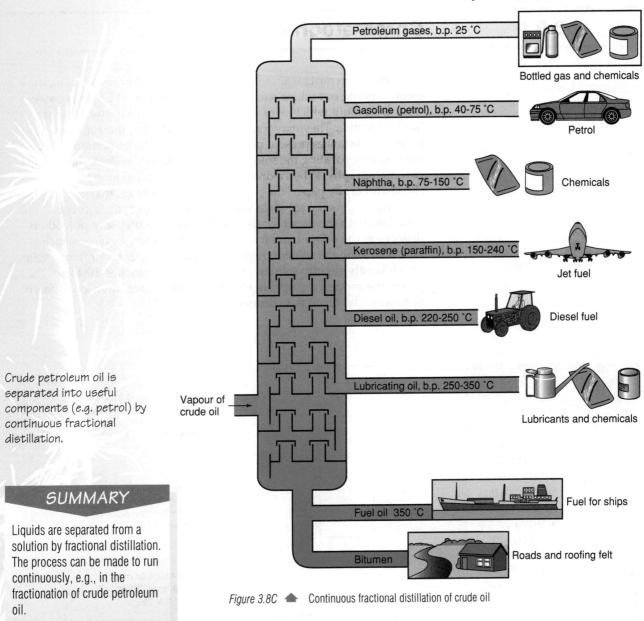

Crude petroleum oil is separated into useful components (e.g. petrol) by continuous fractional distillation.

*Figure 3.8C* ▲ Continuous fractional distillation of crude oil

## SUMMARY

Liquids are separated from a solution by fractional distillation. The process can be made to run continuously, e.g., in the fractionation of crude petroleum oil.

## CHECKPOINT

1. Your 12-year-old sister wants to know what you have been doing in your science lessons. Write a technical report, in words which she can understand, explaining *why* the petroleum industry distils crude oil and *how* fractional distillation works.
2. Your sister pours vinegar into the bottle of cooking oil by mistake. She asks you to help her to put things right. How could you separate the two liquids?
3. In Hammond Innes' novel, *Ice Station Zero*, a vehicle breaks down because the villains have put sugar in the petrol tank. Explain how you could separate the sugar from the petrol.
4. After a collision at sea, thousands of litres of oil escape from a tanker. A salvage ship sucks up a layer of oil mixed with sea water from the surface of the sea.
   (a) Describe how you could separate the oil from the sea water.
   (b) Say why it is important to be able to do this.

# 3.9 Chromatography

Frequently, chemists want to **analyse** a mixture. (**Analysis** of a mixture means finding out which substances are present in it.) Chemists may want to find out which dyes and preservatives have been added to a food substance. They may want to find out whether there are any harmful substances present in drinking water. Chromatography is one method of separating the solutes in a solution. Figure 3.9A shows **paper chromatography**. When a drop of solution is applied to the chromatography paper, the paper absorbs the solutes, that is, binds them to its surface. As the solvent rises through the paper, the solvent competes with the paper for the solutes. Some solutes stay put; others dissolve in the solvent and travel in it up the paper. A solute which is very soluble in the solvent travels through the paper faster than a solute which is only slightly soluble. When the solvent reaches the top of the paper, the process is stopped. Different solutes have travelled different distances. The result is a **chromatogram**.

## KEY SCIENTIST

Analytical chemistry is an important branch of chemistry. A career as an analytical chemist could take you into many different fields. You could be analysing the air we breathe or the water we drink. You could be testing processed foods to find out what colourings and flavour-enhancers have been added to them. You could be in the police force. Many crimes are solved by chemists analysing materials found at the scene of the crime.

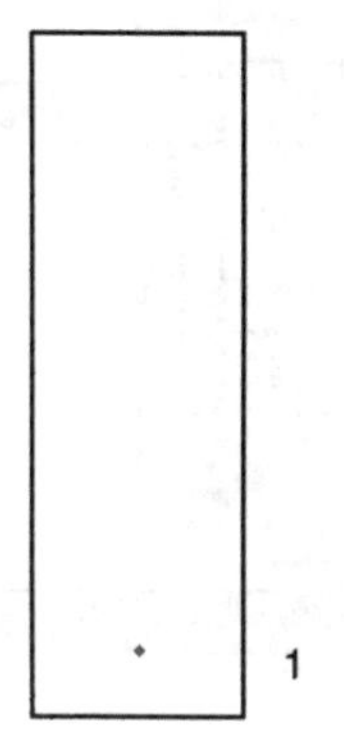


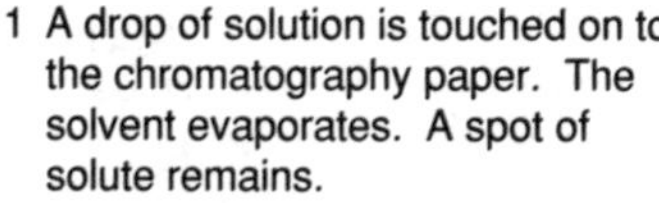

1 A drop of solution is touched on to the chromatography paper. The solvent evaporates. A spot of solute remains.


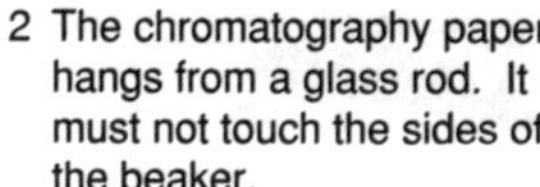

2 The chromatography paper hangs from a glass rod. It must not touch the sides of the beaker.


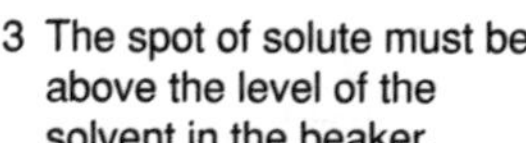

3 The spot of solute must be above the level of the solvent in the beaker.


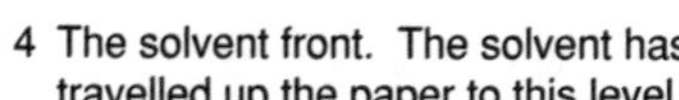

4 The solvent front. The solvent has travelled up the paper to this level.


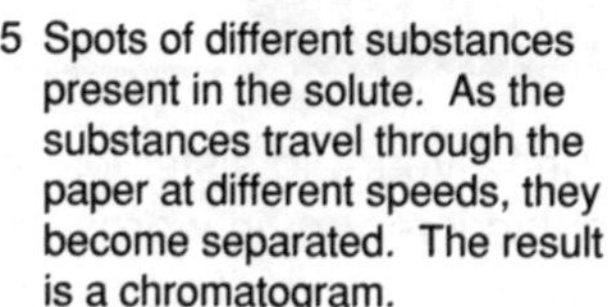

5 Spots of different substances present in the solute. As the substances travel through the paper at different speeds, they become separated. The result is a chromatogram.


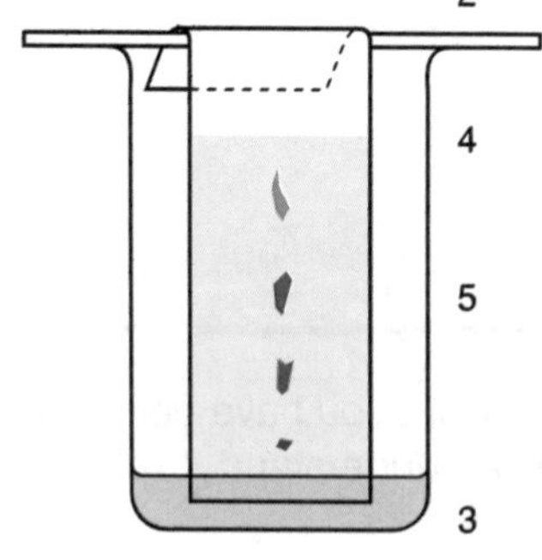


*Figure 3.9A* ▲ Paper chromatography

*Analytical chemists use many techniques. One of them is chromatography.*

Many solvents are used in chromatography. Ethanol (alcohol), ethanoic acid (vinegar) and propanone (a solvent often called acetone) are common. A chemist has to experiment to find out which solvent gives a good separation of the solutes. With a solvent other than water, a closed container should be used so that the chromatography paper is surrounded by the vapour of the solvent (see Figure 3.9B).

## SUMMARY

Chromatography can be used to separate the solutes in a solution. It is used in analysis, that is, in finding out what substances are present in a mixture.

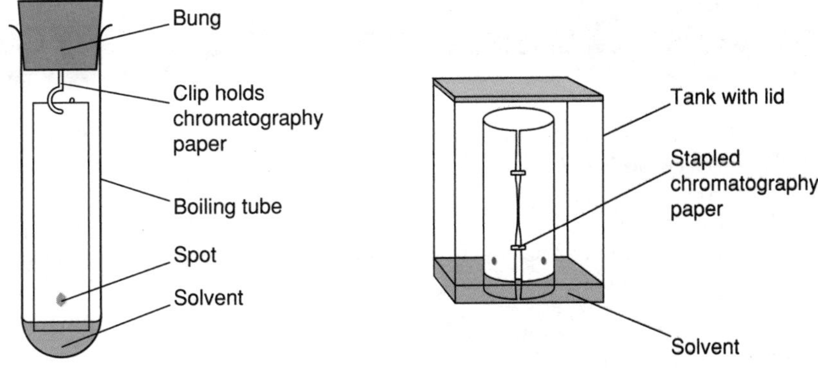


*Figure 3.9B* ▲ Chromatography with a solvent other than water

## CHECKPOINT

▸ **1** A chemist is asked to find out what substances are present in two mixtures, $M_1$ and $M_2$. He makes a chromatogram of the two mixtures and a chromatogram of some substances which he suspects may be present. The figure below shows his results.
Can you interpret his results? What substances are present in $M_1$ and $M_2$?

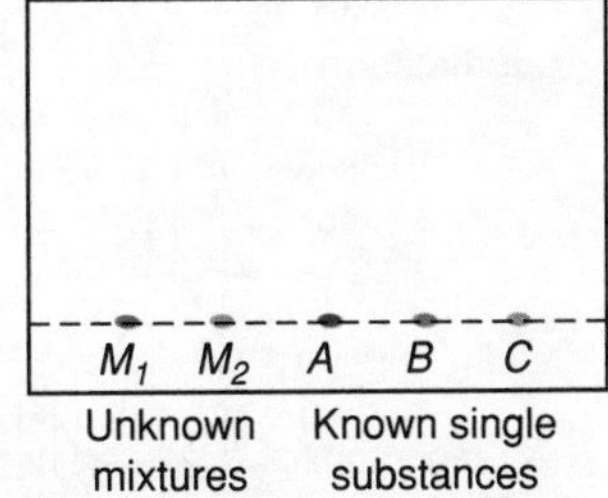


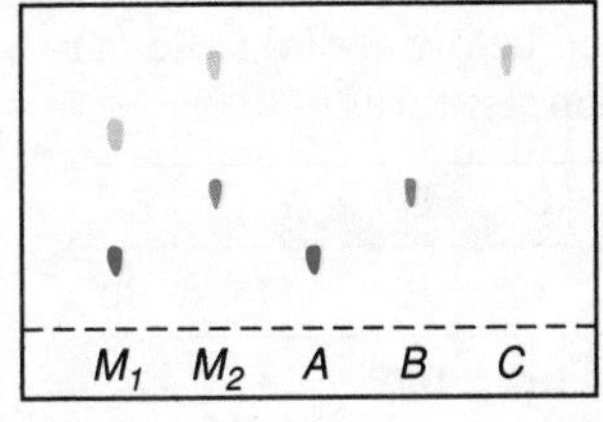

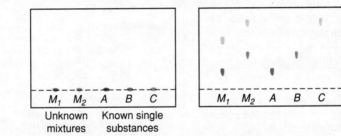

▸ **2** Imagine that you are a detective investigating forged bank notes. Your investigations take you to a house where you find a printing press and some inks. Describe how you would find out whether these inks are the same as those used to make the bank notes.

▸ **3** Dr Ecksplor discovers a strangely coloured orchid in Brazil. Professor Seeker finds an orchid of the same colour in Ecuador. The two scientists wonder whether the two flowers contain the same pigment. They feel sure that the pigment does not dissolve in water.

(a) Why do they feel sure that the pigment does not dissolve in water?

(b) Describe an experiment which they could do to find out whether the two orchids contain the same pigment.

▸ **4** An analytical chemist has the task of finding out whether the red colouring in a new food product contains any dyes which are not permitted in foods. She runs a chromatogram on the food colouring and on some of the permitted red food dyes. The figure opposite shows her results. What conclusion can you draw from these chromatograms?

Note: Dye 1 = E122 Carmoisine, Dye 2 = E162 Beetroot red, Dye 3 = E128 Red 2G, Dye 4 = E120 Cochineal, Dye 5 = E124 Ponceau 4R, Dye 6 = E160 Capsorubin

Dyes which are allowed in foods are given **E numbers**.

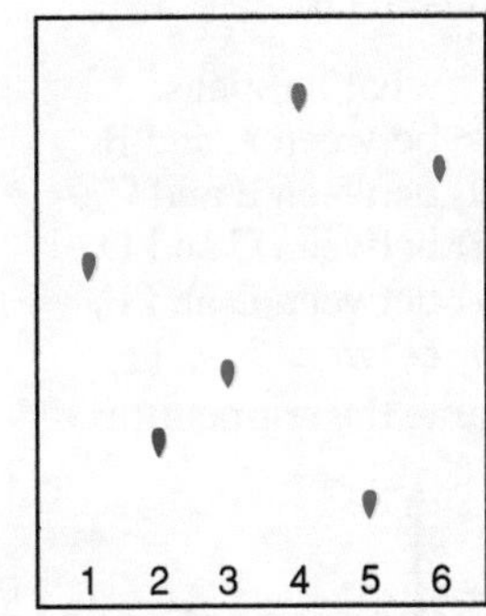


(a) Chromatogram of permitted dyes

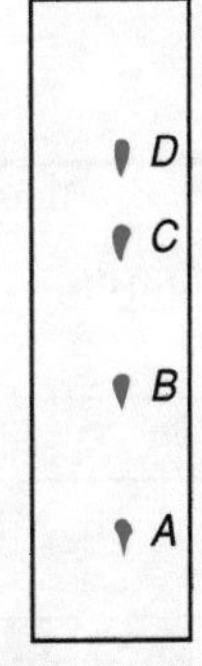


(b) Chromatogram of red food colouring

# Theme Questions

**1** A solid X melts at 58 °C. Which of the following substances could be X?
A Chloroethanoic acid, m.p. 63 °C
B Diphenylamine, m.p. 53 °C
C Ethanamide, m.p. 82 °C
D Trichloroethanoic acid, m.p. 58 °C

**2** An impure sample of a solid Y melts at 77 °C. Which of these solids could be Y?
E Dibromobenzene, m.p. 87 °C
F 1, 4-Dinitrobenzene, m.p. 72 °C
G 3-Nitrophenol, m.p. 97 °C
H Propanamide, m.p. 81 °C

**3** An impure sample of liquid Z boils at 180 °C. Which of these liquids could be Z?
I Benzoic acid, b.p. 249 °C
J Butanoic acid, b.p. 164 °C
K Hexanoic acid, b.p. 205 °C
L Methanoic acid, b.p. 101 °C

**4** Which of the substances listed below are (a) solid (b) liquid (c) gaseous at room temperature?

| *Pure substance* | A | B | C | D | E | F |
|---|---|---|---|---|---|---|
| *Melting point* (°C) | 8 | −92 | 41 | 63 | −111 | −30 |
| *Boiling point* (°C) | 101 | −21 | 182 | 189 | 11 | 172 |

**5** The graph shows how the temperature of a substance rises as it is heated. At A, the substance is a solid.
(a) Say what happens:
(i) between A and B,
(ii) between B and C,
(iii) between C and D,
(iv) between D and E,
(v) between E and F.
(b) Name the temperatures $T_1$ and $T_2$.

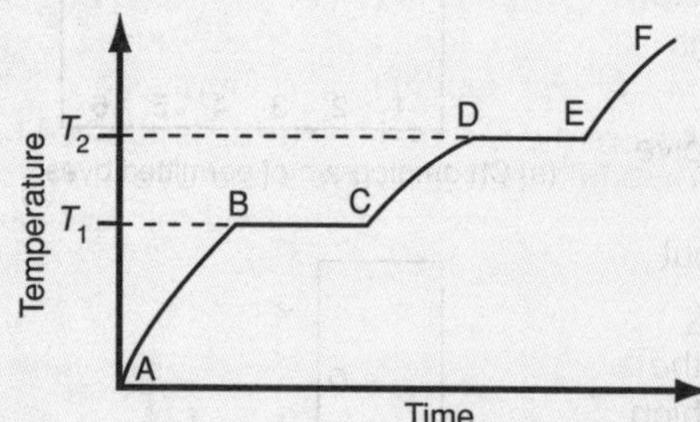


**6** The table gives the solubility of potassium nitrate at various temperatures.

| *Temperature* (°C) | 0 | 10 | 20 | 40 | 60 |
|---|---|---|---|---|---|
| *Solubility* (g per 100 g of water) | 13 | 21 | 32 | 64 | 110 |

(a) On graph paper, plot solubility (on the vertical axis) against temperature (on the horizontal axis). Draw a smooth curve through the points. Use your graph to answer (b) and (c).
(b) What is the solubility of potassium nitrate at 30 °C?
(c) At what temperature is the solubility of potassium nitrate 85 g/100 g?
(d) A 100 g mass of water is saturated with potassium nitrate at 60 °C and cooled at 30 °C. What happens?

**7** The diagram shows what happens when you breathe.

**Inhalation**

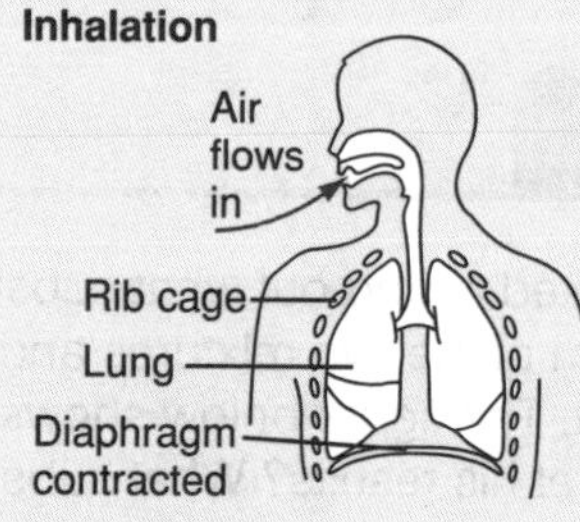


The diaphragm contracts and flattens so increasing the volume of the chest cavity.
What does this do to the air pressure in the cavity? Why does it cause air to flow into the lungs?

**Exhalation**

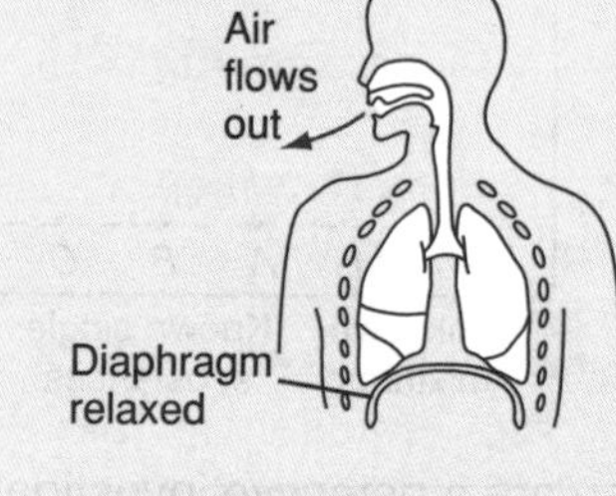


The diaphragm relaxes and pushes up into the chest cavity.
What does this do to the air pressure in the cavity? Why does it causes air to flow out of the lungs?

**8** A king in medieval times asked his goldsmith to make him a new crown, nothing fancy, just plain with no jewels. He gave the goldsmith 2.00 kg of gold. When the crown arrived, the king had it weighed. It was exactly 2.00 kg in mass. He tried it on. The crown did not look exactly the right colour, and the king wondered whether the goldsmith had kept back some of the gold and alloyed the rest with a cheaper metal to make up the mass. What could the king do to find out whether the crown was pure gold? Describe the measurements which he would have to make.

**9** (a) In which state of matter is water in (i) the sea (ii) icicles and (iii) steam?
(b) How does the kinetic theory explain the difference between the three states?
(c) In the Great Salt Lake in the USA, a person can float very easily.
(i) Explain why the high concentration of salt makes it easier to float.
(ii) Describe an experiment which you could do to find out what mass of salt is present in 100 cm³ of the lake water.

**10** (a) What sort of mixture can be separated by (i) filtration, (ii) chromatography, (iii) centrifuging?

(b) A mixture contains two liquids, A and B. A boils at 75 °C, and B boils at 95 °C. Draw an apparatus you could use to separate A and B. Label the drawing.

**11** The police are investigating a case of poison pen letters. The police chemist makes chromatograms from the ink in the letters and the ink in the pens of three suspects. The diagram below shows her results. What conclusion can you reach?

**12** Chlorophyll can be obtained from nettles. These are ground with liquid ethanol until the solution is saturated with chlorophyll. The saturated solution is separated from the rest of the nettles. Ethanol is then evaporated from the solution to leave the solid chlorophyll. Care must be taken because ethanol is flammable.

(a) (i) Suggest why the nettles are ground with ethanol rather than water.

(ii) What is meant by a *saturated solution*?

(b) Name the method used to separate the saturated solution from the rest of the nettles.

(c) The solution of chlorophyll was then evaporated using the apparatus shown below.

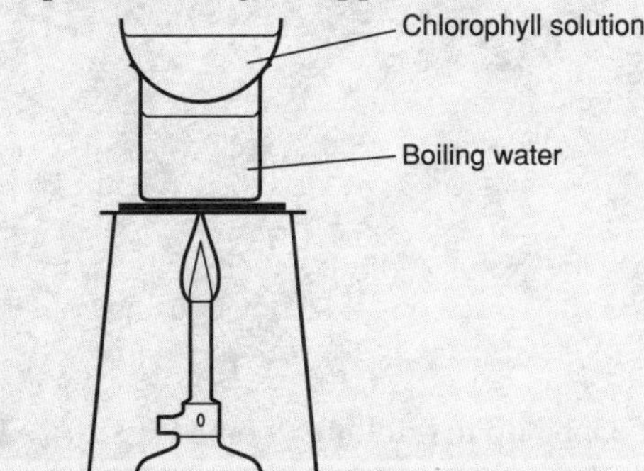


Why is the evaporating dish not heated directly with a Bunsen burner?

(d) The purity of the chlorophyll can be found by testing its melting point.

(i) Draw a diagram to show an apparatus used to find the melting point of chlorophyll.

(ii) The melting point of pure chlorophyll is 125 °C. The chlorophyll obtained in the above experiment melted at about 110 °C. What does this tell you about the purity of the sample of chlorophyll obtained?

(iii) Give a reason for your answer to (ii).

(ULEAC)

**13** Use the information in the table. Say how you could separate:

(a) aluminium and cobalt

(b) chromium and polyurethane.

| *Substance (in powder form)* | *Solubility in water* | *Solubility in ethanol* | *Magnetic properties* |
|---|---|---|---|
| Aluminium | Insoluble | Insoluble | None |
| Cobalt | Insoluble | Insoluble | Magnetic |
| Chromium | Insoluble | Insoluble | None |
| Polyurethane | Insoluble | Soluble | None |

**14** Select the correct endings to the sentences.

(a) A solid can be purified by crystallisation because

(i) it dissolves in cold water but not in hot water
(ii) it is insoluble in hot and cold water
(iii) it is very soluble in hot and cold water
(iv) it is more soluble in hot water than in cold water
(v) it is more soluble in cold water than in hot water.

(b) Two liquids can be separated by distillation if

(i) they have different densities
(ii) they form two layers
(iii) they have different boiling points
(iv) their boiling points are the same
(v) they have different freezing points.

(c) A solvent can be separated from a solution because

(i) the particles of the solvent (liquid) are smaller than those of the solute (dissolved substance)
(ii) the solvent is less dense than the solute
(iii) the solvent has a higher freezing point than the solute
(iv) the solvent evaporates more easily than the solute
(v) the solvent condenses more easily than the solute.

**15** (a) Two bottles have been accidentally knocked off a store room shelf. The two white powders were collected and the broken glass removed. Both chemicals are expensive so it is worth separating them. One, silver nitrate, is soluble in water. The other, silver chloride, is insoluble in water. Describe how you would obtain dry samples of each chemical from the mixture.

(b) Soluble substances can be separated by chromatography. This method is used to check the dyes present in food colourings.

One particular food colouring is allowed to contain only dye A, dye B and dye C. The results of an investigation of this food colouring are shown below:

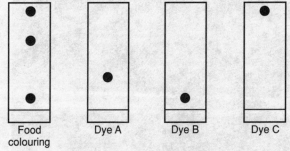

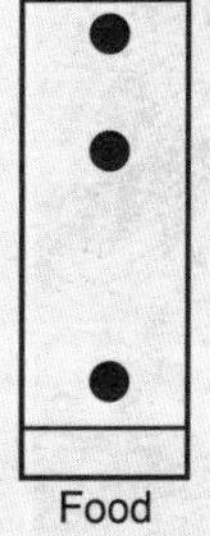
Food colouring

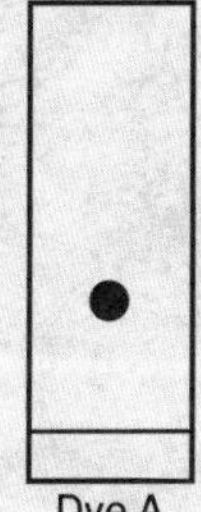
Dye A

Dye B

Dye C

(i) Which dyes are present in the food colouring?
(ii) Does the food colouring break the rules?
(iii) Explain your answer to part (ii). (WJEC)

Theme B

# The atom

'Matter is composed of atoms.'

This is what John Dalton said in 1808. What a simple statement this appears to be! Yet the complex developments that followed from this statement fill the whole of physics and chemistry.

*What are atoms?*

*How many different kinds are there?*

*How do atoms differ from even smaller particles?*

You will find some of the answers to these questions in Theme B.

# Topic 4 Elements and compounds

## 4.1 Silicon

FIRST THOUGHTS

Elements; they're elementary: they are simple substances, and there are 106 of them – all different. You have already met many useful elements, like silicon and gold.

### The tiny chip in the mighty micro

Before 1950, a computer was a massive combination of circuits and valves which took up a whole room. Nowadays, a microcomputer the size of a typewriter can do the same job as the old-style computer. Microcomputers can be fitted in aeroplanes and spacecraft. The size and weight of the old-style computers made this impossible. Many people own a personal computer, a PC, to streamline jobs such as budgeting their expenses. The change in computer size has been brought about by the use of **silicon chips**. Figure 4.1A shows an electronic circuit built on to the surface of a silicon chip. Such circuits are very reliable because they are less affected by age, moisture and vibration than the old-style circuits.

The photograph shows how the size of some silicon chips compares with the eye of a needle.

*Figure 4.1A* ▲ An integrated circuit built on the surface of a silicon chip

Silicon is an **element**. An element is a pure substance which cannot be split up into simpler substances. Elements are classified as metallic elements and non-metallic elements. Silicon is a non-metallic element. Most elements are either electrical conductors (substances which allow an electric current to flow through them) or electrical insulators (substances which do not allow an electric current to pass through them). Silicon is an unusual element in being a **semiconductor**: its behaviour is between that of a conductor and that of an insulator.

Silicon is a semiconductor; treatment with other elements allows it to conduct electricity more easily. This is how electronic circuits are created on a silicon chip.

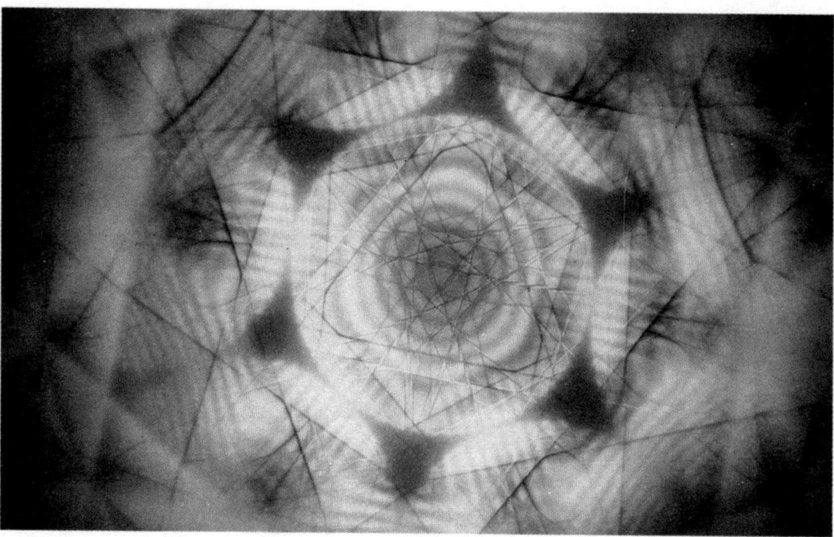

*Figure 4.1B* An electron diffraction micrograph of silicon

The first step in making a silicon chip is to slice large crystals of silicon into wafers. Then tiny areas of the wafer are treated with other elements which make silicon become an electrical conductor. The result is the creation of a thousand electronic circuits on each wafer. A chip 0.5 to 1.0 cm across contains thousands of tiny electrical switches called **transistors**. They allow on-off electric signals, which are the basis of computers, to occur at very high speeds.

The development of silicon chips has revolutionised the computer industry. Computers are now used in every kind of job: in business as word-processors and for keeping accounts and records; in science for recording measurements and doing calculations and for collecting and organising information in a database.

## Number crunching

Very difficult and long calculations can be done on a computer. Computers operate so rapidly that they can solve in minutes problems that would take weeks to compute by hand. This is why computers are used to obtain fast and accurate weather forecasts. Earth scientists use computers to record the readings of their **seismometers** all over the world. The use of computers to process information is called **information technology**.

## Tedious jobs

Keeping track of records is a job for a computer. Debiting and crediting accounts, keeping a list of the stock in a shop or factory and such jobs are handled easily by a computer because a computer can repeat the same procedure over and over without error.

## Planning ahead

Using computers can help businesses to plan ahead. They can try out various plans and see how each will affect profits. In a similar way, computers can be used to try out various approaches to environmental problems to see how each approach will affect the animal and plant populations.

## Data-logging

Computers are used in science for recording and storing measurements. A sensor which measures pH, temperature, humidity, etc., can be connected to a microcomputer. The readings are recorded and displayed as a table or as a chart or as a graph. This can be useful for taking readings over a long period of time, e.g. monitoring water pollution, monitoring weather conditions, recording the temperature in the core of a nuclear reactor, recording the light emitted by a distant star.

### It's a fact!

Plans for the car of the future include a radar set and a computer to tell the driver how far ahead the next car is. It will make fast driving much safer.

## Science at work

Jobs which once used to be done laboriously by hand are now done in a fraction of the time by computer. Computers have taken the drudgery out of a lot of jobs. They have also contributed to safety in aeroplanes and in industrial plants. The use of microcomputers has created jobs in the manufacture of computers (hardware) and the writing of computer programs which tell the computer what to do (software). The new jobs are technical jobs. Skilled people are needed to fill them.

*Figure 4.1C* ▲ Using microcomputers

Computer programs are used as sources of information on a multitude of topics; such programs are called **databases**. In forensic science, materials found at the scene of a crime are analysed. This information can be fed into a computer to become part of a database which could help the police to solve a similar crime.

### Word processing

Computers with a word processor program are used in place of typewriters. Some programs enable the user to do 'desktop publishing'. This is used to present reports for business and scientific purposes in a neat, clear, easy-to-read format.

## SUMMARY

The element silicon is a semiconductor. Silicon is used to make transistors (devices which allow current to flow in one direction only). Circuits built on to silicon chips are the basis of the microcomputer industry.

### Your science lessons

You can use microcomputers in your science lessons for various purposes:

- extracting information from databases or making your own database,
- using a spreadsheet to display results from an experiment in a table or graph,
- using computer programs to test your ideas,
- recording readings from sensors and displaying and analysing the results.

## CHECKPOINT

**1** Can you think of jobs which are done by microcomputers in;
(a) shops (b) the car industry (c) schools (d) homes?
How do you and your family benefit from the use of micros in shops and the car industry? Do you benefit from the use of micros in any other way?

**2** (a) What kinds of jobs are lost through the introduction of computers?
(b) What kinds of jobs are created by the spread of computers?
(c) Can the people who lose their jobs as a result of (a) take the jobs created in (b)? If your answer is *yes*, explain why. If your answer is *no*, explain what you think could be done to solve the problem.

**3** Would space travel have been possible before the age of the microcomputer? Explain your answer.

# 4.2 Gold

## The prospector's dream element

Gold has always held a great fascination for the human race. Gold occurs **native**, that is, as the free element, not combined with other elements. For thousands of years, people have been able to collect gold dust from river beds and melt the dust particles together to form lumps of gold. We can still see some of the objects which the goldsmiths made thousands of years ago because gold never tarnishes. For centuries, people used gold coins.

*Figure 4.2A* ▲ Panning for gold

*Figure 4.2B* ▲ Gold jewellery

Gold is a metallic element. The shine and the ability to be worked into different shapes are characteristic of metals. Unlike gold, most metals tarnish in air. Gold is not attacked by air or water or any of the other chemicals in the environment. It is used in electrical circuits when it is essential that the circuits do not corrode. For example, in microcomputers gold wires connect silicon chips to external circuits. Spacecraft use gold connections in their electrical circuits.

### SUMMARY

Gold is a metallic element. It conducts electricity and is easily worked. Unlike many other metals, gold never becomes tarnished. Gold is used for jewellery and in electrical circuits which must not corrode.

### CHECKPOINT

1. Why has gold always been a favourite metal for jewellers to work with?
2. What use is made of gold in modern technology?
3. Dentists use gold to fill teeth. Why is gold a suitable metal for this job?

# 4.3 Copper and bronze

FIRST THOUGHTS

More metallic elements and some alloys:
- Copper, bronze and the Bronze Age
- Iron, steel and the Industrial Revolution

## A step up from the Stone Age

Thousands of years ago, Stone Age humans found lumps of copper embedded in rocks. Attracted by the colour and shine of copper, they hammered it into bracelets and necklaces. Then they started to use copper to make arrowheads, spears, knives and cooking pans. They found that copper tools did not break like the stone tools they were used to. Copper knives could be ground to a sharper edge than stone, and copper dishes did not crack as pottery bowls did. The shine, the ability to be worked into different shapes and the ability to conduct heat are typical of metals. Copper is a metallic element.

Stone Age people also discovered the alloy of copper and tin called **bronze** (a mixture of metals is called an alloy). Bronze is harder than copper or tin and can be ground to a sharper edge. Bronze weapons and tools made such a difference to the way people lived that they gave their name to the Bronze Age. Thanks to the new tools hunting and farming no longer occupied all the time of all the members of the community. Some people were able to spend time on painting, making pottery and building homes. The arrival of the Bronze Age was the beginning of civilisation.

Copper is still an important metal. Copper is a good electrical conductor – a typical metal. It is easily drawn into wire. Copper wire is used in electrical circuits. Wires, cables, overhead power lines, switches and windings in electrical motors are made of copper. Half the world's production of copper (total 8 million tonnes a year) is used by electrical industries.

It's a fact!

What happened in 1886?
Three thousand miles of copper cable were laid under the Atlantic.
What for?
This was the start of the transatlantic telegraph system.

The discovery of copper and its alloy, bronze, allowed the human race to move out of the Stone Age into the Bronze Age. The science of metallurgy was born.

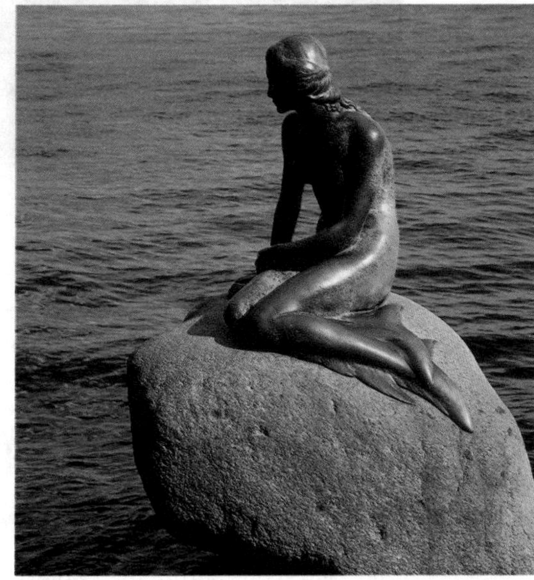

*Figure 4.3A* Bronze in use

CHECKPOINT

1 What advantages did copper tools have over tools made of (a) stone and (b) gold?

2 Explain why the discovery of bronze was so important that it gave its name to the Bronze Age.

3 What is the most important use of copper today?

## 4.4 Iron

### Our most important metal

The discovery of iron led the human race into the Iron Age.

Metallurgists were able to make harder and sharper tools from iron than from bronze.

Steel alloys were used to make the machinery that made the Industrial Revolution possible.

Our ancestors discovered how to extract iron from iron-bearing rocks over 3000 years ago. They used iron to make hammers, axes and knives, which did not break like stone tools and did not bend like bronze tools. Iron tools and weapons made such a difference to the way the human race lived that they gave their name to the Iron Age.

Many centuries later, iron and steel made the Industrial Revolution possible. The various types of **steel** are alloys of iron with carbon and other elements. Iron and steel are hard and strong: they can be hammered into flat blades and ground to a cutting edge. Our way of life in the twentieth century depends on machines made of iron and steel, buildings constructed on a framework of steel girders, and cars, lorries, trains, railways and ships made of steel.

Iron is a typical metallic element. An exceptional characteristic of iron is that it can be magnetised.

*Figure 4.4A* ▲ A steel furnace

**SUMMARY**

Iron is a metallic element. Steel is an alloy. Being hard and strong, iron and steel are used for the manufacture of tools, machinery, motor vehicles, trains and ships.

## 4.5 Carbon

### Diamond and graphite

**FIRST THOUGHTS**

Some non-metallic elements:

- Can brilliant diamond and greasy graphite be the same element?
- Can the killer, chlorine, save lives?

*Why are diamonds so often used in engagement rings?* The sparkle of diamonds comes from their ability to reflect light. The 'fire' of diamonds arises from their ability to split light into flashes of colour. The hardness of diamonds means that they can be worn without becoming scratched. Diamonds are 'for ever'. Diamond is the hardest naturally occurring substance. The only thing that can scratch a diamond is another diamond.

*Figure 4.5A* ▲ Diamonds

## It's a fact!

Diamond knives! Where? In surgery: eye surgeons use diamond knives to remove cataracts. The edges are so sharp and so even that the surgeon can make a clean cut without tearing.

The hardness of diamond finds it many uses in industry. It is able to cut through metals, ceramics, glass, stone and concrete. Diamond-tipped saws slice silicon wafers from large crystals of silicon. As well as for cutting, diamonds are used for grinding, sharpening, etching and polishing. Oil prospectors would not be able to drill through hard rock without the help of drills studded with small diamonds (see Figure 4.5B). A 20 cm bit may be studded with 60 g of small diamonds.

*Figure 4.5B* ▲ Oil rig workers using a diamond studded drill

## It's a fact!

Nineteenth century chemists thought graphite was a source of riches. All they had to do was to find the right conditions and graphite would change into diamond. It was more difficult than they expected. Not until the 1950s was a method found. Now 20 tonnes of manufactured diamonds are produced every year.

Diamond is a strange material, prized for its beauty and its usefulness. It is one form of the non-metallic element carbon.

Graphite is a shiny dark grey solid. It is so soft that it rubs off on your fingers and on paper. Pencil 'leads' contain graphite mixed with clay. Graphite is also used as a lubricant, in cars for example. Graphite is a second form of the element carbon. It is unusual in that it is the only non-metallic element which conducts electricity (see Topic 7.1).

Diamond and graphite are both pure forms of carbon. They are called **allotropes** of carbon. The existence of two or more crystalline forms of an element is called **allotropy**. Small diamonds (micron size, $10^{-6}$ m) can be made by heating graphite to a high temperature (1300 °C) under high pressure (60 000 atmospheres) for a few minutes. You can read more about the difference between the allotropes in Topic 4.8.

## SUMMARY

Diamonds are beautiful jewels. Diamond is the hardest naturally occurring material. Small diamonds are used in industry for cutting, grinding and drilling.

Graphite is a slightly shiny grey solid, which conducts electricity. Graphite is used as a lubricant and as an electrical conductor. Diamond and graphite are allotropes (pure forms) of the element, carbon.

## CHECKPOINT

1. How are small industrial diamonds made?
2. If the process for making industrial diamonds is carried on for a week, large gem-sized diamonds can be obtained. These large diamonds are dearer than natural diamonds. Why do you think this is so?
3. What uses are made of diamonds? What characteristics of diamond make it suitable for the uses you mention?
4. What is graphite used for? Why is graphite a suitable material for the uses you mention?

# 4.6 Chlorine

## It's a fact!

The man who thought up the plan of using chlorine in warfare was the German chemist, Fritz Haber. Before the war, Haber won the Nobel prize for his discovery of a method of manufacturing ammonia. Haber said, 'A man belongs to the world in time of peace but to his country in time of war'. Haber's wife was so distressed by his involvement in the war that in 1916 she committed suicide.

The British scientist, Michael Faraday, was asked to develop poison warfare during the Crimean War, but he refused.

## The life-saver

Chlorine is a killer. Its most infamous use was in the First World War when the German Army released cylinders of chlorine gas. The poisonous green cloud was driven by the wind into the trenches occupied by the British and French forces. Thousands of soldiers died, either choking on the gas or being shot as they retreated.

Chlorine is now used to kill germs. It is the bactericide which the water industry uses to make sure that our water supply is safe to drink. Chlorine has saved more lives than any other chemical. Before chlorine was used in the disinfection of the water supply, deaths from water-borne diseases, such as cholera and dysentery, were common. These diseases are still common in parts of the world which do not have safe water for drinking.

Chlorine is a non-metallic element. Many household bleaches and disinfectants contain chlorine.

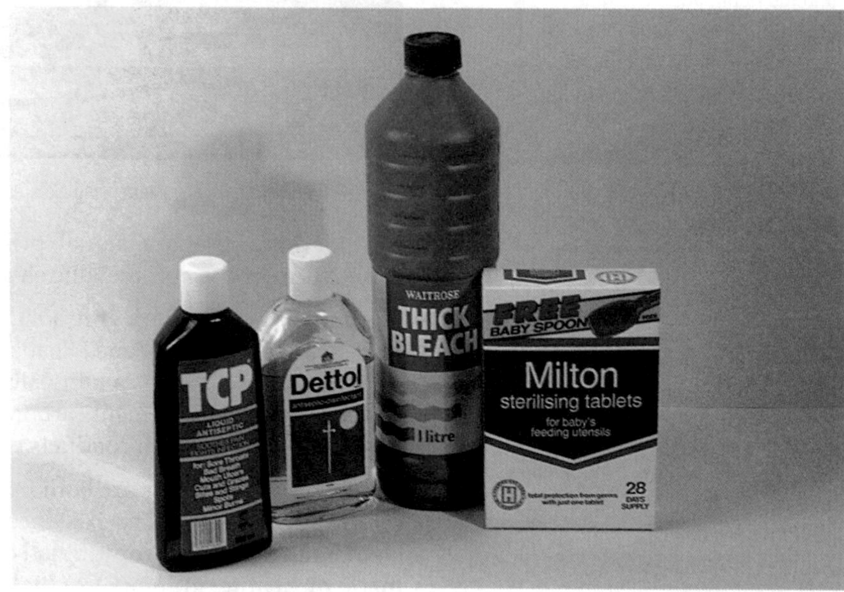


*Figure 4.6A* ▲ Bleach and disinfectant

## It's a fact!

In 1831, an epidemic of cholera hit London, and 50 000 people died. The cholera germs had been carried in the drinking water. That can't happen now that the water supply is disinfected with chlorine

# 4.7 Elements

## FIRST THOUGHTS

You will have noticed that in science there is a need to classify things: to sort them into groups of similar members. Elements can be classified as metallic and non-metallic elements, with some exceptions.

Gold, copper, and iron are typical **metallic elements**. Like all metallic elements, they conduct electricity. Many of the metallic substances we use are not elements; they are **alloys**. Steel, brass, bronze, gunmetal, solder and many others are alloys. An alloy is a combination of two or more metallic elements and sometimes non-metallic elements also. Silicon, carbon and chlorine are **non-metallic elements**. Some of their characteristics are **typical** of non-metallic elements, but some are **atypical** (not typical). Diamond is atypical in being shiny; most non-metallic elements are dull. Graphite is the only non-metallic element that conducts electricity. Silicon is one of the few semiconductors of electricity.

*Do you know how many elements there are, how many exist naturally, and how many are man-made?*

There are 92 elements found on Earth. A further 14 elements have been made by scientists. Table 4.1 summarises the characteristics of metallic elements and non-metallic elements. You will see that there are many differences. Metallic and non-metallic elements also differ in their **chemical reactions**.

SUMMARY

Table 4.1 lists the differences between metallic and non-metallic elements. Alloys are combinations of metallic elements and sometimes non-metallic elements also.

*Table 4.1* Characteristics of metallic and non-metallic elements

| Metallic elements | Non-metallic elements |
|---|---|
| Solids, except for mercury which is a liquid. | Solids and gases, except for bromine which is a liquid. |
| Hard and dense. | Most of the solid elements are softer than metals, but diamond is very hard. |
| A smooth metallic surface is shiny, but many metals tarnish in air, e.g. iron rusts. | Most are dull, but diamond is brilliant. |
| The shape can be changed by hammering: they are **malleable**. They can be pulled out into wire form: they are **ductile**. | Many solid non-metallic elements break easily when you try to change their shape. Diamond is the exception in being hard and strong. |
| Conduct heat, although highly polished surfaces reflect heat. | Poor thermal conductors. |
| Good electrical conductors. | Poor electrical conductors, except for graphite. Some are semiconductors, e.g. silicon. |
| Make a pleasing sound when struck: are **sonorous**. | Are not **sonorous**. |

# 4.8 The structures of some elements

FIRST THOUGHTS

The structure of an element means the arrangement of particles in the element. As you study this section, think about how the structure of an element affects the properties of that element.

*Can you name an element which fits each of these descriptions?*

(a) a solid metallic element
(b) a liquid metallic element
(c) a solid non-metallic element
(d) a gaseous non-metallic element
(e) a hard solid element
(f) a soft solid element
(g) a shiny element
(h) a dull element

*How do these differences arise?* The reason lies in the different arrangements of atoms in the different elements. In many elements the atoms are bonded together in groups called **molecules**.

Oxygen, chlorine and many non-metallic elements consist of individual molecules. There are strong bonds between the atoms in the molecules, but between molecules there is only a very weak attraction. The molecules move about independently, and these elements are gaseous.

Sulphur is a yellow solid. There are two forms of sulphur, which form differently shaped crystals. The crystals of **rhombic sulphur** are octahedral; those of **monoclinic sulphur** are needle-shaped. Rhombic and monoclinic sulphur are **allotropes** of sulphur: the only difference between them is the shape of their crystals (see Figure 4.8A). The reason why the crystals are shaped differently is that the sulphur molecules are packed into different arrangements in the allotropes.

## Did you know?

In 1985, new allotropes of carbon were discovered. They consist of molecules containing 30–70 carbon atoms. The allotropes are known as **fullerenes**. The structure of a fullerene molecule is a closed cage of carbon atoms. The most symmetrical structure is buckminsterfullerene, $C_{60}$, which is a perfect sphere of carbon atoms. It is named after the geodesic domes designed by the architect Buckminster Fuller. The atoms are bonded together in 20 hexagons and 12 pentagons, which fit together like those on the surface of a football. Fullerenes have been nicknamed 'bucky balls'.

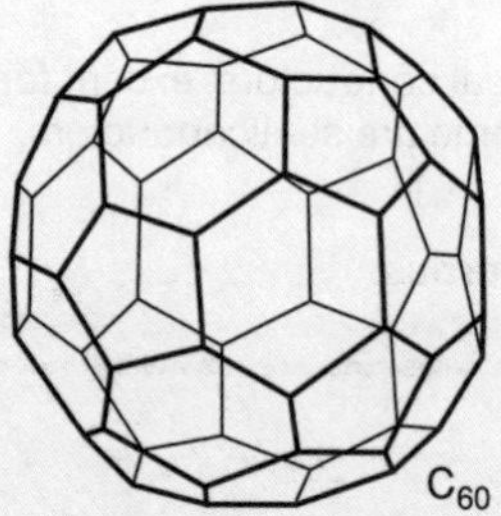
C60

Scientists believe they will find important applications as semiconductors, superconductors, lubricants, catalysts and in batteries.

## SUMMARY

The carbon atoms in diamond are joined by chemical bonds to form a giant molecular structure.

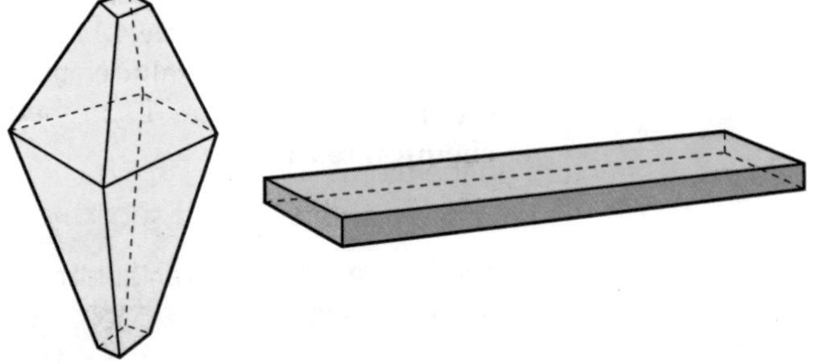

*Figure 4.8A* Allotropes of sulphur (×20)

Figure 4.8B shows a model of diamond. The structure is described as **macromolecular** or **giant molecular**. You can see that the carbon atoms form a regular arrangement. A crystal of diamond contains millions of carbon atoms arranged in this way. Every carbon atom is joined by chemical bonds to four other carbon atoms. It is very difficult to break this structure. The macromolecular structure is the source of diamond's hardness.

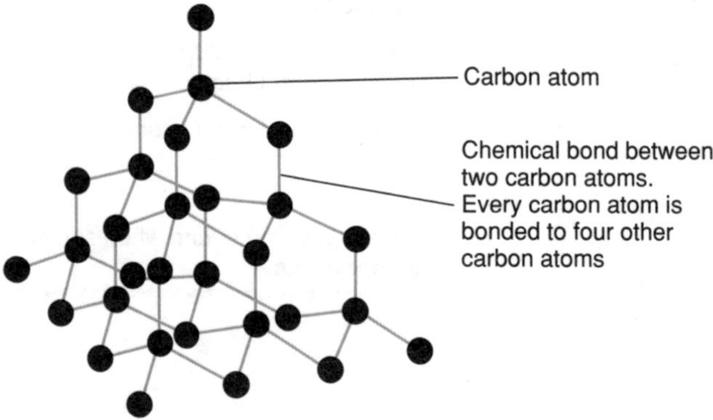
Carbon atom
Chemical bond between two carbon atoms. Every carbon atom is bonded to four other carbon atoms

*Figure 4.8B* The arrangement of carbon atoms in diamonds

Figure 4.8C shows a model of graphite. Like diamond, graphite is a crystalline solid with a macromolecular structure. Graphite has a **layer structure**. Within each layer, the carbon atoms are joined by chemical bonds. Between layers, there are only weak forces of attraction. These weak forces allow one layer to slide over the next layer. This is why graphite is soft and rubs off on your fingers. The structure of graphite enables it to be used as a lubricant and in pencil 'leads' to mark paper.

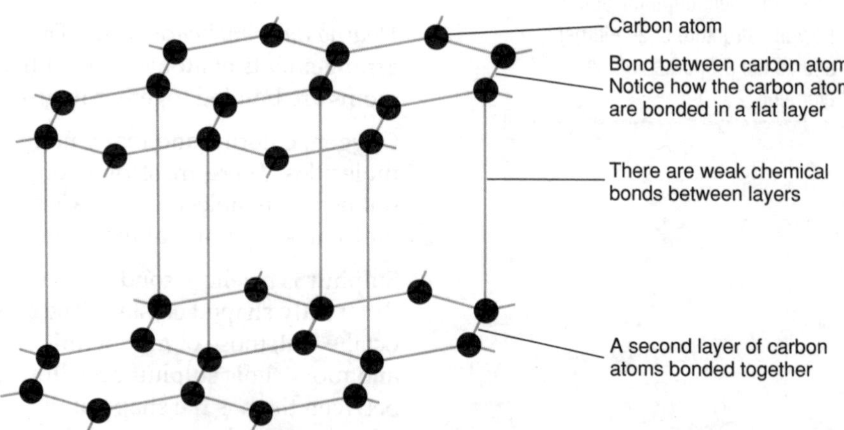
Carbon atom
Bond between carbon atoms. Notice how the carbon atoms are bonded in a flat layer
There are weak chemical bonds between layers
A second layer of carbon atoms bonded together

*Figure 4.8C* The arrangement of carbon atoms in graphite

## CHECKPOINT

1. (a) What kind of atoms are there in a diamond?
   (b) What kind of structure do the atoms form?
   (c) Why does this structure make diamond a hard substance?
   (d) What uses of diamond depend on its hardness?
   (e) Why are diamonds often chosen for engagement rings?
2. (a) Explain how its structure makes graphite less hard than diamond.
   (b) Why are diamond and graphite called allotropes?
   (c) What is graphite used for?
3. Say how the allotropes differ and how they resemble one another.

# 4.9 Compounds

**FIRST THOUGHTS**

Most of the substances you see around you are not elements: they are compounds. Compounds can be made from elements by chemical reactions.

The Verey distress rocket contains the metallic element magnesium. When it is heated – when the fuse is lit – magnesium burns in the oxygen of the air. It burns with a brilliant white flame. A white powder is formed. This powder is the compound, magnesium oxide. A change which results in the formation of a new substance is called a **chemical reaction**. A chemical reaction has taken place between magnesium and oxygen. The elements have combined to form a compound.

Magnesium + Oxygen → Magnesium oxide
Element + Element → Compound

A compound is a pure substance which contains two or more elements chemically combined. A compound of oxygen and one other element is called an **oxide**. Making a compound from its elements is called **synthesis**.

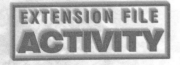

*Have you ever burned magnesium ribbon? You were synthesising magnesium oxide.*

Chemical reactions can synthesise compounds, and chemical reactions can also decompose (split up) compounds. Some compounds can be decomposed into their elements by heat. An example is silver oxide (see Figure 4.9A). The chemical reaction that takes place is

Silver oxide $\xrightarrow{\text{Heat}}$ Silver + Oxygen
Compound → Elements

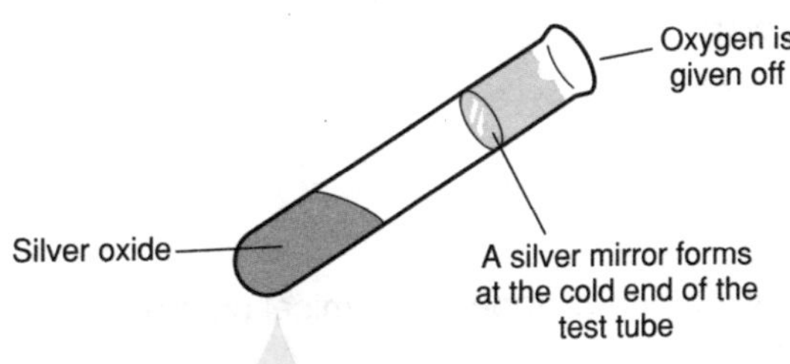


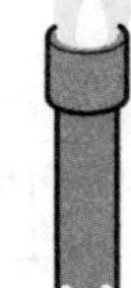

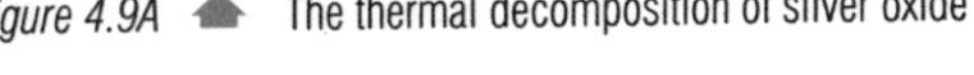

Oxygen is given off

A silver mirror forms at the cold end of the test tube

*Figure 4.9A* ▲ The thermal decomposition of silver oxide

Splitting up a compound by heat is called **thermal decomposition**.

Some compounds can be decomposed into their elements by the passage of a direct electric current. An example is sodium chloride (common salt) which is a compound of sodium and chlorine. A compound of chlorine with one other element is called a **chloride**.

Some compounds can be decomposed by heat, e.g. silver oxide.

Some can be decomposed by a direct electric current, e.g. sodium chloride (when molten) and water.

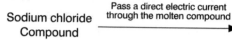

Sodium chloride Compound —Pass a direct electric current through the molten compound→ Sodium + Chlorine Elements

The chemical reaction that occurs when a compound is split up by means of electricity is called **electrolysis**.

Water is a compound. It can be electrolysed to give the elements hydrogen and oxygen (see Figure 4.9B).

Water is a compound of hydrogen and oxygen. You could call it hydrogen oxide. It is possible to make water by a chemical reaction between hydrogen and oxygen.

Water ⇌ Hydrogen + Oxygen

(forward: Electrolysis decomposes the compound into elements; reverse: This compound can be synthesised from its elements)

## SUMMARY

Elements combine to form compounds. Some compounds can be split up by heat in thermal decomposition. Some compounds are decomposed by electrolysis (the passage of a direct electric current through the molten compound or a solution of the compound). Molten common salt can be electrolysed to give the elements sodium and chlorine. Water can be electrolysed to give the elements hydrogen and oxygen.

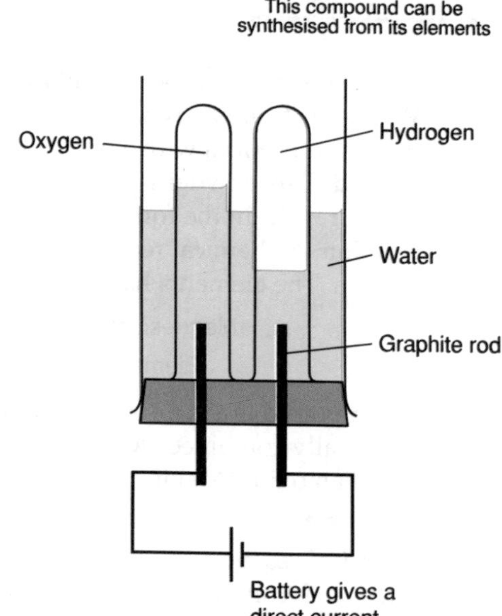


*Figure 4.9B* ▲ The decomposition of water by electrolysis

## CHECKPOINT

1 In a chemical change, a new substance is formed. In a physical change, no new substance is formed. Say which of the following changes are chemical changes.
   (a) Evaporating water.
   (b) Electrolysing water.
   (c) Melting wax.
   (d) Cracking an egg.
   (e) Boiling an egg.

2 Which of the following brings about a chemical change?
   (a) Heating magnesium.
   (b) Heating silver oxide.
   (c) Heating sodium chloride.
   (d) Passing a direct electric current through copper wire.
   (e) Passing a direct electric current through molten salt.

# 4.10 ▶ Mixtures and compounds

**FIRST THOUGHTS**

Try to explain the difference between a mixture and a compound. Then read this section, and see whether you are right.

There are a number of differences between a compound and a mixture of elements. A mixture can contain its components in any proportions. A compound has a **fixed composition**. It always contains the same elements in the same percentages by mass. You can mix together iron filings and powdered sulphur in any proportions, from 1% iron and 99% sulphur to 99% iron and 1% sulphur. When a chemical reaction takes place between iron and sulphur, you get a compound called iron(II) sulphide. It always contains 64% iron and 36% sulphur by mass. Table 4.2 summarises the differences between mixtures and compounds.

*Table 4.2* ▼ Differences between mixtures and compounds

| Mixtures | Compounds |
|---|---|
| A mixture can be separated into its parts by methods such as distillation and dissolving. | A chemical reaction is needed to split a compound into simpler compounds or into its elements. |
| No chemical change takes place when a mixture is made. | When a compound is made, a chemical reaction takes place, and often heat is given out or taken in. |
| A mixture behaves in the same way as its components. | A compound does not have the characteristics of its elements. It has a new set of characteristics. |
| A mixture can contain its components in any proportions. | A compound always contains its elements in fixed proportions by mass; for example, calcium carbonate (marble) always contains 40% calcium, 12% carbon and 48% oxygen by mass. |

**SUMMARY**

A compound is a pure substance which consists of two or more elements chemically combined. The components of a mixture are not chemically combined. Table 4.2 lists the differences between mixtures and compounds.

**CHECKPOINT**

1. Group the following into mixtures, compounds and elements: rain water, sea water, common salt, gold dust, aluminium oxide, ink, silicon, air.
2. Name an element which can be used for each of the following uses: surgical knife, pencil 'lead', wedding ring, saucepan, crowbar, thermometer, plumbing, electrical wiring, microcomputer circuit, disinfecting swimming pools, fireworks, artists' sketching material.
3. What are the differences between a mixture of iron and sulphur and the compound iron sulphide?
4. (a) What happens when you connect a piece of copper wire across the terminals of a battery? What happens when you disconnect the wire from the battery? Is the copper wire the same as before or has it changed?
   (b) What happens when you hold a piece of copper in a Bunsen flame for a minute and then switch off the Bunsen? Is the copper wire the same or different?
   (c) What type of change or changes occur in (a) and (b)?

# Topic 5 Symbols, formulas, equations

## 5.1 Symbols

FIRST THOUGHTS

Molly had a little dog
Her dog don't bark no more
'Cos what she thought was $H_2O$
Was really $H_2SO_4$

*Anon*

This rhyme shows the importance of getting your formulas right!

SUMMARY

The symbol of an element is a letter or two letters which stand for one atom of the element.

For every element there is a **symbol**. For example, the symbol for sulphur is S. The letter S stands for one atom of sulphur. Sometimes, two letters are needed. The letters Si stand for one atom of silicon: the symbol for silicon is Si. The symbol of an element is a letter or two letters which stand for one atom of the element. In some cases the letters are taken from the Latin name of the element, Ag from *argentum* (silver) and Pb from *plumbum* (lead) are examples. Table 5.1 gives a short list of symbols. There is a complete list at the end of the book.

*Table 5.1* ▼ The symbols of some carbon elements

| Element | Symbol | Element | Symbol | Element | Symbol |
|---|---|---|---|---|---|
| Aluminium | Al | Gold | Au | Oxygen | O |
| Barium | Ba | Hydrogen | H | Phosphorus | P |
| Bromine | Br | Iodine | I | Potassium | K |
| Calcium | Ca | Iron | Fe | Silver | Ag |
| Carbon | C | Lead | Pb | Sodium | Na |
| Chlorine | Cl | Magnesium | Mg | Sulphur | S |
| Copper | Cu | Mercury | Hg | Tin | Sn |
| Fluorine | F | Nitrogen | N | Zinc | Zn |

## 5.2 Formulas

*Figure 5.2A* ▲ A model of one molecule of carbon dioxide

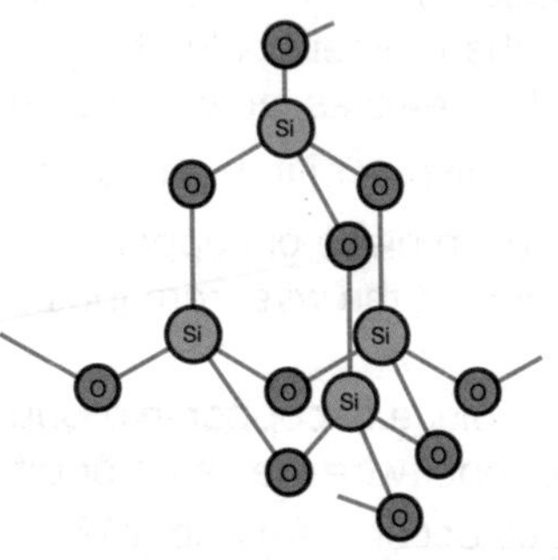


*Figure 5.2B* ▲ The structure of silicon(IV) oxide

For every compound there is a formula. The formula of a compound contains the symbols of the elements present and some numbers. The numbers show the ratio in which atoms are present. The compound carbon dioxide consists of molecules. Each molecule contains one atom of carbon and two atoms of oxygen. The formula of the compound is $CO_2$. The 2 below the line multiplies the O in front of it. To show three molecules of carbon dioxide you write $3CO_2$.

Sand is impure silicon dioxide (also called silicon(IV) oxide). It consists of macromolecules, which contain millions of atoms. There are twice as many oxygen atoms as silicon atoms in the macromolecule. The formula of silicon dioxide is therefore $SiO_2$.

The formulas of some of the compounds mentioned in this chapter are:

- Water, $H_2O$ (two H atoms and one O atom; the 2 multiplies the H in front of it).
- Sodium chloride, NaCl (one Na; one Cl).
- Silver oxide, $Ag_2O$ (two Ag: one O).
- Iron(II) sulphide, FeS (one Fe; one S).

The formula for aluminium oxide is $Al_2O_3$. This tells you that the compound contains two aluminium atoms for every three oxygen atoms. The numbers below the line multiply the symbols immediately in front of them.

The formula for calcium hydroxide is $Ca(OH)_2$. The 2 multiplies the symbols in the brackets. There are 2 oxygen atoms, 2 hydrogen atoms and 1 calcium atom. To write $4Ca(OH)_2$ means that the whole of the formula is multiplied by 4. It means 4Ca, 8O and 8H atoms. Table 5.2 lists the formulas of some common compounds.

Don't be one of those people who say that formulas are difficult!

$H_2O$ = two H atoms + one O atom

$H_2SO_4$ = two H atoms + one S atom + four O atoms

Is that difficult?

*Table 5.2 ▼ The formulas of some common compounds*

| Compound | Formula |
|---|---|
| Water | $H_2O$ |
| Carbon monoxide | CO |
| Carbon dioxide | $CO_2$ |
| Sulphur dioxide | $SO_2$ |
| Hydrogen chloride | HCl |
| Hydrochloric acid | HCl(aq) |
| Sulphuric acid | $H_2SO_4(aq)$ |
| Nitric acid | $HNO_3(aq)$ |
| Sodium hydroxide | NaOH |
| Sodium chloride | NaCl |
| Sodium sulphate | $Na_2SO_4$ |
| Sodium nitrate | $NaNO_3$ |
| Sodium carbonate | $Na_2CO_3$ |
| Sodium hydrogencarbonate | $NaHCO_3$ |
| Calcium oxide | CaO |
| Calcium hydroxide | $Ca(OH)_2$ |
| Calcium chloride | $CaCl_2$ |
| Calcium sulphate | $CaSO_4$ |
| Calcium carbonate | $CaCO_3$ |
| Calcium hydrogencarbonate | $Ca(HCO_3)_2$ |
| Copper(II) oxide | CuO |
| Copper(II) sulphate | $CuSO_4$ |
| Aluminium chloride | $AlCl_3$ |
| Aluminium oxide | $Al_2O_3$ |
| Ammonia | $NH_3$ |
| Ammonium chloride | $NH_4Cl$ |
| Ammonium sulphate | $(NH_4)_2SO_4$ |

SUMMARY

The formula of a compound is a set of symbols and numbers. The symbols show which elements are present in the compound. The numbers give the ratio in which the atoms of different elements are present.

## 5.3 Valency

Some atoms can form only one chemical bond. They can therefore combine with only one other atom. Elements with such atoms are said to have a **valency** of one. Hydrogen has a valency of one, and chlorine has a valency of one.

H—

An atom of hydrogen can form one bond
Hydrogen has a valency of 1

Cl—

An atom of chlorine can form one bond
Chlorine has a valency of 1

—O—

One atom of oxygen can form two bonds
Valency of oxygen = 2

—N— (with one bond above)

One atom of nitrogen can form three bonds
Valency of nitrogen = 3

—C— (with one bond above and one below)

One atom of carbon can form four bonds
Valency of carbon = 4

You need to know how many bonds an atom of an element can form. Then the formulas of its compounds can be worked out.

The formula of a compound depends on the valencies of the elements in the compound. When hydrogen combines with chlorine, oxygen, nitrogen and carbon, the compounds formed have the formulas:

H—Cl    H—O—H    H—N—H (with H above N)    H—C—H (with H above and below C)

These are the compounds hydrogen chloride (HCl), water ($H_2O$), ammonia ($NH_3$), and methane ($CH_4$). They have different formulas because the valencies of the elements Cl, O, N and C are different.

The photograph shows molecules of the compounds formed with hydrogen
by chlorine with 1 bond,
by oxygen with 2 bonds,
by nitrogen with 3 bonds
and by carbon with 4 bonds.

*Figure 5.3A* ▲ Models of HCl, $H_2O$, $NH_3$ and $CH_4$

Some elements have more than one valency. Sulphur, for example, forms compounds in which it has a valency of 2, e.g. $H_2S$, compounds in which it has a valency of 4, e.g. $SCl_4$, and compounds in which it has a valency of 6, e.g. $SF_6$.

SUMMARY

The formula of a compound depends on the valencies of the elements in the compound.

# 5.4 Equations

FIRST THOUGHTS

An equation tells what happens in a chemical reaction. A word equation gives the names of the reactants and the products.

A chemical equation gives their symbols and formulas.

In a chemical reaction, the starting materials, the **reactants**, are changed into new substances, the **products**. The atoms present in the reactants are not changed in any way, but the bonds between the atoms change. Chemical bonds are broken, and new chemical bonds are made. The atoms enter into new arrangements as the products are formed. Symbols and formulas give us a nice way of showing what happens in a chemical reaction. We call this way of describing a chemical reaction a **chemical equation**.

**Example 1** Copper and sulphur combine to form copper sulphide. Writing a **word equation** for the reaction

Copper + Sulphur → Copper sulphide

The arrow stands for **form**.

Writing the symbols for the elements and the formula for the compound gives the chemical equation

$$Cu + S \rightarrow CuS$$

Adding the state symbols

$$Cu(s) + S(s) \rightarrow CuS(s)$$

*Why is this called an equation?* The two sides are equal. On the left hand side, we have one atom of copper and one atom of sulphur; on the right hand side, we have one atom of copper and one atom of sulphur combined as copper sulphide. The atoms on the left hand side and the atoms on the right hand side are the same in kind and in number.

**Example 2** Calcium carbonate decomposes when heated to give calcium oxide and carbon dioxide. The word equation is

Calcium carbonate → Calcium oxide + Carbon dioxide

The chemical equation is

$$CaCO_3(s) \rightarrow CaO(s) + CO_2(g)$$

**Example 3** Carbon burns in oxygen to form the gas carbon dioxide. The word equation is

Carbon + Oxygen → Carbon dioxide

We must use the formula $O_2$ for oxygen because oxygen consists of molecules which contain two oxygen atoms. The chemical equation is

$$C(s) + O_2(g) \rightarrow CO_2(g)$$

## SUMMARY

How to write a balanced chemical equation;

- Write the word equation.
- Put in the symbols of the elements and the formulas of the compounds.
- Add the state symbols.
- Balance the equation. Do this by multiplying symbols or formulas.
  **Never** change a formula.
- Check again;
  no. of atoms of each element on LHS = no. of atoms of each element on RHS

**Example 4** Magnesium burns in oxygen to form the solid magnesium oxide.

Magnesium + Oxygen → Magnesium oxide

$$Mg(s) + O_2(g) \rightarrow MgO(s)$$

There is something wrong here! The two sides are not equal. The left hand side has two atoms of oxygen; the right hand side has only one. Multiplying MgO by 2 on the right hand side should fix it

$$Mg(s) + O_2(g) \rightarrow 2MgO(s)$$

Now there are two oxygen atoms on both sides, but there are two magnesium atoms on the right hand side and only one on the left hand side. Multiply Mg by 2

$$2Mg(s) + O_2(g) \rightarrow 2MgO(s)$$

The equation is now **a balanced chemical equation**. Check up. On the left hand side, number of Mg atoms = 2; number of O atoms = 2. On the right hand side, number of Mg atoms = 2; number of O atoms = 2. The equation is balanced.

## CHECKPOINT

1 Refer to the table of elements at the end of the book (i.e. the Periodic Table). Write down the names and symbols of the elements with atomic numbers 8, 10,20,24,38,47,50,80,82. Say what each of these elements is used for. (The term 'atomic number' will be explained in Topic 6.)

2 Give meanings of the state symbols (s), (l), (g), (aq). (See Topic 1.1 if you need to revise.)

3 Try writing balanced chemical equations for the following reactions.
(a) Zinc and sulphur combine to form zinc sulphide.
(b) Copper reacts with oxygen to form copper(II) oxide.
(c) Sulphur and oxygen form sulphur dioxide.
(d) Magnesium carbonate decomposes to form magnesium oxide and carbon dioxide.
(e) Hydrogen and copper(II) oxide form copper and water.
(f) Carbon and carbon dioxide react to form carbon monoxide.
(g) Magnesium reacts with sulphuric acid to form hydrogen and magnesium sulphate.
(h) Calcium reacts with water to form hydrogen and a solution of calcium hydroxide.
(i) Zinc reacts with steam to form hydrogen and zinc oxide.
(j) Aluminium and chlorine react to form aluminium chloride.

4 Write balanced chemical equations for the following reactions.
(a) Zinc and sulphur combine to form zinc sulphide, ZnS
(b) Copper and chlorine combine to form copper(II) chloride, $CuCl_2$
(c) Sulphur burns in oxygen to form sulphur dioxide, $SO_2$
(d) Magnesium carbonate decomposes to form magnesium oxide and carbon dioxide
(e) Calcium burns in oxygen to form calcium oxide.

5 How many atoms are present in the following?
(a) $CaCl_2$
(b) $3CaCl_2$
(c) $Cu(OH)_2$
(d) $5Cu(OH)_2$
(e) $H_2SO_4$
(f) $3H_2SO_4$
(g) $2NaNO_3$
(h) $3Cu(NO_3)_2$

# Topic 6 Inside the atom

## 6.1 Becquerel's key

FIRST THOUGHTS

This topic will give you a glimpse of the fascinating story of the discovery of radioactivity.

In 1896, a French physicist, Henri Becquerel, left some wrapped photographic plates in a drawer. When he developed the plates, he found the image of a key. The plates were 'fogged' (partly exposed). The areas of the plates which had not been exposed were in the shape of a key. Looking in the drawer, Becquerel found a key and a packet containing some uranium compounds. He did some further tests before coming to a strange conclusion. He argued that some unknown rays, of a type never met before, were coming from the uranium compounds. The mysterious rays passed through the wrapper and fogged the photographic plates. Where the key lay over the plates, the rays could not penetrate, and the image formed on the plates.

*Figure 6.1A* ▲ Becquerel's key

A young research worker called Marie Curie took up the problem in 1898. She found that this strange effect happened with all uranium compounds. It depended only on the amount of uranium present in the compound and not on which compound she used. Madame Curie realised that this ability to give off rays must belong to the 'atoms' of uranium. It must be a completely new type of change, different from the chemical reactions of uranium salts. This was a revolutionary new idea. Marie Curie called the ability of uranium atoms to give off rays **radioactivity**.

Marie Curie's husband, Pierre, joined in her research into this brand new branch of science. Together, they discovered two new radioactive elements. They called one **polonium**, after Madame Curie's native country, Poland. They called the second **radium**, meaning 'giver of rays'. Its salts glowed in the dark.

Many scientists puzzled over the question of why the atoms of these elements, uranium, polonium and radium, give off the rays which Marie Curie named radioactivity. The person who came up with an explanation was the British physicist, Lord Rutherford. In 1902 he suggested that radioactivity is caused by atoms splitting up. This was another revolutionary idea. The word 'atom' comes from the Greek word for 'cannot be divided'. When the British chemist John Dalton put forward his Atomic Theory in 1808, he said that atoms cannot be created or destroyed or split. Lord Rutherford's idea was proved by experiment to be correct. We know now that many elements have atoms which are unstable and split up into smaller atoms.

KEY SCIENTIST

The Curies worked for four years in a cold, ill-equipped shed at the University of Paris. From a tonne of ore from the uranium mine, Madame Curie extracted a tenth of a gram of uranium. The Curies published their work in research papers and exchanged information with leading scientists in Europe. A year later, they were awarded the Nobel prize, the highest prize for scientific achievement. Pierre Curie died in a road accident. Marie Curie went on with their work and won a second Nobel prize. She died in middle-age from leukaemia, a disease of the blood cells. This was caused by the radioactive materials she worked with.

# 6.2 Protons, neutrons and electrons

**FIRST THOUGHTS**

Atoms are made up of even smaller particles: protons, neutrons and electrons. As you read this section try to visualise how these particles are arranged inside the atom.

**KEY SCIENTIST**

Rutherford had an easier life in science than the Curies. He arrived in Britain from New Zealand in 1895. By the age of 28, he was a professor. He worked in the universities of Montreal in Canada, Manchester and Cambridge. He was knighted in 1914 and made Lord Rutherford of Nelson in 1931. His co-worker Otto Hahn described him as 'a very jolly man'. Many stories were told about Rutherford. He would whistle 'Onward, Christian soldiers' when the research work was going well and 'Fight the good fight' when difficulties had to be overcome.

The work of Marie and Pierre Curie, Rutherford and other scientists showed that atoms are made up of smaller particles. These **subatomic particles** differ in mass and in electrical charge. They are called **protons**, **neutrons** and **electrons** (see Table 6.1).

*Table 6.1* Sub-atomic particles

| Particle | Mass (in atomic mass units) | Charge |
|---|---|---|
| Proton | 1 | $+e$ |
| Neutron | 1 | 0 |
| Electron | 0.0005 | $-e$ |

Protons and neutrons both have the same mass. We call this mass one **atomic mass unit**, one u (1.000 u = $1.67 \times 10^{-27}$ kg). The mass of an atom depends on the number of protons and neutrons it contains. The electrons in an atom contribute very little to its mass. The number of protons and neutrons together is called the **mass number**.

Electrons carry a fixed quantity of negative electric charge. This quantity is usually written as $-e$. A proton carries a fixed charge equal and opposite to that of the electron. The charge on a proton can be written as $+e$. Neutrons are uncharged particles. Whole atoms are uncharged because the number of electrons in an atom is the same as the number of protons. The number of protons (which is also the number of electrons) is called either the **atomic number** or the **proton number**. You can see that

Number of neutrons = Mass number − Atomic (proton) number

For example, an atom of potassium has a mass of 39 u and an atomic (proton) number of 19. The number of electrons is 19, the same as the number of protons. The number of neutrons in the atom is

$$39 - 19 = 20$$

## Relative atomic mass

The lightest of atoms is an atom of hydrogen. It consists of one proton and one electron. Chemists compared the masses of other atoms with that of a hydrogen atom. They use **relative atomic mass**. The relative atomic mass, $A_r$, of calcium is 40. This means that one calcium atom is 40 times as heavy as one atom of hydrogen.

$$\text{Relative atomic mass of an element} = \frac{\text{Mass of one atom of the element}}{\text{Mass of one atom of hydrogen}}$$

## CHECKPOINT

1. Some relative atomic masses are:
   $A_r(H) = 1$, $A_r(He) = 4$, $A_r(C) = 12$, $A_r(O) = 16$, $A_r(Ca) = 40$.
   Copy and complete the following sentences.
   (a) A calcium atom is ______ times as heavy as an atom of hydrogen.
   (b) A calcium atom is ______ times as heavy as an atom of helium.
   (c) One carbon atom has the same mass as ______ helium atoms.
   (d) ______ helium atoms have the same mass as two oxygen atoms.
   (e) Two calcium atoms have the same mass as ______ oxygen atoms.
2. Element E has atomic number 9 and mass number 19. Say how many protons, neutrons and electrons are present in one atom of E.
3. State (i) the atomic number and (ii) the mass number of:
   (a) an atom with 17 protons and 18 neutrons,
   (b) an atom with 27 protons and 32 neutrons,
   (c) an atom with 50 protons and 69 neutrons.

# 6.3 The arrangement of particles in the atom

### It's a fact!

If the nucleus of an atom was the size of a cricket ball, the nearest electron would be in the stands.

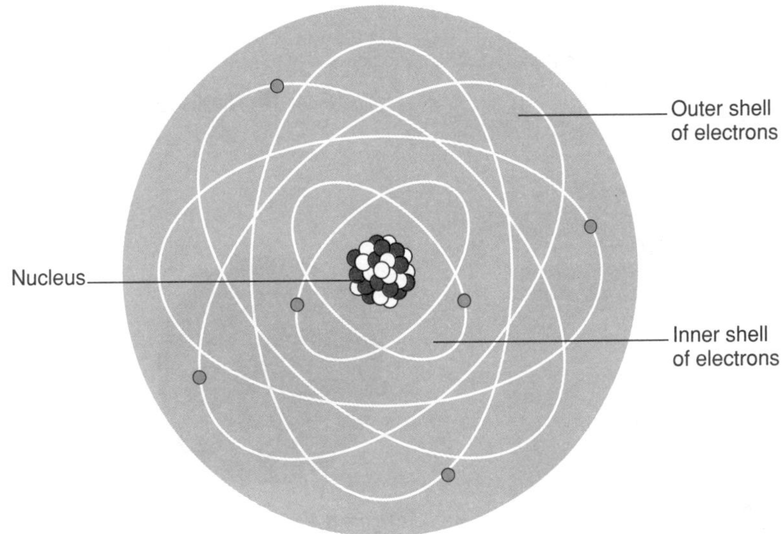


*Figure 6.3A* ▲ The structure of an atom

Lord Rutherford showed, in 1914, that most of the volume of an atom is space. Only protons and electrons were known in 1914; the neutron had not yet been discovered. Rutherford pictured the massive particles, the protons, occupying a tiny volume in the centre of the atom. Rutherford called this the **nucleus**. We now know that the nucleus contains neutrons as well as protons. The electrons occupy the space outside the nucleus. The nucleus is minute in volume compared with the volume of the atom.

The electrons of an atom are in constant motion. They move round and round the nucleus in paths called **orbits**. The electrons in orbits close to the nucleus have less energy than electrons in orbits distant from the nucleus.

## 6.4 How are the electrons arranged?

Electrons move round the nucleus in paths called orbits. A group of orbits of similar energy is called a shell.

Electrons need more energy to move in distant orbits than in orbits close to the nucleus.

The number of electrons in an atom = atomic number

Figure 6.4A illustrates the electrons of an atom in their orbits. The orbits are grouped together in **shells**. A shell is a group of orbits with similar energy. The shells distant from the nucleus have more energy than those close to the nucleus. Each shell can hold up to a certain number of electrons. In any atom, the maximum number of electrons in the outermost group of orbits is eight.

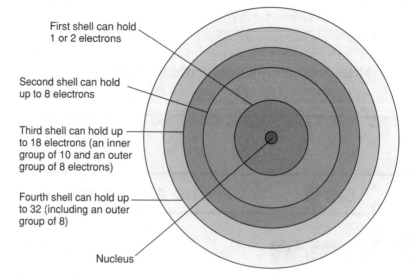

First shell can hold 1 or 2 electrons

Second shell can hold up to 8 electrons

Third shell can hold up to 18 electrons (an inner group of 10 and an outer group of 8 electrons)

Fourth shell can hold up to 32 (including an outer group of 8)

Nucleus

*Figure 6.4A* ▲ Shells of electron orbits in an atom

The diagram shows the arrangement of electrons in the different shells of the oxygen atom and the sodium atom. These arrangements are called electronic configurations.

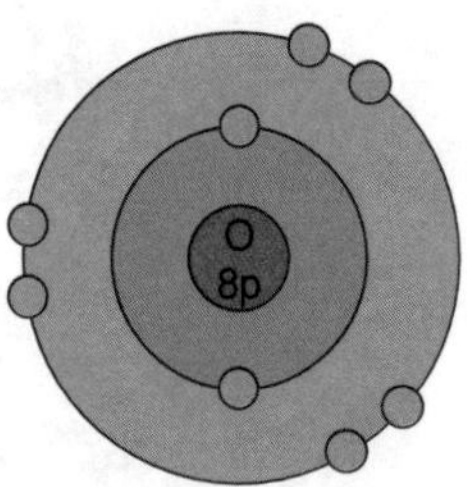


*Figure 6.4B* ▲ The arrangement of electrons in the oxygen atom

The atomic number of an element tells you the number of electrons in an atom of the element. The electrons fill the innermost orbits in the atom first. An atom of oxygen has 8 electrons. Two electrons enter the first shell, which is then full. The other 6 electrons go into the second shell (see Figure 6.4B).

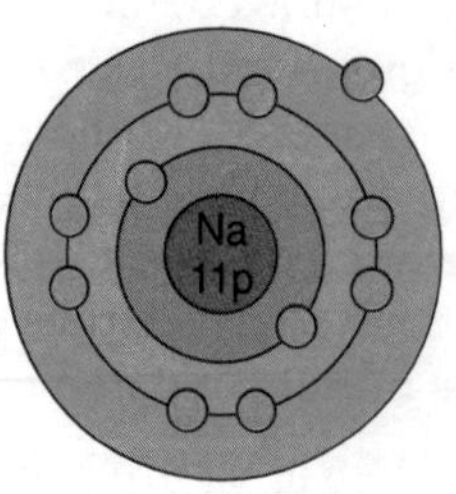


*Figure 6.4C* ▲ The electron configuration of sodium

An atom of sodium has atomic number 11. The first shell is filled by 2 electrons, the second shell is filled by 8 electrons, and 1 electron occupies the third shell. The arrangement of electrons can be written as (2.8.1). It is called the **electron configuration** of sodium. Table 6.2 gives the electron configurations of the first 20 elements.

## SUMMARY

The atoms of all elements are made up of three kinds of particles. These are:

- protons, of mass 1 u and electric charge $+e$
- neutrons, of mass 1 u, uncharged
- electrons, of mass 0.0005 u and electric charge $-e$.

The protons and neutrons make up the nucleus at the centre of the atom. The electrons circle the nucleus in orbits. Groups of orbits with the same energy are called shells. The 1st shell can hold 2 electrons; the 2nd shell can hold 8 electrons; the 3rd shell can hold 18 electrons. The arrangement of electrons in an atom is called the electron configuration.

*Table 6.2* Electron configurations of the atoms of the first twenty elements

| Element | Symbol | Atomic (proton) number | Number of electrons in ... 1st shell | 2nd shell | 3rd shell | 4th shell | Electron configuration |
|---|---|---|---|---|---|---|---|
| Hydrogen | H | 1 | 1 | | | | 1 |
| Helium | He | 2 | 2 | | | | 2 |
| Lithium | Li | 3 | 2 | 1 | | | 2.1 |
| Beryllium | Be | 4 | 2 | 2 | | | 2.2 |
| Boron | B | 5 | 2 | 3 | | | 2.3 |
| Carbon | C | 6 | 2 | 4 | | | 2.4 |
| Nitrogen | N | 7 | 2 | 5 | | | 2.5 |
| Oxygen | O | 8 | 2 | 6 | | | 2.6 |
| Fluorine | F | 9 | 2 | 7 | | | 2.7 |
| Neon | Ne | 10 | 2 | 8 | | | 2.8 |
| Sodium | Na | 11 | 2 | 8 | 1 | | 2.8.1 |
| Magnesium | Mg | 12 | 2 | 8 | 2 | | 2.8.2 |
| Aluminium | Al | 13 | 2 | 8 | 3 | | 2.8.3 |
| Silicon | Si | 14 | 2 | 8 | 4 | | 2.8.4 |
| Phosphorus | P | 15 | 2 | 8 | 5 | | 2.8.5 |
| Sulphur | S | 16 | 2 | 8 | 6 | | 2.8.6 |
| Chlorine | Cl | 17 | 2 | 8 | 7 | | 2.8.7 |
| Argon | Ar | 18 | 2 | 8 | 8 | | 2.8.8 |
| Potassium | K | 19 | 2 | 8 | 8 | 1 | 2.8.8.1 |
| Calcium | Ca | 20 | 2 | 8 | 8 | 2 | 2.8.8.2 |

## CHECKPOINT

1. Silicon has the electron configuration (2.8.4). What does this tell you about the arrangement of electrons in the atom? Sketch the arrangement. (See Figures 6.4B and 6.4C for help.)
2. Sketch the arrangement of electrons in the atoms of (a) He (b) C (c) F (d) Al (e) Mg. (See Table 6.2 for atomic numbers.)
3. Copy this table, and fill in the missing numbers.

| Particle | Mass number | Atomic number | Number of... protons | neutrons | electrons |
|---|---|---|---|---|---|
| Nitrogen atom | 14 | 7 | – | – | – |
| Sodium atom | 23 | – | – | – | 11 |
| Potassium atom | 39 | – | 19 | – | – |
| Uranium atom | 235 | 92 | – | – | – |

# 6.5 Isotopes

FIRST THOUGHTS

There are two sorts of chlorine atom and three sorts of hydrogen atom. You can find out the difference between them in this section.

Atoms of the same element all contain the same number of protons, but the number of neutrons may be different. Forms of an element which differ in the number of neutrons in the atom are called **isotopes**. For example, the element chlorine, with relative atomic mass 35.5, consists of two kinds of atom with different mass numbers.

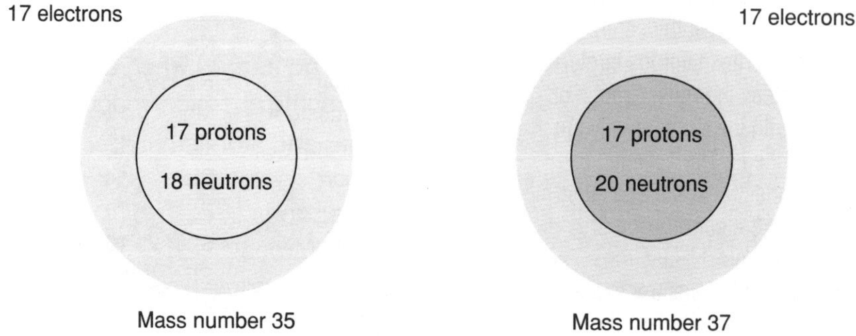


*Figure 6.5A* The isotopes of chlorine

Since the chemical reactions of an atom depend on its electrons, all chlorine atoms react in the same way. The number of neutrons in the nucleus does not affect chemical reactions. The different forms of chlorine are isotopes. Their chemical reactions are the same. In any sample of chlorine, there are three chlorine atoms with mass 35 u for each chlorine atom with mass 37 u so the average atomic mass is

$$\frac{(3 \times 35) + 37}{4} = 35.5\,\text{u}$$

This is why the relative atomic mass of chlorine is 35.5.

Isotopes are shown as:

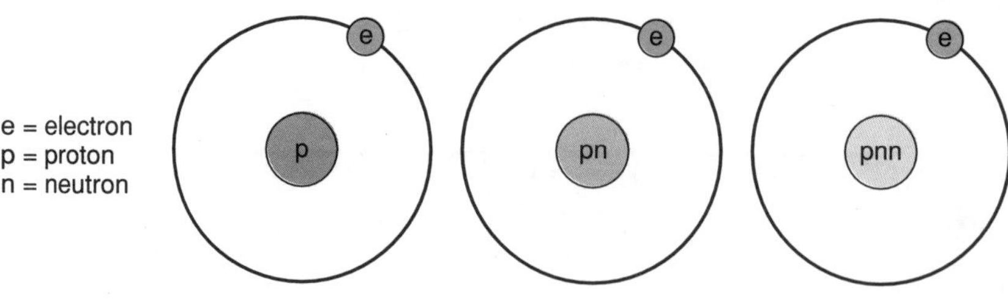


*Figure 6.5B* The isotopes of hydrogen

The isotopes of chlorine are written as $^{35}_{17}Cl$ and $^{37}_{17}Cl$. The isotopes of hydrogen are $^{1}_{1}H$, $^{2}_{1}H$ and $^{3}_{1}H$. They are often referred to as hydrogen-1, hydrogen-2 and hydrogen-3. Hydrogen-2 is also called deuterium, and hydrogen-3 is also called tritium. Carbon has three isotopes, carbon-12, carbon-13 and carbon-14.

SUMMARY

Isotopes are atoms of the same element which differ in the number of neutrons. They contain the same number of protons (and therefore the same number of electrons).

## CHECKPOINT

**1** Hydrogen, deuterium and tritium are isotopes.

(a) Copy and complete this sentence.

Isotopes are _____ of an element which contain the same number of _____ and _____ but different numbers of _____.

(b) Copy and complete the table.

| | *Hydrogen* | *Deuterium* | *Tritium* |
|---|---|---|---|
| *Atomic number* | | | |
| *Mass number* | | | |

(c) Write the formula of the compound formed when deuterium reacts with oxygen.

(d) Explain why isotopes have the same chemical reactions.

**2** Write the symbol with mass number and atomic number (as above) for each of the following isotopes.

(a) oxygen with 8 protons and 8 neutrons

(b) argon with 18 protons and 22 neutrons

(c) bromine with 35 protons and 45 neutrons

(d) chromium with 24 protons and 32 neutrons

**3** Two of the atoms described below have similar chemical properties. Which two are they?

Atom X contains 9 protons and 10 neutrons.

Atom Y contains 13 protons and 14 neutrons.

Atom Z contains 17 protons and 18 neutrons.

Explain your answer.

**4** Strontium-90 is a radioactive isotope formed in nuclear reactors. It can be accumulated in the human body because it follows the same chemical pathway through the body as another element X which is essential for health. After referring to the position of strontium in the Periodic Table, say which element you think is X.

**5** Sodium has the electron arrangement (2.8.1).

(a) Draw and label a diagram to show how the protons, neutrons and electrons are arranged in a sodium atom.

(b) Explain why sodium is electrically uncharged.

# Topic 7 Ions

## 7.1 Which substances conduct electricity?

FIRST THOUGHTS

Some elements and compounds conduct electricity. You can find out how they do it in this topic.

Electrical wires are often made of the metal copper. Copper is an **electrical conductor**: it allows electricity to pass through it. *Are all metals electrical conductors? Are there substances other than metals that conduct?* Figure 7.1A. shows how you can test a solid to find out whether it conducts electricity.

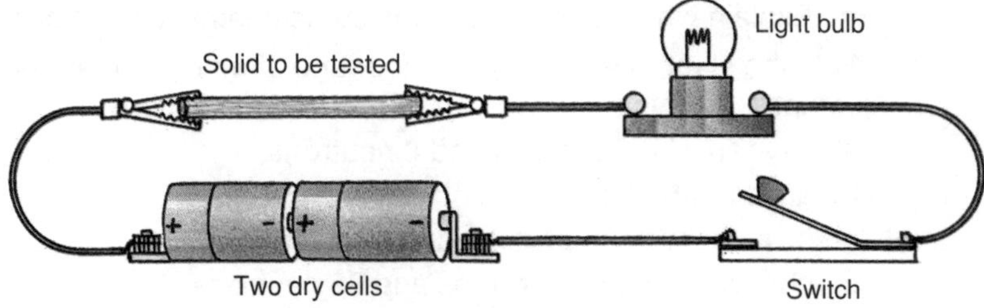


*Figure 7.1A* A testing circuit. If the solid conducts electricity, the bulb lights

The diagrams show circuits which allow electricity to pass and light a bulb ...

... if the solid conducts

... if the liquid conducts.

Figure 7.1B shows a beaker of liquid. The two graphite rods in the liquid are **electrodes**: they can conduct a direct electric current into and out of the liquid. *Draw a **circuit diagram** like Figure 7.1A showing how you could test the liquid to see whether it conducts electricity.*

You should start your study of this topic by doing some experiments on conduction. *Are your results like those in Table 7.1?*

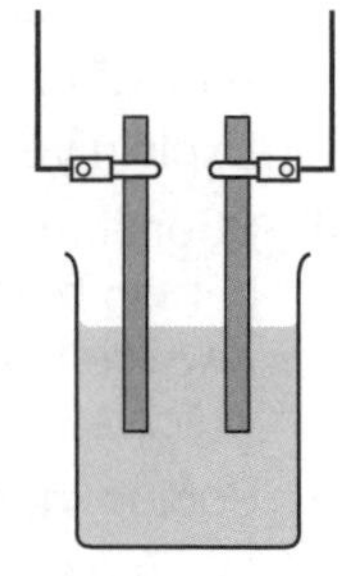

*Figure 7.1B* Test this liquid

*Table 7.1* Electrical conductors

| *Solids* | *Liquids* |
|---|---|
| Metallic elements | Mercury, the liquid metal |
| Alloys (mixtures of metals) | Solutions of acids, bases and salts |
| Graphite (a form of carbon) | Molten salts |
| (Solid compounds do not conduct.) | (Liquids such as ethanol and sugar solution do not conduct.) |

SUMMARY

- **Solids.** Metallic elements, alloys and graphite (a form of carbon) conduct electricity.
- **Liquids.** Solutions of acids, alkalis and salts conduct electricity.

## 7.2 Molten solids and electricity

When a solid, e.g. a **metal**, conducts electricity, the current is carried by electrons. The battery forces the electrons through the conductor. The metal may change, for example it may become hot, but when the current stops flowing, the metal is just the same as before. When a metal conducts electricity, no chemical reaction occurs. When a molten salt conducts electricity, chemical changes occur, and new substances are formed.

When a metal conducts electricity no permanent change occurs.

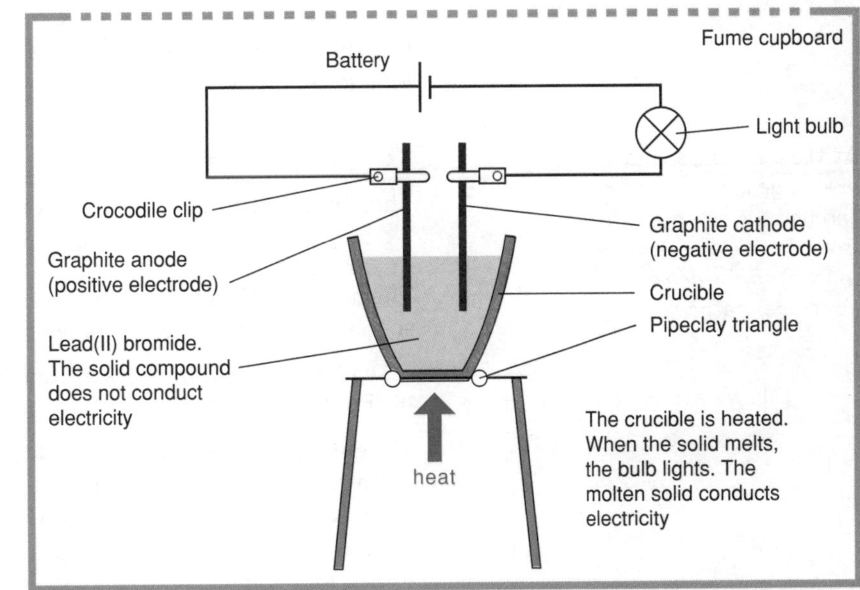


*Figure 7.2A* Passing a direct electric current through lead(II) bromide. This experiment should be done in a fume cupboard. (TAKE CARE: do not inhale the bromine vapour.)

When a molten salt conducts electricity the salt is decomposed – electrolysed – by the current. The diagram shows how molten lead(II) bromide can be electrolysed.

> Which electrode is which?
> **AnoDe**
> AD → ADD → + → positive
> The anode is positive . . .
> and the cathode is negative.

Figure 7.2A shows an experiment to find out what happens when a direct electric current passes through a molten **salt**. A salt is a compound of a metallic element with a non-metallic element or elements. Lead(II) bromide is a salt with a fairly low melting point. The container through which the current passes is called a **cell**. The rods which conduct electricity into and out of the cell are called **electrodes**. The electrode connected to the positive terminal of the battery is called the **anode**. The electrode connected to the negative terminal is called the **cathode**. The electrodes are usually made of elements such as platinum and graphite, which do not react with electrolytes.

When the salt melts, the bulb lights, showing that the molten salt conducts electricity. At the positive electrode (anode), bromine can be detected. It is a non-metallic element, a reddish-brown vapour with a very penetrating smell. (TAKE CARE: do not inhale bromine vapour.) At the negative electrode (cathode), lead is formed. After cooling, a layer of lead can be seen on the cathode.

The experiment shows that lead(II) bromide has been split up by the electric current. It has been **electrolysed**. Compounds which conduct electricity are called **electrolytes**. Remember that all substances consist of particles (see Theme A, Topic 2). Since bromine goes only to the positive electrode, it follows that bromine particles have a negative charge. Since lead appears at the negative electrode only, it follows that lead particles have a positive charge. These charged particles are called **ions**. Positive ions are called **cations** because they travel towards the cathode. Negative ions are called **anions** because they travel towards the anode.

Check that you know the meaning of:
- cell
- electrode
- anode
- cathode
- anion
- cation
- electrolysis
- electrolyte.

*How do ions differ from atoms*? A bromide ion, $Br^-$, differs from a bromine atom, Br, in having one more electron. The extra electron gives it a negative charge.

Bromine atom + Electron → Bromide ion
$$Br + e^- \rightarrow Br^-$$

A lead(II) ion differs from a lead atom by having two fewer electrons. It therefore has a double positive charge, $Pb^{2+}$.

Lead atom → Lead ion + 2 Electrons
$$Pb \rightarrow Pb^{2+} + 2e^-$$

At the electrodes ions are discharged. Electrons travel through the external circuit from the positive electrode (anode) to the negative electrode (cathode).

*What happens when the ions reach the electrodes*? They are **discharged**: they lose their charge. The positive electrode takes electrons from bromide ions so that they become bromine atoms.

Bromide ion → Bromine atom + Electron (taken by positive electrode)

$Br^-(l)$ → $Br(g)$ + $e^-$

Then bromine atoms pair up to form bromine molecules:

$$2Br(g) \rightarrow Br_2(g)$$

The negative electrode gives electrons to the positively charged lead ions so that they become lead atoms.

Lead ion + 2 Electrons (taken from the negative electrode → Lead atom

$Pb^{2+}(l)$ + $2e^-$ → $Pb(l)$

The electrons which are supplied to the anode by the discharge of bromine ions travel round the external circuit to the cathode. At the cathode, they combine with lead(II) ions (see Figure 7.2B).

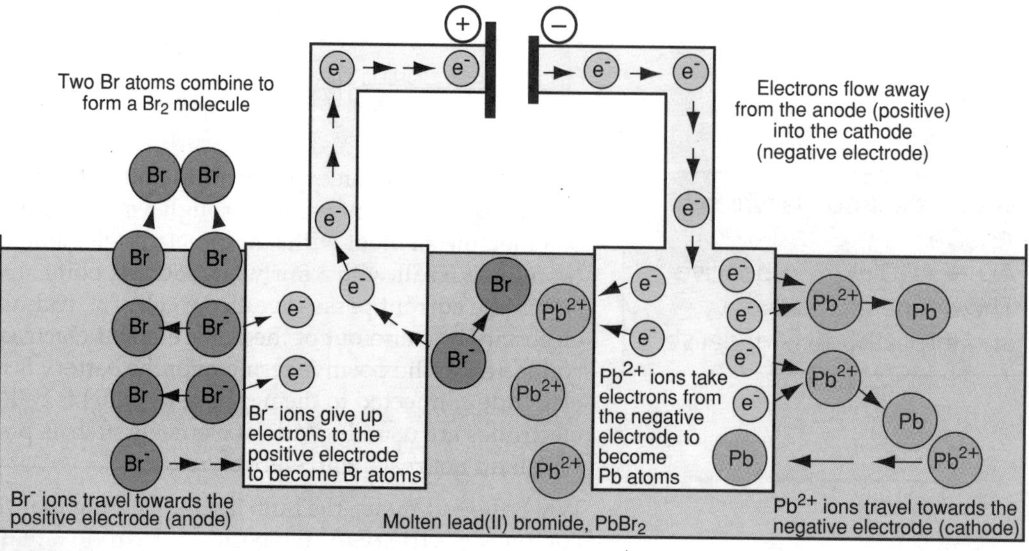
+
–
Two Br atoms combine to form a $Br_2$ molecule
Electrons flow away from the anode (positive) into the cathode (negative electrode)
$Br^-$ ions give up electrons to the positive electrode to become Br atoms
$Pb^{2+}$ ions take electrons from the negative electrode to become Pb atoms
$Br^-$ ions travel towards the positive electrode (anode)
Molten lead(II) bromide, $PbBr_2$
$Pb^{2+}$ ions travel towards the negative electrode (cathode)

*Figure 7.2B* ▲ The flow of electrons

## SUMMARY

Some compounds conduct electric current when they are molten. As they do they are electrolysed, that is, split up by the current. The explanation of electrolysis is that these compounds are composed of positive and negative ions.

Lead(II) bromide is an uncharged substance because there are two $Br^-$ ions for each $Pb^{2+}$ ions. The formula is $PbBr_2$. *Why does solid lead(II) bromide not conduct electricity*?

In the solid salt, the ions cannot move. They are fixed in a rigid three-dimensional structure. In the molten solid, the ions can move and make their way to the electrodes.

You will have noticed that lead(II) ions have two units of charge, whereas bromide ions have a single unit of charge. By experiment, it is possible to find out what charge an ion carries. Table 7.2 shows the results of such experiments.

*Table 7.2* ▼ Some common ions

| Positive ions | | | Negative ions | |
|---|---|---|---|---|
| +1 | +2 | +3 | −1 | −2 |
| Hydrogen, $H^+$ | Copper, $Cu^{2+}$ | Aluminium, $Al^{3+}$ | Bromide, $Br^-$ | Oxide, $O^{2-}$ |
| Sodium, $Na^+$ | Iron(II), $Fe^{2+}$ | Iron(III), $Fe^{3+}$ | Chloride, $Cl^-$ | Sulphide, $S^{2-}$ |
| Potassium, $K^+$ | Lead(II), $Pb^{2+}$ | | Iodide, $I^-$ | Carbonate, $CO_3^{2-}$ |
| | Magnesium, $Mg^{2+}$ | | Hydroxide, $OH^-$ | Sulphate, $SO_4^{2-}$ |
| | Zinc, $Zn^{2+}$ | | Nitrate, $NO_3^-$ | |

# 7.3 ▸ Solutions and conduction

FIRST THOUGHTS

The best way to begin this topic is by doing experiments to find out what types of solutions are electrolysed.

KEY SCIENTIST

The scientist who did the first work on electrolysis was Michael Faraday. He was born in 1791, the son of a blacksmith. He received only a very basic education. When he was apprenticed to a London book binder, books on chemistry and physics came into his hands. He found them fascinating and was led to study science in his spare time. He attended a series of lectures given by Sir Humphry Davy, the director of the Royal Institution. Faraday wrote up the lectures and illustrated them with careful diagrams. He sent the notebook to Davy, asking for some kind of work in the laboratory. Davy agreed, and Faraday soon became a capable research assistant. On Davy's retirement, Faraday became director of the Royal Institution. Faraday is famous for his discovery of electromagnetic induction and for his work on the chemical effects of electric currents. In 1834, he put forward the theory that electrolysis could be explained by the existence of charged particles of matter. We now call them *ions*.

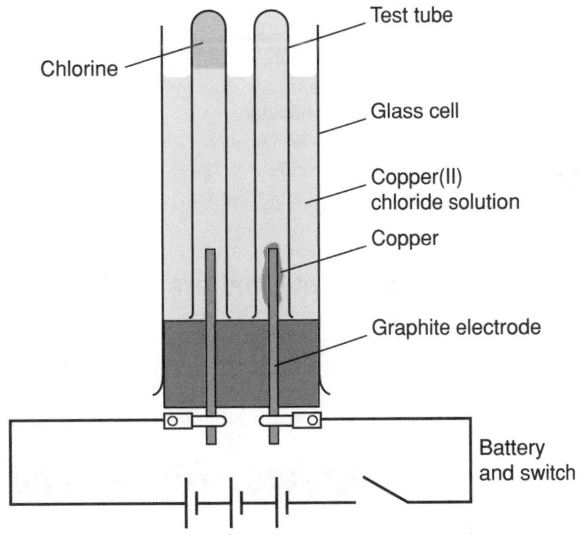


*Figure 7.3A* ▲ Electrolysis of copper(II) chloride

In Figure 7.3A a solution of the salt copper(II) chloride is being electrolysed. At the positive electrode (anode), bubbles of the gas chlorine can be seen. The negative electrode (cathode), becomes coated with a reddish brown film of copper.

*Look at Figure 7.3A. At which electrode is copper deposited? Which kind of charge must copper ions carry? At which electrode is chlorine given off? Which kind of charge must chloride ions carry?*

When they reach the electrodes, the ions are discharged.

At the negative electrode

Copper(II) ion + 2 Electrons (taken from the cathode) → Copper atom

$Cu^{2+}(aq) + 2e^- \rightarrow Cu(s)$

At the positive electrode

Chloride ion → Chlorine atom + Electron (given up at the anode)

$Cl^-(aq) \rightarrow Cl(g) + e^-$

Pairs of chlorine atoms then join to form molecules.

$2Cl(g) \rightarrow Cl_2(g)$

The electrons given to the anode by the discharge of chloride ions travel round the external circuit to the cathode. At the cathode, they combine with copper(II) ions.

To make sure you understand it, draw a diagram showing what happens to the ions and electrons in the electrolysis of copper(II) ch'oride solution. You can refer to Figure 7.2B for help.

Solid copper(II) chloride does not conduct electricity, but a solution of the salt in water does conduct. It is not the water in the solution that makes it conduct: experiment shows that water is a very poor electrical conductor. *What is the reason for the difference in behaviour between the solid and the solution?* In a solid, the ions are fixed in position, held together by strong attractive forces between positive and negative ions. In a solution of a salt, the ions are free to move (see Figure 7.3B).

The ions... which is which?
How to remember:
Current carries copper cations to cathode
**Cat**ions travel to the **cat**hode...
**An**ions travel to the **an**ode.

## SUMMARY

When a direct electric current is passed through solutions of some compounds, they are electrolysed, that is, split up by the current. Chemical changes occur in the electrolyte, and, as a result, new substances are formed.

Compounds which are electrolytes consist of positively and negatively charged particles called ions. In molten **ionic** solids and in solutions, the ions are free to move.

Water is a non-conductor because it is not ionic

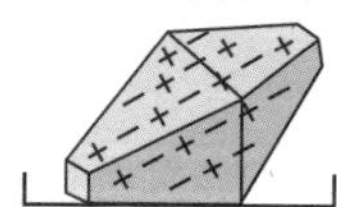
The crystalline solid is a non-conductor because the ions are held in a rigid structure

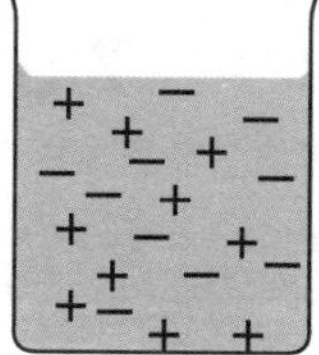
The solution is a good conductor because the ions are now free to move

*Figure 7.3B* The ions must be free to move

### Safety matters

Tap water conducts electricity. Never handle electrical equipment with wet hands. If the equipment is faulty, you run a much bigger risk of getting a lethal shock if you have wet hands.

## Non-electrolytes and weak electrolytes

Some liquids and solutions do not conduct electricity. It follows that these **non-electrolytes** consist of molecules, not ions. Some substances conduct electricity to a very slight extent. They are **weak electrolytes.** These compounds consist mainly of molecules, which do not conduct. A small fraction of the molecules split up to form ions which do conduct. Such compounds are **partially ionised**.

## CHECKPOINT

**1** A sodium atom can be written as $^{23}_{11}Na$.
The number of protons in a sodium atom is _____ and the number of electrons is _____. The charge on a sodium atom is therefore _____.
A sodium ion can be written as $^{23}_{11}Na^+$
The number of protons in a sodium ion is _____. The charge on the ion is _____ and the number of electrons is therefore _____.
The word equation for the formation of a sodium ion is:
sodium atom _____ electron _____ sodium atom
The symbol equation is: __________

**2** An atom of chlorine can be shown as $^{35}_{17}Cl$.
(a) How many (i) protons and (ii) electrons are there in a chlorine atom?
(b) What is the overall charge on a chlorine atom?
A chloride ion can be written as $^{35}_{17}Cl^-$
(c) What is the overall charge on a chloride ion?
(d) How many (i) protons and (ii) electrons does it contain?
Copy and complete the word equation for the formation of a chloride ion:
chlorine atom _____ electron _____ chloride ion
The symbol equation is: __________

**3** (a) Divide the following list into (i) electrical conductors, (ii) non-conductors.
copper, ethanol (alcohol), mercury, limewater, sugar solution, molten copper chloride, distilled water, sodium chloride crystals, molten sodium chloride, sodium chloride solution, dilute sulphuric acid

**4** Explain the words: electrolysis, electrolyte, electrode, anode, cathode, ion, anion, cation.

**5** Why are sodium chloride crystals not able to conduct electricity?

# 7.4 More examples of electrolysis

FIRST THOUGHTS

Sometimes the products formed when solutions are electrolysed are difficult to predict, but there are rules to help you.

## Sodium chloride solution

When molten sodium chloride (common salt) is electrolysed, the products are sodium and chlorine. When the aqueous solution of sodium chloride is electrolysed, the products are hydrogen (at the negative electrode) and chlorine (at the positive electrode). To explain how this happens, we have to think about the water present in the solution. Water consists of molecules, but a very small fraction of the molecules ionise into hydrogen ions and hydroxide ions.

$$\text{Water} \rightarrow \text{Hydrogen ions} + \text{Hydroxide ions}$$
$$H_2O(l) \rightarrow H^+(aq) + OH^-(aq)$$

Hydrogen ions are attracted to the negative electrode as well as sodium ions. Sodium ions are more stable than hydrogen ions. It is easier for the negative electrode to give an electron to a hydrogen ion than it is for it to give an electron to a sodium ion. Sodium ions remain in solution while hydrogen ions are discharged to form hydrogen atoms. These atoms join in pairs to form hydrogen molecules.

$$H^+(aq) + e^- \rightarrow H(g)$$
$$2H(g) \rightarrow H_2(g)$$

Can you explain how these products are formed?

- hydrogen and chlorine when sodium chloride solution is electrolysed
- copper and oxygen when copper(II) sulphate solution is electrolysed

Although the concentration of hydrogen ions in the solution is very low, it is kept topped up by the ionisation of more water molecules.

At the positive electrode, there are chloride ions and also hydroxide ions. The hydroxide ions have come, in very low concentration, from the ionisation of water molecules. Chloride ions are discharged while hydroxide ions remain in solution.

## Copper(II) sulphate solution

When copper(II) sulphate is electrolysed, copper is deposited on the cathode, and oxygen is evolved at the anode. The oxygen comes from the water in the solution. At the positive electrode, there are hydroxide ions, $OH^-$, as well as sulphate ions, $SO_4^{2-}$. The hydroxide ions have come in low concentration from the ionisation of water molecules (see above). It is easier for the positive electrode to take electrons away from hydroxide ions than from sulphate ions, and hydroxide ions are discharged. The OH groups which are formed exist for only a fraction of a second before rearranging to give oxygen and water.

$$OH^-(aq) \rightarrow OH(aq) + e^-$$
$$4OH(aq) \rightarrow 2H_2O(l) + O_2(g)$$

Although only a tiny fraction of water molecules is ionised, once hydroxide ions have been discharged, more water molecules ionise to replace them with fresh hydroxide ions.

## Dilute sulphuric acid

Figure 7.4A shows the electrolysis of dilute sulphuric acid.

The electrolysis of dilute sulphuric acid produces hydrogen and oxygen.

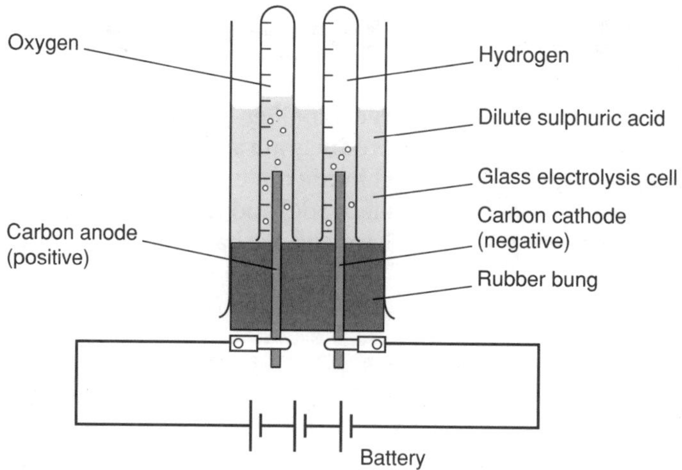


*Figure 7.4A* ▲ The electrolysis of dilute sulphuric acid

# 7.5 Which ions are discharged?

You will have seen that some ions are easier to discharge at an electrode than others.

## Cations

Some metals are more **reactive** than others. The ions of very reactive metals are difficult to discharge. Sodium is a very reactive metal. When it reacts, sodium atoms form sodium ions. It is difficult to force a sodium ion to accept an electron and turn back into a sodium atom. Hydrogen ions are discharged in preference to sodium ions. The ions of less reactive metals, such as copper and lead, are easy to discharge.

*Figure 7.5A* ▲ Sodium ions do not want to accept electrons

## Anions

Sulphate ions and nitrate ions are very difficult to discharge. When solutions of sulphates and nitrates are electrolysed, hydroxide ions are discharged instead, and oxygen is evolved.

### SUMMARY

Water is ionised to a very small extent into $H^+(aq)$ ions and $OH^-(aq)$ ions. These ions are discharged when some aqueous solutions are electrolysed. The ions of very reactive metals, e.g. $Na^+$, are difficult to discharge. In aqueous solution, hydrogen ions are discharged instead. Hydrogen is evolved at the negative electrode. The anions $SO_4^{2-}$ and $NO_3^-$ are difficult to discharge. In aqueous solution, $OH^-$ ions are discharged instead, and oxygen is evolved at the positive electrode.

## CHECKPOINT

1 Copy and complete this passage.

When molten sodium chloride is electrolysed, ______ is formed at the positive electrode and ______ is formed at the negative electrode.

When aqueous sodium chloride is electrolysed, ______ is formed at the positive electrode and ______ is formed at the negative electrode. The reaction that takes place at the positive electrode is ______________________ and the reaction that takes place at the negative electrode is ______________________.

2 A solution of potassium bromide is electrolysed. At one electrode a colourless gas is given off. At the other a brown vapour appears.

(a) At which electrode does the brown vapour appear? Which ions have been discharged?

(b) Which gas is produced at the other electrode? Which ions have been discharged?

(c) Write equations for the discharge of the ions at each electrode.

# 7.6 Applications of electrolysis

### FIRST THOUGHTS

Electrolysis is useful for:

- electroplating,
- extracting metals from their ores,
- manufacturing chemicals.

### Science at work

Do you know what happens when a music group makes a recording? Sound waves activate a machine which cuts a groove in a soft plastic disc. This disc is coated with graphite so that it will conduct. Then it is electroplated with a thick layer of metal. The metal plating is prised off the plastic disc and used to stamp out copies of the original.

## Electroplating

Some metals are prized for their strength and others for their beauty. Beautiful metals like silver and gold are costly. Often, objects made from less expensive metals are given a coating of silver or gold. The coating layer must stick well to the surface. It must be even and, to limit the cost, it should be thin. Depositing the metal by electrolysis is ideal. The technique is called **electroplating**.

Electroplating is used for protection as well as for decoration. You may have a bicycle with chromium-plated handlebars. The layer of chromium protects the steel underneath from rusting. Chromium does not stick well to steel. Steel is first electroplated with copper, which adheres well to steel, then it is nickel-plated and finally chromium-plated. The result is an attractive bright surface which does not corrode.

The rusting of iron is a serious problem. One solution to the problem is to coat iron with a metal which does not corrode. Food cans must not rust. They are made of iron coated with tin. Tin is an unreactive metal, and the juices in foods do not react with it.

*Figure 7.6A* ▲ Chromium plate

Electroplating is used to give a thin even layer of tin. A layer of zinc is applied to iron in the manufacture of 'galvanised' iron. Often, electroplating is employed.

SUMMARY

Electroplating is used to coat a cheap metal with an expensive metal. It is used to coat a metal like iron, which rusts, with a metal which does not corrode, e.g. tin or zinc.

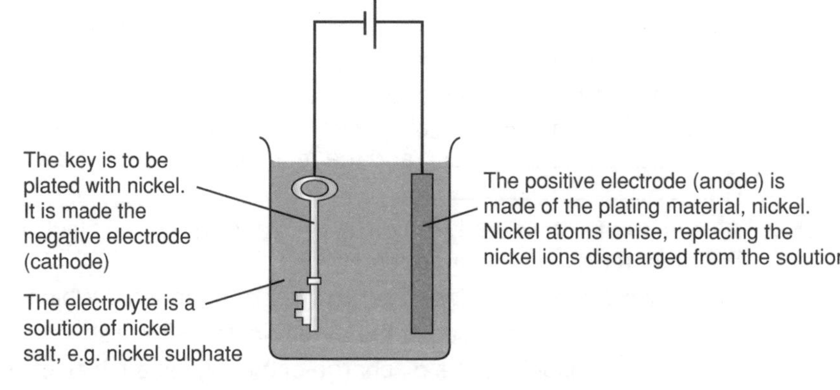

The key is to be plated with nickel. It is made the negative electrode (cathode)
The electrolyte is a solution of nickel salt, e.g. nickel sulphate
The positive electrode (anode) is made of the plating material, nickel. Nickel atoms ionise, replacing the nickel ions discharged from the solution

*Figure 7.6B* Electroplating

## Extraction of metals from their ores

Reactive metals are difficult to extract from their ores. For some of them, electrolysis is the only method which works. Sodium is obtained by the electrolysis of molten dry sodium chloride, and aluminium is obtained by the electrolysis of molten aluminium oxide.

Electrolysis can also be used as a method of purifying metals. Pure copper is obtained by an electrolytic method.

## Manufacture of sodium hydroxide

The three important chemicals, sodium hydroxide, chlorine and hydrogen are all obtained from the plentiful starting material, common salt. The method of manufacture is to electrolyse a solution of sodium chloride (common salt) in a diaphragm cell.

SUMMARY

Electrolysis is used in the extraction of some metals from their ores, e.g. sodium, aluminium, copper.

*Figure 7.6C* A room of diaphragm cells for the electrolysis of sodium chloride

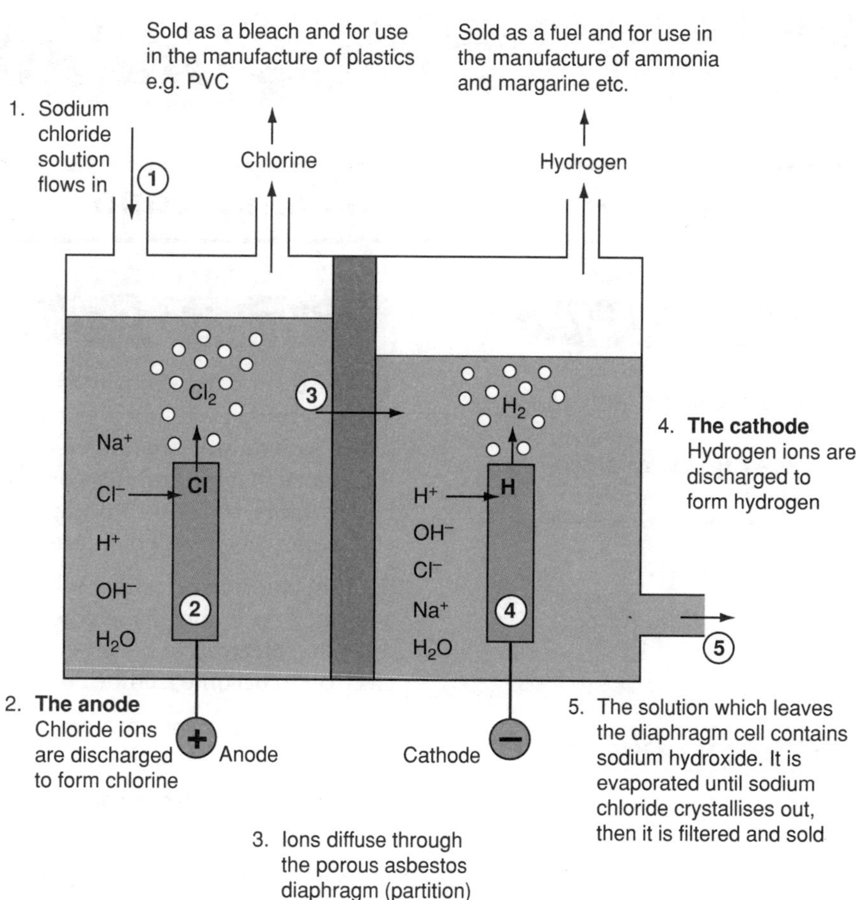

Sold as a bleach and for use in the manufacture of plastics e.g. PVC
Sold as a fuel and for use in the manufacture of ammonia and margarine etc.
1. Sodium chloride solution flows in
Chlorine
Hydrogen
$Cl_2$
$H_2$
Na+
Cl−
H+
OH−
$H_2O$
Cl
H
4. **The cathode** Hydrogen ions are discharged to form hydrogen
2. **The anode** Chloride ions are discharged to form chlorine
Anode
Cathode
5. The solution which leaves the diaphragm cell contains sodium hydroxide. It is evaporated until sodium chloride crystallises out, then it is filtered and sold
3. Ions diffuse through the porous asbestos diaphragm (partition)

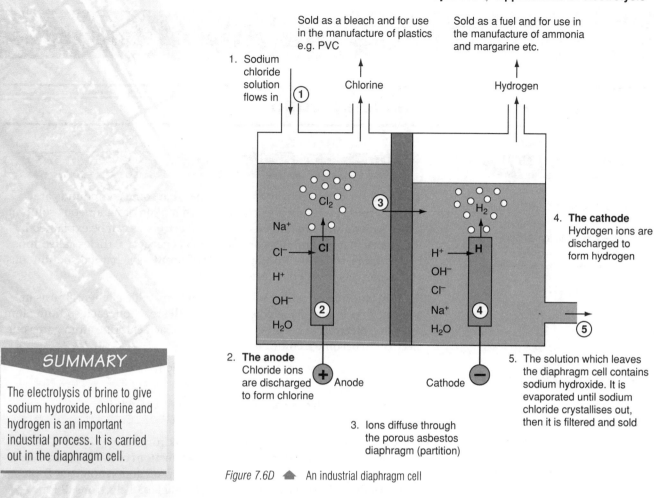

*Figure 7.6D* ▲ An industrial diaphragm cell

## SUMMARY

The electrolysis of brine to give sodium hydroxide, chlorine and hydrogen is an important industrial process. It is carried out in the diaphragm cell.

## CHECKPOINT

**1** (a) You are asked to electroplate a nickel spoon with silver. Draw the apparatus and the circuit you would use. Say what the electrodes are made of and what charge they carry. Say what electrolyte you could use.
(b) Explain why silver plating is popular.

**2** Both paint and chromium plating are used to protect parts of a car body.
(a) What advantages does paint have over chromium plating?
(c) What advantages does chromium plating have over paint?
(c) Why is chromium plating, rather than paint, used on the door handles?

**3** Many people like gold-plated watches and jewellery.
(a) What two advantages do gold-plated articles have over cheaper metals?
(b) Why do people choose gold plate rather than pure gold?
(c) What is the advantage of solid gold over gold plate?

# Topic 8 The chemical bond

## 8.1 The ionic bond

FIRST THOUGHTS

What holds the atoms in a compound together? What holds the ions in a compound together? You can find out in this topic.

### Sodium chloride

Topic 7 dealt with electrolysis. You found that some compounds are electrolytes: they conduct electricity when molten or in aqueous solution. Other compounds are non-electrolytes. Electrolysis can be explained by the theory that electrolytes consist of small charged particles called **ions**. For example, sodium chloride consists of positively charged sodium ions and negatively charged chloride ions, $Na^+ Cl^-$.

Sodium chloride is formed when sodium (a metallic element) burns in chlorine (a non-metallic element). During the reaction, each sodium atom loses one **electron** to become a sodium ion. Each chlorine atom gains one electron to become a chloride ion.

$$\text{Sodium atom} \rightarrow \text{Sodium ion} + \text{Electron}$$
$$Na \rightarrow Na^+ + e^-$$

$$\text{Chlorine atom} + \text{Electron} \rightarrow \text{Chloride ion}$$
$$Cl + e^- \rightarrow Cl^-$$

The noble gases are unreactive – possibly because each has a full outer shell of electrons.

A sodium atom reacts by giving one electron to leave a full outer shell.

A chlorine atom can accept one electron to attain a full outer shell.

Look at the arrangement of electrons in a sodium atom (Figure 8.1A). There is one electron more than there is in an atom of neon. Neon is one of the noble gases, helium, neon, argon, krypton and xenon (see Topic 11.3). All the noble gases have a full outer shell of 8 electrons (2 for helium). They are very unreactive elements; until 1965 they were thought to take part in no chemical reactions at all. It is believed that the lack of reactivity of the noble gases is due to the stability of the full outer shell of electrons. When atoms react they gain or lose or share electrons to attain an outer shell of 8 electrons.

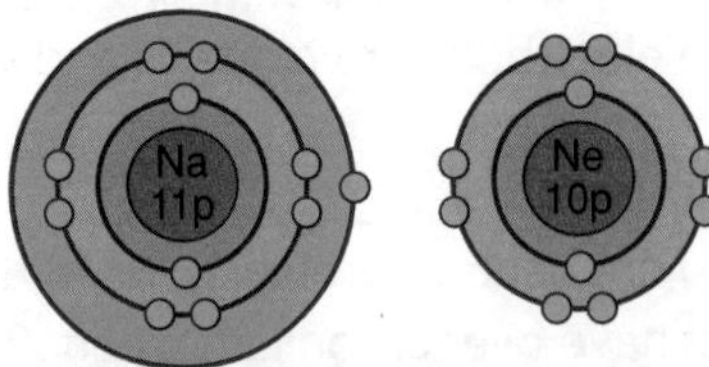


*Figure 8.1A* ▲ Atoms of sodium and neon

When a sodium atom loses the lone electron from its outermost shell, the outer shell that remains contains 8 electrons. A sodium atom cannot just shed an electron. The attraction between the nucleus and the electrons prevents this. It can, however, give an electron to a chlorine atom. A chlorine atom is prepared to accept an electron because this gives it the same electron arrangement as argon: a full outer shell (Figure 8.1B).

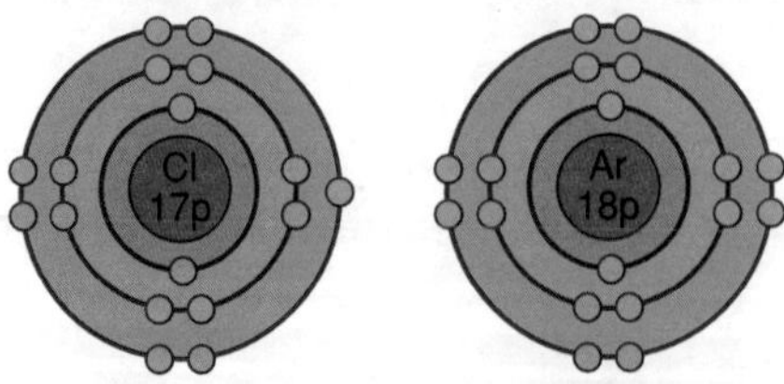


*Figure 8.1B* ▲ The arrangement of electrons in chlorine and argon

Figure 8.1C shows what happens when an atom of sodium gives an electron to an atom of chlorine. A full outer shell is left behind in sodium, and a full outer shell is created in chlorine. In the diagram, sodium electrons have been shown as x and chlorine electrons as o. This 'dot-and-cross diagram' does not imply that the electrons are different!

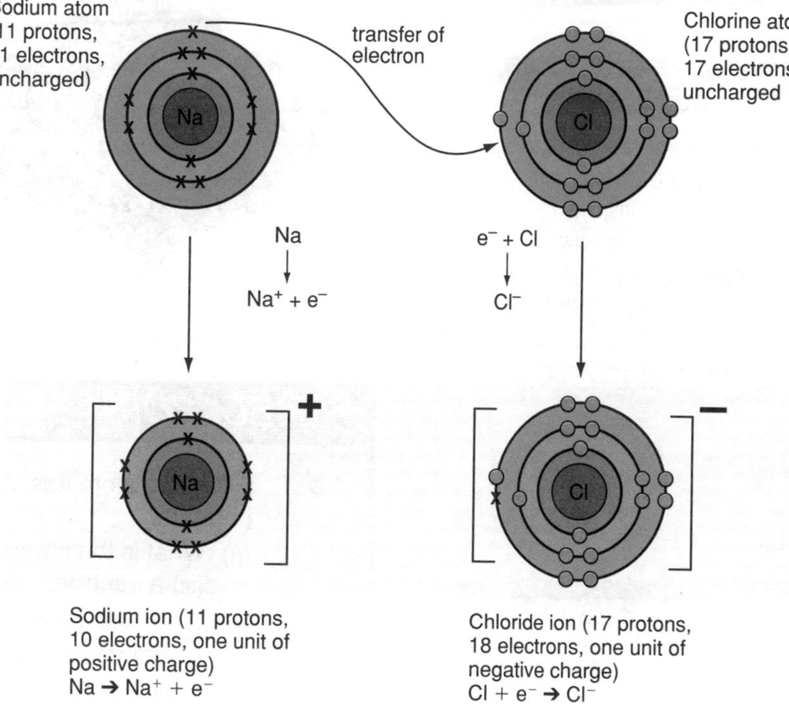


*Figure 8.1C* ▲ The formation of sodium chloride

Ions with opposite charges are formed. They are held together by electrostatic attraction in a crystal structure. If the solid is melted the ions move and conduct electricity.

When particles have opposite electric charges, a force of attraction exists between them. It is called **electrostatic attraction**. In sodium chloride, the sodium ions and chloride ions are held together by electrostatic attraction. The electrostatic attraction is the **chemical bond** in the compound, sodium chloride. It is called an **ionic bond** or **electrovalent bond**. Sodium chloride is an ionic or electrovalent compound. The compounds which conduct electricity when they are melted or dissolved are electrovalent compounds.

*Figure 8.1D* ▲ There is an attraction between sodium ions and chloride ions

A pair of ions, $Na^+ Cl^-$, does not exist by itself. It attracts other ions. The sodium ion attracts chloride ions, and the chloride ion attracts sodium ions. The result is a three-dimensional structure of alternate $Na^+$ and $Cl^-$ ions (Figure 8.1E). This is a **crystal** of sodium chloride. The crystal is uncharged because the number of sodium ions is equal to the number of chloride ions. The forces of attraction between the ions hold them in position in the structure. Since they cannot move out of their positions, the ions cannot conduct electricity. When the salt melts, the three-dimensional structure breaks down, the ions can move towards the electrodes and the molten solid can be electrolysed. There are millions of ions in even the tiniest crystal. The structures of ionic solids, such as sodium chloride, are described as **giant ionic structures**.

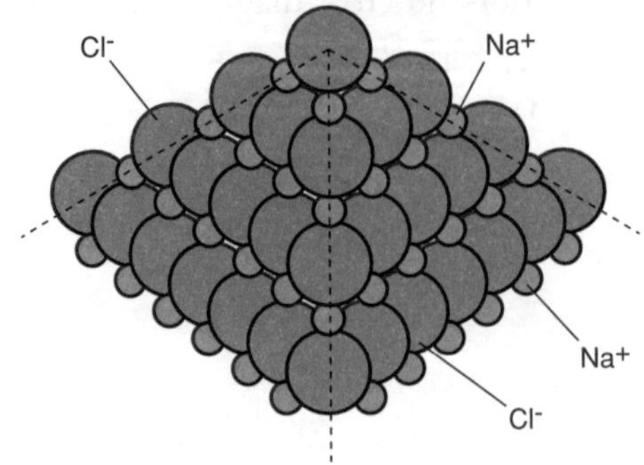


*Figure 8.1E* ▲ The structure of sodium chloride

## SUMMARY

Sodium and chlorine react to form positive sodium ions and negative chloride ions. The electrostatic attraction between these oppositely charged ions is an ionic bond or electrovalent bond.

## CHECKPOINT

1. A sodium atom has 11 electrons, 11 protons and 12 neutrons.
   (a) What is the charge on (i) an electron (ii) a proton and (iii) a neutron?
   (b) What is the overall charge on a sodium atom?
   A sodium ion has 10 electrons, 11 protons and 12 neutrons.
   (c) What is the type of charge on a sodium ion?
2. A chlorine atom has 17 protons and 17 electrons.
   (a) What is the overall charge on a chlorine atom?
   A chloride ion has 17 protons and 18 electrons.
   (b) What is the type of charge on a chloride ion?

Why is the formula of magnesium fluoride $MgF_2$? An Mg atom can lose 2 electrons to form an $Mg^{2+}$ ion. An F atom can gain one electron to form an $F^-$ ion. Therefore 2F combine with one Mg to form $Mg^{2+}\ 2F^-$ $MgF_2$.

## 8.2 Other ionic compounds

There are thousands of ionic compounds. In general, metallic elements give away electrons to form positive ions (cations). Non-metallic elements take up electrons to form negative ions (anions). Ionic compounds are formed when metallic elements give electrons to non-metallic elements.

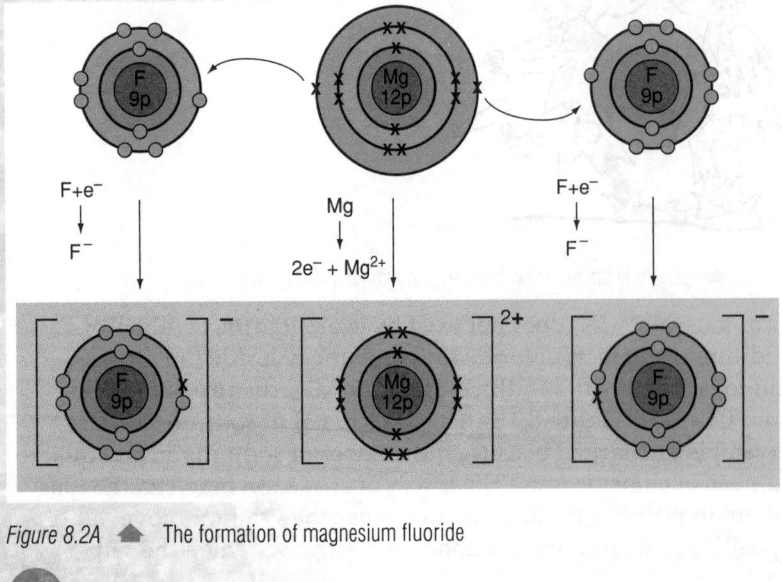

*Figure 8.2A* ▲ The formation of magnesium fluoride

### Magnesium fluoride

Magnesium (2.8.2) has two electrons in its outermost shell. It needs to lose these to gain a stable, full outer shell of 8 electrons. Fluorine (2.7) needs to gain one electron. One magnesium atom needs two fluorine atoms to accept two electrons (see Figure 8.2A). The formula of magnesium fluoride is $Mg^{2+}2F^-$ or $MgF_2$.

## Magnesium oxide

One atom of oxygen (2.6) can accept two electrons to give it the electron arrangement of neon. One atom of magnesium (2.8.2) therefore combines with one atom of oxygen (Figure 8.2B). The ions $Mg^{2+}$ and $O^{2-}$ are formed. The attraction between them is an ionic bond. The formula of magnesium oxide is $Mg^{2+}O^{2-}$ or MgO.

Why is the formula of magnesium oxide MgO? One O atom can gain 2 electrons to form $O^{2-}$. One O atom can therefore combine with one Mg atom to form $Mg^{2+}\ O^{2-}$, MgO.

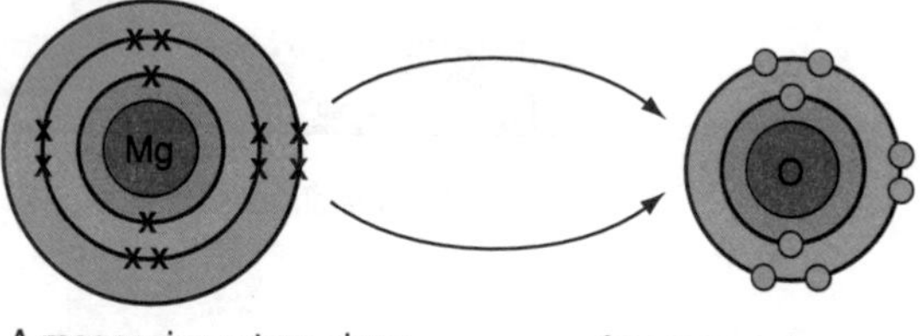


A magnesium atom gives 2 electrons to form a $Mg^{2+}$ ion.
$Mg \rightarrow Mg^{2+} + 2e^-$

An oxygen atom accepts 2 electrons to form an oxide ion, $O^{2-}$.
$O + 2e^- \rightarrow O^{2-}$

*Figure 8.2B* ▲ The formation of magnesium oxide

### CHECKPOINT

1. Copy and complete the following passage (lithium is an alkali metal in Group 1 and fluorine is a halogen in Group 7).

   Lithium and fluorine combine to form ______ fluoride. When this happens, each lithium atom gives ______ to a fluorine atom. Lithium ions with a ______ charge and fluoride ions with a ______ charge are formed. The ______ ions and ______ ions are held together by a strong ______ attraction. This ______ attraction is a chemical bond. This type of chemical bond is called the ______ bond or ______ bond. A ______ dimensional structure of ions is built up.
2. Sodium is a silvery-grey metal which reacts rapidly with water. Chlorine is a poisonous green gas. Sodium chloride is a white, crystalline solid, which is used as table salt. Explain how the sodium in sodium chloride differs from sodium metal. Explain how the chlorine in sodium chloride differs from chlorine gas.
3. Lithium has atomic number 3. Fluorine has atomic number 9. Draw a dot-and-cross diagram to show the bonding in lithium fluoride.
4. There are two differences between
   (a) a sodium ion, $Na^+$ (2.8), and a neon atom, Ne (2.8);
   (b) a chloride ion, $Cl^-$ (2.8.8), and an argon atom, Ar (2.8.8).
   What are the differences?
5. Potassium has atomic number 19. State the number of protons and electrons in one atom of potassium. Sketch the arrangement of electrons in shells in (a) an atom of potassium (b) a potassium ion.

## 8.3 ▸ Formulas of ionic compounds

### FIRST THOUGHTS

How do you work out the formula of an ionic compound? You balance the charges on the ions, and suddenly formulas make sense!

Ions are charged, but ionic compounds are uncharged. A compound has no overall charge because the sum of positive charges is equal to the sum of negative charges. In magnesium chloride, $MgCl_2$, every magnesium ion, $Mg^{2+}$, is balanced in charge by two chloride ions, $2Cl^-$.

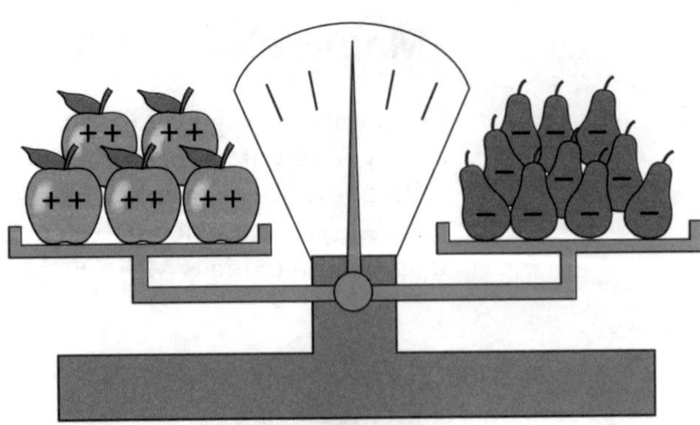


*Figure 8.3A* ▲ The charges balance

This is how you can work out the formulas of electrovalent compounds.

*Compound: Magnesium chloride*

| | |
|---|---|
| The ions present: | $Mg^{2+}$ $Cl^-$ |
| The charges must balance: | One $Mg^{2+}$ ion needs two $Cl^-$ ions. |
| Ions in the formula: | $Mg^{2+}$ and $2Cl^-$ ions |
| The formula is: | $MgCl_2$ |

*Compound: Sodium sulphate*

| | |
|---|---|
| The ions present: | $Na^+$ $SO_4^{2-}$ |
| The charges must balance: | Two $Na^+$ are needed to balance one $SO_4^{2-}$ |
| Ions in the formula: | $2Na^+$ and $SO_4^{2-}$ |
| The formula is: | $Na_2SO_4$ |

*Compound: Calcium hydroxide*

| | |
|---|---|
| The ions present: | $Ca^{2+}$ $OH^-$ |
| The charges must balance: | Two $OH^-$ ions balance one $Ca^{2+}$ ion |
| Ions in the formula: | $Ca^{2+}$ and $2OH^-$ |
| The formula is: | $Ca(OH)_2$ |

The brackets tell you that the 2 multiplies everything inside them. There are 2O atoms and 2H atoms, in addition to the 1Ca ion.

Have you grasped the idea? This exercise will help you to check.

***Ionic compounds***

Copy and complete these formulas:

*Compound: Iron(II) sulphate*

| | |
|---|---|
| The ions present: | $Fe^{2+}$ $SO_4^{2-}$ |
| The charges must balance: | One $Fe^{2+}$ ion balances __ $SO_4^{2-}$ ions |
| Ions in the formula: | ____ and ____ |
| The formula is: | ____ |

*Compound: Iron(III) sulphate*

| | |
|---|---|
| The ions present: | $Fe^{3+}$ $SO_4^{2-}$ |
| The charges must balance: | __ $Fe^{3+}$ balance __ $SO_4^{2-}$ ions |
| Ions in the formula: | __ $Fe^{3+}$ and __ $SO_4^{2-}$ |
| The formula is: | $Fe_2(SO_4)_3$ |
| | The formula contains __ Fe, __ S and __ O. |

What is the difference between iron(II) and iron(III) and between copper(I) and copper(II)?

You will notice that the sulphates of iron are named iron(II) sulphate and iron(III) sulphate. The roman numerals, II and III, show whether a compound contains iron(II) ions, $Fe^{2+}$, or iron(III) ions, $Fe^{3+}$.

Table 8.1 shows the symbols and charges of some ions. From this table, you can work out the formula of any compound containing these ions.

**SUMMARY**

The formula of an ionic compound is worked out by balancing the charges on the ions.

Table 8.1 ▼ The symbols of some common ions

| Name | Symbol | Name | Symbol |
|---|---|---|---|
| Aluminium | $Al^{3+}$ | Bromide | $Br^-$ |
| Ammonium | $NH_4^+$ | Carbonate | $CO_3^{2-}$ |
| Barium | $Ba^{2+}$ | Chloride | $Cl^-$ |
| Calcium | $Ca^{2+}$ | Hydrogencarbonate | $HCO_3^-$ |
| Copper(I) | $Cu^+$ | Hydroxide | $OH^-$ |
| Copper(II) | $Cu^{2+}$ | Iodide | $I^-$ |
| Hydrogen | $H^+$ | Nitrate | $NO_3^-$ |
| Iron(II) | $Fe^{2+}$ | | |
| Iron(III) | $Fe^{3+}$ | Oxide | $O^{2-}$ |
| Lead(II) | $Pb^{2+}$ | Phosphate | $PO_4^{3-}$ |
| Magnesium | $Mg^{2+}$ | | |
| Mercury(II) | $Hg^{2+}$ | Sulphate | $SO_4^{2-}$ |
| Potassium | $K^+$ | | |
| Silver | $Ag^+$ | Sulphide | $S^{2-}$ |
| Sodium | $Na^+$ | | |
| Zinc | $Zn^{2+}$ | Sulphite | $SO_3^{2-}$ |

**CHECKPOINT**

1 Write the formulas of the following ionic compounds.
(a) potassium chloride
(b) potassium sulphate
(c) ammonium chloride
(d) magnesium bromide
(e) copper(II) hydroxide
(f) zinc sulphate
(g) calcium carbonate
(h) aluminium chloride
(i) sodium hydrogencarbonate

2 Write the formulas of the following ionic compounds.
(a) sodium hydroxide
(b) calcium hydroxide
(c) iron(II) hydroxide
(d) iron(III) hydroxide
(e) aluminium oxide
(f) iron(III) oxide

3 Name the following.
(a) $AgBr$
(b) $AgNO_3$
(c) $Cu(NO_3)_2$
(d) $Al_2(SO_4)_3$
(e) $FeBr_2$
(f) $FeBr_3$
(g) $PbO_2$
(h) $PbO$
(i) $Zn(OH)_2$
(j) $NH_4Br$
(k) $Ca(HCO_3)_2$

# 8.4 The covalent bond

**FIRST THOUGHTS**

Some compounds are non-electrolytes because they consist of molecules, not ions. In this section, we think about the type of chemical bond in these molecular compounds.

Many compounds are non-electrolytes. They do not conduct electricity; therefore they cannot consist of ions. They contain a different kind of chemical bond: the covalent bond.

## Chlorine $Cl_2$

Two atoms of chlorine combine to form a molecule $Cl_2$. Both chlorine atoms have the electron arrangement Cl (2.8.7). Both chlorine atoms want to gain an electron to attain a full octet. They do this by sharing electrons (see Figure 8.4A). The atoms have come close enough for the outer shells to overlap. Then the shared pair of electrons can orbit round both chlorine atoms. The shared pair of electrons is a **covalent bond**.

How to attain a full outer shell of electrons; this is the problem.

Two Cl atoms manage it by sharing a pair of electrons between them in the molecule $Cl_2$.

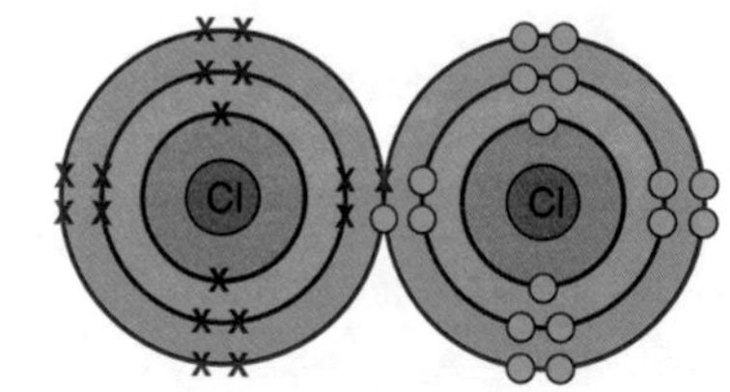


The pair of electrons ●x is shared between the 2Cl atoms. One electron comes from each Cl.

Figure 8.4A ▲ The chlorine molecule

## Hydrogen chloride HCl

In hydrogen chloride HCl two electrons are shared between the hydrogen and chlorine atoms to form a covalent bond (see Figure 8.4B).

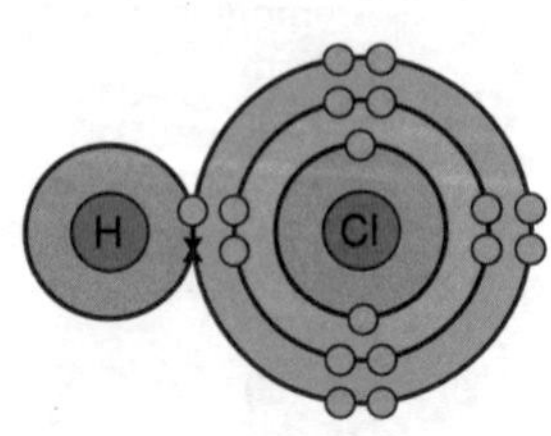
The hydrogen atom shares its electron with the chlorine atom. H now has a full shell of 2 electrons, the same arrangement as helium.
H
Cl
The chlorine atom shares one of its electrons with the hydrogen atom. Cl now has a full shell of 8 electrons, the same arrangement as argon.

*Figure 8.4B* ▲ A molecule of hydrogen chloride

The shared pair of electrons is attracted to the hydrogen nucleus and to the chlorine nucleus and bonds the two nuclei together. Other covalent molecules are illustrated in Figure 8.4C.

## Water $H_2O$

Two hydrogen atoms each share an electron with an oxygen atom. Each hydrogen atom has an outer shell of 2 electrons, and the oxygen atom has an outer shell of 8 electrons.

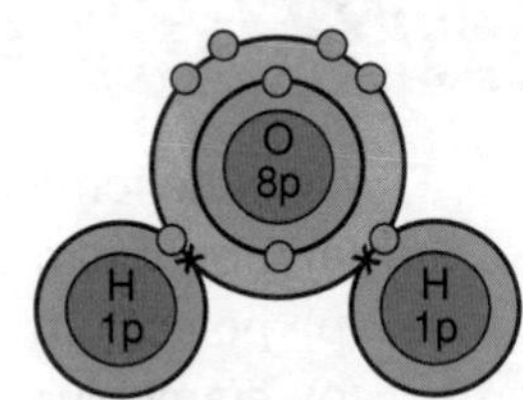
O
8p
H
1p
H
1p

## Ammonia $NH_3$

In ammonia, N shares 3 of its electrons, one with each of 3 H atoms. N now has 8 electrons in its outer shell (like Ne) and H has 2 electrons in its outer shell (like He).

Other examples of molecules with covalent bonds – shared pairs of electrons – are $H_2O$, $NH_3$, $CH_4$.

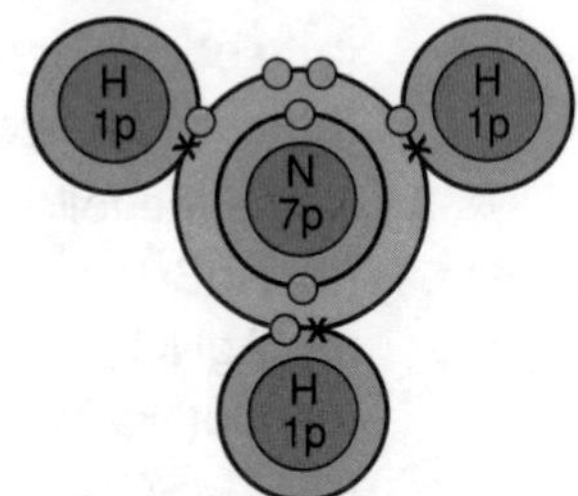
H
1p
H
1p
N
7p
H
1p

## Methane $CH_4$

The C atom shares 4 electrons, one with each of 4 H atoms.

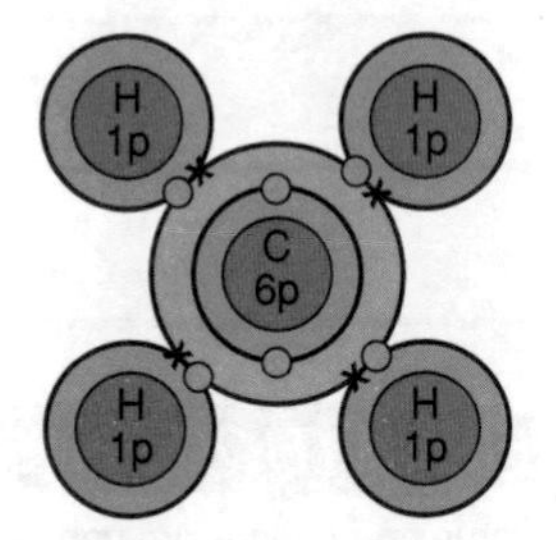
H
1p
H
1p
C
6p
H
1p
H
1p

*Figure 8.4C* ▲ $H_2O$, $NH_3$, $CH_4$

# Double covalent bonds

If two pairs of electrons are shared the bond is a **double bond** (see Figure 8.4D).

## Carbon dioxide $CO_2$

There is a double covalent bond in the $CO_2$ molecule.

The C atom shares 2 electrons with each of the 2 O atoms. Each O atom shares 2 of its electrons with the C atom. In this way 2 pairs of electrons are shared between the C atom and each O atom. Each C=O bond is called a double bond.

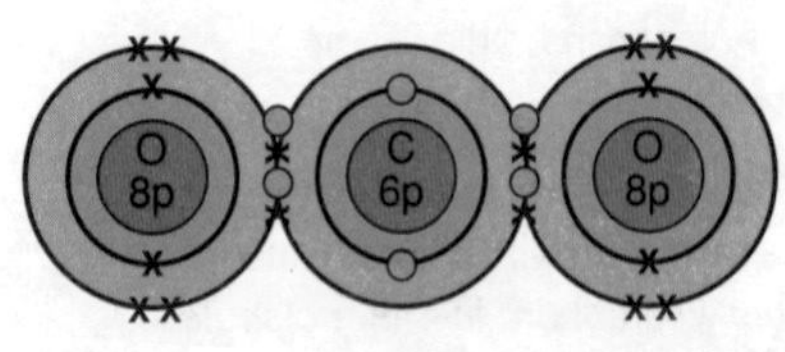
O
8p
C
6p
O
8p

*Figure 8.4D* ▲ $CO_2$

The shapes of some covalent molecules are shown in Figure 8.4E

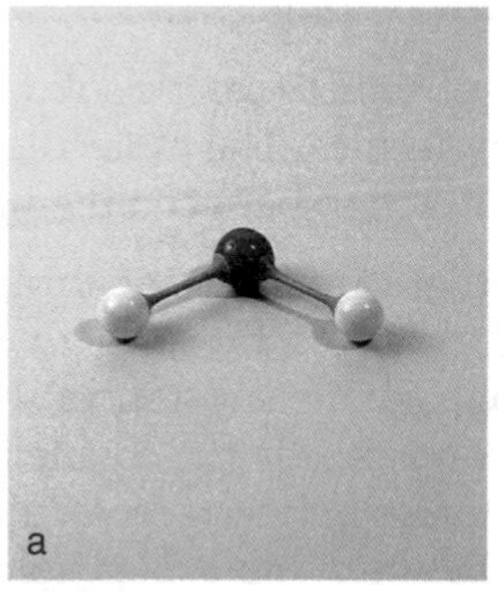

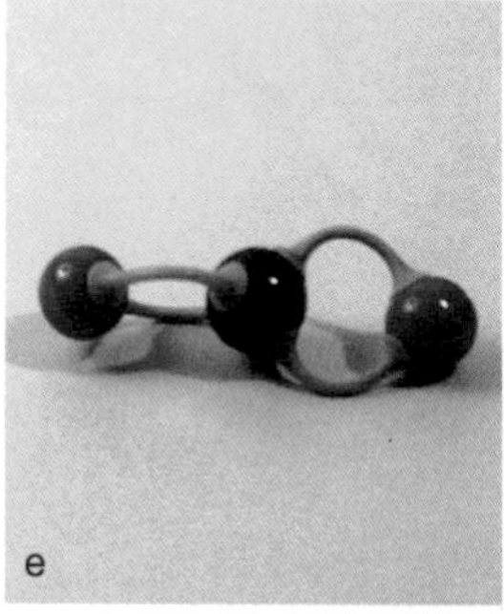

*Figure 8.4E* Covalent molecules (a) water, $H_2O$ (b) ammonia, $NH_3$ (c) methane, $CH_4$ (d) oxygen, $O_2$ (e) carbon dioxide, $CO_2$ (f) ethene, $C_2H_4$

**SUMMARY**

Compounds which are non-electrolytes contain covalent bonds. Each of the atoms in a covalent bond completes its outer shell of electrons by sharing a pair of electrons. The shared electrons are attracted to both atomic nuclei and therefore bond the atoms.

**CHECKPOINT**

1. Sketch the arrangement of electrons in H (1) and F (2.7). Show by means of a 'dot-and-cross diagram' how a covalent bond is formed in HF.
2. Sketch the arrangement of electrons in the covalent compounds $BF_3$ [ B (2.3), F (2.7)], $PH_3$ [P (2.8.5), H (1)] and $H_2S$ [S (2.8.6), H (1)].

## 8.5 Formulas of covalent compounds

**SUMMARY**

The formula of a covalent compound follows from a 'dot-and-cross diagram'.

The formula of a covalent compound depends on the sort of 'dot-and-cross diagram' that can be drawn for the compound. This shows the number of each type of atom in a molecule. This is the formula of the compound. Some covalent compounds you will meet on your course are sulphur dioxide, $SO_2$, sulphur trioxide, $SO_3$, ethane, $C_2H_6$, ethanol (alcohol), $C_2H_5OH$.

## 8.6 Forces between molecules

Covalent substances can be solids, liquids or gases. Gases consist of separate molecules. The molecules are far apart, and there are almost no forces of attraction between them.

In liquids there are forces of attraction between the molecules. In some covalent substances, the attractive forces between molecules are strong enough to make the substances solids. Ice is a solid. There are strong covalent bonds inside the covalent water molecules and weaker attractive forces between molecules (see Figure 8.6A). The forces of attraction between molecules hold them in a three-dimensional structure. It is described as a **molecular structure**.

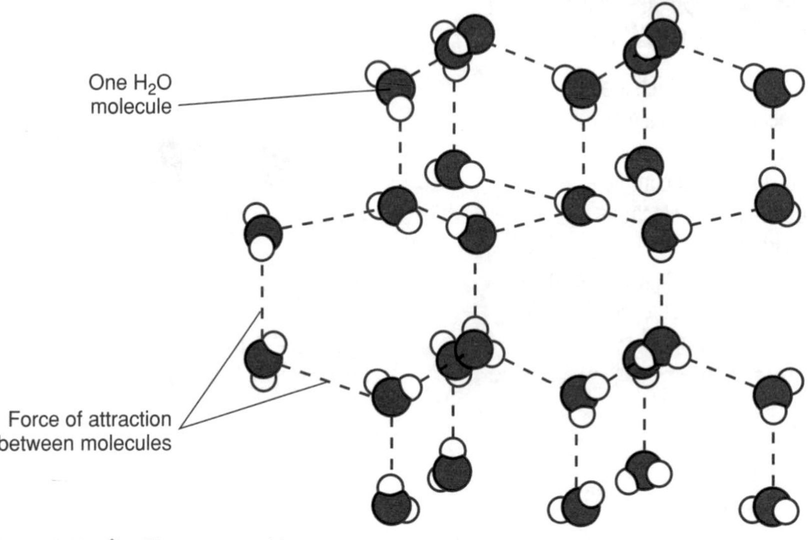


*Figure 8.6A* ▲ The structure of ice

Diamond is an **allotrope** of carbon. Diamond has a structure in which every carbon atom is joined by covalent bonds to four other carbon atoms (see Figure 4.8B). Millions of carbon atoms form a giant molecule. The structure is giant molecular or macromolecular.

Graphite, the other allotrope of carbon, has a layer structure (see Figure 4.8C). Strong covalent bonds join the carbon atoms within each layer. The layers are held together by weak forces of attraction.

### SUMMARY

Some covalent substances are composed of individual molecules. Others are composed of larger units. These may be chains or layers or macromolecular structures.

### CHECKPOINT

1 (a) Why does diamond have a high melting point?
(b) Why is it very difficult to scratch a diamond?
(c) Why does graphite rub off on to your hands?

2 Solid iodine consists of shiny black crystals. Iodine vapour is purple. What is the difference in chemical bonding between solid and gaseous iodine?

## 8.7 Ionic and covalent compounds

What difference does the type of bond make to the properties of ionic and covalent compounds?

The physical and chemical characteristics of substances depend on the type of chemical bonds in the substances.

*Table 8.7* ▼ Differences between ionic and covalent substances

| ***Ionic bonding*** | ***Covalent bonding*** |
|---|---|
| Ionic compounds are formed when a metallic element combines with a non-metallic element. An **ionic bond** is formed by **transfer of electrons** from one atom to another to form ions. | Atoms of non-metallic elements combine with other non-metallic elements by **sharing pairs of electrons** in their outer shells. A shared pair of electrons is a **covalent bond**. |
| Atoms of metallic elements form positive ions (cations), e.g. $Na^+$, $Mg^{2+}$, $Al^{3+}$.<br>Atoms of non-metallic elements form negative ions (anions), e.g. $O^{2-}$ and $Cl^-$. | The maximum number of covalent bonds that an atom can form is equal to the number of electrons in the outer shell. An atom may not use all its electrons in bond formation. |
| Ionic compounds are **electrolytes**; they conduct electricity when molten or in solution and are split up in the process – electrolysed. | Covalent compounds are **non-electrolytes**. |
| The strong electrostatic attraction between ions of opposite charge is an **ionic bond**. An ionic compound is composed of a giant regular structure of ions (see Figure 8.6A). This regular structure makes ionic compounds **crystalline**. The strong forces of attraction between ions make it difficult to separate the ions, and ionic compounds therefore have **high melting and boiling points**. | There are three **types of covalent substances**:<br>(a) Many covalent substances are composed of small individual molecules with only very small forces of attraction between molecules. Such covalent substances are gases, e.g. HCl, $SO_2$, $CO_2$, $CH_4$.<br>(b) Some covalent substances consist of small molecules with weak forces of attraction between the molecules. Some such covalent substances are low boiling point liquids, e.g. ethanol, $C_2H_5OH$<br>Others are low melting point solids, e.g. iodine and solid carbon dioxide, which consist of molecular crystals (see Figure 8.7A).<br>(c) Some covalent substances consist of giant molecules, e.g. quartz (silicon(IV) oxide, Figure 8.7B). These substances have high melting and boiling points. Atoms may link in chains or sheets, e.g. graphite (Figure 4.8C) or in 3-dimensional structures, e.g. diamond (Figure 4.8B) and quartz (Figure 8.7B). Substances with giant molecular structures have high melting and boiling points.<br>**Organic solvents**, e.g. ethanol and propanone, have covalent bonds. They dissolve covalent compounds but not ionic compounds. |

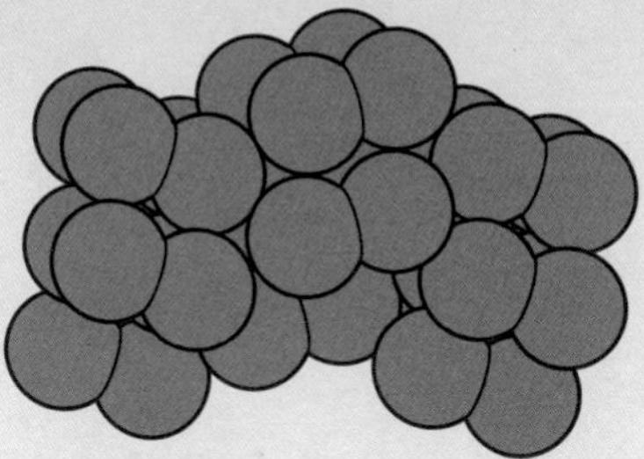

*Figure 8.7A* ▲ The structure of an iodine crystal

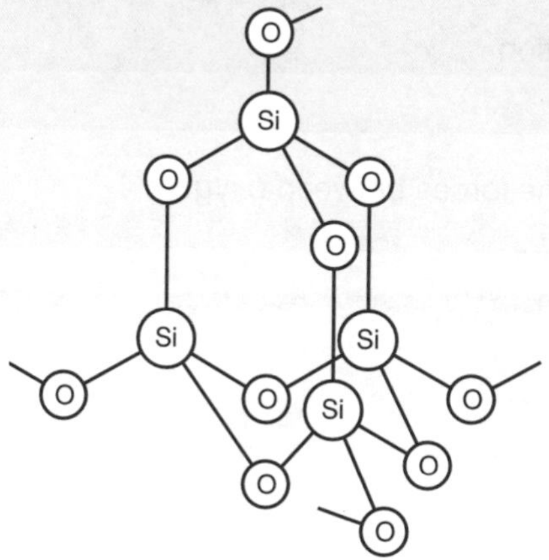


*Figure 8.7B* ▲ The structure of silicon(IV) oxide (quartz)

*Can you tell from its properties whether a substance has ionic bonds or covalent bonds? The table will help you.*

## SUMMARY

The type of chemical bonds present, ionic or covalent, decides the properties of a compound, e.g. its physical state (s, l, g), boiling and melting points, electrolytic conductivity.

## CHECKPOINT

1 Explain why the covalent compound $CH_2Br_2$ is a volatile liquid whereas the ionic compound $PbBr_2$ is a solid.

2 Explain why the compound $CHCl_3$, chloroform, has a powerful smell whereas sodium chloride, NaCl, has none.

3 Astatine, At, is an element with 7 electrons in its outer shell. Caesium, Cs, is an element with one electron in its outer shell. What type of bonding would you predict in caesium astatide? Write its formula.

4 Radium, Ra, is an element with 2 electrons in its outer shell, and iodine, I, is an element with 7 electrons in the outer shell. What type of bonding would you predict in radium iodide? Write its formula.

5 Oxygen and chlorine combine to form a number of compounds, e.g. $Cl_2O$ and $ClO_2$. What type of bonding would you expect in these compounds?

6 Name covalent substances which consist of (a) individual molecules, (b) a molecular structure, (c) a giant molecular structure.

7 A girl gets some copper(II) sulphate solution on her shirt. Which will dissolve the blue stain out better, water or alcohol?

8

| *Solid* | *State* | *Melting point* | *Does it conduct electricity?* |
|---|---|---|---|
| A | Solid | 650 °C | Conducts when molten |
| B | Liquid | −20 °C | Does not conduct |
| C | Solid | 700 °C | Does not conduct when molten |
| D | Solid | 85 °C | Does not conduct when molten |
| E | Gas | −100 °C | Does not conduct |

From the information in the table, say what you can about the chemical bonds in A, B, C, D and E.

9 Below are five statements. Give a piece of evidence (e.g. a physical or chemical property of the substance) in support of each statement.

(a) An aqueous solution of sodium chloride contains ions.

(b) Copper exists as positive ions in copper(II) sulphate solution.

(c) Ethanol (alcohol) is a covalent compound.

(d) The forces between oxygen molecules are very weak.

(e) The forces between iodine molecules are stronger than the forces between oxygen molecules.

# Theme Questions

| *Element* | Argon | Bromine | Calcium | Carbon | Chlorine | Gold | Hydrogen | Mercury | Phosphorus | Sulphur |
|---|---|---|---|---|---|---|---|---|---|---|
| *Metal or non-metal* | N-M | N-M | M | N-M | N-M | M | N-M | M | N-M | N-M |
| *Melting point (°C)* | −189 | −7 | 850 | 3730 (sublimes) | −101 | 1060 | −259 | −39 | 44 (white) 590 (red) | 113 (rh) 119 (mono) |
| *Boiling point (°C)* | −186 | 59 | 1490 | 4830 | −35 | 2970 | −252 | 357 | 280 | 445 |
| *Density (g/cm³)* | 0.0017 | 3.1 | 1.5 | 2.3 (gr) 3.5 (di) | 0.003 017 | 19.3 | 0.000 083 | 13.6 | 1.8 (wh) 2.3 (red) | 2.1 (rh) 2.0 (mono) |

For questions 1 – 10 refer to the table of elements above.

**1** Name the element with (a) the highest melting point (b) the lowest melting point.

**2** Name the element with (a) the greatest density and (b) the lowest density.

**3** Name the metal with (a) the lowest melting point and (b) the highest melting point.

**4** Name the element which is liquid at room temperature and is (a) a metal (b) a non-metal.

**5** Write down the names of all the elements which are (a) metals (b) non-metals.

**6** Write down the names of all the elements which are (a) solids (b) liquids (c) gases.

**7** Name the element which has the smallest temperature range over which it exists as a liquid.

**8** Name the gaseous element which is (a) the most dense (b) the least dense.

**9** Name the solid element which is (a) the most dense (b) the least dense.

**10** Why are two sets of values given for (a) sulphur (b) phosphorus?

**11** Copy out these equations into your book, and then balance them.

(a) $H_2O_2(aq) \rightarrow H_2O(l) + O_2(g)$

(b) $Fe(s) + O_2(g) \rightarrow Fe_3O_4(s)$

(c) $Mg(s) + N_2(g) \rightarrow Mg_3N_2(s)$

(d) $P(s) + Cl_2(g) \rightarrow PCl_3(s)$

(e) $P(s) + Cl_2(g) \rightarrow PCl_5(s)$

(f) $SO_2(g) + O_2(g) \rightarrow SO_3(g)$

(g) $Na_2O(s) + H_2O(l) \rightarrow NaOH(aq)$

(h) $KClO_3(s) \rightarrow KCl(s) + O_2(g)$

(i) $NH_3(g) + O_2(g) \rightarrow N_2(g) + H_2O(g)$

(j) $Fe(s) + H_2O(g) \rightarrow Fe_3O_4(s) + H_2(g)$

**12** The element X has atomic number 11 and mass number 23. State how many protons and neutrons are present in the nucleus. Sketch the arrangement of electrons in an atom of X.

**13** Atom A has atomic number 82 and mass number 204. Atom B has atomic number 80 and mass number 204. How many protons has atom A? How many neutrons has atom B? Are atoms A and B isotopes of the same element? Explain your answer.

**14** The structure of one molecule of an industrial solvent, dichloromethane, is shown below.

```
     H
     |
Cl — C — Cl
     |
     H
```

(a) How many atoms are there in one molecule of dichloromethane?

(b) How many bonds are there in one molecule of dichloromethane?

**15** Describe how you could use a battery and a torch bulb to test various materials to find out whether they are electrical conductors.
Divide the following list into conductors and non-conductors:
silver, steel, polythene, PVC, brass, candle wax, lubricating oil, dilute sulphuric acid, petrol, alcohol, sodium hydroxide solution, sugar solution, limewater

**16** (a) Name three types of substance that conduct electricity.

(b) Why can molten salts conduct electricity while solid salts cannot?

(c) Why can solutions of salts conduct electricity while pure water cannot?

(d) Explain why metal articles can be electroplated but plastic articles cannot.

**17** Name the substances A to H in the table. Write equations for the discharges of the ions which form these substances.

| *Electrolyte* | *Anode* | *Cathode* |
|---|---|---|
| Copper(II) chloride solution | A | B |
| Sodium chloride solution | C | D |
| Dilute sulphuric acid | E | F |
| Sodium hydroxide solution | G | H |

**18** Your aunt runs a business making souvenirs. She asks your advice on how to electroplate a batch of small brass medallions with copper. Describe how this could be done. Draw a diagram of the apparatus and the circuit she could use.

**19** In the electrolysis of potassium bromide solution, what element is formed (a) at the positive electrode (b) at the negative electrode? Why does the solution around the negative electrode become alkaline?

**20** (a) Explain why the ionic compound $PbBr_2$ melts at a higher temperature than the covalent compound $CH_2Br_2$.

(b) You can smell the compound $CHCl_3$, chloroform. You cannot smell the compound KCl. What difference in chemical bonding is responsible for the difference?

**21** The diagram shows apparatus that could be used in a laboratory to find out the effect of an electric current on an aqueous solution of sodium chloride.

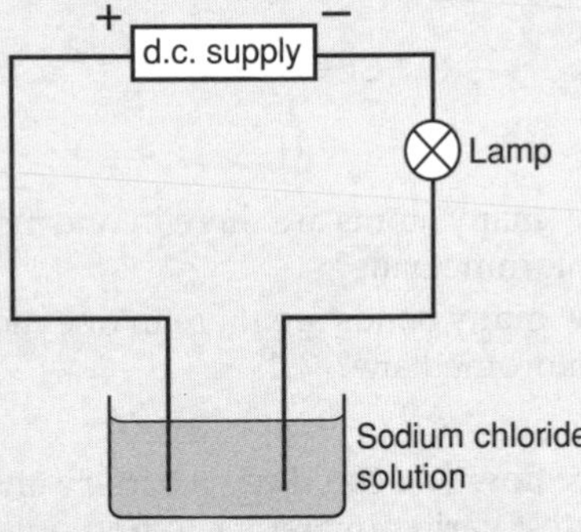


(a) On a copy of the diagram, (i) label the cathode, (ii) show the direction of flow of the electrons.

(b) What **two** observations would show that the solution conducts an electric current and that a chemical reaction is taking place?

(c) For the product formed at the positive electrode (i) give its name (ii) write the word equation for its formation.

(d) Give **one** reason why sodium is **not** a product when an electric current is passed through aqueous sodium chloride solution. (ULEAC)

**22** Sodium chloride, chlorine and diamond are three very different substances.

(a) State the type of bond present in (i) sodium chloride, (ii) chlorine, (iii) diamond.

(b) What is the difference in structure which explains why diamond is a solid up to a very high temperature, but chlorine is a gas at room temperature?

(c) Explain why sodium chloride will conduct electricity when it is molten, but not when it is solid.

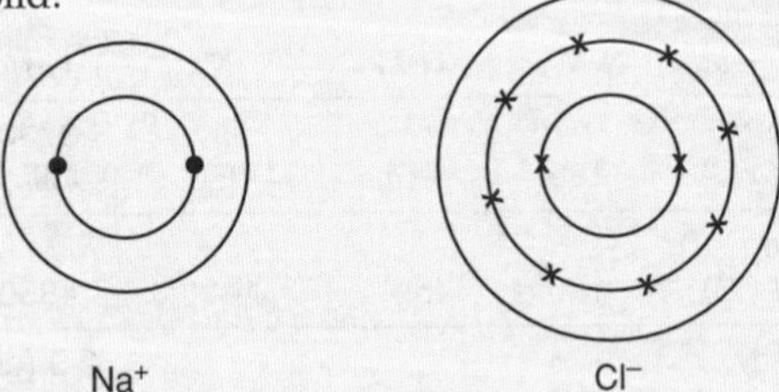


(d) Complete the diagrams to show the electronic structure of sodium ions and chloride ions.

(e) Draw a diagram to show the electronic structure in a chlorine molecule. (For example a dot-and-cross diagram.) (ULEAC)

**23** (*a*) From the Periodic Table at the back of this book deduce the electronic arrangement of:
(i) lithium; (ii) oxygen; (iii) fluorine.

(*b*) When lithium reacts with fluorine, lithium fluoride, LiF, is formed. It is made up of positive and negative ions.
(i) How are the positive lithium ions formed from lithium atoms?
(ii) How are the negative fluoride ions formed from fluorine atoms?
(iii) **Explain** how the ions are held together in lithium fluoride.
(iv) Name this type of bonding.

(*c*) Fluorine also forms a *gaseous* compound with oxygen, with the formula $F_2O$. Explain how these two elements bond to form $F_2O$ and name the type of bonding used.

(*d*) **Explain** why the bonding in lithium fluoride, LiF, produces a high melting point solid. (WJEC)

**24** (a) Some people sprinkle salt on their food.
Too much of the salt called sodium chloride is bad for our health.
Some people now buy "Low-Salt" which contains mainly potassium chloride.

(i) Which ion in salt is bad for our health if we eat too much of it?
(ii) The elements sodium and potassium have similar chemical properties. Explain why.

(b) Brine is a solution of sodium chloride in water. Brine is used in the manufacture of chlorine, hydrogen and sodium hydroxide by electrolysis.

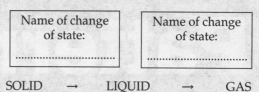

| sodium chloride | + | water | → | chlorine | + | hydrogen | + | sodium hydroxide |
|---|---|---|---|---|---|---|---|---|
| [compound] | | [compound] | | [element] | | [element] | | [compound] |

What are the differences between an element, a mixture and a compound? You must use the idea of atoms, ions and molecules. Use these substances as examples to explain your answer.

(c) Hydrogen is used in weather balloons.

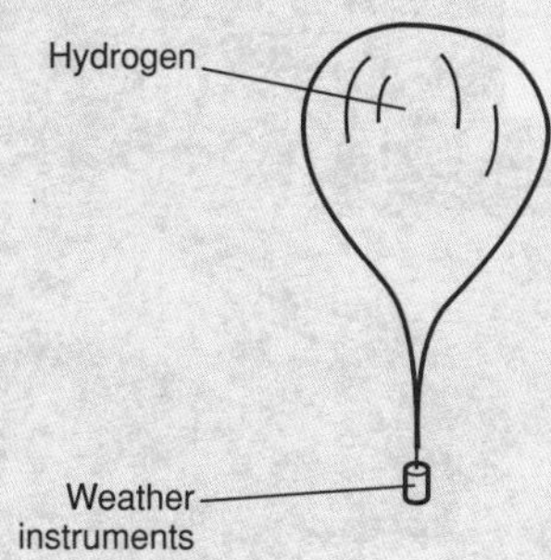


(i) Explain what happens to the volume enclosed by the balloon if you add more hydrogen.

(ii) Explain what happens to the volume of gas in the balloon if the air temperature falls.

(d) Molten sodium chloride is used in the manufacture of sodium by electrolysis. In sodium street lamps the metal is used in the form of gas.

(i) Copy and complete the diagram by naming each change of state as sodium changes from a solid to a vapour (gas).

| Name of change of state: ................................ | | Name of change of state: ................................ |
|---|---|---|

SOLID → LIQUID → GAS

(ii) Give **two** different ways of making liquid sodium change into sodium vapour.

(iii) Copy and complete the following sentences.

When a gas becomes liquid the change of state is called ____________

The opposite of melting is ____________

(SEG)

# Patterns of behaviour

Metals

Non-metals

Transition metals

| 1 | 2 | | | | | | | | | | | 3 | 4 | 5 | 6 | 7 | 0 |
|---|---|---|---|---|---|---|---|---|---|---|---|---|---|---|---|---|---|
| | | | H | | | | | | | | | | | | | | He |
| Li | Be | | | | | | | | | | | B | C | N | O | F | Ne |
| Na | Mg | | | | | | | | | | | Al | Si | P | S | Cl | Ar |
| K | Ca | Sc | Ti | V | Cr | Mn | Fe | Co | Ni | Cu | Zn | Ga | Ge | As | Se | Br | Kr |
| Rb | Sr | Y | Zr | Nb | Mo | Tc | Ru | Rh | Pd | Ag | Cd | In | Sn | Sb | Te | I | Xe |
| Cs | Ba | La | Hf | Ta | W | Re | Os | Ir | Pt | Au | Hg | Tl | Pb | Bi | Po | At | Rn |

There are thousands of chemicals and hundreds of thousands of chemical reactions. Fortunately for us there is order, not chaos, in the universe. There are patterns of behaviour. All acids have reactions in common, bases have reactions in common, and acids and bases react together. Oxidising agents and reducing agents have complementary properties, and they react together in oxidation-reduction reactions.

The Periodic Table is a remarkable pattern of behaviour. The properties of all the elements, 105 of them, fall into place in this excellent system of classification.

# Topic 9 Acids and bases

## 9.1 ▶ Acids

**FIRST THOUGHTS**

Can you say what the elements in each of these sets of materials have in common?
**List A:** lime juice, vitamin C, stomach juice, vinegar
**List B:** toothpaste, household ammonia, Milk of Magnesia, lime

If you have remembered, well done! You will learn more about acids and bases in this topic.

Cells in the linings of our stomachs produce hydrochloric acid. This is a powerful acid. If a piece of zinc was dropped into hydrochloric acid of the concentration which the stomach contains, it would dissolve. In the stomach, hydrochloric acid works to kill bacteria which are present in foods and to soften foods. It also helps in **digestion** by providing the right conditions for the enzyme pepsin to begin the digestion of proteins.

Sometimes the stomach produces too much acid. Then the result is the pain of 'acid indigestion' and 'heartburn'. The many products which are sold as 'antacids' and indigestion remedies all contain compounds that can react with hydrochloric acid to neutralise (counteract) the excess of acid. In this chapter, we shall look at the properties (characteristics) of acids and the substances which react with them.

A sour taste usually shows that a substance contains an acid. The word 'acid' comes from the Latin word for 'sour'. Vinegar contains ethanoic acid, sour milk contains lactic acid, lemons contain citric acid and rancid butter contains butanoic acid. For centuries, chemists have been able to extract acids such as these from animal and plant material. They call these acids **organic acids**. As chemistry has advanced, chemists have found ways of making sulphuric acid, hydrochloric acid, nitric acid and other acids from minerals. They call these acids **mineral acids**. Mineral acids in general react much more rapidly than organic acids. We describe mineral acids as **strong acids** and organic acids as **weak acids**. Table 9.1 lists some common acids.

Solutions of acids (and other substances) can be **dilute solutions** or **concentrated solutions**. A dilute solution contains a small amount of acid per litre of solution. A concentrated solution contains a large amount of acid per litre of solution. Solutions of acids are always called, say, dilute hydrochloric acid or concentrated hydrochloric acid. Concentrated acids are very corrosive, and you need to know which type of solution you are dealing with.

This symbol means **corrosive**. Concentrated acids always carry this symbol.

The table lists
- organic acids, many of which are weak acids
- mineral acids, many of which are strong acids

*Table 9.1* ▼ Some common acids

| Acid | Strong or weak | Where you find it |
|---|---|---|
| Ascorbic acid | Weak | In fruits. Also called Vitamin C |
| Carbonic acid | Weak | In fizzy drinks: these contain the gas carbon dioxide which reacts with water to form carbonic acid. |
| Citric acid | Weak | In fruit juices, e.g. lemon juice |
| Ethanoic acid | Weak | In vinegar |
| Hydrochloric acid | Strong | In digestive juices in the stomach; also used for cleaning ('pickling') metals before they are coated |
| Lactic acid | Weak | In sour milk |
| Nitric acid | Strong | Used for making fertilisers and explosives |
| Phosphoric acid | Strong | In anti-rust paint; used for making fertilisers |
| Sulphuric acid | Strong | In car batteries; used for making fertilisers |

Check that you know the difference between
- a strong acid and a concentrated acid
- a weak acid and a dilute acid

*Figure 9.1A* ▲ Some common acids

*Figure 9.1B* ▲ The acid taste

## SUMMARY

Acids
- have a sour taste,
- change the colour of indicators,
- react with many metals to give hydrogen and a salt,
- react with carbonates and hydrogencarbonates to give carbon dioxide, a salt and water,
- react with bases (see Topic 9.2).

# What do acids do?

1 **Acids have a sour taste.**
You will know the taste of lemon juice (which contains citric acid) and vinegar (which contains ethanoic acid). **Do not taste any of the strong acids.**

2 **Acids change the colour of substances called indicators.**
For example, acids turn blue litmus red (see Table 9.4).

3 **Acids react with many metals to produce hydrogen and a salt of the metal.**
Hydrogen is an element. It is a colourless, odourless (without smell) gas and is the least dense of all the elements. With air, hydrogen forms an explosive mixture, which is the basis of a test for hydrogen. If you put a lighted splint into hydrogen, you hear an explosive 'pop' (see Figure 9.1C).

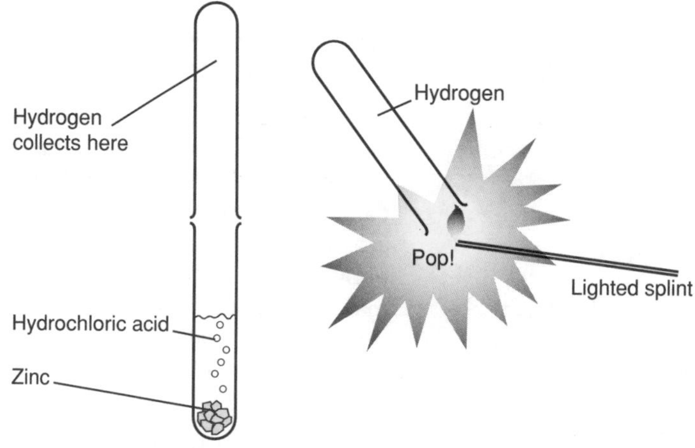


*Figure 9.1C* ▲ (a) Making hydrogen (b) Testing for hydrogen

Some metals, e.g. copper, react very slowly with acids. Some metals, e.g. sodium, react dangerously fast. Much heat is given out, and the hydrogen which is formed may explode. Examples of metals which react at a safe speed are magnesium, zinc and iron.

Zinc + Sulphuric acid → Hydrogen + Zinc sulphate

$$Zn(s) + H_2SO_4(aq) \rightarrow H_2(g) + ZnSO_4(aq)$$

4 **Acids react with carbonates to give carbon dioxide, a salt and water.**
Acid indigestion is caused by an excess of hydrochloric acid in the stomach. Some indigestion tablets contain magnesium carbonate. The reaction that takes place when magnesium carbonate reaches the stomach is

Magnesium carbonate + Hydrochloric acid → Carbon dioxide + Magnesium chloride + Water

$$MgCO_3(s) + 2HCl(aq) \rightarrow CO_2(g) + MgCl_2(aq) + H_2O(l)$$

Other anti-acid tablets contain sodium hydrogencarbonate. In this case, the reaction that happens in the stomach is

Sodium hydrogencarbonate + Hydrochloric acid → Carbon dioxide + Sodium chloride + Water

$$NaHCO_3(s) + HCl(aq) \rightarrow CO_2(g) + NaCl(aq) + H_2O(l)$$

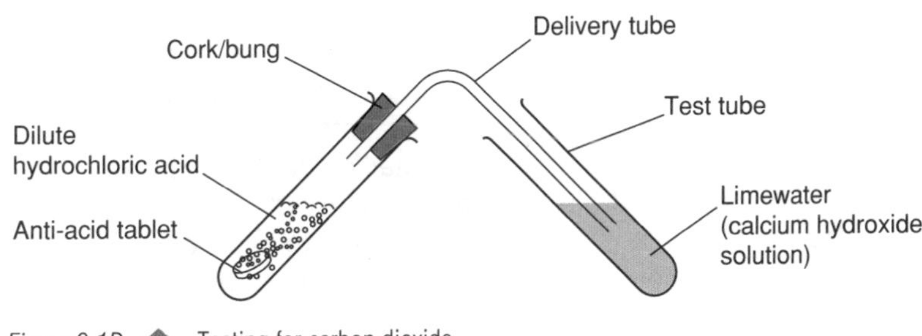


*Figure 9.1D* ▲ Testing for carbon dioxide

**5 Acids neutralise bases**
(see Topic 9.4.)

## What is an acid?

A Swedish chemist called Svante Arrhenius explained why different acids have so much in common. He put forward the theory that aqueous solutions of all acids contain hydrogen ions, $H^+(aq)$, in high concentration. By 'high' concentration, he meant a concentration much higher than that in water. This theory explains many reactions of acids, including **neutralisation, electrolysis of solutions** of acids and the reaction of metals with acids

| Metal atoms + Hydrogen ions | → | Metal ions + Hydrogen molecules |
|---|---|---|
| $M(s) + 2H^+(aq)$ | → | $M^{2+}(aq) + H_2(g)$ |

Arrhenius gave this definition of an acid:

> An acid is a substance that releases hydrogen ions when dissolved in water.

A solution of a strong acid has a much higher concentration of hydrogen ions than a solution of a weak acid. A solution of a strong acid therefore reacts much more rapidly than a solution of a weak acid.

### SUMMARY

Different acids have similar reactions. The reason is that solutions of all acids contain hydrogen ions in high concentration. The hydrogen ions are responsible for the typical reactions of acids.

### CHECKPOINT

1 Where in the kitchen could you find
(a) a weak acid with a pleasant taste?
(b) a weak acid with an unpleasant taste?
(c) a weak acid with a very sour taste?

2 What is the difference between a concentrated solution of a weak acid and a dilute solution of a strong acid?

3 What is the difference between a concentrated solution of sulphuric acid and a dilute solution of sulphuric acid? Why do road tankers of sulphuric acid carry the sign shown above?

4 *Toffee Recipe 1* Boil sugar and water with a little butter until the mixture thickens. Pour into a greased tray to set.
*Toffee Recipe 2* Repeat Recipe 1. When the mixture thickens, add vinegar and 'bicarbonate of soda' (sodium hydrogencarbonate). Pour into a greased tray to set.
One of these recipes gives solid toffee. The other gives a honeycomb of toffee containing bubbles of gas. Which recipe gives the honeycomb? Why?

5 Rosie and Luke visited the underground caves in the limestone rock at Wookey Hole in Somerset. A guide told them that the chemical name for limestone is calcium carbonate. The children took a small piece of limestone rock to school and did an experiment with it. They put a piece of rock in a beaker and added dilute hydrochloric acid. Immediately, bubbles of gas were given off. (See the figures opposite.)

(a) Why must you always wear safety glasses when working with acids?
(b) What is the name of the gas in the bubbles?
(c) What chemical test could they do to 'prove' what the gas was?

Rosie lit a candle, and then tilted the beaker so that gas poured on to it (see lower figure). The candle went out.

(d) What two things about the gas did this experiment tell the children?
(e) What could the gas be used for?

Their teacher prepared a gas jar of the gas. Luke added distilled water and shook the gas jar. Then he added litmus solution.

(f) What colour did the indicator turn? (See Table 9.1 and *What do acids do?* for help.)

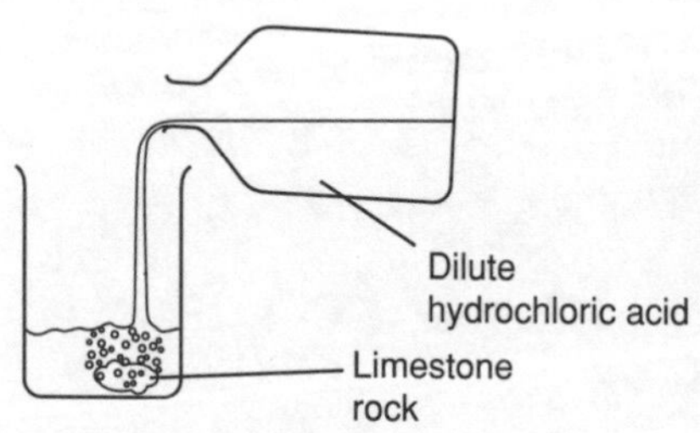

(a) Adding acid to limestone

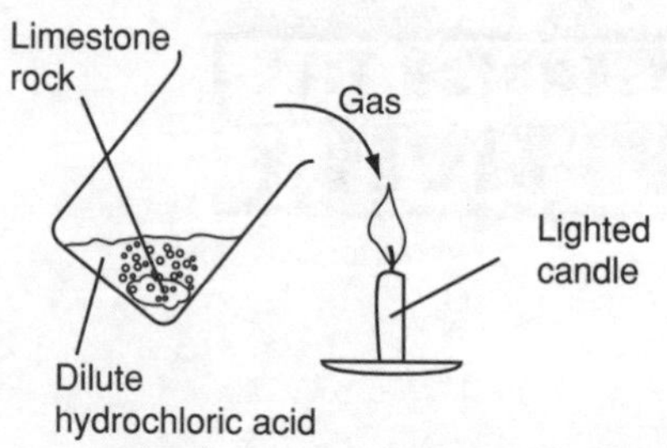

(b) Testing the gas on a lighted candle

## 9.2 Bases

Figure 9.2A Spreading lime

Bases are substances that neutralise (counteract) acids. Figure 9.2A shows a farmer spreading the base called 'lime' (calcium hydroxide) on a field. Some soils are too acidic to grow good crops. 'Liming' neutralises some of the acid and increases the **soil fertility**.

The product of the reaction between an acid and a base is a neutral substance, neither an acid nor a base, which is called a **salt**. Lime (calcium hydroxide) is a base. It reacts with nitric acid in the soil to form the salt calcium nitrate and water

Calcium hydroxide + Nitric acid → Calcium nitrate + Water

A definition of a base is:

> A base is a substance that reacts with an acid to form a salt and water only.

Acid + Base → Salt + Water

Soluble bases are called **alkalis**. Sodium hydroxide is an alkali; it reacts with an acid to form a salt and water.

Sodium hydroxide + Hydrochloric acid → Sodium chloride + Water

$NaOH(aq) + HCl(aq) \rightarrow NaCl(aq) + H_2O(l)$

Can you explain these terms:

- acid,
- base,
- salt,
- alkali?

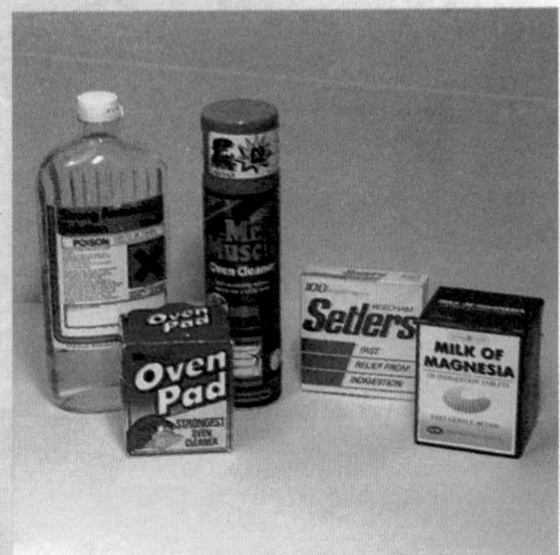


*Figure 9.2B* ▲ Some common bases

Limewater (calcium hydroxide solution) is an alkali. A test for carbon dioxide is that it turns limewater cloudy. The reaction is an acid-base reaction to form a salt and water.

$$\text{Carbon dioxide} + \text{Calcium hydroxide} \rightarrow \text{Calcium carbonate} + \text{Water}$$
$$CO_2(g) + Ca(OH)_2(aq) \rightarrow CaCO_3(s) + H_2O(l)$$

The salt, calcium carbonate, appears as a cloud of insoluble white powder.

*Table 9.2* ▼ Some common bases

| State | Strong or weak | Where you find it |
|---|---|---|
| Ammonia | Weak | In cleaning fluids for use as a degreasing agent; also used in the manufacture of fertilisers |
| Calcium hydroxide | Strong | Used to treat soil which is too acidic |
| Calcium oxide | Strong | Used in the manufacture of cement, mortar and concrete |
| Magnesium hydroxide | Strong | In anti-acid indigestion tablets and Milk of Magnesia |
| Sodium hydroxide | Strong | In oven cleaners as a degreasing agent; also used in soap manufacture |

Table 9.2 lists some common bases. Different bases have a number of reactions in common.

1 Bases neutralise acids (see previous page).
2 Soluble bases can change the colour of **indicators**, e.g. turn red litmus blue (see Table 9.4).
3 Soluble bases feel soapy to your skin. The reason is that soluble bases convert some of the oil in your skin into **soap**. Some household cleaning solutions, e.g. ammonia solution, use soluble bases as degreasing agents. They convert oil and grease into soluble soaps which are easily washed away.
4 A solution of an alkali in water contains hydroxide ions, $OH^-(aq)$. This solution will react with a solution of a metal salt. Most metal hydroxides are insoluble. When a solution of an alkali is added to a solution of a metal salt, an insoluble metal hydroxide is **precipitated** from solution. (A precipitate is a solid which forms when two liquids are mixed.) For example

## SUMMARY

Bases neutralise acids to form salts. Soluble bases are called alkalis. Metal oxides and hydroxides are bases. Alkalis change the colours of indicators. They are degreasing agents, and have a soapy 'feel'. Solutions of alkalis contain hydroxide ions, $OH^-(aq)$ in high concentration.

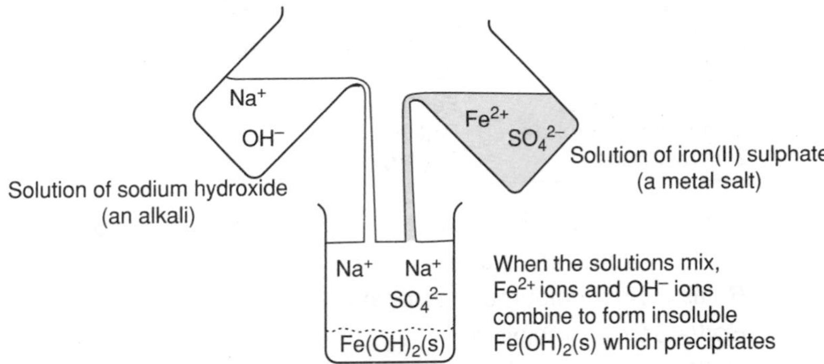


*Figure 9.2C* ▲ Precipitating an insoluble hydroxide

$$\text{Iron(II) Sulphate (solution)} + \text{Sodium hydroxide (solution)} \rightarrow \text{Iron(II) hydroxide (precipitate)} + \text{Sodium sulphate (solution)}$$
$$FeSO_4(aq) + 2NaOH(aq) \rightarrow Fe(OH)_2(s) + Na_2SO_4(aq)$$

A solution of a strong base contains a higher concentration of hydroxide ions than a solution of a weak base.

Check that you know the difference between
- a strong base and a weak base
- a weak base and a dilute base
- a base and an alkali.

*Table 9.3* Examples of bases

| Metal oxides | Metal hydroxides | Alkalis (soluble bases) |
|---|---|---|
| Copper(II) oxide, $CuO$ | Sodium hydroxide, $NaOH$ | Sodium hydroxide, $NaOH$ |
| Zinc oxide, $ZnO$ | Magnesium hydroxide, $Mg(OH)_2$ | Potassium hydroxide, $KOH$ |
| | | Calcium hydroxide, $Ca(OH)_2$ |
| (*Most metal oxides and hydroxides are insoluble*) | | Ammonia solution, $NH_3(aq)$ |

## Ionic equations

An ionic equation shows the essentials.

There is another way of writing the equation of the reaction between iron(II) sulphate and sodium hydroxide. The reaction is the combination of iron(II) ions and hydroxide ions. The sodium ions and sulphate ions take no part in the reaction, and the equation can be written without them.

$$Fe^{2+}(aq) + 2OH^-(aq) \rightarrow Fe(OH)_2(s)$$

This is called an **ionic equation**.

## CHECKPOINT

1 (a) Write the names of the four common alkalis.
(b) Write the names of four insoluble bases.

2 Four bottles of solution are standing on a shelf in the prep room. Their labels have come off and are lying on the floor. They read, Copper(II) sulphate, Iron(II) sulphate, Iron(III) sulphate and Zinc sulphate. The lab assistant knows that some insoluble metal hydroxides are coloured (see the table below).

| *Hydroxide* | *Formula* | *Colour* |
|---|---|---|
| Copper(II) hydroxide | $Cu(OH)_2(s)$ | Blue |
| Iron(II) hydroxide | $Fe(OH)_2(s)$ | Green |
| Iron(III) hydroxide | $Fe(OH)_3(s)$ | Rust |
| Magnesium hydroxide | $Mg(OH)_2(s)$ | White |
| Zinc hydroxide | $Zn(OH)_2(s)$ | White |

She adds sodium hydroxide solution to a sample of each solution. Her results are:

*Bottle 1* White precipitate

*Bottle 2* Rust-coloured precipitate

*Bottle 3* Blue precipitate

*Bottle 4* Green precipitate

Say which label should be stuck on each bottle.

3 Write the formulas for the bases: calcium oxide, copper(II) oxide, zinc oxide, magnesium hydroxide, iron(II) hydroxide, iron(III) hydroxide.

# 9.3 Summarising the reactions of acids and bases

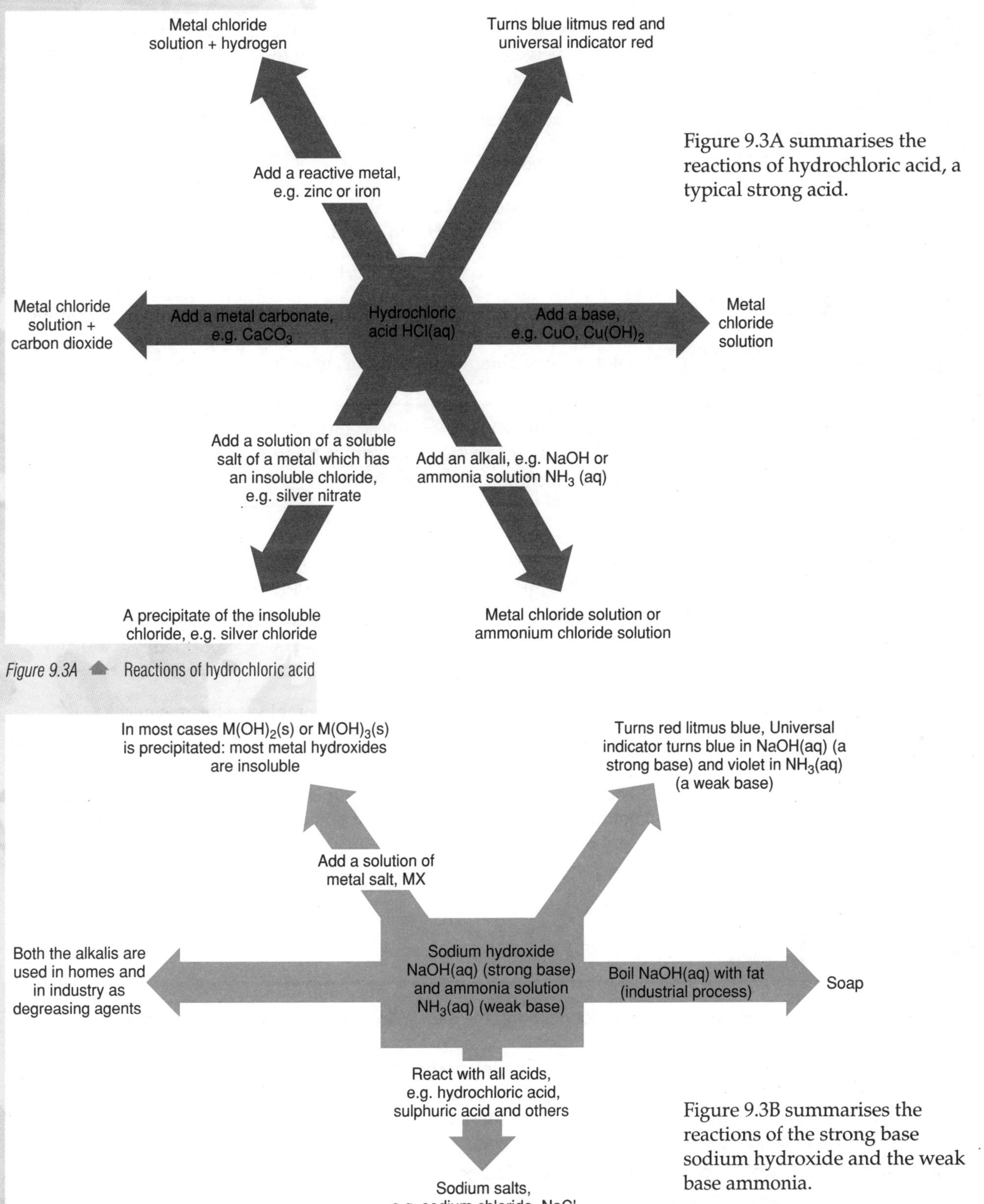
Metal chloride solution + hydrogen
Turns blue litmus red and universal indicator red
Add a reactive metal, e.g. zinc or iron
Metal chloride solution + carbon dioxide
Add a metal carbonate, e.g. $CaCO_3$
Hydrochloric acid HCl(aq)
Add a base, e.g. CuO, $Cu(OH)_2$
Metal chloride solution
Add a solution of a soluble salt of a metal which has an insoluble chloride, e.g. silver nitrate
Add an alkali, e.g. NaOH or ammonia solution $NH_3$ (aq)
A precipitate of the insoluble chloride, e.g. silver chloride
Metal chloride solution or ammonium chloride solution
In most cases $M(OH)_2$(s) or $M(OH)_3$(s) is precipitated: most metal hydroxides are insoluble
Turns red litmus blue, Universal indicator turns blue in NaOH(aq) (a strong base) and violet in $NH_3$(aq) (a weak base)
Add a solution of metal salt, MX
Both the alkalis are used in homes and in industry as degreasing agents
Sodium hydroxide NaOH(aq) (strong base) and ammonia solution $NH_3$(aq) (weak base)
Boil NaOH(aq) with fat (industrial process)
Soap
React with all acids, e.g. hydrochloric acid, sulphuric acid and others
Sodium salts, e.g. sodium chloride, NaCl sodium sulphate, $Na_2SO_4$
Ammonium salts, e.g. ammonium chloride, $NH_4Cl$ ammonium sulphate, $(NH_4)_2SO_4$

Figure 9.3A summarises the reactions of hydrochloric acid, a typical strong acid.

*Figure 9.3A* Reactions of hydrochloric acid

Figure 9.3B summarises the reactions of the strong base sodium hydroxide and the weak base ammonia.

*Figure 9.3B* Some reactions of alkalis (soluble bases)

# 9.4 Neutralisation

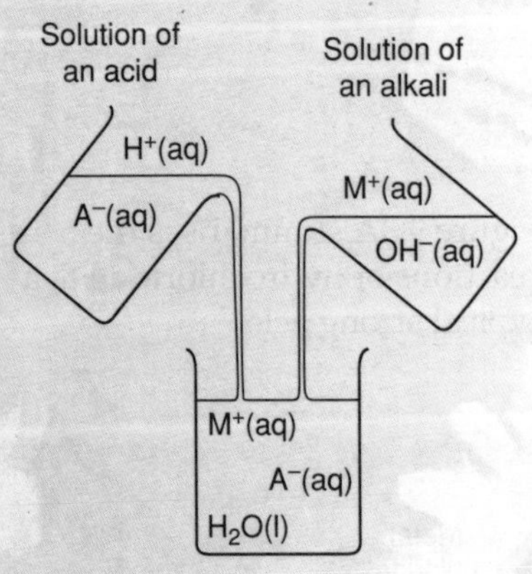


$H^+(aq)$ ions and $OH^-(aq)$ ions have combined to form water molecules, $H_2O(l)$

The solution is a solution of the salt MA

*Figure 9.4A* ▲ Acid + alkali → salt + water

## SUMMARY

Neutralisation is the combination of hydrogen ions (from an acid) with hydroxide ions (from an alkali or an insoluble base) or oxide ions (from an insoluble base) to form water and a salt.

*What takes place when an acid neutralises a soluble base?* An example is the reaction between hydrochloric acid and the alkali (soluble base) sodium hydroxide solution.

| Hydrochloric acid | + | Sodium hydroxide | → | Sodium chloride | + Water |
|---|---|---|---|---|---|
| $HCl(aq)$ | + | $NaOH(aq)$ | → | $NaCl(aq)$ | + $H_2O(l)$ |
| *acid* | + | *alkali* | → | *salt* | + *water* |

When the solutions of acid and alkali are mixed, hydrogen ions, $H^+(aq)$, and hydroxide ions, $OH^-(aq)$, combine to form water molecules.

$$H^+(aq) + OH^-(aq) \rightarrow H_2O(l)$$

Sodium ions, $Na^+(aq)$, and chloride ions, $Cl^-(aq)$, remain in the solution, which becomes a solution of sodium chloride. If you evaporate the solution, you obtain solid sodium chloride.

> Neutralisation is the combination of hydrogen ions from an acid and hydroxide ions from a base to form water molecules. In the process, a salt is formed.

*What happens when an acid neutralises an insoluble base?* An example is the reaction between sulphuric acid and copper(II) oxide

| Sulphuric acid | + | Copper(II) oxide | → | Copper(II) sulphate | + Water |
|---|---|---|---|---|---|
| $H_2SO_4(aq)$ | + | $CuO(s)$ | → | $CuSO_4(aq)$ | + $H_2O(l)$ |
| *acid* | + | *base* | → | *salt* | + *water* |

Hydrogen ions and oxide ions, $O^{2-}$, combine to form water

$$2H^+(aq) + O^{2-}(s) \rightarrow H_2O(l)$$

The resulting solution contains copper(II) ions and sulphate ions: it is a solution of copper(II) sulphate. If you evaporate it, you will obtain copper(II) sulphate crystals.

## CHECKPOINT

1. Bee stings hurt because bees inject acid into the skin. Wasp stings hurt because wasps inject alkali into the skin.
   Your little brother is stung by a bee. You are in charge. What do you use to treat the sting: 'bicarbonate of soda' (sodium hydrogencarbonate), calamine lotion (zinc carbonate), vinegar (ethanoic acid) or Milk of Magnesia (magnesium carbonate)? Would you use the same treatment for a wasp sting? Give reasons for your answers.
2. 'Acid drops' which you buy from a sweet shop contain citric acid.
   - Put an acid drop in your mouth.
   - Put some baking soda (sodium hydrogencarbonate) on your hand.
   - Lick some of the baking soda into your mouth.

   What happens to the taste of the acid drop? Why does this happen?

# 9.5 Indicators and pH

Table 9.4 ▾ The colours of some common indicators

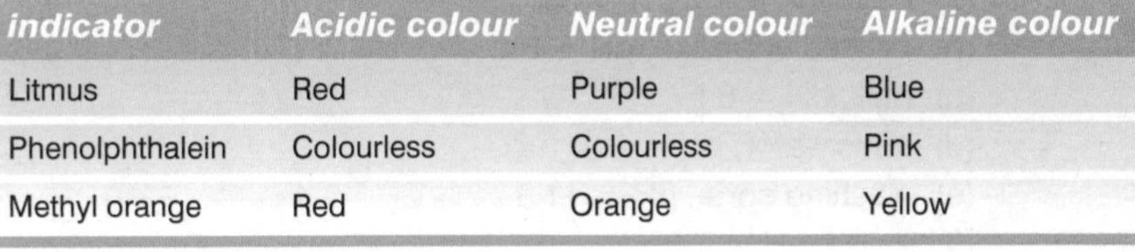

| indicator | Acidic colour | Neutral colour | Alkaline colour |
|---|---|---|---|
| Litmus | Red | Purple | Blue |
| Phenolphthalein | Colourless | Colourless | Pink |
| Methyl orange | Red | Orange | Yellow |

## SUMMARY

In general, mineral acids are stronger than organic acids. Some bases are stronger than others. Ammonia is a weak base. The pH of a solution is a measure of its acidity or alkalinity. Universal indicator turns different colours in solutions of different pH.

EXTENSION FILE ACTIVITY

Universal indicator turns different colours in strongly acidic and weakly acidic solutions. It can also distinguish between strongly basic and weakly basic solutions (see Figure 9.5A). Each universal indicator colour is given a **pH number**. The pH number measures the acidity or alkalinity of the solution. For a neutral solution, pH = 7; for an acidic solution, pH < 7; for an alkaline solution, pH > 7.

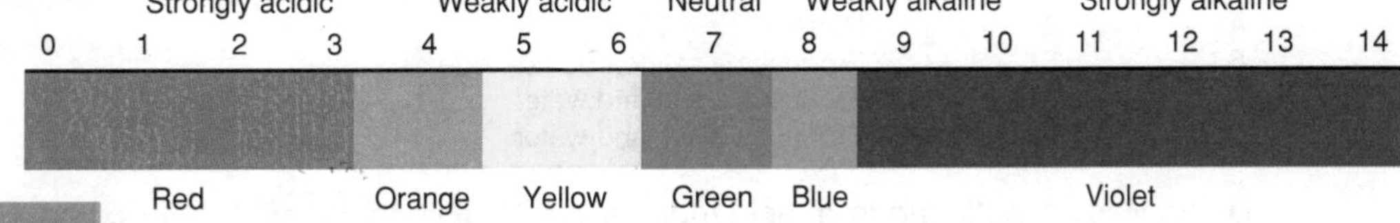


Figure 9.5A ▴ The colour of universal indicator in different pH solutions

## It's a fact!

Why must you never use both ammonia as a degreaser and chlorine as a bleach on the same cleaning job? **Don't do it! You could get a dangerous reaction between them.**

Figure 9.5B shows the pH values of solutions of some common acids and alkalis. Some salts do not have a pH value of 7. Carbonates and hydrogencarbonates are alkaline in solution.

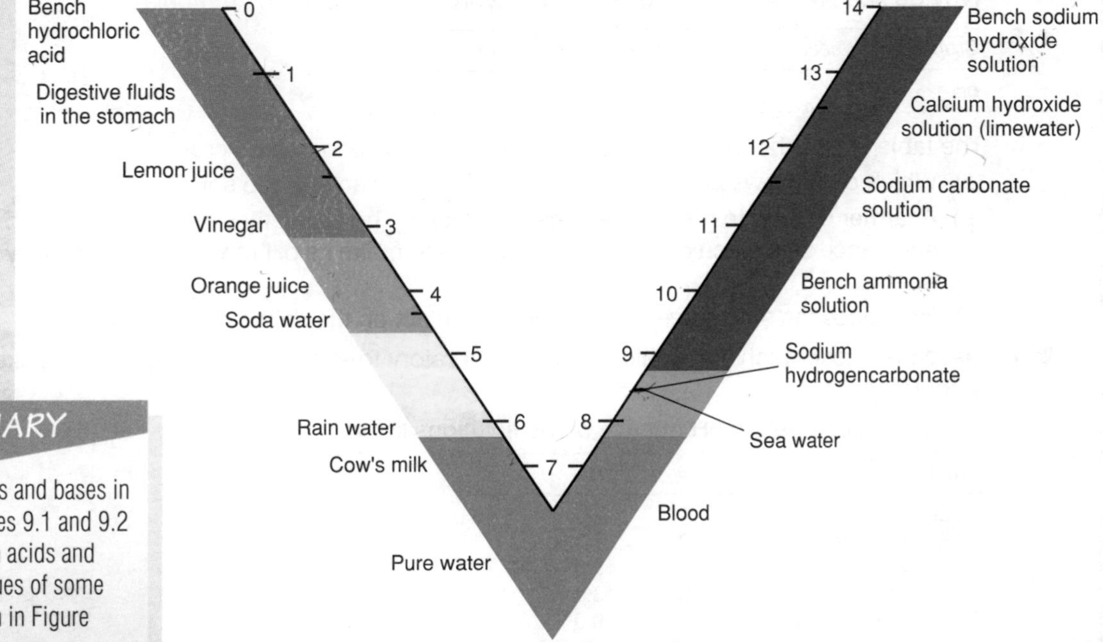


Figure 9.5B ▴ The pH values of some different solutions

## SUMMARY

We use many acids and bases in everyday life. Tables 9.1 and 9.2 list some common acids and bases. The pH values of some solutions are given in Figure 9.5B.

## CHECKPOINT

**1** Say whether these substances are strongly acidic, weakly acidic, neutral, weakly basic or strongly basic:

(a) cabbage juice, pH = 5.0
(b) pickled cabbage, pH = 3.0
(c) milk, pH = 6.5
(d) tonic water, pH = 8.2
(e) washing soda, pH = 11.5
(f) saliva, pH = 7.0
(g) blood, pH = 7.4

**2** You know what it feels like to be stung by a nettle. Is the substance which the nettle injects an acid or an alkali or neither? Think up a method of extracting some of the substance from a nettle and testing to see whether the extract is acidic or alkaline or neutral. If your teacher approves of your plan, try it out.

**3** Figure 9.5A shows the colours of universal indicator. What colour would you expect universal indicator to be when added to each of the following? (a) distilled water (b) lemon juice (c) household ammonia (d) battery acid (e) oven cleaner

**4**

| *Liquid* | *pH value* | *Reaction with acid* |
|---|---|---|
| A | 1.0 | None |
| B | 8.5 | Produces a salt, carbon dioxide and water |
| C | 8.5 | Produces a salt and water |
| D | 13 | Produces a salt and water |

Uncle Harry is suffering from acid indigestion. Explain to him why it would not be a good idea to drink either Liquid A or Liquid D. Would you advise him to take Liquid B or Liquid C? Explain your choice.

**5** The oven sprays which are sold for cleaning greasy ovens contain a concentrated solution of sodium hydroxide.
Why does sodium hydroxide clean the greasy oven?
Why does sodium hydroxide work better than ammonia?
What two safety precautions should you take to protect yourself when using an oven spray?
Why do domestic cleaning fluids contain ammonia, rather than sodium hydroxide?
Why do soap manufacturers use sodium hydroxide, rather than ammonia?

**6**

| *Crop* | *Wheat* | *Potatoes* | *Sugar beet* |
|---|---|---|---|
| *pH* | 6 | 9 | 7 |

The table shows the most suitable values of pH for growing some crops.

(a) Which crop grows best in (i) an acidic soil and (ii) an alkaline soil?
(b) A farmer wanted to grow sugar beet. On testing, he found that his soil had a pH of 5. Name a substance which could be added to the soil to make its pH more suitable for growing sugar beet.
(c) How does the substance you mention in (b) act on the soil to change its pH?

**7** Describe a test which you could do in the laboratory to show that citric acid is a weaker acid than sulphuric acid.

**8** Pair them up. Give the pH of each of the solutions listed.

| *Solution* | | *pH* |
|---|---|---|
| 1 Ethanoic acid | A | 7.0 |
| 2 Sodium chloride | B | 1.0 |
| 3 Sulphuric acid | C | 5.0 |
| 4 Ammonia | D | 13.0 |
| 5 Sodium hydroxide | E | 9.0 |

# Topic 10 Salts

## 10.1 Sodium chloride

**FIRST THOUGHTS**

Sodium chloride, NaCl, is the salt which we call 'common salt' or simply 'salt'. The average human body contains about 250 g of sodium chloride.

Sodium chloride is essential for life: it enables muscles to contract, it enables nerves to conduct nerve impulses, it regulates osmosis and it is converted into hydrochloric acid, which helps digestion to take place in the stomach. Deprived of sodium chloride, the body goes into convulsions; then paralysis and death may follow. When we sweat, we lose both water and sodium chloride. We also **excrete** sodium chloride in urine. Our kidneys control the quantity of sodium chloride which we excrete. If we eat too much salt, our kidneys excrete sodium chloride; if we eat too little salt, our kidneys excrete water but no sodium chloride.

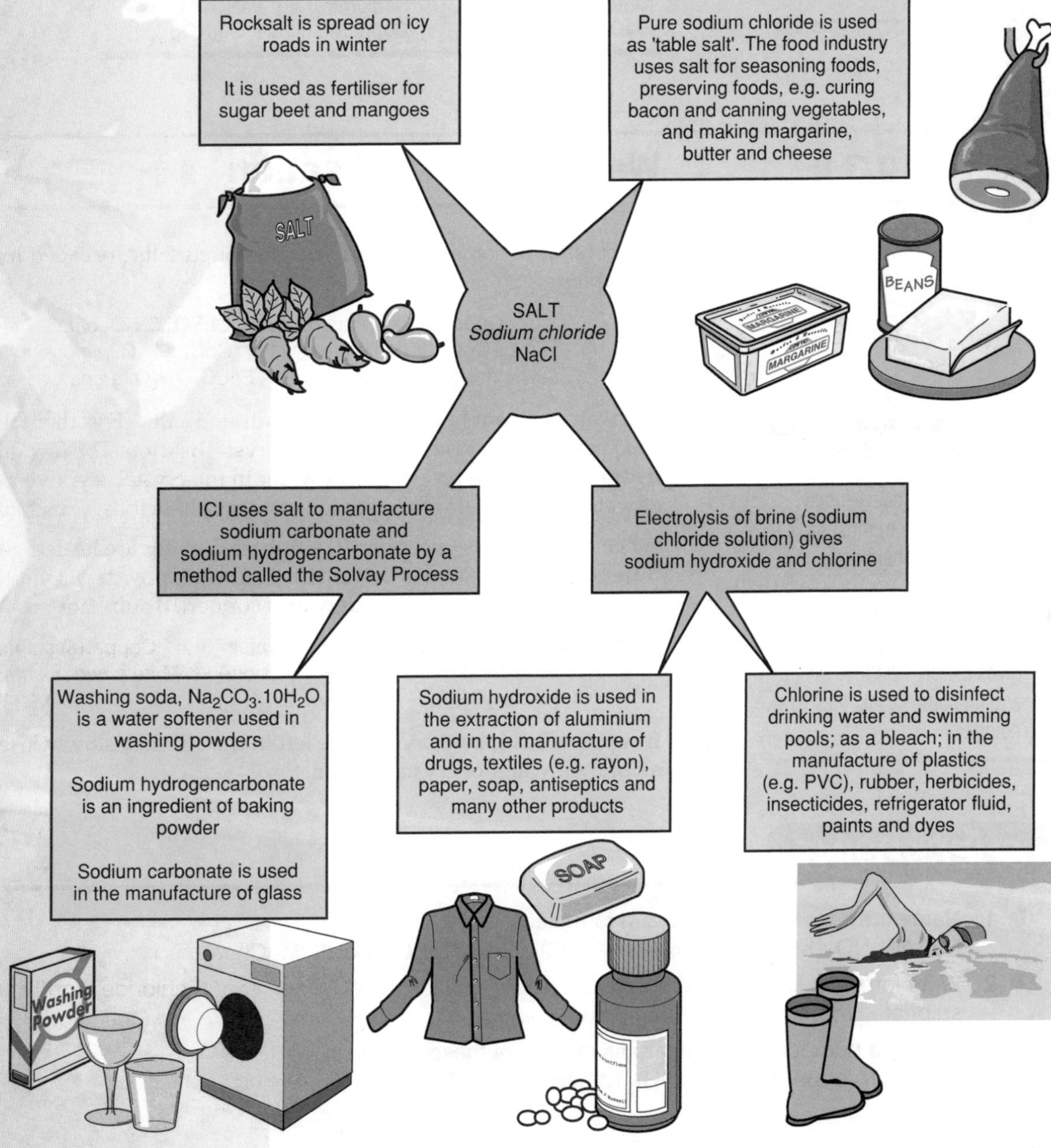

Rocksalt is spread on icy roads in winter
It is used as fertiliser for sugar beet and mangoes
Pure sodium chloride is used as 'table salt'. The food industry uses salt for seasoning foods, preserving foods, e.g. curing bacon and canning vegetables, and making margarine, butter and cheese
SALT
SALT
Sodium chloride
NaCl
MARGARINE
BEANS
ICI uses salt to manufacture sodium carbonate and sodium hydrogencarbonate by a method called the Solvay Process
Electrolysis of brine (sodium chloride solution) gives sodium hydroxide and chlorine
Washing soda, $Na_2CO_3.10H_2O$ is a water softener used in washing powders
Sodium hydrogencarbonate is an ingredient of baking powder
Sodium carbonate is used in the manufacture of glass
Sodium hydroxide is used in the extraction of aluminium and in the manufacture of drugs, textiles (e.g. rayon), paper, soap, antiseptics and many other products
Chlorine is used to disinfect drinking water and swimming pools; as a bleach; in the manufacture of plastics (e.g. PVC), rubber, herbicides, insecticides, refrigerator fluid, paints and dyes
SOAP
Washing Powder


*Figure 10.1A* ▲ Some of the uses of sodium chloride

## 10.2 Salts of some common acids

### It's a fact!

A French army under Napoleon I invaded Russia in 1812. The harsh Russian winter forced the French army to retreat. Salt starvation was one of the hardships which Napoleon's soldiers endured on their retreat from Moscow. Shortage of salt lowered their resistance to disease and epidemics spread; it prevented their wounds from healing and led to infection and death.

Many salts can be made by neutralising acids. In a salt, the hydrogen ions in the acid have been replaced by metal ions or by the ammonium ion. Salts of hydrochloric acid are called **chlorides**. Salts of sulphuric acid are called **sulphates**. Salts of nitric acid are called **nitrates**. See Table 10.1.

*Table 10.1* Salts of some common acids

| Acids | Salts |
|---|---|
| Hydrochloric acid<br>$HCl(aq)$ | Sodium chloride, $NaCl$<br>Calcium chloride, $CaCl_2$<br>Iron(II) chloride, $FeCl_2$<br>Ammonium chloride, $NH_4Cl$ |
| Sulphuric acid<br>$H_2SO_4(aq)$ | Sodium sulphate, $Na_2SO_4$<br>Zinc sulphate, $ZnSO_4$<br>Iron(III) sulphate, $Fe_2(SO_4)_3$ |
| Nitric acid<br>$HNO_3$ | Sodium nitrate, $NaNO_3$<br>Copper(II) nitrate, $Cu(NO_3)_2$ |

## 10.3 Water of crystallisation

The crystals of some salts contain water. Such salts are called **hydrates**. Examples are:

Copper(II) sulphate-5-water, $CuSO_4.5H_2O$, *blue*
Cobalt(II) chloride-6-water, $CoCl_2.6H_2O$, *pink*
Iron(II) sulphate-7-water, $FeSO_4.7H_2O$, *green*

The water present in the crystals of hydrated salts gives them their shape and their colour. It is called **water of crystallisation**. The formula of the hydrate shows the proportion of water in the crystal, e.g. five molecules of water to each pair of copper ions and sulphate ions: $CuSO_4.5H_2O$.

When blue crystals of copper(II) sulphate-5-water are heated gently, the water of crystallisation is driven off. The white powdery solid that remains is anhydrous (without water) copper(II) sulphate.

| | | | |
|---|---|---|---|
| Copper(II) sulphate-5-water | → | Water + | Copper(II) sulphate |
| (blue crystals) | → | (vapour) + | (white powder, *anhydrous*) |
| $CuSO_4.5H_2O(s)$ | → | $5H_2O(g)$ + | $CuSO_4(s)$ |

If copper(II) sulphate crystals are left in the air, they slowly lose some or all of their water of crystallisation.

### SUMMARY

- Some salts occur naturally, but most of them must be made by the chemical industry.
- Some salts crystallise with water of crystallisation. This gives the crystals their shape and, in some cases, their colour.

### CHECKPOINT

1. Name the salts with the following formulas: $NaI$, $NH_4NO_3$, $KBr$, $BaCO_3$, $Na_2SO_4$, $ZnCl_2$, $CrCl_3$, $NiSO_4$, $CaCl_2$, $MgCl_2$.
2. Write formulas for the salts: potassium nitrate, potassium bromide, iron(II) chloride, iron(II) sulphate, iron(III) bromide, ammonium chloride.
3. State the number of oxygen atoms in each of these formulas: (a) $Pb_3O_4$ (b) $3Al_2O_3$ (c) $Fe(OH)_3$ (d) $MgSO_4.7H_2O$ (e) $Co(NO_3)_3.9H_2O$

4 The table shows how samples of four substances change in mass when they are left in the air.

| *Substance* | *Mass of sample fresh from bottle (g)* | *Mass of sample after 1 week in the air (g)* |
|---|---|---|
| A | 13.10 | 13.21 |
| B | 15.25 | 15.25 |
| C | 11.95 | 5.01 |

Which of the three substances in the table (a) loses water of crystallisation on standing (b) absorbs water from the air and (c) is unchanged by exposure to air?

5 *Barbara:* Bother! The holes in the salt cellar are blocked again. I wonder why that happens.
*Razwan:* It happens because sodium chloride absorbs water from the air.
*Gwynneth:* My mum puts a few grains of rice in the salt cellar. That stops it getting clogged.
(a) Explain why the absorption of water by sodium chloride will block the salt cellar.
(b) Explain how rice grains stop the salt cellar clogging.
(c) Describe an experiment which you could do to show that what you say in (b) is correct.

6 When a tin of biscuits is left open, do the biscuits absorb water vapour from the air or give out water vapour? Describe an experiment which you could do to find out.

## 10.4 ▶ Useful salts

**FIRST THOUGHTS**

Salts – who needs them? Farmers, doctors, dentists, industrial manufacturers, photographers: they all use salts in their work.

### Agricultural uses

**NPK fertilisers** are mixtures of salts. They contain ammonium nitrate and ammonium sulphate to supply nitrogen to the soil. They contain phosphates to supply phosphorus and potassium chloride as a source of potassium.

Potato blight used to be a serious problem. It destroyed the potato crop in Ireland in 1846. Thousands of people starved and many more were forced to leave the country in search of food. Now, the fungus which causes potato blight can be killed by spraying with a solution of a simple salt, copper(II) sulphate. This fungicide is also used on vines to protect the grape harvest.

*Figure 10.4A* ▲ NPK fertiliser

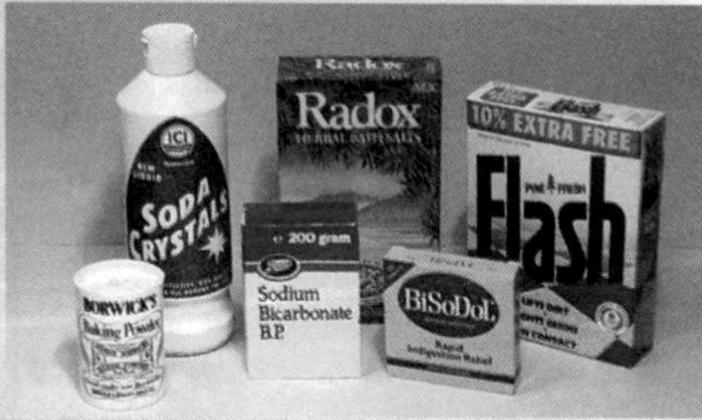


*Figure 10.4B* ▲ Washing soda, bath salts, baking powder

## Domestic uses

Sodium carbonate-10-water is known as 'washing soda'. It is used as a **water softener**. It is an ingredient of washing powders and is also sold as bath salts.

Sodium hydrogencarbonate is known as 'baking soda'. It is added to self-raising flour. When heated at a moderate oven temperature, it decomposes.

$$\text{Sodium hydrogencarbonate} \rightarrow \text{Sodium carbonate} + \text{Carbon dioxide} + \text{Steam}$$

$$2NaHCO_3(s) \rightarrow Na_2CO_3(s) + CO_2(g) + H_2O(g)$$

The carbon dioxide and steam which are formed make bread and cakes 'rise'.

## Medical uses

Figure 10.4C shows someone's leg being set in a plaster cast. Plaster of Paris is the salt calcium sulphate-$\frac{1}{2}$-water, $CaSO_4.\frac{1}{2}H_2O$. It is made by heating calcium sulphate-2-water, $CaSO_4.2H_2O$, which is mined. When plaster of Paris is mixed with water, it combines and sets to form a strong 'plaster cast'. It is also used for plastering walls.

People who are always tired and lacking in energy may be suffering from anaemia. This illness is caused by a shortage of haemoglobin (the red pigment which contains iron) in the blood. It can be cured by taking iron compounds in the diet. The 'iron tablets' which anaemic people may be given often contain iron(II) sulphate-7-water.

Patients who are suspected of having a stomach ulcer may be given a 'barium meal'. It contains the salt barium sulphate. After a barium meal, an X-ray photograph of the body shows the path taken by the salt. Being large, barium ions show up well in X-ray photographs.

### Science at work

Liquid crystals are a form of matter which has a regular structure like a solid and which flows like a liquid. Many liquid crystals change their colour with the temperature. A recent invention uses the behaviour of liquid crystals to diagnose appendicitis. A thin plastic film coated with a suitable liquid crystal is placed on the patient's abdomen. Since an inflamed appendix produces heat, it shows up immediately as a change in the colour of the film. Appendicitis can be difficult to diagnose by other methods because the pain is often felt some distance from the appendix. Can you think of some further uses of liquid crystal temperature strips?

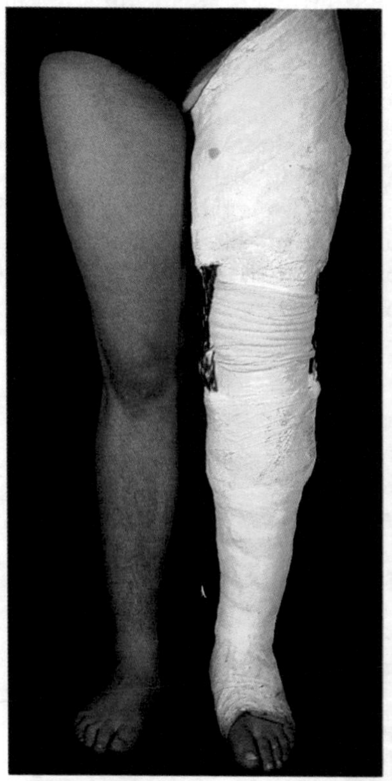

*Figure 10.4C* ▲ A patient having his leg set in a plaster cast

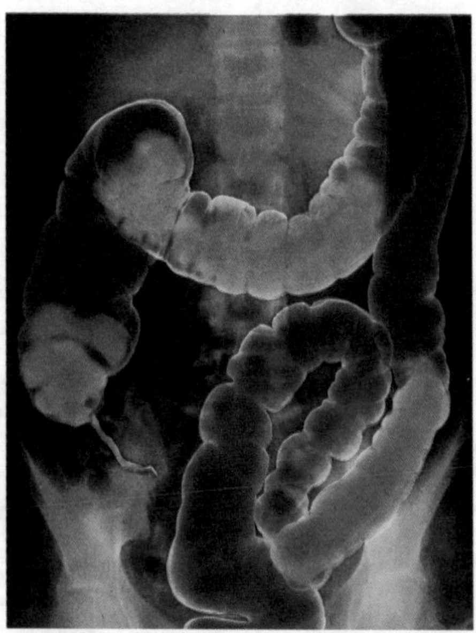

*Figure 10.4D* ▲ X-ray photograph of a patient after consuming a barium meal

## KEY SCIENTIST

The fluoride story started in 1901 when Fred McKay began work as a dentist in Colorado, USA. Many of his patients had mottled teeth or dark brown stains on their teeth. He could find no information about this condition and called it 'Colorado Brown Stain'. He became convinced that the stain was caused by something in the water. After Oakley in Idaho changed its source of drinking water, the children there no longer developed mottled or stained teeth. The drinking water in a number of towns was analysed and showed that teeth were affected when there were over 2 p.p.m (parts per million) of fluoride in the drinking water. It affected children's teeth more than adults' teeth. McKay also noticed that people with 'Colorado Brown Stain' had less dental decay than other patients!

## EXTENSION FILE ASSIGNMENT

What is the evidence that fluorides promote healthy teeth?

Analyse the data and see what you think.

You could use a computer program to display the data as a bar graph.

## SUMMARY

Salts are used
- in the home, e.g. for cleaning and in baking
- in medicine, e.g. plaster casts, 'iron' tablets, 'barium meals'
- in dentistry, e.g. fluorides to protect against decay
- in photography – silver bromide and 'fixer' and 'developer'.

Some salts occur naturally; others do not. Methods of making salts are therefore important.

## Fluorides for healthy teeth

*Why do dentists recommend that people use toothpastes containing the salt calcium fluoride*? Tooth enamel reacts with calcium fluoride to form a harder enamel which is better at resisting attack by mouth acids. To make sure that everyone gets protection from tooth decay, many water companies add a small amount of calcium fluoride to drinking water. The concentration of fluoride ions must not rise above 1 p.p.m. Spending 7p per person each year on fluoridation saves the National Health Service £6 per person each year in dentistry. Some people are violently opposed to the fluoridation of water supplies. This is because they are worried about the effects of drinking too much fluoride. Worldwide, 230 million people drink fluoridated water.

## Industrial uses

Sodium chloride has many uses in industry; see Figure 10.1A.

## Photography

Silver bromide, AgBr, is the salt which is used in black and white photography. Light affects silver bromide. A photographic film is a piece of celluloid covered with a thin layer of gelatin containing silver bromide. When you take a photograph, light falls on to some areas of the film. In exposed areas of film, light converts some silver ions, $Ag^+$, into silver atoms, which are black. These black atoms form a *latent image* of the object.

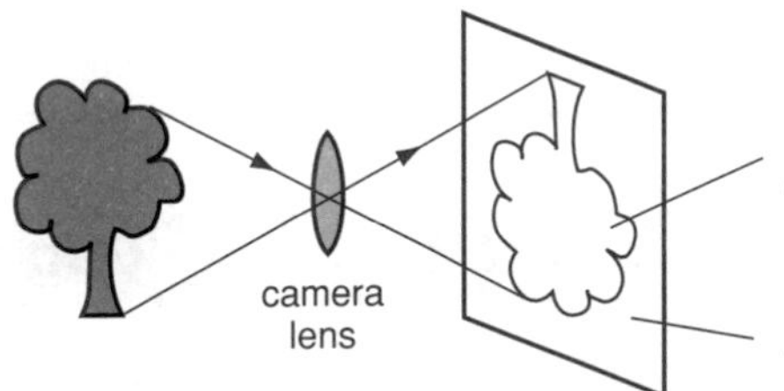
camera lens
Light from the tree falls onto the film here
More light, coming from the sky, falls on the film here

1 The film is **exposed**. The light converts some silver ions, $Ag^+$, into silver atoms, which are black. The more light that falls on the film, the more silver ions are converted. These black atoms form a **latent image** of the tree ('latent' means hidden)

2 The film is **developed**. It is placed in a 'developer' solution whivh continues the process started by light, converting more silver ions in the areas affected by light into silver atoms

3 The film is placed into a solution of **fixer** which removes unchanged silver bromide

4 The result is a **negative**. The tree appears light against a dark background

5 To make a **print** the negative is placed on a piece of light-sensitive printing paper and exposed to light. The pattern of dark and light areas in the print is the reverse of that of the negative

*Figure 10.4E* ▲ Black and white photography

Some of these useful salts are mined, e.g. sodium chloride, magnesium sulphate and calcium sulphate. Others must be made by the chemical industry, e.g. silver bromide and copper(II) sulphate.

# 10.5 Methods of making salts

FIRST THOUGHTS

Since salts are so useful, chemists have found methods of making them. This topic describes the methods.

The method which you choose for making a salt depends on whether it is soluble or insoluble. Soluble salts are made by neutralising an acid. Insoluble salts are made by adding two solutions. Table 10.4 summarises the facts about the solubility of salts.

*Table 10.2* Soluble and insoluble salts

| Salts | Soluble | Insoluble |
|---|---|---|
| Chlorides | Most are soluble | Silver chloride<br>Lead(II) chloride |
| Sulphates | Most are soluble | Barium sulphate<br>Calcium sulphate<br>Lead(II) sulphate |
| Nitrates | All are soluble | None |
| Carbonates | Sodium and potassium carbonates | Most are insoluble |
| Ethanoates | All are soluble | None |
| Sodium salts | All are soluble | None |
| Potassium salts | All are soluble | None |
| Ammonium salts | All are soluble | None |

SUMMARY

Soluble salts are made by the reactions:

- Acid + Reactive metal → Salt + Hydrogen
- Acid + Metal oxide → Salt + Water
- Acid + Metal carbonate → Salt + Water + Carbon dioxide

In these three methods, an excess of the solid reactant is used, and unreacted solid is removed by filtration.

## Methods for making soluble salts

An acid is neutralised by adding a metal, a solid base or a solid metal carbonate or a solution of an alkali.

*Method 1:* Acid + Metal → Salt + Hydrogen
*Method 2:* Acid + Metal oxide → Salt + Water
*Method 3:* Acid + Metal carbonate → Salt + Water + Carbon dioxide
*Method 4:* Acid + Alkali → Salt + Water

The practical details of Methods 1–3 are as follows.

**Step one** Add an excess (more than enough) of the solid to the acid (see Figure 10.5A).

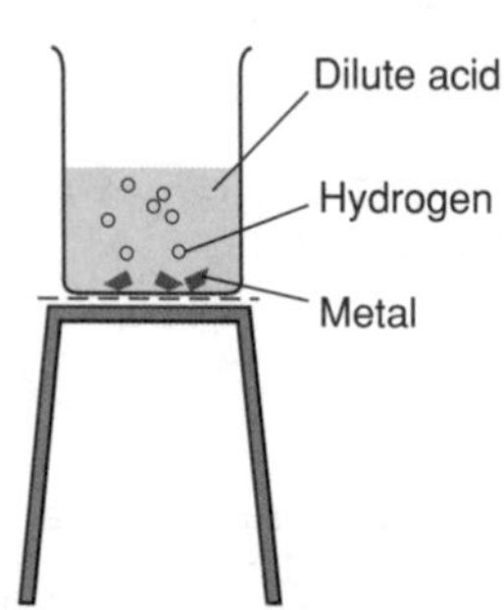


**Method 1**
Warm the acid.
Switch off the Bunsen.
Add an excess of the metal to the acid.
Wait until no more hydrogen is evolved.
The reaction is then complete.

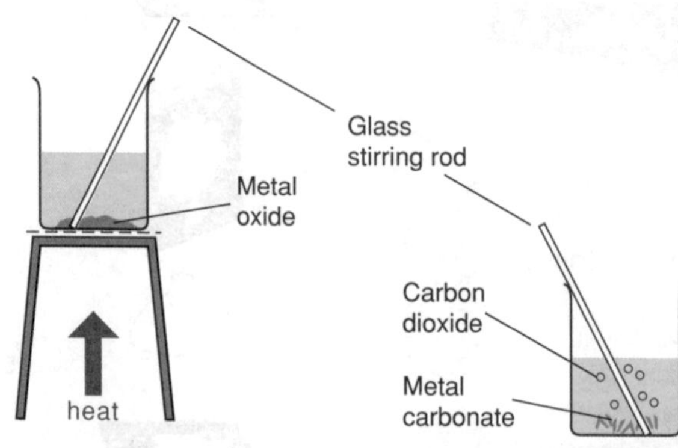


**Method 2**
Add an excess of the metal oxide to the acid. Wait until the solution no longer turns blue litmus red. The reaction is then over.

**Method 3**
Add an excess of the metal carbonate to the acid. Wait until no more carbon dioxide is evolved. The reaction is then over.

*Figure 10.5A* Adding an excess of the solid reactant to the acid

**Step two** Filter to remove the excess of solid (Figure 10.5B).

**Step three** Gently evaporate the filtrate (Figure 10.5C).

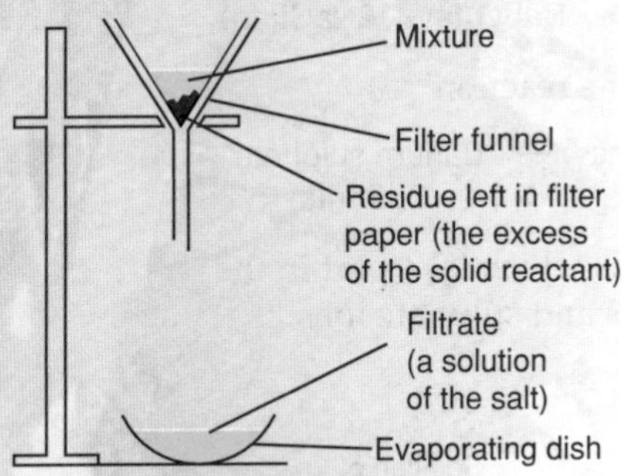


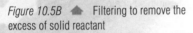

*Figure 10.5B* ▲ Filtering to remove the excess of solid reactant

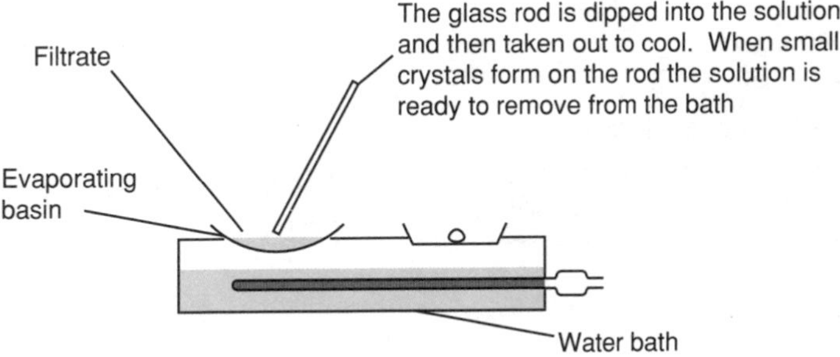


*Figure 10.5C* ▲ Evaporating the filtrate

**Step four** As the solution cools, crystals of the salt form. Separate the crystals from the solution by filtration. Use a little distilled water to wash the crystals in the filter funnel. Then leave the crystals to dry.

## ■ Acid + alkali

*Method 4* Reaction between an acid and an alkali

Ammonia is an alkali. The preparation of ammonium salts is important because they are used as fertilisers. A laboratory method for making ammonium sulphate is shown in Figure 10.5D.

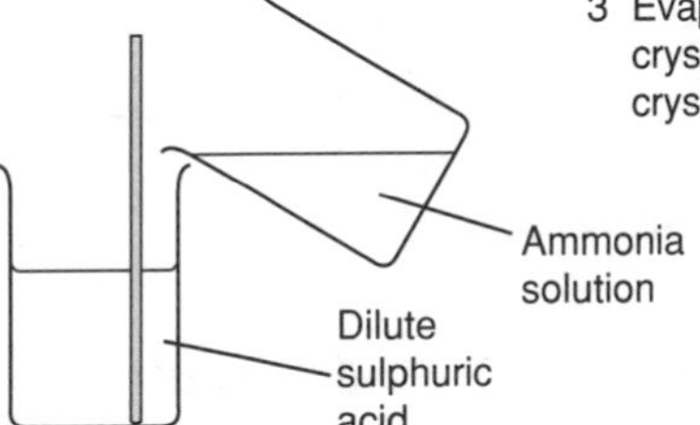

1 Add the ammonia solution to the dilute sulphuric acid, stirring constantly
2 From time to time, remove a drop of acid on the end of a glass rod. Spot it onto a strip of indicator paper on a white tile. When the test shows that the solution has become akaline, stop adding ammonia. You have now added an excess of ammonia
3 Evaporate the solution until it begins to crystallise. Leave it to stand. Filter to obtain crystals of ammonium sulphate
Ammonia solution
Dilute sulphuric acid

*Figure 10.5D* ▲ Making ammonium sulphate

### SUMMARY

Soluble salts can be made by the reaction:
Acid + Alkali → Salt + Water

This method is used for ammonium salts. An excess of ammonia is added. The excess is removed when the solution is evaporated.

## Method for insoluble salts

Insoluble salts are made by mixing two solutions. For example, you can make the insoluble salt barium sulphate by mixing a solution of the soluble salt barium chloride with a solution of the soluble salt sodium sulphate. When the two solutions are mixed, the insoluble salt barium sulphate is precipitated (thrown out of solution). What do you think remains in solution? The method is called **precipitation**.

## SUMMARY

Insoluble salts are made by precipitation. A solution of a soluble salt of the metal is added to a solution of a soluble salt of the acid.

Remember

- All nitrates are soluble.
- All sodium, potassium and ammonium salts are soluble.

## Precipitation

The equation for the reaction is

| Barium chloride | + | Sodium sulphate | → | Barium sulphate | + | Sodium chloride |
|---|---|---|---|---|---|---|
| $BaCl_2(aq)$ | + | $Na_2SO_4(aq)$ | → | $BaSO_4(s)$ | + | $2NaCl(aq)$ |

An ionic equation can be written for the reaction

| Barium ions | + | Sulphate ions | → | Barium sulphate |
|---|---|---|---|---|
| $Ba^{2+}(aq)$ | + | $SO_4^{2-}(aq)$ | → | $BaSO_4(s)$ |

The ionic equation shows only the ions which take part in the precipitation reaction: the barium ions and sulphate ions.

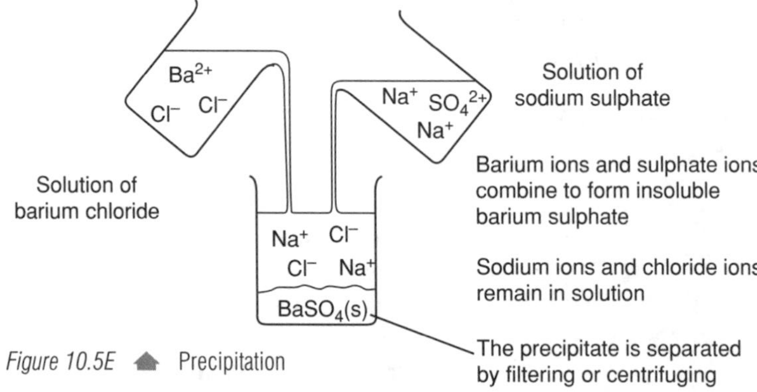

$Ba^{2+}$
$Cl^-$ $Cl^-$
$Na^+$ $SO_4^{2+}$
$Na^+$
Solution of sodium sulphate
Solution of barium chloride
Barium ions and sulphate ions combine to form insoluble barium sulphate
$Na^+$ $Cl^-$
$Cl^-$ $Na^+$
$BaSO_4(s)$
Sodium ions and chloride ions remain in solution
The precipitate is separated by filtering or centrifuging

*Figure 10.5E* ▲ Precipitation

## CHECKPOINT

1 Refer to Figure 10.5A.
   (a) How do you know that all the acid has been used up
      (i) in the reaction with a metal?
      (ii) in the reaction with a metal oxide?
      (iii) in the reaction with a metal carbonate?
   (b) Why is it important to make sure that all the acid is used up?
   (c) If some acid were left unneutralised in Step 1, what would happen to it in Step 3? How would it affect the crystals of salt formed in Step 4?
   (d) Why is it easier to remove an excess of base than an excess of acid?

2 Complete the following word equations
   (a) magnesium + sulphuric acid → _____ sulphate + _____
   (b) zinc oxide + hydrochloric acid → _____ chloride + _____
   (c) calcium carbonate + hydrochloric acid → _____ + _____ + _____
   (d) nickel oxide + hydrochloric acid → _____ + _____
   (e) chromium oxide + sulphuric acid → _____ + _____
   (f) magnesium oxide + nitric acid → _____ nitrate + _____

3 Write chemical equations for the reactions (a), (b) and (c) in Question 2.

4 Refer to Table 10.4 for solubility
   (a) Strontium sulphate, $SrSO_4$, is insoluble. Which soluble strontium salt and which soluble sulphate could you use to make strontium sulphate?
   (b) What would you do to obtain a dry specimen of strontium sulphate?
   (c) Write a word equation for the reaction.
   (d) Write an ionic equation for the reaction (the strontium ion is $Sr^{2+}$).

5 Barium carbonate, $BaCO_3$, is insoluble.
   (a) Name two solutions which you could mix to give a precipitate of barium carbonate.
   (b) Say what you would do to obtain barium carbonate from the mixture.
   (c) Write a word equation for the reaction.
   (d) Write an ionic equation (the barium ion is $Ba^{2+}$).

# Topic 11 The Periodic Table

## 11.1 Arranging elements in order

FIRST THOUGHTS

The physical and chemical properties of elements and the arrangement of electrons in atoms of the elements; they all fall into place in the Periodic Table.

The atom is composed of protons, neutrons and electrons; see Topic 6.2. The number of protons in an atom is the same as the number of electrons and is called the **atomic number**. Electrons are arranged in groups at different distances from the nucleus called shells; see Topic 6.4. The arrangement of electrons in the atom of an element is called the **electron configuration**. For example calcium with 20 electrons in the atom has the electron configuration 2.8.8.2, meaning that there are 2 electrons in the first shell (nearest to the nucleus), 8 in the second shell, 8 in the third shell and 2 electrons in the fourth shell (furthest away from the nucleus). In this topic we shall see how the electron configurations of elements tie in with their chemical reactions. Some interesting patterns emerge when the elements are taken in the order of their atomic numbers and then arranged in rows. A new row is started after each noble gas (see Table 11.1). The arrangement is called the Periodic Table. You will have seen copies of the Periodic Table on the walls of chemistry laboratories. It simplifies the job of learning about the chemical elements.

Check that you remember the meanings of these terms:

- atomic number
- shell of electrons
- electron configuration.

If not, look at Topic 6 again!

*Table 11.1* A section of the Periodic Table

| | Group 1 | Group 2 | Group 3 | Group 4 | Group 5 | Group 6 | Group 7 | Group 0 |
|---|---|---|---|---|---|---|---|---|
| Period 1 | H (1) | | | | | | | He (2) |
| Period 2 | Li (2.1) | Be (2.2) | B (2.3) | C (2.4) | N (2.5) | O (2.6) | F (2.7) | Ne (2.8) |
| Period 3 | Na (2.8.1) | Mg (2.8.2) | Al (2.8.3) | Si (2.8.4) | P (2.8.5) | S (2.8.6) | Cl (2.8.7) | Ar (2.8.8) |
| Period 4 | K (2.8.8.1) | Ca (2.8.8.2) | | | | | | |

You can see that the arrangement in Table 11.1 has the following features:

- The elements are listed in order of increasing atomic number.
- The eight vertical columns are called **groups**. Elements which have the same number of electrons in the outermost shell fall into the same group of the Periodic Table.
- The noble gases are in Group 0. For the rest of the elements, the group number is the number of electrons in the outermost shell.
- The horizontal rows are called **periods**. The first period contains only hydrogen and helium. The second period contains the elements lithium to neon. The third period contains the elements sodium to argon.

The complete Periodic Table is shown in Figure 11.1A and at the back of this book.

The Periodic Table lists the elements in order of a certain property. What is it?

What can you say about

- all the members of a group of the Periodic Table?
- all the members of a period of the Periodic Table?

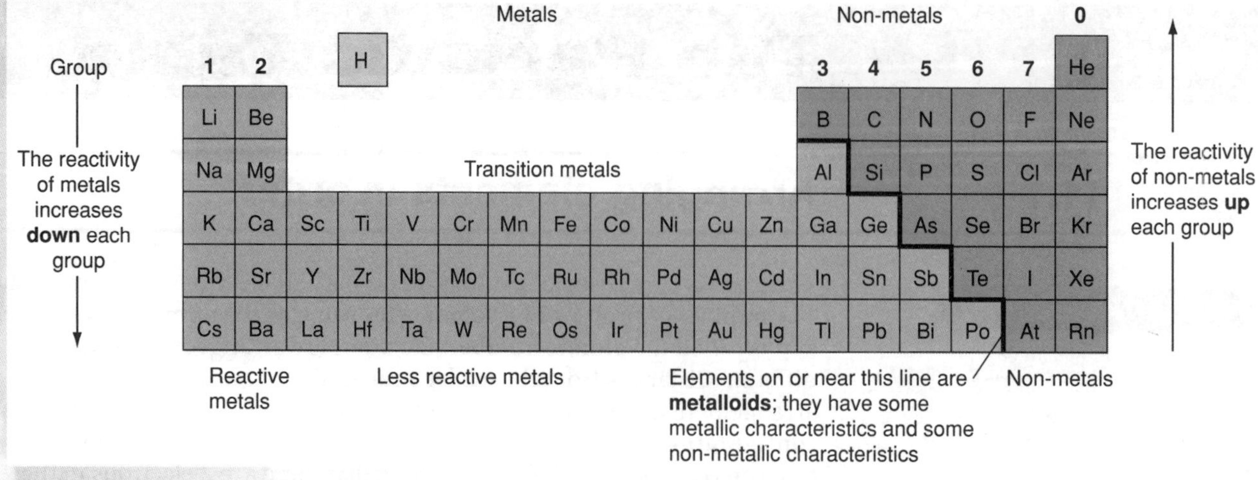


*Figure 11.1A* ▲ The Periodic Table

## 11.2 Patterns in the Periodic Table

Could you list the differences between the physical properties of metallic and non-metallic elements? If not, refer to Table 4.1.

Patterns can be seen in the arrangement of the elements in the Periodic Table. The metallic elements are on the left-hand side and in the centre block and the non-metallic elements on the right-hand side.

Table 11.2 summarises the differences between the chemical properties of metallic and non-metallic elements.

*Table 11.2* ▼ Chemical properties of metallic and non-metallic elements

| Metallic elements | Non-metallic elements |
|---|---|
| Metals which are high in the reactivity series react with dilute acids to give hydrogen and a salt of the metal | Non-metallic elements do not react with dilute acids |
| Metallic elements form positive ions, e.g. $Na^+$, $Zn^{2+}$, $Fe^{3+}$ | Non-metallic elements form negative ions, e.g. $Cl^-$, $O^{2-}$ |
| Metal oxides and hydroxides are bases, e.g. $Na_2O$, CaO, NaOH. If they dissolve in water they give alkaline solutions, e.g. NaOH | Many oxides are acids and dissolve in water to give acidic solutions, e.g. $CO_2$, $SO_2$. Some oxides are neutral and insoluble, e.g. CO |
| The chlorides of metals are ionic crystalline solids, e.g. NaCl | The chlorides of the non-metals are covalent liquids or gases, e.g. HCl(g), $CCl_4$(l) |

**SUMMARY**

Metallic and non-metallic elements differ in:

- their reactions with acids,
- the acid–base nature of their oxides,
- the nature of their chlorides,
- the type of ions they form.

### Metals

Table 11.2 compares

- the alkali metals in Group 1,
- the alkaline earths in Group 2 and
- the transition metals in between Groups 2 and 3.

The reactive metals are at the left-hand side of the table, less reactive metallic elements in the middle block and non-metallic elements at the right-hand side.

The differences between the metals in Group 1, those in Group 2 and the transition metals are summarised in Table 11.3 (M stands for the symbol of a metallic element). Dilute sulphuric acid reacts with metals in the same way as dilute hydrochloric acid. Dilute nitric acid is an oxidising agent and attacks metals, e.g. copper, which are not sufficiently reactive to react with other dilute acids.

*Table 11.3* Some reactions of metals

**Group 1 • The alkali metals**

| *Element* | *Symbol* | *Reaction with ...* air | water | non-metallic elements | dilute hydrochloric acid | *Trend* |
|---|---|---|---|---|---|---|
| Lithium<br>Sodium<br>Potassium<br>Rubidium<br>Caesium | Li<br>Na<br>K<br>Rb<br>Cs | All burn vigorously to form an oxide of formula $M_2O$ (M=symbol of metal) | All are stored under oil<br>They react vigorously with cold water to give hydrogen and the hydroxide MOH.<br>The hydroxides are all strong alkalis. | All combine with non-metals to form salts (and oxides).<br>The salts are crystalline ionic solids. The alkali metals are the cations (positive ions) in the salts. | The reaction is dangerously violent. | The vigour of all these reactions increases down the group. |

**Group 2 • The alkaline earths**

| *Element* | *Symbol* | *Reaction with ...* air | water | dilute hydrochloric acid | *Trend* |
|---|---|---|---|---|---|
| Beryllium<br>Magnesium<br>Calcium<br>Strontium<br>Barium | Be<br>Mg<br>Ca<br>Sr<br>Ba | Burn to form the strongly basic oxides, MO, which are sparingly soluble or insoluble. | Be reacts very slowly.<br>Mg burns in steam.<br>Ca, Sr, Ba react readily to form hydrogen and the alkali $M(OH)_2$. | React readily to give hydrogen and a salt, e.g. $MgCl_2$. | The vigour of all these reactions increases down the group.<br>Group 2 elements are less reactive than Group 1. |

**Transition metals • The block of elements between Group 2 and Group 3**

| *Element* | *Symbol* | *Reaction with ...* air | water | dilute hydrochloric acid | *Trend* |
|---|---|---|---|---|---|
| Iron<br>Zinc | Fe<br>Zn | When heated, form oxides without burning. The oxides and hydroxides are weaker bases than those of Groups 1 and 2 and are insoluble | Iron rusts slowly.<br>Iron and zinc react with steam to form hydrogen and the oxide. | Iron and zinc react to give hydrogen and a salt. | Transition metals are less reactive than Groups 1 and 2.<br>In general, their compounds are coloured.<br>They are used as catalysts. |
| Copper | Cu | | Copper does not react. | Copper does not react. | |

*Table 11.4* Physical properties of Group 1

| *Element* | *Symbol* | *m.p. (°C)* | *b.p. (°C)* | *Density (g/cm³)* |
|---|---|---|---|---|
| Lithium | Li | 180 | 1336 | 0.53 |
| Sodium | Na | 98 | 883 | 0.97 |
| Potassium | K | 64 | 759 | 0.86 |
| Rubidium | Rb | 39 | 700 | 1.53 |
| Caesium | Cs | 29 | 690 | 1.9 |

## SUMMARY

The alkali metals of Group 1 are very reactive.
Their oxides are strong bases.
Their hydroxides are strong alkalis.
Reactivity increases with the size of the atom (down the group).
The alkaline earths of Group 2 are less reactive than Group 1.
Reactivity increases with the size of the atom.

## The alkali metals

From Table 11.3, you can see how the Periodic Table makes it easier to learn about all the elements. Look at the elements in Group 1: lithium, sodium, potassium, rubidium and caesium. They are all very reactive metals. Their oxides and hydroxides have the general formulas $M_2O$ and MOH (where M is the symbol of the metallic element) and are strongly basic. The oxides and hydroxides dissolve in water to give strongly alkaline solutions. The reactivity of the alkali metals increases as you pass down the group. If you know these facts, you do not need to learn the properties of all the metals separately. If you know the properties of sodium, you can predict those of potassium and lithium. The Periodic Table saves you from having to learn the properties of 106 elements separately!

## The alkaline earths

The metals in Group 2 are less reactive than those in Group 1. They form basic oxides and hydroxides with the general formulas MO and $M(OH)_2$. Their oxides and hydroxides are either sparingly soluble or insoluble. The reactivity of the alkaline earths increases as you pass down the group. Again, if you know the chemical reactions of one element, you can predict the reactions of other elements in the group.

# The transition metals

The transition metals in the block between Group 2 and Group 3 are a set of similar metallic elements. They are less reactive than those in Groups 1 and 2. Their oxides and hydroxides are less strongly basic and are insoluble.

## Physical properties

Transition metals are hard and dense. They are good conductors of heat and electricity (e.g. copper wire in electrical circuits). Their melting points, boiling points and heats of melting are all higher than those for Group 1 and Group 2 metals. These are all a measure of the strength of the metallic bond. Iron, cobalt and nickel are strongly magnetic. Some of the other transition metals are weakly magnetic.

## Extraction

Transition metals are less reactive than those in Groups 1 and 2. They can be extracted from their ores by means of chemical reducing agents. The method of extraction follows the steps listed below.

1 The ore is concentrated, e.g. by flotation.
2 Sulphide ores are roasted to convert them into oxides.
3 The oxide is reduced by heating with coke to form the metal and carbon monoxide.
4 The metal is purified.

**SUMMARY**

Transition metals compared with Groups 1 and 2 are
- less reactive
- harder, denser and stronger
- use more than one valency (except zinc)
- have coloured ions (except zinc)
- form oxo-ions, e.g. $MnO_4^-$
- are catalysts.

## Valency

Transition metals characteristically use more than one valency.

*Table 11.5* Some transition metals, ions and compounds

| Metal | Ions | Compounds |
|---|---|---|
| Copper | $Cu^+$, $Cu^{2+}$ | $Cu_2O$, $CuSO_4$ |
| Iron | $Fe^{2+}$, $Fe^{3+}$ | $FeSO_4$, $Fe_2(SO_4)_3$ |
| Cobalt | $Co^{2+}$, $Co^{3+}$ | $CoCl_2$, $CoCl_3$ |
| Chromium | $Cr^{2+}$, $Cr^{3+}$<br>$Cr_2O_7^{2-}$ | $CrCl_2$, $CrCl_3$<br>$K_2Cr_2O_7$ |
| Manganese | $Mn^{2+}$, $MnO_4^-$ | $MnSO_4$, $MnO_2$, $KMnO_4$ |

## Colour

Transition metals have coloured ions, except for zinc (see Table 11.6).

In addition to forming simple cations, transitions metals form oxo-ions in combination with oxygen. Examples are:

chromate(VI), $CrO_4^{2-}$ (yellow), dichromate(VI), $Cr_2O_7^{2-}$ (orange)
manganate(VII), $MnO_4^-$ (purple), zincate, $ZnO_2^{2-}$ (colourless)

*Table 11.6* The colours of some ions

| Ion | Colour |
|---|---|
| $Cu^{2+}(aq)$ | blue |
| $Fe^{2+}(aq)$ | green |
| $Fe^{3+}(aq)$ | rust |
| $Ni^{2+}(aq)$ | green |
| $Mn^{2+}(aq)$ | pink |
| $Cr^{3+}(aq)$ | blue |
| $Zn^{2+}(aq)$ | colourless |

## Catalysis

Many transition metals and their compounds are important catalysts. Examples are:

- iron and iron(III) oxide in the Haber process for making ammonia (Topic 23.2)
- platinum in the oxidation of ammonia during the manufacture of nitric acid (Topic 23.3)
- vanadium(V) oxide in the Contact process for making sulphuric acid (Topic 23.4)
- nickel in the hydrogenation of oils to form fats (Topic 29.1).

## Silicon and germanium

Silicon and germanium are semiconductors.

On the borderline between metals and non-metals in the Periodic Table (Figure 11.1A) are silicon and germanium. These elements are semiconductors, intermediate between metals, which are electrical conductors, and non-metals, which are non-conductors of electricity. They are vital to the computer industry.

## The halogens

The elements in Group 7 are fluorine, chlorine, bromine and iodine. These elements are a set of very reactive non-metallic elements (see Table 11.7). They are called **the halogens** because they react with metals to form salts. (Greek: halogen = salt-former)

*Table 11.7* The halogens

| Element | Symbol | m.p. (°C) | b.p. (°C) | Colour | Reaction with metals to form salts | Reaction with hydrogen to form the compound HX | |
|---|---|---|---|---|---|---|---|
| Fluorine | F | −223 | −188 | Pale yellow | Dangerously reactive | Explodes | Reactivity decreases down the group ↓ |
| Chlorine | Cl | −103 | −35 | Yellow-green | Readily combines to form chlorides. | Explodes in sunlight | |
| Bromine | Br | −7 | 59 | Red-brown | Combines when heated to form bromides | Reacts when heated | |
| Iodine | I | 114 | 184 | Purple-black | Combines when heated to form iodides. The halogens are the anions (negative ions) in their salts. | Reaction is not complete | |

### Reactions of the halogens

1 Sodium burns in halogens vigorously to form the halide, sodium fluoride, NaF, sodium chloride, NaCl, sodium bromide, NaBr, or sodium iodide, NaI. The reaction with fluorine is dangerously explosive.

2 When heated, iron reacts vigorously with chlorine to form iron(III) chloride, $FeCl_3$. With bromine the product is iron(III) bromide, $FeBr_3$. With iodine the reaction is less vigorous, and the product is iron(II) iodide, $FeI_2$. Iodine is a less powerful oxidising agent than the other halogens; see Topic 12.5.

3 The halogens are oxidising agents; see Topic 12.5.

4 The halogen halides (hydrogen fluoride, hydrogen chloride, hydrogen bromide and hydrogen iodide) are gases. They are covalent compounds. They dissolve in and react with water to form solutions of strong acids. The acids consist of hydrogen ions and halide ions, e.g.

Hydrogen cloride + Water → Hydrochloric acid

$$HCl(g) + aq \rightarrow H^+(aq) + Cl^-(aq)$$

**SUMMARY**

The halogens in Group 7 are a reactive group of non-metallic elements.
They react with metals to form salts.
Reactivity decreases with the size of the atom (down the group).
The halogens are oxidising agents.
Hydrogen halides react with the water to form strong acids.

SUMMARY

The noble gases in Group 0 are very unreactive.
Each atom of a noble gas has a full outer shell of electrons

## The noble gases

The elements in group 0 are helium, neon, argon, krypton, xenon and radon. These elements are called the noble gases. They are present in air (see Topic 14). The noble gases exist as single atoms, e.g. He, Ne. Their atoms do not combine in pairs to form molecules as do the atoms of most gaseous elements, e.g. $O_2$, $H_2$. For a long time, no-one was able to make the noble gases take part in any chemical reactions. In 1960, however, two of them, krypton and xenon, were made to combine with the very reactive element, fluorine. *Why are the noble gases so exceptionally unreactive?* Chemists came to the conclusion that it is the full outer shell of electrons that makes the noble gases unreactive.

CHECKPOINT

1 Which of the alkali metals (a) float on water? (b) can be cut with a knife made of iron? (c) melt at the temperature of boiling water?
   (d) Why would it be very dangerous to put an alkali metal into boiling water?
   (e) What is another name for sodium chloride?
   (f) What would you expect rubidium chloride to look like?
   (g) The alkali metals are kept under oil to protect them from the air. Which substances in the air would attack them?
   (h) What pattern can you see in (i) the melting points (ii) the boiling points of the alkali metals?

2 (a) Which of the halogens is (i) a liquid and (ii) a solid at room temperature (20 °C)?
   (b) Why is fluorine not studied in school laboratories?
   (c) Write the formulas for (i) sodium iodide (ii) potassium fluoride.

3 Magnesium chloride, $MgCl_2$, is a solid of high melting point and carbon tetrachloride, $CCl_4$, is a volatile liquid. Explain how the differences in chemical bonding account for these differences.

4 Choose from the elements: Na, Mg, Al, Si, P, S, Cl, Ar.
   (a) List the elements that react readily with cold water to form alkaline solutions.
   (b) List the elements that form sulphates.
   (c) Name the elements which exist as molecules containing (i) 1 atom (ii) 2 atoms.
   (d) Which element has both metallic and non-metallic properties?

5 The elements sodium and potassium have the electron arrangements Na (2.8.1), K (2.8.8.1). How does this explain the similarity in their reactions?

# 11.3 Electron configuration and chemical reactions

## Group 0: helium, neon, argon, krypton and xenon

The noble gases have a full outer shell of electrons, e.g. helium (2), neon (2.8) and argon (2.8.8). For many years no reactions of the noble gases were known, until in 1960 krypton and xenon were made to combine with fluorine. They exist as single atoms: their atoms do not combine to form molecules as do the atoms of other gaseous elements, e.g. $O_2$, $N_2$.

SUMMARY

The chemical properties of elements depend on the electron configurations of their atoms. The unreactive noble gases all have a full outer shell of 8 electrons.

The reactive alkali metals have a single electron in the outer shell.

The halogens are reactive non-metallic elements which have 7 electrons in the outer shell: they are 1 electron short of a full shell.

## Group 1: lithium, sodium, potassium, rubidium, caesium

The alkali metals (see Table 11.3) all have one electron in the outer shell, e.g. lithium (2.1), sodium (2.8.1) and potassium (2.8.8.1). They form ionic compounds. The outer electrons are involved in the formation of ionic bonds; see Topic 8.1–2. This is why it is believed that the similar electron configuration is the reason why the metals behave in a very similar way.

## Group 2: beryllium, magnesium, calcium, strontium, barium

The alkaline earth metals (see Table 11.3) all have two electrons in the outer shell, e.g. beryllium (2.2), magnesium (2.8.2) and calcium (2.8.8.2). It is the outer electrons that are involved in the formation of bonds. This is why it is believed that the similar electron configuration is the reason why the metals all behave in a very similar way.

## Group 7: fluorine, chlorine, bromine, iodine

The halogens all have 7 electrons in the outer shell, e.g. fluorine (2.7) and chlorine (2.8.7). They all have atoms that want to acquire an extra electron to form a halide ion and form an ionic compound. This is why the halogens all behave in a similar way; see Table 11.7.

# 11.4 Ions in the Periodic Table

The Periodic Table helps when it comes to remembering the charge on an ion.

***Investigating ions***

| *Element* | *Symbol for ion* | *Group of Periodic Table* |
|---|---|---|
| Sodium | $Na^+$ | |
| Potassium | $K^+$ | |
| Calcium | $Ca^{2+}$ | |
| Magnesium | $Mg^{2+}$ | |
| Aluminium | $Al^{3+}$ | |
| Nitrogen | $N^{3-}$ | |
| Oxygen | $O^{2-}$ | |
| Sulphur | $S^{2-}$ | |
| Chlorine | $Cl^-$ | |
| Bromine | $Br^-$ | |

- Copy the table. Fill in the number of the Group in the Periodic Table to which each of the elements belongs.
- What is the connection between the charge on a cation and the number of the Group in the Periodic Table to which the element belongs?
- What is the connection between the charge on an anion and the number of the Group in the Periodic Table to which the element belongs?
- What would you expect to be the charge on (a) a barium ion (Group 2) and (b) a fluoride ion (Group 7)?

This exercise will show you how useful the Periodic Table is. One of the things it helps you to remember is the charge on an ion.

Take the number of the group in the Periodic Table to which an element belongs. How is this related to the charge on the ions of the element?

The charge on the ions formed by an element is called the valency of that element. Magnesium form $Mg^{2+}$ ions: magnesium has a valency of 2. Sodium forms $Na^+$ ions: sodium has a valency of 1. Oxygen forms $O^{2-}$ ions: oxygen has a valency of 2. Iodine forms $I^-$ ions: iodine has a valency of 1. Some elements have a variable valency. Iron forms $Fe^{2+}$ ions and $Fe^{3+}$ ions: iron has a valency of 2 in some compounds and 3 in others.

Non-metallic elements often combine with oxygen to form anions, e.g. sulphate, $SO_4^{2-}$ and nitrate, $NO_3^-$.

- **Metallic elements**

Charge on cation = Number of the Group in the Periodic Table to which the element belongs = Valency of element

- **Non-metallic elements**

Charge on anion = 8 − Number of Group in Periodic Table to which the element belongs = Valency of element

## 11.5 The history of the Periodic Table

**FIRST THOUGHTS**

John Newlands had the idea of the Periodic Table, but it didn't catch on. Dmitri Mendeleev took over and improved on Newlands' work. He was able to predict the properties of new elements that had not yet been discovered

The person who has the credit for drawing up the Periodic Table is a Russian scientist called Dmitri Mendeleev. He extended the work of a British chemist called John Newlands. In 1864, 63 elements were known to Newlands. He arranged them in order of relative atomic mass. When he started a new row with every eighth element, he saw that elements with similar properties fell into vertical groups. He spoke of 'the regular periodic repetition of elements with similar properties'. This gave rise to the name 'periodic table'. Sometimes, there were misfits in Newlands' table, for example, iron did not seem to belong where he put it with oxygen and sulphur. Newlands' ideas were not accepted.

Mendeleev had an idea about the troublesome misfits. He realised that many elements had yet to be discovered. Instead of slotting elements into positions where they did not fit he left gaps in the table. He expected that when further elements were discovered they would fit the gaps. He predicted that an element would be discovered to fit into the gap he had left in the table below silicon.

*Table 11.8* The predictions which Mendeleev made for the undiscovered element which he called 'eka-silicon' (below silicon) compared with the properties of germanium

| *Mendeleev's predicted properties for eka-silicon, Ek (1871)* | | *The properties of germanium Ge (discovered in 1886)* |
|---|---|---|
| *Appearance* | Grey metal | Grey-white metal |
| *Density* | ~5.5 g/cm³ | 5.47 g/cm³ |
| *Relative atomic mass* | 73.4 | 72.6 |
| *Melting point* | ~800 °C | 958 °C |
| *Reaction with oxygen* | Forms the oxide $EkO_2$. The oxide EkO may also exist. | Forms the oxide $GeO_2$ GeO also exists |

**SUMMARY**

In the Periodic Table, elements are arranged in order of increasing atomic (proton) number in 8 vertical groups. The horizontal rows are called periods.

By the time Mendeleev died in 1907, many of the gaps in the table had been filled by new elements. The noble gases were unknown to Mendeleev when he drew up his table. The first of them was discovered in 1894. As the noble gases were discovered, they all fell into place between the halogens and the alkali metals. They formed a new group, Group 0. This was a spectacular success for the Periodic Table.

## CHECKPOINT

1 (a) When was germanium discovered?
(b) How do the properties of germanium compare with Mendeleev's prediction?
(c) How long did Mendeleev have to wait to see if he was right?
(d) In which group of the Periodic Table does germanium come?
(e) What important articles are manufactured from germanium and silicon?

2 Radium, Ra, is a radioactive element of atomic number 88 which falls below barium in Group 2. What can you predict about
(a) the nature of radium oxide
(b) the reaction of radium and water
(c) the reaction of radium with dilute hydrochloric acid?
State the physical state and type of bonding in any compounds you mention. Give the names and formulas of any compounds formed.

3 Astatine, At, is a radioactive element of atomic number 85 which follows fluorine in Group 7. What can you predict about
(a) the nature of its compound with hydrogen
(b) the reaction of astatine with sodium?
State the physical state and type of bonding in any compounds you mention. Give the names and formulas of any compounds formed.

# Topic 12 Oxidation-reduction reactions

## 12.1 Oxidation and reduction

**FIRST THOUGHTS**

The rocket launch shown in Figure 12.1A takes place thanks to oxidation–reduction reactions. Kerosene burns in oxygen to give the power required for lift-off. In stage 2 and stage 3 of the rocket's journey into space, hydrogen burns in oxygen. These are oxidation–reduction reactions.

**SUMMARY**

Oxidation is the gain of oxygen or loss of hydrogen by a substance. Reduction is the loss of oxygen or gain of hydrogen by a substance. An oxidising agent gives oxygen to or takes hydrogen from another substance. A reducing agent takes oxygen from or gives hydrogen to another substance.

You have already met both chemical reactions that are described as oxidation reactions and also reduction reactions. In this topic, we shall take a closer look at some of these reactions.

**Example 1**:

Copper + Oxygen → Copper(II) oxide

$2Cu(s) + O_2(g) \rightarrow CuO(s)$

In this reaction, copper has gained oxygen, and we say that copper has been **oxidised**. This reaction is an **oxidation**, and oxygen is an **oxidising agent**.

**Example 2**:

Lead(II) oxide + Hydrogen → Lead + Water

$PbO(s) + H_2(g) \rightarrow Pb(s) + H_2O(l)$

In this reaction, lead(II) oxide has lost oxygen, and we say that lead(II) oxide has been **reduced**. Hydrogen has taken oxygen from another substance, and we therefore describe hydrogen as a **reducing agent**. The reaction is a **reduction**. On the other hand, notice that hydrogen has gained oxygen: hydrogen has been oxidised. You can also describe this reaction as an oxidation. Since lead(II) oxide has given oxygen to another substance, in this reaction lead(II) oxide is an oxidising agent. Oxidation and reduction are occurring together. One reactant is the oxidising agent and the other reactant is the reducing agent.

This is reduction (PbO → Pb)

$PbO(s) + H_2(g) \rightarrow Pb(s) + H_2O(l)$

This is oxidation ($H_2$ → $H_2O$)

Oxidising agent (PbO) Reducing agent ($H_2$)

**Example 3**:

Zinc oxide + Carbon → Zinc + Carbon monoxide

$ZnO(s) + C(s) \rightarrow Zn(s) + CO(g)$

Again, you see in this example that oxidation and reduction are occurring together. This is always true, however many examples you consider. Since oxidation never occurs without reduction, it is better to call these reactions **oxidation–reduction reactions** or **redox reactions**.

**CHECKPOINT**

1 State which is the oxidising agent and which is the reducing agent in each of these reactions:
  (a) the thermit reaction: Aluminium + Iron oxide → Iron + Aluminium oxide
  (b) roasting the ore tin sulphide in air: Tin sulphide + Oxygen → Tin oxide + Sulphur dioxide
  (c) 'smelting' tin oxide with coke: Tin oxide + Carbon → Tin + Carbon monoxide
  (d) roasting the ore copper sulphide in air: $2CuS(s) + 3O_2(g) \rightarrow 2CuO(s) + 2SO_2(g)$

# 12.2 Gain or loss of electrons

**SUMMARY**

Oxidation and reduction occur together in oxidation–reduction reactions or redox reactions.

Are you ready for another way of looking at oxidation–reduction reactions?

In a redox reaction

- the reducing agent gives electrons to the oxidising agent,
- and the oxidising agent accepts them.

**How to remember**

Which is which: oxidation or reduction – loss or gain of electrons?

OIL Oxidation is loss.

RIG Reduction is gain.

Metals are reducing agents. They react by giving electrons to e.g. the hydrogen ions of an acid or the molecules of a halogen.

It is not obvious in Example 1 in Topic 12.1 that reduction accompanies oxidation.

$$\text{Copper} + \text{Oxygen} \rightarrow \text{Copper(II) oxide}$$
$$2Cu(s) + O_2(g) \rightarrow 2CuO(s)$$

Atoms of copper, Cu, have been converted into copper(II) ions, $Cu^{2+}$. Molecules of oxygen, $O_2$, have been converted into oxide ions, $O^{2-}$. This means that copper atoms have lost electrons:

$$Cu(s) \rightarrow Cu^{2+}(s) + 2e^-$$

and oxygen molecules have gained electrons:

$$O_2(g) + 4e^- \rightarrow 2O^{2-}(s)$$

This picture holds for all oxidation–reduction reactions.

- The substance which is oxidised loses electrons.
- The substance which is reduced gains electrons.
- An oxidising agent accepts electrons.
- A reducing agent gives electrons.

Oxidation is the gain of oxygen or the loss of hydrogen or the loss of electrons.

Reduction is the loss of oxygen or the gain of hydrogen or the gain of electrons.

Metals are reducing agents. Metals react (see Topics 12.1, 19.3) by losing electrons to form positive ions, $M^+$, $M^{2+}$ and $M^{3+}$. When a metal reacts with an acid:

$$\text{Metal atom} \rightarrow \text{Metal ion} + \text{electrons}$$
$$M(s) \rightarrow M^{2+}(aq) + 2e^-$$

$$\text{Hydrogen ions} + \text{electrons} \rightarrow \text{Hydrogen molecule}$$
$$2H^+(aq) + 2e^- \rightarrow H_2(g)$$

The metal atoms have supplied electrons, and hydrogen ions have accepted electrons. The metal atoms have been oxidised, and hydrogen ions have been reduced.

$$M(s) + 2H^+(aq) \rightarrow M^{2+}(aq) + H_2(g)$$

Some metals form more than one type of ion. When iron(II) compounds are converted into iron(III) compounds, $Fe^{2+}$ ions lose electrons: they are oxidised.

$$Fe^{2+}(aq) \rightarrow Fe^{3+}(aq) + e^-$$

The oxidation is brought about by a substance which can accept electrons – an oxidising agent, e.g. chlorine. Chlorine is reduced to chloride ions.

$$Cl_2(aq) + 2e^- \rightarrow 2Cl^-(aq)$$

The complete reaction is:

$$2Fe^{2+}(aq) + Cl_2(aq) \rightarrow 2Fe^{3+}(aq) + 2Cl^-(aq)$$

# 12.3 Redox reactions and cells

## Electrolysis cells

Redox reactions involve the gain and the loss of electrons. The reactions that happen at the electrodes in electrolysis involve either the gain or the loss of electrons. Let us look at the connection between electrode reactions and redox reactions.

In the electrolysis of copper(II) chloride, the electrode reactions are as follows.

$$\text{Cathode (negative electrode):} \quad Cu^{2+}(aq) + 2e^- \rightarrow Cu(s)$$

Here copper ions, $Cu^{2+}$, gain electrons: they are reduced to copper atoms, Cu.

$$\text{Anode (positive electrode):} \quad Cl^-(aq) \rightarrow Cl(g) + e^-$$

Here chloride ions, $Cl^-$ give up electrons: they are oxidised to chlorine atoms, Cl. These immediately combine to form chlorine molecules, $Cl_2$. The cathode reaction is reduction; the anode reaction is oxidation.

**SUMMARY**

In an electrolysis cell, the anode reaction is oxidation, and the cathode reaction is reduction.

## Chemical cells

A simple chemical cell can be made by immersing two different metals in an electrolyte and connecting them with a conducting wire. A current flows through the wire (see Figure 12.3A).

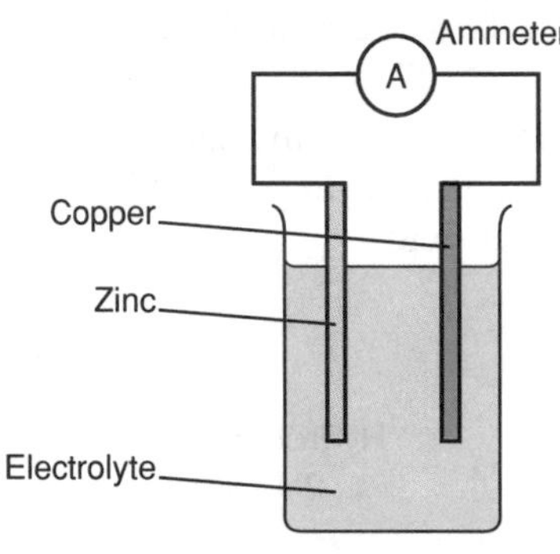


*Figure 12.3A* ▲ A chemical cell

**Example 1**: The zinc–copper chemical cell

The half-equations for the two electrode reactions are:

$$\text{Zinc electrode:} \quad Zn(s) \rightarrow Zn^{2+}(aq) + 2e^-$$

Oxidation (loss of electrons) occurs at this electrode; therefore this is the anode.

$$\text{Copper electrode:} \quad Cu^{2+}(aq) + 2e^- \rightarrow Cu(s)$$

Reduction (gain of electrons) occurs at this electrode; therefore this is the cathode. Notice that, in the chemical cell, the anode is negative and the cathode is positive.

The equation for the whole cell reaction is:

$$Zn(s) + Cu^{2+}(aq) \rightarrow Zn^{2+}(aq) + Cu(s)$$

**SUMMARY**

- The reactions which take place in chemical cells are redox reactions.
- Oxidation takes place at the anode.
- Reduction takes place at the cathode.
- Half-equations are written for the electrode reactions.
- The half-equations can be added to give the overall cell reaction.

# 12.4 ▶ Tests for oxidising agents and reducing agents

## Test for reducing agents

Acidified potassium manganate(VII) is a powerful oxidising agent. A reducing agent will change the colour from the purple of $MnO_4^-(aq)$ to the pale pink of $Mn^{2+}(aq)$. Since potassium manganate(VII) is a strong oxidising agent, it can oxidise – and therefore test for – many reducing agents.

For example, (see Figure 12.7A),

Soak a piece of filter paper in acidified potassium manganate(VII) solution, and hold it in a stream of sulphur dioxide.

The paper turns from purple to a very pale pink.

## Test for oxidising agents

Many oxidising agents will oxidise the iodide ion, $I^-$, to iodine, $I_2$. The presence of iodine can be detected because it forms a dark blue compound with starch. Potassium iodide solution and starch can therefore be used to test for oxidising agents.

| $2I^-(aq)$ | → | $I_2(aq) + 2e^-$ |
|---|---|---|
| colourless | | brown, forms a dark blue compound with starch |

For example,

Wet a filter paper with a solution of potassium iodide and starch, and hold it in a stream of chlorine.

The colour changes from white to dark blue.

**SUMMARY**

Reducing agents decolourise acidified potassium manganate(VII) solution. Oxidising agents give a dark blue colour with a solution of potassium iodide and starch.

**CHECKPOINT**

▶ 1 (a) Say which of the following will decolourise acidified potassium manganate(VII): $FeSO_4(aq)$, $ZnSO_4(aq)$, $SO_2(g)$

(b) Say which of the following will turn starch-iodide paper blue: $Br_2(aq)$, $Fe_2(SO_4)_3$, $Cl_2$, $I_2$, $HCl(aq)$

# 12.5 ▶ Chlorine as an oxidising agent

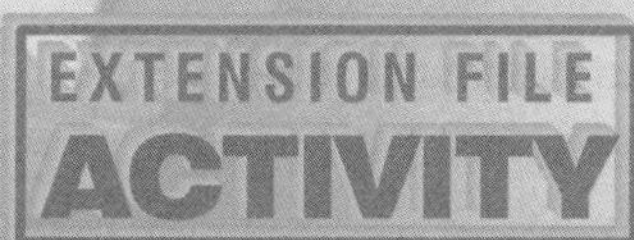


The reactions of chlorine are dominated by its readiness to act as an oxidising agent: to gain electrons and form chloride ions.

| Chlorine molecules | + | Electrons | → | Chloride ions |
|---|---|---|---|---|
| $Cl_2(aq)$ | + | $2e^-$ | → | $2Cl^-(aq)$ |

### Water treatment

Of major importance is the role played by chlorine in making our tap water safe to drink. The ability of chlorine to kill bacteria is due to its oxidising power.

## Metals

Sodium and chlorine react to form sodium chloride. Sodium atoms are oxidised (give electrons), becoming sodium ions, $Na^+$. Chlorine molecules are reduced (gain electrons) to become chloride ions, $Cl^-$.

$$Na(s) \rightarrow Na^+(s) + e^-$$

So the full equation for sodium and chlorine is:

$$2Na(s) + Cl_2(g) \rightarrow 2Na^+(s) + 2Cl^-(s)$$

## Ions of metals of variable valency

The oxidation of iron(II) salts to iron(III) salts by chlorine has been mentioned (Topic 12.2).

## Halogens

The reactions of chlorine with bromides and iodides are oxidation–reduction reactions. Chlorine displaces bromine from bromides. This happens because chlorine is a stronger oxidising agent than bromine is. Chlorine takes electrons away from bromide ions: bromide ions are oxidised to bromine molecules. Chlorine molecules are reduced to chloride ions.

REDUCTION ($Cl_2 \rightarrow 2Cl^-$)

$$Cl_2(aq) + 2Br^-(aq) \rightarrow 2Cl^-(aq) + Br_2(aq)$$

OXIDATION ($2Br^- \rightarrow Br_2$)

Chlorine molecules are ready to accept electrons to form chloride ions.

This is the basis of chlorine's reactivity as an oxidising agent, with, e.g.

- metals
- metal ions
- other halogens.

The bromine formed can be detected more easily if a small volume of an organic solvent, e.g. trichloroethene, is added. Bromine dissolves in the organic solvent to form an orange solution.

Similarly, chlorine displaces iodine from iodides because chlorine is a stronger oxidising agent than iodine. Again, iodine can be detected readily by adding a small volume of an organic solvent. Iodine dissolves to give a purple solution.

Bromine displaces iodine from iodides because bromine is a stronger oxidising agent than iodine.

Fluorine is the most powerful oxidising agent of the halogens. It is dangerously reactive and is not used in schools and colleges. In order of oxidising power, the halogens rank:

$$F_2 > Cl_2 > Br_2 > I_2$$

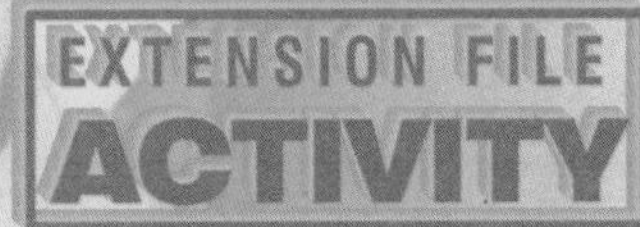


## Oxidation–reduction

*Oxidation is*

- the gain of oxygen or
- the loss of hydrogen or
- the loss of electrons by a substance.

*Reduction is*

- the loss of oxygen or
- the gain of hydrogen or
- the gain of electrons by a substance.

*An oxidising agent*

- gives oxygen to or
- takes hydrogen from or
- takes electrons from a substance.

*A reducing agent*

- takes oxygen from or
- gives hydrogen to or
- gives electrons to a substance.

### SUMMARY

Chlorine is an oxidising agent. It is used in water treatment to kill bacteria. It oxidises metals, metal ions, bromides and iodides. In order of oxidising power, the halogens rank

$F_2 > Cl_2 > Br_2 > I_2$

# 12.6 Bromine from sea water

The oxidising power of the halogens decreases as you descend Group 7. Chlorine is a more powerful oxidising agent than bromine; therefore chlorine will oxidise bromide ion to bromine while chlorine is reduced to chloride ion.

$$\text{chlorine} + \text{bromide ion} \rightarrow \text{chloride ion} + \text{bromine}$$
$$Cl_2(aq) + 2Br_2(aq) \rightarrow 2Cl^-(aq) + Br_2(l)$$

Sea water is a plentiful source of bromide ion. The extraction of bromine from sea water is carried out in Anglesey, Wales and in the Dead Sea, Israel. Israel has the advantage that the sea water can be concentrated by evaporation by the sun's heat before treatment. During evaporation, sodium chloride, potassium chloride and magnesium chloride crystallise. These salts are electrolysed to give chlorine. The process is illustrated in Figure 12.6A.

The oxidising power of chlorine is used to convert bromide ions in sea water into bromine.

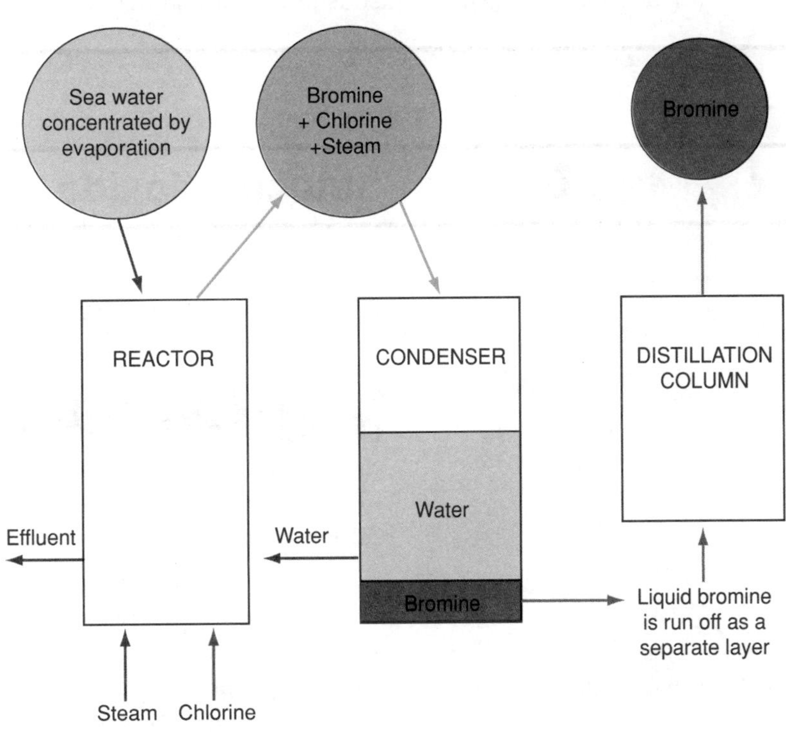


*Figure 12.6A* ▲ Flow diagram for the extraction of bromine from sea water

Bromine is a corrosive liquid with a poisonous vapour. Extreme care has to be taken in transporting bromine from the producer to the user.

Bromine is shipped from Israel to the petrochemicals plant at Fawley near Southampton. Bromine is a corrosive liquid which vaporises easily (b.p. 58 °C). The vapour is poisonous, with a penetrating smell, and inhaling it makes you choke. Both liquid and vapour are caustic to the skin, producing serious burns. Transporting this dangerous liquid from Israel to Britain is a difficult business, but the safety record is good. It is shipped in lead-lined steel tanks which hold several tonnes of bromine. At Fawley the tanks must be emptied. Nitrogen is led into the tank to pump bromine out through a plastic pipe into a storage cylinder of glass-lined steel. When required bromine is pumped from the storage tank through plastic pipes into the plant. A bromine detector in the area ensures the safety of operators.

## CHECKPOINT

1 (a) Write a word equation for the oxidation of sodium iodide by chlorine.
(b) Write a balanced chemical equation for the reaction in (a).
(c) Write a word equation for the oxidation of potassium iodide by bromine.
(d) Write a balanced chemical equation for the reaction in (c).

2 The lab. technician has taken delivery of three bottles containing crystalline white solids. Unfortunately, their labels have come off.

| Potassium chloride | Potassium bromide | Potassium iodide |
|---|---|---|

The technician has to decide which label to stick on each of the three bottles. All she has to work with is a bottle of chlorine water and an organic solvent. How can she solve the problem?

3 With which one of the following pairs of reagents would a displacement reaction take place?
(a) aqueous bromine and aqueous potassium chloride
(b) aqueous bromine and aqueous sodium chloride
(c) aqueous chlorine and aqueous potassium iodide
(d) aqueous iodine and aqueous potassium bromide.

# 12.7 Sulphur dioxide as a reducing agent

Sulphur dioxide is a gas with a very unpleasant, penetrating smell. It is poisonous, causing congestion followed by choking, and at sufficiently high concentrations it will kill.

## Sulphurous acid

Sulphur dioxide is extremely soluble in water. It 'fumes' in moist air as it reacts with water vapour to form a mist. It reacts with water to form sulphurous acid, $H_2SO_3$.

Sulphur dioxide + Water → Sulphurous acid
$SO_2(g) + H_2O(l) \rightarrow H_2SO_3(aq)$

Sulphur dioxide reacts with water to form sulphurous acid ...
... which is oxidised to sulphuric acid ...
... by oxygen in the air ...
... and by bromine water, potassium dichromate(VI) and potassium manganate(VII).

Sulphurous acid is a weak acid which forms salts called sulphites, e.g. sodium sulphite $Na_2SO_3$. Sulphurous acid is slowly oxidised by oxygen in the air to sulphuric acid, $H_2SO_4$. Sulphuric acid is formed whenever sulphurous acid accepts oxygen from an oxidising agent.

Sulphurous acid + Oxygen from the air or from an oxidising agent → Sulphuric acid
$H_2SO_3(aq) + (O) \rightarrow H_2SO_4(aq)$

For example,

Sulphurous acid + Bromine water → Sulphuric acid + Hydrogen bromide
$H_2SO_3(aq) + Br_2(aq) + H_2O(l) \rightarrow H_2SO_4(aq) + 2HBr(aq)$

## Tests for sulphur dioxide

1 Sulphur dioxide reduces potassium dichromate(VI) from orange dichromate ions, $Cr_2O_7^{2-}$, to blue chromium(III) ions, $Cr^{3+}$, passing through the intermediate colour of green, which results from a mixture of orange ions and blue ions.

2 It reduces potassium manganate(VII) from purple manganate(VII) ions, $MnO_4^-(aq)$ to pale pink, almost colourless manganese(II) ions, $Mn^{2+}$.

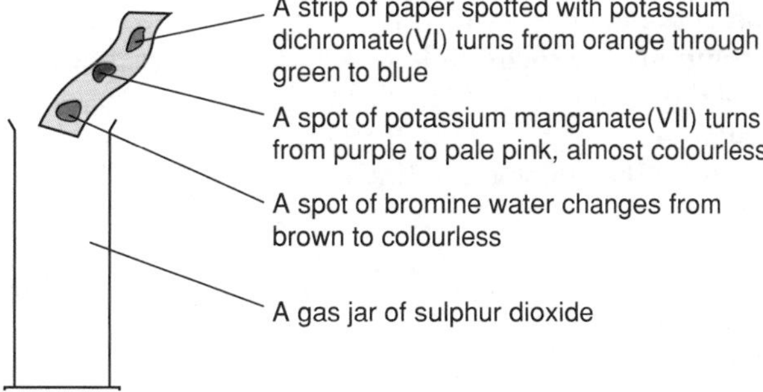
A strip of paper spotted with potassium dichromate(VI) turns from orange through green to blue
A spot of potassium manganate(VII) turns from purple to pale pink, almost colourless
A spot of bromine water changes from brown to colourless
A gas jar of sulphur dioxide

*Figure 12.7A* ▲ Testing for sulphur dioxide

## SUMMARY

Sulphur dioxide is an acidic, poisonous gas with a pungent smell. It dissolves in water to form sulphurous acid, which is a reducing agent.

## CHECKPOINT

▶ 1 Makers of home-brewed beer and wine rinse their bottles with a solution of sodium sulphite before filling them. Explain why they do this.

▶ 2 Write the symbol for the reducing agent in each of the following reactions.

$Na(s) + 2H_2O(l) \rightarrow H_2(g) + 2NaOH(aq)$

$O_2(g) + S(s) \rightarrow SO_2(g)$

$2Al(s) + 3Cl_2(g) \rightarrow 2AlCl_3(s)$

$Cu(s) + S(s) \rightarrow CuS(s)$

# Theme Questions

1 The indicator phenolphthalein is colourless in neutral and acidic solutions and turns pink in alkaline solutions. Of the solutions A, B and C, one is acidic, one is alkaline, and one is neutral. Explain how, using only phenolphthalein, you can find out which solution is which.

2 Seven steel bars were placed in solutions of different pH for the same time. The table shows the percentage corrosion of the steel bars.

| *pH of solution* | 1 | 2 | 3 | 4 | 5 | 6 | 7 |
|---|---|---|---|---|---|---|---|
| *Percentage corrosion of steel bar* | 65 | 60 | 55 | 50 | 20 | 15 | 10 |

(a) On graph paper, plot the percentage corrosion against the pH of the solution.
(b) Read from your graph the percentage corrosion at pH 4.5.

3 Look at the information from the label of a bottle of concentrated orange squash.

**Concentrated orange squash**
Ingredients: Sugar, Water, Citric acid, Flavourings, Preservative E250, Artificial sweetener, Yellow colourings E102 and E103

(a) Explain what is meant by 'concentrated'.
(b) A sample of concentrated orange squash is mixed with water. A piece of universal indicator paper is dipped in. What colour will the paper turn?
(c) Why does using universal indicator paper give a better result than adding universal indicator solution to the orange squash?
(d) Describe a simple experiment you could do to prove that **only two** yellow substances are present in the concentrated orange squash.

4 Zinc sulphate crystals, $ZnSO_4 . 7H_2O$, can be made from zinc and dilute sulphuric acid by the following method.
*Step 1* Add an excess of zinc to dilute sulphuric acid. Warm.
*Step 2* Filter.
*Step 3* Partly evaporate the solution from Step 2. Leave it to stand.
(a) Explain why an excess of zinc is used in Step 1.
(b) How can you tell when Step 1 is complete?
(c) Name the residue and the filtrate in Step 2.
(d) Explain why the solution is partly evaporated in Step 3.
(e) Would you dry the crystals by strong heating or by gentle heating? Explain your answer.
(f) Write a word equation for the reaction. Write a symbol equation.

5 What method would you use to make the insoluble salt lead(II) carbonate? Explain why you have chosen this method. Say what starting materials you would need, and say what you would do to obtain solid lead(II) carbonate. (For solubility information, see Theme C, Table 10.4)

6 Explain the following:
(a) Tea changes colour when lemon juice is added.
(b) Sodium sulphate solution is used as an antidote to poisoning by barium compounds.
(c) Washing soda takes some of the pain out of bee stings.
(d) Toothpastes containing aluminium hydroxide fight tooth decay.
(e) Calcium fluoride is added to some toothpastes.
(f) Scouring powders often contain sodium hydroxide and powdered stone.

7 A student receives the following instructions: 'Make a solution of copper sulphate by neutralising dilute sulphuric acid, using the base copper oxide.'
(a) Explain the meanings of the terms: solution, neutralising, dilute, base.
(b) Draw an apparatus which the student could use for carrying out the instructions.
(c) Describe exactly how the student should carry out the instructions.

8 The table shows the colours of some indicators.

| *Indicator* | *pH* 1 2 3 4 5 6 7 8 9 10 11 12 13 14 |
|---|---|
| Methyl orange | ←Red→ ←——— Yellow ———→ |
| Bromocresol green | ←Yellow→ ←——— Blue ———→ |
| Phenol red | ←Yellow→ ←——— Red ———→ |
| Phenolphthalein | ←Colourless→ ←——— Red ———→ |

(a) What colour is a solution of pH 10 when a few drops of bromocresol green are added?
(b) What colour is a solution of pH 3.5 when a few drops of methyl orange are added?
(c) What colour is a solution of pH 6.5 when a few drops of phenol red are added?
(d) A solution turns yellow when either methyl orange or phenol red is added. What is the approximate pH of the solution?
(e) A solution is colourless when phenolphthalein is added and red when phenol red is added. What is the pH of the solution?
(f) A mixture of bromocresol green, phenol red and phenolphthalein is added to a solution of pH 10. What is the colour?
(g) A mixture of all four indicators is added to a strong acid. What is the colour?

**9** (a) Use the Periodic Table to answer the following questions:

(i) Give the symbols for the elements carbon and sodium.

(ii) Give the symbol for any inert gas.

(iii) Give the symbol for any element in Group 7 (the halogens)

(iv) Give **one** reason why the symbol for helium is He and not H.

(b) One molecule of carbon dioxide contains one atom of carbon and two atoms of oxygen. Its formula is written as $CO_2$.

Write the formula of:

(i) a molecule of sulphur dioxide, which has one atom of sulphur and two atoms of oxygen.

(ii) a molecule of sulphuric acid, which contains two atoms of hydrogen, one atom of sulphur and four atoms of oxygen.

**10** Explain why Mendeleev fitted the elements into vertical groups in his Periodic Table. Why did he fit lithium, sodium, potassium, rubidium and caesium into the same group? Why did he leave some gaps in his table? The noble gases were discovered after Mendeleev had written his Periodic Table. What do they have in common (a) in their chemical reactions, (b) in their electronic configurations and (c) in their positions in the Periodic Table?

**11** The table below shows the formulae, melting points and pH values of aqueous solutions (when soluble) of the oxides of some elements.

| Atomic (proton) number of element | 11 | 12 | 13 | 14 | 15 | 16 | 17 | 18 |
|---|---|---|---|---|---|---|---|---|
| Formula of oxide | $Na_2O$ | MgO | $Al_2O_3$ | $SiO_2$ | $P_4O_6$ | $SO_2$ | $Cl_2O$ | no oxide |
| Melting point (K) | 1193 | 3173 | 2313 | 1883 | 297 | 198 | 253 | – |
| pH of aqueous solution (if soluble) | 14 | 11 | insol | insol | 3 | 3 | 3 | – |

(a) (i) How many of these oxides are solid at a room temperature of 20 °C?

(ii) Why does the element with atomic number 18 not have an oxide?

(iii) When sodium oxide is added to water, the solution formed has a pH of 14. Explain the reason for this.

(b) (ii) Give the following information for an atom of the chlorine isotope $^{35}_{17}Cl$; number of protons; number of neutrons; number of electrons; electron structure.

(ii) Draw a dot and cross diagram to show the electronic structure of a molecule of chlorine(I) oxide, $Cl_2O$. Show only the electrons in the outermost shell (highest energy level), and use o for electrons from the oxygen atom and x for electrons from the chlorine atoms.

(iii) Explain, in terms of the forces present, why chlorine(I) oxide has a low melting point.

(c) Magnesium is manufactured by the electrolysis of molten magnesium chloride.

(i) Use the table at the beginning of the question to suggest why molten magnesium chloride is used as the electrolyte, rather than molten magnesium oxide.

(ii) Name the product at the cathode, and that at the anode.

(iii) Write an ionic equation for the reaction that takes place at the cathode of the cell during the electrolysis. (ULEAC)

**12** Use the Periodic Table at the back of the book to help you to answer this question.

(a) State **one** similarity and **one** difference in the electronic structure of the elements

(i) across the Period from sodium to argon.

(ii) down Group 7 from fluorine to astatine.

(b) (i) State the trend in reactivity of the Group 1 elements.

(ii) Explain this trend in terms of atomic structure.

(c) Hydrogen is an element which is difficult to fit into a suitable position in the Periodic Table. Give reasons why hydrogen could be placed in either Group 1 or Group 7. (NEAB)

**13** Chlorine is manufactured by the electrolysis of brine (saturated sodium chloride solution). One type of cell uses membranes which allow positive ions to pass through but not negative ions or water molecules. The anode and cathode are on opposite sides of the membrane.

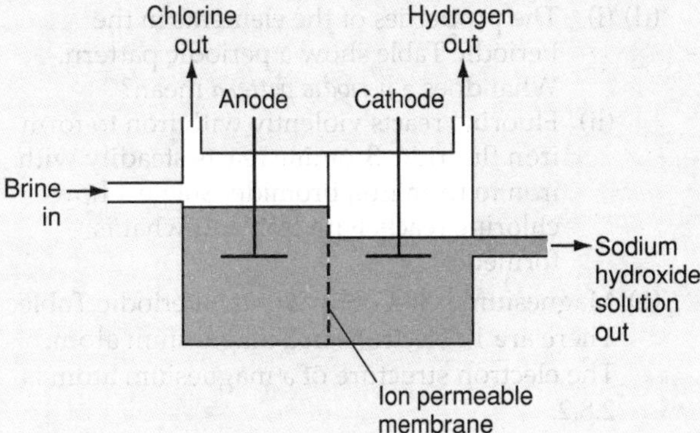


(a) Write the ionic equation for the electrolytic reaction at the anode.

(b) Explain why fresh brine is pumped into the **anode** side of the cell.

Chlorine is used in the production of bromine from sea water. There are four key stages in the industrial process.

**A** In the sea water bromide ions are changed into bromine by passing chlorine into the water.

**B** Bromine vapour is removed from this mixture by blowing air through.

**C** The bromine vapour is passed into a mixture of sulphur dioxide and water vapour where it forms hydrobromic acid (HBr).

**D** Fairly concentrated bromine is obtained from this by blowing in a mixture of steam and chlorine.

(c) Give the letter of the stage or stages in which
  (i) bromine is reduced;
  (ii) chlorine is reduced.

(d) Name the reagent which reduces bromine.

(e) Write a balanced ionic equation for the reaction of bromide ions with chlorine molecules.

(ULEAC)

**14** The diagram shows the position of some of the elements in the Periodic Table.

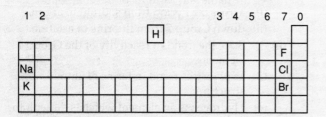

| 1 | 2 | | 3 | 4 | 5 | 6 | 7 | 0 |
|---|---|---|---|---|---|---|---|---|
| | | H | | | | | | |
| | | | | | | | F | |
| Na | | | | | | | Cl | |
| K | | | | | | | Br | |

(a) Copy and complete this sentence.
The elements in the Periodic Table are arranged in order of their ________.

(b) Name an element shown in this diagram which is:
  (i) a metal;
  (ii) a non-metal.

(c) Name **two** elements in the diagram which react with each other to form an ionic compound.

(d) (i) The properties of the elements in the Periodic Table show a periodic pattern. What does a *periodic pattern* mean?
  (ii) Fluorine reacts violently with iron to form iron fluoride. Bromine reacts steadily with iron to form iron bromide. Suggest how chlorine reacts with iron and what is formed.

(e) Magnesium is in Group 2 of the Periodic Table. There are 12 electrons in a magnesium atom. The electron structure of a magnesium atom is 2.8.2.
  (i) Write down what this information tells you about the arrangement of the electrons in a magnesium atom.
  (ii) Write down the electronic structure of calcium, the next element in Group 2, which contains 20 electrons.

(f) When the elements of Group 2 react with water, the atoms each lose two electrons to form positive ions.
Use ideas about electronic structure to suggest why the reaction of calcium and water is faster than the reaction of magnesium with water.

(MEG)

**15** Use the Periodic Table at the back of this book to help you answer this question.
The table below gives information about the elements in one row of the Periodic Table.

| *Group* | I | II | III | IV | V | VI | VII | 0 |
|---|---|---|---|---|---|---|---|---|
| *element* | Li | Be | B | C | N | O | F | Ne |
| *boiling point/°C* | 1331 | 2487 | 3927 | 4827 | −196 | −183 | −188 | −246 |

(a) Describe how the boiling point changes across this row of elements.

(b) The next row of the Periodic Table includes the elements **sodium**, **silicon**, **sulphur** and **argon**.
  (i) Which of these four elements is likely to have the highest boiling point?
  (ii) Explain your choice.

(c) The elements of Group 0 are called the noble gases.
  (i) What is the main characteristic of the chemistry of these elements?
  (ii) Use your knowledge of atomic structure to explain why these elements behave in this way.

(d) The table shows information about the noble gases.

| *element* | *relative atomic mass* | *boiling point/ °C* |
|---|---|---|
| helium | 4.0 | −269 |
| neon | 20.2 | −246 |
| argon | 39.9 | −186 |
| krypton | 83.8 | −152 |
| xenon | 131.3 | −108 |

  (i) Describe the pattern shown in the boiling points of these elements.
  (ii) Use the information in the table to suggest an explanation for the pattern you have noticed.

(MEG)

**16** (a) Copy and complete the table (refer to Figure 9.5A if necessary).

| *Colour of universal indicator* | *pH of solution* |
|---|---|
| red | ______ |
| ______ | 11 |
| green | ______ |

(b) Copy and complete the following word equation:
acid + alkali → ______ + ______

(c) Name this type of reaction.

(d) Name the products formed when the acid is dilute hydrochloric acid and the alkali is potassium hydroxide.

(e) Name the acid and alkali needed to give ammonium nitrate.

**17** The graph shows the average number of fillings per child plotted against the fluoride content of the water in different regions.

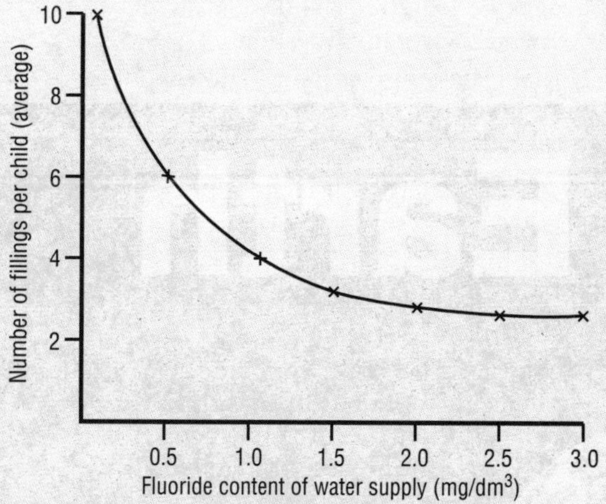


(a) Say what you can deduce from the graph about tooth decay.
(b) Give one reason why some people are in favour of adding fluoride to the water supply.
(c) Give one reason why some people are against adding fluoride to the water supply.

**18** Choose from the following list a substance which fits each of the descriptions below.

fluorine  hydrogen  sodium hydroxide
potassium  chromium  copper(II) sulphate
potassium chloride  chlorine  calcium  silver

(a) a halogen used to treat drinking water
(b) a transition element
(c) a compound which dissolves to form a coloured solution
(d) an alkali
(e) a metallic element which reacts vigorously with water
(f) a metallic element which does not react with water
(g) an alkaline earth
(h) the salt of a Group 1 element and a Group 7 element.

**19** (a) In the Periodic Table the elements are arranged in an increasing order. What is it that increases?
(b) (i) How does the reactivity of the elements change as you pass down Group 1?
(ii) What happens to the electrons in the outer shell of the atom when elements of Group 1 react?
(iii) How does this explain the change in reactivity down the group?
(c) (i) How does the reactivity of the elements change as you pass down Group 7?
(ii) What happens to the electrons in the outer shell of the atom when elements of Group 7 react?
(iii) How does this explain the change in reactivity down the group?

**20** Refer to the Periodic Table (Figure 11.1A) and to Tables 11.3 and 11.7.
(a) Rubidium, Rb, is in Group 1.
(i) Name two products you would expect when rubidium reacts with water.
(ii) Describe what you would expect to see during the reaction.
(b) Iodine, I, is in Group 7. It reacts with hydrogen to form hydrogen iodide, a soluble gas.
(i) Name a common chemical which you would expect to behave like hydrogen iodide solution.
(ii) Name two products which you would expect when hydrogen iodide solution reacts with magnesium.
(iii) Describe what you would expect to see during the reaction.
(c) Give the name and formula of the compound formed between rubidium and iodine.

**21**

(a) Copy and complete the table by using information in the diagrams.

| *Substance* | *pH* | *Acidic or alkaline?* | *Strength* |
|---|---|---|---|
| Toilet cleaner | 2 | ____ | ____ |
| ____ | ____ | acidic | weak |
| Water | ____ | ____ | |
| Soap | 9 | alkaline | ____ |
| ____ | 12 | ____ | ____ |

(b) How could you test soap to show that it is alkaline?
(c) (i) What effect would toilet cleaner have on your skin if you spilt some?
(ii) Why is it better to wash toilet cleaner from your skin with soap and water than with water alone?
(d) A deposit of calcium carbonate (limescale) forms in kettles. Would you recommend using toilet cleaner to get rid of limescale? Is there something else in the table that would be better? Explain your answer.
(e) Mrs Ahmed has a sink blocked by grease. She puts some washing soda into the sink. Explain how it works to get rid of the grease.

## Theme D

# Planet Earth

Our planet provides us with the materials which we use to make the food, clothing, housing and all the possessions we need and prize. The plenty we enjoy has led to pollution. The air is polluted by objectionable gases, dust and smoke from our cars and factories. Rivers and lakes are polluted by waste from factories and fields. The land is spoiled by rubbish tips. People are awakening to the need to combat pollution and save our planet. Steps are being taken to reduce the emission of pollutants into the air and into rivers and lakes. A start has been made in recycling rubbish to preserve the landscape and to conserve Earth's resources.

The Earth's atmosphere has evolved slowly over billions of years to the composition which it has today. The rocks in the Earth's crust are constantly undergoing slow geological changes. The continents we know today are vastly different from those that first emerged on Earth. This theme tells the story of these changes.

# Topic 13 Air

## 13.1 ▶ Oxygen the life-saver

**FIRST THOUGHTS**

As you study this topic, think about the vital importance of air as a source of:

- oxygen – which supports the life of plants and animals
- carbon dioxide – which supports plant life
- nitrogen – the basis of natural and synthetic fertilisers
- the noble gases – filling our light bulbs

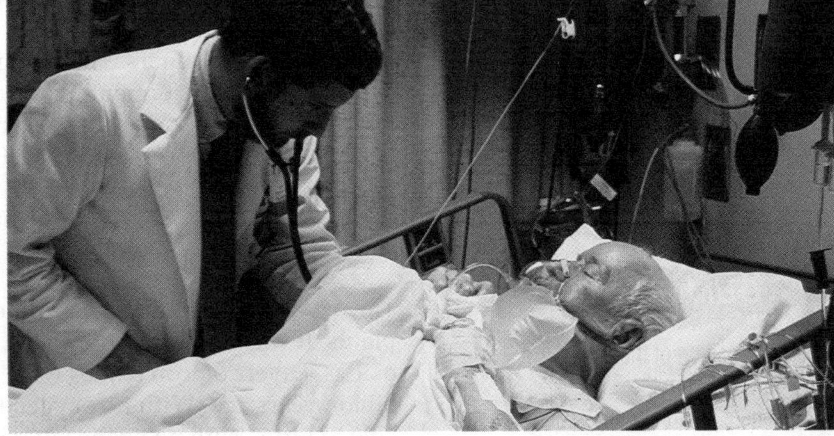

*Figure 13.1A* ▲ Oxygen in hospitals

Hospitals need oxygen for patients who have difficulty in breathing. Tiny premature babies often need oxygen. They may be put into an oxygen high pressure chamber. This chamber contains oxygen at 3–4 atmospheres pressure, which makes 15–20 times the normal concentration of oxygen dissolve in the blood. People who are recovering from heart attacks and strokes also benefit from being given oxygen. Patients who are having operations are given an anaesthetic mixed with oxygen.

In normal circumstances, we can obtain all the oxygen we need from the air. Figure 13.1B shows the percentages of oxygen, nitrogen and other gases in pure, dry air. Water vapour and pollutants may also be present in air.

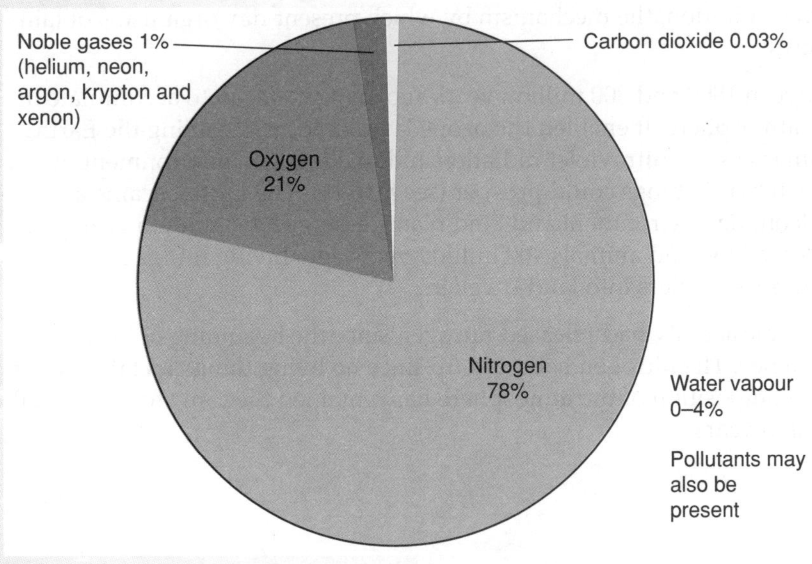


*Figure 13.1B* ▲ Composition of clean, dry air in percentage by volume

## 13.2 ▶ The evolution of Earth's atmosphere

The present-day atmosphere has evolved from a mixture of gases released from the interior of the Earth about 4.6 billion years ago. The mixture was probably similar to that released from volcanoes today: water vapour 64% by mass, carbon dioxide 24%, sulphur dioxide 10%, nitrogen 1.5%. This mixture has evolved into today's atmosphere which is mainly nitrogen and oxygen with only a trace of carbon dioxide.

The Earth is far enough from the Sun to allow the primordial water vapour to condense to make oceans. Carbon dioxide dissolved in the oceans in large quantities and was eventually converted into limestone beds. Since Venus is closer to the Sun than Earth is, oceans never formed and carbon dioxide remains in the atmosphere.

When living things emerged on Earth, they altered the composition of the atmosphere. The first primitive living things evolved in the ocean about 3.5 billion years ago. The atmosphere was irradiated with ultraviolet light which provided the energy needed to bring about chemical reactions that resulted in the formation of substances such as amino acids and sugars. The first living things were aquatic. They derived energy by fermenting the organic matter that had been formed. Eventually they became able to carry out photosynthesis to produce organic matter (represented as $CH_2O$).

$$CO_2 + H_2O + h\upsilon \rightarrow \underset{\text{organic matter}}{(CH_2O)} + O_2$$

The oxygen formed in photosynthesis would have poisoned primitive plants. It was not released into the air but was converted into iron oxides. Layers of iron oxides were laid down between 3 billion and 1.5 billion years ago. With no oxygen in the atmosphere, no ozone, $O_3$, was able to form. With no ozone layer in the atmosphere to absorb ultraviolet (UV) radiation from the Sun, the surface of the Earth was too hot to support life. However, bacteria were able to live in the ocean, protected from UV radiation by water. They eventually acquired enzymes that enabled them to use the oxygen in the atmosphere to oxidise organic matter in the ocean with the liberation of energy. The process evolved into respiration, the mechanism by which present-day organisms obtain energy.

Between 1800 and 800 million years ago, oxygen began to accumulate in the atmosphere. It enabled the ozone layer to form, shielding the Earth from excessive ultraviolet radiation and making it an environment in which living things could prosper (see § 16.11). The Earth became a welcoming environment and land plants emerged 450 million years ago, followed by land animals 400 million years ago. Living things evolved from sea-dwellers into land-dwellers.

Volcanic activity had released nitrogen since the beginning of Earth's existence. The nitrogen had built up since no living things metabolised it. The composition of the atmosphere has remained the same for about 300 million years.

**SUMMARY**

The composition of the air is very different from the atmosphere that originally surrounded the Earth. Changes happened gradually over millions of years. Water vapour condensed to make oceans ... carbon dioxide dissolved in the oceans ... primitive plants evolved and carried on photosynthesis ... bacteria evolved and began to use oxygen in respiration ... oxygen accumulated and the ozone layer formed ... making Earth an environment in which living things could prosper.

## 13.3 How oxygen and nitrogen are obtained from air

The method used by industry to obtain oxygen and nitrogen from air is **fractional distillation** of liquid air. Air must be cooled to −200 °C before it liquefies. It is very difficult to get down to this temperature. However, one way of cooling a gas is to compress it and then allow it to expand suddenly. Figures 13.3A and 13.3B show how air is first liquefied and then distilled.

Compression of a gas causes a rise in temperature; expansion causes a cooling. This cooling effect is used to liquefy air.

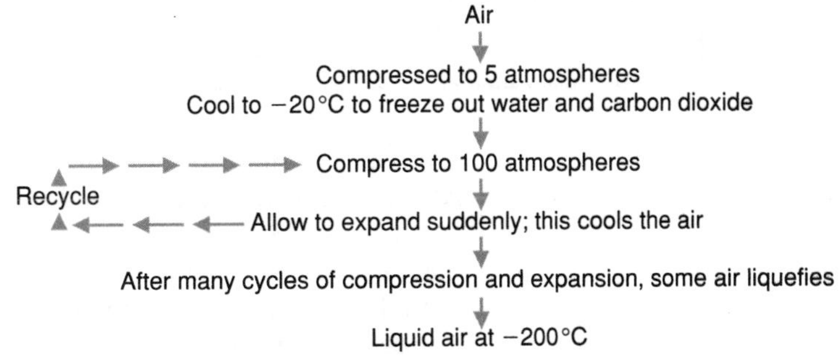
Air
Compressed to 5 atmospheres
Cool to −20 °C to freeze out water and carbon dioxide
Compress to 100 atmospheres
Recycle
Allow to expand suddenly; this cools the air
After many cycles of compression and expansion, some air liquefies
Liquid air at −200 °C

*Figure 13.3A* ▲ Liquefaction of air

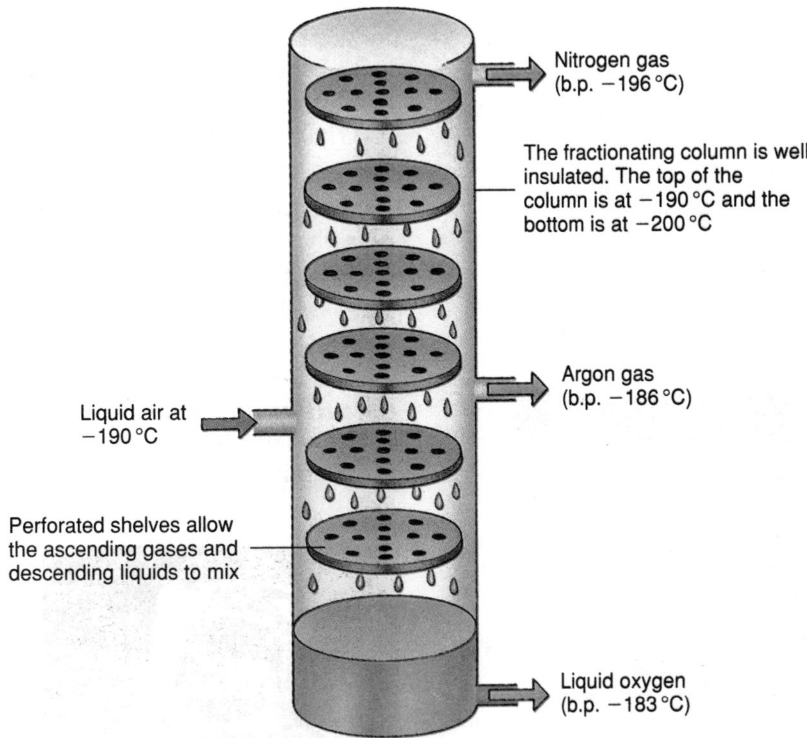
Nitrogen gas
(b.p. −196 °C)
The fractionating column is well insulated. The top of the column is at −190 °C and the bottom is at −200 °C
Argon gas
(b.p. −186 °C)
Liquid air at
−190 °C
Perforated shelves allow the ascending gases and descending liquids to mix
Liquid oxygen
(b.p. −183 °C)

*Figure 13.3B* ▲ Fractional distillation of air

Oxygen, nitrogen and argon are redistilled. They are stored under pressure in strong steel cylinders. Industry finds many important uses for them.

## SUMMARY

Air is a mixture of nitrogen, oxygen, noble gases, carbon dioxide, water vapour and pollutants. The fractional distillation of liquid air yields oxygen, nitrogen and argon.

## CHECKPOINT

1. The top of the fractionating column in Figure 13.3B is at −190 °C. Explain why nitrogen is a gas at the top of the column but oxygen is a liquid.
2. The table lists nitrogen, oxygen and the noble gases.

| *Gas* | *Boiling point* (°C) |
|---|---|
| Argon | −186 |
| Helium | −269 |
| Krypton | −153 |
| Neon | −246 |
| Nitrogen | −196 |
| Oxygen | −183 |
| Xenon | −108 |

(a) Which gas has (i) the lowest boiling point and (ii) the highest boiling point?
(b) List the gases which would still be gaseous at −200 °C.

## 13.4 Uses of oxygen

Oxygen is a colourless, odourless gas which is slightly soluble in water. All living things need oxygen for respiration. Breathing becomes difficult at a height of 5 km above sea level, where the air pressure is only half that at sea level. Climbers who are tackling high mountains take oxygen with them. Aeroplanes which fly at high altitude carry oxygen. Astronauts must carry oxygen, and even unmanned space flights need oxygen. Deep-sea divers carry cylinders which contain a mixture of oxygen and helium.

Industry has many uses for oxygen. Figure 13.4B shows an oxyacetylene torch being used to weld metal at a temperature of about 4000 °C. The hot flame is produced by burning the gas ethyne (formerly called acetylene) in oxygen. Substances burn faster in pure oxygen than in air.

The steel industry converts brittle cast **iron** into strong **steel**. Cast iron is brittle because it contains impurities such as carbon, sulphur and phosphorus. These impurities will burn off in a stream of oxygen. Steel plants use one tonne of oxygen for every tonne of cast iron turned into steel. Many steel plants make their own oxygen on site.

The proper treatment of sewage is important for public health. Air is used to help in the decomposition of sewage. Without this treatment, sewage would pollute many rivers and lakes. One method of treating polluted lakes and rivers to make them fit for plants and animals to live in is to pump in oxygen.

### It's a fact!

The introduction of the oxy-acetylene flame brought about a revolution in metal working. Industry moved out of the blacksmith era into the twentieth century with gas-welding and flame-cutting techniques.

*Figure 13.4A* Astronaut using oxygen

*Figure 13.4B* Oxyacetylene welding being used on an automated production line

### SUMMARY

Uses for pure oxygen are:

- treating patients who have breathing difficulties
- supporting high-altitude pilots, mountaineers, deep-sea divers and space flights
- in steel making and other industries
- treating polluted water

### CHECKPOINT

1 Explain why oxygen is used in (a) steelworks (b) metal working (c) sewage treatment (d) space flights and (e) hospitals. Say what advantage pure oxygen has over air for each purpose.

## 13.5 Nitrogen

FIRST THOUGHTS

So far, oxygen seems to be the important part of the air, but nitrogen has its uses too, as you will find out in this section.

Nitrogen is a colourless, odourless gas which is slightly soluble in water. It does not readily take part in chemical reactions. Many uses of nitrogen depend on its unreactive nature. Liquid nitrogen (below −196 °C) is used when an inert (chemically unreactive) refrigerant is needed. The food industry uses it for the fast freezing of foods.

Vets use the technique of artificial insemination to enable a prize bull to fertilise a large number of dairy cows. They carry the semen of the bull in a type of vacuum flask filled with liquid nitrogen.

Many foods are packed in an atmosphere of nitrogen. This prevents the oils and fats in the foods from reacting with oxygen to form rancid products. As a precaution against fire, nitrogen is used to purge oil tankers and road tankers. The silos where grain is stored are flushed out with nitrogen because dry grain is easily ignited.

*Figure 13.5A* ▲ Oil tanker delivering oil to a refinery. Once it is empty, the tanks will be purged with nitrogen

*Figure 13.5B* ▲ Grain silos

SUMMARY

Nitrogen is a rather unreactive gas. It is used to provide an inert (chemically unreactive) atmosphere.

## 13.6 The nitrogen cycle

Nitrogen is an essential element in **proteins**. Some plants have nodules on their roots which contain nitrogen-fixing bacteria. These bacteria **fix** gaseous nitrogen, that is, convert it into nitrogen compounds. From these nitrogen compounds, the plants can synthesise proteins. Members of the legume family, such as peas, beans and clover, have nitrogen-fixing bacteria.

Plants other than legumes synthesise proteins from nitrates. *How do nitrates get into the soil?* Nitrogen and oxygen combine in the atmosphere during lightning storms and in the engines of motor vehicles during combustion. They form nitrogen oxides (compounds of nitrogen and oxygen). These gases react with water to form nitric acid. Rain showers bring nitric acid out of the atmosphere and wash it into the ground, where it reacts with minerals to form nitrates. Plants take in these nitrates through their roots. They use them to synthesise proteins. Animals obtain the proteins they need by eating plants or by eating the flesh of other animals.

*Figure 13.6A* ▲ Nodules containing nitrogen fixing bacteria on the roots of a legume

In the excreta of animals and the decay products of animals and plants, **ammonium salts** are present. Nitrifying bacteria in the soil convert ammonium salts into nitrates. Both nitrates and ammonium salts can be removed from the soil by **denitrifying bacteria**, which convert the compounds into nitrogen. To make the soil more fertile, farmers add both nitrates and ammonium salts as fertilisers. The balance of processes which put nitrogen into the air and processes which remove nitrogen from the air is called the **nitrogen cycle** (see Figure 13.6B).

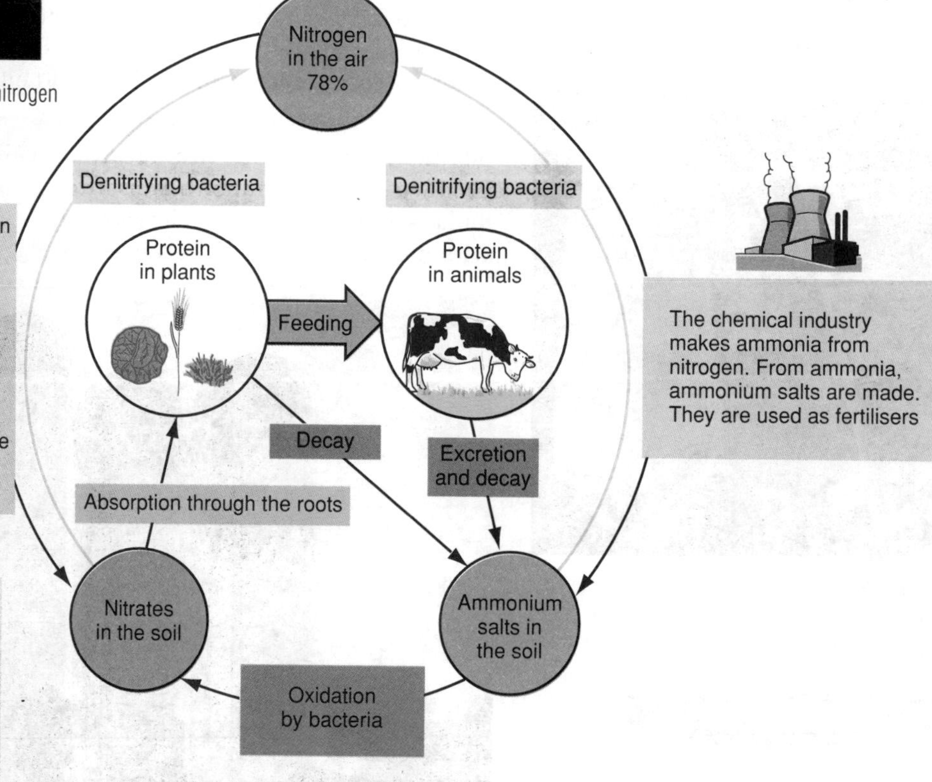

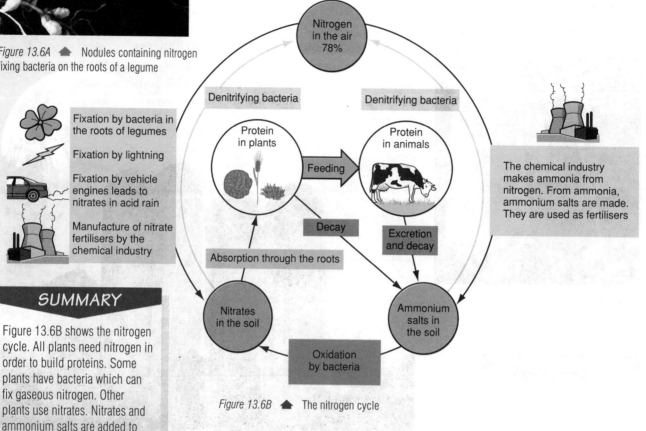


*Figure 13.6B* ▲ The nitrogen cycle

## SUMMARY

Figure 13.6B shows the nitrogen cycle. All plants need nitrogen in order to build proteins. Some plants have bacteria which can fix gaseous nitrogen. Other plants use nitrates. Nitrates and ammonium salts are added to soil to make it more fertile.

## CHECKPOINT

**1** *Process A* Nitrogen-fixing bacteria convert nitrogen gas into nitrates.
*Process B* Nitrifying bacteria use oxygen to convert ammonium compounds (from decaying plant and animal matter) into nitrates.
*Process C* Denitrifying bacteria turn nitrates into nitrogen.

(a) Say where *Process A* takes place.
(b) Say what effect the presence of air in the soil will have on *Process B*.
(c) Say what effect waterlogged soil which lacks air will have on *Process C*.
(d) Explain why plants grow well in well-drained, aerated soil.
(e) A farmer wants to grow a good crop of wheat without using a fertiliser. What could he plant in the field the previous year to ensure a good crop?
(f) Explain why garden manure and compost fertilise the soil.

**2** Explain why nitrogen is used in (a) food packaging (b) oil tankers (c) hospitals and (d) food storage.

# 13.7 ▸ Carbon dioxide and the carbon cycle

## FIRST THOUGHTS

The percentage by volume of carbon dioxide in clean, dry air is only 0.03%. Perhaps you think this makes carbon dioxide sound rather unimportant. Through studying this section, you may change your mind about the importance of carbon dioxide!

## SUMMARY

Figure 13.7A shows the carbon cycle. Photosynthesis is the process in which green plants use sunlight, carbon dioxide and water to make sugars and oxygen. Respiration is the process in which animals and plants oxidise carbohydrates to carbon dioxide and water with the release of energy.

Plants need carbon dioxide, and animals, including ourselves, need plants. Plants take in carbon dioxide through their leaves and water through their roots. They use these compounds to **synthesise** (build) sugars. The reaction is called **photosynthesis** (photo means light). It takes place in green leaves in the presence of sunlight. Oxygen is formed in photosynthesis.

Photosynthesis (in plants)

$$\text{Sunlight} + \text{Carbon dioxide} + \text{Water} \xrightarrow{\text{catalysed by chlorophyll in green leaves}} \text{Glucose (a sugar)} + \text{Oxygen}$$

The energy of sunlight is converted into the energy of the chemical bonds in glucose.

Animals eat foods which contain starches and sugars. They inhale (breathe in) air. Inhaled air dissolves in the blood supply to the lungs. In the cells, some of the oxygen in the dissolved air oxidises sugars to carbon dioxide and water and energy is released. This process is called **respiration**. Plants also respire to obtain energy.

Respiration (in plants and animals)

$$\text{Glucose} + \text{Oxygen} \rightarrow \text{Carbon dioxide} + \text{Water} + \text{Energy}$$

The processes which take carbon dioxide from the air and those which put carbon dioxide into the air are balanced so that the percentage of carbon dioxide in the air stays at 0.03%. This balance is called the **carbon cycle** (see Figure 13.7A).

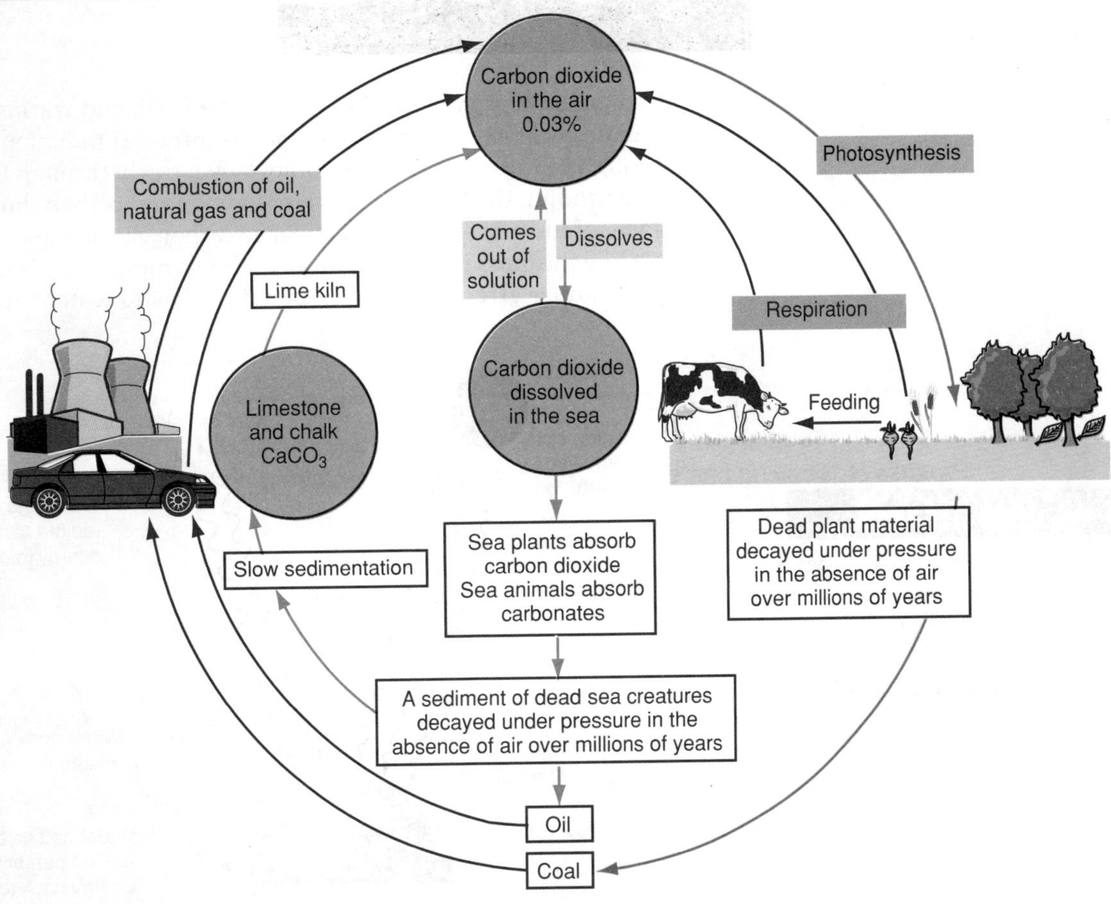


*Figure 13.7A* ▲ The carbon cycle

## CHECKPOINT

1. Name two processes which add carbon dioxide to the atmosphere.
2. Suggest a place where you would expect the percentage of carbon dioxide in the atmosphere to be lower than average.
3. Suggest two places where you would expect the percentage of carbon dioxide in the atmosphere to be higher than average.

## 13.8 The greenhouse effect

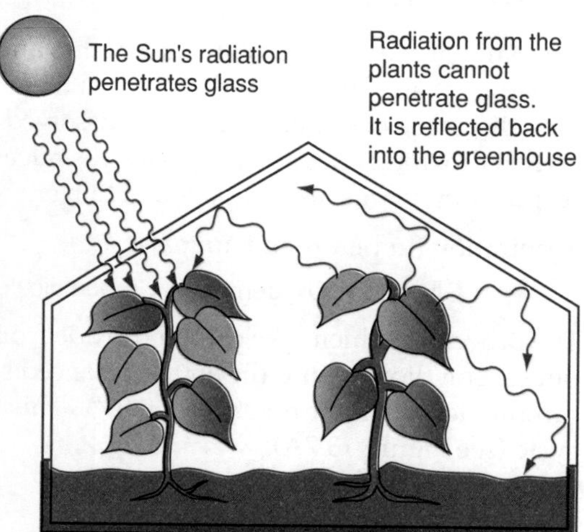

The Sun's radiation penetrates glass
Radiation from the plants cannot penetrate glass. It is reflected back into the greenhouse

Figure 13.8A A greenhouse

The sun is so hot that it emits high-energy radiation. The Sun's rays can pass easily through the glass of a greenhouse. The plants in the greenhouse are at a much lower temperature. They send out infra-red radiation which cannot pass through the glass. The greenhouse therefore warms up (Figure 13.8A).

Radiant energy from the Sun falls on the Earth and warms it. The Earth radiates heat energy back into space as infra-red radiation. Unlike sunlight, infra-red radiation cannot travel freely through the air surrounding the Earth. Both water vapour and carbon dioxide absorb some of the infra-red radiation. Since carbon dioxide and water vapour act like the glass in a greenhouse, their warming effect is called the **greenhouse effect**. Without carbon dioxide and water vapour, the

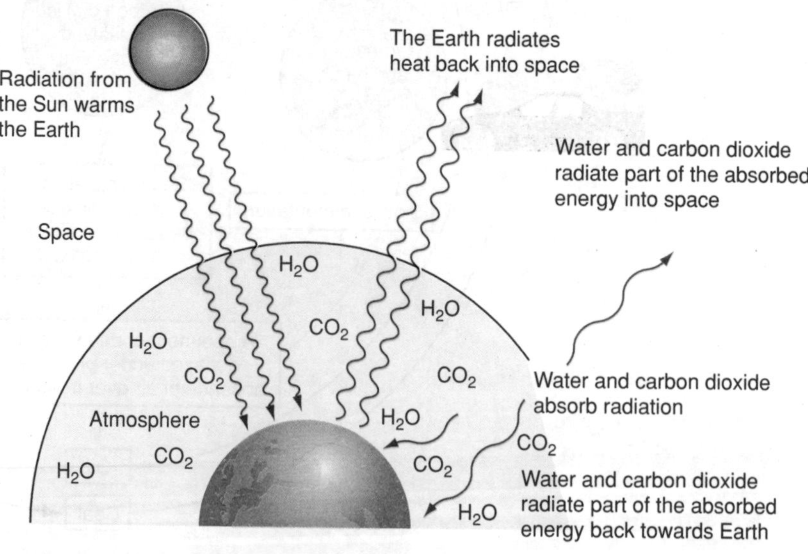


Figure 13.8B The greenhouse effect

### It's a fact!

Worldwide, 16 thousand million tonnes of carbon dioxide are formed each year by the combustion of fossil fuels!

## It's a fact!

Every month, an area of tropical forest equal to the size of Belgium is felled and burned in South America.

surface of the Earth would be at −40 °C. Most of the greenhouse effect is due to water vapour. The Earth does, however, radiate some wavelengths which water vapour cannot absorb. Carbon dioxide is able to absorb some of the radiation which water vapour lets through.

The surface of the Earth has warmed up by 0.75 °C during the last century. The rate of warming up is increasing. Unless something is done to stop the temperature rising, there is a danger that the temperature of the Arctic and Antarctic regions might rise above 0 °C. Then, over the course of a century or two, polar ice would melt and flow into the oceans. If the level of the sea rose, low-lying areas of land would disappear under the sea.

One reason for the increase in the Earth's temperature is that we are putting too much carbon dioxide into the air. The combustion of coal and oil in our power stations and factories sends carbon dioxide into the air. The second reason is that we are felling too many trees. In South America, huge areas of tropical forest have been cut down to make timber and to provide land for farming. In many Asian countries, forests have been cut down for firewood. The result is that worldwide there are fewer trees to take carbon dioxide from the air by photosynthesis. The percentage of carbon dioxide is increasing, and some scientists calculate that it will double by the year 2000 (see Figure 13.8C). This would raise the Earth's temperature by 2 °C.

## SUMMARY

Carbon dioxide and water vapour reduce the amount of heat radiated from the Earth's surface into space and keep the Earth warm. Their action is called the greenhouse effect. The percentage of carbon dioxide in the atmosphere is increasing, and the temperature of the Earth is rising. If it continues to rise, the polar ice caps could melt.

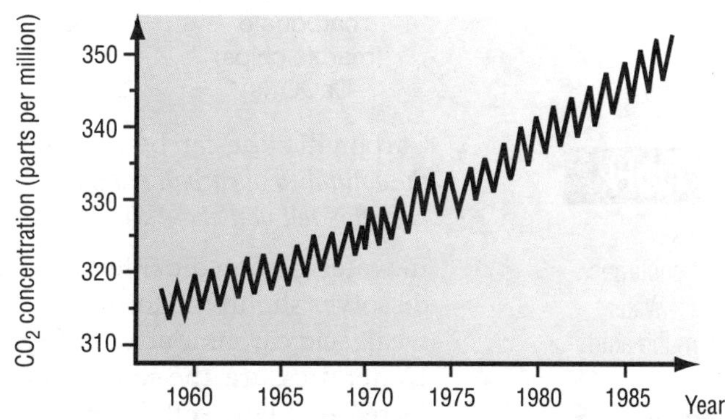


*Figure 13.8C* ▸ Atmospheric concentration of carbon dioxide (These measurements were made at the Mauna Loa Observatory in Hawaii)

## CHECKPOINT

**1** The amount of carbon dioxide in the atmosphere is slowly increasing.

(a) Suggest two reasons why this is happening.

(b) Explain why people call the effect which carbon dioxide has on the atmosphere the 'greenhouse effect'.

(c) Why are some people worried about the greenhouse effect?

(d) Suggest two things which could be done to stop the increase in the percentage of carbon dioxide in the atmosphere.

**2** *Selima* Did you hear what Miss Sande said about the greenhouse effect making the temperature of the Earth go up?

*Joshe* I don't know what she's worried about. We wouldn't be here at all if it weren't for the greenhouse effect.

(a) What does Joshe mean by what he says? What would the Earth be like without the greenhouse effect?

(b) Is Joshe right in thinking there is no cause for worry?

**3** The burning of fossil fuels produces 16 000 million tonnes of carbon dioxide per year. Carbon dioxide is thought to increase the average temperature of the air. It is predicted that the effect of this increase in temperature will be to melt some of the ice at the North and South Poles. Describe the effects which this could have on life for people in other parts of the world.

# 13.9 Carbon dioxide

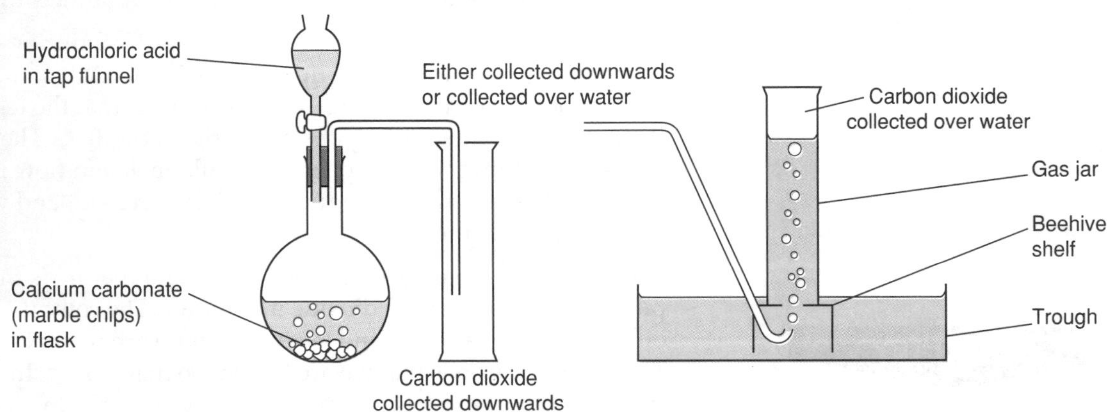


*Figure 13.9A* ▲ A laboratory preparation of carbon dioxide

Do not try using dilute sulphuric acid. A layer of insoluble calcium sulphate forms on the outside of each marble chip and brings the reaction to a stop.

Carbon dioxide can be made in the laboratory. Figure 13.9A shows one method. You will remember from Topic 9 that a carbonate reacts with an acid to form carbon dioxide, a salt and water. In this case

| Calcium carbonate (marble chips) | + | Hydrochloric acid | → | Carbon dioxide | + | Calcium chloride | + | Water |
|---|---|---|---|---|---|---|---|---|
| $CaCO_3(s)$ | + | $2HCl(aq)$ | → | $CO_2(g)$ | + | $CaCl_2(aq)$ | + | $H_2O(l)$ |

Carbon dioxide can be collected over water. (*What does this tell you about the solubility of carbon dioxide?*) It can also be collected downwards. (*What does this tell you about the density of carbon dioxide?*)

In water, carbon dioxide dissolves slightly to form the weak acid carbonic acid, $H_2CO_3$. Under pressure, the solubility increases. Many soft drinks are made by dissolving carbon dioxide under pressure and adding sugar and flavourings. When you open a bottle of fizzy drink, the pressure is released and bubbles of carbon dioxide come out of solution. People like the bubbles and the slightly acidic taste.

When carbon dioxide is cooled, it turns into a solid. This is known as **dry ice** and as **Dricold**. It is used as a refrigerant for icecream and meat. When it warms up, dry ice sublimes (turns into a vapour without melting first).

*Figure 13.9B* ▲ Subliming carbon dioxide being poured from a gas jar

## Science at work

Pop stars sometimes use Dricold on stage. As it sublimes, it cools the air on stage. Water vapour condenses from the cold air in swirling clouds.

## SUMMARY

Carbon dioxide:

- is a colourless, odourless gas
- is denser than air
- dissolves slightly in water to form carbonic acid
- does not burn
- allows few materials to burn in it

## CHECKPOINT

1. Why do icecream sellers prefer dry ice to ordinary ice?
2. Why is carbon dioxide used by the soft drinks industry?

## 13.10 Testing for carbon dioxide and water vapour

**It's a fact!**

Lake Nyos is a beautiful lake in Cameroon, West Africa. On 21 August 1986, a loud rumbling noise came from the lake. One billion cubic metres of gas burst out of the lake. The dense gas rushed downward to the village of Lower Nyos. The 1200 villagers were suffocated in their sleep. The gas was carbon dioxide. It had been formed by volcanic activity beneath the lake.

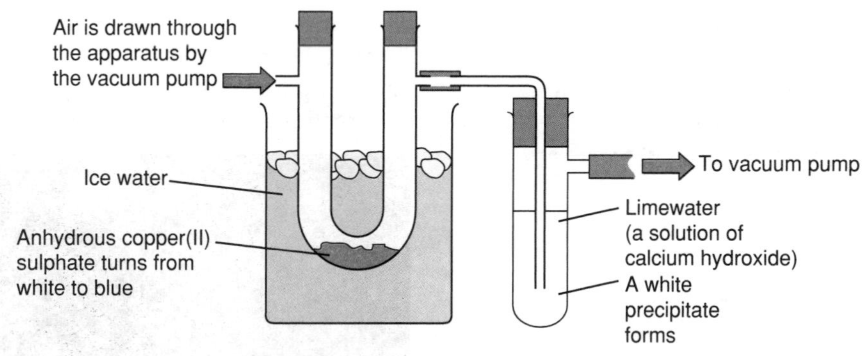
Air is drawn through the apparatus by the vacuum pump
Ice water
Anhydrous copper(II) sulphate turns from white to blue
To vacuum pump
Limewater (a solution of calcium hydroxide)
A white precipitate forms

*Figure 13.10A* ▲ Testing for carbon dioxide and water vapour in air

Figure 13.10A shows how you can test for the presence of carbon dioxide and water vapour in air.

### Test for carbon dioxide

A white precipitate is formed when carbon dioxide reacts with a solution of calcium hydroxide, **limewater**.

| Carbon dioxide | + | Calcium hydroxide (limewater) | → | Calcium carbonate (white precipitate) | + | Water |
|---|---|---|---|---|---|---|
| $CO_2(g)$ | + | $Ca(OH)_2(aq)$ | → | $CaCO_3(s)$ | + | $H_2O(l)$ |

### Test for water vapour

Water turns anhydrous copper(II) sulphate from white to blue.

| Copper(II) sulphate (anhydrous, a white solid) | + | Water | → | Copper(II) sulphate-5-water (blue crystals) |
|---|---|---|---|---|
| $CuSO_4(s)$ | + | $5H_2O(l)$ | → | $CuSO_4.5H_2O(s)$ |

**SUMMARY**

The carbon dioxide in the air turns limewater cloudy. The water vapour in the air turns anhydrous copper(II) sulphate from white to blue.

**CHECKPOINT**

1. Malachite is a green mineral found in many rocks. How could you prove that malachite is a carbonate?
2. (a) How could you prove that whisky contains water?
   (b) How could you show that whisky is not pure water? (Don't say 'Taste it': tasting chemicals is often dangerous.)

## 13.11 The noble gases

Deep-sea divers have to take their oxygen with them (see Figure 13.11A). If they take air in their cylinders, nitrogen dissolves in their blood. Although the solubility of nitrogen is normally very low, at the high pressures which divers experience the solubility increases. When the divers surface, their blood can dissolve less nitrogen than it can at high pressure. Dissolved nitrogen leaves the blood to form tiny bubbles. These cause severe pains, which divers call 'the bends'. The solution to this problem is to breathe a mixture of oxygen and **helium**. Since helium

### It's a fact!

Why are the noble gases so called?

They don't take part in many chemical reactions. Like the nobility of old, they don't seem to do much work.

is much less soluble than nitrogen, there is much less danger of the bends. Helium is a safe gas to use because it takes part in no chemical reactions. It is one of the **noble gases** (see Figure 13.1B).

For a long time, it seemed that the noble gases (helium, neon, argon, krypton and xenon) were unable to take part in any chemical reactions. They were called the 'inert gases'. Argon, the most abundant of them, makes up 0.09% of the air. Figure 13.3B shows how argon is obtained by the fractional distillation of liquid air.

What are 'the bends'? Why does breathing a mixture of oxygen and helium stop divers from getting 'the bends'?

*Figure 13.11A* ▲ A diver carries a mixture of oxygen and helium

Neon, argon, krypton and xenon are used in display lighting (see Figure 13.11B). The discharge tubes which are used as strip lights contain these gases at low pressure. When the gases conduct electricity, they glow brightly. Most electric light bulbs are filled with argon. The filament of a light bulb is so hot that it would react with other gases. Krypton and xenon are also used to fill light bulbs.

The low density of helium makes it useful. It is used to fill balloons and airships.

*Figure 13.11B* ▲ Neon lights

### SUMMARY

The noble gases take part in few chemical reactions. They are present in the air. Argon is used to fill light bulbs. Neon and other noble gases are used in illuminated signs. Helium is used to fill airships.

## CHECKPOINT

**1** The first airships contained hydrogen. Modern airships use helium. The table gives some information about the two gases and air.

| | *Hydrogen* | *Helium* | *Air* |
|---|---|---|---|
| *Density* (g/cm$^3$) | $8.3 \times 10^{-5}$ | $1.66 \times 10^{-4}$ | $1.2 \times 10^{-3}$ |
| *Chemical reactions with air* | Forms an explosive mixture with air | No known chemical reactions | |

(a) What advantage does helium have over hydrogen for filling airships?

(b) What advantage does hydrogen have compared with helium?

(c) Why is air not used for 'airships'?

# Topic 14 Oxygen

## 14.1 Blast-off

**FIRST THOUGHTS**

In this topic, you can find out about some of the elements and compounds that react with oxygen. Oxygen is a very reactive element, and many substances burn in it.

Why do rocket launches use pure oxygen? After all, burning a tonne of kerosene can supply only a certain amount of heat. Burning the fuel in oxygen does not produce more heat than burning it in air.

The reason is that fuels burn faster in pure oxygen and release heat faster and can therefore deliver more power.

**It's a fact!**

The mass of oxygen in the atmosphere is 12 hundred million million tonnes!

On 16 July 1969, ten thousand people gathered at the Kennedy Space Centre in Florida, USA. They had come to watch the spacecraft *Apollo 11* lift off on its journey to the moon. While the *Saturn* rocket stood on the launch pad, its roaring jet engines burned 450 tonnes of kerosene in 1800 tonnes of pure oxygen. The thrust from the jets shot the rocket through the lower atmosphere, trailing a jet of flame. At a height of 65 km the first stage of the rocket separated. For six minutes, the second stage burned hydrogen in pure oxygen, taking the spacecraft to a height of 185 km. Then the second stage separated. The third stage, burning hydrogen in oxygen, put *Apollo 11* into an orbit round the Earth at a speed of 28 000 km/hour. From this orbit, *Apollo 11* headed for the moon. The energy needed to lift *Apollo 11* into space came from burning fuels (kerosene and hydrogen) in pure oxygen. Fuels burns faster in oxygen than they do in air. They therefore deliver more power. We shall return to this important reaction of burning in Topic 14.7.

*Figure 14.1A* ▲ Rocket launch

## 14.2 Test for oxygen

**KEY SCIENTIST**

A new gas was discovered by the British chemist, Joseph Priestley, on 1 August 1786. Heating mercury in air he found that it combined with part of the air to form a solid which he called 'red calx of mercury'. When Priestley heated 'red calx of mercury', he obtained mercury and a new gas, which when he breathed it produced a 'light and easy sensation in his chest'. The gas was oxygen.

Oxygen is a colourless, odourless gas, which is only slightly soluble in water. It is neutral. Oxygen allows substances to burn in it: it is a good **supporter of combustion**. One test for oxygen is to lower a glowing wooden splint into the gas. If the splint starts to burn brightly, the gas is oxygen.

**Oxygen relights a glowing splint** (see Figure 14.2A).

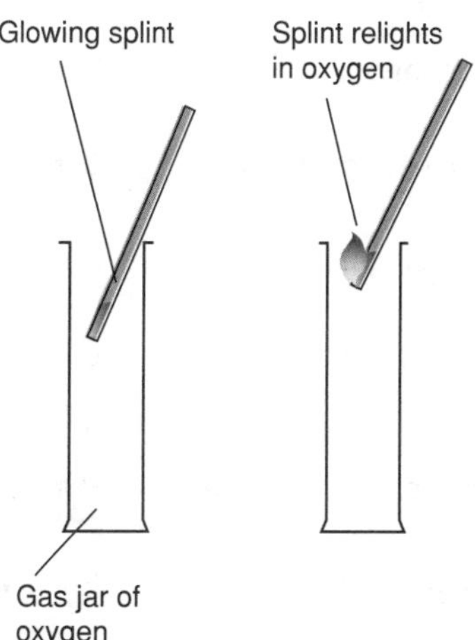


*Figure 14.2A* ▲ Testing for oxygen

## 14.3 A method of preparing oxygen in the laboratory

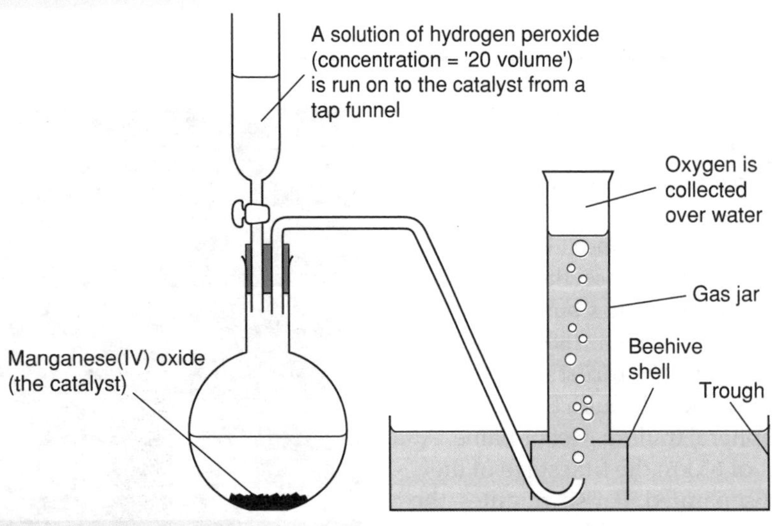
A solution of hydrogen peroxide (concentration = '20 volume') is run on to the catalyst from a tap funnel
Oxygen is collected over water
Gas jar
Beehive shell
Trough
Manganese(IV) oxide (the catalyst)

*Figure 14.3A* ▲ A laboratory preparation of oxygen

Hydrogen peroxide is a colourless liquid which decomposes to give oxygen and water.

Hydrogen peroxide → Oxygen + Water

$2H_2O_2(aq) \rightarrow O_2(g) + 2H_2O(l)$

Solutions of hydrogen peroxide are kept in stoppered brown bottles to slow down the rate of decomposition. If you want to speed up the formation of oxygen, you can add a **catalyst**. A substance which speeds up a reaction without being used in the reaction is called a catalyst. The catalyst manganese(IV) oxide, $MnO_2$, is often used.

## 14.4 The reaction of oxygen with some elements

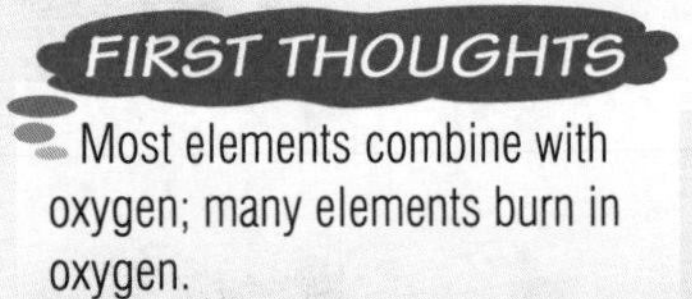

FIRST THOUGHTS

Most elements combine with oxygen; many elements burn in oxygen.

Table 14.1 shows how some elements react with oxygen. The products of the reactions are **oxides**. An oxide is a compound of oxygen with one other element.

*Table 14.1* ▼ How some elements react with oxygen

| Element | Observation | Product | Action of product on water |
|---|---|---|---|
| Calcium (metal) | Burns with a red flame | Calcium oxide, $CaO$ (a white solid) | Dissolves to give a strongly alkaline solution |
| Copper (metal) | Does not burn; turns black | Copper(II) oxide, $CuO$ (a black solid) | Insoluble |
| Iron (metal) | Burns with yellow sparks | Iron oxide, $Fe_3O_4$ (a blue-black solid) | Insoluble |
| Magnesium (metal) | Burns with a bright white flame | Magnesium oxide, $MgO$ (a white solid) | Dissolves slightly to give an alkaline solution, pH = 9 |
| Sodium (metal) | Burns with a yellow flame | Sodium oxide, $Na_2O$ (a yellow-white solid) | Dissolves readily to form a strongly alkaline solution, pH = 10 |
| Carbon (non-metal) | Glows red | Carbon dioxide, $CO_2$ (an invisible gas) | Dissolves slightly to give a weakly acidic solution, pH = 4 |
| Phosphorus (non-metal) | Burns with a yellow flame | Phosphorus(V) oxide, $P_2O_5$ (a white solid) | Dissolves to give a strongly acidic solution, pH = 2 |
| Sulphur (non-metal) | Burns with a blue flame | Sulphur dioxide, $SO_2$ (a fuming gas with a choking smell) | Dissolves readily to form a strongly acidic solution, pH = 2 |

# 14.5 Oxides

A pattern can be seen in the characteristics of oxides. The oxides of metallic elements are **bases**. The bases which dissolve in water are called **alkalis**. Most of the oxides of non-metallic elements are acids, but some are neutral. Acids react with bases to form **salts**.

## Amphoteric oxides

Some oxides and hydroxides are both basic (they react with acids to form salts) and acidic (they react with bases to form salts). Such oxides and hydroxides are **amphoteric**. They include the oxides and hydroxides of zinc, aluminium and lead.

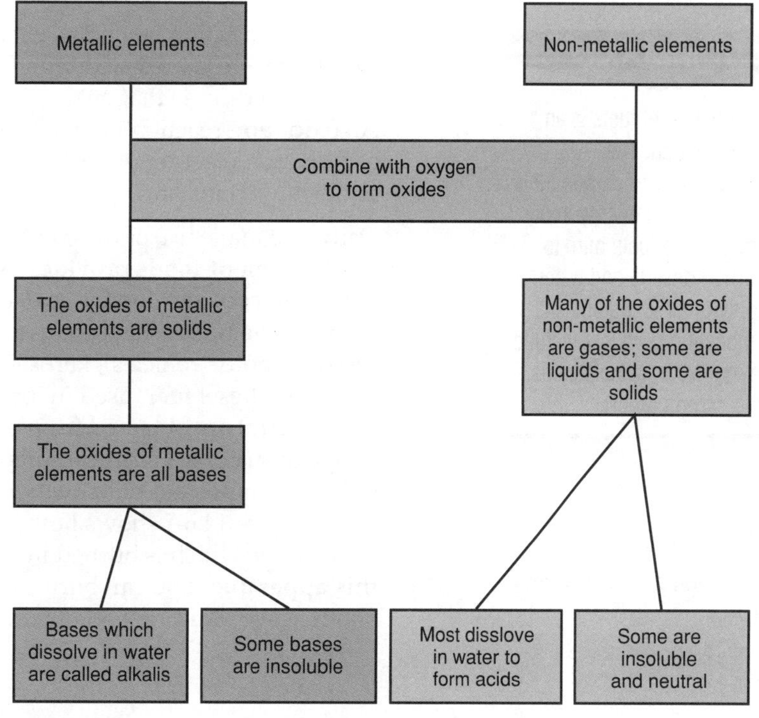


*Figure 14.5A* ▲ Oxides

When an element combines with oxygen, we say it has been **oxidised**. Oxygen **oxidises** copper to copper(II) oxide. This reaction is an **oxidation**.

Copper is oxidised

Copper + Oxygen → Copper(II) oxide

$2Cu(s) + O_2(g) \rightarrow 2CuO(s)$

This reaction is oxidation

The opposite of oxidation is **reduction**. When a substance loses oxygen, it is **reduced**. Copper(II) oxide can be reduced by heating it and passing hydrogen over it.

Copper(II) oxide is reduced

Copper(II) oxide + Hydrogen → Copper + Water

$CuO(s) + H_2(g) \rightarrow Cu(s) + H_2O(l)$

Hydrogen is oxidised

You can see from the equation that oxidation and reduction occur together. Copper(II) oxide is reduced to copper while hydrogen is oxidised to water. Copper(II) oxide is called an **oxidising agent** because it gives oxygen to hydrogen. Hydrogen is called a **reducing agent** because it takes oxygen from copper(II) oxide.

### SUMMARY

Most elements combine with oxygen to form oxides. Many elements burn in oxygen. The oxides of metals are basic. The oxides of non-metallic elements are acidic or neutral. An oxidising agent gives oxygen to another substance. A reducing agent takes oxygen from another substance.

### CHECKPOINT

1 (a) Describe two differences in the physical characteristics of metallic and non-metallic elements (see Topic 4.7 also).
(b) State two differences between the chemical reactions of metallic and non-metallic elements (see Topic 9.1 also).

2 Write word equations and balanced chemical equations for the combustion in oxygen of (a) sulphur (b) carbon (c) magnesium (d) sodium.

# 14.6 Combustion

SUMMARY

The combustion of fuels is an oxidation reaction. The combustion of hydrocarbon fuels is a vital source of energy in our economy. These fuels burn to form carbon dioxide and water. If the supply of air is insufficient, the combustion products include carbon monoxide (a poisonous gas) and carbon (soot).

As you read, check your understanding of these terms:

- oxidation,
- combustion,
- burning,
- fuel,
- respiration.

In many oxidation reactions, energy is given out. The fireworks called 'sparklers' are coated with iron filings. When the iron is oxidised to iron oxide, you can see that energy is given out in the form of heat and light. An oxidation reaction in which energy is given out is called a combustion reaction. A combustion in which there is a flame is described as burning. Substances which undergo combustion are called fuels.

Daily, we make use of the combustion of fuels. In respiration, the **combustion of foods** provides us with energy. We use fuels to heat our homes, to cook our food, to run our cars and to generate electricity. Many of the fuels which we use are derived from petroleum oil. Petrol (used in motor vehicles), kerosene (used in aircraft and as domestic paraffin), diesel fuel (used in lorries and trains) and natural gas (used in gas cookers) are obtained from **petroleum oil**. These fuels are mixtures of **hydrocarbons**. Hydrocarbons are compounds of carbon and hydrogen only. It is important to know what products are formed when these fuels burn. Figure 14.6A shows how you can test the products of combustion of kerosene which is burned in paraffin heaters. **Do not use petrol in this apparatus**. You can burn a candle instead of kerosene. Candle wax is another hydrocarbon fuel obtained from crude oil.

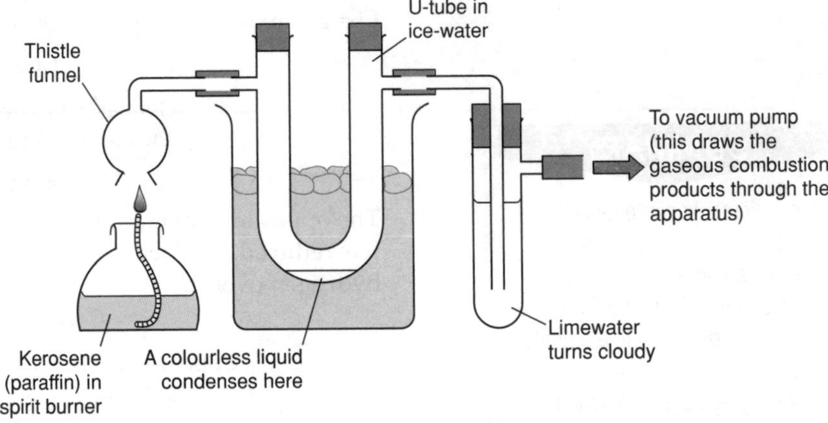


*Figure 14.6A* ▲ How to test the combustion products of a hydrocarbon fuel, e.g. kerosene or candle wax (*NOTE: do not use petrol*)

SUMMARY

Oxidation is the addition of oxygen to a substance. Combustion is oxidation with the release of energy. Burning is combustion accompanied by a flame. Respiration is combustion which takes place in living tissues. In respiration, food materials are oxidised in the cells with the release of energy.

The combustion products are carbon dioxide and water. Other hydrocarbon fuels give the same products.

Hydrocarbon + Oxygen → Carbon dioxide + Water vapour

If you burn a candle in this apparatus, you will see a deposit of carbon (soot) in the thistle funnel. This happens when the air supply is insufficient to oxidise all the carbon in the hydrocarbon fuel to carbon dioxide, $CO_2$. Another product of incomplete combustion is the poisonous gas carbon monoxide, CO. Because you cannot see or smell carbon monoxide, it is doubly dangerous. Many times, people have been poisoned by carbon monoxide while running a car engine in a closed garage. The engine could not get enough oxygen for complete combustion to occur. The exhaust gases from petrol engines always contain some carbon monoxide, some unburnt hydrocarbons and some soot, in addition to the harmless products: carbon dioxide and water.

## CHECKPOINT

1. (a) What type of compound is present in petrol?
   (b) What products are formed in combustion (i) if there is plenty of air and (ii) if there is a limited supply of air?
2. In February 1988 newspapers carried a report of a woman who fell asleep in front of a fire and never woke up. Later, workmen removed three buckets full of birds' nesting materials from the chimney. What do you think had caused the woman's death?
3. Why should you make sure the window is open if you use a gas heater in the bathroom?

# 14.7 Rusting

**FIRST THOUGHTS**

Rusting is a costly nuisance. Methods of slowing down the process can save a lot of money.

These experiments explore the conditions which favour rusting. The point is that if you know the conditions that promote rusting you can avoid these conditions and prolong the life of iron objects.

Many metals become corroded by exposure to the air. The corrosion of iron and steel is called rusting. Rust is the reddish brown solid, hydrated iron(III) oxide, $Fe_2O_3.nH_2O$. (The number of water molecules, $n$, varies.)

Rusting is an oxidation reaction:

$$\text{Iron} + \text{Oxygen} \rightarrow \text{Iron(III) oxide}$$
$$4Fe(s) + 3O_2(g) \rightarrow 2Fe_2O_3(s)$$

Rusting is a nuisance. Cars, ships, bridges, machines and other costly items made from iron and steel rust. To prevent rusting, or at least to slow it down, saves a lot of money. Before you can prevent rusting, you first have to know what conditions speed up rusting. Figure 14.7A shows some experiments on the rusting of iron nails.

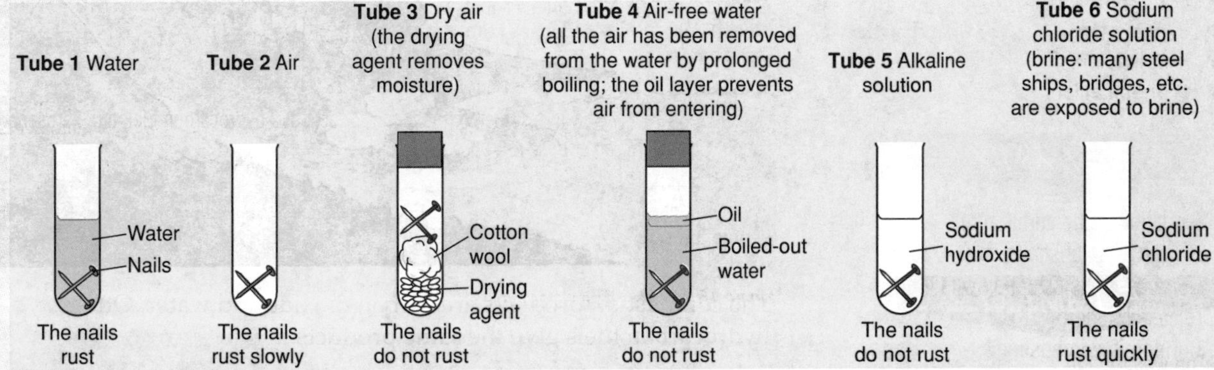


*Figure 14.7A* ▲ Experiments on the rusting of iron nails.

**SUMMARY**

Rusting is the oxidation of iron and steel to iron(III) oxide.

The experiments show that in order to rust, iron needs air and water and a trace of acid. The carbon dioxide in the air provides sufficient acidity. Rusting is accelerated by salt. Bridges and ships are exposed to brine, and cars are exposed to the salt that is spread on the roads in winter. It is obviously very important to find ways of rust-proofing these objects.

# Topic 15 Water

## 15.1 The water cycle

FIRST THOUGHTS

The first living things evolved in water. As more complex plants and animals evolved, water remained essential for life.

*Where does all the rain come from? Why does the atmosphere never run out of water?* Four-fifths of the world's surface is covered by water. From oceans, rivers and lakes, water evaporates into the atmosphere. Plants give out water vapour in **transpiration.** As it rises into a cooler part of the atmosphere, water vapour condenses to form clouds of tiny droplets. If the clouds are blown upward and cooled further, larger drops of water form and fall to the ground as rain (or snow). *Where does the rain go?* Rain water trickles through soil, where some is taken up by plants. The rest passes through porous rocks to become part of rivers, lakes, ground water and the sea. This chain of events is called the **water cycle** (see Figure 15.1A).

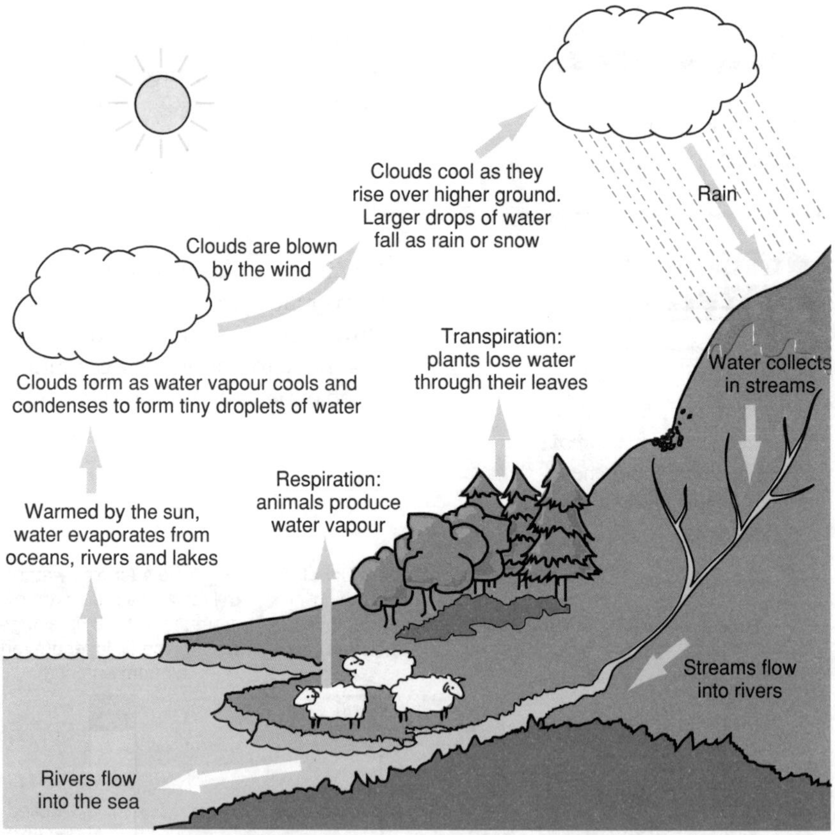

It's a fact!

A large tree can lose 300 litres of water vapour in an hour by transpiration.



*Figure 15.1A* ▲ The water cycle

Check that you understand the meaning of the terms:

- condensation,
- evaporation,
- respiration,
- transpiration.

As rain water falls through the air, it dissolves oxygen, nitrogen and carbon dioxide. The dissolved carbon dioxide forms a solution of the weak acid, carbonic acid. Natural rain water is therefore weakly acidic. In regions where the air is polluted, rain water may dissolve sulphur dioxide and oxides of nitrogen, which make it strongly acidic; it is then called **acid rain**. As rain water trickles through porous rocks, it dissolves salts from the rocks. The dissolved salts are carried into the sea. When sea water evaporates, the salts remain behind.

SUMMARY

The water cycle:

- Evaporation, transpiration and respiration send water vapour into the atmosphere.
- Condensation forms clouds which return water to the Earth as rain, hail or snow.

## 15.2 ▶ Dissolved oxygen

FIRST THOUGHTS

Oxygen sensors connected to a computer are used to monitor the concentration of dissolved oxygen in commercial fish tanks. Fish farmers and breeders are warned when the concentration starts dropping and can take preventive action before any damage is done.

The fact that oxygen dissolves in water is vitally important. The solubility is low: water can dissolve no more than 10 g oxygen per tonne of water, that is 10 p.p.m. (parts per million). This is high enough to sustain fish and other water-living animals and plants. When the level of dissolved oxygen falls below 5 p.p.m. aquatic plants and animals start to suffer.

Water is able to purify itself of many of the pollutants which we pour into it. Bacteria which are present in water feed on plant and animal debris. These bacteria are **aerobic** (they need oxygen). They use dissolved oxygen to oxidise organic material (material from plants and animals) to harmless products, such as carbon dioxide and water. This is how the bacteria obtain the energy which they need to sustain life. If a lot of untreated sewage is discharged into a river, the dissolved oxygen is used up more rapidly than it is replaced, and the aerobic bacteria die. Then **anaerobic** bacteria (which do not need oxygen) attack the organic matter. They produce unpleasant-smelling decay products.

Some synthetic (manufactured) materials, e.g. plastics, cannot be oxidised by bacteria. These materials are nonbiodegradable, and they last for a very long time in water.

SUMMARY

Air dissolves in water. The dissolved oxygen in water keeps fish alive. If too much organic matter, e.g. sewage, is discharged into a river, the dissolved oxygen is used up in oxidising the organic matter, and the fish die.

## 15.3 ▶ Water treatment

FIRST THOUGHTS

The earliest human settlements were always beside rivers. The settlers needed water to drink and used the river to carry away their sewage and other waste. Obtaining clean water is more difficult now.

The water that we use is taken mainly from lakes and rivers. Water treatment plants purify the water to make it safe to drink. They do this by:

- filtration to remove solid matter followed by...
- bacterial oxidation to get rid of organic matter and...
- treatment with chlorine to kill germs.

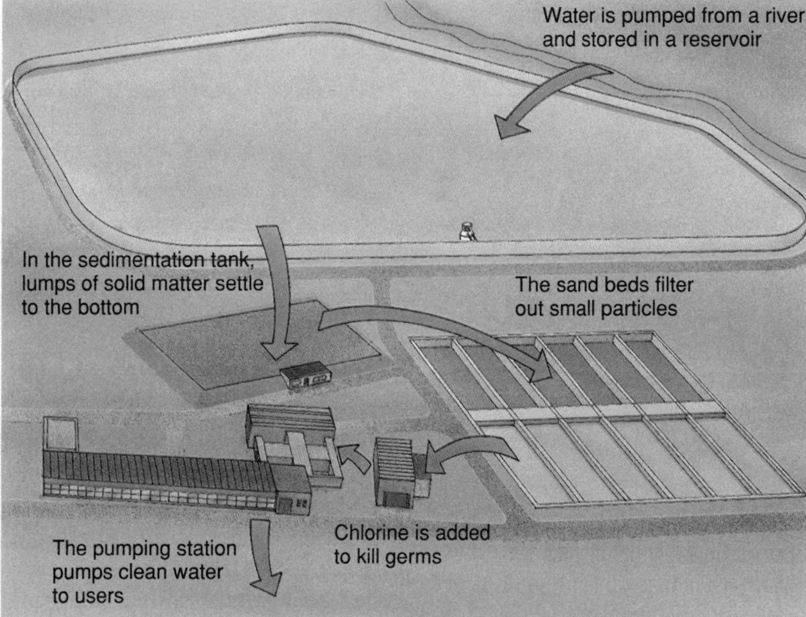
Water is pumped from a river and stored in a reservoir
In the sedimentation tank, lumps of solid matter settle to the bottom
The sand beds filter out small particles
Chlorine is added to kill germs
The pumping station pumps clean water to users

*Figure 15.3A* ▲ A water treatment works

Science at work

The water industry makes use of computers. Sensors detect the pH, oxygen concentration and other qualities of the water and relay the measurements to a computer. This constant monitoring enables the industry to control the quality of the water it provides.

SUMMARY

Water treatment works take water from lakes and rivers. After filtration, followed by chlorination, the water is safe to drink.

In some areas, the water supply comes from ground water (water held underground in porous layers of rock). As rain water trickles down from the surface through porous rocks, the solid matter suspended in it is filtered out. Ground water therefore does not need the complete treatment. It is pumped out of the ground and chlorinated before use.

KEY SCIENTIST

In 1854, 50 000 people died of cholera in London. Dr John Snow did some scientific detective work to find the source of the disease. He marked the deaths from cholera and the positions of the street pumps from which people obtained their water on a map of London. Dr Snow came to the conclusion that one of the pumps was supplying contaminated water. Which was it? He tested his theory by removing the handle of the pump. There was a sudden fall in the number of people getting cholera. The pump water had been contaminated by sewage leaking into it.

## 15.4 Sewage works

FIRST THOUGHTS

Rivers carry away our sewage, our industrial grime, the waste chemicals from our factories and the waste heat from our power stations. Sewage works try to ensure that rivers are not overloaded with waste.

Homes, factories, businesses and schools all discharge their used water into sewers which take it to a sewage works. There, the dirty water is purified until it is fit to be discharged into a river (see Figure 15.4A). The river dilutes the remaining pollutants and oxidises some of them. The digested sludge obtained from a sewage works can be used as a fertiliser. Raw sewage cannot be used as fertiliser because it contains harmful bacteria.

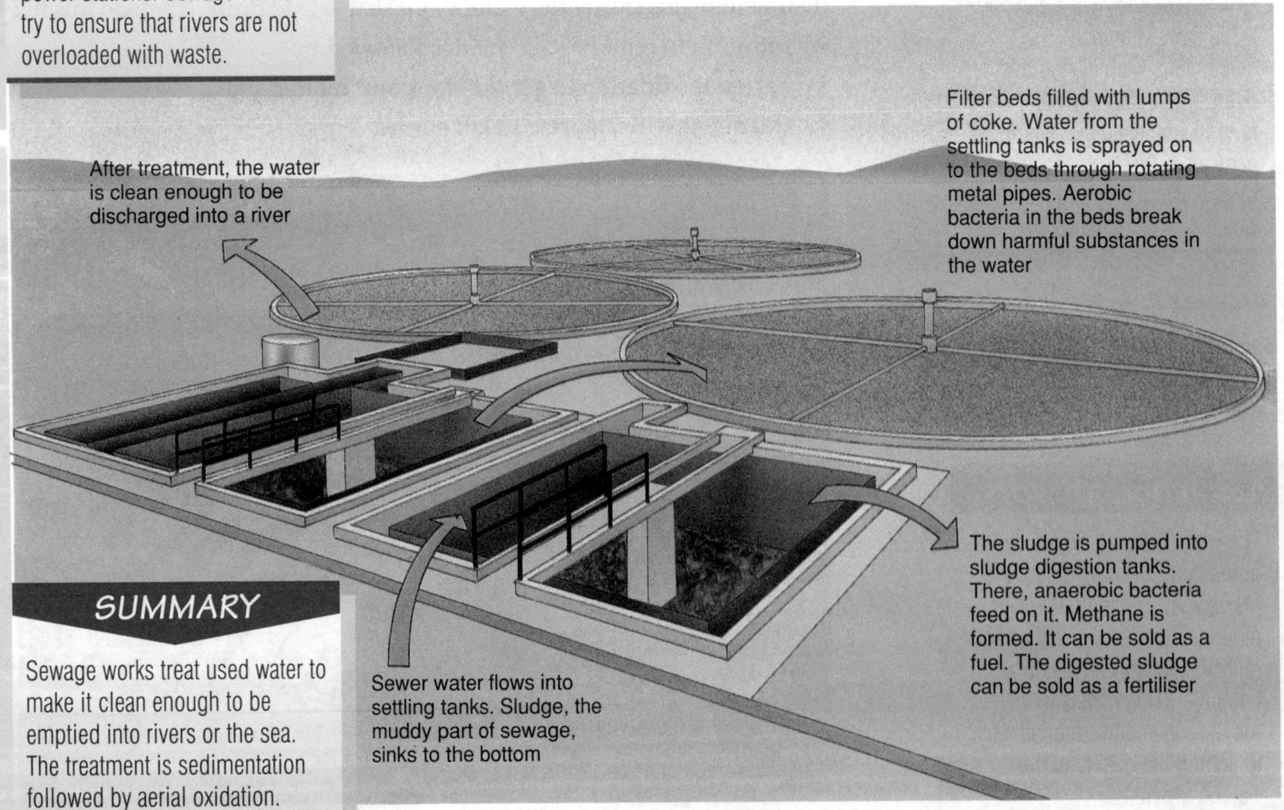

Filter beds filled with lumps of coke. Water from the settling tanks is sprayed on to the beds through rotating metal pipes. Aerobic bacteria in the beds break down harmful substances in the water
After treatment, the water is clean enough to be discharged into a river
The sludge is pumped into sludge digestion tanks. There, anaerobic bacteria feed on it. Methane is formed. It can be sold as a fuel. The digested sludge can be sold as a fertiliser
Sewer water flows into settling tanks. Sludge, the muddy part of sewage, sinks to the bottom

SUMMARY

Sewage works treat used water to make it clean enough to be emptied into rivers or the sea. The treatment is sedimentation followed by aerial oxidation.

*Figure 15.4A* ▲ A sewage works

## 15.5 ▶ Uses of water

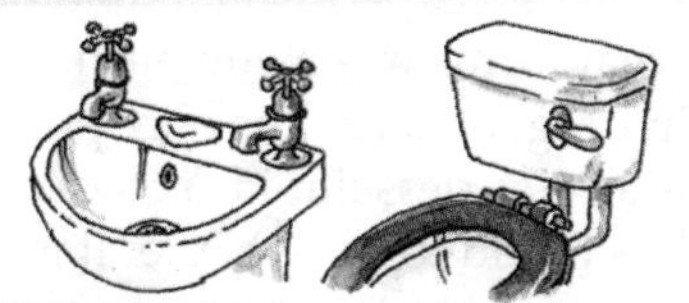

*Figure 15.5A* ▲ Water: 160 litres a day

The uses shown in Figure 15.5A show only 10% of the total amount of water you use. The other 90% is used:

- to grow your food (agricultural use),
- to make your possessions (industrial use as a solvent, for cleaning and for cooling),
- to generate electricity (used as a coolant in power stations).

The total water consumption in an industrialised country amounts to around 80 000 litres (80 tonnes) a year per person. The manufacture of:

- 1 tonne of steel uses 45 tonnes of water,
- 1 tonne of paper uses 90 tonnes of water,
- 1 tonne of nylon uses 140 tonnes of water,
- 1 tonne of bread uses 4 tonnes of water,
- 1 motor car uses 450 tonnes of water,
- 1 litre of beer uses 10 litres of water.

Water used for many industrial purposes is purified and recycled.

**It's a fact!**

An industrial country uses about 80 tonnes of water per person per year.

**CHECKPOINT**

▶ 1 The table shows the world consumption of water over the past 30 years.

| *Year* | *World consumption of water (millions of tonnes per day)* |
|---|---|
| 1960 | 10.0 |
| 1970 | 11.5 |
| 1975 | 13.0 |
| 1980 | 15.0 |
| 1985 | 17.0 |
| 1990 | 20.0 |

(a) On graph paper, plot the consumption (on the vertical axis) against the year (on the horizontal axis).
(b) Say what has happened to the demand for water over the past 30 years.
(c) Suggest three reasons for the change.
(d) From your graph, predict what the consumption of water will be in the year 2000.

## 15.6 ▶ Water: the compound

The one thing that everyone knows – the formula of water!

Water is a compound. When a direct electric current passes through it, water splits up: it is electrolysed. The only products formed in the electrolysis of water are the gases hydrogen and oxygen. The volume of hydrogen is twice that of oxygen. From this result, chemists have calculated that the formula for water is $H_2O$.

$$\text{Water} \xrightarrow{\text{electrolyse}} \text{Hydrogen} + \text{Oxygen}$$
$$2H_2O(l) \rightarrow 2H_2(g) + O_2(g)$$

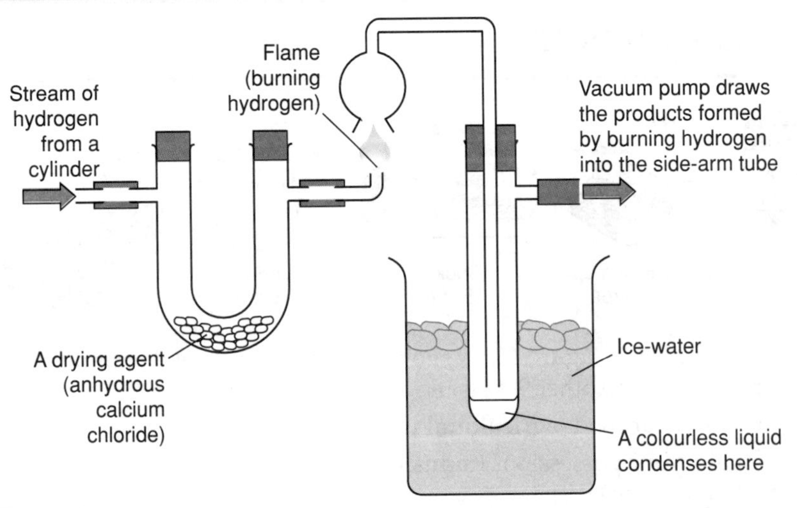


*Figure 15.6A* What is formed when hydrogen burns in air?

Water is the oxide of hydrogen. *Can it be made by the combination of hydrogen and oxygen*? Figure 15.6A shows an experiment to find out what forms when you burn hydrogen in air. The only product is a colourless liquid. You can test this liquid to see whether it is water.

Since hydrogen burns with the evolution of much heat to form a harmless product, water, hydrogen is used as a fuel.

## SUMMARY

Tests for water:

- Turns anhydrous copper(II) sulphate from white to blue.
- Turns anhydrous cobalt(II) chloride from blue to pink.

Tests for pure water:

- Boiling point = 100 °C at 1 atm
- Freezing point = 0 °C at 1 atm

Water is formed when hydrogen burns in air.

### Tests for water

- Water turns white anhydrous copper(II) sulphate blue.

| Copper(II) sulphate | + Water | → | Copper(II) sulphate-5-water |
|---|---|---|---|
| $CuSO_4(s)$ | $+ 5H_2O(l)$ | → | $CuSO_4.5H_2O(s)$ |
| (white solid) | | | (blue solid) |

- Water turns blue anhydrous cobalt(II) chloride pink.

| Cobalt(II) chloride | + Water | → | Cobalt(II) chloride-6-water |
|---|---|---|---|
| $CoCl_2(s)$ | $+ 6H_2O(l)$ | → | $CoCl_2.6H_2O(s)$ |
| (blue solid) | | | (pink solid) |

Any liquid which contains water will give positive results in these tests. To find out whether a liquid is pure water, you can find its boiling point and freezing point. At 1 atm, pure water boils at 100 °C and freezes at 0 °C.

The tests show that the liquid formed when hydrogen burns in air is in fact water.

| Hydrogen + Oxygen | → | Water |
|---|---|---|
| $2H_2(g) + O_2(g)$ | → | $2H_2O(l)$ |

# 15.7 Pure water

## SUMMARY

Water is a good solvent. The presence of a solute raises the boiling point and lowers the freezing point. Pure water is obtained by distillation.

Almost all substances dissolve in water to some extent: that is, water is a good **solvent**. Since water is such a good solvent, it is difficult to obtain pure water. Distillation is one method of purifying water. In some countries, distillation is used to obtain drinking water from sea water. The technique is called desalination (desalting). Hong Kong has a large desalination plant which has never been used because the cost of importing the oil needed to run it is so high. Saudi Arabia and Bahrain operate desalination plants. *Why do you think they need the plants and can afford to run them?*

When chemists describe water as pure, they mean that the water contains no dissolved material. This is different from what a water company means by pure water: they mean that the water contains no harmful substances. Safe drinking water contains dissolved salts. Water which contains substances that are bad for health is **polluted** water.

## CHECKPOINT

These questions will enable you to revise solubility curves. Use the figure opposite to help you.

▶ 1 One kilogram of water saturated with potassium chloride is cooled from 80 °C to 20 °C. What mass of potassium chloride crystallises out?

▶ 2 One kilogram of water saturated with sodium chloride is cooled from 80 °C to 20 °C. What mass of sodium chloride crystallises out?

▶ 3 Dissolved in 100 g of water at 100 °C are 30 g of sodium chloride and 50 g of potassium chloride. What will happen when the solution is cooled to 20 °C?

▶ 4 Dissolved in 100 g of water at 80 °C are 10 g of potassium sulphate and 70 g of potassium bromide. What will happen when the solution is cooled to 20 °C?

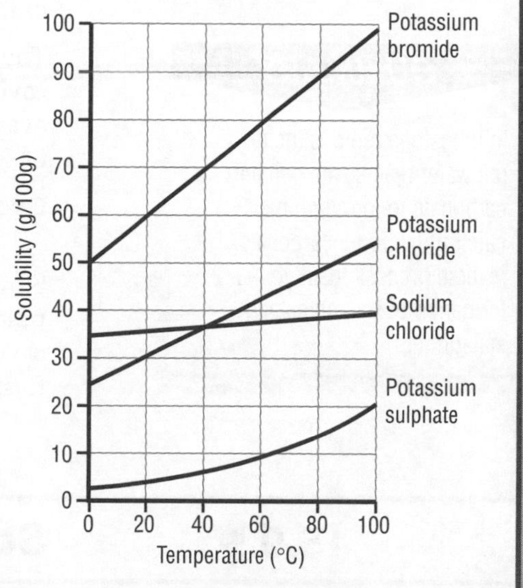


# 15.8 ▶ Underground caverns

In limestone regions, rain water trickles over rocks composed of calcium carbonate (limestone) and magnesium carbonate. These carbonates do not dissolve in pure water, but they react with acids. The carbon dioxide dissolved in rain water makes it weakly acidic. It reacts with the carbonate rocks to form the soluble salts, calcium hydrogencarbonate and magnesium hydrogencarbonate.

Calcium carbonate (limestone) + Water + Carbon dioxide → Calcium hydrogencarbonate solution

$$CaCO_3(s) + H_2O(l) + CO_2(g) \rightarrow Ca(HCO_3)_2(aq)$$

What caused the formation of this cavern at Wookey Hole?

A well-known reaction between calcium carbonate and a dilute acid!

*Figure 15.8A* ▲ A cavern at Wookey Hole (note the stalactites and stalagmites)

This chemical reaction is responsible for the formation of the underground **caves** and potholes which occur in limestone regions. Over thousands of years, large masses of carbonates have been dissolved out of the rock (see Figure 15.8A).

The reverse reaction can take place. Sometimes, in an underground cavern, a drop of water becomes isolated. With air all round it, water will evaporate. The dissolved calcium hydrogencarbonate turns into a grain of solid calcium carbonate.

Calcium hydrogencarbonate → Calcium carbonate + Water + Carbon dioxide

$$Ca(HCO_3)_2(aq) \rightarrow CaCO_3(s) + H_2O(l) + CO_2(g)$$

Slowly, more grains of calcium carbonate are deposited. Eventually, a pillar of calcium carbonate may have built up from the floor of the cavern. This is called a **stalagmite**. The same process can lead to the formation of a **stalactite** on the roof of the cavern.

SUMMARY

In limestone regions, acidic rainwater reacts with calcium carbonate to form soluble calcium hydrogencarbonate. The reverse process leads to the formation of stalactites and stalagmites.

## 15.9 Soaps

FIRST THOUGHTS

Have you ever tried to wash greasy hands without using soap? The problem is that grease and water do not mix. You can find out how soap solves the problem in this section.

Soaps are able to form a bridge between grease and water. They are the sodium and potassium salts of organic acids. One soap is sodium hexadecanoate, $C_{15}H_{31}CO_2Na$. A model of the soap is shown in Figure 15.9A.

(Hexadecane means sixteen. Count up. *Are there 16 carbon atoms?*)

It consists of a sodium ion and a hexadecanoate ion, which we will call a *soap* ion for short. The *soap* ion has two parts (see Figure 15.9B). The head, which is attracted to water, is a—$CO_2^-$ group (a **carboxylate** group). The tail, which is repelled by water and attracted by grease, is a long chain of —$CH_2$— groups. Figure 15.9C shows how *soap* ions wash grease from your hands.

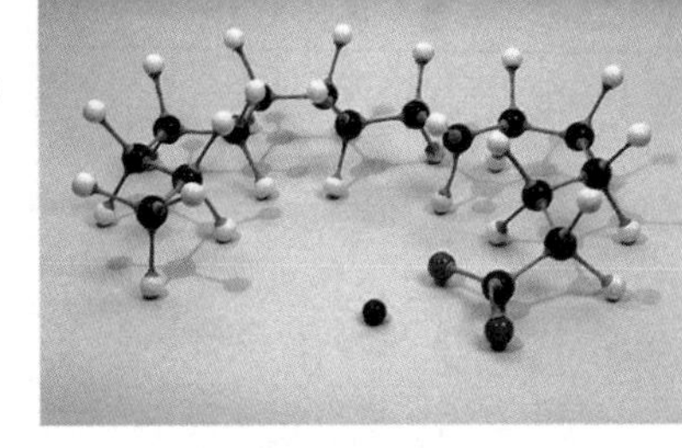

Figure 15.9A ▲ A model of the soap, sodium hexadecanoate

$$CH_3\,CH_2\,CH_2\,CH_2\,CH_2\,CH_2\,CH_2\,CH_2\,CH_2\,CH_2\,CH_2\,CH_2\,CH_2\,CH_2\,CH_2{-}\overset{\overset{\displaystyle O}{\|}}{C}{-}O^-\,Na^+$$

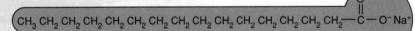

Figure 15.9B ▲ A *soap* ion

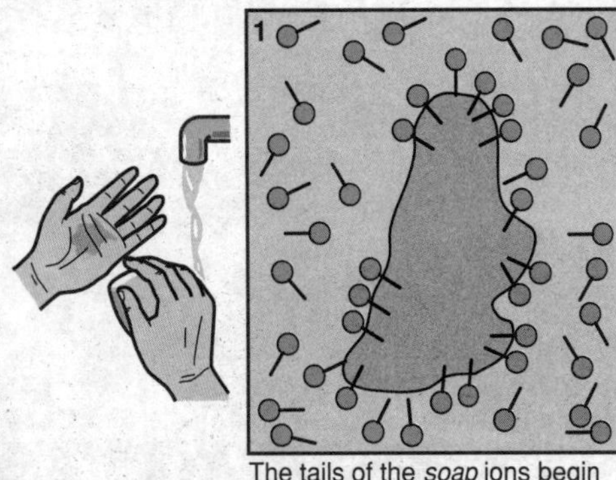


The tails of the *soap* ions begin to dissolve in the grease. The heads remain dissolved in the water

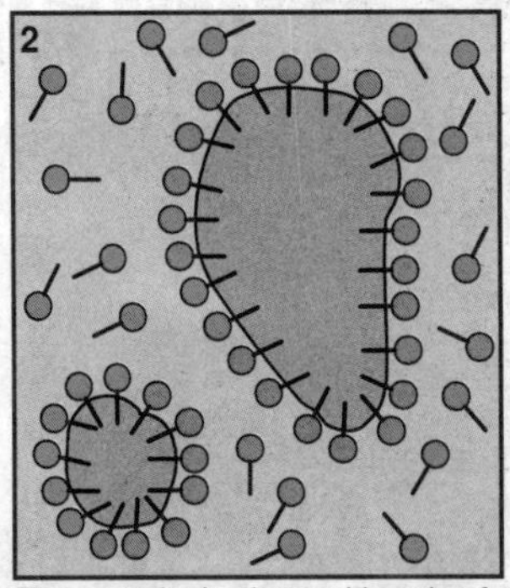


The negatively charged heads of the *soap* ions repel one another; this repulsion makes the grease break up into small droplets, which are suspended in water. The soap has **emulsified** the grease and water (made them mix)

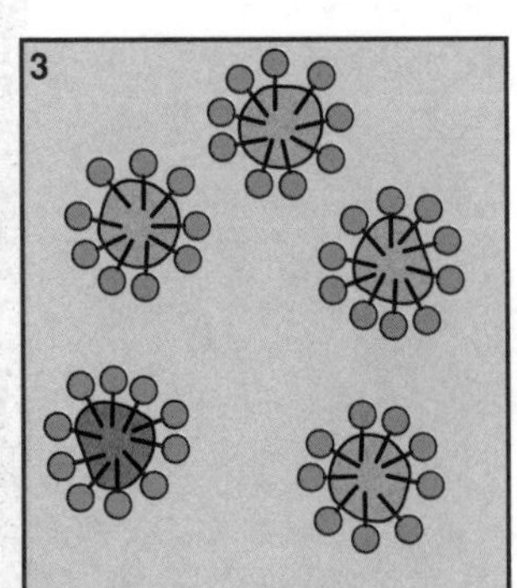


The emulsified grease is washed away by water

Figure 15.9C ▲ The cleansing action of soap

## Manufacture of soap

SUMMARY

Soaps are able to emulsify oil and water.

Soaps are made by boiling together animal fats or vegetable oils and a strong alkali, e.g. sodium hydroxide. The reaction is called **saponification** (soap-making).

Fat (or oil) + Sodium hydroxide → Soap + Glycerol

When sodium chloride is added to the mixture, soap solidifies. The product is purified to remove alkali from it. Perfume and colouring are added before the soap is formed into bars.

# 15.10 ▶ Soapless detergents

Soaps are one type of detergent (cleaning agent). There is another type of detergent known as **soapless detergents**. Many washing powders and household cleaning fluids are soapless detergents. Often they are referred to simply as 'detergents'. Soapless detergents are made from petroleum oil. They are the sodium salts of sulphonic acids (see Figure 15.10A).

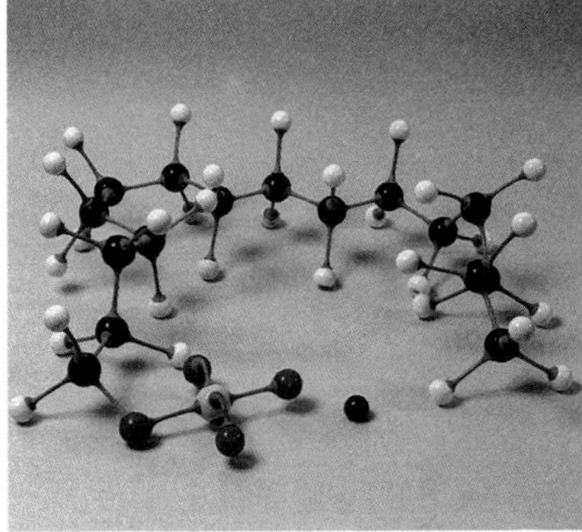

*Figure 15.10A* ▲ A model of a soapless detergent (note the tail, a chain of $—CH_2—$ groups which dissolves in grease, and the head, a sulphate group, $—SO_4^-$, which dissolves in water)

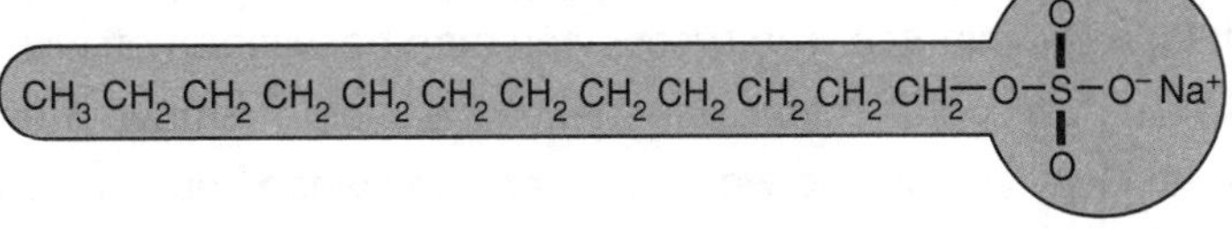


*Figure 15.10B* ▲ A soapless detergent

SUMMARY

Soapless detergents have many advantages over soaps and are used in washing powders. Soaps are better for use on the skin.

Soapless detergents are very good at removing oil and grease. They are too powerful for use on the skin, and the gentler action of a soap is better. Shampoos are mild detergents.

## CHECKPOINT

1. Figure 15.10B shows the formula of a soapless detergent. Say which part of the ion will attach itself to grease and which part will remain in the water.
2. Explain how a soapless detergent is able to dislodge grease from dirty clothes.
3. Why is it important that the water in a washing machine is agitated?
4. Why is it important to rinse clothes well after washing them?

## 15.11 Bleaches and alkaline cleaners

### It's a fact!

In March 1988, 15 workers at a laundry were taken to hospital after inhaling chlorine. The laundry said that suppliers had delivered the wrong chemicals, a mixture of hydrochloric acid and bleach.

### SUMMARY

Household bleaches are chlorine compounds. They work by oxidising dirt and germs. It is not safe to use a bleach together with an acid. Alkaline cleaners work by saponifying grease and oil. It is not safe to use ammonia and a bleach together.

EXTENSION FILE ACTIVITY

### Bleaches

Many household bleaches contain chlorine compounds, e.g. sodium chlorate(I), NaClO. They are powerful oxidising agents, killing germs and oxidising dirt. You should not use bleaches together with other cleaning agents. An acid will react with sodium chlorate(I) to liberate the poisonous gas chlorine.

### Alkaline cleaners

Many household cleaners are alkalis. They react with grease and oil to form an emulsion of glycerol and soap, which can be washed away. The reaction is saponification. Sodium hydroxide, NaOH, is used in oven-cleaners; sodium carbonate, $Na_2CO_3$, is used in washing powders, and ammonia is used in solution as a household cleaner. You should not use ammonia together with a bleach because they can react to form poisonous chloroamines.

### Dry cleaning

Many covalent substances do not dissolve in water but dissolve in covalent solvents such as white spirit (a petroleum fraction), ethanol and propanone. Dry cleaning is carried out with such non-aqueous, covalent solvents.

### CHECKPOINT

Oven-cleaners contain sodium hydroxide. A greasy oven is wiped with a pad of oven-cleaner and left for a few minutes. The grease can then be washed off with water.

1 Explain how sodium hydroxide makes it easier to remove grease.

2 Explain why you should wear rubber gloves when you use this kind of oven-cleaner.

3 What effect does it have on the cleaning job if you warm the oven first?

## 15.12 Hard water and soft water

Something went wrong with the names: hard water and soft water. Should we call them difficult water and easy water – meaning difficult or easy to get a lather with soap?

In some parts of the country, the tap water is described as **hard water**. This means that it is hard to get a lather with soap. Instead of forming a lather, soap forms an insoluble scum. Water in which soap lathers easily is **soft water**. Hard water contains soluble calcium and magnesium salts. They combine with *soap* ions to form insoluble calcium and magnesium compounds. These compounds are the insoluble scum that floats on the water.

*Soap* ions (in solution) + Calcium ions (in solution) → Scum (insoluble solid)

SUMMARY

Hard water contains calcium ions and magnesium ions. They combine with soap ions to form an insoluble scum. Detergents work well even in hard water because their calcium and magnesium salts are soluble.

If you go on adding soap, eventually all the calcium ions and magnesium ions will be precipitated as scum. After that, the soap will be able to work as a cleaning agent.

Soapless detergents are able to work in hard water because their calcium and magnesium salts are soluble. For many purposes, people prefer soapless detergents to soaps. Sales of soapless detergents are four times as high as those of soaps.

# 15.13 Methods of softening hard water

## Temporary hardness

*Figure 15.13A* ▲ Scale on a kettle element

Hardness which can be removed by boiling is called temporary hardness. Temporarily hard water contains dissolved calcium hydrogencarbonate and magnesium hydrogencarbonate, and these compounds decompose when the water is boiled. The resulting water is soft water.

EXTENSION FILE ACTIVITY

Calcium hydrogencarbonate → Calcium carbonate + Carbon dioxide + Water

$$Ca(HCO_3)_2(aq) \rightarrow CaCO_3(s) + CO_2(g) + H_2O(l)$$

A deposit of calcium carbonate and magnesium carbonate forms. This is the **scale** which is deposited in kettles and water pipes.

## Permanent hardness

Hardness which cannot be removed by boiling is called permanent hardness. It is caused by dissolved chlorides and sulphates of calcium and magnesium. These compounds are not decomposed by heat.

## Washing soda

Washing soda is sodium carbonate-10-water. It can soften both temporary and permanent hardness. Washing soda precipitates calcium ions and magnesium ions as insoluble carbonates.

Calcium ions + Carbonate ions → Calcium carbonate

$$Ca^{2+}(aq) + CO_3^{2-}(aq) \rightarrow CaCO_3(S)$$

## Exchange resins

Ion exchange resins are substances which take ions of one kind out of aqueous solution and replace them with ions of a different kind. Permutits are manufactured ion exchange resins. They replace calcium and magnesium ions in water by sodium ions.

Calcium ions + Sodium permutit → Sodium ions + Calcium permutit

SUMMARY

- Temporary hardness is removed by boiling.
- Permanent hardness is removed by adding sodium carbonate (washing soda) or by running water through an exchange resin.

# 15.14 Advantages of hard water

SUMMARY

- Hard water is better than soft water for drinking ...
  ... and preferred by some industries.

Hard water has some advantages over soft water for health reasons. The **calcium** compounds in hard water strengthen bones and teeth. The calcium content is also beneficial to people with a tendency to develop heart disease.

Some industries prefer hard water. The leather industry prefers to cure leather in hard water. The brewing industry likes hard water for the taste which the dissolved salts give to the beer.

## CHECKPOINT

1 Gwen washes her hair in hard water. Which kind of shampoo would you advise her to choose: a mild soapless detergent or a soap? Explain your advice.

2 (a) Explain the difference between hard and soft water.
(b) Why is drinking hard water better for health than drinking soft water?
(c) Which solutes make water hard? Explain how the substances you mention get into tap water.
(d) Name a use for which soft water is preferred to hard water. Explain why.
(e) Describe one method of softening hard water. Explain how it works.
(f) Why are detergents preferred to soaps for use in hard water?
(g) Why is it better to use distilled water rather than tap water in a steam iron?

3 The table gives some information on three brands of shampoo.

| *Brand* | *Price of bottle (p)* | *Volume (cm$^3$)* |
|---|---|---|
| Soffen | 40 | 204 |
| Sheeno | 50 | 350 |
| Silken | 60 | 480 |

(a) Which brand is sold in the smallest bottle?
(b) Calculate what volume of shampoo (in cm$^3$) you get for 1 p if you buy (i) Soffen (ii) Sheeno and (iii) Silken. Say which shampoo is the cheapest.
(c) Suggest three things which a person might consider, other than price, when choosing a shampoo.
(d) Describe an experiment you could do to find out which of the shampoos is best at producing a lather. Mention any steps you would take to make sure the test was fair.

4 Some washing powders contain enzymes. Zenab decides to test whether the washing powder Biolwash, which contains an enzyme, washes better than Britewash, which does not. Zenab decides to use 1 g of washing powder in 100 cm$^3$ of warm water and to do her tests on squares of cotton fabric.
(a) Suggest some everyday substances which stain cloth and which would be interesting to experiment on.
(b) Describe how Zenab could do a fair test to compare the washing action of Biolwash and Britewash. What factors must be kept the same in the two experiments?
(c) Zenab finds that Biolwash washes better than Britewash on many stains. Another student, Ahmed, did his tests at 80 °C and found that Britewash gave a cleaner result than Biolwash. Can you explain the difference between Zenab's and Ahmed's results? For enzymes, see Topic 26.8.

# 15.15 Colloids

**FIRST THOUGHTS**

The bluish haze of tobacco smoke and the brilliant sunsets in deserts – what do they have in common? The answer is the scattering of light by **colloidal particles** suspended in air. In this topic you will find out what colloidal particles are.

Sand is largely silica, $SiO_2$. Silica does not dissolve in water. Why is it, then, that when an aqueous solution of a silicate is acidified, silica is not precipitated? Instead there forms either a clear solution with a slight pearly sheen or a jelly-like precipitate or a solid-like gel in which liquid is trapped. The form of the product depends on the pH. The solution and the jelly-like precipitate and the gel are all **colloids**, also called **colloidal dispersions** and **colloidal suspensions**. The particles of silica are **suspended** or **dispersed** (spread) through the liquid. The solid (silica) is the **disperse phase** and the liquid (water) is the **dispersion phase** or **dispersion medium**. The silica content may be as high as 30% by mass. Measurements show that the silica particles contain between 2000 and 20 000 $SiO_2$ units.

*How do colloids differ from solutions and suspensions?*

**Solutions** are homogeneous (the same all through) mixtures of two or more substances. The particles of the solute (the dissolved substance) are of atomic or molecular size (about one nanometre long, 1 nm; 1 nm = $10^{-9}$ m).

**Suspensions** are heterogenous (not the same all through) mixtures. They contain relatively large particles (over 1000 nm long) of insoluble solid or liquid suspended in (scattered throughout) a liquid. In time, the particles settle out.

**Colloids** are heterogenous mixtures whose particles are larger than the molecules and ions which form solutions but smaller than the particles which form suspensions (between 1 nm and 1000 nm long). The particles cannot be separated by filtration through filter paper.

Colloidal particles are smaller than the particles of solutes and larger than the particles of suspensions.

Colloidal particles are invisible but scatter light.

## Optical properties

The particles of a colloid are too small to be visible, but they are large enough to scatter light. The effect resembles the scattering of light in dust-laden air. It is named the Tyndall effect after its discoverer (see Figure 15.15A).

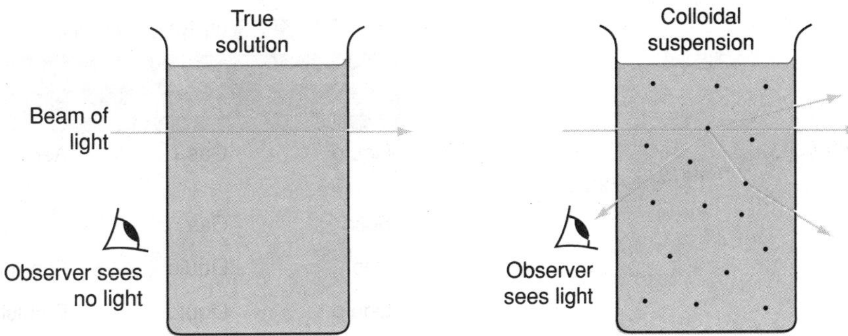


*Figure 15.15A* ▲ The scattering of light by a colloid

## Charge

The reason colloidal particles do not clump together to form a precipitate is that they are charged. In order to precipitate a colloid, the charge on the particles must be neutralised. This can be done by adding ions of opposite charge. When water is purified for drinking, clay particles and other colloidal suspensions must be removed. The impure water is treated with, for example, aluminium sulphate. Aluminium ions, $Al^{3+}$, are small in size and highly charged. They neutralise the negative charges on the clay particles, which are then able to clump together and settle out of solution.

The proteins in blood are colloidal and are negatively charged. Small cuts can be treated with styptic pencils. These contain $Al^{3+}$ or $Fe^{3+}$ ions which neutralise the charges on the colloidal particles of protein and help the blood to clot.

## Electrostatic precipitation

When gases and air are fed into an industrial process, they often contain colloidal particles. To remove these particles, **electrostatic precipitation** is used. The charge on the particles is used to attract the particles to charged metal plates (see Figure 15.15B).

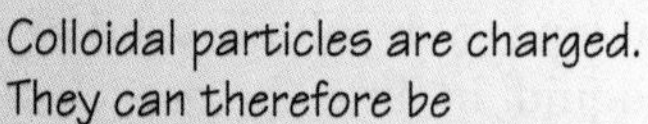
Colloidal particles are charged. They can therefore be

- precipitated by neutralising the charge
- separated by electrophoresis.

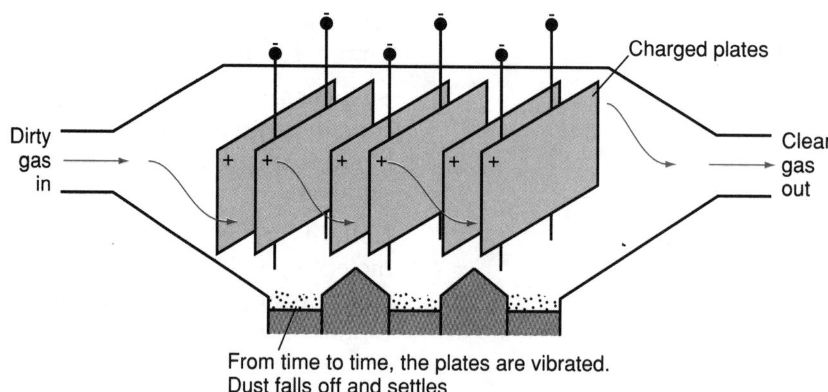


*Figure 15.15B* ▲ Electrostatic precipitation

## Classification

Some colloids of different kinds are tabulated in Table 15.1.

*Table 15.1* ▼ Some types of colloids

| Dispersed phase | Dispersed medium | Type | Example |
|---|---|---|---|
| Liquid | Gas | Aerosol | Fog, mist from aerosol spray can, clouds |
| Solid | Gas | Aerosol | Smoke, dust-laden air |
| Gas | Liquid | Foam | Soap suds, whipped cream |
| Liquid | Liquid | Emulsion | Oil in water, milk, mayonnaise, protoplasm |
| Solid | Liquid | Sol | Clay, starch in water, protein in water, gelatin in water |
| Gas | Solid | Solid foam | Lava, pumice, styrofoam, marshmallows |
| Liquid | Solid | Solid emulsion | Pearl, opal, jellies, butter, cheese |
| Solid | Solid | Solid sol | Some gems, e.g. black diamond, coloured glass, some alloys |

## Electrophoresis

Colloidal particles migrate in an electric field as ions do. They are attracted to one electrode and repelled by the other. Colloidal particles of different types migrate at different rates. This movement under the influence of an electric field is called **electrophoresis**. Electrophoresis can be used to separate different substances. Blood contains a number of proteins in colloidal suspension.

## Emulsifiers

An emulsion is a colloidal mixture of oil and water. The oil is dispersed through the water as a suspension of tiny drops. Examples are milk, cream, mayonnaise and many sauces. The natural tendency is for oil and water to separate. An emulsifier keeps the two together as a colloidal dispersion. An emulsifier ion has two parts. One is a polar group, with a negative charge, which is attracted to water (the water-loving group). The other is a hydrocarbon chain which is attracted to fat and oil (the fat-loving group) (see Figure 15.15C(a)). When the emulsifier is added to a mixture of oil and water, the emulsifier ions orient themselves so that the water-loving group dissolves in the water and the fat-loving group dissolves in the oil. Emulsifier ions arrange themselves round each droplet of oil (Figure 15.15C(b)). As the surface of each droplet is negatively charged, the drops repel one another and do not run together.

Emulsifiers are important
- in salad dressings
- as soaps
- as soapless detergents.

They are able to make oil and water mix.

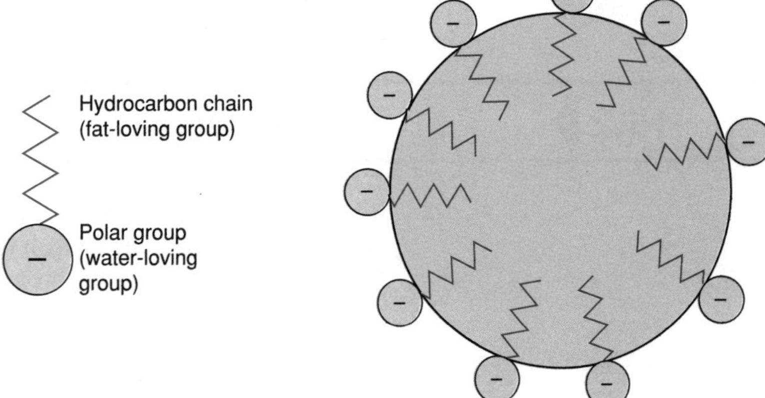


*Figure 15.15C* ▲ (a) An emulsifier ion; (b) A drop of oil surrounded by emulsifier ions

### SUMMARY

Colloidal particles are larger than those in solutions but smaller than those in suspensions. They scatter light. They are charged, and if they lose their charge they settle out of the colloidal dispersion. Colloidal particles move under the influence of an electric field in electrophoresis.

### CHECKPOINT

1. How can you tell the difference between:
   (a) a solution and a suspension,
   (b) a suspension and a colloid,
   (c) a solution and a colloid?
2. How can solid colloidal particles be made to separate:
   (a) from a sol,
   (b) from an aerosol?
3. Explain:
   (a) why a salad dressing contains an emulsifier,
   (b) how the emulsifier works.

# Topic 16 Air pollution

## 16.1 Smog

FIRST THOUGHTS

As you read through this topic, think about the 15 000 or 20 000 litres of air that you breathe in each day. Obviously you hope that it is clean air. Unfortunately the air that most of us breathe is polluted. In this topic, we shall look into what can be done to reduce the pollution of air.

Four thousand people died in the great London smog of December 1952. Smog is a combination of smoke and fog. Fog consists of small water droplets. It forms when warm air containing water vapour is suddenly cooled. The cool air cannot hold as much water vapour as it held when it was warm, and water condenses. When smoke combines with fog, fog prevents smoke escaping into the upper atmosphere. Smoke stays around, and we inhale it. Smoke contains particles which irritate our lungs and make us cough. Smoke also contains the gas sulphur dioxide. This gas reacts with water and oxygen to form sulphuric acid, $H_2SO_4$. This strong acid irritates our lungs, and they produce a lot of mucus which we cough up.

The Government did very little about the cause of smog until 1956. Then there was another killer smog. A private bill brought by a Member of Parliament (the late Mr Robert Maxwell, the newspaper owner) gained such widespread support that the Government was forced to act. The Government introduced its own bill, which became the Clean Air Act of 1956. The Act allowed local authorities to declare smokeless zones. In these zones, only low-smoke and low-sulphur fuels can be burned. The Act banned dark smoke from domestic chimneys and industrial chimneys.

## 16.2 The problem

All the dust and pollutants in the air pass over the sensitive tissues of our lungs. Any substance which is bad for health is called a **pollutant**. The lung diseases of cancer, bronchitis and emphysema are common illnesses in regions where air is highly polluted. From our lungs, pollutants enter our bloodstream to reach every part of our bodies. The main air pollutants are shown in Table 16.1.

Can you spot from the table the three main sources of pollutants?

*Table 16.1* The main pollutants in air (Emissions are given in millions of tonnes per year in the UK.)

| Pollutant | Emission | Source |
|---|---|---|
| Carbon monoxide, CO | 100 | Vehicle engines and industrial processes |
| Sulphur dioxide, $SO_2$ | 33 | Combustion of fuels in power stations and factories |
| Hydrocarbons | 32 | Combustion of fuels in factories and vehicles |
| Dust | 28 | Combustion of fuels; mining; factories |
| Oxides of nitrogen, NO and $NO_2$ | 21 | Vehicle engines and fuel combustion |
| Lead compounds | 0.5 | Vehicle engines |

In this topic, we shall look at where these pollutants come from, what harm they do and what can be done about them.

## 16.3 ▶ Dispersing air pollutants

*Fortunately there is a mechanism for carrying away many of our air pollutants.*

The surface of the Earth absorbs energy from the Sun and warms up. The Earth warms the lower atmosphere. The air in the upper atmosphere is cooler than the air near the Earth. **Convection currents** carry warm air upwards. Cold air descends to take its place (see Figure 16.3A). In this way, the warm dirty air from factories and motor vehicles is carried upwards and spread through the vast upper atmosphere.

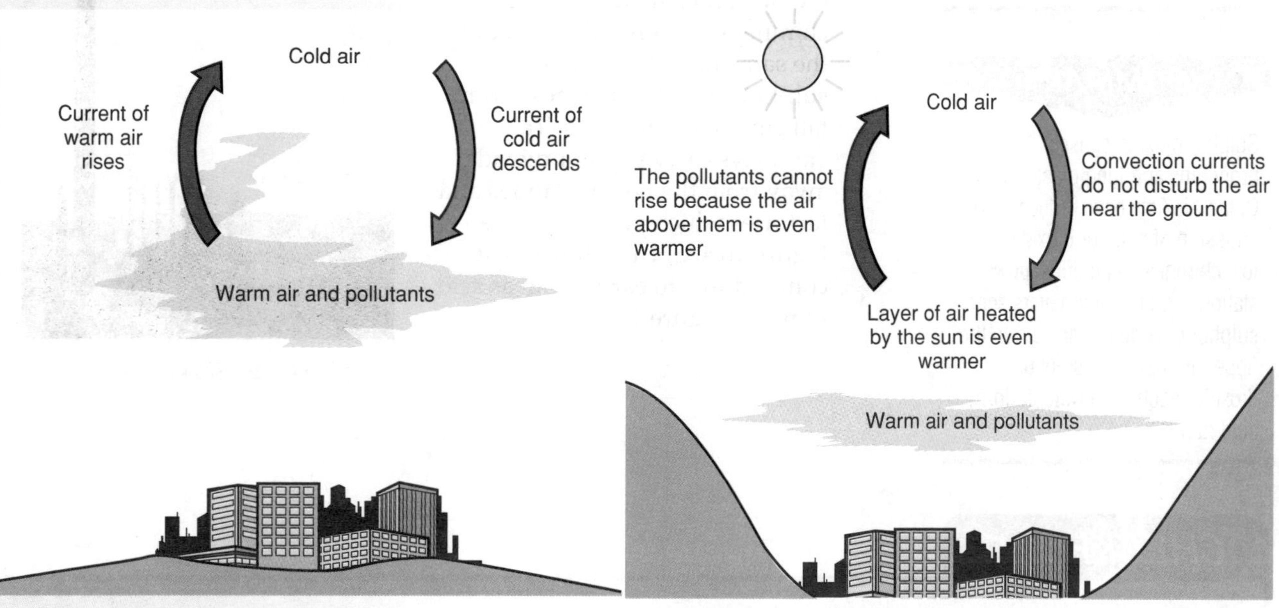
Cold air
Current of warm air rises
Current of cold air descends
Warm air and pollutants
Cold air
The pollutants cannot rise because the air above them is even warmer
Convection currents do not disturb the air near the ground
Layer of air heated by the sun is even warmer
Warm air and pollutants

*Figure 16.3A* ▲ Convection currents disperse pollutants

*Figure 16.3B* ▲ A temperature inversion traps pollutants

### SUMMARY

Pollutants are carried upwards by rising currents of warm air. A temperature inversion stops the dispersal of pollutants. Temperature inversions occur in places with a hot climate and still air.

A low-lying area surrounded by higher ground tends to have still air. If an area like this has a hot climate, it is possible for the Sun to warm a layer of air in the upper atmosphere (Figure 16.3B). If the Sun is very hot, this layer of air may become warmer than that near the ground. There is a **temperature inversion**. The air near the ground is no longer carried upwards and dispersed. Pollutants accumulate in the layer of still air at ground level, and the city dwellers are forced to breathe them.

## 16.4 ▶ Sulphur dioxide

*The root of the sulphur dioxide problem is that coal and oil contain sulphur.*

### Where does sulphur dioxide come from?

Worldwide, 150 million tonnes of sulphur dioxide a year are emitted. Almost all the sulphur dioxide in the air comes from industrial sources. The emission is growing as countries become more industrialised. Half of the output of sulphur dioxide comes from the burning of coal. Most of the coal is burned in power stations. All coal contains between 0.5 and 5 per cent sulphur.

$$\underset{\text{(coal)}}{\text{Sulphur}} + \text{Oxygen} \rightarrow \underset{\text{(air)}}{\text{Sulphur dioxide}}$$

$$S(s) + O_2(g) \rightarrow SO_2(g)$$

EXTENSION FILE
ACTIVITY

Industrial smelters, which obtain metals from sulphide ores, also produce tonnes of sulphur dioxide daily.

## It's a fact!

In Canada, the countryside surrounding the nickel smelter in Sudbury, Ontario, was so devastated by sulphur dioxide emissions that in 1968 US astronauts practised moon-walking there before attempting the first lunar landing.

## SUMMARY

Sulphur dioxide causes bronchitis and lung diseases. The Clean Air Acts have reduced the emission of sulphur dioxide from low chimneys. Factories, power stations and metal smelters send sulphur dioxide into the air. In the upper atmosphere, sulphur dioxide reacts with water to form acid rain.

### What harm does sulphur dioxide do?

Sulphur dioxide is a colourless gas with a very irritating smell. Inhaling sulphur dioxide causes coughing, chest pains and shortness of breath. It is poisonous; at a level of 0.5%, it will kill. Sulphur dioxide is thought to be one of the causes of bronchitis and lung diseases.

### What can be done about it?

After the Clean Air Acts of 1956 and 1968, the emission of sulphur dioxide and smoke from the chimneys of houses decreased. At the same time, the emission of sulphur dioxide and smoke from tall chimneys increased. Tall chimneys carry sulphur dioxide away from the power stations and factories which produce it (see Figure 16.4A). Unfortunately it comes down to earth again as acid rain (see Figure 16.5B).

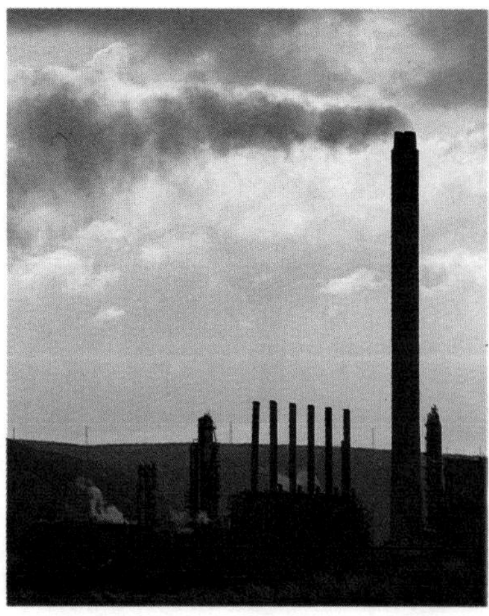

*Figure 16.4A* ▲ Smoking chimney

## CHECKPOINT

1 The emission of sulphur dioxide from low chimneys decreased between 1955 and 1975 from 1.7 million tonnes a year to 0.6 million tonnes a year. In the same period, the emission of sulphur dioxide from tall chimneys increased from 1.4 million tonnes to 3.0 million tonnes.

(a) Explain why there was a decrease in sulphur dioxide emission from low chimneys.

(b) Who benefited from the decrease in emission from low chimneys?

(c) Explain why tall chimneys are not a complete answer to the problem of sulphur dioxide emission.

# 16.5 Acid rain

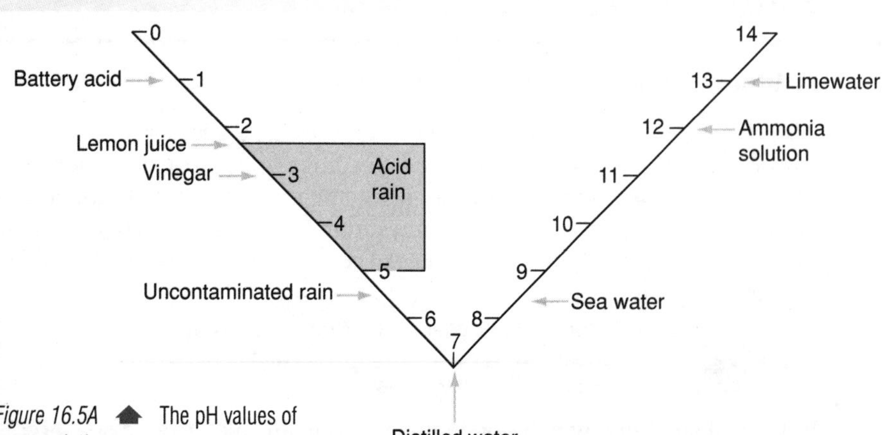


*Figure 16.5A* ▲ The pH values of some solutions

Rain water is naturally weakly acidic. It has a pH of 5.4. Carbon dioxide from the air dissolves in it to form the weak acid, carbonic acid, $H_2CO_3$. What we mean by acid rain is rain which contains the strong acids, sulphuric acid and nitric acid. Acid rain has a pH between 2.4 and 5.0 (see Figure 16.5A).

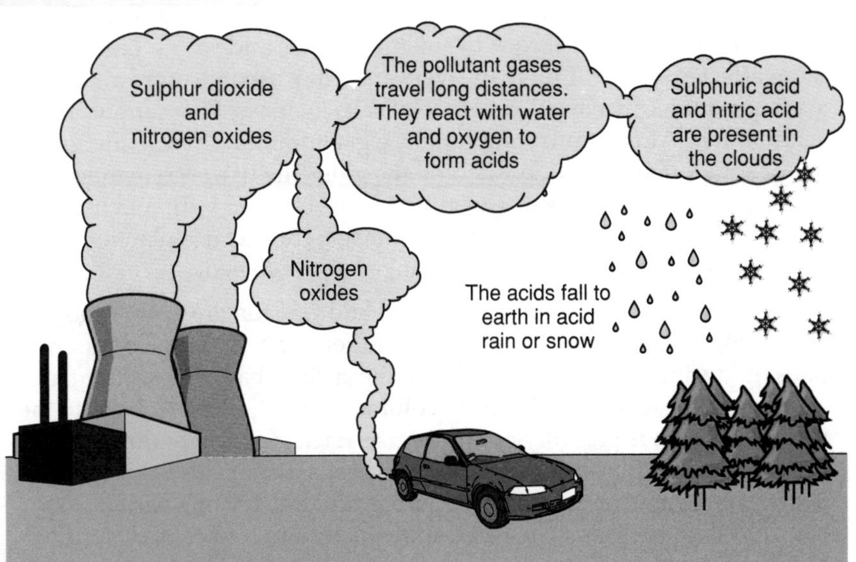


Figure 16.5B ▲ Where acid rain comes from

*How do sulphuric acid and nitric acid get into rain water?* Tall chimneys emit sulphur dioxide and other pollutant gases, such as oxides of nitrogen. Air currents carry the gases away. Before long, the gases react with water vapour and oxygen in the air. Sulphuric acid, $H_2SO_4$, and nitric acid, $HNO_3$, are formed. The water vapour with its acid content becomes part of a cloud. Eventually it falls to earth as acid rain or acid snow which may turn up hundreds of miles away from the source of pollution (see Figure 16.5C).

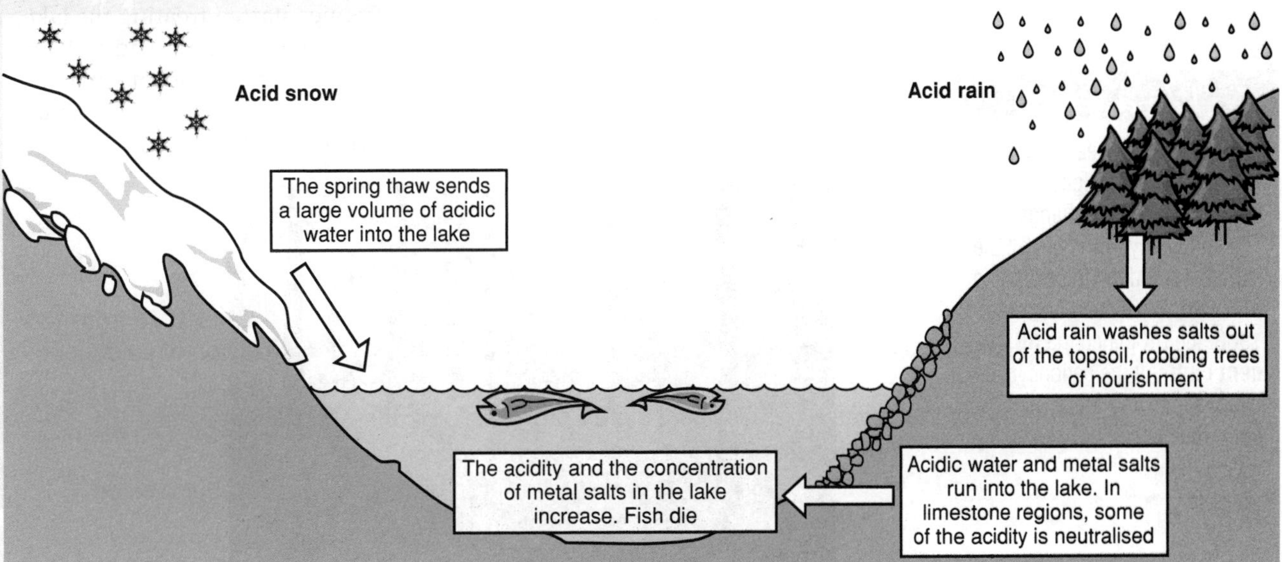


Figure 16.5C ▲ Where acid rain goes

Acid rain which falls on land is absorbed by the soil. At first, the nitrates in the acid rain fertilise the soil and encourage the growth of plants. But acid rain reacts with minerals, converting the metals in them into soluble salts. The rain water containing these soluble salts of calcium, potassium, aluminium and other metals trickles down through the soil into the subsoil where plant roots cannot reach. In this way, salts are leached out of the topsoil, and crops are robbed of nutrients. One of the salts formed by acid rain is aluminium sulphate. This salt damages the roots of trees. The damaged roots are easily attacked by viruses and bacteria, and the trees die of a combination of malnutrition and disease. The Black Forest is a famous beauty spot in Germany which makes money from tourism. About half the trees there are now damaged or dead. Pollution is an important issue in Germany. The Green Party is a major political party. It is compaigning for the reduction of pollution. There are dead forests also in Czechoslovakia and Poland. In 1987, the UK Forestry Commission reported that damage to spruce and pine trees is as widespread in the UK as in Germany. One third of British trees are damaged, but not everyone agrees that the cause of the damage is acid rain.

**It's a fact!**

The most acidic rain ever recorded fell on Pitlochry in Scotland; it had a pH of 2.4.

## Science at work

It is possible to monitor the emission from the chimneys of factories and power stations from a distance. A van can carry equipment using infra-red radiation to detect sulphur dioxide. The readings are automatically recorded by a computer.

The acidic rain water trickles through the soil until it meets rock. Then it travels along the layer of rock to emerge in lakes and rivers. Lakes are more affected by acid rain than rivers are. They become more and more acidic, and the concentrations of metal salts increase. Fish cannot live in acidic water. Aluminium compounds, e.g. aluminium hydroxide, come out of solution and are deposited on the gills. The fish secrete mucus to try to get rid of the deposit. The gills become clogged with mucus, and the fish die. An acid lake is perfectly transparent because plants, plankton, insects and other living things have perished.

Thousands of lakes in Norway, Sweden and Canada are now 'dead' lakes. One reason why these countries suffer badly is that acidic snow piles up during the winter months. In the spring thaw, the accumulated snow melts suddenly, and a large volume of acidic water flows into the lakes. Acid rain is partially neutralised as it trickles slowly through soil and over rock. Limestone, in particular, keeps damage to a minimum by neutralising some of the acidity. There is not time for this partial neutralisation to occur when acid snow melts and tonnes of water flow rapidly down the hills and into the lakes.

The UK is affected too. In 1982, lakes and rivers in south-west Scotland had become so acidic that the water companies started treating the lakes with calcium hydroxide (lime). The aim is to neutralise the acidic water and revive stocks of fish. In Wales, the water company has for some years poured tonnes of powdered limestone into acidic lakes. A number of lakes are 'dead' and the fish in many others are threatened.

## It's a fact!

In 1979, 31 European countries, including the UK, signed the Convention on Long Range Transboundary Pollution. All the nations agreed to stop exporting pollution across their borders. In 1986, the UK joined the **30 per cent club**. All the nations in the 'club' agreed to reduce their emissions of sulphur dioxide by 30 per cent.

*Figure 16.5D* ▲ The effects of acid rain

**EXTENSION FILE ASSIGNMENT**

## What can be done about acid rain?

There are three main methods of attacking the problem of acid rain. They all cost money, but then the damage done by acid rain costs money too.

1 Low-sulphur fuels can be used. Crushing coal and washing it with a suitable solvent reduces the sulphur content by 10 to 40 per cent. The dirty solvent must be disposed of without creating pollution on land or in rivers. Oil refineries could refine the oil which they sell to power stations. The cost of the purified oil would be higher, and the price of electricity would increase.

2 Flue gas desulphurisation, FGD, is the removal of sulphur from power station chimneys after the coal has been burnt and before the waste gases leave the chimneys. As the combustion products pass up the chimney, they are bombarded by jets of wet powdered limestone. The acid gases are neutralised to form a sludge. The method will remove 95 per cent of the acid combustion products. FGD can be fitted to existing power stations. One of the products is calcium sulphate, which can be sold to the plaster board industry and to cement manufacturers.

## It's a fact!

Together, the USA, Canada and Europe send 100 million tonnes of sulphur dioxide into the air each year. Nine tenths of this comes from burning coal and oil.

**Science at work**

The CEGB fitted FGD to Drax power station in Yorkshire and Fidlers' Ferry in Wales. At a cost of £40 million, the FGD plant at Drax power station started operating in 1993.

3 Pulverised fluidised bed combustion, PFBC, uses a new type of furnace. The furnace burns pulverised coal (small particles) in a bed of powdered limestone. An upward flow of air keeps the whole bed in motion. The sulphur is removed during burning. The PFBC uses much more limestone than the FGD method: one power station needs 1 million tonnes of limestone a year (4 times as much as the FGD method). The PFBC method also produces a lot more waste material, which has to be dumped.

**CHECKPOINT**

➧ 1 What is the advantage of building a power station close to a densely populated area? What is the disadvantage?

➧ 2 Why do power stations and factories have tall chimneys? Are tall chimneys a solution to the problem of pollution? Explain your answer.

➧ 3 Why does acid rain attack (a) iron railings (b) marble statues and (c) stone buildings?

➧ 4 (a) Why does Sweden suffer badly from acid rain?
(b) Why do lakes suffer more than rivers from the effects of acid rain?

➧ 5 A country decides to increase the price of electricity so that the power stations can afford to use refined low-sulphur fuel oil. In what ways will the country actually save money by reducing the emission of sulphur dioxide?

## 16.6 ➧ Carbon monoxide

### Where does it come from?

Worldwide, the emission of carbon monoxide is 350 million tonnes a year. Most of it comes from the exhaust gases of motor vehicles. Vehicle engines are designed to give maximum power. This is achieved by arranging for the mixture in the cylinders to have a high fuel to air ratio. This design leads to incomplete combustion. The result is the discharge of carbon monoxide, carbon and unburnt **hydrocarbons**.

### What harm does carbon monoxide do?

The root of the carbon monoxide problem is incomplete combustion of hydrocarbon fuels. The poisonous effect of carbon monoxide is due to combination with haemoglobin.

Oxygen combines with **haemoglobin**, a substance in red blood cells. Carbon monoxide is 200 times better at combining with haemoglobin than oxygen is. Carbon monoxide is therefore able to tie up haemoglobin and prevent it combining with oxygen. A shortage of oxygen causes headache and dizziness, and makes a person feel sluggish. If the level of carbon monoxide reaches 0.1% of the air, it will kill. Carbon monoxide is especially dangerous in that, being colourless and odourless, it gives no warning of its presence. Since carbon monoxide is produced by motor vehicles, it is likely to affect people when they are driving in heavy traffic. This is when people need to feel alert and to have quick reflexes.

### What can be done?

Soil contains organisms which can convert carbon monoxide into carbon dioxide or methane. This natural mechanism for dealing with carbon monoxide cannot cope in cities, where the concentration of carbon monoxide is high and there is little soil to remove it. People are trying out a number of solutions to the problem.

SUMMARY

Carbon monoxide is emitted by vehicle engines. It is poisonous. Catalytic converters fitted in the exhaust pipes of cars reduce the emission of carbon monoxide.

- Vehicle engines can be tuned to take in more air and produce only carbon dioxide and water. Unfortunately, this increases the formation of oxides of nitrogen (see Topic 16.7).
- Catalytic converters are fitted to the exhausts of many cars. The catalyst helps to oxidise carbon monoxide in the exhaust gases to carbon dioxide (see Topic 16.7).
- New fuels may be used in the future. Some fuels, e.g. alcohol, burn more cleanly than hydrocarbons (see Topic 24.1).

CHECKPOINT

1 (a) What are the products of complete combustion of petrol?
(b) What harm do these products do?
(c) What conditions lead to the formation of carbon monoxide?
(d) What harm does it do?

2 How does carbon monoxide act on the body?

3 Which types of people are likely to breathe in carbon monoxide? Is there anything they can do to avoid it?

4 A family was spending the weekend in their caravan. At night, the weather turned cold, so they shut the windows and turned up the paraffin heater. In the morning, they were all dead. What had gone wrong? Why did they have no warning that something was wrong?

## 16.7 Oxides of nitrogen

When fuels are burned in air, nitrogen is present. Combustion temperatures are high enough to make some nitrogen combine with oxygen. As a result, the gases nitrogen monoxide, NO, and nitrogen dioxide, $NO_2$, are formed. These gases enter the air from the chimneys of power stations and factories and from the exhausts of motor vehicles. This mixture of gases is sometimes shown as $NO_x$ or even NOX.

SUMMARY

Oxides of nitrogen, nitrogen monoxide, NO, and nitrogen dioxide, $NO_2$ ($NO_x$), are formed when oxygen and nitrogen combine at high temperature

- in vehicle engines
- in power stations
- in factory furnaces.

They react with air and water to form nitric acid. This strong acid is very toxic and corrosive. Catalytic converters can be fitted in cars to convert nitrogen oxides into nitrogen.

### What harm do they do?

Nitrogen monoxide, NO, is not a very dangerous gas. However, it quickly reacts with air to form nitrogen dioxide, $NO_2$. Nitrogen dioxide is highly toxic and irritates the breathing passages. It reacts with oxygen and water to form nitric acid, an ingredient of acid rain.

### What can be done?

A reaction which can be used to reduce the quantity of nitrogen monoxide in exhaust gases is

$$\text{Nitrogen monoxide} + \text{Carbon monoxide} \rightarrow \text{Nitrogen} + \text{Carbon dioxide}$$
$$2NO(g) + 2CO(g) \rightarrow N_2(g) + 2CO_2(g)$$

This reaction takes place in the presence of a catalyst. A metal cylinder containing the catalyst is fitted in the vehicle exhaust (see Figure 16.8B). All new UK cars are now fitted with these **catalytic converters**.

# 16.8 Hydrocarbons

FIRST THOUGHTS

The source of most of the hydrocarbons in air is the natural decay of plant and animal material.

They are also emitted by vehicle engines.

## Where do they come from?

Hydrocarbons are present naturally in air. Methane, $CH_4$, is one of the products of decay of plant material. Only 15 per cent of the hydrocarbons in the air come from human activities. They affect our health because they are concentrated in city air.

## What harm do they do?

Hydrocarbons by themselves cause little damage. In intense sunlight, **photochemical reactions** occur ('photo' means light). Hydrocarbons react with oxygen and oxides of nitrogen to form irritating and toxic compounds.

SUMMARY

Hydrocarbons from vehicle exhausts take part in photochemical reactions to form irritating and toxic compounds. The emission of hydrocarbons may be reduced by running the engine at a lower temperature with a catalyst.

## What can be done?

The hydrocarbons in the exhausts of petrol engines can be reduced by increasing the oxygen supply so as to burn the petrol completely. This also decreases the formation of carbon monoxide. There is a snag, however. Increasing the oxygen supply increases the formation of oxides of nitrogen in the engine. The problem may have a solution. Research workers are trying the idea of running the engine at a lower temperature (to reduce the combination of oxygen and nitrogen) with a catalyst (to assist complete combustion of hydrocarbons at the lower temperature).

Science at work

A microcomputer in a vehicle engine can reduce the emission of pollutants by adjusting the fuel to air ratio to the speed of the vehicle.

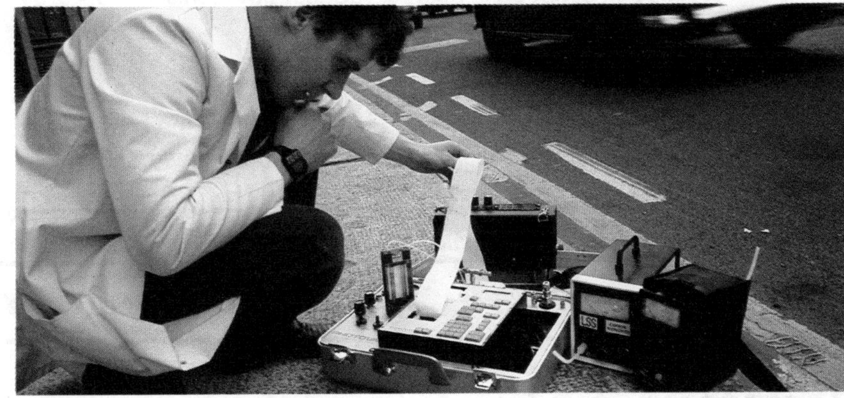

*Figure 16.8A* ▲ Testing the emission from a vehicle exhaust

SUMMARY

Catalytic converters reduce carbon monoxide, oxides of nitrogen and hydrocarbons in vehicle exhausts.

## Catalytic converters

Pollution by carbon monoxide, oxides of nitrogen and hydrocarbons is reduced when catalytic converters are fitted to vehicle exhausts. Figure 16.8B shows the structure of a catalytic converter.

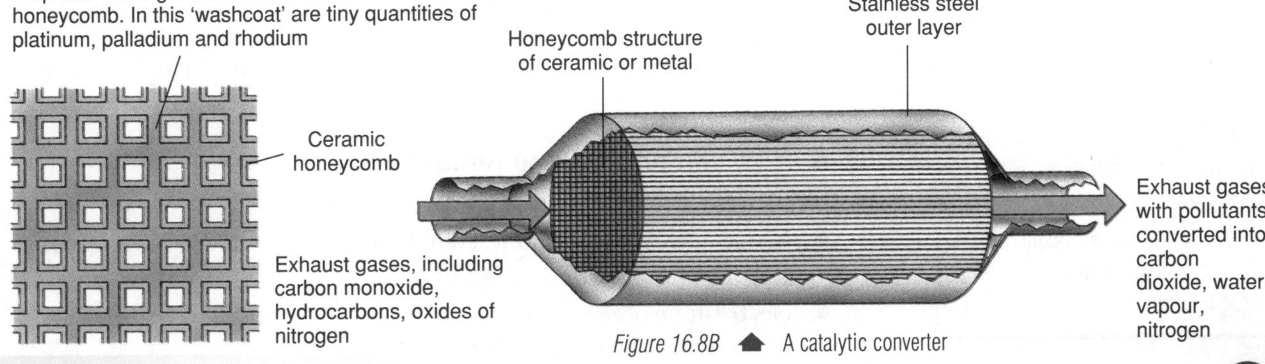


*Figure 16.8B* ▲ A catalytic converter

## CHECKPOINT

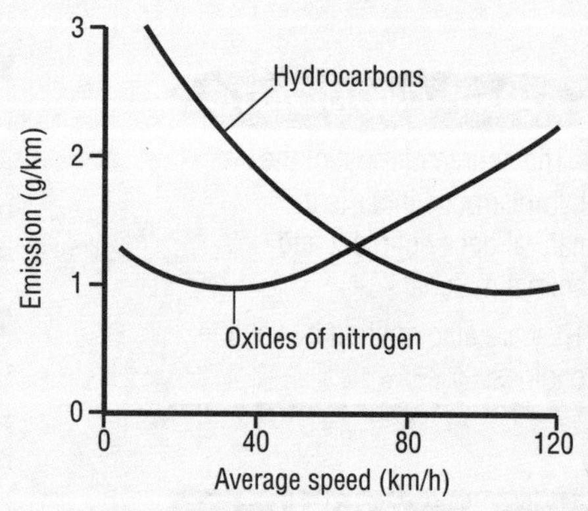


1. What products are formed by the combustion of hydrocarbons in petrol engines?
2. How does the supply of air affect the course of combustion?
3. What is the advantage of increasing the air to fuel ratio in the combustion chamber?
4. What are the pollutants that form when the air to fuel ratio is high? What can be done about them?
5. Copy and complete this summary.
   In internal combustion engines, a high air to fuel ratio:
   decreases the emission of unburnt _____ A
   decreases the emission of _____ B
   increases the emission of _____ C
   A way of reducing the emission of C would be to run the engine at a lower temperature. A _____ would be needed to promote _____ combustion and reduce the emission of A and B.

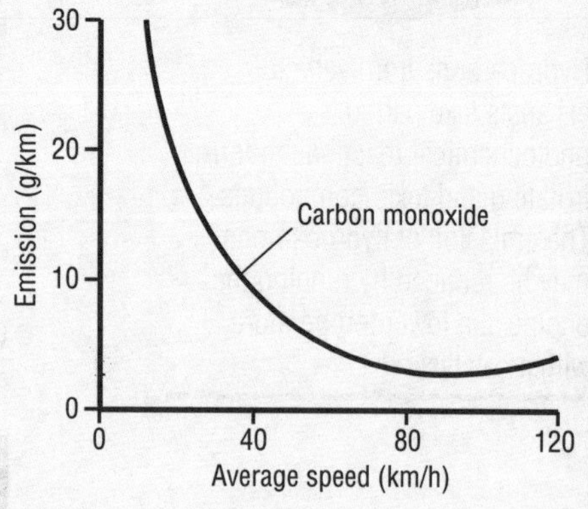


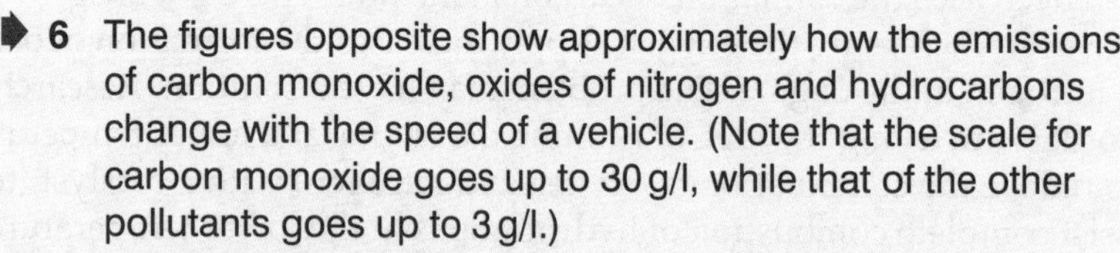

6. The figures opposite show approximately how the emissions of carbon monoxide, oxides of nitrogen and hydrocarbons change with the speed of a vehicle. (Note that the scale for carbon monoxide goes up to 30 g/l, while that of the other pollutants goes up to 3 g/l.)
   (a) Say what speed is best for reducing the emission of
      (i) carbon monoxide
      (ii) oxides of nitrogen
      (iii) hydrocarbons.
   (b) (i) What speed would you recommend as the best to reduce overall pollution?
      (ii) What is this speed in miles per hour (5 mile = 8 km)?

## 16.9 Smoke, dust and grit

Millions of tonnes of smoke, dust and grit are present in the atmosphere. Dust storms, forest fires and volcanic eruptions send matter into the air. Human activities such as mining, land-clearing and burning coal and oil add to the solid matter in the air.

### What harm do particles do?

Particles darken city air by scattering light. Smoke increases the danger of smog. Solid particles fall as grime on people, clothing, buildings and plants.

Sunlight which meets dust particles is reflected back into space and prevented from reaching the Earth. Some scientists believe that the increasing amount of dust in the atmosphere is serious. A fourfold increase in the amount of dust would make the Earth's temperature fall by about 3 °C. This would affect food production.

### How can particles be removed?

Industries use a number of methods. These include:

- using sprays of water to wash out particles from their waste gases
- passing waste gases through filters,
- electrostatic precipitators, which remove dust particles from waste gases by **electrostatic attraction**

### SUMMARY

Particles of smoke and dust and grit are sent into the air by factories, power stations and motor vehicles. Dirt damages buildings and plants. It pollutes the air we breathe; mixed with fog, it forms smog.

Dust is removed by washing, filtration, electrostatic precipitation.

# 16.10 ▶ Metals

Mercury and lead enter the air from the smelting of metal ores and the combination of fuels.

Much of the lead in the air is present because lead compounds are added to petrol. The amount is decreasing as sales of unleaded petrol increase.

Many heavy metals and their compounds are serious air pollutants. 'Heavy' metals are metals with a density greater than 5 g/cm$^3$.

## Mercury

Earth-moving activities, such as mining and road-making, disturb soil and rock and allow the mercury which they contain to escape into the air. Mercury vapour is also released into the air during the smelting of many metal ores and the combustion of coal and oil. Both mercury and its compounds cause kidney damage, nerve damage and death.

*Figure 16.10A* ▲ City dwellers breathe exhaust gases

## Lead

### ■ Where does it come from?

The lead compounds in the air all come from human activity. Vehicle engines, the combustion of coal and the roasting of metal ores send lead and its compounds into the air. Unlike the other pollutants in exhaust gases, lead compounds have been purposely added to the fuel. Tetraethyl lead, TEL, is added to improve the performance of the engine.

### ■ What harm does it do?

Lead compounds settle out of the air on to plant crops, and contaminate our food. The level of lead in our environment is high: some areas still have lead plumbing; old houses may have peeling lead-based paint. City dwellers take in lead from many sources. Many people have blood levels of lead which are nearly high enough to produce the symptoms of lead poisoning. Symptoms of mild lead poisoning are headache, irritability, tiredness and depression. Higher levels of lead cause damage to the brain, liver and kidneys. Scientists have suggested that behaviour disorders such as hooliganism and vandalism may be due in part to lead poisoning.

### ■ What can be done?

This type of pollution can be remedied. We can stop adding lead compounds to petrol. Research chemists have found other compounds which can be used to improve engine performance. Vehicles made in the UK after autumn 1990 are adjusted to run on unleaded petrol. Petrol stations now stock unleaded petrol. This must be used in vehicles fitted with catalytic converters because they are 'poisoned' by lead compounds.

**SUMMARY**

Heavy metals are serious air pollutants. Levels of mercury and lead and their compounds in the air are increasing.

**CHECKPOINT**

1. Name the pollutants which come from motor vehicles.
2. Name the pollutants which can be reduced by fitting catalytic converters into vehicle exhausts. What effect will this modification have on the price of cars?
3. Catalytic converters will only work with unleaded petrol. When TEL is no longer added to petrol, motorists will have to use high octane (4 star) fuel. What effect will this have on the cost of motoring?

4 What effect does the use of TEL have on the air, apart from its effect on catalytic converters?

5 In which ways will the control of pollution from vehicles cost money? In which ways will a reduction in the level of pollutants in the air save money? (Consider the effects of pollution on people and materials.) Will the expense be worthwhile?

# 16.11 Chlorofluorohydrocarbons

*Ozone, $O_3$, is an allotrope of oxygen, $O_2$. The ozone layer in the upper atmosphere protects us from receiving too much ultraviolet radiation from the Sun.*

## The ozone layer

Ozone is an allotrope of oxygen, $O_2$. There is a layer of ozone, $O_3$, surrounding the Earth. It is 5 km thick at a distance of 25–30 km from the Earth's surface. The ozone layer cuts out some of the ultraviolet light coming from the Sun. Ultraviolet light is bad for us and for crops. Long exposure to ultraviolet light can cause skin cancer. This complaint is common in Australia among people who spend a lot of time out of doors. If anything happens to decrease the ozone layer, the incidence of skin cancer from exposure to ultraviolet light will increase. An excess of ultraviolet light kills **phytoplankton**, the minute plant life of the oceans which are the primary food on which the life of an ocean depends.

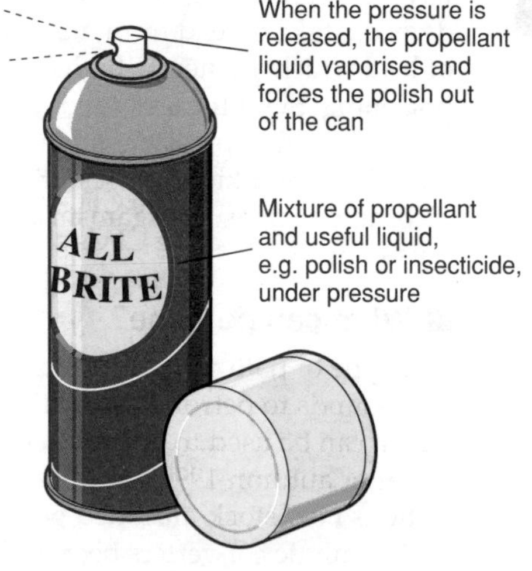
When the pressure is released, the propellant liquid vaporises and forces the polish out of the can
Mixture of propellant and useful liquid, e.g. polish or insecticide, under pressure
ALL BRITE

*Figure 16.11A* ▲ An aerosol can

### The problem

Ozone is a very reactive element. If the upper atmosphere becomes polluted, ozone will oxidise the pollutants. Two pollutants are accumulating in the upper atmosphere. One is chlorofluorohydrocarbons (CFCs). They are very unreactive compounds. They spread through the atmosphere without reacting with other substances and drift into the upper atmosphere. There they meet ozone, which oxidises CFCs and in doing so is converted into oxygen.

Ozone + CFC → Oxygen + Oxidation products

CFCs were formerly used as the propellants in aerosol cans (Figure 16.11A) and are used as refrigerant liquids in fridges, freezers and air conditioners.

Another pollutant found at this height is nitrogen monoxide, NO. It comes from the exhausts of high-altitude aircraft, such as Concorde. Ozone oxidises nitrogen monoxide to nitrogen dioxide:

Ozone + Nitrogen monoxide → Oxygen + Nitrogen dioxide

$$O_3(g) + NO(g) \rightarrow O_2(g) + NO_2(g)$$

### Science at work

Microcomputers and sensors are taken up in aircraft to keep watch on the ozone layer. The sensor detects the thickness of the ozone layer and the microcomputer records the measurements.

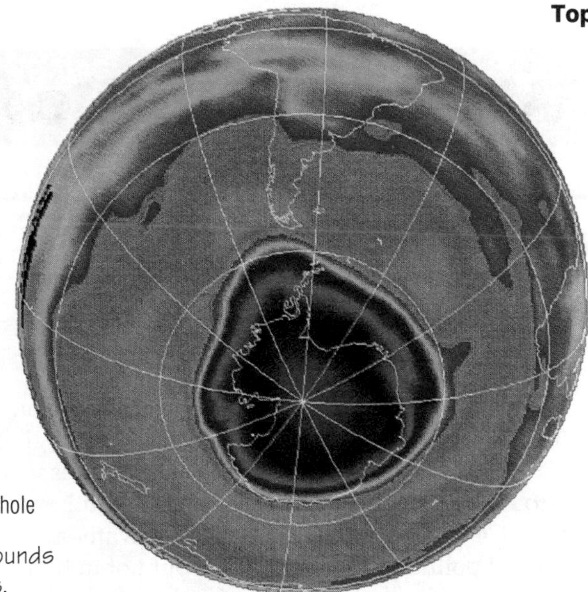

Figure 16.11B ▲ The ozone hole

The very unreactive compounds chlorofluorocarbons, CFCs, react with nothing in the lower atmosphere. When they reach the upper atmosphere they react with ozone. Oxides of nitrogen also react with ozone. CFCs and $NO_x$ decrease the thickness of the ozone layer.

## What should be done?

*Is it happening? Is the ozone layer becoming thinner?* In June 1980 the British Antarctic Expedition discovered that there was a gap in the ozone layer over Antarctica during certain months. In 1987, research workers in the US confirmed that there was a thinning of the ozone layer which was 'large, sudden and unexpected ... far worse than we thought'. In 1988 a team of scientists working in the Arctic Ocean discovered that the ozone layer over Northern Europe was thinner than it had been.

In 1990 a US plane flying at high altitude took samples which showed that levels of ozone-destroying gases are 50 times higher than expected over the Arctic. An ozone 'hole' over the Arctic would be even more serious than the 'hole' over the Antarctic because more people live in the northern hemisphere.

Knowing that the danger had appeared over more populated regions of the globe spurred many countries to take action. At a meeting in Montreal in 1987 many countries agreed to reduce their use of CFCs by 50% by the year 2000. Since that date, many countries have agreed to speed up their programme of phasing out CFCs. Aerosols containing CFCs have been banned in the USA since 1988. In 1988 many makers of toiletries in the UK agreed to stop using CFCs by the end of 1989. They are now using spray cans with different propellants, which they label 'ozone-friendly', or pump-action cans. There is more of a problem with the CFCs used as refrigerants, in air conditioners, in the manufacture of polyurethane foam and as solvents. Chemists are now finding stable compounds to replace CFCs. In the USA, Du Pont Chemicals have agreed to stop using CFCs after the year 2000. In the UK, ICI chemists have found substitutes which will enable ICI to do the same.

### SUMMARY

The ozone layer protects animals and plants from ultraviolet radiation. As it reacts with pollutants in the upper atmosphere, the ozone layer is becoming thinner. CFCs and nitrogen monoxide from high altitude planes are the culprits. The use of CFCs is being reduced.

### CHECKPOINT

1. Look round your kitchen, bathroom and garage. How many products in aerosol cans do you buy? How convenient is it to have each of these products in an aerosol can? What inconvenience would you suffer if aerosol cans were banned? How many of the aerosol cans are labelled 'ozone-friendly'? What does this mean?
2. Speaking on 23 February 1988, the Prince of Wales announced that he had banned aerosols from his household. He said that some members of his household had difficulty in finding a suitable alternative hairspray.
   (a) What concern led the Prince of Wales to take this step?
   (b) What properties must the propellant in the hairspray possess to work effectively and to be safe in use?
   (c) What substitute can you suggest for an aerosol hairspray?
3. How does their lack of chemical reactivity make CFCs (a) useful and (b) dangerous?

# Topic 17 Water pollution

## 17.1 Pollution by industry

FIRST THOUGHTS

What's the problem? We need clean drinking water – nothing could be more important. We need industry, and industry needs to dispose of waste products; in the process our water is polluted.

### Controls

You will notice that many industrial firms are on river banks. These firms can get rid of waste products by discharging them into rivers. Until 1989, the quantities of waste which industries were allowed to discharge into rivers were controlled by the water authority of each region. Under the 1974 Control of Pollution Act, the water authorities had power to control pollution in inland rivers but not in tidal rivers, estuaries and the sea (except for the discharge of radioactive waste: see Topic 8.8). In spite of the Act, more than 2800 km of Britain's largest rivers are too dirty and lacking in oxygen to keep fish alive.

In 1989 the UK Water Privatisation Bill became law. The water authorities were sold to private companies, and are now run for profit as other industries are. The Government set up a National Rivers Authority to watch over the quality of water and prosecute polluters.

SUMMARY

The National Rivers Authority controls pollution of inland rivers. It does not regulate the discharge of pollutants into tidal rivers, estuaries and the sea. The estuaries in the UK are heavily polluted by industry and by sewage.

#### Estuaries

Many of the worst polluters discharge into coastal waters and estuaries. The oil refineries, chemical works, steel plants and paper mills on coasts and estuaries can pour all the waste they want into estuaries and the sea. In the 1930s, fishermen could make a living in the Mersey. Now, it is too foul to keep fish alive. One reason is the discharge of raw sewage into the Mersey. The other is that too many firms pour waste into the estuary. There is unemployment in Merseyside, and the Government does not want to make life difficult for industry in the area. The industries on the banks of the Mersey have been given permission to fall below the standards of the Control of Pollution Act.

Other estuaries, such as the Humber, the Tees, the Tyne and the Clyde, are also polluted by industry.

The photograph shows the industries on the banks of the Mersey. Since this is a tidal water, the industries can discharge all their wastes into it. The tides can no longer keep up with carrying all the pollutants away to be diluted at sea.

*Figure 17.1A* ▲ The Mersey

SUMMARY

Many industrial firms do not keep their discharges of wastes within the limits set by law.

FIRST THOUGHTS

Why has it taken so long for industry to react to the tragedy of Minamata?

## Mercury and its compounds

A well-known case of industrial pollution is the tragedy of Minamata, a fishing village on the shore of Minamata Bay in Japan. A plastics factory started discharging waste into the bay in 1951. By 1953, a thousand people in Minamata were seriously ill. Some were crippled, some were paralysed, some went blind, some became mentally deranged, and some died. The cause of the disease was found to be the mercury compounds which the plastics factory discharged into the Bay. Although the level of mercury compounds in the Bay water was low, mercury was concentrated by a **food chain** (see Figure 17.1B). The level of mercury in the fish in the Bay was high, and fishers and their families became ill through eating the fish.

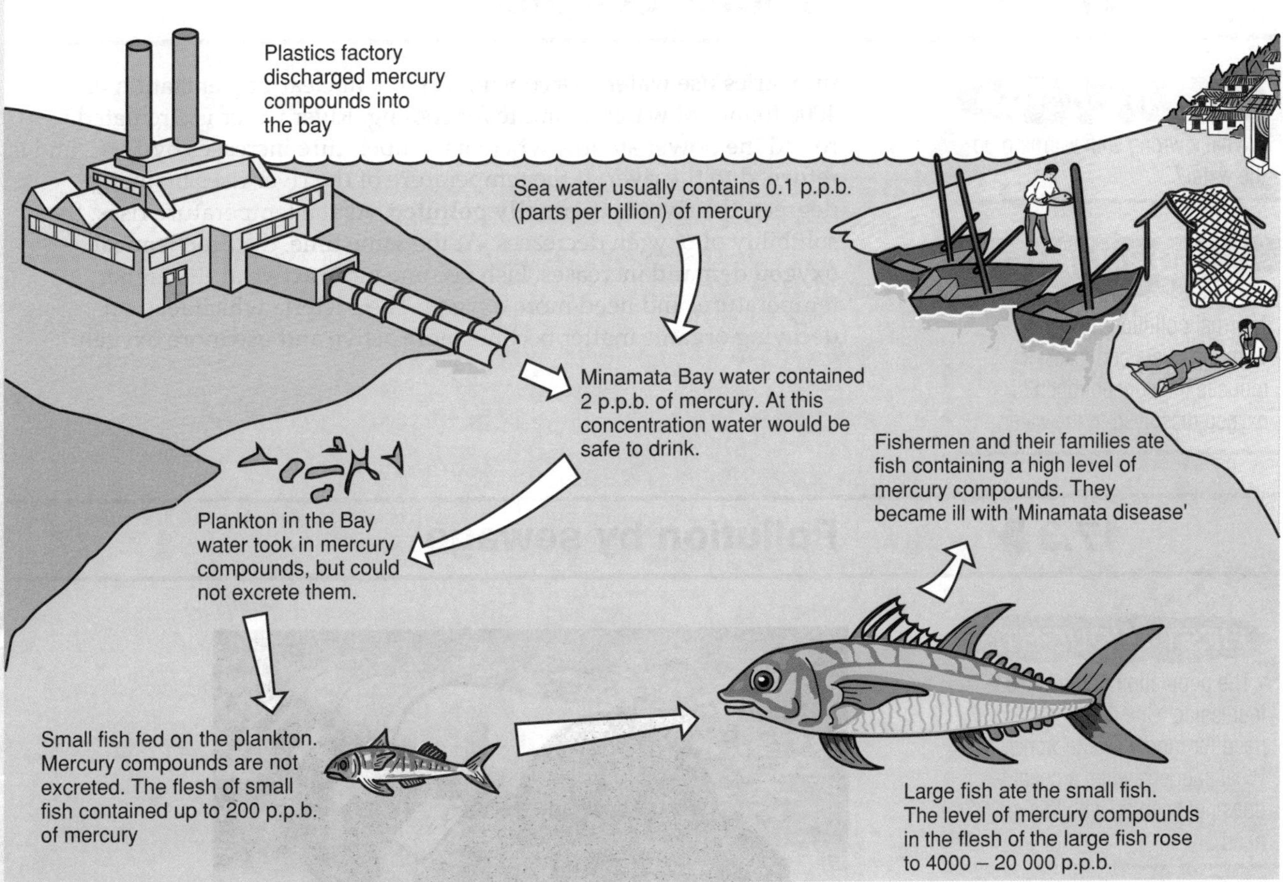

Plastics factory discharged mercury compounds into the bay
Sea water usually contains 0.1 p.p.b. (parts per billion) of mercury
Minamata Bay water contained 2 p.p.b. of mercury. At this concentration water would be safe to drink.
Plankton in the Bay water took in mercury compounds, but could not excrete them.
Small fish fed on the plankton. Mercury compounds are not excreted. The flesh of small fish contained up to 200 p.p.b. of mercury
Large fish ate the small fish. The level of mercury compounds in the flesh of the large fish rose to 4000 – 20 000 p.p.b.
Fishermen and their families ate fish containing a high level of mercury compounds. They became ill with 'Minamata disease'

*Figure 17.1B* ▲ The food chain which led to the Minamata disease

SUMMARY

Mercury and its compounds are poisonous. If mercury gets into a lake or river, it is converted slowly into soluble compounds. These are likely to accumulate in fish and may be eaten by humans.

Other countries have experienced the results of mercury pollution. In 1967, many lakes and rivers in Sweden were found to be so contaminated by mercury that fishing had to stop. In 1970, high mercury levels were found in hundreds of lakes in Canada and the USA. As late as 1988, the ICI plant on Merseyside discharged more mercury than the permitted level. Now that the danger is known, the polluting plants have taken care to reduce spillage of mercury. The danger is still there, however. Mercury from years of pollution lies in the sediment at the bottom of lakes. Slowly it is converted by bacteria into soluble mercury compounds. These may get into a food chain.

## CHECKPOINT

1 When a car engine has an oil change, the waste oil is sometimes poured down the drain. What is wrong with doing this?

2 Does it matter whether rivers are clean and stocked with fish or foul and devoid of life? Explain your answer.

3 The Minamata tragedy happened when Japan was building up its industry after the war. In spite of Japan's experience, Sweden, Canada and the USA found an excess of mercury in their lakes twenty years later. Why had they not learned from Japan's mistake? (You will not find the answer in the back!)

## 17.2 Thermal pollution

**FIRST THOUGHTS**

What's wrong with warming up the water?

**SUMMARY**

Thermal pollution means warming rivers and lakes. It reduces the concentration of oxygen dissolved in the water.

Industries use water as a coolant. A large nuclear power station uses 4000 tonnes of water a minute for cooling. River water is circulated round the power station, where its temperature increases by 10 °C, and is returned to the river. If the temperature of the river rises by many degrees, the river is **thermally polluted**. As the temperature rises, the solubility of oxygen decreases. At the same time, the **biochemical oxygen demand** increases. Fish become more active at the higher temperature, and need more oxygen. The bacteria which feed on decaying organic matter become more active and use more oxygen.

## 17.3 Pollution by sewage

**FIRST THOUGHTS**

The population of the UK is increasing. One result is the need for more sewage works. What happens when a country does not keep up with this need?

*Figure 17.3A* ▲ British beaches and the EC standard

Estuaries and bathing beaches suffer when untreated sewage is discharged into estuaries and the sea.

In Topic 15.4, you read how sewage is treated before it is discharged into rivers or the sea. Unfortunately, as some water companies do not have enough plants to treat all their area's sewage, they discharge some raw sewage into rivers and estuaries. The Mersey receives raw sewage from Liverpool and other towns. In Sussex, sewage treatment works are inadequate and sewage is discharged into the sea. This creates some nasty results at several bathing beaches in the county.

SUMMARY

During the 1980s, the United Nations set a target of safe water and sanitation for all by 1990. The aim was to provide wells and pumps, kits for disinfecting water and hygienic toilets. The sum needed was £25 billion, slightly more than the world spends on its armies in one month. The target was not reached by 1990, but the work is continuing.

The quality of the water at dozens of Britain's bathing beaches fails to meet standards set by the European Community (EC). Many British beaches have more coliform bacteria and faecal bacteria in the water than the EC standard.

## The Third World

Of the four billion people in the world, two billion have no toilets, and one billion have unsafe drinking water. In Third World countries (the developing countries) three out of five people have difficulty in obtaining clean water. Some Third World communities have to use a river as a source of drinking water as well as for disposal of their sewage. Bacteria are present in faeces, and they infect the water. Many diseases are spread by contaminated water. They include cholera, typhoid, river blindness, diarrhoea and schistosomiasis. Four-fifths of the diseases in the Third World are linked to dirty water and lack of sanitation. Five million people each year are killed by water-borne disease.

*Figure 17.3B* ▲ Their water supplies

# 17.4 ▸ Pollution by agriculture

FIRST THOUGHTS

Farmers need to use fertilisers. What happens when a crop does not use all the fertiliser applied to it? There can be pollution, as this section explains.

## Fertilisers

A lake has a natural cycle. In summer, algae grow on the surface, fed by nutrients which are washed into the lake. In autumn the algae die and sink to the bottom. Bacteria break down the algae into nutrients. Plants need the elements carbon, hydrogen, oxygen, nitrogen and phosphorus. Water always provides enough carbon, hydrogen and oxygen; plant growth is limited by the supply of nitrogen and phosphorus. Sometimes farm land surrounding a lake receives more fertiliser than the crops can absorb. Then the unabsorbed nitrates and phosphates in the fertiliser wash out of the soil into the lake water. When fertilisers wash into a lake, they upset the natural cycle. The algae multiply rapidly to produce an **algal bloom**. The lake water comes to resemble a cloudy greenish soup. When the algae die, bacteria feed on the dead material and multiply. The increased bacterial activity consumes much of the dissolved oxygen. There is little oxygen left in the water, and fish die from lack of oxygen. The lake becomes difficult for boating because masses of algae snag the propellers. The name given to this accidental fertilisation of lakes and rivers is **eutrophication**.

Many parts of the Norfolk Broads are now covered with algal bloom. The tourist industry centred on the Broads would like to see them restored to their former condition.

## Science at work

Research chemists are working on a microbiological method for stripping nitrates from drinking water. A bacterium converts nitrates and hydrogen into nitrogen and water. The bacterium is bound to beads of calcium alginate. Hydrogen and water containing nitrates are passed through a tube containing the beads. Nitrogen is formed, and nitrate-free water flows out of the tube.

Lough Neagh in Northern Ireland is the UK's biggest inland lake. It supplies Belfast's water and it also supports eel-fishing. Algae now block the filters through which water flows to the water treatment plant. Eels and other fish are in danger as the level of dissolved oxygen falls. The problem is being tackled by removing phosphates from the treated sewage which enters the lough. Treatment with a solution containing aluminium ions and iron(III) ions precipitates phosphates. Each year, this stops 60 tonnes of phosphorus in the form of phosphates from entering Lough Neagh. This pollution is unnecessary. Detergents without phosphates would leave laundry only a little less sparkling white, but would not pollute our rivers and lakes.

Fertiliser which is not absorbed by crops can be carried into the ground water (the water in porous underground rock). Ground water provides one third of Britain's drinking water. The EC has set a maximum level of nitrates in drinking water at 50 mg/l (12 p.p.m. of nitrogen in the form of nitrate). Four out of the ten water companies in England and Wales have drinking water which exceeds this nitrate level. In 1989, the EC decided to prosecute the UK for falling below EC water standards.

EXTENSION FILE ACTIVITY

*Figure 17.4A* ▲ Algal bloom

There are two health worries over nitrates. Nitrates are converted into nitrites (salts containing the $NO_2^-$ ion). Some chemists think that nitrites are converted in the body into nitrosoamines. These compounds cause cancer. The other worry is that nitrites oxidise the iron in haemoglobin. The oxidised form of haemoglobin can no longer combine with oxygen. The extreme form of nitrite poisoning is 'blue baby' syndrome, in which the baby turns blue from lack of oxygen. Babies are more at risk than adults because babies' stomachs are less acidic and assist the conversion of nitrates into nitrites.

The level of nitrites in drinking water permitted by the EC is 0.1 mg/l. Some parts of London have nitrite levels which are higher than this. The UK Government has agreed to bring the UK into line with the rest of Europe. To install nitrate-stripping equipment would cost £200 million. *Should the Government reduce the use of fertilisers? How? Should the Government introduce a tax on fertilisers or ration fertilisers?*

## SUMMARY

When a crop receives more fertiliser than it can use, nitrates and phosphates wash into lakes and rivers. There, they stimulate the growth of weeds and algae. When the plants die, bacterial decay of the dead material uses oxygen. The resulting shortage of dissolved oxygen kills fish. The level of nitrates in ground water, from which we obtain much of our drinking water, is rising. Pollution can be reduced by reducing the application of fertilisers and by omitting phosphates from detergents.

## Pesticides

### FIRST THOUGHTS

What are the 'drins'? Why is the EC worried about the level of drins in UK water?

Other pollutants which must worry us are the pesticides dieldrin, endrin and aldrin (sometimes called the 'drins'). They cause liver cancer and affect the central nervous system. The EC sets a maximum level of $5 \times 10^{-9}$ g/l for 'drins'. Half the water in the UK exceeds this level. The danger with the 'drins' is that fish take them in and do not excrete them. The level of 'drins' in fish may build up to 6000 times the level in water. Figure 17.4B shows what happened when DDT, another powerful insecticide, was used to spray Clear Lake in California to get rid of mosquitoes. It is another example of pollutants being concentrated by a food chain.

## SUMMARY

Pesticides are serious pollutants, especially when they can be concentrated through a food chain.

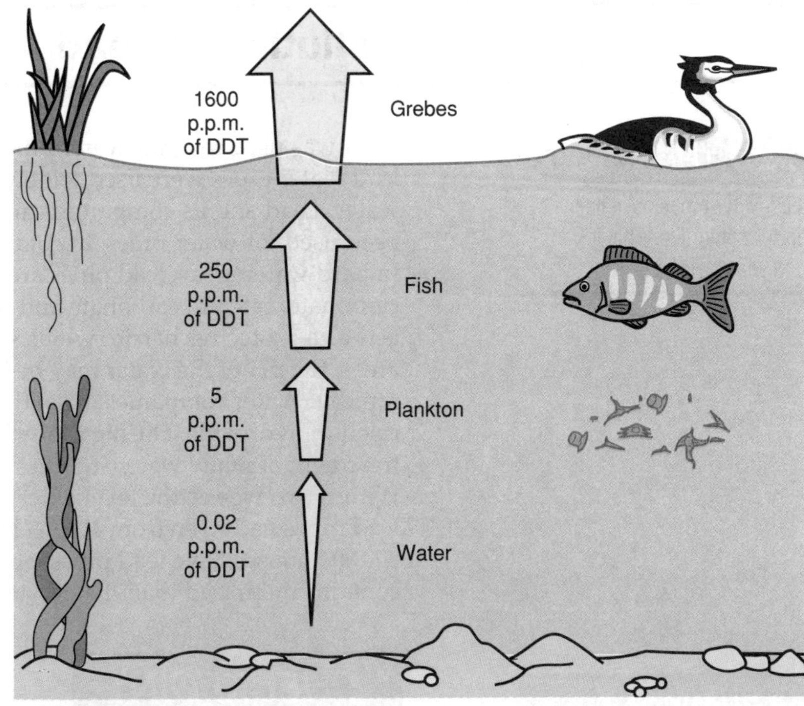


*Figure 17.4B* ▲ A food chain in Clear Lake, California

## CHECKPOINT

▶ **1** Groups of settlers in North America always built their villages on river banks, and discharged their sewage into the river. How did the river dispose of the sewage? Why can this method of sewage disposal not be used for larger settlements?

▶ **2** British Tissues make toilet paper, paper towels, paper handkerchiefs, etc. They use a lot of bleach on the paper, and this bleach is one of the chemicals which the firm has to dispose of. Can you suggest how the firm could reduce the problem of bleach disposal?

▶ **3** (a) Why do some lakes develop a thick layer of algal bloom?

(b) Why is algal bloom less likely to occur in a river?

(c) What harm does algal bloom do to a lake that is used as

(i) a reservoir

(ii) a fishing lake

(iii) a boating lake?

▶ **4** The concentration of nitrates in ground water is rising. Explain:

(a) why this is happening,

(b) why some people are worried about the increase.

▶ **5** Water companies can tackle the problem of high nitrate levels by:

- blending water from high-nitrate sources with water from low-nitrate sources
- closing some sources of water
- treating the water with chemicals
- ion exchange
- microbiological methods

(a) Say what you think are the advantages and disadvantages of each of these methods.

(b) Which do you think would be the most expensive treatments? How will water companies be able to pay for the treatment?

(c) Suggest a different method of reducing the level of nitrates in ground water.

(d) Say who would pay for the method which you mention in (c) and how they would find the money.

## 17.5 Pollution by lead

FIRST THOUGHTS

Lead water pipes have been used for centuries. What is wrong with them?

Lead was used to make water pipes in Roman times. Until the 1950s, lead water pipes were used widely in the UK. Slowly, lead dissolves in water. Lead and its compounds are poisonous. Since 1950, copper has been used for water pipes. In cold water systems, plastic pipes are used. In hard water areas, lead pipes are safe because a layer of insoluble lead carbonate, calcium carbonate and magnesium carbonate builds up and acts as a protective barrier which stops lead dissolving. In soft water areas, the pH of the water may be less than 5, and lead dissolves more rapidly. Water companies make the water more alkaline by adding calcium hydroxide. During the 1983 strike of UK water workers, the treatment of many water supplies with calcium hydroxide stopped. Within two weeks, the level of lead in the tap water of some houses with lead pipes had risen from 40 mg/l to over 800 mg/l. In the North-West, 600 000 houses have lead pipes. Some parts of Scotland have water which contains more lead than the EC standard.

SUMMARY

Some parts of the UK have lead water pipes. Lead compounds are toxic. In hard water areas, the deposition of insoluble scale in the pipes stops lead dissolving. In other areas, the solution of lead is reduced by keeping the water alkaline.

CHECKPOINT

 1 (a) How does lead get into our drinking water?
(b) What harm does it do?
(c) In what ways can the amount of lead in drinking water be reduced?
(d) Pollution of air by lead is another problem. How does lead get into the air? (See Topic 16 if you need to revise.)

## 17.6 Pollution by oil

FIRST THOUGHTS

Huge oil tankers sail the sea. Sometimes one has an accident – not often, but often enough to cause a pollution disaster. The first big oil spill off the UK coast was when the oil tanker, the Torrey Canyon, sank off Cornwall in 1967. The pollution fouled beaches in Cornwall and killed thousands of sea birds. Now, tankers of up to 550 000 tonnes are in use, and accidents can happen on a much larger scale.

Prince William Sound was a beautiful unspoiled bay in Alaska. It was home to a huge variety of sea animals. Death struck over an area of 1300 square kilometres in 1989. The supertanker *Exxon Valdez* left Valdez with a cargo of oil from Alaska. Only 40 km out of port, the tanker hit a submerged reef and 60 million litres of crude oil leaked from her tanks. Fish, sea mammals and migrating birds from all parts of the American continent perished in the giant oil slick.

Sea birds dive to obtain food. If the sea is polluted by a discharge of oil, sea birds may find themselves in an oil slick when they surface. Then oil sticks to their feathers and they cannot fly. They drift on the surface, becoming more and more waterlogged until they die of hunger and exhaustion. Thirty five thousand sea birds died in the *Exxon Valdez* disaster.

The *Exxon Valdez* disaster was by no means the biggest ever. The increase in the size of tankers since 1945 has been spectacular. In 1945 the deadweight of the largest tanker was 16 500 tonnes, whereas the *Seawise Giant* built in 1980 has a deadweight of 565 000 tonnes. The very large crude carriers are difficult to stop and to change direction, and when accidents happen the results are more serious. The *Exxon Valdez* accident promoted legislation to prevent repetition of such disasters. In 1990 the Oil Pollution Act was passed in the USA, specifying that in future tankers that want to operate in US waters must have a double hull, with a 3 m gap between the outer hull and the oil tanks inside.

A double hull is not the answer to all accidents, but it would have saved the *Sea Empress* when she grounded off the UK in 1996. In February 1996, the oil tanker *Sea Empress* ran on to rocks outside Milford Haven. A stretch of Welsh coastline 190 km long was contaminated by oil. Oil slicks hit the marine nature reserves of Skomer and Lundy Islands and six sites of special scientific interest (SSSIs, on the map). These include major bird colonies and the country's only coastal national park. Within a month, over 1000 oiled birds had been counted, and 300 dead birds washed ashore.

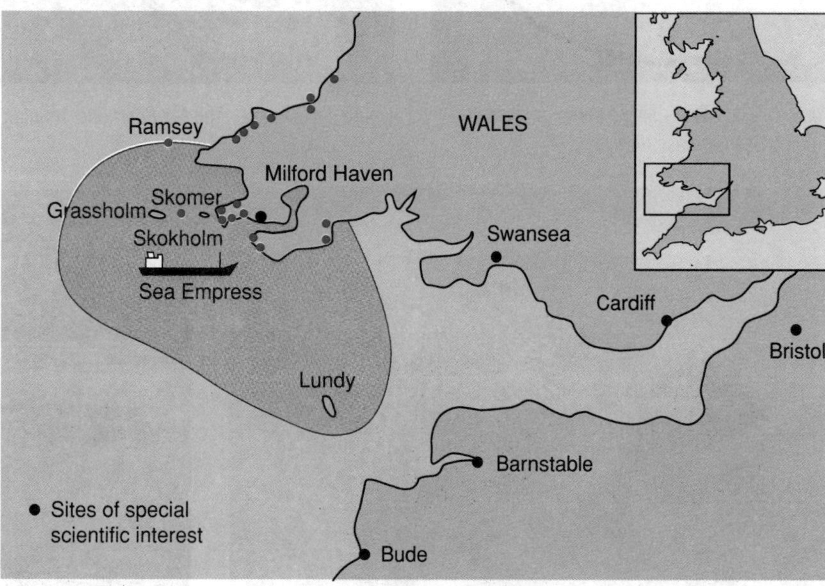


*Figure 17.6A* ▲ Milford Haven in South Wales

More recent accidents involved the *Exxon Valdez* 1988 (60 000 tonnes of oil lost) the *Sea Empress* in 1996 (70 000 tonnes of oil lost).

A double hull is a protection against accidents

The *Sea Empress* was carrying 130 000 tonnes of crude oil on her way to Milford Haven refinery. After losing 6000 tonnes of oil she was refloated by tugs. Two days later, in rough weather, the ship grounded again. The following day stormy weather made it necessary to evacuate the ship, and the ship grounded yet again, this time on rocks. Leakage of oil continued for 9 days until the salvage crews were able to pump the oil from the ship. The leakage of oil was estimated as 70 000 tonnes. At the time of writing the damage to wildlife and to the coastline has not been assessed.

Oil spills at sea are the results of capsizings, collisions and accidental spills during loading and unloading at oil terminals. There is another source. After a large tanker has unloaded, it may have 200 tonnes of oil left in its tanks. While in port, the tanker is flushed out with water sprays, and the cleaning water is collected in a special tank, where the oil separates. Some captains save time by flushing out their tanks at sea and pumping the wash water overboard. This is illegal. Maritime nations have tried to set up standards to stop pollution of the seas, but several nations have not signed the agreements. Enforcing agreements is very difficult as it is impossible to detect everything that happens at sea.

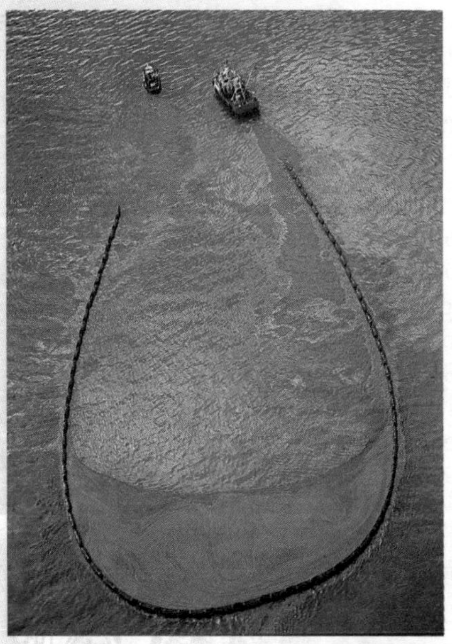

*Figure 17.6A* ▲ Using a boom to clean up after the *Sea Empress* disaster

*Figure 17.6B* ▲ The *MV Braer* after running aground on the Shetland Islands 1993

*Figure 17.6C* ▲ Ten thousand sea otters perished in the *Exxon Valdez* spill. Some ate poisoned fish. Others drowned when their fur became clogged with oil

*Figure 17.6D* ▲ Oil spill in Alaska

*Figure 17.6E* ▶ More victims

## Science at work

Bacteria can be used to clean out a tanker's storage compartment. The empty tank is filled with sea water, nutrient, air and bacteria. When the tanker reaches its destination, the tank contains clean water, a small amount of recoverable oil and an increased number of bacteria. The bacteria can be used as animal feed.

## SUMMARY

Spillage of oil from large tankers is a source of pollution at sea. It kills marine animals and washes ashore to pollute beaches.

Various methods have been tried for the removal of oil from the surface of the sea.

- **Dispersal** Chemicals are added to emulsify the oil. The danger is that they may be toxic to marine life.
- **Sinking** Oil may be treated with sand and other fine materials to make it sink. A danger is that the sunken oil may cover and destroy the feeding areas of marine creatures.
- **Burning** Burning oil is dangerous as a fire can spread rapidly over the sea. Research has been done on safe methods of burning oil, but they leave 15 per cent of the oil behind as lumps of tar.
- **Absorbing** Absorbents do not work well in the open sea. They provide the best way of cleaning a beach or preventing an oil spill from reaching the shore.
- **Skimming off** The method of surrounding an oil spill with a line of booms to prevent it spreading and then pumping oil off the surface has been used with some success.
- **Solidifying** Scientists at British Petroleum have discovered chemicals which will solidify oil spills. The chemicals must be sprayed on to the oil slick from the air. They convert the oil into a rubber-like solid which can be skimmed off the surface in nets.
- **Bacteria** There are bacteria which will decompose petroleum. A mixture of bacteria (of the correct strain) and nutrients is sprinkled on to the spill from the air.

## CHECKPOINT

▶ 1 (a) What are the causes of oil spills at sea?
(b) What damage do they do?
(c) Who pays to clean up the mess?
(d) Suggest what can be done to stop pollution of the sea by oil.

# Topic 18 Earth

## 18.1 The structure of the Earth

**FIRST THOUGHTS**

What is Earth like deep down, many kilometres below the surface? It sounds an impossible question to answer, but scientists have found methods of investigating the deep structure of the Earth.

The study of the Earth is called **geology**, and a person who works in this branch of science is called a **geologist**. The research work of geologists has enabled them to construct a model of Earth's structure (see Figure 18.1A).

*How was Earth formed?* A molten mass cooled down over millions of years. Dense materials sank deeper into the centre to form a core of dense molten rock. Less dense material remained on the surface to form a **crust** of solid rock (50 km thick). Gaseous matter outside the crust is the **atmosphere**. Earth's atmosphere is chiefly oxygen and nitrogen.

As Earth cooled, water vapour condensed to form rivers, lakes and oceans on the surface of Earth. No other planet has oceans and lakes, though Mars has some water vapour and polar ice caps.

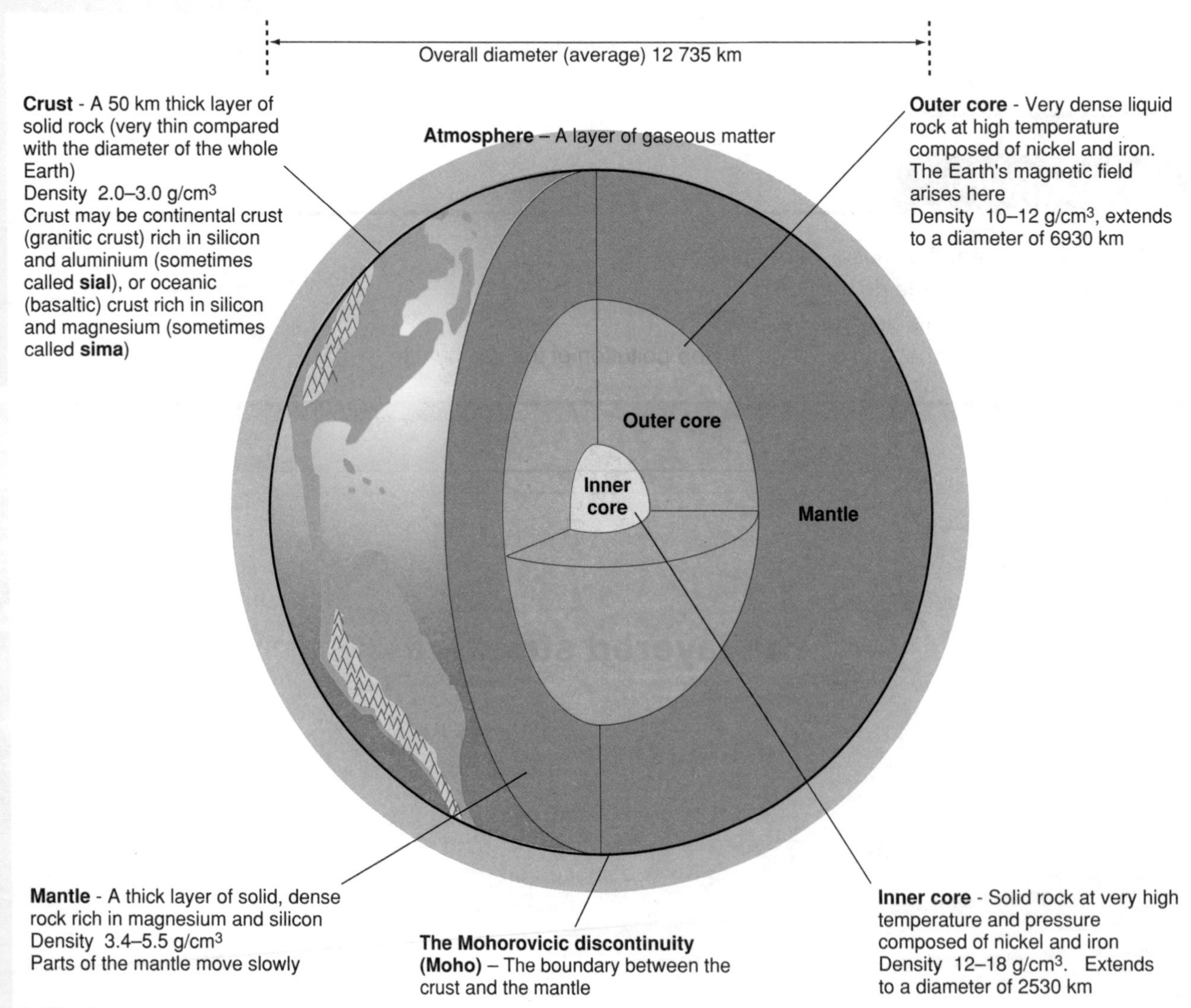

Overall diameter (average) 12 735 km
Crust - A 50 km thick layer of solid rock (very thin compared with the diameter of the whole Earth)
Density 2.0–3.0 g/cm³
Crust may be continental crust (granitic crust) rich in silicon and aluminium (sometimes called sial), or oceanic (basaltic) crust rich in silicon and magnesium (sometimes called sima)
Atmosphere – A layer of gaseous matter
Outer core - Very dense liquid rock at high temperature composed of nickel and iron. The Earth's magnetic field arises here
Density 10–12 g/cm³, extends to a diameter of 6930 km
Outer core
Inner core
Mantle
Mantle - A thick layer of solid, dense rock rich in magnesium and silicon
Density 3.4–5.5 g/cm³
Parts of the mantle move slowly
The Mohorovicic discontinuity (Moho) – The boundary between the crust and the mantle
Inner core - Solid rock at very high temperature and pressure composed of nickel and iron
Density 12–18 g/cm³. Extends to a diameter of 2530 km


*Figure 18.1A* ▲ The structure of the Earth

Check that you know the meaning of:

- crust
- continental crust/granit crust
- oceanic crust/basaltic crust
- mantle
- outer core
- inner core
- Moho

## Earth's crust

Earth's crust is composed of **rocks** (see Topic 19) and **soils**. Rocks are composed of compounds, e.g. carbonates, oxides and silicates, and some free elements, e.g. sulphur and copper. The elements and compounds which occur naturally in Earth's crust are called **minerals**. Soils have been formed by the breakdown of rocks and vegetation. The crust is divided into continental and oceanic crust.

| Continental crust | Oceanic crust |
|---|---|
| • Forms continents and their shelves<br>• Up to 70km thick in mountain ranges<br>• Density ~ 2.7 g/cm$^3$<br>• Age: up to 3700 million years<br>• Same composition as granite rock<br>• Often called granitic crust<br>• Rich in silicon and aluminium<br>• The deeper parts of continental crust are of a denser material similar to oceanic crust. | • Beneath deep sea floors<br>• Average thickness 6 km<br>• Density ~ 3.0 g/cm$^3$<br>• Age: up to 220 million years<br>• Same composition as basalt rock<br>• Often called basaltic crust<br>• Rich in silicon and magnesium<br>• Material similar to oceanic crust is thought to lie beneath the continents. |

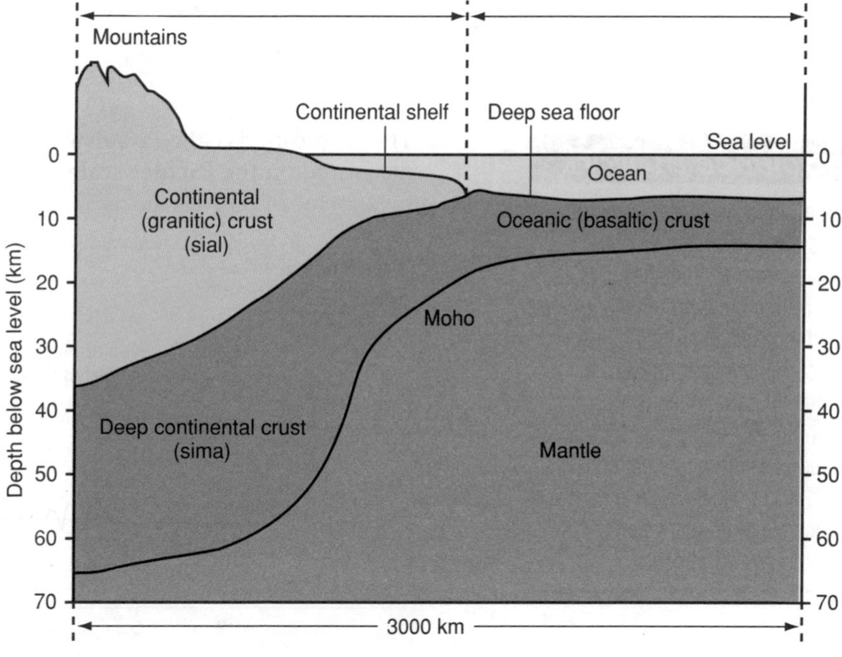

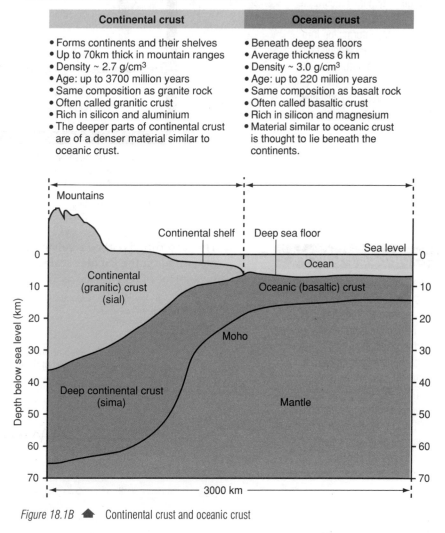

*Figure 18.1B* ▲ Continental crust and oceanic crust

## SUMMARY

The Earth has a layered structure: inner core, outer core, mantle, crust and atmosphere. The crust is composed of oceanic (basaltic) crust beneath the ocean floors and continental (granitic) crust, which forms the Earth's land masses. The boundary between the crust and the mantle is called the Moho. The study of the Earth is called geology.

# 18.2 ▶ The layered structure of the Earth

## FIRST THOUGHTS

We cannot obtain information about the deep structure of Earth by mining or by drilling. The world's deepest mine is 3.5 km in depth, and the radius of Earth is over 6000 km. Evidence about the interior of Earth comes from the study of earthquakes and volcanoes.

## Earthquakes

The study of earthquakes is called **seismology**. About 500 000 earthquakes occur every year. Only about 1000 of these are strong enough to cause damage, and only a few are serious. An earthquake occurs when forces inside Earth become strong enough to fracture large masses of rock and make them move. The energy which is released travels through the Earth as a series of shock waves. Earthquakes are limited to the rigid part of the crust. They cannot occur in the molten part of the mantle. Most earthquakes are generated within 600 km of Earth's surface. The point where an earthquake originates is called

## Science at work

Seismic waves are used by geologists in the search for oil. They are shock waves, like earthquake waves. Prospectors produce seismic waves by detonating small explosive charges. Then they record the wave patterns on seismometers and feed the information into computers. The computers record the position, thickness and density of rock layers. This is very much more convenient than drilling holes to find out!

**focus**. The nearest point on Earth's surface directly above it is the **epicentre**. Shock waves are felt most strongly at the epicentre and then spread out from it. Earthquake shocks are recorded by an instrument called a **seismometer** (see Figure 18.2A).

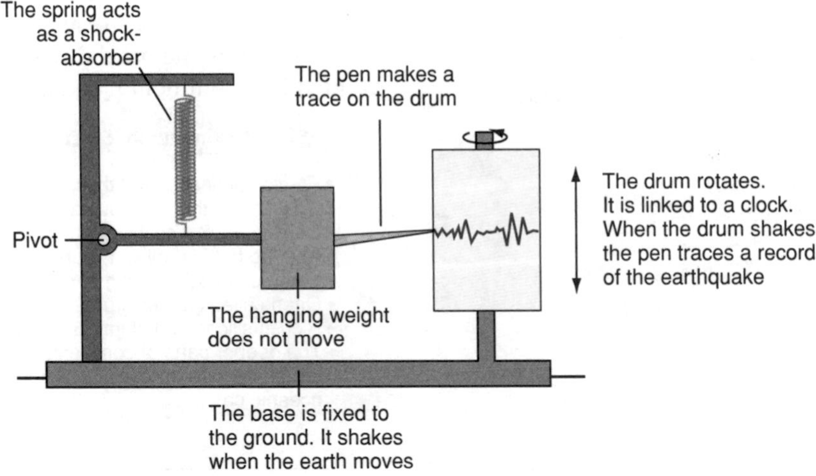


*Figure 18.2A* ▲ A seismometer

## Science at work

Computers are used to record and analyse the patterns of waves from earthquakes. By recording the form of different earthquakes and logging the occurrence of earthquakes in different parts of the world, computers enable scientists to predict where future earthquakes will occur. However, it is not yet possible to predict when an earthquake will occur.

The recording is called a **seismogram**. The energy of the earthquake is measured on the **Richter scale**. Each point on the scale means an increase by a factor of ten: a scale of 5 is ten times as powerful as a scale of 4.

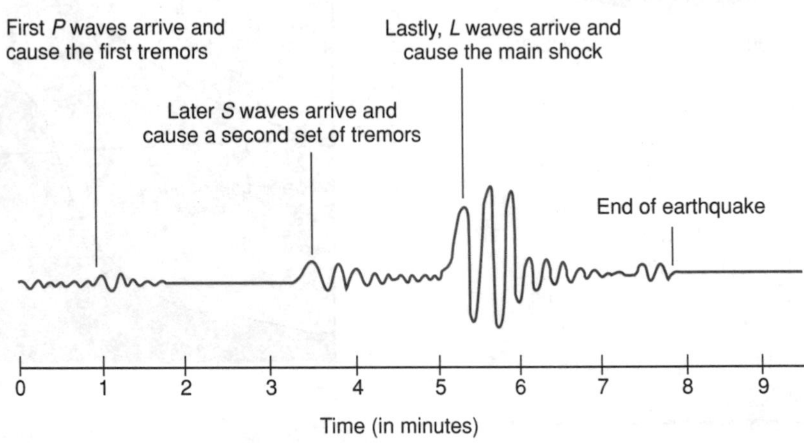


*Figure 18.2B* ▲ A seismogram

The seismogram in Figure 18.2B records three types of shock waves. *P* waves – **primary waves**, *S* waves – **secondary waves** and *L* waves – **long waves**. Figure 18.2C shows the difference between them.

The pattern of waves received by a seismometer depends on what the waves have passed through inside the Earth. Waves are either reflected (bounced back) or refracted (bent) when they travel from one type of material into another.

The positions of boundaries, e.g. the Moho, and the thicknesses and densities of the zones have all been worked out from the way that earthquake shock waves have been affected by passing through them. *S* waves do not pass through the outer core at all. Since it is known that *S* waves do not travel through liquids, this is evidence that the outer core is in the liquid state. This shows that its temperature must be very high.

Check that you know the meaning of

- earthquake

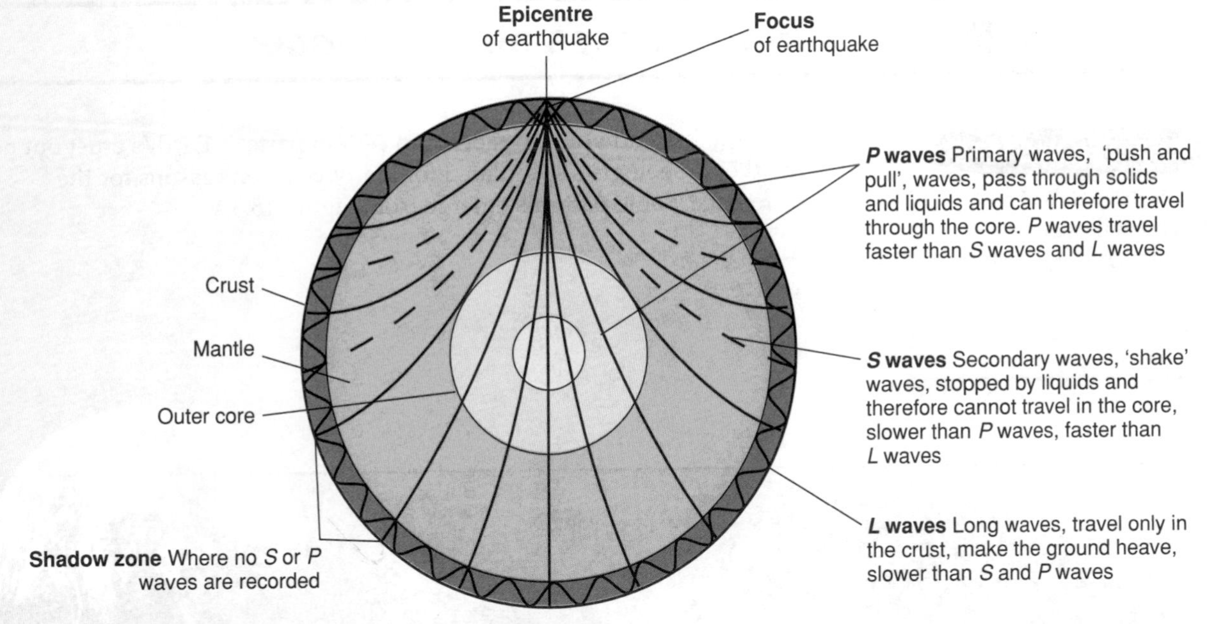


*Figure 18.2C* ▲ Shock waves from an earthquake

## Volcanoes

The lava erupted by volcanoes gives information about the crust and upper mantle, where lava is produced, but not about deeper layers. Plotting on a map the places where volcanoes have occurred gives information about the regions of Earth where heat is being generated and causing volcanic activity.

## Meteorites

Meteorites reach Earth from space. They are pieces of rock and dust which have been attracted towards Earth by Earth's gravity. Most meteorites burn up when they reach Earth's atmosphere, but some fall to Earth's surface. Geologists believe that meteorites may be samples of planetary material dating from the time of formation of the solar system.

## Magnetism

Earth's magnetic field is evidence for the presence of iron in the core.

### SUMMARY

Evidence for the structure of the Earth comes from:

- the patterns of shock waves produced by earthquakes
- material erupted by volcanoes
- the positions of earthquakes and volcanoes on the map
- meteorites
- the Earth's magnetic field.

### CHECKPOINT

➧ **1** The overall density of Earth is 5.5 g/cm³. The rocks in the Earth's crust have densities of 2.5 to 3.0 g/cm³. How can you explain the difference between these values and the much higher density of the whole Earth?

➧ **2** Take a piece of string 3 m long. Imagine that this length represents the 3000 million years that have passed since the first living things appeared on Earth. Mark on the string the length that represents the 2 million years since the human race appeared.

# 18.3 Earthquakes and volcanoes

Why do many parts of Earth experience earthquakes and volcanoes? Why do some parts of Earth have neither? Geologists have found answers to this puzzle.

Earthquakes and volcanoes occur in certain parts of Earth's crust but not in others. Geologists speculated for many years on reasons for the difference. Some patterns emerge from Figure 18.3A.

Areas where earthquakes occur

Active volcanoes

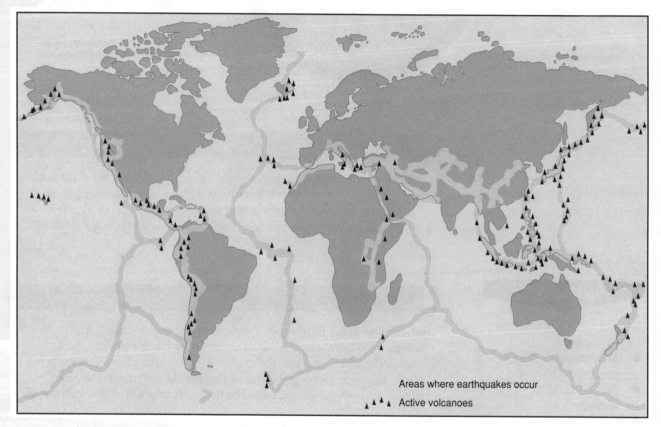

*Figure 18.3A* ▲ Areas of earthquake and volcanic activity

- Earthquakes and volcanoes occur in belts of activity. The belts are hundreds of kilometres wide and thousands of kilometres long. In places, belts join up.
- On land, belts occur along chains of high mountains like the Alps.
- Beneath the sea, belts pass through the centres of oceans and through chains of volcanic islands like the Philippines.
- Surveys of the mid-ocean belts show that the sea floor rises to form chains of huge mountains beneath the sea. These mountain chains are called **oceanic ridges**. The mid-Atlantic ridge rises above sea-level to form Iceland.
- Surveys show that belts which pass through chains of islands are close to deep **oceanic trenches** on the sea floor. The same is true of mountain ranges near the edges of continents. A trench in the Pacific called the Peru-Chile trench runs parallel to the Andes.

## SUMMARY

Earthquakes and volcanoes occur in belts of activity. These belts run:

- along mountain ranges
- along oceanic ridges (mountains on the sea floor)
- along oceanic trenches (deep channels in the sea floor)

# 18.4 ▸ Plate tectonics

**FIRST THOUGHTS**

Is Earth's crust really composed of separate moving pieces? It sounds amazing, but this is one of the newest scientific theories.

*Figure 18.4A* ▲ The result of an earthquake in Los Angeles

Geologists believe that the outer layer of Earth is made up of separate pieces called **plates**. Each plate is a piece of **lithosphere** (crust and uppermost layer of mantle) of 80–120 km in thickness. Movements in the mantle beneath make the plates move very slowly, a few centimetres per year. As a result, plates sometimes rub against each other. If stress builds up to a large extent, the plates may bend. When they spring back into shape, the ground shakes violently: there is an **earthquake**. There has to be a source of energy to produce the movement of plates. Many geologists believe that it is the heat given out when radioactive elements decay (see Figure 18.4L).



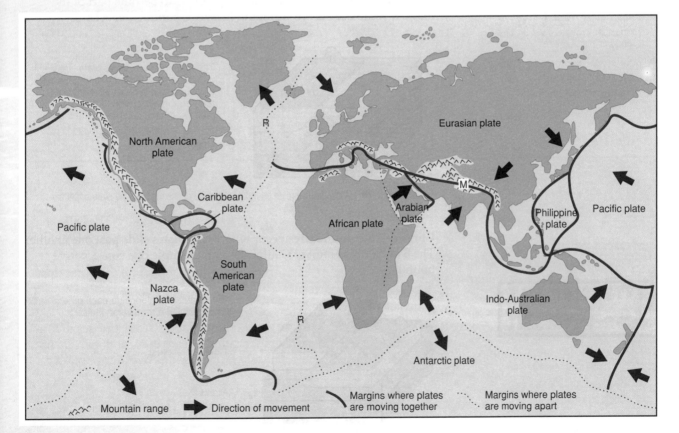

*Figure 18.4B* ▲ The plates which make up Earth's crust

**It's a fact!**

In April 1990 the second largest earthquake in Britain this century hit Clun in Shropshire. It registered 5.2 on the Richter scale.

## Boundaries

Boundaries between plates may be **constructive, destructive** or **conservative**.

**Constructive boundaries** occur where the plates are moving apart (see Figure 18.4C). Evidence for plate movement was found in the 1960s when the volcanoes along oceanic ridges were studied. Lava erupts from these volcanoes and cools to form new oceanic crust along the edges of the plates on either side. The plates move away from the ridge and the width of the ocean floor is increased. The process is called **sea-floor spreading**. An example is the mid-Atlantic trench.

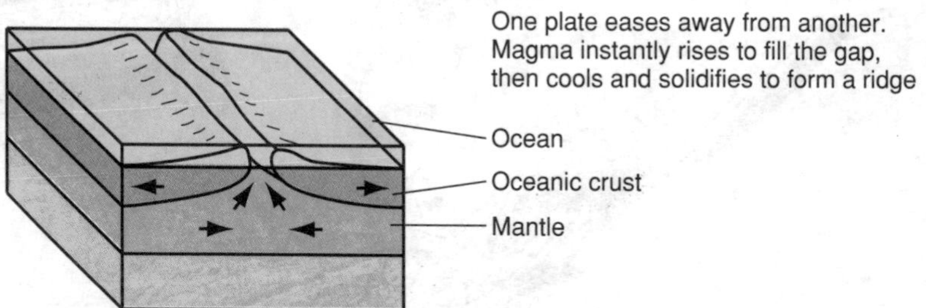

One plate eases away from another. Magma instantly rises to fill the gap, then cools and solidifies to form a ridge
Ocean
Oceanic crust
Mantle

*Figure 18.4C* ▲ A constructive plate boundary. Material is added at the plate boundary and the plates move apart

Boundaries between plates may be

- constructive – where new crust is formed between plates moving apart
- destructive – where crust is forced downwards as plates collide
- conservative – where plates slide past one another with no gain or loss of crust

**Destructive boundaries** occur where the plates are in collision (see Figure 18.4D). Oceanic trenches are regions where plates meet as they move together. When they meet, the edge of one plate is forced to slide beneath the other and move down into the mantle. This process is called **subduction**. The descending plate edge melts to become part of the mantle. Some oceans are shrinking in size as subduction occurs.

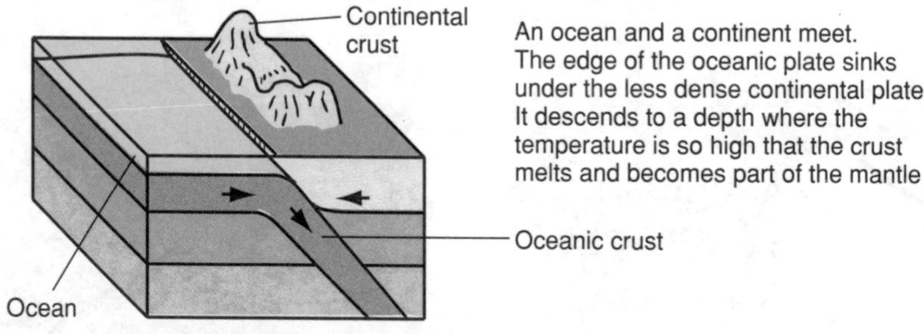


*Figure 18.4D* ▲ A destructive plate boundary. Material descends from the plate boundary

**Conservative boundaries** occur where two plates slide past one another. The San Andreas fault in California is an example of a conservative boundary.

**EXTENSION FILE ASSIGNMENT**

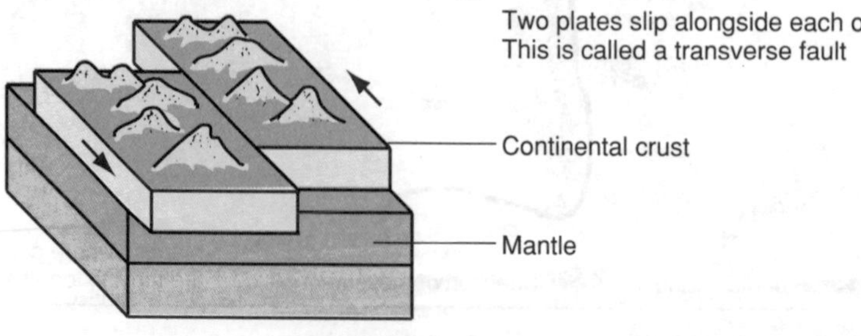

Two plates slip alongside each other. This is called a transverse fault
Continental crust
Mantle

*Figure 18.4E* ▲ A conservative plate boundary. No material is gained or lost

## SUMMARY

Earth's crust and the upper part of the mantle are together called the lithosphere. Geologists believe that the lithosphere consists of separate plates and that these plates are moving slowly. At oceanic ridges, the plates are moving apart. At oceanic trenches, the plates are moving together. When plates meet, continental crust is pushed up to form mountain ranges. Plate movement gives rise to earthquakes and volcanoes.

## The conveyor belt

As material is subducted at an oceanic trench and added at an oceanic ridge, the net effect is to convey material from one edge of a plate to another. Since the mass of material taken away at oceanic trenches is equal to the mass of material added at oceanic ridges, the plate remains the same size.

This movement of plates has been called a **conveyor belt**. Continents ride the conveyor belt beneath them: as the plates move, the continents on them move. This has been happening for thousands of millions of years. Although the rate is slow, only a few centimetres a year, the continents have already travelled thousands of kilometres. About 300 million years ago, northern Europe was near the equator, and tropical forests grew there. These later decayed to form coal deposits.

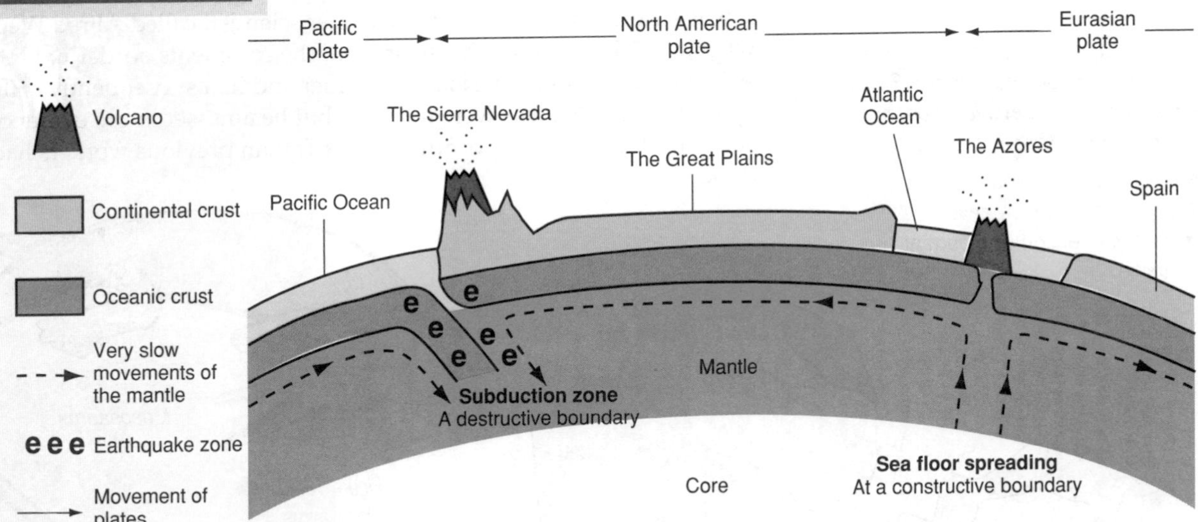


*Figure 18.4F* ▲ Movement of plates on the 'conveyor belt'

Figure 18.4G shows what happens when oceanic crust meets continental crust at a destructive plate boundary.

The oceanic plate sinks below the less dense continental plate. Some of the sediment on the surface of the oceanic plate is scraped off. It piles up on the landward side

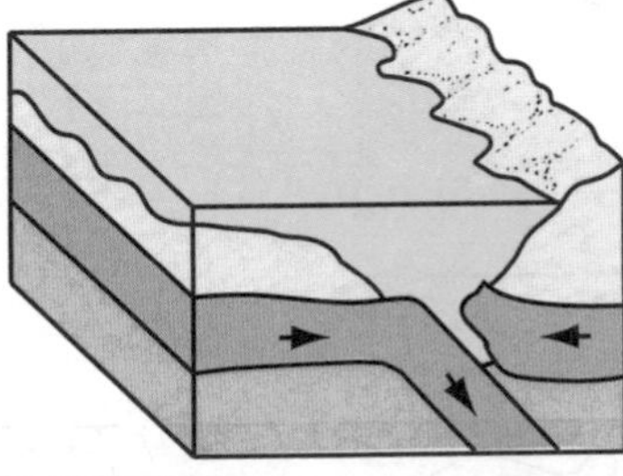

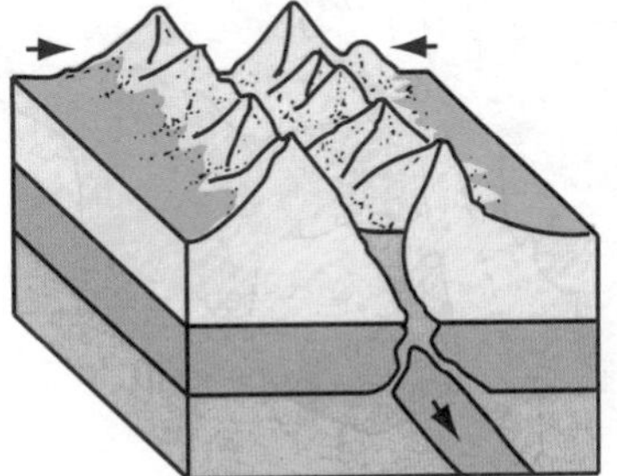

The sediment is compressed by the continental plate. It folds to form a mountain range. The Andes, on the west coast of South America, formed this way

*Figure 18.4G* ▲ Collision between an oceanic plate and a continental plate: formation of a mountain range. (When two continents collide the sediments from both are squeezed up to form mountains.)

## The theory of plate tectonics

You will remember the theory that Earth originated from a cloud of hot gas which condensed to form a ball of hot liquid. The densest materials sank into the **core** of the liquid while the less dense materials rose to the surface. There they cooled and solidified to form a **crust**. Until the twentieth century, scientists considered that Earth went on cooling until the present day. They thought that as the interior cooled and contracted, the crust folded to fit it and as a result mountain ranges were formed. Scientists compared the folding crust with the wrinkling of the skin of an apple as the inner part of the fruit dries out and shrinks.

Continents travel on a 'conveyor belt'. Although they travel at only a few centimetres a year they have moved thousands of kilometres since they were formed.

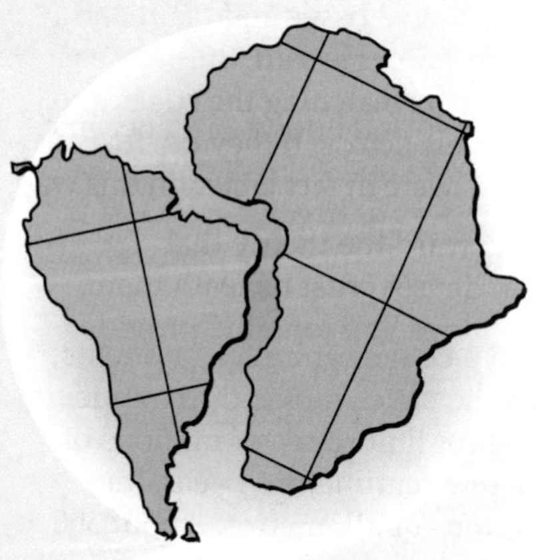

*Figure 18.4H* ▲ The 'jigsaw' fit of South America and Africa

Can you appreciate the evidence for continental drift? Look at the 'fit' between South America and Africa

- in the type of rocks
- in the types of fossils
- in the position of mountain belts.

There were some mysteries which this view of Earth could not explain. For example, it is possible in the UK to find rocks and fossils which can only be formed in desert conditions. Mountains occur in ranges, not as isolated peaks. Earthquakes and volcanoes are collected in belts of activity and not scattered over Earth's surface.

Geologists noticed the similarity between the east coastline of South America and the west coastline of Africa (see Figure 18.4H). The coastlines look as though the continents could have been joined together in a previous era.

*If Africa and South America were once joined, how did the Atlantic come to be formed?* In the eighteenth and nineteenth centuries, it was widely believed that a vast flood, such as Noah's flood in the Old Testament, had forced the continents apart. *And how did the Pacific ocean originate?* The nineteenth century belief was that a huge section of continental land mass was gouged out to form the moon and the hole that was left became the Pacific.

It was a German meteorologist (weather scientist) called Alfred Wegener who, in 1915, promoted the theory that the continents bordering the Atlantic were at one time joined together and had subsequently drifted apart. Wegener's ideas were not new, but he amassed more evidence in support of the theory of **continental drift** than previous workers had done.

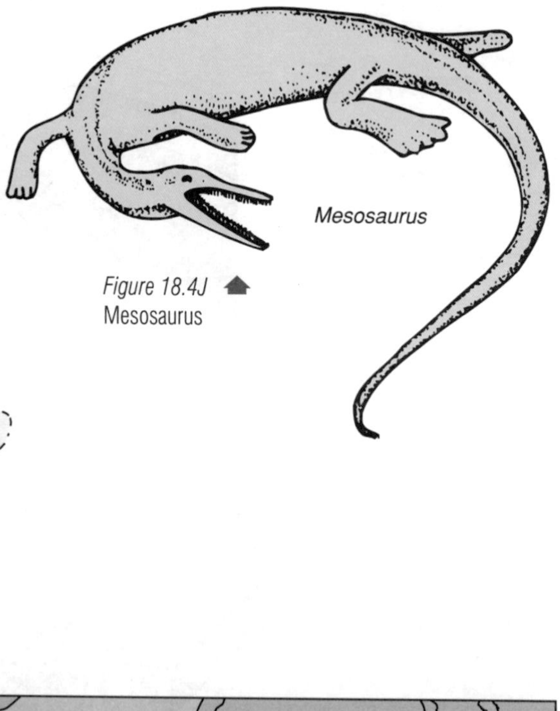

*Mesosaurus*

*Figure 18.4J* ▲ Mesosaurus

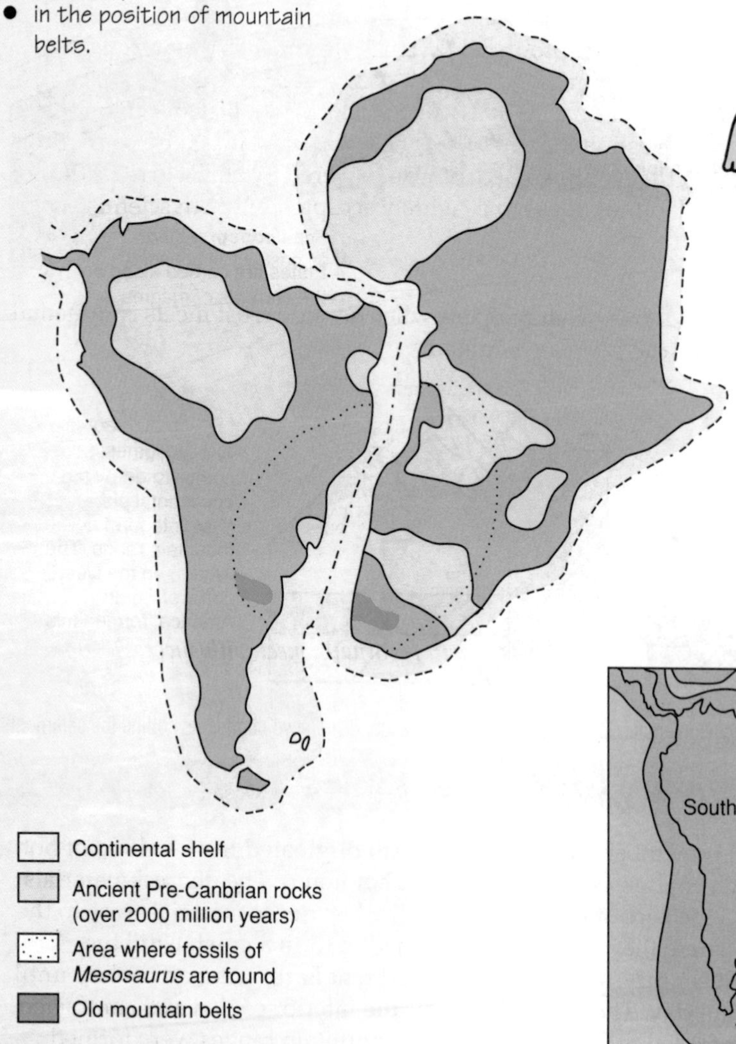

Continental shelf

Ancient Pre-Canbrian rocks (over 2000 million years)

Area where fossils of *Mesosaurus* are found

Old mountain belts

*Figure 18.4I* ▲ Evidence that South America and Africa were joined

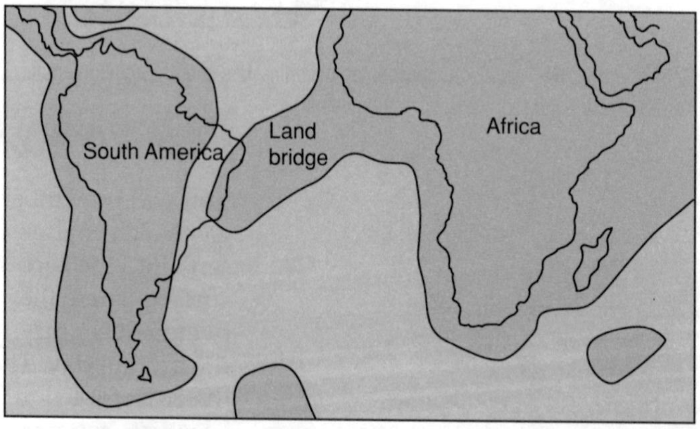

South America

Land bridge

Africa

*Figure 18.4K* ▲ The land bridge theory

Wegener described the matching up of geology, fossils and plant and animal populations on both sides of the Atlantic. He said, 'It is just as if we were to refit the torn pieces of a newspaper by matching the edges and check whether the lines of print run smoothly across. If they do, there is nothing left but to conclude that the pieces were in fact joined in this way.'

*How can continents drift?* Wegener put forward the theory that continental land masses float on a fluid, denser crust beneath them. According to his theory, continents can move horizontally, provided that there are forces acting on them, forces which last for geological eras. Wegener was unable to come up with a convincing idea to explain the force which must have driven continents apart.

The theory of continental drift is not restricted to South America and Africa. It is believed that long ago all the southern continents were joined together as one land mass, called **Gondwanaland** and the northern continents were joined to form a supercontinent called **Laurasia**. It is believed that at a still earlier time all the present-day continents were part of a single land mass, known as **Pangaea**, which began to break up and spread about 200 million years ago (see Figure 18.4O).

Continents need energy in order to drift. Where does this energy come from?

Where were the continents of Gondwanaland, Laurasia and Pangaea?

What evidence is there that these continents drifted apart?

Wegener's theory was derided in 1915, but is accepted today. *Why did it take so long?* People's ideas were dominated by the belief that Earth was cooling, shrinking and folding. Some physicists considered that Earth's crust was too rigid for sideways movements of the kind Wegener was describing. A turning point came when geologists came to realise the magnitude of the heat produced by Earth's **radioactive materials**. It provides for all volcanic activity with plenty of heat to spare. *What happens to the excess heat?* In 1931, Arthur Holmes, a British geologist, put forward a suggestion that the excess heat was discharged by convection currents and that continental drift was powered by such currents. It took 30 years for Holmes' ideas to be widely accepted by earth scientists.

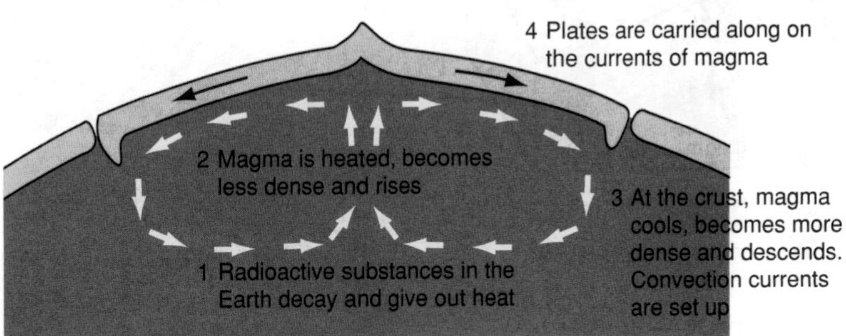

4 Plates are carried along on the currents of magma
2 Magma is heated, becomes less dense and rises
3 At the crust, magma cools, becomes more dense and descends. Convection currents are set up
1 Radioactive substances in the Earth decay and give out heat


*Figure 18.4L* ▲ Convection currents of magma

*Why did the theory of continental drift eventually meet with success?* Some of the evidence is summarised below.

## Evidence of sea-floor spreading

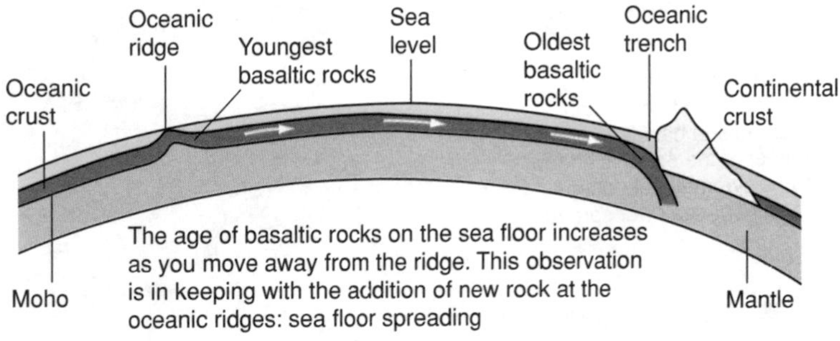


*Figure 18.4M* ▲ Sea-floor spreading

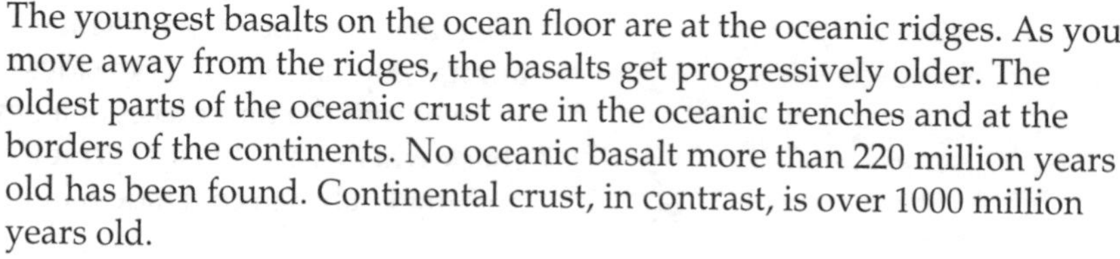

The youngest basalts on the ocean floor are at the oceanic ridges. As you move away from the ridges, the basalts get progressively older. The oldest parts of the oceanic crust are in the oceanic trenches and at the borders of the continents. No oceanic basalt more than 220 million years old has been found. Continental crust, in contrast, is over 1000 million years old.

Every few thousand years Earth's magnetism reverses its polarity. That is to say that over certain periods of time in the past the compass needle would have pointed to the South Pole instead of the North Pole. Oceanic floor basalts include minerals which contain iron and are therefore magnetic. As they crystallise, particles of iron line up with the magnetic field which is operating at the time.

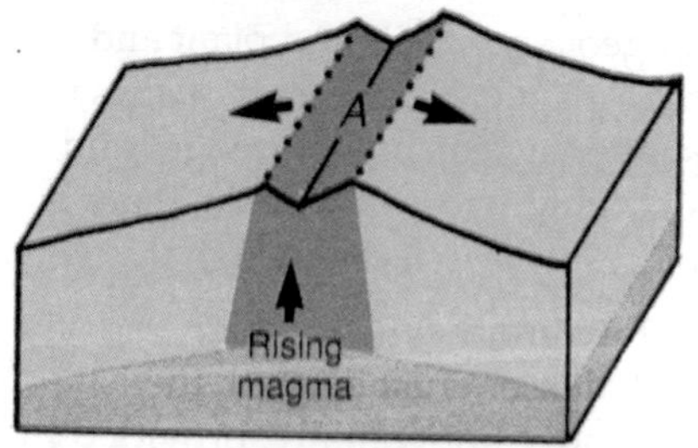


**Band A** The basalt which solidifies as the magma cools is magnetised normally with the North Pole as we know it

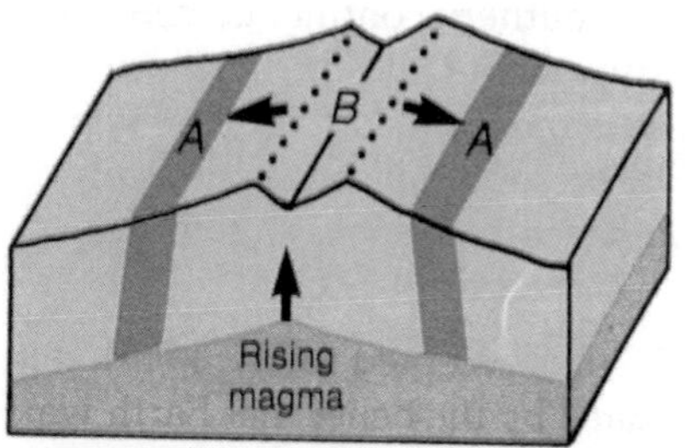


**Band B** The sea-floor has spread. The Earth's magnetic field has reversed. The basalt which now solidifies shows reversed magnetism. As band *B* solidifies, it pushes the two sides of band *A* apart

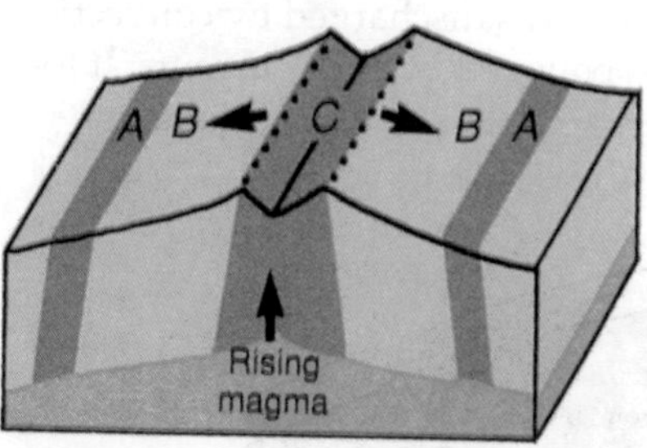


**Band C** The polarity of the Earth's magnetic field has changed again. Normally magnetised basalt solidifies at the centre of the ridge

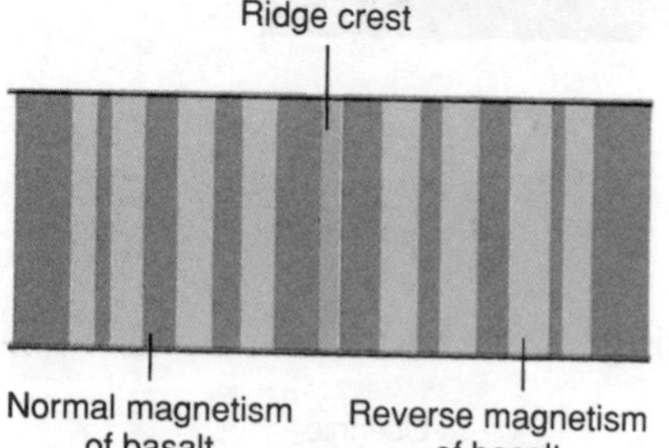


Note the symmetrical pattern of bands of normally magnetised basalt and reversely magnetised basalt on either side of the crest of the oceanic ridge. How could this symmetrical pattern have arisen without sea-floor spreading?

*Figure 18.4N* ▲ Pattern of magnetic stripes at an oceanic ridge

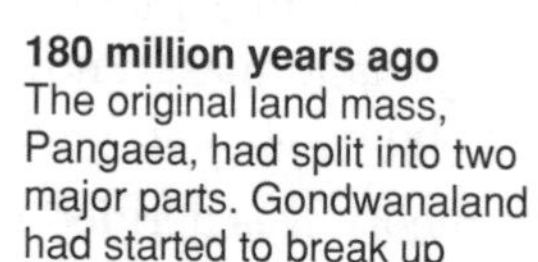

**180 million years ago**
The original land mass, Pangaea, had split into two major parts. Gondwanaland had started to break up

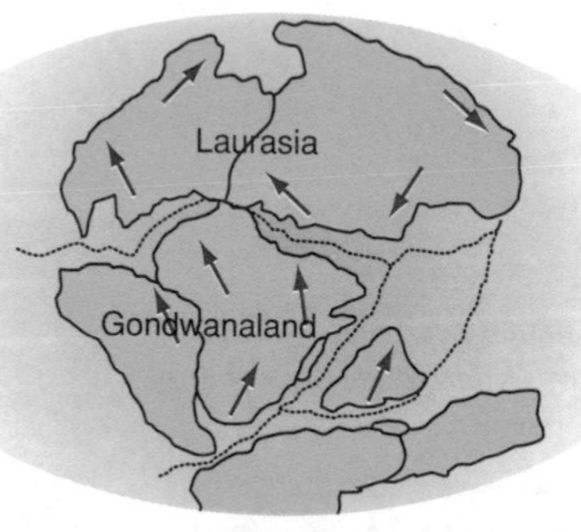


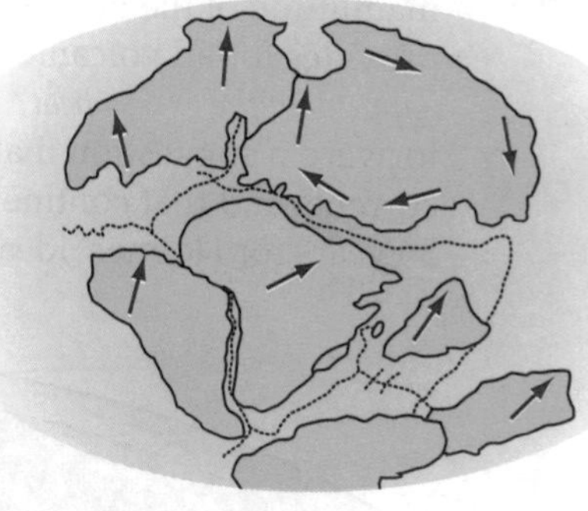

**135 million years ago**
Gondwanaland and Laurasia drifted northwards. The North Atlantic and Indian Oceans widened. The South Atlantic rift lengthened

**65 million years ago**
South America had separated from Africa. Australia and Antarctica were still combined. The Mediterranean Sea had appeared. India was moving towards Asia

**Today**
South America has connected with North America. Australia has separated from Antarctica. India has collided with Asia.

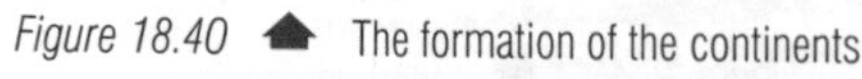

*Figure 18.4O* ▲ The formation of the continents

## SUMMARY

According to the theory of sea-floor spreading, new oceanic crust is being formed along oceanic ridges. The new crust pushes older crust away from the ridges in mid-ocean towards the continents. In this way the sea floor is constantly spreading.

Sea-floor spreading is very uneven. As different lengths of a ridge spread by different amounts, cracks appear between them. These cracks are called transform faults. It is possible to calculate the rate of sea-floor spreading from magnetic data. It comes to 2–10 cm per year for different parts of ridges in different oceans. Figure 18.4O summarises the way continents have drifted apart due to sea-floor spreading.

This model of the Earth, in which rigid slabs of the crust jostle with one another on the surface of a sphere, is called **plate tectonics** (tectonics = construction). From the theory, it has been possible to predict accurately where earthquakes will occur. Unfortunately, earth scientists cannot yet predict the time when a future earthquake will occur.

## CHECKPOINT

1. Refer to Figure 18.4B on p 183.
   (a) Name a plate that is surrounded on all sides by subduction zones.
   (b) Name the mountain range, M.
   (c) The line R–R is the mid-Atlantic ridge. Lava erupts along this ridge. What type of lava is it? Explain how eruptions arise from the movement of plates.
   (d) The San Andreas fault passes through California. What is happening to the plates along this fault?
   (e) Why does the UK experience few earthquakes and no serious quakes?
   (f) Rocks on the Isle of Skye in Scotland show that volcanoes erupted there about 50 million years ago. Explain how this could have happened.
2. Refer to the figure below.
   (a) Name the features A, B and C.
   (b) What type of rock has formed the mountain range, A?
   (c) State the direction of movement in (i) Plate P, (ii) Plate Q and (iii) Plate R.
   (d) Name the zone labelled E. Say what part this zone plays in plate movement.
   (e) Explain what is happening at D.
   (f) Explain how the mountain range A has been formed.

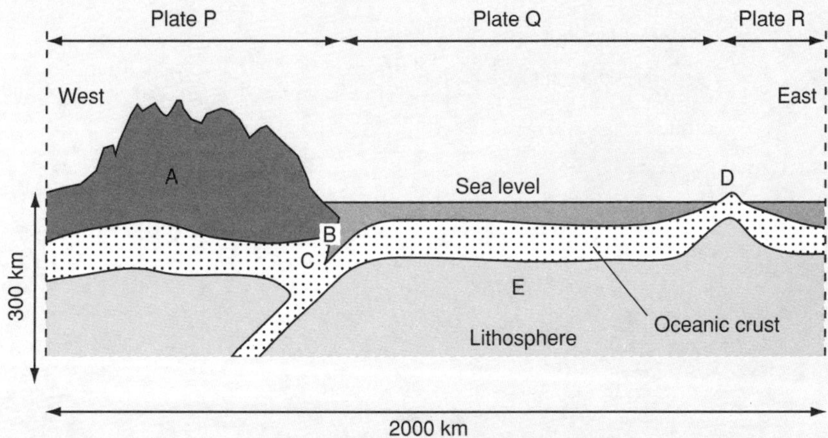


3. The North Atlantic is spreading at a rate of about 5 cm/year.
   (a) How far will it spread during your lifetime (say 80 years)?
   (b) How tall are you?
   (c) How does your answer to (a) compare with your answer to (b)?

**4** Outline two theories which explain how the Pacific Ocean was formed: the plate tectonic theory and a historical theory.

**5** What are the three different types of boundary between plates?

**6** Refer to the figure below.

(a) How old is the ocean floor basalt at (i) the western edge of the ocean, (ii) 500 km east of the centre of the ocean and (iii) at the centre of the ocean?

(b) How do the measurements shown in the graph agree with the theory of sea-floor spreading?

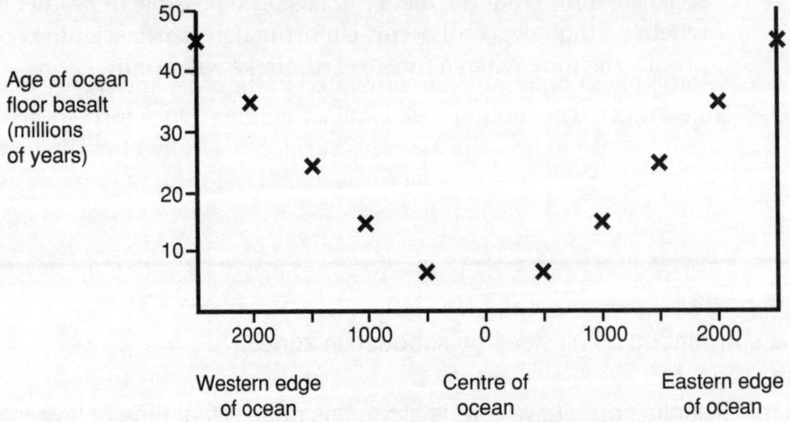


**7** Imagine that you are a journalist who has just attended the 1926 conference at which Wegener put forward his theory of plate tectonics. Write an account for your magazine, *Science Weekly*, describing how other scientists reacted to Wegener's theory.

# Topic 19 Rocks

## 19.1 Types of rock

**FIRST THOUGHTS**

There are three main types of rock: igneous, sedimentary and metamorphic. In this section, you can find out which is which.

### Igneous rocks

Sometimes enough heat is generated in the crust and upper mantle to melt rocks. The molten rock is called **magma**. Once formed, magma tends to rise. If it reaches Earth's surface, it is called **lava**. When cracks appear in Earth's crust, magma is forced out from the mantle on to the surface of Earth. It erupts as a **volcano**, a shower of burning liquid, smoke and dust. When lava crystallises above the surface of the Earth, **extrusive igneous rock** is formed. When magma crystallises below Earth's surface, **intrusive igneous rock** is formed.

The rocks that are formed when volcanic lava cools on the surface of the Earth are called **extrusive igneous rocks**.

The rocks that are formed when magma crystallises inside the Earth are called **intrusive igneous rocks**.

Extrusive igneous rocks formed when the lava erupted from a volcano cools are:

- **basalt**, from free-flowing mobile lava,
- **rhyolite**, from slow-moving lava,
- **pumice**, from a foam of lava and volcanic gases.

Types of rock erupted by a volcano are:

- **agglomerate**, the largest rock fragments which settle close to the vent,
- **volcanic ash**, finer fragments of rock,
- **tuff**, compacted volcanic ash,
- **dust**, which may be carried over great distances by the wind. Sometimes, dust rises high into the atmosphere and affects the weather. This is what happened at Mount St Helens in the USA in 1980.

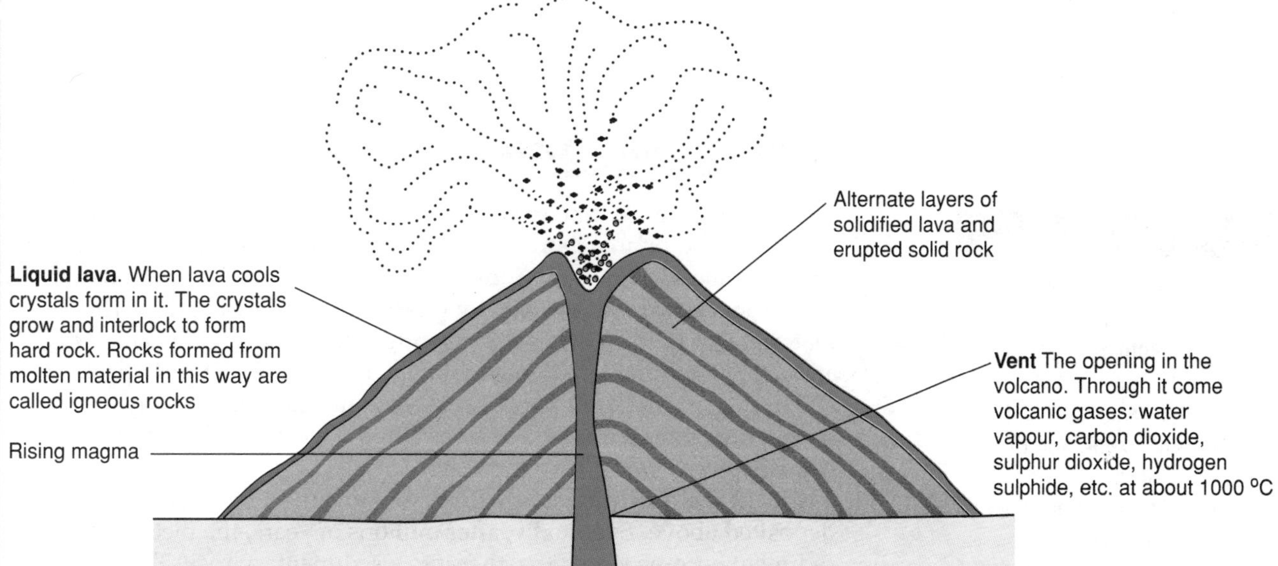

Alternate layers of solidified lava and erupted solid rock
**Liquid lava**. When lava cools crystals form in it. The crystals grow and interlock to form hard rock. Rocks formed from molten material in this way are called igneous rocks
Rising magma
**Vent** The opening in the volcano. Through it come volcanic gases: water vapour, carbon dioxide, sulphur dioxide, hydrogen sulphide, etc. at about 1000 °C
If the lava solidifies in the vent, gas pressure builds up and there is likely to be a violent eruption. If this happens, lava and rock are forced out of the vent in a jet of volcanic gas. The mixture can travel rapidly down the side of the volcano causing death and destruction in its path.


*Figure 19.1A* ▲ A volcano

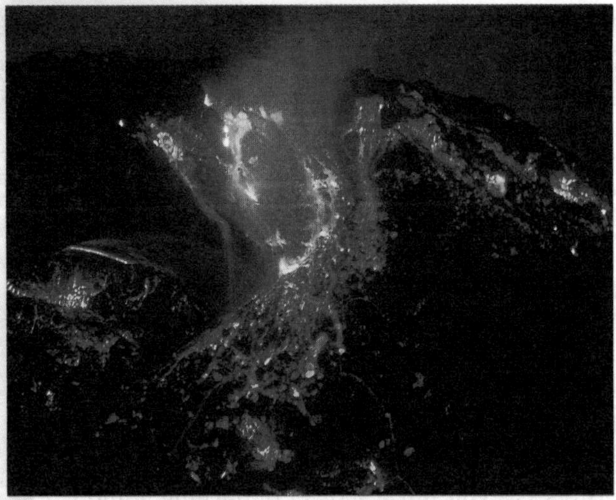

*Figure 19.1B* ▲ Lava rolling down the slopes of Mount Etna, one of the most active volcanoes

*Figure 19.1C* ▲ Mount St. Helens during its second eruption. Dust and ash were spread for many miles around and stayed in the atmosphere for months

*Figure 19.1D* ▲ In 79AD, the city of Pompeii was destroyed when a mixture of lava, rock and ash travelled quickly down the side of the volcano Vesuvius

*Figure 19.1E* ▲ The Giant's Causeway in Northern Ireland is made from basalt, solidified lava

The rocks formed by compressing solid particles are called **sedimentary rocks**.

## Sedimentary rocks

The formation of sedimentary rocks begins when solid particles settle out of a liquid or an air stream to form a **sediment**. The solid material comes from older rocks or from living organisms. All rocks exposed on Earth's surface are worn away by **weathering** and by **erosion**. The material that is worn away is transported by gravity, wind, ice, rivers and seas. The transported material may be fragments of rock, pebbles and grains of sand, or it may be dissolved in water. Eventually the transported material is deposited as a **bed** (layer) of sediment. It may be deposited on a sea bed, on a sea shore or in a desert. The beds of sediment are slowly compacted (pressed together) as other material is deposited above. Eventually, after millions of years, the pieces of sediment become joined together into a sedimentary rock. This process is called **lithification**. Examples of sedimentary rocks are:

- **limestone**, formed from the shells of dead animals,
- **coal**, formed from the remains of dead plants,
- **sandstone**, compacted grains of sand.

### It's a fact!

There are 540 active volcanoes, including 80 on the sea bed. In addition to **active** volcanoes (which have erupted in the past 80 years) there are **dormant** (resting) and **extinct** (dead) volcanoes. Rocks found in the Lake District and in North Wales prove that volcanoes were erupting there 450–500 million years ago, showing that Britain must have been in contact with plate margins in the past.

Rocks formed by the action of high pressure and high temperature on other types of rock are called **metamorphic rocks.**

## Metamorphic rocks

Igneous and sedimentary rocks can be changed by high temperature or high pressure into harder rocks. The new rocks are called **metamorphic rocks** (from the Greek for 'change of shape'). Examples of metamorphic rocks are:

- **marble**, formed when limestone is close to hot igneous rocks,
- **slate**, formed from clay, mud and shale at high pressure,
- **metaquartzite**, from metamorphism of sandstone.

The composition of the Earth's crust is: igneous rocks 65%, sedimentary rocks 8% and metamorphic rocks 27%. The differences are summarised in Table 19.1. Rocks are composed of compounds, e.g. carbonates, oxides and silicates, and some free elements, e.g. sulphur and copper. The elements and compounds which occur naturally in Earth's crust are called **minerals**.

*Table 19.1* ▼ Types of rock

| | *Igneous* | *Sedimentary* | *Metamorphic* |
|---|---|---|---|
| *Type of grain* | Crystalline | Fragmental: grains do not usually interlock (They do in some limestones) | Crystalline |
| *Direction of grain* | Grains usually not lined up | Grains usually not lined up | Grains usually lined up |
| *Mode of formation* | Crystallisation of magma | Deposition of particles | Recrystallisation of other rocks |
| *Fossil remains* | Absent | May be present | Absent |
| *Appearance when broken* | Shiny | Usually dull | Shiny |
| *Ease of breaking* | Hard, not easily split, may crumble if weathered | May be soft and crumble, but some are hard to break | Hard, but may split in layers, may crumble if weathered |
| *Examples* | Basalt, granite, rhyolite, pumice | Limestone, clay, sandstone, mudstone | Marble, hornfels, slate, schist |

## The rock cycle

Only igneous rocks are formed from new material brought into the crust. The original crust of Earth must have been made entirely from igneous rocks. The slow processes by which metamorphic and sedimentary rocks are formed from igneous rock and also converted back into igneous rock is called the rock cycle (see Figure 19.1G).

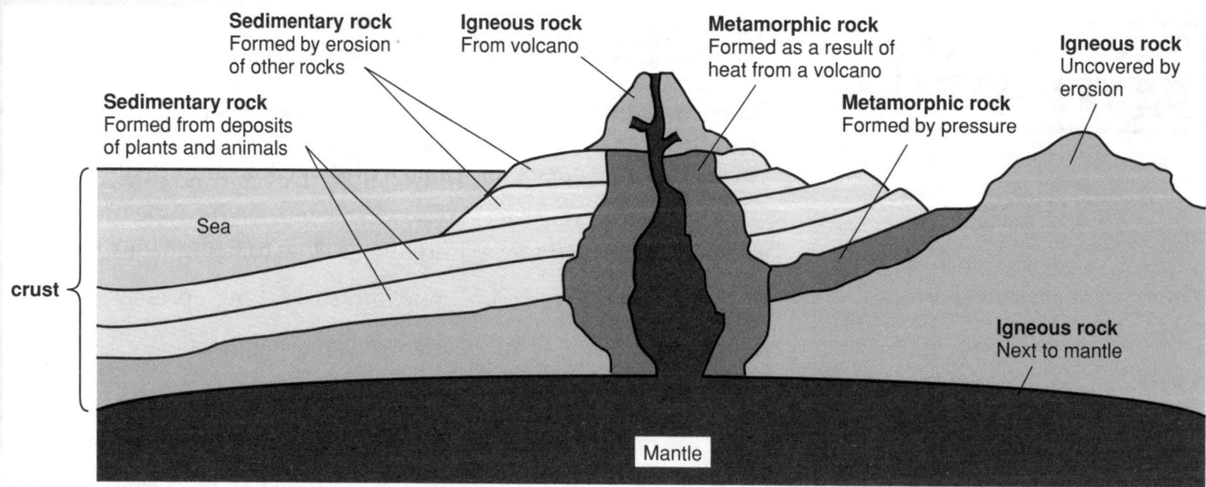


*Figure 19.1F* ▲ Types of rock

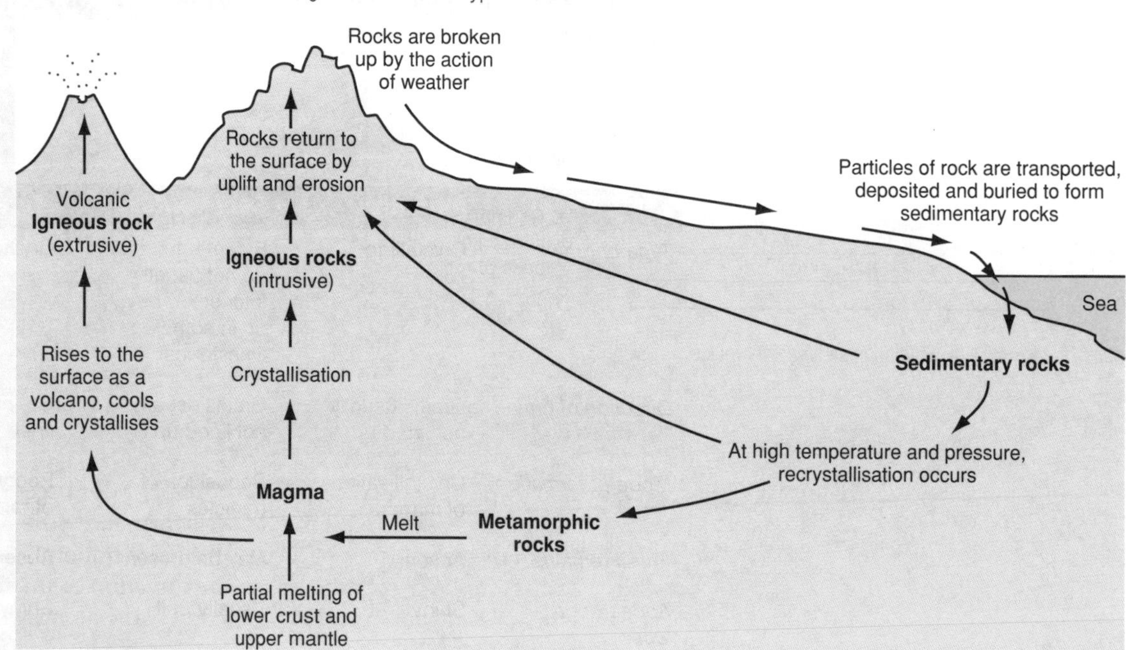


*Figure 19.1G* ▲ The rock cycle

## SUMMARY

When volcanoes erupt, they emit volcanic gases, lava, ash and pieces of solid rock. Lava cools and solidifies to form igneous rock. The solid rock settles as agglomerate. Ash is compacted to form tuff. Rocks are weathered into smaller particles. These particles are deposited as a sediment, which becomes compacted to form sedimentary rock. Igneous and sedimentary rocks may be changed by high temperature and pressure into metamorphic rock. The slow processes by which rock material is recycled are called the rock cycle.

## CHECKPOINT

1 Copy and complete this passage.

When volcanoes erupt, molten rock called _____ streams out of the Earth. It solidifies to form _____ rocks, e.g. _____. When deposits of solid materials are compressed to form rocks, _____ rocks are formed, e.g. _____. The action of heat and pressure can turn _____ rocks and _____ rocks into _____ rocks.

# 19.2 ▶ Deformation of rocks

The deformation of rocks is caused by forces acting within the crust. It results in the formation of **folds**, **faults**, **cleavage** and **joints**. The extent of deformation produced by a force depends on the type of rock: brittle rocks may fracture to produce a fault, while soft rocks may crumple to produce a fold.

(a) A fold in limestone rock

## Folding

When rocks are compressed (squeezed) they may become folded. Sedimentary rocks are often folded. The beds of rock which have been laid down are no longer horizontal; they are folded to bulge upwards (an **anticline**) or downwards (a **syncline**).

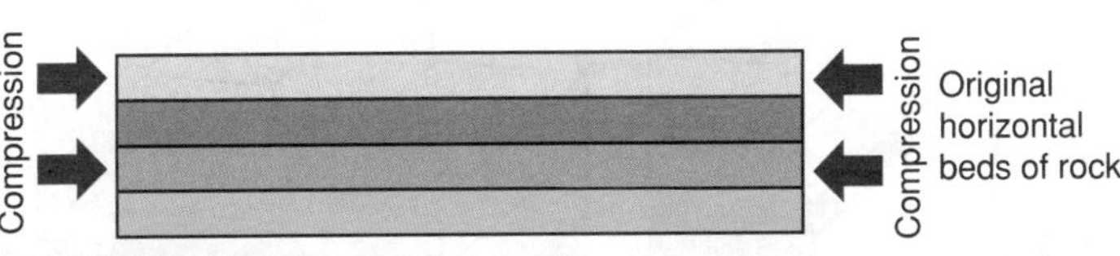


*Figure 19.2A* ▲ (b) Folding in sedimentary rock

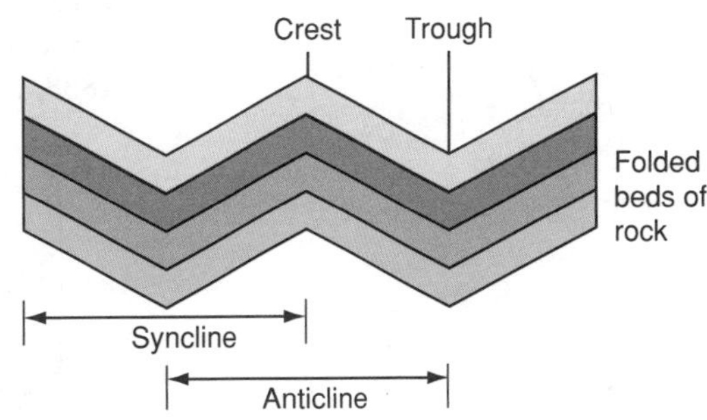


(c) The formation of a fold

## Cleavage

Cleavage is the splitting into thin sheets of rock under pressure. Slate is easily broken along **cleavage planes**. Slate is formed by the metamorphism of shales. Clay minerals and flaky minerals, such as mica, are recrystallised to lie perpendicular to the direction of maximum stress. The slate which results has a weakness in one plane along which it can be easily broken.

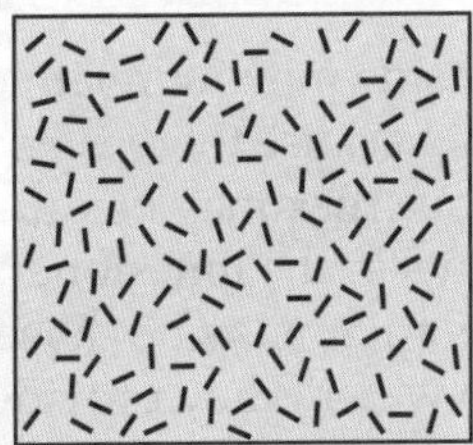


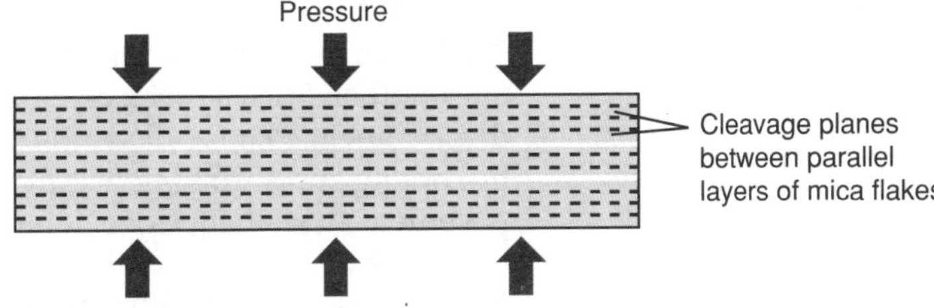


*Figure 19.2B* ▲ The formation of slate

## Faults

A powerful force can break rocks and cause faults (breaks) in the rock. A break allows rocks on one side of a fault to move against rocks on the other side of the fault. As long as forces keep acting, rocks will continue to keep moving against each other along the line of the fault. Since there is friction between the edges of rocks, the movement takes place in jerks. Each of these jerks may cause an earthquake.

Figure 19.2C ▲ (a) A vertical fault

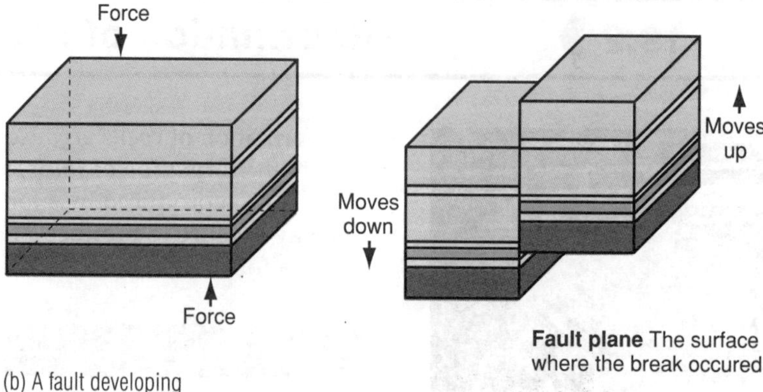


(b) A fault developing

Forces acting within the crust deform rocks.

The result may be

- a fold (an anticline or a syncline)
- a fault
- a cleavage
- a joint.

Check that you know what each of these terms means.

In Figure 19.2C the faults are vertical displacements. Horizontal displacements occur at faults where two of the Earth's plates slide past each other. The San Andreas fault which lies beneath California in the USA is of this type. There were severe earthquakes in California in 1838, 1906 and 1989.

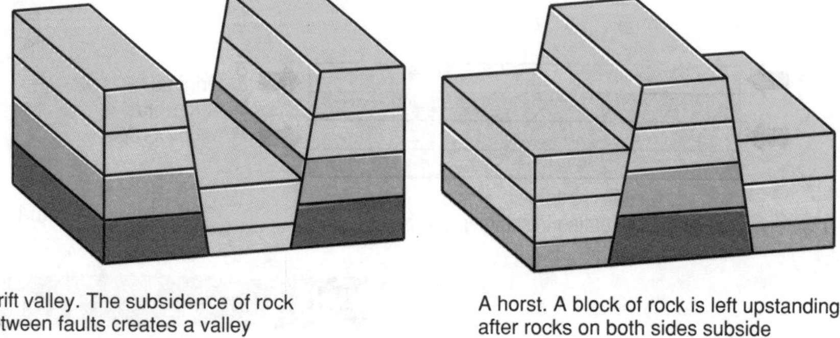

A rift valley. The subsidence of rock between faults creates a valley
A horst. A block of rock is left upstanding after rocks on both sides subside


Figure 19.2D ▲ (a) A rift valley. The valley has subsided between faults. It may become the course of a river or the site of a lake. (b) A horst left standing when rocks on both sides subside

## It's a fact!

Ever since 'the trembler' of 1906, California has been waiting for 'the big one'. The earthquake of 1906 registered 8.3 on the Richter scale and flattened three-quarters of San Francisco leaving 700 dead. Scientists predicted that there would be another earthquake. They predicted the site and the size of the future earthquake but could not predict the date. In 1989 an earthquake, measuring 6.9 on the Richter scale hit San Francisco killing 300 people. This time the damage was limited because engineers had designed buildings to withstand an earthquake. Scientists predict that another earthquake, of magnitude 8, is still to come.

## Joints

A fracture may occur without the rocks on either side moving relative to one another. Such a break cannot be called a fault because there is no displacement so it is called a **joint**. Joints in igneous rock may be caused by cooling and shrinking. Figure 19.2F shows joints in the limestone pavement at Malham Cove. Joints in sedimentary rocks may be caused by loss of water. Joints make a rock permeable to water. They provide weaknesses which may be affected by weathering.

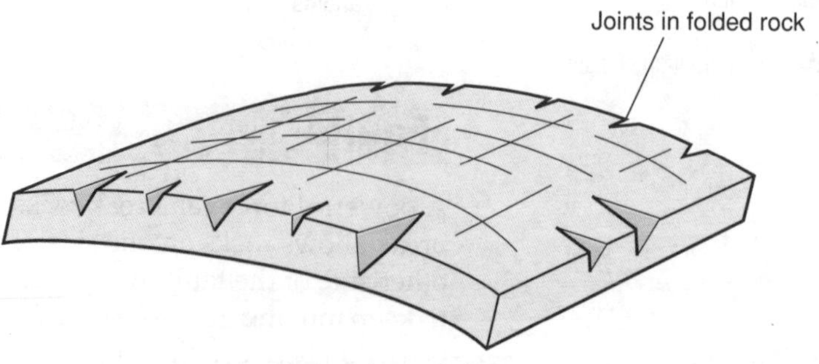


Figure 19.2E ▲ Joints in folded rock

## SUMMARY

Sedimentary rock is laid down in horizontal layers. These may be deformed by:

- folding, with the formation of synclines and anticlines
- cleavage, splitting into thin sheets
- faulting, developing breaks which allow slabs of rock to slide against one another
- jointing, developing breaks without any movement of rock.

*Figure 19.2F* ▲ Limestone pavement at Malham Cove

## CHECKPOINT

▸ 1 The figure shows a section through some layers of rock.

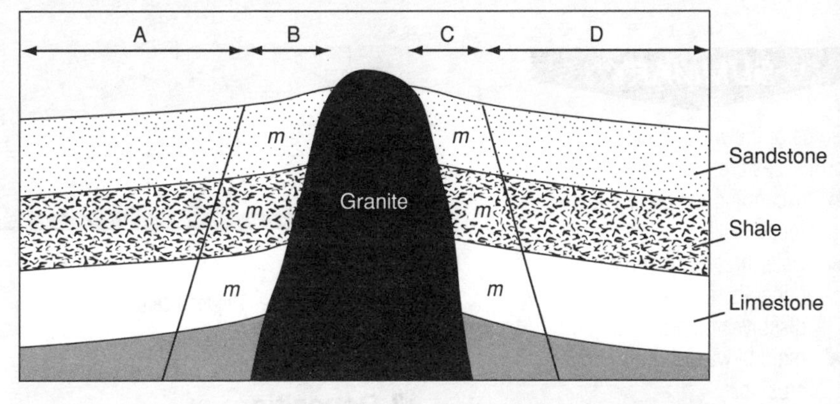


(a) Explain the statement: *Limestone, shale and sandstone are **sedimentary** rocks.*

(b) What type of rock is granite?

(c) Explain how the granite could have pushed through the layers of sedimentary rock.

(d) Explain the statement: *The layers of sedimentary rock in regions B and C have been **metamorphosed**.*

(e) Why have the sedimentary rocks at *A* and *D* not been metamorphosed?

# 19.3 ▸ The forces which shape landscapes

## FIRST THOUGHTS

What forces shaped the varied landscape that we see around us? What forces pushed up the mountain ranges, smoothed the plains and carved out the valleys?

Rocks are continually being broken down into smaller particles by forces in the environment. These processes are called **weathering.** Weathering may be brought about by:

- **physical forces**, especially in deserts and high mountains,
- **chemical reactions**, especially in warm, wet climates.

## Rain

Water is an important weathering agent. Water expands when it freezes. If water enters a crack in a rock and then freezes, it will force the crack to open wider. When the ice thaws, water will penetrate further into the rock. After cycles of freeze and thaw, pieces of rock will break off.

Water reacts with some minerals, like mica. The reaction produces tiny particles which are easily transported away and deposited as a sediment of mud or clay.

EXTENSION FILE
ASSIGNMENT

## Rivers and streams

Rivers and streams carry water back to the oceans as part of the **water cycle**. A fast-flowing stream can carry a lot of particles in suspension (see Figure 19.3A). A very fast stream can push sand and pebbles along with it.

Running water causes erosion. The bed load and the suspension load rub against the bed and sides of the river channel. In addition, there are chemical reactions between water and rocks.

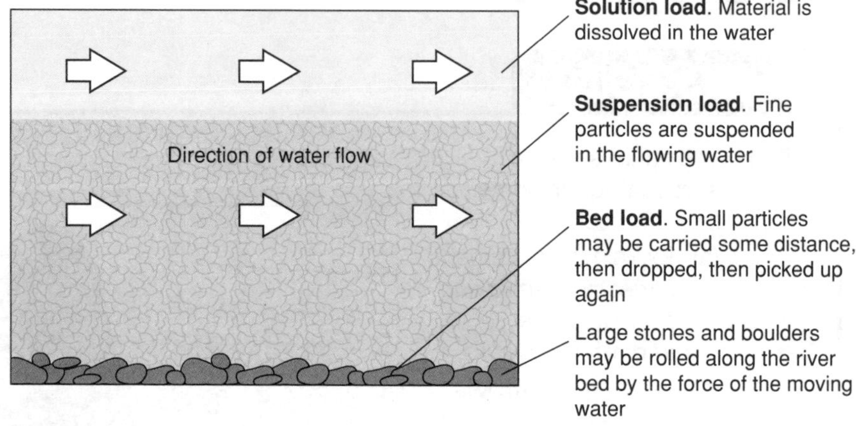
Direction of water flow
**Solution load**. Material is dissolved in the water
**Suspension load**. Fine particles are suspended in the flowing water
**Bed load**. Small particles may be carried some distance, then dropped, then picked up again
Large stones and boulders may be rolled along the river bed by the force of the moving water

*Figure 19.3A* ▲ The load carried by a stream

### SUMMARY

Some of the weathering processes are:

- expanding ice breaks up rock
- water breaks up some minerals, e.g. mica, and dissolves others
- running water brings about erosion.

Sediments are deposited when rivers lose speed: on the inside curve of river bends and where rivers flow into lakes and seas.

### Deposition

Rivers deposit the sand and gravel which they carry. The deposits form when rivers lose speed:

- on the inside curves of river bends,
- when a river flows into the sea or a lake.

## Underground water

Water can pass through certain kinds of rocks. It can seep through joints in beds of sedimentary rock. Such rocks, like limestone, are described as **permeable**. Other rocks, like sandstone, have tiny spaces between their mineral grains which allow water to enter. These rocks are **porous**.

Water held in rocks below the surface is called **ground water**. Some of it finds a way back to the surface to become a **spring**. Some of it remains as an underground reserve. About one third of the UK water supply is drawn from ground water.

Rain water causes weathering of rocks. This happens also below ground. Rain water permeates the ground and dissolves some minerals; gypsum is dissolved by rain. The acid in rain water reacts with limestone to form soluble compounds. Small openings in the rock become wider and in time form large underground passages which carry underground rivers and streams. When you visit **underground caves** you walk along dried-up river beds.

### SUMMARY

Underground water reacts with limestone and other rocks to form soluble compounds. In time, underground caverns are formed.

## The sea

Erosion by the sea is illustrated in Figure 19.3B The cave has been gouged out by waves at a weak point in the cliff, e.g. a joint. The arch has been formed by two caves meeting back to back. The sea stack has been left where the top of an arch has fallen away.

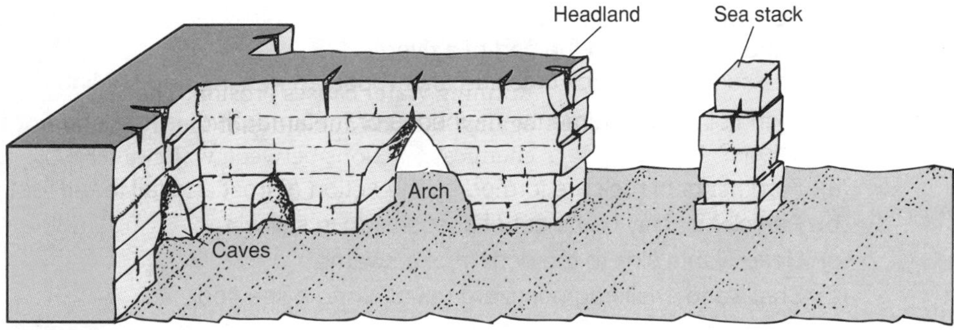


*Figure 19.3B* ▲ Landscaping by the sea

## Glaciers

*Figure 19.3C* ▲ Glacier

In some regions the temperature is low enough for snow to exist all the year round. As layers of snow build up, the lower levels become compressed into a mass of solid ice. When the ice begins to move under the influence of gravity, a **glacier** develops. Glaciers move slowly downhill, usually at less than 1 m/day. As glacial ice moves over a land surface, it wears down rocks by:

- **plucking**, freezing round pieces of rock and carrying them along,
- **grinding**, wearing down the rocks over which it is moving by means of the sharp rocks which become attached to the bottom of the glacier.

When a glacier melts, all the material which it carries is deposited.

### SUMMARY

Forces which mould the landscape are:

- Rain, causing the formation of cracks in rock when it freezes
- Rivers and streams, transporting and depositing rock fragments
- Underground water, dissolving soluble minerals, e.g. gypsum, and reacting with limestone and other rocks to form soluble substances
- The sea, eroding rocks
- Glaciers, plucking rocks from the landscape and grinding land surfaces
- Wind, eroding sand and soil, especially in dry regions.

## Wind

In dry desert regions, wind is a landscaping agent, for example in shaping sand dunes. Moisture holds particles of sand and soil together and makes it much more difficult for the wind to remove them. In moist regions, plants grow, and their roots bind the soil, making it much more difficult to erode.

*Figure 19.3D* ▲ Sand dunes

## CHECKPOINT

1 Which of the following are needed for the formation of a glacier?
(a) mountains (b) steep-sided valleys (c) heavy snowfall (d) heavy rainfall (e) low temperatures

2 Which of the following are examples of (a) erosion and (b) weathering?
(i) Waves breaking against a cliff.
(ii) Rocks splitting after a cold winter.
(iii) Soil carried by the wind.
(iv) Sand carried along the bed of a river.
(v) The surface of a rock cracking after repeated heating and cooling.

3 What kind of rock (sedimentary, igneous or metamorphic) is formed under each of the following conditions?
(a) Fragments of rock are formed by the action of frost and fall to the foot of a mountain.
(b) Particles of clay come out of suspension in still water.
(c) Dead plants sink to the bottom of a swamp.
(d) Shells and shell fragments are rolled along a sea floor.

## 19.4 Soil

As a result of weathering, rocks break up into small particles. These particles become part of the soil. Soil also contains water and **humus**: matter that was formed by the decay of dead plants and animals. Soils differ in the type of rock from which they were formed, the size of the particles, the amounts of water, salts and humus which they contain and the pH. The way in which the size of particles affects the properties of soils is shown in Figure 19.4A and Table 19.2. Most soils are mixtures of particles of different sizes and have properties in between those of sandy soils and clay soils. Such a mixture is called **loam**.

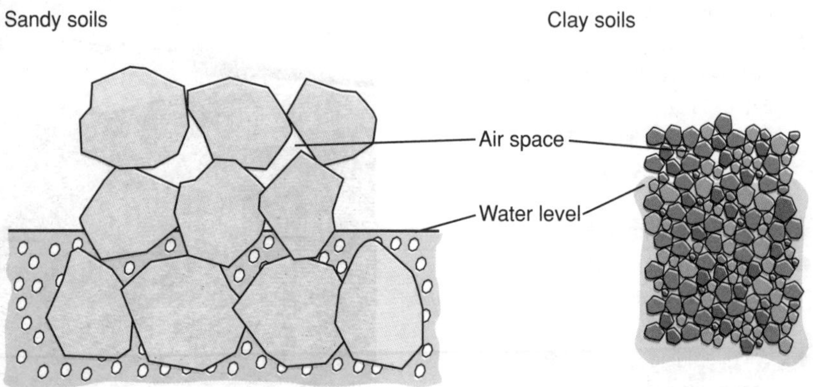


*Figure 19.4A* ▲ Clay soils and sandy soils

Clay soils and sandy soils both have advantages. The table lists some of them.

*Table 19.2* ▼ The properties of sandy soils and clay soils

| | ***Sandy soils*** | ***Clay soils*** |
|---|---|---|
| *Particle size* | Large, >0.2 mm | Small, <0.002 mm |
| *Air spaces* | Large | Small |
| *Drainage* | Rapid, leaving a dry soil | Slow, leaving a wet soil |
| *Temperature* | Fluctuates, tending to be higher | More consistent, tending to be lower |
| *Cultivation* | Easy to dig and plough because they are dry and loose | Difficult to dig and plough because they are wet and sticky |
| *For plant growth ...* | Plants may suffer from lack of water. Minerals may be leached (washed) from the soil by rain | Plant roots may lack oxygen if the soil becomes waterlogged. The mineral content is high since minerals tend to stick to clay particles. |

Humus improves soils by the following means:

- It improves the texture of clay soils by helping the particles to stick together to form crumbs.
- It improves the texture of sandy soils by increasing the soil's ability to hold water.
- Humus reduces the leaching of minerals.
- By absorbing water, humus makes soil more fertile. This is especially important in sandy soils.
- Humus provides food for detritus feeders, e.g. woodlice and earthworms, which in turn fertilise the soil.
- Its high water content enables humus to absorb heat and warm the soil.

**SUMMARY**

Soil contains small particles of rock, water, humus and salts. Sandy soils have larger particles than clay soils. Loams are mixtures of sandy soils and clay soils. Humus improves all soils.

## 19.5 ▶ The geological time scale

**FIRST THOUGHTS**

Geologists can date the different layers of rocks which they unearth. Inspecting the fossils which they find helps, and the radioactivity of the rock tells a story.

Figure 19.5A shows the **geological column**. It divides Earth's history (4600 million years) into **eras**. Each era is divided into **periods**. Human life evolved in the Quaternary period. When geologists describe a rock as being of the Silurian age, they mean that the rock was formed between 435 and 395 million years ago. It was during the Pre-Cambrian period that the Earth's crust solidified, oceans and atmospheres developed, and the first living organisms appeared.

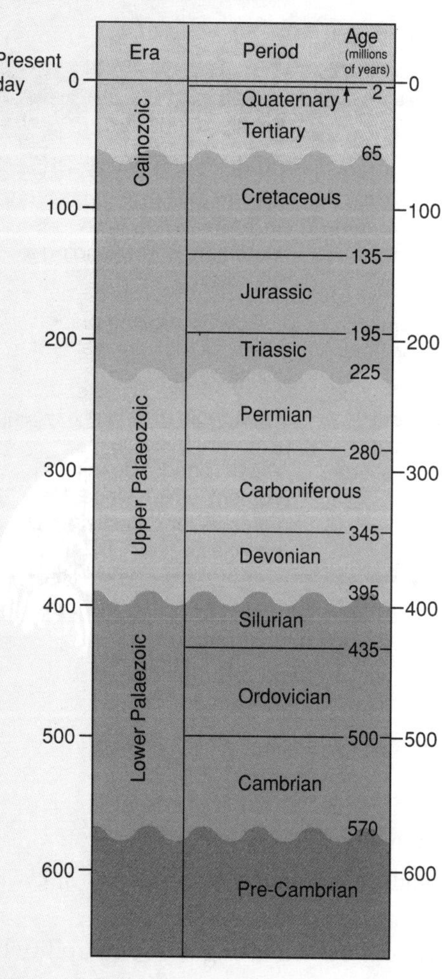


Figure 19.5A ▲ The geological column

## Fossils

Geologists are able to say what period a rock dates from by examining the fossils that the rock contains. Fossils are the preserved remains of, or marks made by, dead plants and animals (see Figure 19.5B). If a rock contains the imprints of the shells of creatures known to have been living 300 million years ago, the rock may be Carboniferous.

Figure 19.5B ▲ Carboniferous fossils

## Relative dating

Relative dating does not give the age of rocks but enables you to classify them (arrange them) in order of age. If one sedimentary rock lies above another, it is very likely that the upper rock is younger than the lower one (although folding of the rocks can reverse the order). Fragments of rock included in another rock must be older than the rock that surrounds them.

## Dating from radioactivity

Some elements are **radioactive**. They have unstable atoms which split up (decay) to form atoms of stable elements. Imagine that a rock contains the radioactive element A, which decays very slowly to form element B. Then the ratio of B to A in the rock increases as the years go by. A measurement of the ratio of B to A will give the length of time for which A has been decaying, that is, the age of the rock. It is by radioactive dating that the age of Earth has been established as 4600 million years.

EXTENSION FILE
ASSIGNMENT

## CHECKPOINT

1 A geologist made a sketch of the beds of rock in a quarry (see the figure below). He tabulated the fossils which he found in the four layers of rock.

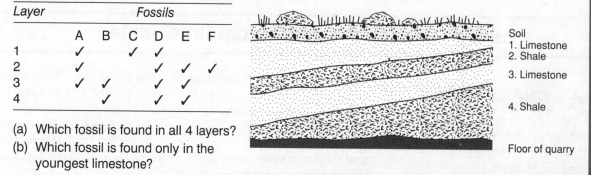

| *Layer* | *Fossils* | | | | | |
|---|---|---|---|---|---|---|
| | A | B | C | D | E | F |
| 1 | ✓ | | ✓ | ✓ | | |
| 2 | ✓ | | | ✓ | ✓ | ✓ |
| 3 | ✓ | ✓ | | ✓ | ✓ | |
| 4 | | ✓ | | ✓ | ✓ | |

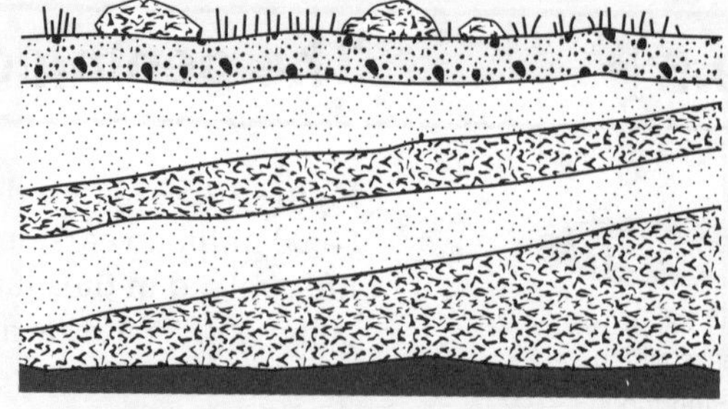


(a) Which fossil is found in all 4 layers?

(b) Which fossil is found only in the youngest limestone?

(c) Which fossil can be used to give the age of layer 2? Explain your answer.

# Theme Questions

1 (a) What happens to the temperature of a gas if the gas is compressed suddenly?

(b) What happens to the temperature of a gas if the gas is allowed to expand suddenly? How is this effect used to liquefy air? Why is this method chosen for the liquefaction of air?

(c) Boiling points are nitrogen, −196 °C; oxygen, −183 °C.

Explain how the difference in boiling points makes it possible to separate oxygen and nitrogen from liquid air.

(d) Give one-large scale industrial use for (i) oxygen (ii) nitrogen and (iii) another gas which is obtained from liquid air. Say why each gas is chosen for that particular use.

2 Three of the gases in air dissolve in water.

(a) Which of them dissolves to give an acidic solution? What use is made of this solution?

(b) Which of the three gases is a nuisance to deep-sea divers? Explain why, and say how the problem has been solved.

(c) The life processes of plants and animals depend on the solubility of two of these gases. Explain why.

3 The table shows the composition of inhaled air and exhaled air, excluding water vapour, and a comment on the content of water vapour.

| | *Percentage by volume* | |
|---|---|---|
| | *Inhaled air* | *Exhaled air* |
| Oxygen | 21 | 17 |
| Nitrogen | 78 | 78 |
| Carbon dioxide | 0.03 | 4 |
| Noble gases | 1 | 1 |
| Water vapour | Variable | Saturated |

(a) Describe the differences between inhaled air and exhaled air.

(b) Briefly give the cause of each of these differences.

4 A classroom contains 36 pupils. The doors and windows are closed for half an hour. Answer these questions about the air at the end of the half hour.

(a) Will the air temperature be higher or lower? Explain your answer.

(b) Will the air be more or less humid (moist)? Explain your answer.

(c) Will the percentage of carbon dioxide in the air be higher or lower? Explain your answer. Say how the change in carbon dioxide content will affect the class.

5 You are given four gas jars. One contains oxygen, one nitrogen, one carbon dioxide and one hydrogen. Describe how you would find out which is which.

6 The diagram shows an apparatus which is being used to pass a sample of air slowly over heated copper.

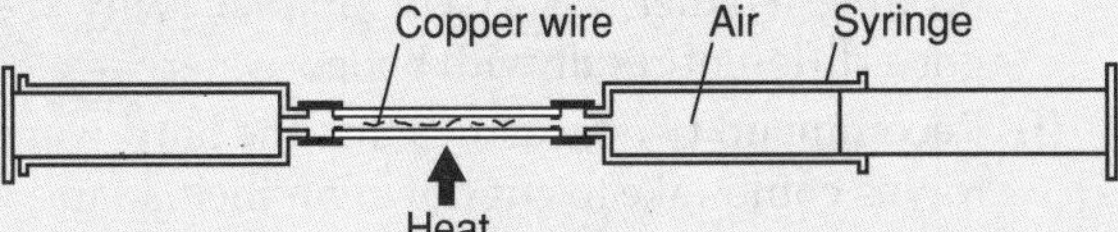


(a) Describe how the appearance of the copper changes.

(b) Which gas is removed from the air by copper? Name the solid product formed.

(c) If 250 $cm^3$ of air are treated in this way, what volume of gas will remain?

(d) Name the chief component of the gas that remains.

(e) Name two other gases that are present in air.

7 When petrol burns in a car engine, carbon dioxide and carbon monoxide are two of the products.

(a) Write the formula of (i) carbon dioxide (ii) carbon monoxide.

(b) Explain the statement 'Carbon monoxide is a product of incomplete combustion.'

(c) Red blood cells contain haemoglobin. What vital job does haemoglobin do in the body?

(d) If people breathe in too much carbon monoxide, it may kill them. How does carbon monoxide cause death?

(e) Explain how people can be poisoned accidentally by carbon monoxide.

(f) What precautions can people take to make sure that carbon monoxide is not formed in their homes?

(g) The blood of people who smoke contains more carbon monoxide than the blood of non-smokers. Can you explain why?

8 Ruth carried out an experiment to compare the hardness of the water from three towns. She measured 50 $cm^3$ of each water sample into separate conical flasks. She added soap solution gradually to each flask, shaking them until a lather was formed. Her results are shown in the table.

| *Water sample* | *Volume of soap solution* ($cm^3$) |
|---|---|
| Distilled water | 2.0 |
| Johnstown water | 7.5 |
| Mansville water | 10.0 |
| Rumchester water | 4.0 |

(a) Say what piece of apparatus Ruth could use for measuring (i) the 50 $cm^3$ samples of water and (ii) the volume of soap solution added.

(b) Explain why she did a test on distilled water.

(c) Why does distilled water require the smallest volume of soap solution to form a lather?

(d) Which town has the hardest water? Explain your answer.

(e) When Ruth boiled a 50 $cm^3$ sample of Rumchester water before testing it, she found that the volume of soap solution needed to produce a lather was 2.0 $cm^3$. Explain why she got a different result with boiled water.

(f) Recommend two measures that the hard water towns could take to cut down on their soap consumption.

**9** The diagram shows rain falling on the ground and trickling over underground rocks.

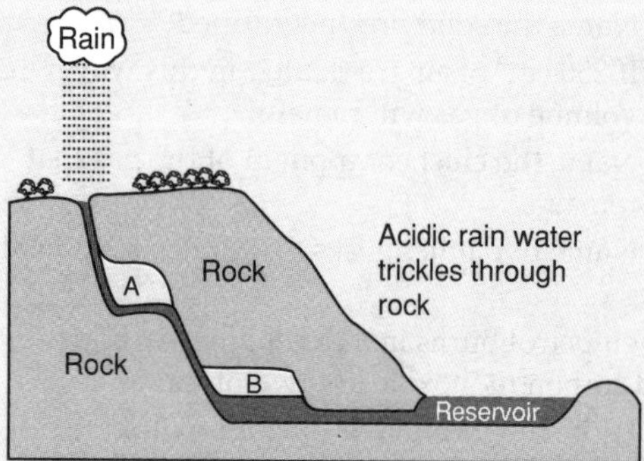


(a) Explain why natural rain water is weakly acidic.

(b) Name a type of rock which will be attacked by acidic rain water.

(c) Name the type of cavity, A, that is formed as a result of the action of rain water on the rock.

(d) Name the type of water that accumulates at B.

(e) Water flows from B into a reservoir. What treatment does the water need before it is fit to drink? Explain why this water receives a different treatment from river water.

(f) Will the water in the reservoir be hard water or soft water? Explain your answer.

(g) State one advantage and one disadvantage of hard water compared with soft water.

**10** (a) What is the difference between raw sewage and digested sludge?

(b) Why can raw sewage not be used as a fertiliser?

(c) The digested sludge which sewage works produce is sold to farmers. Who benefits from this sale?

(d) Why do water treatment works treat drinking water with chlorine?

(e) How and why does the treatment of ground water differ from that of river water?

**11** (a) Name three pollutants that are produced by power stations.

(b) For one of the pollutants, describe the kind of cleaning system that can be used to stop the pollutant being discharged into the air.

(c) How will the cost of electricity be affected by (i) installing the cleaning system and (ii) stocking the chemicals consumed in running the cleaning system?

**12** (a) What is the difference between oxygen and ozone?

(b) What converts oxygen into ozone?

(c) What converts ozone into oxygen?

(d) What is the ozone layer? Where is it?

(e) Why is the ozone layer becoming thinner?

(f) Why does the decrease in the ozone layer make people worry?

**13** The nitrogen monoxide content of the atmosphere is $5 \times 10^6$ tonnes. One supersonic transport (SST) flies on average 2500 hours per year and emits 3 tonnes of nitrogen monoxide per hour.

(a) How much nitrogen monoxide will be emitted in one year by a fleet of 50 SSTs?

(b) Calculate the ratio

$$\frac{\text{Nitrogen monoxide emission by fleet of SSTs}}{\text{Quantity of nitrogen monoxide already in the atmosphere}}$$

**14** Cleopatra's needle has corroded more in London since 1878 than it did in 30 centuries in the Egyptian desert. Can you explain this?

**15** The diagram shows acid rain falling on the shores of three lakes.

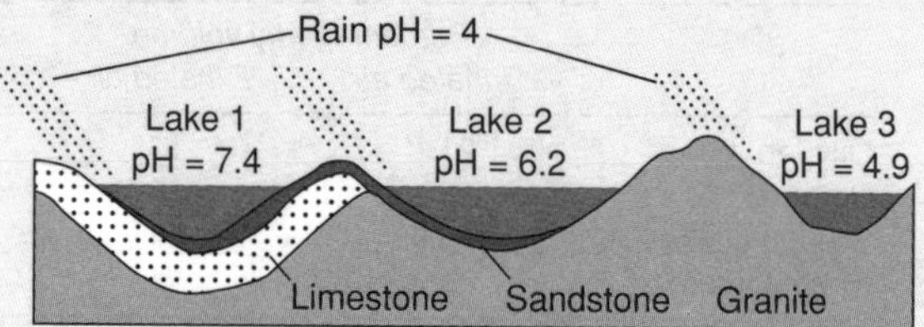


(a) Unpolluted rain water has a pH of 6.8. What gives it this weak acidity?

(b) By acid rain, we mean rain with a pH below 5.6. Name two substances that react with rain to make it strongly acidic.

(c) Explain why Lake 3 is more acidic than Lake 2.

(d) Explain why Lake 2 is more acidic than Lake 1.

(e) Lakes in Sweden become more acidic in the spring. Suggest an explanation.

(f) Acidic lakes in Sweden are treated with crushed limestone. Explain how this reduces the acidity. Give two disadvantages of this solution to the problem.

**16** Many industrial plants take water from a river and then return it at a higher temperature. What harm can this do? What name is given to this practice?

**17** Bacteria in river water are able to convert many pollutants into harmless products. What, then, is the harm in dumping waste into rivers?

**18** What happens when plastics are dumped in lakes and rivers?

**19** It takes 60 g of oxygen per day to oxidise the sewage from one person.
Water contains 10 p.p.m. of dissolved oxygen. What mass of water is robbed of its dissolved oxygen by oxidising the sewage from a village of 400 people in one day? (1 tonne = 1000 kg)

**20** Water normally contains 10 p.p.m. of oxygen. It takes 4 g of oxygen to oxidise completely 1 g of oil. A garage does an oil change on a car, and pours the 4 kg of dirty oil down the drain. What mass of water will be stripped of its dissolved oxygen by oxidising the oil from this car?

**21** Why is the pollution of estuaries so common? Who gains from being able to pollute estuaries? Who loses from this pollution?

**22** A river carries material along with it, as large particles, as small particles and in solution.
(a) How does the river transport large particles, e.g. pebbles?
(b) How does the river transport fine particles, e.g. clay?
(c) When the river flows into a lake, what happens to (i) the large particles and (ii) the fine particles?
(d) Explain how (c) leads to the formation of sedimentary rocks.

**23** The table below shows the substances present in the atmosphere at various stages in the history of the Earth.

| *Time* | *Atmosphere* |
|---|---|
| 4000 million years ago | No atmosphere |
| 3800 million years ago | Carbon dioxide, steam, hydrogen, small amounts of methane and ammonia |
| 2700 million years ago | Carbon dioxide begins to decrease |
| 2400 million years ago | Oxygen starts to appear |
| 600 million years ago | Larger amounts of oxygen present |

(a) What was the origin of the gases present in the atmosphere 3800 million years ago?
(b) Using the table, suggest, with a reason, an approximate period when the oceans formed.
(c) The first life on Earth may have consisted of very simple molecules that could reproduce. These molecules may have been formed by the action of lightning and ultraviolet radiation on substances in the Earth's atmosphere.
Using the table, suggest the four elements which were most likely to be present in these first simple molecules.
(d) The amount of oxygen gas in the atmosphere increased as plants began to evolve.
(i) Write a balanced equation to show the overall process by which oxygen is formed by plants.
(ii) State the type of energy change involved in the reaction when plants form oxygen.
(iii) What is the function of chlorophyll in this reaction?
(iv) Explain the overall energy change in terms of breaking and making chemical bonds.
(ULEAC)

**24** The theory of plate tectonics suggests that: the crust of the Earth is made up of plates; the plates move at various speeds and in various directions.
(a) Give **three** features that may be observed at the plate boundaries.
(b) What causes the Earth's plates to move? Explain your answer.
(c) Geologists studied the rocks either side of the plate boundary in the middle of the Atlantic. They found that the rocks had an interesting magnetic pattern.

White stripes: reversely magnetised (S)
Black stripes: normally magnetised (N)

This magnetic pattern gave good evidence for the theory that the continents do move apart. Explain how. (SEG)

**25** (a) It is suggested that, millions of years ago, South America and Africa were part of the same land mass.
(i) What name is given to the phenomenon of the separation of a land mass?
(ii) Look at the diagram. What evidence is there that separation of a land mass may have occurred?

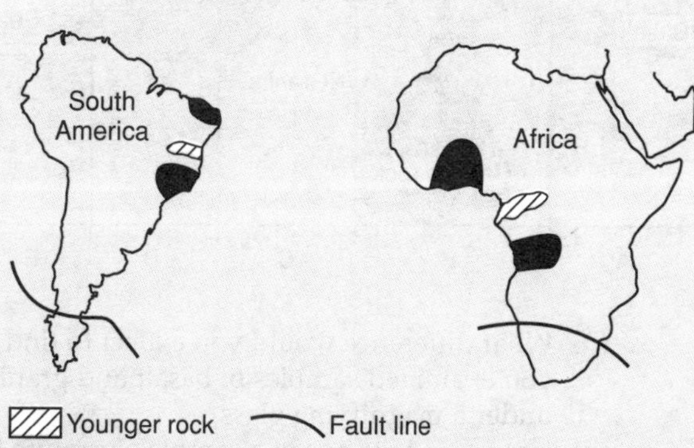

(iii) Fossil analysis in South America and Africa produces the following results.

| *Period* | *When present /millions of years ago* | *Fossils found in South America* | *Africa* |
|---|---|---|---|
| Quaternary | 0–2 | llamas<br>no camels | no llamas<br>camels |
| Tertiary | 2–65 | no horses | horses |
| Cretaceous | 65–135 | dinosaurs | dinosaurs |
| Permian | 225–280 | mesosauri<br>lizards | mesosauri<br>lizards |
| Carboniferous | 280–345 | palm trees | palm trees |

From the data, suggest when separation of South America from Africa may have occurred, and give your reason.

(b) What kind of information does the magnetic record of rocks give?

(c) The diagram shows **magma** creating a mid-ocean ridge.

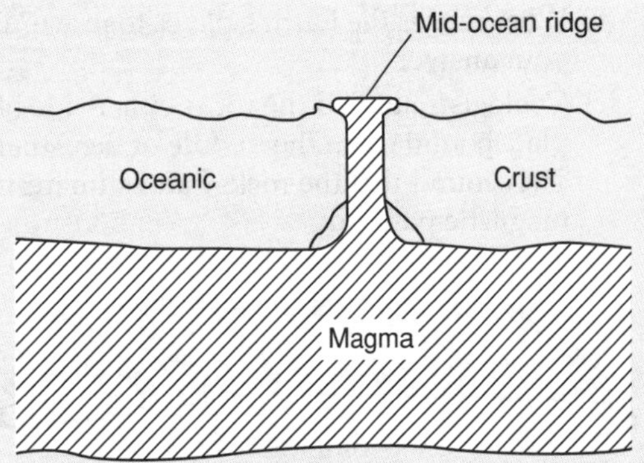


(i) What is meant by the term magma?

(ii) Explain how mid-ocean ridges are linked with movement of land masses such as South America and Africa. (MEG)

**26** (a) The diagram below represents some of the natural processes at work in forming rocks.

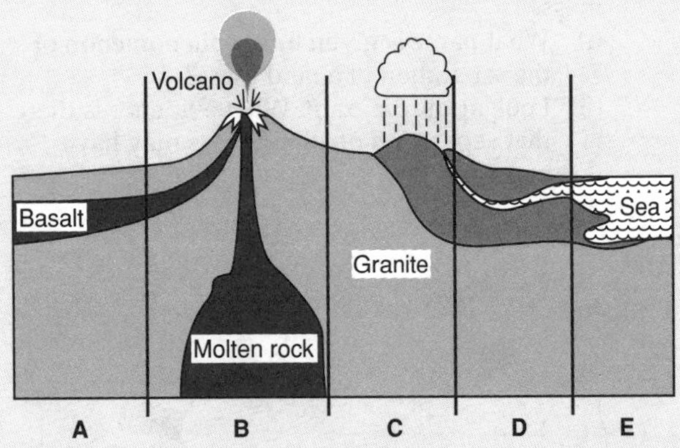


(i) What difference would you expect to find if you examined samples of basalt and granite under a magnifying glass?

(ii) How could the age of a sample of granite be determined?

(b) When sedimentary rocks are subjected to high pressures and temperatures, the minerals in them change into new ones with new structures. The rocks formed are described as metamorphic. This is a continuous process and occurs particularly at the margins of tectonic plates.

(i) Explain what is meant by a tectonic plate.

(ii) In which region, **A**, **B**, **C**, **D** or **E**, of the diagram might there be a plate margin?

(iii) Explain what type of margin you would expect to find there.

(iv) Use the information in this question to explain what is meant by the 'rock cycle' and how plate movements are thought to contribute to it. Use a diagram as part of your explanation. (MEG)

**27** The diagram represents part of the rock cycle.

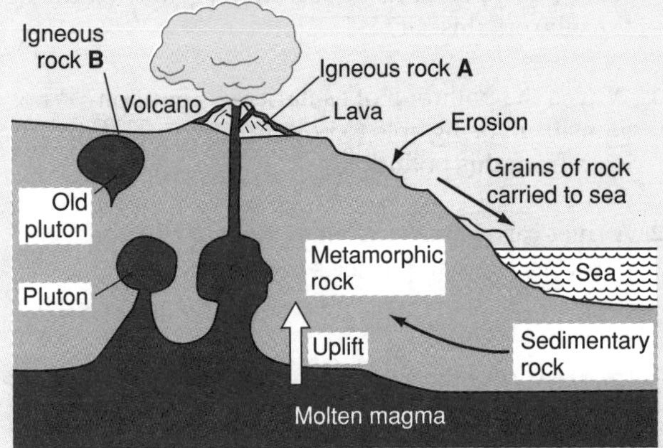


(a) (i) Explain how the grains of eroded rock which are carried into the sea are changed into sedimentary rock.

(ii) Why does chalk, which is a sedimentary rock, often contain fossils?

(iii) Igneous rock **A** contains small crystals of basalt. Why are these crystals small?

(iv) The diagram shows two **plutons**. A pluton is rock formed when molten magma is trapped **before it reaches** the surface. Igneous rock **B** is found in a pluton and contains large crystals of granite. Why are these crystals large?

(b) What are the two conditions required to convert sedimentary rock into metamorphic rock?

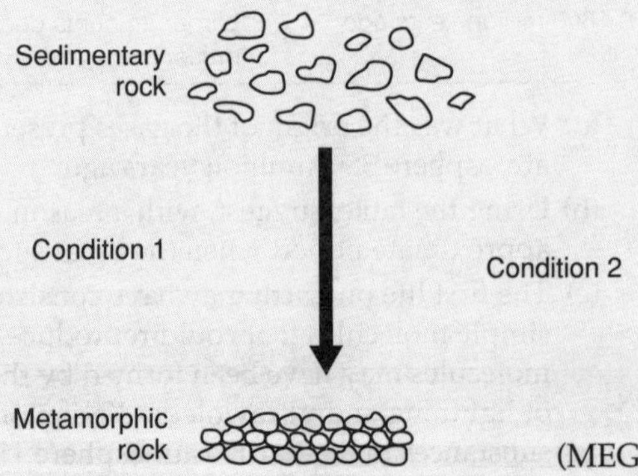


(MEG)

**28** Coal is a complex mixture of carbon and hydrocarbons. It also contains sulphur.

Carbon dioxide, water vapour, sulphur dioxide and dust particles are formed when coal burns.

(a) Five years after a coal-burning factory was built, the stonework on a building nearby started to crumble.

Suggest how the factory might have caused this.

(b) Sulphur dioxide needs to be removed. The waste gases pass through a mixture of powdered calcium carbonate and water.

Why would you expect a reaction to take place between dissolved sulphur dioxide and calcium carbonate?

(c) Some power stations fit electrostatic precipitators so that dust particles do not escape into the atmosphere.

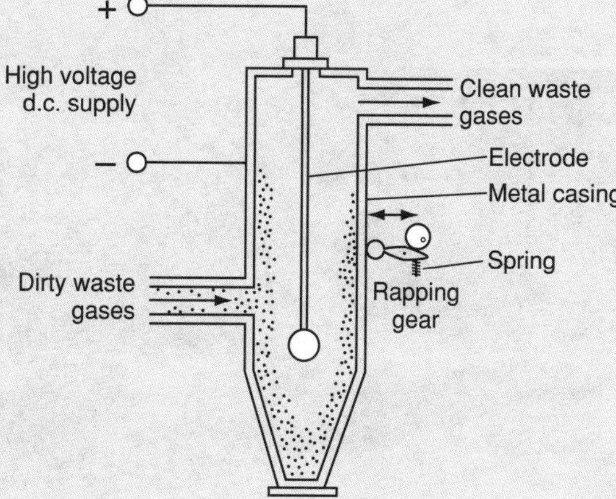


Use the diagram to help explain how electrostatic precipitation removes dust particles from the dirty waste gases.

(d) Hydrocarbons such as methane may contribute to global warming.

The diagram below shows the direction of some energy transfers in the atmosphere.

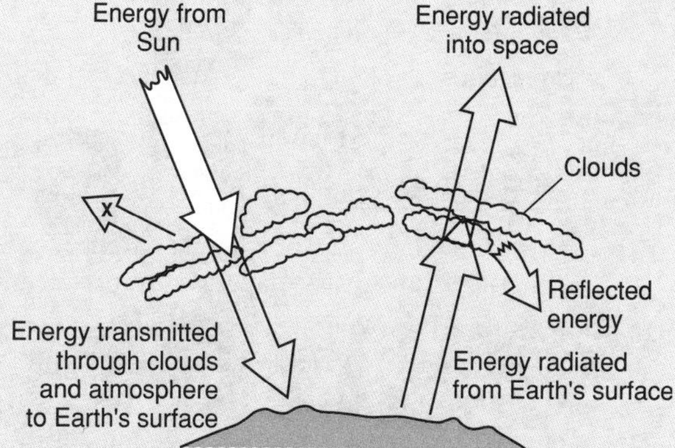


(i) The label for arrow **X** is missing from the diagram. Suggest what the label should be.

(ii) What is the most common type of radiation leaving the Earth's surface?

(iii) In which region of the electromagnetic spectrum must methane absorb radiation when it behaves as a 'greenhouse gas'?

(iv) Use the diagram to explain why an increased concentration of methane in the atmosphere might contribute to increased global warming.

(v) Suggest and explain what may happen to the temperature of the lower atmosphere if the cloud cover over the whole Earth is reduced. (MEG(N))

## Theme E

# Using Earth's resources

The Earth's crust and atmosphere supply us with many different materials. The study of the way materials behave is the science of chemistry. Chemists devise ways of changing the substances which are found in nature into new substances. Rocks in the Earth's crust supply us with metal ores, and chemists have worked out ways of extracting metals from these ores. The structure and chemical reactions of metals explain why metals play such a unique role in the daily life of a civilised society. Metals can be combined in chemical cells which supply us with power.

Rocks are also the source of the essential building materials limestone, concrete and glass. Rock salt is the starting point in the manufacture of many important chemicals.

Nitrogen from the air and natural gas are the starting materials in the manufacture of fertilisers.

# Topic 20 Metals

## 20.1 Metals and alloys

FIRST THOUGHTS

Metals and alloys have played an important part in history. The discovery of bronze made it possible for the human race to advance out of the Stone Age into the Bronze Age. Centuries later, smiths found out how to extract iron from rocks, and the Iron Age was born. In the nineteenth century, the invention of steel made the Industrial Revolution possible.

Check that you know the difference between a metal and an alloy.

Metals are strong materials. They are used for purposes where strength is required. Metals can be worked into complicated shapes. They can be ground to take a cutting edge. They conduct heat and electricity. As science and technology advance, metals are put to work for more and more purposes.

### Alloys

An alloy is a mixture of metallic elements and in some cases non-metallic elements also. Many metallic elements are not strong enough to be used for the manufacture of machines and vehicles which will have to withstand stress. Alloying a metallic element with another element is a way of increasing its strength. Steel is an alloy of iron with carbon and often other metallic elements. Duralumin is an alloy of aluminium with copper and magnesium. This alloy is much stronger than pure aluminium and is used for aircraft manufacture (Figure 20.1A).

Alloys have different properties from the elements of which they are composed. Brass is made from copper and zinc. It has a lower melting point than either of these metals. This makes it easier to cast, that is, to melt and pour into moulds. Brass musical instruments have a more pleasant and sonorous sound than instruments made from either of the two elements copper or zinc.

SUMMARY

An alloy is a mixture of metals, with, in some cases, non-metallic elements also. Alloys have different properties from the elements of which they are composed.

The differences between the physical properties of metals and non-metallic elements are summarised in Table 4.1.

*Figure 20.1A* ▲ Concorde

# 20.2 The structure of metals

FIRST THOUGHTS

Metals possess their remarkable and useful properties because of the type of chemical bond between the atoms: the metallic bond.

## The metallic bond

A piece of metal consists of positive metal ions and free electrons (see Figure 20.2A). The free electrons are the outermost electrons which break free when the metal atoms form ions. Free electrons move about between the metal ions. This is what prevents the metal ions from being driven apart by repulsion between their positive charges.

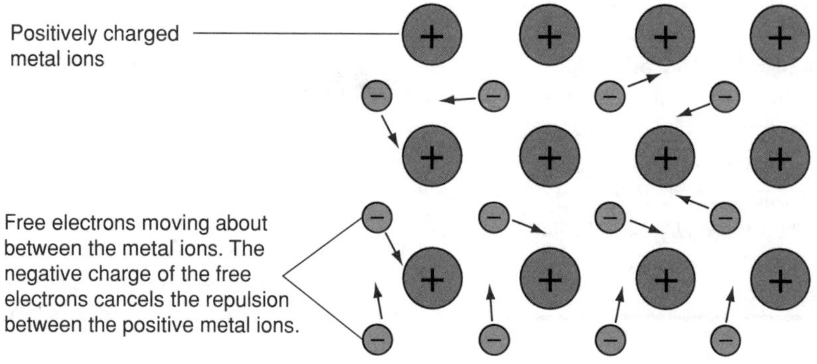

Positively charged metal ions
Free electrons moving about between the metal ions. The negative charge of the free electrons cancels the repulsion between the positive metal ions.

*Figure 20.2A* The metallic bond

SUMMARY

The metallic bond is a strong bond. It gives metals their strength and allows them to conduct electricity.

The metallic bond explains how metals **conduct electricity**. It also explains how metals can change their shape without breaking. When a metal is bent, the shape changes but the free electrons continue to hold the metal ions together.

## The crystalline structure of metals

### Grains

When a metal solidifies from the molten state, many crystals start to grow at the same time in the liquid, and they produce a solid mass of small crystals (see Figure 20.2B). These individual crystals are called '**grains**' by metallurgists. Each grain grows independently, and the directions in which the atoms are arranged are not related to those in neighbouring grains (Figure 20.2B). Between the grains are atoms which have not fitted into the crystal structure. These are often atoms of impurities. The crystal structure of the metal grains is imperfect. There are faults in the orderly arrangement of atoms called **dislocations** (see Figure 20.2C).

Metals solidify as a mass of small crystals called **grains**. There are boundaries between the grains.

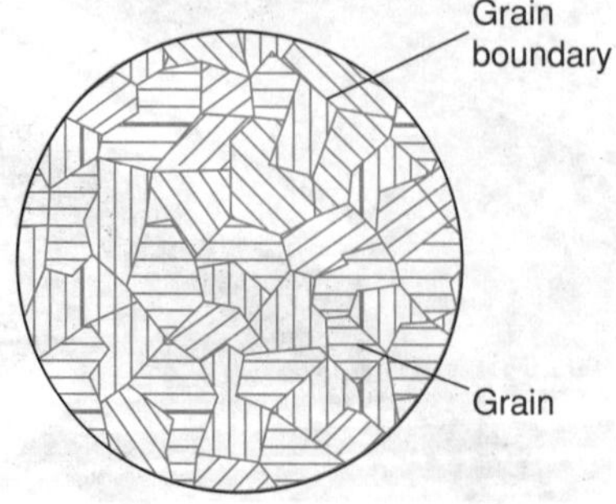


*Figure 20.2B* Metal grains

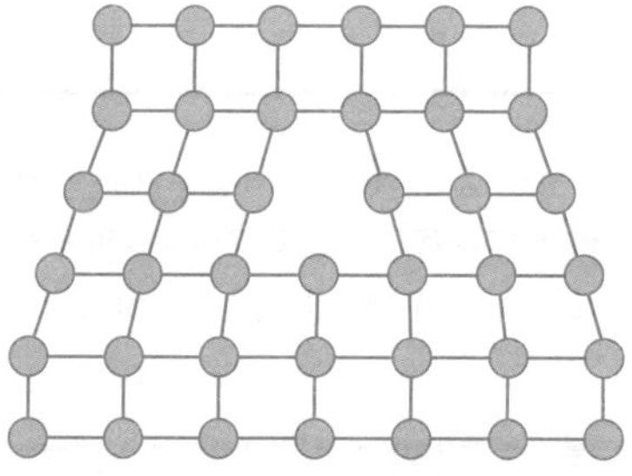

Figure 20.2C ▲ A dislocation

The presence of dislocations causes weakness. When the structure is stressed, the dislocation may be forced to take up another position. As the process is repeated, the dislocation may travel along to the crystal boundary (Figure 20.2D). The mobility of dislocations explains the malleability of metals.

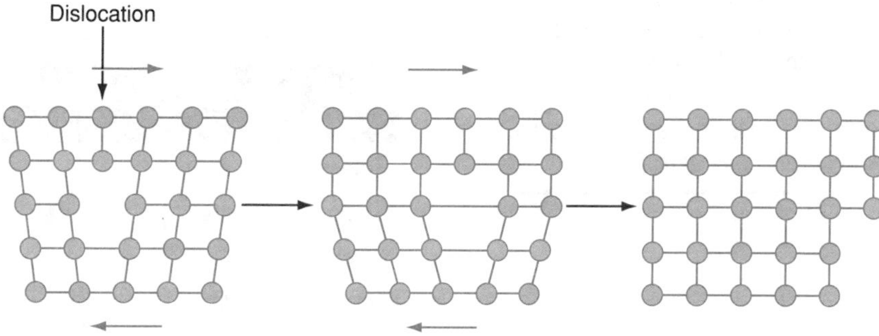
Dislocation

Figure 20.2D ▲ Dislocations tend to move towards a grain boundary

Within the grains are imperfections: dislocations. They make metals much more malleable than they would be otherwise.

Other imperfections in the structure are caused by the presence of extra atoms and by atoms being missing from a particular position (see Figure 20.2E).

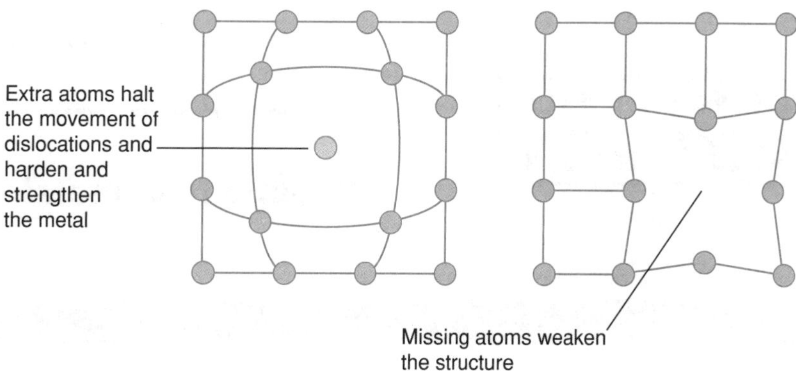
Extra atoms halt the movement of dislocations and harden and strengthen the metal
Missing atoms weaken the structure

Figure 20.2E ▲ Imperfections in the structure of grains

Alloying is used to harden a metal by providing additional atoms which will be trapped between the grains. The greater the number of grains, the stronger is the metal. Therefore 'grain refining agents' are sometimes added to molten metal to provide nuclei from which crystals grow to produce 'fine-grained' metal, which contains a large number of small grains.

## SUMMARY

A metal consists of a mass of tiny crystals called grains. Faults in the crystal structure of a grain are called dislocations. The presence of dislocations weakens a metal. The mobility of dislocations under stress gives metals their ability to change shape without breaking. A piece of metal with a large number of small grains is stronger than a piece with a smaller number of larger grains.

## CHECKPOINT

1 (a) Why is it appropriate to describe a metal as a polycrystalline material?
(b) How does it come about that a molten metal crystallises in this way?

2 Explain what a dislocation is and how the malleability of a metal depends on the presence of dislocations.

# 20.3 The chemical reactions of metals

## FIRST THOUGHTS

You have already met the reaction between metals and acids in Topic 9.1 and the reaction between metals and air in Topic 14.4. In this section, you will find out how to classify metals on the basis of the vigour of their chemical reactions.

## Reactions of metals with air

Many metals react with the oxygen in the air (see Table 20.1).

*Table 20.1* The reactions between metals and oxygen

| Metal | Symbol | Reaction when heated in air | Reaction with cold air |
|---|---|---|---|
| Potassium | K | Burn in air to form oxides | React slowly with air to form a surface film of the metal oxide. This reaction is called **tarnishing** |
| Sodium | Na | | |
| Calcium | Ca | | |
| Magnesium | Mg | | |
| Aluminium | Al | | |
| Zinc | Zn | | |
| Iron | Fe | | |
| Tin | Sn | When heated in air, these metals form oxides without burning | |
| Lead | Pb | | |
| Copper | Cu | | |
| Silver | Ag | Do not react | Silver tarnishes in air |
| Gold | Au | | Do not react |
| Platinum | Pt | | |

EXTENSION FILE ACTIVITY

## Reactions of metals with water

*Table 20.2* Reactions of metals with cold water and steam

| Metal | Reaction with water |
|---|---|
| **Potassium**<br>The hydrogen that is formed burns<br>Potassium reacts violently with water | A violent reaction occurs. Hydrogen and potassium hydroxide solution, a strong alkali, are formed. The reaction is so exothermic that the hydrogen burns. The flame is coloured lilac by potassium vapour.<br>Potassium + Water → Hydrogen + Potassium hydroxide<br>$2K(s) + 2H_2O(l) \rightarrow H_2(g) + 2KOH(aq)$<br>Potassium is kept under oil to prevent water vapour and oxygen in the air from attacking it. |

*Table 20.2* ▼ Reactions of metals with cold water and steam (continued)

| Metal | Reaction with water |
|---|---|
| **Sodium** 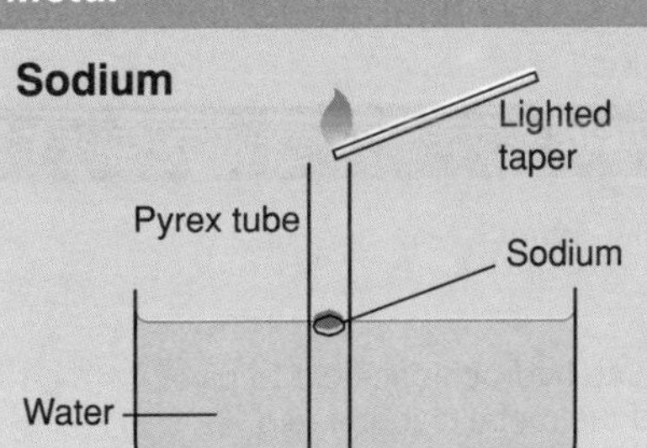  | Reacts slightly less violently than potassium does. Hydrogen and sodium hydroxide solution, a strong alkali, are formed. The hydrogen formed burns with a yellow flame. The flame colour is due to the presence of sodium vapour.<br>Sodium + Water ➔ Hydrogen + Sodium hydroxide<br>$2Na(s) + 2H_2O(l) \rightarrow H_2(g) + 2NaOH(aq)$<br>Sodium is kept under oil to prevent water vapour and oxygen in the air attacking it. |
| **Calcium** 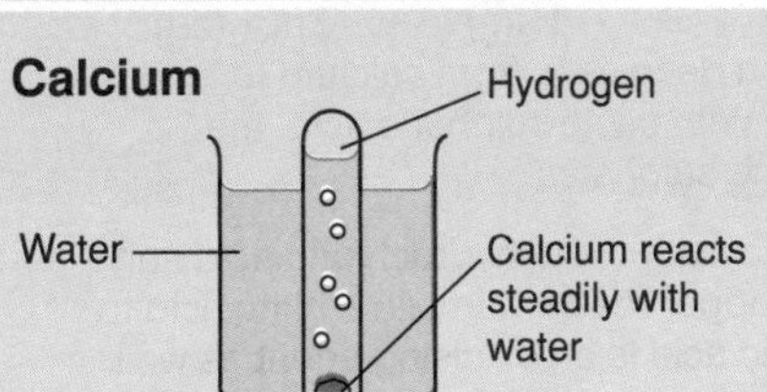  | Reacts readily but not violently with cold water to form hydrogen and calcium hydroxide solution, the alkali limewater.<br>Calcium + Water ➔ Hydrogen + Calcium hydroxide<br>$Ca(s) + 2H_2O(l) \rightarrow H_2(g) + Ca(OH)_2(aq)$ |
| **Magnesium** 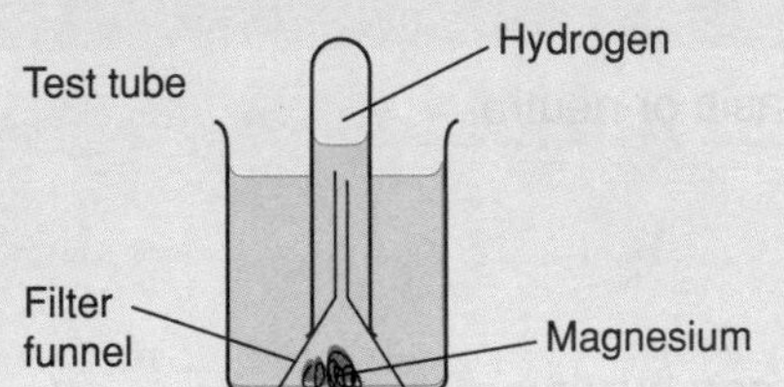  | Reacts slowly with cold water to form hydrogen and magnesium hydroxide.<br>Magnesium + Water ➔ Hydrogen + Magnesium hydroxide<br>$Mg(s) + 2H_2O(l) \rightarrow H_2(g) + Mg(OH)_2(aq)$ |
| 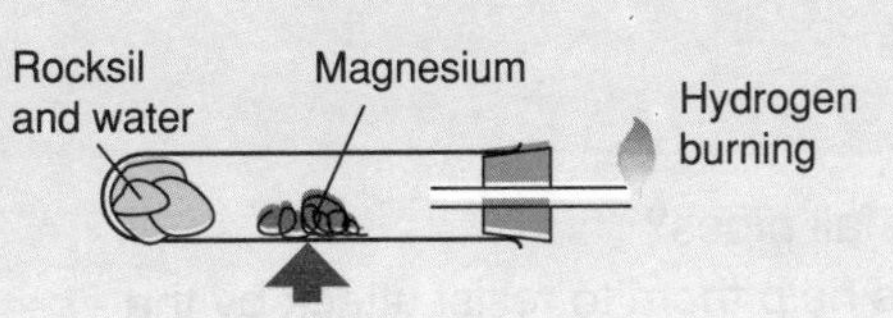  | Burns in steam to form hydrogen and magnesium oxide.<br>Magnesium + Steam ➔ Hydrogen + Magnesium oxide<br>$Mg(s) + H_2O(g) \rightarrow H_2(g) + MgO(s)$ |
| **Aluminium** | Aluminium has a surface layer of aluminium oxide which is unreactive. When the oxide layer is removed, aluminium reacts readily with water to form hydrogen and aluminium oxide. |
| **Zinc** 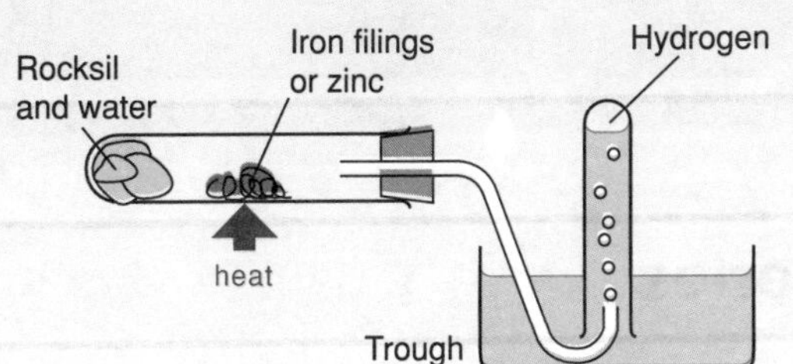  | Reacts with steam to form hydrogen and zinc oxide.<br>Zinc + Steam ➔ Hydrogen + Zinc oxide<br>$Zn(s) + H_2O(g) \rightarrow H_2(g) + ZnO(s)$ |
| **Iron** | Reacts with steam to form hydrogen and the oxide, $Fe_3O_4$, tri-iron tetraoxide, which is blue-black in colour.<br>Iron + Steam ➔ Hydrogen + Tri-iron tetraoxide<br>$3Fe(s) + 4H_2O(g) \rightarrow 4H_2(g) + Fe_3O_4(s)$ |
| **Tin**<br>**Lead**<br>**Copper**<br>**Silver**<br>**Gold**<br>**Platinum** | Do not react |

## Reactions of metals with dilute acids

*Table 20.3* The reactions of metals with dilute acids

| Metal | Reaction with dilute acid |
|---|---|
| Potassium<br>Sodium<br>Lithium | The reaction is dangerously violent<br>**Do not try it** |
| Calcium<br>Magnesium<br>Aluminium<br>Zinc<br>Iron<br>Tin<br>Lead | These metals react with dilute hydrochloric acid to give hydrogen and a solution of the metal chloride, e.g.<br>Zinc + Hydrochloric acid → Hydrogen + Zinc chloride<br>$Zn(s) + 2HCl(aq) \rightarrow H_2(g) + ZnCl_2(aq)$<br>The vigour of the reaction decreases from calcium to lead. Lead reacts very slowly. With dilute sulphuric acid, the metals give hydrogen and sulphates. |
| Copper<br>Silver<br>Gold<br>Platinum | These metals do not react with dilute hydrochloric acid and dilute sulphuric acid. (Copper reacts with dilute nitric acid to form copper nitrate. Nitric acid is an oxidising agent as well as an acid.) |

### CHECKPOINT

See Theme D, Topic 14.5 if you need to revise.

1 State whether the oxides of the following elements are acidic or basic or neutral.
(a) iron
(b) carbon
(c) copper
(d) sulphur
(e) zinc

2 Write (a) word equations (b) balanced chemical equations for the combustion of the following elements in oxygen to form oxides.
(i) zinc
(ii) magnesium
(iii) sodium
(iv) carbon

3 (a) Which are attacked more by acid rain: lead gutters or iron fall pipes?
(b) Food cans made of iron are coated with tin. How does this help them to resist attack by the acids in foods?

4 Write word equations and symbol equations for the reactions between:
(a) magnesium and hydrochloric acid,
(b) iron and hydrochloric acid to form iron(II) chloride,
(c) zinc and sulphuric acid,
(d) iron and sulphuric acid to form iron(II) sulphate.

## 20.4 The reactivity series

### FIRST THOUGHTS

There are over seventy metallic elements. One way of classifying them is to arrange them in a sort of league table, with the most reactive metals at the top of the league and the least reactive metals at the bottom. This section shows how it is done.

There are over 70 metals in the Earth's crust. Table 20.4 summarises the reactions of metals with oxygen, water and acids. You will see that the same metals are the most reactive in the different reactions.

Table 20.4 ▼ Reactions of metals

<table>
<tr><th>Metal</th><th>Reaction when heated in oxygen</th><th>Reaction with cold water</th><th>Reaction with dilute hydrochloric acid</th></tr>
<tr><td>Potassium</td><td rowspan="8">Burn to form oxides</td><td rowspan="4">Displace hydrogen; form alkaline hydroxides</td><td rowspan="3">React dangerously fast</td></tr>
<tr><td>Sodium</td></tr>
<tr><td>Lithium</td></tr>
<tr><td>Calcium</td><td rowspan="5">Displace hydrogen; form metal chlorides</td></tr>
<tr><td>Magnesium</td><td>Slow reaction</td></tr>
<tr><td>Aluminium</td><td rowspan="3">No reaction, except for slow rusting of iron; all react with steam</td></tr>
<tr><td>Zinc</td></tr>
<tr><td>Iron</td></tr>
<tr><td>Tin</td><td rowspan="4">Oxides form slowly without burning</td><td rowspan="6">Do not react, even with steam</td><td rowspan="2">React very slowly</td></tr>
<tr><td>Lead</td></tr>
<tr><td>Copper</td><td rowspan="4">Do not react</td></tr>
<tr><td>Silver</td></tr>
<tr><td>Gold</td><td rowspan="2">Do not react</td></tr>
<tr><td>Platinum</td></tr>
</table>

The metals can be listed in order of **reactivity**, in order of their readiness to take part in chemical reactions. This list is called the **reactivity series** of the metals. Table 20.5 shows part of the reactivity series.

Table 20.5 ▼ Part of the reactivity series

<table>
<tr><th>Metal</th><th>Symbol</th><th></th></tr>
<tr><td>Potassium</td><td>K</td><td rowspan="13">Reactivity decreases from top to bottom ↓</td></tr>
<tr><td>Sodium</td><td>Na</td></tr>
<tr><td>Calcium</td><td>Ca</td></tr>
<tr><td>Magnesium</td><td>Mg</td></tr>
<tr><td>Aluminium</td><td>Al</td></tr>
<tr><td>Zinc</td><td>Zn</td></tr>
<tr><td>Iron</td><td>Fe</td></tr>
<tr><td>Tin</td><td>Sn</td></tr>
<tr><td>Lead</td><td>Pb</td></tr>
<tr><td>Copper</td><td>Cu</td></tr>
<tr><td>Silver</td><td>Ag</td></tr>
<tr><td>Gold</td><td>Au</td></tr>
<tr><td>Platinum</td><td>Pt</td></tr>
</table>

If you have used aluminium saucepans, you may be surprised to see aluminium placed with the reactive metals high in the reactivity series. In fact, aluminium is so reactive that, as soon as aluminium is exposed to the air, the surface immediately reacts with oxygen to form aluminium oxide. The surface layer of aluminium oxide is unreactive and prevents the metal from showing its true reactivity.

## SUMMARY

- Many metals react with oxygen; some metals burn in oxygen. The oxides of metals are bases.
- A few metals (for example sodium) react with cold water to give a metal hydroxide and hydrogen. Some metals react with steam to give a metal oxide and hydrogen.
- Many metals react with dilute acids to give hydrogen and the salt of the metal.

Metals fall into the same order of reactivity in all these reactions. This order is called the reactivity series of the metals.

## CHECKPOINT

1 In some parts of the world, copper is found 'native'. Why is zinc never found native?

2 The ancient Egyptians put gold and silver objects into tombs. Explain why people opening the tombs thousands of years later find the objects still in good condition. Why are no iron objects found in the tombs?

3 The following metals are listed in order of reactivity:

sodium > magnesium > zinc > copper

Describe the reactions of these metals with (a) water, (b) dilute hydrochloric acid. Point out how the reactions illustrate the change in reactivity.

# 20.5 Predictions from the reactivity series

**FIRST THOUGHTS**

The classifications you have been learning about are useful: they enable you to make predictions about chemical reactions.

Where in the Periodic Table do metals occur?

Revise Topic 11.2 if you are in doubt!

## Competition between metals for oxygen

Aluminium is higher in the reactivity series than iron. When aluminium is heated with iron(III) oxide, a very vigorous, exothermic reaction occurs

| Aluminium + | Iron(III) oxide | → | Iron + | Aluminium oxide |
|---|---|---|---|---|
| $2Al(s)$ + | $Fe_2O_3(s)$ | → | $2Fe(s)$ + | $Al_2O_3(s)$ |

This reaction is called the **thermit reaction** (therm = heat). It is used to mend railway lines because the iron formed is molten and can weld the broken lines together (see Figure 20.5A).

*Figure 20.5A* Using the thermit reaction

SUMMARY

Reactive metals displace metals lower down the reactivity series from their compounds.

## Displacement reactions

Metals can displace other metals from their salts. A metal which is higher in the reactivity series will displace a metal which is lower in the reactivity series from a salt. Zinc will displace lead from a solution of lead nitrate.

CHECKPOINT

1. Which metal is best for making saucepans: zinc, iron or copper? Explain your choice.
2. Gold is used for making electrical contacts in space capsules. Explain (a) why gold is a good choice and (b) why it is not more widely used.
3. A metal, romin, is displaced from a solution of one of its salts by a metal, sarin, A metal, tonin, displaces sarin from a solution of one of its salts. Place the metals in order of reactivity.
4. The following metals are listed in order of reactivity, with the most reactive first:

   Mg, Zn, Fe, Pb, Cu, Hg, Au

   List the metals which will:

   (a) occur 'native',
   (b) react with cold water,
   (c) react with steam,
   (d) react with dilute acids,
   (e) displace copper from copper(II) nitrate solution.

# 20.6 Uses of metals and alloys

FIRST THOUGHTS

As you read through this section, think about the reasons which lead to the choice of a metal or alloy for a particular purpose.

Metals and alloys have thousands of uses. The purposes for which a metal is used are determined by its physical and chemical properties. Sometimes a manufacturer wants a material for a particular purpose and there is no metal or alloy which fits the bill. Then metallurgists have to invent a new alloy with the right characteristics. Table 20.6 gives some examples.

*Table 20.6* What are metals and alloys used for?

| Metal/Alloy | Characteristics | Uses |
|---|---|---|
| Aluminium (Duralumin is an important alloy) | Low density<br>Never corroded<br>Good electrical conductor<br>Good thermal conductor<br>Reflector of light | Aircraft manufacture (Duralumin)<br>Food wrapping<br>Electrical cable<br>Saucepans<br>Car headlamps |
| Brass (alloy of copper and zinc) | Not corroded<br>Easy to work with<br>Sonorous<br>Yellow colour | Ships' propellers<br>Taps, screws<br>Trumpets<br>Ornaments |
| Bronze (alloy of copper and tin) | Harder than copper<br>Not corroded<br>Sonorous | Coins, medals<br>Statues, springs<br>Church bells |

It's a fact!

One $cm^3$ of gold can be hammered into enough thin gold leaf to cover a football pitch. Gold foil is used for decorating books, china, etc.

## It's a fact!

At the launch of the 'Platinum 1990' exhibition, the Johnson Matthey precious metals group presented a £300,000 wedding dress made of platinum. They had made it by lining super-thin platinum foil with paper, shredding the platinum-paper combination into strands and weaving the strands into a fabric. The Japanese designer who made the dress sent instructions to the exhibition, *Ironing the dress is strictly forbidden.*

## SUMMARY

Metals and alloys are essential for many different purposes. Metals and alloys are chosen for particular uses because of their physical properties and their chemical reactions.

*Table 20.6* What are metals and alloys used for? (continued)

| Metal/Alloy | Characteristics | Uses |
|---|---|---|
| Copper | Good electrical conductor<br>Not corroded | Electrical circuits<br>Water pipes and tanks |
| Gold | Beautiful colour<br>Never tarnishes<br>Easily worked | Jewellery<br>Electrical contacts<br>Filling teeth |
| Iron | Hard, strong, inexpensive, rusts | Motor vehicles, trains, ships, buildings |
| Lead | Dense<br>Unreactive | Protection from radioactivity<br>Was used for all plumbing (Lead is no longer used for water pipes as it reacts very slowly with water) |
| Magnesium | Bright flame | Distress flares, flash bulbs |
| Mercury | Liquid at room temperature | Thermometers<br>Electrical contacts<br>Dental amalgam for filling teeth |
| Silver | Good electrical conductor<br>Good reflector of light<br>Beautiful colour and shine (tarnishes in city air) | Electrical contacts<br>Mirrors<br>Jewellery |
| Sodium | High thermal capacity | Coolant in nuclear reactors |
| Solder (alloy of tin and lead) | Low melting point | Joining metals, e.g. in an electrical circuit |
| Steel (alloy of iron) | Strong | Construction, tools, ball bearings, magnets, cutlery, etc. |
| Tin | Low in reactivity series | Coating 'tin cans' |
| Titanium | Low in density<br>Stays strong at high and low temperatures | Supersonic aircraft |
| Zinc | High in reactivity series | Protection of iron and steel; see Table 20.11. |

## CHECKPOINT

**1** Name the metal which is used for each of these purposes. Explain why that metal is chosen.
(a) Thermometers
(b) Window frames
(c) Sinks and draining boards
(d) Radiators
(e) Water pipes
(f) Household electrical wiring
(g) Scissor blades

**2** Explain the following:
(a) Some mirrors have aluminium sprayed on to the back of the glass instead of silver, which was used previously.
(b) Although brass is a colourful, shiny material, it is not used for jewellery.
(c) Titanium oxide has replaced lead carbonate as the pigment in white paint.

# 20.7 Compounds and the reactivity series

## The stability of metal oxides

- Reactive metals form compounds readily.
- The compounds of reactive metals are difficult to split up.

Magnesium is a reactive metal, and magnesium oxide is difficult to reduce to magnesium.

Copper is an unreactive metal, and copper(II) oxide is easily reduced to copper. Hydrogen will reduce hot copper(II) oxide to copper.

$$\text{Copper(II) oxide} + \text{Hydrogen} \xrightarrow{\text{Heat}} \text{Copper} + \text{Water}$$
$$CuO(s) + H_2(g) \rightarrow Cu(s) + H_2O(l)$$

Carbon is another reducing agent. When heated, it will reduce the oxides of metals which are fairly low in the reactivity series. Reduction by carbon is often the method employed to obtain metals from their ores.

$$\text{Zinc oxide} + \text{Carbon} \xrightarrow{\text{Heat}} \text{Zinc} + \text{Carbon monoxide}$$
$$ZnO(s) + C(s) \rightarrow Zn(s) + CO(g)$$

Carbon and hydrogen are reducing agents. They can reduce many metal oxides.

Neither hydrogen nor carbon will reduce the oxides of metals which are high in the reactivity series. Aluminium is high in the reactivity series; its oxide is difficult to reduce. The method used to obtain aluminium from aluminium oxide is electrolysis.

If a metal is very low in the reactivity series, its oxide will decompose when heated. The oxides of silver and mercury decompose when heated.

SUMMARY

The oxides of metals low in the reactivity series (e.g. Cu) are easily reduced by carbon and hydrogen. The oxides of metals high in the reactivity series are difficult to reduce.

Compounds of metals low in the reactivity series are decomposed by heat. Compounds of metals high in the reactivity series are stable to heat.

## The stability of other compounds

When compounds are described as stable it means that they are difficult to decompose (split up) by heat. Table 20.7 shows how the stability of the compounds of a metal is related to the position of the metal in the reactivity series.

The sulphates, carbonates and hydroxides of the most reactive metals are not decomposed by heat. Those of other metals decompose to give oxides.

*Table 20.7* ▾ Action of heat on compounds

| Cation | Anion | | | | |
|---|---|---|---|---|---|
| | **Oxide** | **Chloride** | **Sulphate** | **Carbonate** | **Hydroxide** |
| Potassium, Sodium | No decomposition | No decomposition | No decomposition | No decomposition | No decomposition |
| Calcium, Magnesium, Aluminium, Zinc, Iron, Lead, Copper | No decomposition | No decomposition | Oxide and sulphur trioxide $MO + SO_3$ Some also give $SO_2$ | Oxide and carbon dioxide $MO + CO_2$ | Oxide and water $MO + H_2O$ |
| Silver | Metal + oxygen | No decomposition | Metal + $O_2 + SO_3$ | Metal + $O_2 + CO_2$ | Do not form hydroxides |
| Gold | Not formed | No decomposition | Metal + $O_2 + SO_3$ | Metal + $O_2 + CO_2$ | Do not form hydroxides |

# 20.8 Extraction of metals from their ores

FIRST THOUGHTS

One of the important jobs which chemists do is to find ways of extracting metals from the rocks of the Earth's crust.

This section tells you
- what is meant by a metal occurring 'native',
- when a rock is called an ore,
- how chemists decide on the method to use for extracting a metal from its ore.

Metals are found in rocks in the Earth's crust. A few metals, such as gold and copper, occur as the free metal, uncombined. They are said to occur 'native'. Only metals which are very unreactive can withstand the action of air and water for thousands of years without being converted into compounds. Most metals occur as compounds.

Rock containing the metal compound is mined. Then machines are used to crush and grind the rock. Next a chemical method must be found for extracting the metal. All these stages cost money. If the rock contains enough of the metal compound to make it profitable to extract the metal, the rock is called an ore.

*Figure 20.8A* ▲ 'Native' copper

The method used to extract a metal from its ore depends on the position of the metal in the reactivity series (see Table 20.8).

*Table 20.8* ▼ Methods used for the extraction of metals

| Metal | Method |
|---|---|
| Potassium<br>Sodium<br>Calcium<br>Magnesium | The anhydrous chloride is melted and electrolysed |
| Aluminium | The anhydrous oxide is melted and electrolysed |
| Zinc<br>Iron<br>Lead<br>Copper | Found as sulphides and oxides. The sulphides are roasted to give oxides; the oxides are reduced with carbon |
| Silver<br>Gold | Found 'native' (as the free metal) |

## Sodium

Sodium is obtained by the electrolysis of molten dry sodium chloride. The same method is used for potassium, calcium and magnesium. This is an expensive method of obtaining metals because of the cost of the electricity consumed.

## Aluminium – the 'newcomer' among metals

Aluminium is mined as **bauxite**, an ore which contains aluminium oxide, $Al_2O_3.2H_2O$. This ore is very plentiful, yet aluminium was not extracted from it until 1825. Another 60 years passed before a commercial method of extracting the metal was invented. *What was the problem?*

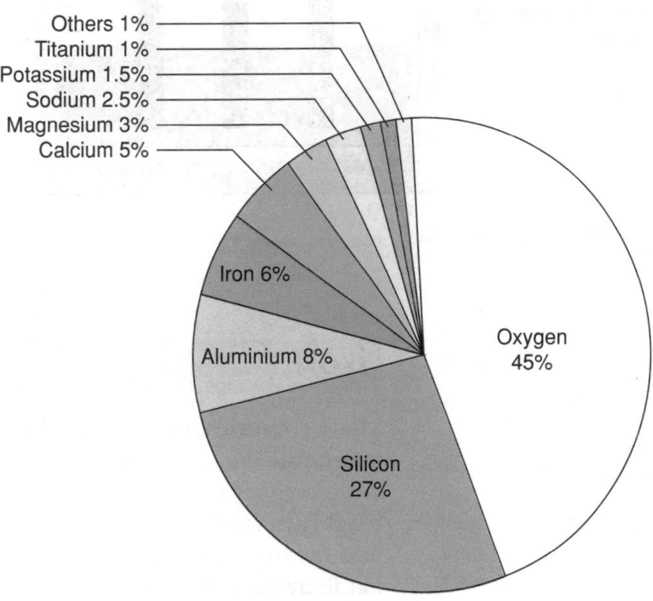


*Figure 20.8B* ▲ Elements in the Earth's crust

### KEY SCIENTISTS

The story of how aluminium came into use tells of contributions from a number of scientists of different ages and different nationalities over a period of 60 years. They were: Hans Christian Oersted, the Danish academic; Henri Sainte-Claire Déville, the French industrial chemist; Charles Martin Hall, an American student; Paul Héroult, a French student.

A Danish scientist called Hans Christian Oersted succeeded in obtaining aluminium from aluminium oxide in 1825. First he made aluminium chloride from aluminium oxide. Then he used potassium amalgam (an alloy of potassium and mercury) to displace aluminium from aluminium chloride.

The German chemist Friedrich Wöhler altered the method somewhat. He used potassium (instead of potassium amalgam).

A French chemist, Henri Sainte-Claire Déville, tackled the problem of scaling up the reaction. He used sodium in the displacement reaction. In 1860 he succeeded in making the reaction yield aluminium on a large scale. The price of aluminium tumbled. Instead of being an expensive curiosity, it became a useful commodity. With its exceptional properties, aluminium soon found new uses, and the demand for the new metal increased.

Aluminium was expensive because of the cost of the sodium used in its extraction. Chemists were keen to find a less costly method of extracting the new metal. Many thought it should be possible to obtain aluminium by electrolysing molten aluminium oxide. The difficulty was the high melting point, 2050 °C, which made it impossible to keep the compound molten while a current was passed through it.

The big breakthrough came in 1886. Two young chemists working thousands of miles apart made the discovery at the same time. An American called Charles Martin Hall, aged 21, and a Frenchman called Paul Héroult, aged 23, discovered that they could obtain a solution of aluminium oxide by dissolving it in molten cryolite, $Na_3AlF_6$, at 700 °C. On passing electricity through the melt, the two men succeeded in obtaining aluminium. Their method is still used. The Hall-Héroult cell is shown in Figure 20.8C. Anhydrous pure aluminium oxide must be obtained from the ore and then electrolysed in the cell.

### Science at work

In 1985, a firm in Northampton had the idea of buying empty aluminium drink cans for recycling. They installed machines which pay for aluminium cans fed in but not for steel cans. *How can the machines tell the difference between aluminium cans and steel cans?*

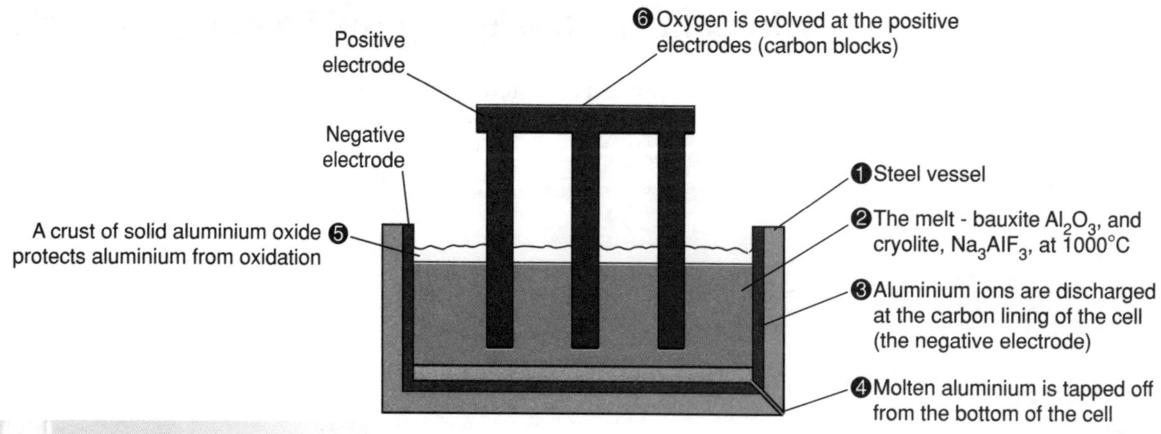


*Figure 20.8C* Electrolysis of aluminium oxide

## Iron

Many countries have plentiful resources of iron ores, **haematite**, $Fe_2O_3$, **magnetite**, $Fe_3O_4$ and **iron pyrites**, $FeS_2$. The sulphide ore is roasted in air to convert it into an oxide. The oxide ores are reduced to iron in a **blast furnace** (Figure 20.8D). Iron ore and coke and limestone are fed into the furnace. Iron ores and limestone are plentiful resources. Coke is made by heating coal.

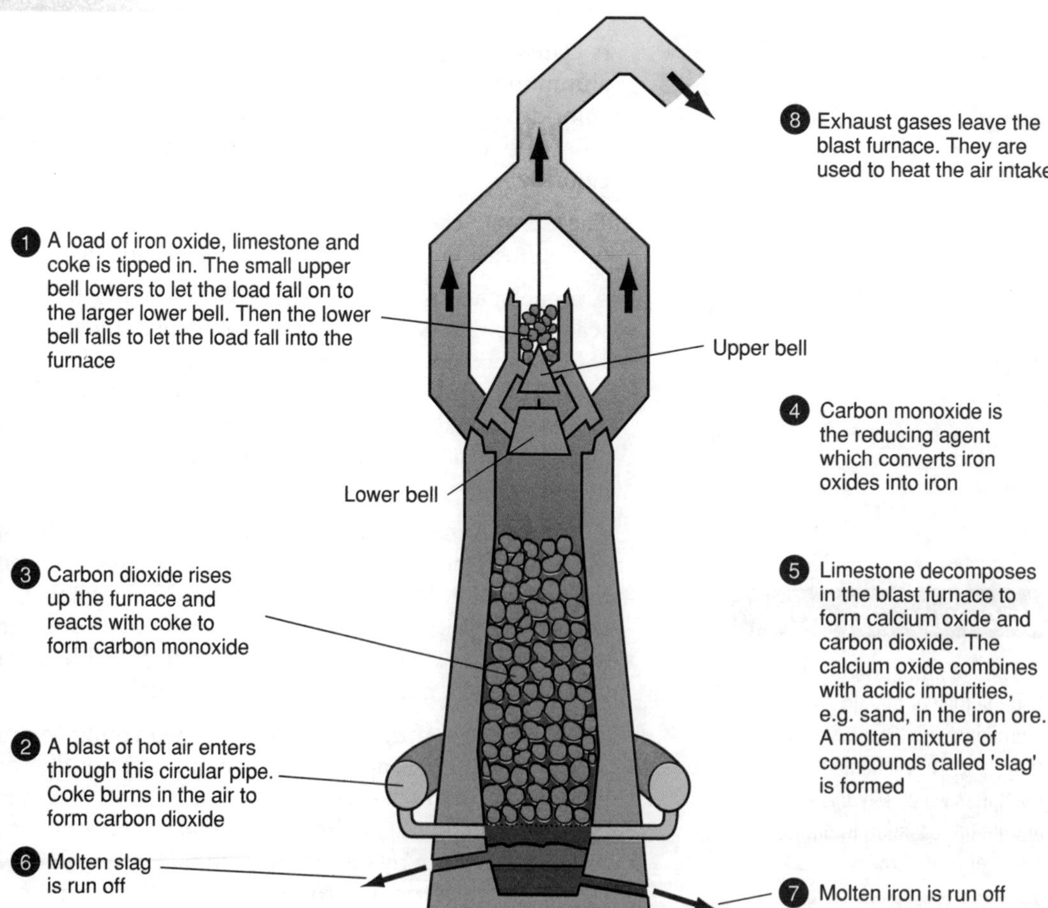

8 Exhaust gases leave the blast furnace. They are used to heat the air intake
1 A load of iron oxide, limestone and coke is tipped in. The small upper bell lowers to let the load fall on to the larger lower bell. Then the lower bell falls to let the load fall into the furnace
Upper bell
4 Carbon monoxide is the reducing agent which converts iron oxides into iron
Lower bell
3 Carbon dioxide rises up the furnace and reacts with coke to form carbon monoxide
5 Limestone decomposes in the blast furnace to form calcium oxide and carbon dioxide. The calcium oxide combines with acidic impurities, e.g. sand, in the iron ore. A molten mixture of compounds called 'slag' is formed
2 A blast of hot air enters through this circular pipe. Coke burns in the air to form carbon dioxide
6 Molten slag is run off
7 Molten iron is run off

*Figure 20.8D* A blast furnace is a tower of steel plates lined with heat-resistant bricks. It is about 50 m high

> **Science at work**
>
> A few years ago, explorers discovered that there was a rich source of minerals on the sea bed. Lumps or 'nodules' of metal ore were found. They consist chiefly of manganese ores but contain other metals such as zinc and nickel also. Methods of scooping the nodules off the sea bed are being developed. There are political problems as well as technical problems. The industralised countries can afford to develop the technology for recovering the nodules. Sometimes nodules are discovered on the floor of the ocean within the territorial limits of a poor country, which cannot afford to develop the technology. A fair arrangement must be worked out to split the profits from recovering the nodules between the country which owns the nodules and the country which provides the technology.

The chemical reactions which take place in the blast furnace are:

**Step 2**

Carbon (coke) + Oxygen → Carbon dioxide

$$C(s) + O_2(g) \rightarrow CO_2(g)$$

**Step 3**

Carbon dioxide + Carbon (coke) → Carbon monoxide

$$CO_2(g) + C(s) \rightarrow 2CO(g)$$

**Step 4**

Iron(III) oxide + Carbon monoxide → Iron + Carbon dioxide

$$Fe_2O_3(s) + 3CO(g) \rightarrow 2Fe(s) + 3CO_2(g)$$

**Step 5**

(a) Calcium carbonate (limestone) → Calcium oxide + Carbon dioxide

$$CaCO_3(s) \rightarrow CaO(s) + CO_2(g)$$

(b) Calcium oxide + Silicon(IV) oxide (sand) → Calcium silicate(slag)

$$CaO(s) + SiO_2(s) \rightarrow CaSiO_3(l)$$

The blast furnace runs continuously. The raw materials are fed in at the top, and molten iron and molten slag are run off separately at the bottom. The slag is used in foundations by builders and road-makers. The process is much cheaper to run than an electrolytic method. With the raw materials readily available and the cost of extraction low, iron is cheaper than other metals.

## Copper

Copper is low in the reactivity series. It is found 'native' (uncombined) in some parts of the world. More often, it is mined as the sulphide. This is roasted in air to give impure copper. Pure copper is obtained from this by the **electrolytic** method shown in Figure 20.8E.

After the cell has been running for a week, the negative electrode becomes very thick. It is lifted out of the cell, and replaced by a new thin sheet of copper. When all the copper has dissolved out of the positive electrode, a new piece of impure copper is substituted.

Other metals are present as impurities in copper ores. Iron and zinc are more reactive than copper, and their ions therefore stay in solution while copper ions are discharged. Silver and gold are less reactive than copper. They do not dissolve and are therefore present in the anode sludge. They can be extracted from this residue.

1 The electrolyte is copper(II) sulphate solution

2 The negative electrode is a strip of pure copper. Copper ions are discharged and copper atoms are deposited on the electrode. The strip of pure copper becomes thicker $Cu^{2+}(aq) + 2e^- \rightarrow Cu(s)$

3 The positive electrode is a lump of impure copper. Copper atoms supply electrons to this electrode and become copper ions which enter the solution. The lump of impure copper becomes smaller $Cu(s) \rightarrow Cu^{2+}(aq) + 2e^-$

4 Anode sludge, this is the undissolved remains of the lump of impure copper

Electrons flow through the external circuit from the positive electrode to the negative electrode

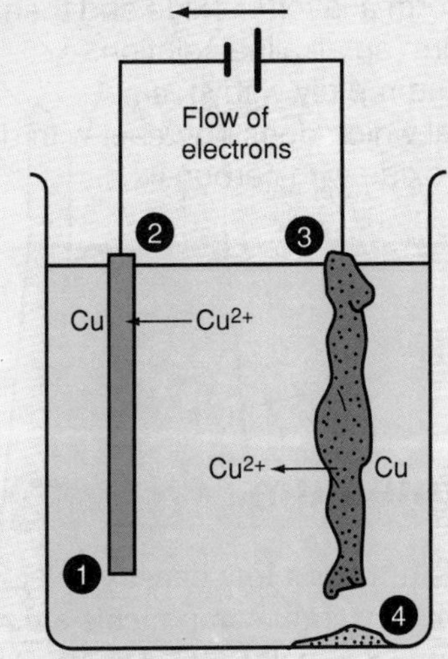


*Figure 20.8E* ▲ Purification of copper

SUMMARY

The method used for extracting a metal from its ores depends on the position of the metal in the reactivity series. Very reactive metals, such as sodium, are obtained by electrolysis of a molten compound. Less reactive metals, such as iron, are obtained by reducing the oxide with elements such as carbon or hydrogen. The metals at the bottom of the reactivity series occur 'native'.

## Silver and gold

Silver and gold are found 'native'. A new deposit of gold was discovered in the Sperrin mountains of Northern Ireland in 1982.

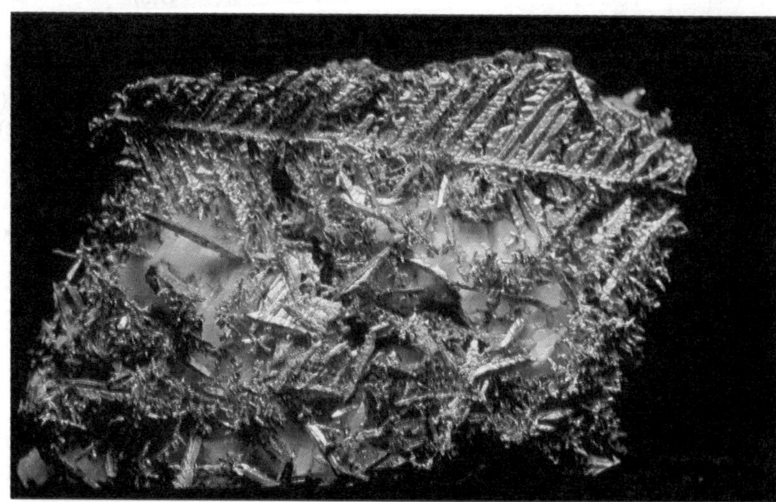

*Figure 20.8F* Natural silver crystals

CHECKPOINT

1 The Emperor Napoleon invested in the French research on new methods of obtaining aluminium. He was interested in the possibility of aluminium suits of armour for his soldiers. What advantage would aluminium armour have had over iron or steel?

2 (a) Arrange in order of the reactivity series the metals copper, iron, zinc, gold and silver.
(b) Arrange the same metals in order of their readiness to form ions, that is, the ease with which the reaction $M(s) \rightarrow M^{n+}(aq) + ne^-$ takes place.
(c) Arrange the same metals in order of the ease of discharging their ions, that is, the ease with which the reaction $M^{n+}(aq) + ne^- \rightarrow M(s)$ takes place.
(d) Refer to Figure 20.8E showing the electrolytic purification of copper. Use your answers to (b) and (c) to explain why copper is deposited on the negative electrode but iron, zinc, gold and silver are not.

3 Explain why (a) the human race started using copper, silver and gold long before iron was known and (b) iron tools were a big improvement on tools made from other metals.

4 Give the names of metals which fit the descriptions A, B, C, D and E.
A Reacts immediately with air to form a layer of oxide and then reacts no further.
B Reacts violently with water to form an alkaline solution.
C Reacts slowly with cold water and rapidly with steam.
D Is a reddish-gold coloured metal which does not react with dilute hydrochloric acid.
E Can be obtained by heating its oxide with carbon.

It's a fact!

All the magnesium that has been produced so far could have been extracted from 4 cubic kilometres of sea water. The sea is an almost inexhaustible reserve of magnesium.

## Magnesium

Magnesium has a low density (1.7 g/cm$^3$) and its alloys are widely used where lightweight components are required, e.g. the aircraft industry. The alloy Dowmetal, 89% Mg, 9% Al, 2% Zn, has a tensile strength which is close to steel. Magnesium is present to the extent of 0.14% in sea water. Figure 20.8G (opposite) illustrates the extraction of magnesium by the Dow process.

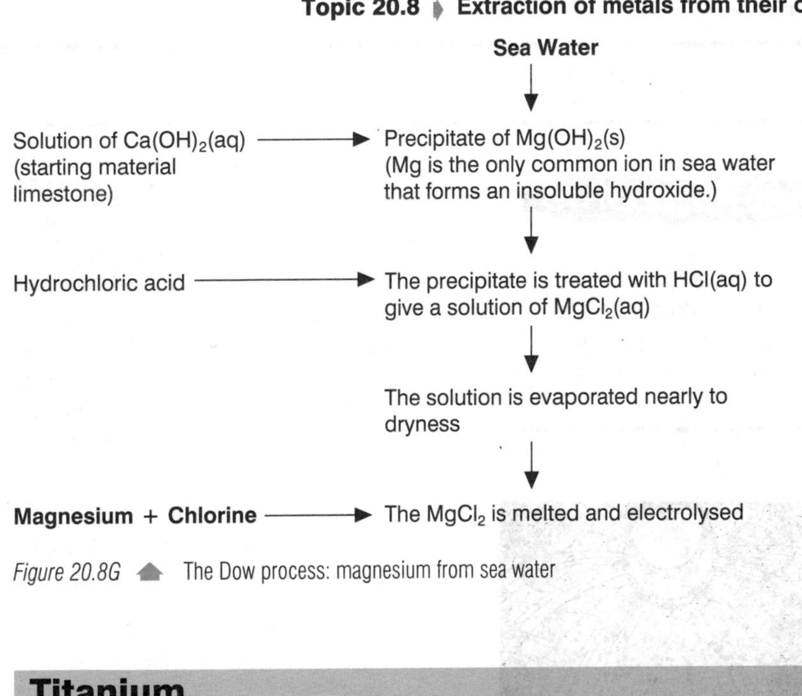

Sea Water
Solution of $Ca(OH)_2(aq)$ (starting material limestone)
Precipitate of $Mg(OH)_2(s)$ (Mg is the only common ion in sea water that forms an insoluble hydroxide.)
Hydrochloric acid
The precipitate is treated with HCl(aq) to give a solution of $MgCl_2(aq)$
The solution is evaporated nearly to dryness
Magnesium + Chlorine
The $MgCl_2$ is melted and electrolysed


*Figure 20.8G* ▲ The Dow process: magnesium from sea water

## Titanium

Titanium is sometimes described as the **aerospace metal**. It has a unique set of properties: low density, high strength, high resistance to corrosion and high melting point. The low density and the ability to retain its strength at high temperatures make titanium and its alloys valuable in the manufacture of spacecraft and aircraft designed to fly at high speeds.

Many transition metals are obtained by reducing the metal oxides. Carbon and carbon monoxide are often used as reducing agents. In the blast furnace for the reduction of iron oxides, carbon monoxide is the reducing agent.

Titanium is mined as rutile, titanium(IV) oxide, $TiO_2$. Titanium(IV) oxide is more difficult to reduce than iron oxides, and carbon monoxide cannot be used. The Kroll process is used. Titanium(IV) chloride, $TiCl_4$, is reduced with magnesium. Magnesium, being higher than titanium in the reactivity series, can displace titanium from its compound.

The steps in the extraction are:

1 Rutile is mixed with coke and ground. It is heated in a furnace, through which chlorine passes.

titanium(IV) oxide + chlorine + carbon → titanium(IV) choride + carbon monoxide

$$TiO_2(s) + 2Cl_2(g) + 2C(s) \rightarrow TiCl_4(g) + 2CO(g)$$

2 The titanium(IV) chloride formed is purified by fractional distillation.

3 The Kroll process. Gaseous titanium(IV) chloride is passed into a reactor containing molten magnesium. Argon is passed through the reactor. Reduction takes place.

titanium(IV) chloride + magnesium → titanium + magnesium chloride

$$TiCl_4(g) + 2Mg(l) \rightarrow Ti(s) + 2MgCl_2(s)$$

4 A spongy deposit of titanium forms on the walls of the reactor. A batch process is employed because the reactor must be allowed to cool so that the deposit of titanium can be scraped off. The metal is distilled and then melted to form ingots.

5 Magnesium chloride is run off from the reactor and electrolysed to give magnesium and chlorine for recycling.

Titanium and its alloys are used in spacecraft and high-speed aircraft. Titanium is mined as the oxide $TiO_2$. This is converted into the chloride $TiCl_4$. Magnesium displaces titanium from its chloride.

# 20.9 Iron and steel

FIRST THOUGHTS

The machines that manufacture our possessions, our means of transport and the frameworks of our buildings: all these depend on the strength of steel.

The iron that comes out of the blast furnace is called **cast iron** (or pig iron). It contains three to four per cent carbon. The carbon content makes it brittle, and cast iron cannot be bent without snapping. The impurity makes the melting point lower than that of pure iron so that cast iron is easier to cast – to melt and mould – than pure iron. Cast iron expands slightly as it solidifies. This helps it to flow into all the corners of a mould and reproduce the shape exactly. By casting, objects with complicated shapes can be made, for example the cylinder block of a car engine (which contains the cylinders in which the combustion of petrol vapour in air takes place).

Iron which contains less than 0.25% carbon is called wrought iron. Wrought iron is strong and easily worked (Figure 20.9A). Nowadays, mild steel has replaced wrought iron.

*Figure 20.9A* ▲ Wrought iron

## Steel

Steel is made by reducing the carbon content of cast iron, which makes it brittle, to less than one per cent. Carbon is burnt off as its oxides, the gases carbon monoxide, CO, and carbon dioxide, $CO_2$. Iron is less easily oxidised. The sulphur, phosphorus and silicon in the iron are also converted into acidic oxides. These are not gases, and a base, such as calcium oxide, must be added to remove them. The base and the acidic oxides combine to form a slag (a mixture of compounds of low melting point).

Figure 20.9B shows a basic oxygen furnace. In it, cast iron is converted into steel. One converter can produce 150–300 tonnes of steel in an hour.

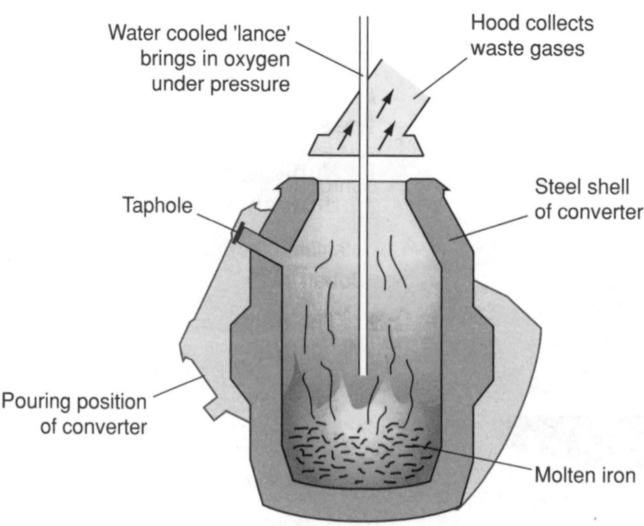


*Figure 20.9B* ▲ A basic oxygen furnace holding 150–300 tonnes of steel

There are various types of steel. They differ in their carbon content and are used for different purposes. Low-carbon steel (mild steel) is pliable; high-carbon steel is hard.

SUMMARY

Cast iron (from the blast furnace) is easy to mould, but is brittle. Wrought iron (the purest form of iron) is easily worked without breaking. Alloy steels contain other elements in addition to iron and carbon. Different steels are suited to different uses.

## Alloy steels

Many elements are alloyed with iron and carbon to give alloy steels with different properties. They all have different uses, for example nickel and chromium give stainless steel, manganese and molybdenum increase strength, and vanadium increases springiness.

### CHECKPOINT

▶ 1 What is the difference between cast iron and wrought iron in (a) composition (b) strength and (c) ease of moulding? Name two objects made from cast iron and two objects made from wrought iron.

## The uses of alloy steels

Some steels are alloys of iron and carbon. They have more uses than cast iron (see Table 20.9).

*Table 20.9* ▼ Cast iron and steel and their uses

| Type of steel | Description | Uses |
|---|---|---|
| Mild steel (<0.25%) carbon) | Pliable (can be bent without breaking) | Chains and pylons |
| Medium steel (0.25–0.45% carbon) | Tougher than mild steel, more springy than high carbon steel | Nuts and bolts<br>Car springs and axles<br>Bridges |
| High carbon steel (0.45–1.5% carbon) | The carbon content makes it both hard and brittle | Chisels, files, razor blades, saws, cutting tools |
| Cast iron (2.5–4.5% carbon) | Cheaper than steel, easily moulded into complicated shapes | Fire grates, gear boxes, brake discs, engine blocks. These articles will break if they are hammered or dropped |

### SUMMARY

Steels contain carbon. Low-carbon steel is pliable; high-carbon steel is hard. Cast iron contains up to 4% carbon.

Many elements are used for alloying with steel. The different alloy steels have different properties which fit them for different uses (see Table 20.10).

*Table 20.10* ▼ Some alloy steels

| Element | Properties of alloy | Uses |
|---|---|---|
| Chromium | Prevents rusting if >10% | Stainless steels, acid-resisting steels, cutlery, car accessories, tools |
| Cobalt | Takes a sharp cutting edge<br>Can be strongly magnetised | High-speed cutting tools<br>Permanent magnets |
| Manganese | Increases strength and toughness | Some is used in all steels; steel in railway points and safes contains a high percentage of manganese |
| Molybdenum | Strong even at high temperatures | Rifle barrels, propeller shafts |
| Nickel | Resists heat and acids | Stainless steel cutlery, industrial plants which must withstand acidic conditions |
| Tungsten | Stays hard and tough at high temperatures | High-speed cutting tools |
| Vanadium | Increases springiness | Springs, machinery |

### SUMMARY

Alloy steels contain added metals. The different elements give a range of steels with different characteristics suited to different uses.

## CHECKPOINT

1. Think what is the most important characteristic of the steel that must be used for making the articles listed below. Should the steel be pliable or springy or hard? Then say whether you would use mild steel or medium steel or high-carbon steel or cast iron for making the following articles. (a) car axles (b) axes (c) car springs (d) ornamental gates (e) drain pipes (f) chisels (g) sewing needles (h) picks (i) saws (j) food cans.
2. Suggest which type of alloy steel could be chosen for: (a) an electric saw (b) a car radiator (c) the suspension of a car (d) a gun barrel (e) a set of steak knives.

# 20.10 Rusting of iron and steel

Iron and most kinds of steel rust. Rust is hydrated iron(III) oxide $Fe_2O_3.nH_2O$ ($n$, the number of water molecules in the formula, varies).

Some of the methods which are used to protect iron and steel against rusting are listed in Table 20.11. Some methods use a coating of some substance which excludes water and air. Other methods work by sacrificing a metal which is more reactive than iron.

*Table 20.11* Rust prevention

| Method | Where it is used | Comment |
|---|---|---|
| A coat of paint | Ships, bridges, cars, other large objects (see Figure 20.10A(a)) | If the paint is scratched, the exposed iron starts to rust. Corrosion can spread to the iron underneath the paintwork which is still sound. |
| A film of oil or grease | Moving parts of machinery, e.g. car engines | The film of oil or grease must be renewed frequently. |
| A coat of plastic | Kitchenware, e.g. draining rack | If the plastic is torn, the iron starts to rust. |
| Chromium plating | Kettles, cycle handlebars (see Figure 20.10A(b)) | The layer of chromium protects the iron beneath it and also gives a decorative finish. It is applied by electroplating. |
| Galvanising (zinc plating) | Galvanised steel girders are used in the construction of buildings and bridges (see Figure 20.10A(c)) | Zinc is above iron in the reactivity series: zinc will corrode in preference to iron. Even if the layer of zinc is scratched, as long as some zinc remains, the iron underneath does not rust. Zinc cannot be used for food cans because zinc and its compounds are poisonous. |
| Tin plating | Food cans | Tin is below iron in the reactivity series. If the layer of tin is scratched, the iron beneath it starts to rust. |
| Stainless steel | Cutlery, car accessories, e.g. radiator grille | Steel containing chromium (10–25%) or nickel (10–20%) does not rust. |
| Sacrificial protection | Ships (see Figure 20.10A(d)) | Blocks of zinc are attached to the hulls of ships below the waterline. Being above iron in the reactivity series, zinc corrodes in preference to iron. The zinc blocks are sacrificed to protect the iron. As long as there is some zinc left, it protects the hull from rusting. The zinc blocks must be replaced. |

(a) Paint protects the car

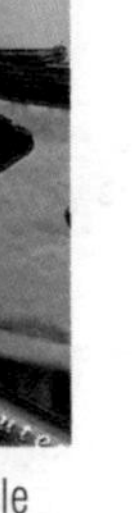
(b) Chromium plating protects the bicycle handlebars

(c) Galvanised steel girders

(d) Zinc bars protect the ship's hull

*Figure 20.10A* Rust prevention

## SUMMARY

Exposure to air and water in slightly acidic conditions makes iron and steel rust. Some of the treatments for protecting iron and steel from rusting are:

- oil or grease,
- a protective coat of another metal, e.g. chromium, zinc, tin,
- alloying with nickel and chromium,
- attaching a more reactive metal, e.g. magnesium or zinc, to be sacrificed.

## It's a fact!

The rusting of iron is an expensive nuisance. Replacing rusted iron and steel structures costs the UK £500 million a year.

The Earth contains huge deposits of iron ores. We need iron for our machinery and for our means of transport. During the twentieth century, we have used more metal than in all the previous centuries put together. If we keep on mining iron ores at the present rate, the Earth's resources may one day be exhausted. We allow tonnes of iron and steel to rust every year. We throw tonnes of used iron and steel objects on the scrap heap. The Earth's iron deposits will last longer if we take the trouble to collect scrap iron and steel and recycle it, that is, melt it down and reuse it.

*Figure 20.10B* ▲ Scrap iron dump

## CHECKPOINT

1 Say how the rusting of iron is prevented:
  (a) in a bicycle chain,
  (b) in a food can,
  (c) in parts of a ship above the water line,
  (d) in parts of a ship below the water line,
  (e) in a galvanised iron roof.

2 'The Industrial Revolution would not have been possible without steel.' Say whether or not you agree with this statement. Give your reasons.

3 The map opposite shows possible sites, A, B, C and D, for a steelworks. Say which site you think is the best, and explain your choice. You will have to consider the need for:
  - iron ore, coke (from coal), limestone,
  - a work-force,
  - transporting iron and steel to customers,
  - removing slag.

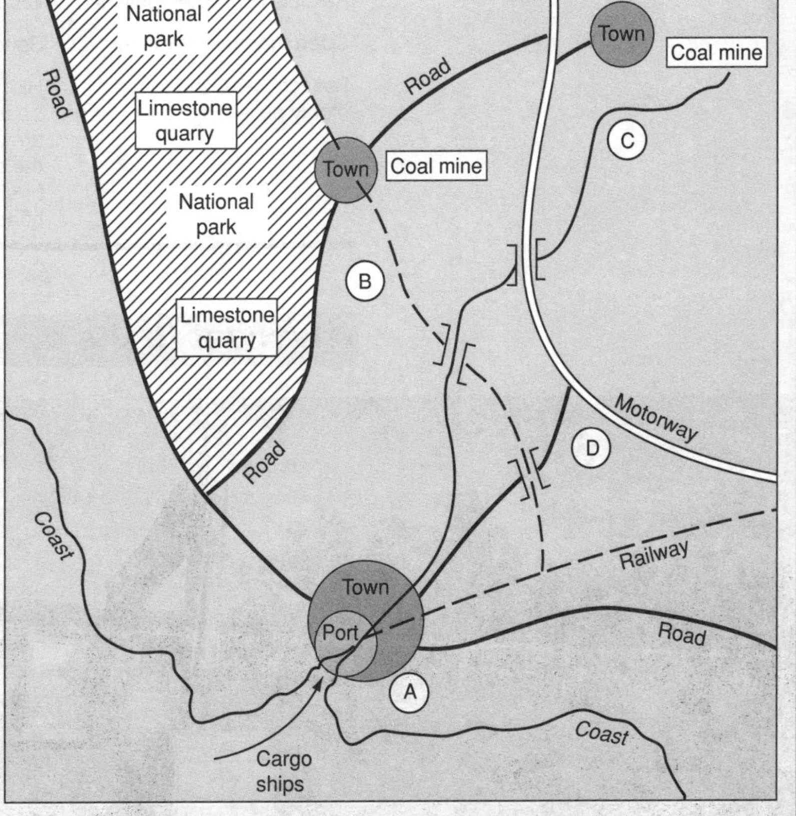

# 20.11 Aluminium

FIRST THOUGHTS

Aluminium is the most plentiful metal in the Earth's crust, yet aluminium was not manufactured until the nineteenth century. The twentieth century has seen aluminium finding more and more vitally important uses.

## Uses of aluminium

Aluminium is a metal with thousands of applications. Some of these are listed in Table 20.12. Pure aluminium is not a very strong metal. Its alloys, such as duralumin (which contains copper and magnesium), are used when strength is needed.

Note how the different uses of aluminium and its alloys depend on their various properties.

*Table 20.12* Some uses of aluminium and its alloys

| Property | Uses for aluminium which depends on this property |
|---|---|
| Never corroded (except by bases) | Door frames and window frames are often made of aluminium. 'Anodised aluminium' is used. The thickness of the protective layer of aluminium oxide has been increased by anodising (making it the positive electrode in an electrolytic cell). |
| Low density | Packaging food: milk bottle tops, food containers, baking foil.<br>The low density and resistance to corrosion make aluminium ideal for aircraft manufacture. Alloys such as duralumin are used because they are stronger than aluminium. |
| Good electrical conductor | Used for overhead cables. The advantage over copper is that aluminium cables are lighter and need less massive pylons to carry them. |
| Reflects light when polished | Car headlamp reflectors |
| Good thermal conductor | Saucepans, etc. |
| Reflects heat when highly polished | Highly polished aluminium reflects heat and can be used as a thermal blanket. Aluminium blankets are used to wrap premature babies. They keep the baby warm by reflecting heat back to the body. Firefighters wear aluminium fabric suits to reflect heat away from their bodies. |

## The cost to the environment

Bauxite is found near the surface in Australia, Jamaica, Brazil and other countries. The ore is obtained by open cast mining (Figure 20.11A). A layer of earth 1 m to 60 m thick is excavated, and the landscape is devastated. In some places, mining companies have spent money on restoring the landscape after they have finished working a deposit.

*Figure 20.11A* An open cast bauxite mine

Pure aluminium oxide must be extracted from bauxite before it can be electrolysed. Iron(III) oxide, which is red, is one of the impurities in bauxite. The waste produced in the extraction process is an alkaline liquid containing a suspension of iron(III) oxide. It is pumped into vast red mud ponds.

Jamaica has a land area of 11 000 km$^2$. Every year, 12 km$^2$ of red mud are created. The Jamaicans are worried about the loss of land. Even when it dries out, red mud is not firm enough to build on. They also worry about the danger of alkali seeping into the water supply. The waste cannot be pumped into the sea because it would harm the fish.

Purified aluminium oxide is shipped to an aluminium plant. The electrolytic method of extracting aluminium is expensive to run because of the electricity it consumes. Aluminium plants are often built in areas which have hydroelectric power (electricity from water-driven generators). This is relatively cheap electricity. The waterfalls and fast-flowing rivers which provide hydroelectric power are found in areas of natural beauty. Local residents often object to the siting of aluminium plants in such beauty spots.

We want to use aluminium – thousands of tonnes a year – and we also want to retain the beauty of the countryside. Sometimes there is a conflict.

There may be other difficulties over hydroelectric power. Purified aluminium oxide has to be transported to a remote area and aluminium has to be transported away. If there are not enough local workers, a workforce may have to be brought into the area and provided with housing. Often it pays to build an aluminium plant in an area which has a big population and good transport, even if the cost of electricity is higher.

The exhaust gases from aluminium plants contain fluorides from the electrolyte. Before leaving the chimney, the exhaust gases are 'scrubbed' with water. The waste water, which contains fluorides, is discharged into rivers. The remaining exhaust gases are discharged into the atmosphere through tall chimneys. In the past, fluoride emissions have been known to pollute agricultural land, killing grass and causing lameness in cattle. Farmers sued for the damage to their cattle. Aluminium plants now take more care to control fluoride emission.

## SUMMARY

Aluminium has brought us many benefits. There is, however, a cost to the environment. The open cast mining of bauxite spoils the landscape. The purification creates unsightly red mud ponds. The extraction can cause pollution through fluoride emission.

Factors which decide the siting of aluminium plants include

- the cost of electricity,
- the cost of transporting the raw material and the product,
- the availability of a workforce.

Figure 20.11B ▲ Recycling aluminium

## CHECKPOINT

1 The percentages of some metals recycled in the UK are shown below.
   (a) Explain what 'recycled' means.
   (b) Plot the figures as a bar chart or as a pie chart.
   (c) Why is lead easy to recycle?
   (d) How can iron be separated from other metals for recycling?
   (e) What resources are saved by recycling aluminium?

| *Metal* | *Aluminium* | *Zinc* | *Iron* | *Tin* | *Lead* | *Copper* |
|---|---|---|---|---|---|---|
| *Percentage recycled* | 28 | 30 | 50 | 30 | 56 | 19 |

## 20.12 Recycling

Few materials are destroyed during their use. Recycling is a possible method of saving resources and energy. Metals and alloys are prime candidates for recycling (see Table 20.13). Other materials are wood, paper, plastics and textiles. Some objects are easier to recycle than others. An iron machine has a high scrap value, but iron bars embedded in concrete are more difficult to reuse. In many cases a number of different metals are used in the manufacture of an item, e.g. a motor vehicle. A motor vehicle contains about 1% by mass of copper and this must be reduced to 0.1% before the scrap can go to the steelworks.

*Table 20.13* Recovery of metals

| Metal | Recovery |
|---|---|
| Aluminium | About 50% is recycled. Recycling uses only 5% of the energy need to make aluminium from its ore |
| Iron | About 50% of the feedstock for steel furnaces is scrap |
| Copper | Pipes, vehicle radiators, etc are recycled |
| Lead | Batteries, pipes, sheet metal, type metal are recycled |
| Zinc | Recovery is from alloys |
| Tin | Recovery is from alloys and from tin-plate |
| Mercury | A large percentage is recovered from mercury cells, instruments and apparatus |
| Gold, silver and platinum | Recovery is high from jewellery, watches and chemical plants |

### SUMMARY

Recycling metals saves Earth's resources of metal ores and saves energy. Do you play your part in recycling?

## CHECKPOINT

1 Why is the recovery of gold and silver high?
2 Why do people find it more convenient to recycle mercury than to dump it?
3 Why is copper an easy metal to recycle?
4 How can other metals be separated from scrap iron?

# Topic 21 Chemical cells

## 21.1 Simple chemical cells

**FIRST THOUGHTS**

A car needs a battery to start the motor and to power the instruments. A battery is a series of chemical cells. A chemical cell is a system for converting the energy of a chemical reaction into electricity. This topic will tell you more about how chemical cells work.

You have studied metals in this theme. You have learned that metals are made up of positive ions and a cloud of moving electrons (Figure 20.2A). What happens when a strip of metal is put into water or a solution of an electrolyte? Figure 21.1A(a) shows what happens when the metal is a reactive metal such as zinc. Some zinc ions pass into solution, leaving their electrons behind on the metal:

$$Zn(s) \rightarrow Zn^{2+}(aq) + 2e^-$$

As a result, the strip of zinc becomes negatively charged. After a while the negative charge on the zinc builds up to a level which prevents any more $Zn^{2+}$ ions from leaving the metal.

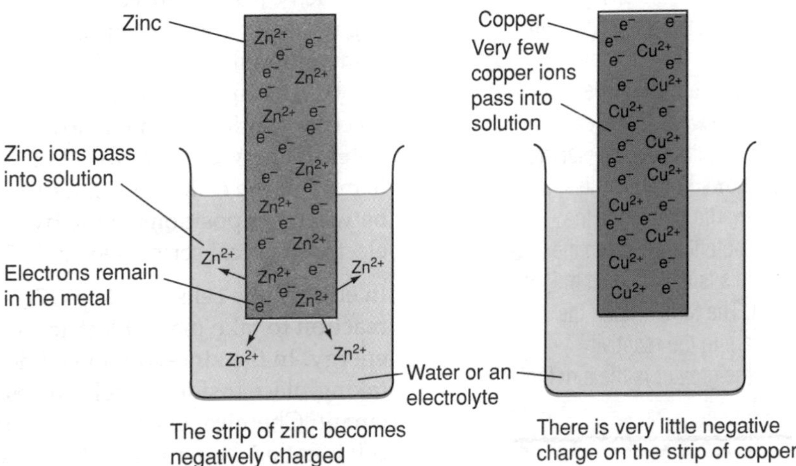


*Figure 21.1A* ▲ (a) Zinc (b) Copper

The reactivity series of metals (see Topic 20.4) is based on the observation that some metals are more ready to form ions (and are therefore more reactive) than others. Copper is one of the less reactive metals. A strip of copper placed in water or a solution of an electrolyte has very little tendency to form ions and to acquire a negative charge (see Figure 21.1A(b)).

Perhaps you can predict what will happen if you immerse a strip of zinc and a strip of copper in a solution and then connect the two metals. Electrons flow through the external circuit from zinc, which is negatively charged, to copper, which has very little charge (see Figure 21.1B).

Electrons flow from the negative electrode to the positive electrode. Conventional electricity flows from the positive electrode to the negative electrode

*Figure 21.1B* ▲ A zinc–copper chemical cell

## SUMMARY

When a metal is placed in a solution of an electrolyte, it acquires an electric charge. When two metals which dip into an electrolyte are connected, there is a difference in electric potential between them, called the e.m.f. of the cell. An electric current flows from one metal to the other. The difference in electric potential between the two metals is called the e.m.f. of the cell. The further apart the metals are in the reactivity series, the greater is the e.m.f. of the cell.

In Figure 21.1B, electrons flow from zinc to copper through the external circuit. Before the nature of electricity was understood, the flow of electricity was regarded as taking place from the positive electrode to the negative electrode, in this case from copper to zinc. The flow of 'conventional' electricity is from copper to zinc (right to left in Figure 21.1B). The direction of flow of electrons is from a metal higher in the reactivity series to a metal lower in the series. The difference in electric potential between the two electrodes is called the electromotive force (e.m.f.) of the cell. The e.m.f. of the cell depends on the difference between the positions of the two metals in the reactivity series (and the electrochemical series; see Table 21.1).

In electrolytic cells (see Topic 7), an electric current causes a chemical reaction to take place. Electrical energy is converted into chemical energy. In the zinc–copper cell described above, the chemical reaction taking place inside the cell causes a current to flow through the external circuit. Chemical energy is converted into electrical energy. This kind of cell is called a **chemical cell** (or a **galvanic cell** or a **voltaic cell**).

## Arranging metals in order of their tendency to lose electrons

In the zinc–copper cell (Figure 21.1A), zinc is the negative electrode and copper is the positive electrode. Copper is not always the positive electrode. If copper is paired with silver in a chemical cell, the reaction:

$$Cu(s) \rightarrow Cu^{2+}(aq) + 2e^-$$

happens to a greater extent than the reaction:

$$Ag(s) \rightarrow Ag^{+}(aq) + e^-$$

The build-up of electrons on copper is greater than the build-up of electrons on silver. Copper is the negative electrode of the cell. Paired with a more reactive metal, e.g. zinc, copper is the positive electrode; paired with a less reactive metal, e.g. silver, copper is the negative electrode.

It is possible to construct a 'league table' of metals, arranging them in order of their tendency to lose electrons. The method is to choose one metal as a reference electrode, say copper, and measure the e.m.f. of a number of different metal–copper cells. The values of e.m.f. place the metals in order of their readiness to give electrons and form cations in aqueous solution. The order is called the **electrochemical series**. Table 21.1 shows a section of the electrochemical series including hydrogen. It also puts cations in the reverse order of the ease with which they accept electrons and are discharged in electrolysis (see Topic 7.5).

## SUMMARY

When you place metals in order of their tendency to lose electrons and gain a positive charge, you obtain the electrochemical series. This is similar to the reactivity series of metals. Some metals occupy different positions in the two series. This happens when an oxide coating or a film of insoluble salt decreases the true reactivity of the metal.

*Table 21.1* Part of the electrochemical series

| Metal | |
|---|---|
| Potassium, K | ↑ |
| Calcium, Ca | |
| Sodium, Na | Increasing ease |
| Magnesium, Mg | of losing electrons |
| Aluminium, Al | to form |
| Zinc, Zn | positive ions |
| Iron, Fe | (cations) |
| Tin, Sn | Decreasing ease |
| Lead, Pb | of discharging ions |
| Hydrogen, H | in electrolysis |
| Copper, Cu | |
| Silver, Ag | |
| Gold, Au | |

Two metals far apart in the electrochemical series give a cell with a bigger e.m.f. than two metals close together.

## The reactivity series

The electrochemical series will remind you of the reactivity series of metals (Topic 20.4). The reactivity series does not always show the true ability of a metal to form ions. Some metals, e.g. aluminium, acquire a coating of the metal oxide which prevents the metal from showing its true reactivity. Other metals, e.g. calcium, form insoluble salts and this slows down their reactions with solutions, e.g. with water and acids. The electrochemical series, on the other hand, shows the true ability of a metal to form ions.

Many metals can be paired up in chemical cells. The further apart the metals are in the electrochemical series, the bigger is the e.m.f. produced by the cell. Cells can be combined in series to form a battery. Three cells each with an e.m.f. of 1.5 V combine to make a battery with an e.m.f. of 4.5 V. The word 'battery' is sometimes used incorrectly for a single cell. For example, what we call a torch 'battery' is often a single cell.

## CHECKPOINT

1. Name the particles that conduct electricity through (a) a chemical cell and (b) an electrical circuit outside the cell.
2. The following pairs of metals are joined to form chemical cells. Place the pairs in list (a) and the pairs in list (b) in order of the e.m.f. of the cells:
   (a) zinc–copper, magnesium–copper, iron–copper
   (b) magnesium–zinc, iron–magnesium, magnesium–tin
3. Zinc and iron are connected to form a cell.
   (a) Which of the two metals is the more able to form ions?
   (b) Which metal will become the negative electrode of the cell and which the positive?
   (c) Sketch a zinc–iron cell showing the direction of flow of electrons in the external circuit.
4. Copper and lead are both low in the electrochemical series. When they are paired up to form a chemical cell, which of the two metals will become the negative electrode?

## The lead–acid accumulator

The type of battery commonly used in motor vehicles is the lead–acid accumulator. The battery delivers 6 V or 12 V, depending on the number of 2 V cells which it contains (see Figure 21.1A). In each cell, one pole is a lead plate and the other is a lead grid packed with lead(IV) oxide, $PbO_2$. The electrolyte is sulphuric acid. When the battery supplies a current, the reactions which occur convert lead and lead(IV) oxide into lead(II) ions. When the reactants have been used up, the battery can no longer supply a current; it is 'flat'. It can, however, be recharged. When the vehicle engine is running, the dynamo (the alternator) rotates and generates electricity which recharges the battery. If the dynamo is faulty, the battery will become flat. It can be recharged by connecting it to a battery charger, which is a transformer connected to the mains. This reverses the sign of each electrode and reverses the chemical reactions that have occurred at the electrodes. The battery is again able to supply a current.

*Figure 21.1C* ▲ A car battery

### SUMMARY

A vehicle battery consists of a number of chemical cells. Chemical reactions in the cells supply an electric current. When the chemicals have been used up they can be reformed by recharging the battery.

- During discharge, chemical energy is converted into electrical energy.
- During charge, electrical energy is converted into chemical energy.

### CHECKPOINT

1. What is the advantage of the lead–acid accumulator over the zinc–copper cell (Figure 21.1B)?
2. What does the alternator in a vehicle engine do?
3. (a) Why do car batteries sometimes become flat?
   (b) How can they be given new life?

## 21.2 Dry cells

*Do you use a portable radio or cassette player, a pocket calculator, a wrist watch or a camera? All these depend on dry cells.*

Dry cells are chemical cells which are used in everyday life. For use in batteries for torches, radios etc., a dry cell is used, one which has a damp paste instead of a liquid electrolyte.

## The zinc–carbon dry cell

You will already be familiar with the zinc–carbon type of dry cell (see Figure 21.2B).

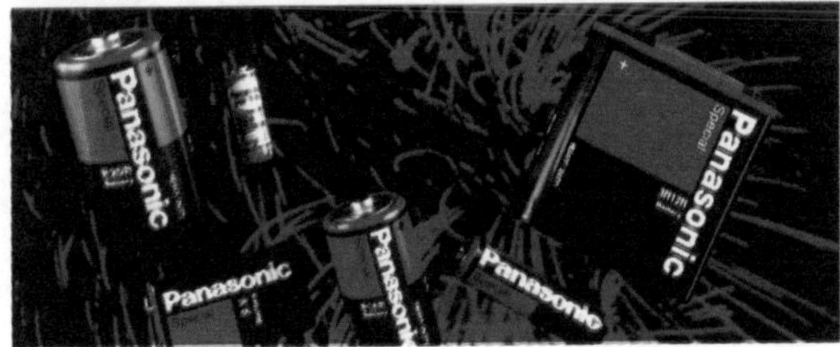


*Figure 21.2B* ▲ Dry cells

The zinc–carbon cell has an e.m.f. of 1.5 V.

The chemical reactions that take place are complex. A summary is:
*Anode*: Zinc atoms are oxidised to zinc ions.
*Cathode*: Manganese(IV) oxide is reduced to manganese(III) oxide.

When the chemicals have been used up, the cell can operate no longer: it is flat.

Cells of this type are 1–2 cm in diameter and 3–10 cm in length. They are used in flashlights, portable radios, toys etc. A disadvantage is that with heavy use the cell quickly becomes flat. Zinc–carbon dry cells cannot be recharged and have a short shelf life.

*Figure 21.2A* ▲ A pocket calculator uses a dry cell

## The alkaline manganese cell

The alkaline manganese cell also uses zinc as anode and manganese(IV) oxide as cathode. The electrolyte contains potassium hydroxide. The zinc anode is slightly porous, giving it a larger surface area. The alkaline manganese cell can therefore deliver more current than the zinc–carbon cell; it has an e.m.f. of 1.5 V. The alkaline manganese cells do not become flat as quickly in use and have a longer shelf life than zinc–carbon cells.

EXTENSION FILE ACTIVITY

## The silver oxide cell

Another type of dry cell is the silver oxide cell. Silver oxide cells measure 0.5–1 cm in diameter and 0.25–0.5 cm in height. These tiny 'batteries' are used in electronic wrist watches, in cameras and in electronic calculators. They have a long life because little current is drawn from them. They are rather expensive.

## The nickel–cadmium cell

The dry cells described above have a limited life. Once the chemicals have reacted, there is no further source of energy. The nickel–cadmium cell (see Figure 21.2C) need not be thrown away when the chemical reaction that gives rise to the e.m.f. is finished. The cell can be recharged. This is done by connecting the cell to a source of direct current. The chemical reaction is reversed, and the cell has a new source of e.m.f. As the cell operates, cadmium is oxidised to cadmium(II) ions, and nickel(IV) oxide is reduced to nickel(II) hydroxide. During recharging, these reactions are reversed.

The nickel–cadmium cell has an e.m.f. of 1.4 V. It has a longer life than the lead storage battery. It can be packaged in a sealed unit ready for use in rechargeable calculators and photographic flash units etc.

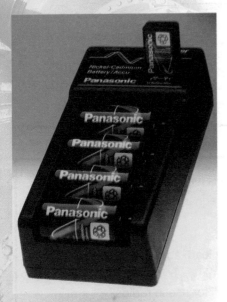

*Figure 21.2C* ▲ Nickel–cadmium cells and chargers

*Table 21.2* Dry cells

| Type | Electrodes | Electrolyte | Voltage | Use | Characteristics |
|---|---|---|---|---|---|
| Zinc–carbon cell | Carbon (+) Zinc (−) | Ammonium chloride | 1.5 V | Radios, cassette players, electrical toys, torches | Cheap. Relatively short life. Not rechargeable. Leakage of electrolyte occurs in the unsealed types of cell. Operates on low current; current drops gradually during discharge. |
| Alkaline manganese cell | Manganese(IV) oxide (+) Zinc (−) | Potassium hydroxide | 1.5 V | Cassette players, electrical toys, appliances which are in use for long periods | Costs about twice as much as the zinc–carbon cell and lasts twice as long. Not rechargeable. Long shelf-life. Gives a large current; voltage remains steady during discharge. |
| Silver oxide cell | Silver oxide (+) Zinc (−) | Potassium hydroxide | 1.5 V | Watches, calculators, cameras, hearing aids | High cost. Not rechargeable. Small in size; light-weight. Voltage remains constant in operation. |
| Nickel–cadmium cell | Nickel oxide (+) Cadmium (−) | Potassium hydroxide | 1.5 V | Cassette players, electrical toys, radios | Expensive. Rechargeable. Gives a large current. |

## SUMMARY

Chemical cells transform the energy of chemical reactions into electrical energy. Examples of dry cells are: the carbon–zinc dry cell, the alkaline manganese cell, the silver oxide dry cell, the nickel–cadmium dry cell.

Fuel cells operate continuously. One kind uses the reaction between hydrogen and oxygen to make water.

## Science at work

Space craft use **fuel cells**. A fuel cell is another means of converting chemical energy into electrical energy. The fuel cell shown in Figure 21.2D uses the reaction:

Hydrogen + Oxygen → Water

The big advantage of a fuel cell over other cells is that it operates continuously with no need for recharging. As long as reactants are fed in, the cell can supply energy.

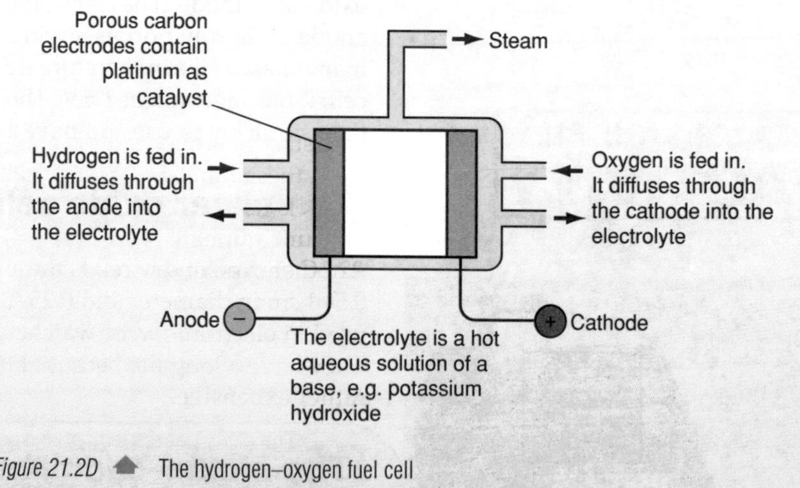
Porous carbon electrodes contain platinum as catalyst
Steam
Hydrogen is fed in. It diffuses through the anode into the electrolyte
Oxygen is fed in. It diffuses through the cathode into the electrolyte
Anode
Cathode
The electrolyte is a hot aqueous solution of a base, e.g. potassium hydroxide

*Figure 21.2D* The hydrogen–oxygen fuel cell

## CHECKPOINT

1. Why is the silver oxide 'battery' suitable for use in a camera? Why is it expensive?
2. Both the lead–acid accumulator and the nickel–cadmium cell are rechargeable.
   (a) What does 'rechargeable' mean?
   (b) What is the advantage of rechargeable cells over others?
   (c) For what uses is the nickel–cadmium cell more suited than the lead–acid accumulator?
   (d) What is the most widespread use of the lead–acid accumulator?
3. Why is the alkaline manganese cell preferred to the silver oxide cell for use in radios?

# Topic 22 Building materials

## 22.1 ▶ Concrete

FIRST THOUGHTS

The UK quarries 90 million tonnes of limestone and chalk each year. Most of it is used in the building industry.

EXTENSION FILE ACTIVITY

*Figure 22.1A* ◀ Concrete and glass

All around you, you can see buildings made of concrete, steel and glass. Concrete is used in the construction of all kinds of homes, schools, factories, skyscrapers, power stations and reservoirs. All our forms of transport depend on the concrete which is used in roads and bridges, docks and airport runways. For extra strength, for example in bridges, reinforced concrete is strengthened by steel supports.

The starting point in the manufacture of concrete is limestone or chalk. Both these minerals are forms of calcium carbonate. First, limestone or chalk is used with shale or clay to make cement. A small percentage of calcium sulphate (gypsum) is also needed (see Figure 22.1B). Figure 22.1C shows how concrete is made from cement.

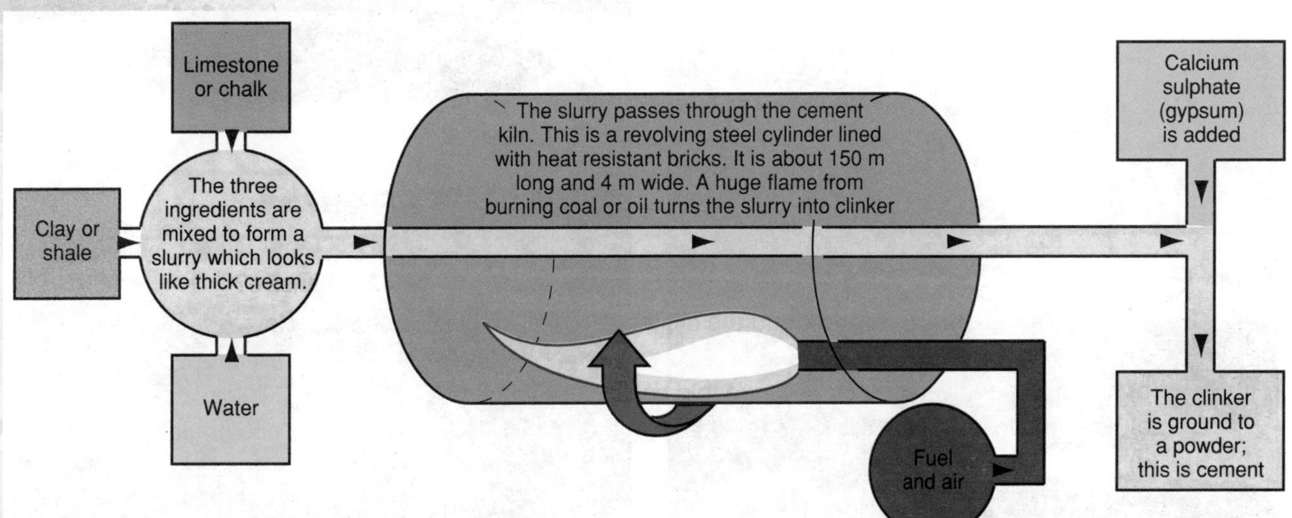

Limestone or chalk
Clay or shale
Water
The three ingredients are mixed to form a slurry which looks like thick cream.
The slurry passes through the cement kiln. This is a revolving steel cylinder lined with heat resistant bricks. It is about 150 m long and 4 m wide. A huge flame from burning coal or oil turns the slurry into clinker
Fuel and air
Calcium sulphate (gypsum) is added
The clinker is ground to a powder; this is cement


*Figure 22.1B* ▲ The manufacture of cement (after Blue Circle)

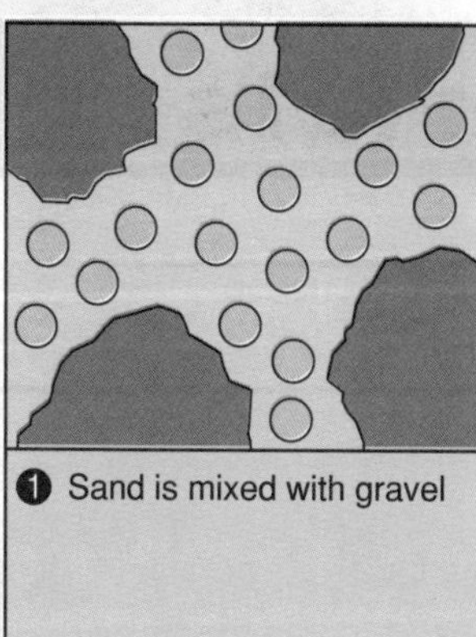


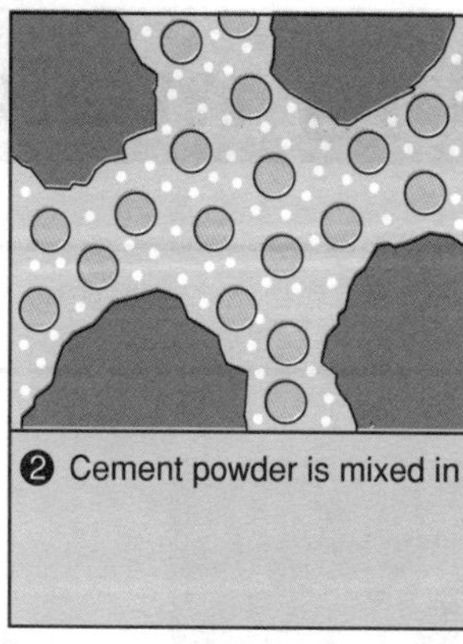


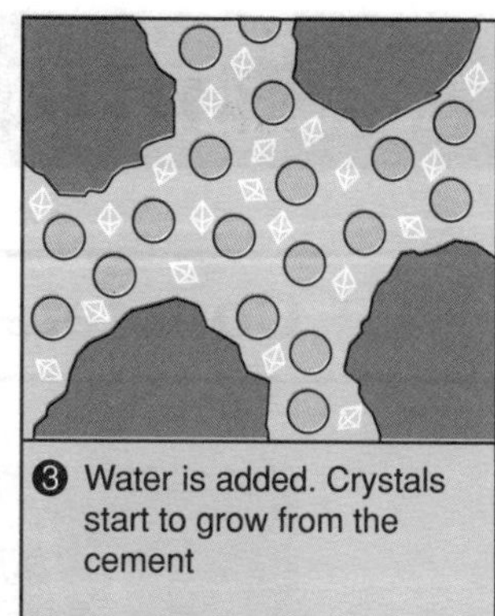

❷ Cement powder is mixed in

❸ Water is added. Crystals start to grow from the cement


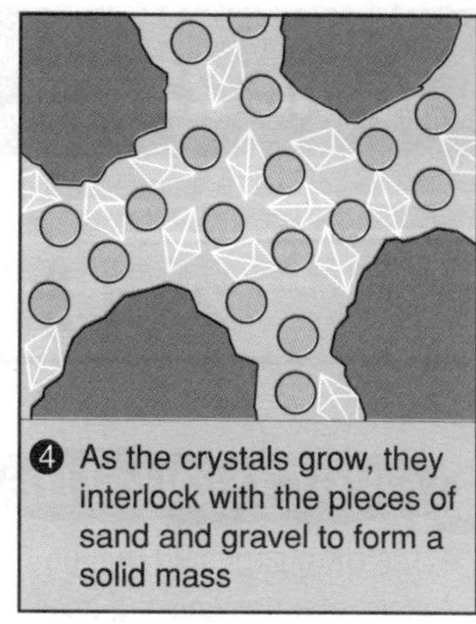

❹ As the crystals grow, they interlock with the pieces of sand and gravel to form a solid mass


*Figure 22.1C* ▲ Making concrete (after Blue Circle)

## SUMMARY

Concrete is a strong and versatile construction material. It is made from cement, sand, gravel and water. Cement is made from chalk or limestone and clay or shale. The quarrying of vast quantities of these raw materials creates environmental problems.

Chalk and limestone are calcium carbonate, $CaCO_3$. Clay and shale consist largely of silicon(IV) oxide, $SiO_2$, and aluminium oxide, $Al_2O_3$, with some other compounds. The chemical reactions which occur between them produce cement, which consists chiefly of calcium silicate, $CaSiO_3$, and calcium aluminate, $CaAl_2O_4$. A little calcium sulphate (gypsum) is added to slow down the rate at which concrete sets.

The UK has large deposits of both limestone and chalk. Limestone is quarried by blasting a hillside with an explosive. Chalk is dug out by mechanical excavators. Both methods devastate the landscape. It happens that these minerals often occur in regions of great natural beauty (Figure 22.1D). We have to balance the damage to our countryside against the useful materials which industry can obtain from limestone and chalk. Mining companies are required to restore the countryside after they exhaust a deposit of limestone or chalk (Figure 22.1E). Limestone is used in the manufacture of iron in blast furnaces, in the manufacture of glass and in the manufacture of lime.

*Figure 22.1D* ▲ A limestone quarry

*Figure 22.1E* ▲ Restoring the scenery

*Figure 22.1G* ▲ Enjoying a 'carbonated' drink

EXTENSION FILE
ACTIVITY

## Lime

When calcium carbonate is heated strongly, it dissociates (splits up) to give calcium oxide and carbon dioxide.

$$\text{Calcium carbonate} \rightleftharpoons \text{Calcium oxide} + \text{Carbon dioxide}$$
$$CaCO_3(s) \rightleftharpoons CaO(s) + CO_2(g)$$

The reaction is a **reversible** reaction: it can go from left to right and also from right to left, depending on the temperature and the pressure. The reaction is carried out industrially in towers called **lime kilns** (see Figure 22.1F). At the temperature of a lime kiln, 1000 °C, calcium carbonate dissociates. The through draft of air carries away carbon dioxide as it is formed and prevents it from recombining with calcium oxide. Otherwise there would be an acid-base reaction between the acid gas, carbon dioxide, and the base, calcium oxide.

Calcium oxide is called **lime** or **quicklime**. It reacts with water to form calcium hydroxide, which is called **slaked lime**.

$$\text{Calcium oxide} + \text{Water} \rightarrow \text{Calcium hydroxide}$$
$$CaO(s) + H_2O(l) \rightarrow Ca(OH)_2(s)$$

On building sites, calcium hydroxide, slaked lime, is mixed with sand to give **mortar**. Mortar is used to hold bricks together. As mortar is exposed to the air, it becomes gradually harder as it reacts with carbon dioxide in the air to form calcium carbonate.

Calcium oxide, lime, is used in agriculture. Farmers spread it on fields to neutralise excess acid in soils.

The carbon dioxide produced in lime kilns is also useful. Carbon dioxide dissolves to a slight extent in water to give a solution of the weak acid, carbonic acid, $H_2CO_3$. Under pressure, the solubility of carbon dioxide increases. The basis of the soft drinks industry is dissolving carbon dioxide in water under pressure, and adding flavourings. When the cap is taken off a bottle, the pressure is decreased, and carbon dioxide comes out of solution.

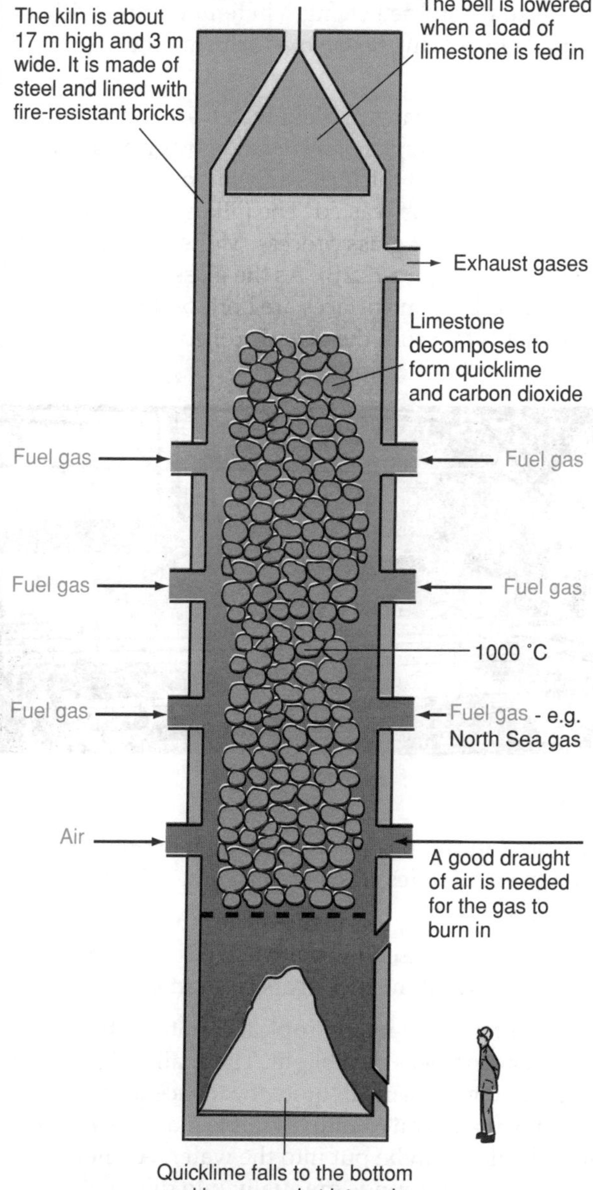
The kiln is about 17 m high and 3 m wide. It is made of steel and lined with fire-resistant bricks
The bell is lowered when a load of limestone is fed in
Exhaust gases
Limestone decomposes to form quicklime and carbon dioxide
Fuel gas
Fuel gas
Fuel gas
Fuel gas
1000 °C
Fuel gas
Fuel gas - e.g. North Sea gas
Air
A good draught of air is needed for the gas to burn in
Quicklime falls to the bottom and is removed at intervals

*Figure 22.1F* ▲ A lime kiln

### SUMMARY

When calcium carbonate (limestone) is heated in lime kilns, calcium oxide (quicklime) and carbon dioxide are produced. On building sites, calcium hydroxide is used to make mortar. On farms, it is used to neutralise excessive acidity of soils. Carbon dioxide is used in the soft drinks industry and as a fire extinguisher.

# 22.2 Glass

FIRST THOUGHTS

The recipe for glass is 4500 years old. Egyptians discovered it when they melted sand with limestone and sodium carbonate. To their surprise, they obtained a transparent material, glass. Egypt is one of the few countries where sodium carbonate occurs naturally.

Glass is a mixture of calcium silicate, $CaSiO_3$, and sodium silicate, $Na_2SiO_3$. In structure, it resembles silicon(IV) oxide, $SiO_2$ (sand), which is a crystalline substance with a macromolecular structure (see Figure 22.2A). In glass, many of the Si—O bonds have been broken, and the structure is less regular. X-rays show that glass does not have the orderly packing of atoms found in other solids. Glass is neither a liquid nor a crystalline solid. It is a **supercooled liquid**: it appears to be solid, but has no sharp melting point.

## Some new uses for glass

Craftsmen worked with glass for thousands of years without knowing anything about its structure. When chemists began to study glass, they made many advances in glass technology. New types of glass were invented. Some of these are described in this section.

Si atom attached to 4 O atoms

O atom attached to 2 Si atoms

*Figure 22.2A* ▲ Bonding in crystalline silica, $SiO_2$

**Pyrex® glass** will stand sudden changes in temperature without cracking. It is made by adding boron oxide during the manufacture of glass.

**Window glass** consists of plates of glass which are the same thickness all over. Formerly, plates were made by grinding a sheet of glass to the required thickness and smoothness. In the process, 30% of the sheet was wasted. The Pilkington Glass Company invented the **float glass** process. Molten glass flows on to a bath of molten tin (Figure 22.2B). As the glass cools and solidifies, the top and bottom surfaces are both perfectly smooth and planar. While it is still soft, the glass is rolled to the required thickness and then cut into sections.

*Figure 22.2B* ▲ Float glass

**Photosensitive glass** is used in sunglasses which darken in bright light and lighten again when the light fades. The glass includes silver chloride.

**Glass ceramic** is almost unbreakable. It is made by heating photosensitive glass in a furnace. Glass ceramic is used in ovenware, electrical insulation, the nose cones of space rockets and the tiles of space shuttles.

**Soluble glass** has important uses. In many tropical countries, the disease, **bilharzia** (or schistosomiasis) is a blight. The snails which carry the disease can be killed by copper salts in quite low concentrations. The copper compounds can be incorporated into a soluble glass. Pellets of the copper-containing soluble glass can be put into the water. As they dissolve, they release copper compounds gradually into the snail-infested water. To make a soluble glass, phosphorus(V) oxide, $P_2O_5$, is used instead of silicon(IV) oxide.

SUMMARY

The glass industry makes:

- Pyrex® glass,
- plate glass,
- light-sensitive glass,
- glass ceramic,
- soluble glass.

## CHECKPOINT

1. (a) Why must a bottle of a carbonated soft drink have a well-fitting cap?
   (b) Why can you not see bubbles inside the closed bottle?
   (c) Why can you see bubbles when the bottle is opened?
2. 'Spreading calcium hydroxide on soil reduces the acidity of the soil.' Describe an experiment which you could do to check whether this statement is true.
3. 'Mortar reacts with the air to form calcium carbonate.' Describe experiments which you could do on new mortar and old mortar to find out whether this statement is true.

# 22.3 Ceramics

*Figure 22.3A* Traditional ceramics

Ceramics are crystalline compounds which consist of an ordered arrangement of atoms. Many are compounds of metallic and non-metallic elements, e.g. silicon or aluminium combined with oxygen or nitrogen. Pottery is a ceramic. Brick is a ceramic. Traditional ceramics such as these are derived from the raw materials clay and silica. They are formed on, for example, a potter's wheel and hardened in a fire kiln. Examples of modern ceramics are silica glass, soda glass, aluminium oxide, $Al_2O_3$, silicon carbide, SiC, silicon nitride, SiN, zirconium oxide, $ZrO_2$, titanium carbide, TiC, zirconium carbide, ZrC, tungsten carbide, WC. They are made by grinding the components to fine powders, mixing them and heating to high temperatures under high pressure in electric kilns to cause 'densification'. New uses are being found for ceramics all the time. They are used in spacecraft, artificial bones, cutting tools, engine parts, turbine blades and electronic components (see next page).

The bonding in ceramics may be ionic or covalent. The arrangement of particles is more complicated than in metals. The bonds are directed in space and the structure is therefore more rigid than a metallic structure.

This structure makes ceramics

- crystalline
- harder than metals (able to cut steel and glass)
- of higher melting point than metals
- poor conductors of heat and electricity because they lack free electrons
- usually opaque because light is reflected by grain boundaries in the crystal structure
- not ductile or malleable or plastic (except at high temperature)
- rather brittle because the bonds are rigid, and dislocation movements are complicated so that it is difficult to change the shape without breaking

Ceramics can be made to behave like metals. Ceramic semiconductors and even superconductors can be made from suitable mixtures of metal oxides. Superplastic ceramics are more plastic than most metals. They are based on grains of silicon nitride and silicon carbide in a 'glue' of silicon(IV) oxide.

## Applications of ceramics

How do ceramics compare with metals and plastics? Different properties mean different uses for these materials; see Topic 29.4.

*Vehicle components* A ceramic coating, e.g. $ZrO_2$, on cast iron improves wear and tear and heat-resistance. It extends the life of diesel engine parts. Ceramic bearings can operate at high speeds without lubricants. Research is going on into the use of ceramics to allow diesel engines and jet engines to run at higher temperatures. The higher the temperature at which an engine operates, the more efficiently it runs and the less fuel it consumes.

*Gas turbine blades* made of ceramics, e.g. silicon nitride, can run at higher temperatures than those made of alloys.

*Machine tools* made of ceramics can rotate twice as fast as metal tools without deforming or wearing out. *Sialon* is a ceramic made of silicon, aluminium, oxygen and nitrogen. It is almost as hard as diamond, and is as strong as steel and as low in density as aluminium. It can be used at up to 1300 °C and needs no lubricants.

*Bioceramics* Recently ceramics, e.g. aluminium oxide and silicon nitride, have been considered as implant materials. Alloys and polymers are used for replacement devices, e.g. hips, knees, teeth etc. Ceramics have a high strength to weight ratio, would wear better than alloys or polymers and would never corrode. For replacing damaged bone, a porous ceramic implant could be used, so as to allow bone to grow into and bond with the implant.

*Space craft* Ceramic tiles protect a space shuttle during re-entry into the Earth's atmosphere. The outside temperature can reach 1500 °C. Each tile consists of a cellular structure of very fine fibres coated with silica. The fibres are loosely packed: 95% of the tiles is air, which gives them their insulating property.

Figure 22.3B ▲ Space-age ceramics

### SUMMARY

Ceramics are crystalline compounds of oxygen or nitrogen with other elements, e.g. silicon, aluminium and some transition elements. Their structure makes them hard-wearing, resistant to corrosion, and of high melting point. Their resistance to heat and corrosion finds them many uses.

### CHECKPOINT

1. What is the advantage of a ceramic over (i) an alloy and (ii) a hard plastic for use as (a) a tooth implant (b) an elbow implant (c) a middle ear prosthesis?
2. Explain why ceramic tiles are fitted to space shuttles.
3. What is the chief difference between the methods used to fire traditional ceramics and modern ceramics?
4. Why is glass transparent and china opaque?
5. Say what structural features of ceramics make them (a) more brittle than metals (b) harder than metals (c) poor electrical conductors.

# Topic 23 Agricultural chemicals

## 23.1 ▶ Fertilisers

FIRST THOUGHTS

The world's population keeps on growing. Many people in the world are short of food. A major contribution to increasing the size of food crops is the use of manufactured fertilisers to supplement natural fertilisers

As the world population has increased, the demand for food has increased. In 1965 it was 3.5 billion, and by 2000 it will have reached 6.5 billion. In many parts of the world, people are either hungry or undernourished. Farming has to be intensive, that is, able to grow large crops on the available land. Crops take nutrients from the soil, and these must be replaced before the next crop is sown (see the nitrogen cycle, Topic 13.6).

Before the nineteenth century, farmers relied on the natural fertilisers, manure and compost, to replace nutrients in the soil. When the world population began to rise sharply in the nineteenth century, farmers needed more fertilisers. They asked the chemists to make some artificial fertilisers. Before the chemists could do this, they had to find out which chemicals in manure fertilised the soil. Then they would be able to make chemical fertilisers to do the same job.

The chemists' work showed that plants need these three groups of chemicals.

1 Large amounts of carbon, hydrogen and oxygen. There is no shortage of these elements. Carbon comes from carbon dioxide in the air; hydrogen and oxygen from water.
2 Small amounts of 'trace elements' (see Table 23.1). So little is needed that there is no need to add these elements to the soil.
3 Larger quantities of the other elements shown in Table 23.1.

The elements which plants obtain from soil are listed in Table 23.1.

*Table 23.1* ▼ Elements obtained by plants from soil

| Major elements | Importance | Trace elements |
|---|---|---|
| Nitrogen | Needed for protein synthesis | Manganese |
| Phosphorus | Needed for development of roots, energy-transfer reactions, nucleic acids | Copper<br>Iron<br>Zinc |
| Potassium | Needed in photosynthesis | Chlorine |
| Magnesium | Present in chlorophyll | Boron |
| Sulphur | Needed for synthesis of some proteins | Molybdenum |
| Calcium | Needed for transport | |

It's a fact!

With 2–3 million tonnes of nitrogen (in the form of nitrates) going on to the soil of the UK each year, 600 000 tonnes of nitrate ion leach out every year.

### How much fertiliser?

Most fertilisers concentrate on supplying the necessary nitrogen (N), phosphorus (P) and potassium (K). **NPK fertilisers** are consumed in huge quantities: in 1990, the world consumption was over 20 million tonnes. In the UK, the consumption of NPK fertilisers is about seven million tonnes a year. The fertilisers cost about £60 a tonne. Farmers and market gardeners need to invest a lot of money in fertilisers. They want good crops, but they do not want to spend more than they need on fertilisers. They can obtain expert advice from the Ministry of

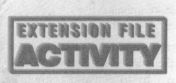

Agriculture, Fisheries and Food. Agricultural chemists at the Ministry will advise them on the type and quantity of fertiliser to apply and the best season of the year for applying it. Every farmer and grower has a different problem. The agricultural chemists must weigh up the type of crop and the type of soil before they can recommend the most suitable treatment.

## Nitrogen

**It's a fact!**

There are islands off the coast of South America which are inhabited by large flocks of sea birds. Mounds of their droppings, called guano, accumulate. Since sea birds eat a fish diet, which is high in protein, guano is rich in nitrogen compounds. For a century, European farmers imported guano from South America to use as a fertiliser.

Nitrogen can be absorbed by plants when it is combined as nitrates. Ammonia and ammonium salts can also be used as fertilisers because ammonium salts are converted into nitrates by organisms which live in the soil. All nitrates and all ammonium salts are soluble so they are easily absorbed. Sometimes concentrated ammonia is used as a fertiliser. Solid fertilisers, ammonium nitrate, ammonium sulphate and urea, ($CON_2H_4$) are applied in pellet form (see Figure 23.1A). The manufacture of nitrogenous fertilisers is described below. The use of more fertiliser than plants can absorb causes eutrophication of lakes and rivers and pollution of ground water (see Topic 17.4).

Manufactured fertilisers contain nitrogen, phosphorus and potassium. Nitrogenous fertilisers include ammonia, urea and ammonium salts, which you see being applied in the photograph.

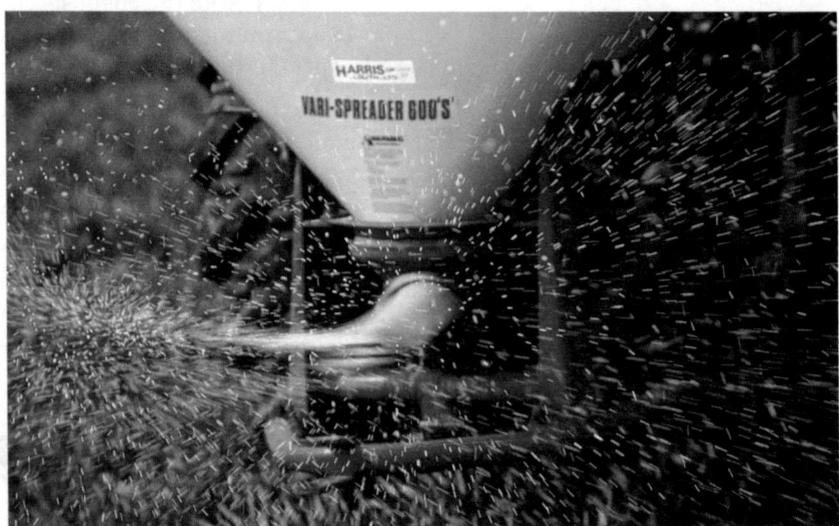


*Figure 23.1A* ▲ Applying pellets of ammonium salts to a field

## 23.2 Manufacture of ammonia

Ammonium salts and nitrates are used as fertilisers. The first step in making them is to make ammonia.

There is plenty of nitrogen in the air. Making it combine with hydrogen to make ammonia was a difficult problem. Fritz Haber solved it.

The first nitrogen-containing fertiliser which the farmers of Europe used was sodium nitrate. They had to import it across the Atlantic from Chile. There were disadvantages to this practice. European chemists tackled the problem of making nitrogenous fertilisers. Nitrogen was the obvious starting material because every country has plenty of it in the air. Making nitrogen combine with other elements proved to be a problem. The problem was solved by a German chemist called Fritz Haber. In 1908, he succeeded in combining nitrogen with hydrogen to form ammonia

$$\text{Nitrogen} + \text{Hydrogen} \rightleftharpoons \text{Ammonia}$$
$$N_2(g) + 3H_2(g) \rightleftharpoons 2NH_3(g)$$

This reaction is reversible: it takes place from right to left as well as from left to right. Some of the ammonia formed dissociates (splits up) into nitrogen and hydrogen. A mixture of nitrogen, hydrogen and ammonia

Conditions which the Haber Process uses to favour the formation of ammonia are high pressure, moderate temperature and a catalyst.

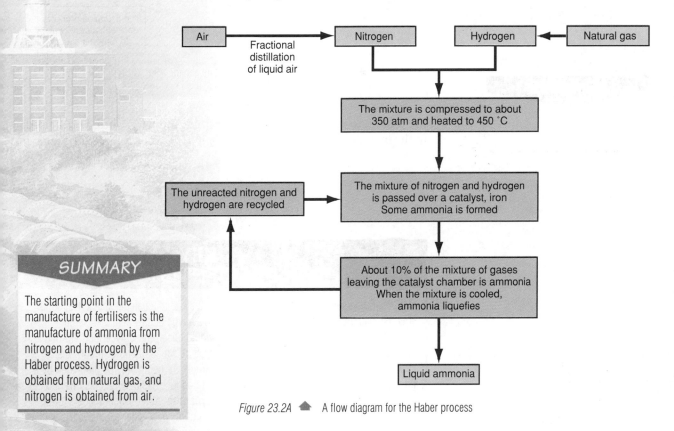

is formed. To increase the percentage of ammonia formed in the reaction, a high pressure and a low temperature are used. A low temperature reduces the dissociation of ammonia, but it also makes the reaction very slow. Modern industrial plants use a compromise temperature and speed up the reaction by means of a catalyst. (The buildings and equipment in which a manufacturing process is carried out are called a **plant**.) The yield of ammonia is about 10%. Ammonia is condensed out of the mixture. With a boiling point of −33 °C, which is higher than that of most gases, ammonia is easily liquefied. The nitrogen and hydrogen which have not reacted are recycled through the plant (see Figure 23.2A).

The hydrogen for the Haber process is obtained from natural gas. The process takes place in a number of stages. The overall reaction is:

Natural gas (methane) + Steam → Hydrogen + Carbon monoxide

$$CH_4(g) + H_2O(g) \rightarrow 3H_2(g) + CO(g)$$

Carbon monoxide is oxidised to carbon dioxide and removed to leave hydrogen.

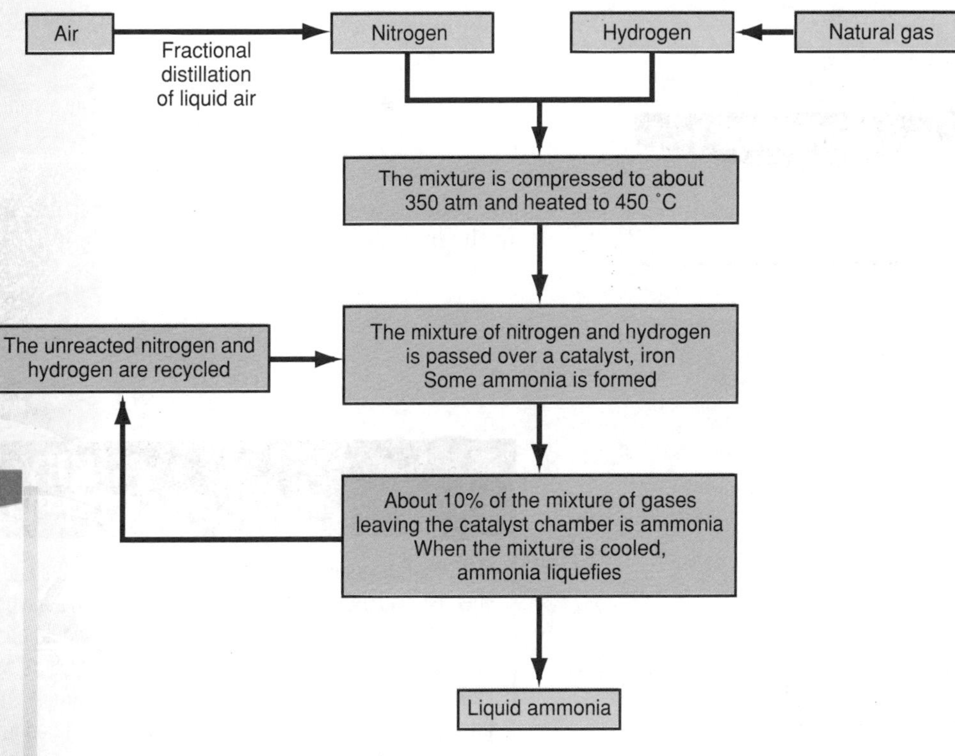


*Figure 23.2A* ▲ A flow diagram for the Haber process

SUMMARY

The starting point in the manufacture of fertilisers is the manufacture of ammonia from nitrogen and hydrogen by the Haber process. Hydrogen is obtained from natural gas, and nitrogen is obtained from air.

## Making fertilisers from ammonia

A concentrated solution of ammonia can be used as a fertiliser. It is easier to store solid fertilisers, such as ammonium salts.

Being a base, ammonia is neutralised by acids to yield ammonium salts

Ammonia + Nitric acid → Ammonium nitrate

$$NH_3(aq) + HNO_3(aq) \rightarrow NH_4NO_3(aq)$$

Ammonia + Sulphuric acid → Ammonium sulphate

$$2NH_3(aq) + H_2SO_4(aq) \rightarrow (NH_4)_2SO_4(aq)$$

Ammonia + Phosphoric acid → Ammonium phosphate

$$3NH_3(aq) + H_3PO_4(aq) \rightarrow (NH_4)_3PO_4(aq)$$

SUMMARY

Ammonia is used as a fertiliser, but ammonium nitrate, sulphate and phosphate are preferred..

# 23.3 Manufacture of nitric acid

Nitric acid is made by the oxidation of ammonia (see Figure 23.3A).

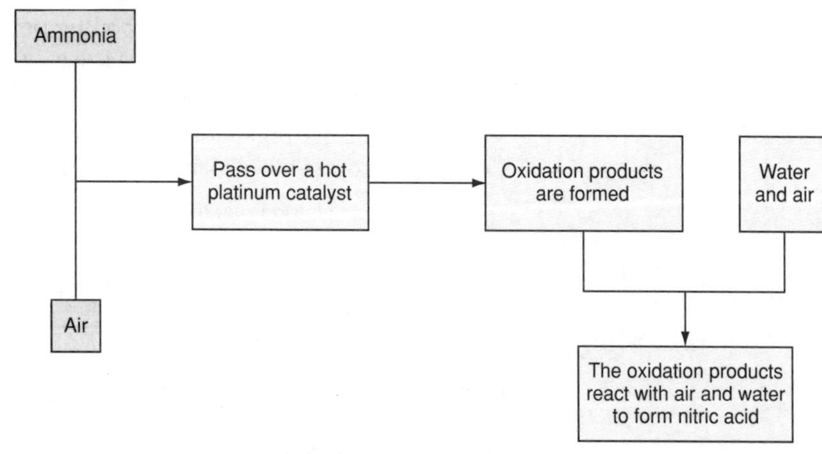


*Figure 23.3A* A flow diagram for the manufacture of nitric acid

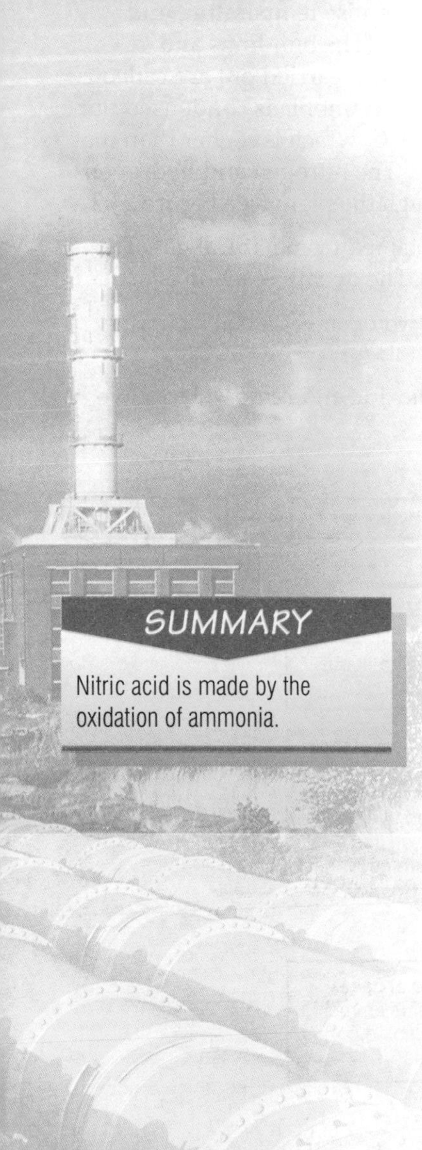

**SUMMARY**

Nitric acid is made by the oxidation of ammonia.

In addition to its importance in the fertiliser industry, nitric acid is used in the manufacture of explosives, such as TNT (trinitrotoluene), and dyes.

## KEY SCIENTIST

In 1846, Professor Sobrero, an Italian chemist, had the idea of reacting concentrated nitric acid with glycerol. He obtained a liquid which explodes when ignited by a lighted fuse. He called it **nitroglycerine**. The new compound was welcomed by miners who started to use it for blasting away rock. Unfortunately, there were accidents because nitroglycerine can explode if it receives a sudden blow, and this sometimes happened in transport.

This drawback did not stop Immanuel Nobel from beginning to manufacture nitroglycerine in 1863 in Sweden. A year later, an explosion killed four people in the factory, including his son, Emil. The other son, Alfred, continued manufacturing, but he looked for a method of stablising nitroglycerine to withstand mechanical shock. He found that a type of clay called kieselguhr would absorb nitroglycerine to form a mixture which did not explode when struck and could only be detonated by a lighted fuse. He patented his invention as **dynamite**. Alfred Nobel made a fortune out of his new, safer explosive. As well as being used to make tunnels, roads and railways and in mining, dynamite was sometimes used in warfare. Alfred Nobel was distressed to see his invention being misused in this way, and he became interested in promoting peace between nations. He financed the Nobel Foundation in Stockholm. This awards prizes each year to the people who have done the most effective work for peace and the most outstanding work in medicine, physics, chemistry and literature.

# 23.4 ▶ Manufacture of sulphuric acid

Sulphuric acid is made by the **contact process**. Sulphur dioxide and oxygen combine in contact with a catalyst, vanadium(V) oxide. Air is used as a source of oxygen. Sulphur dioxide is obtained by:

- Burning sulphur (deposits of sulphur occur in many countries).
- As a by-product of the extraction of metals from sulphide ores.
- As a by-product of the removal of unwanted, unpleasant-smelling sulphur compounds from petroleum oil and natural gas.

A flow diagram for the industrial process is shown in Figure 23.4A.

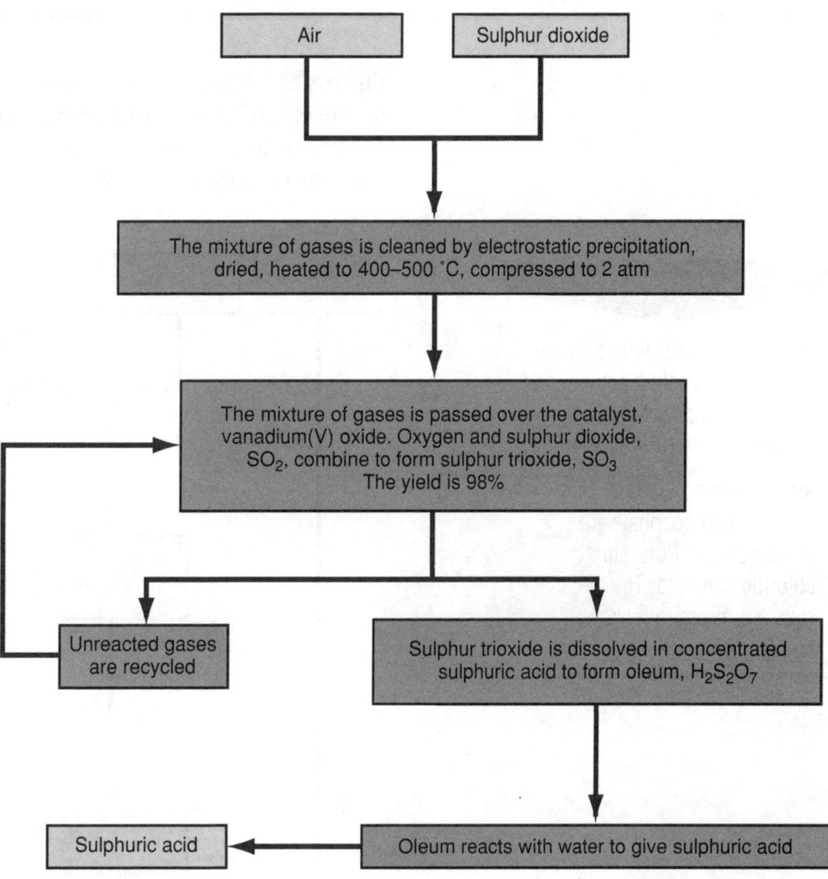


*Figure 23.4A* ▲ The contact process

## SUMMARY

Sulphuric acid is made by the contact process. Sulphur or a metal sulphide is oxidised to sulphur dioxide. Some sulphur dioxide is obtained from natural gas. Sulphur dioxide is oxidised to sulphur trioxide. This reacts with concentrated sulphuric acid to form oleum, which reacts with water to form sulphuric acid. The flow diagram shows the steps involved.

Sulphur trioxide reacts with water to give sulphuric acid

Sulphur trioxide + Water → Sulphuric acid

$$SO_3(s) + H_2O(l) \rightarrow H_2SO_4(l)$$

It is dangerous for industry to carry out the reaction in this way. As soon as sulphur trioxide meets water vapour, it reacts to form a mist of sulphuric acid. It is safer to absorb the sulphur trioxide in sulphuric acid to form oleum, $H_2S_2O_7$, and then convert the oleum into sulphuric acid.

Sulphuric acid is important in the fertiliser industry because it is needed for the manufacture of phosphoric acid. It has many other uses, such as the manufacture of paints, pesticides and plastics.

## 23.5 Manufacture of phosphoric acid

**SUMMARY**

Phosphoric acid is made from calcium phosphate.

Phosphoric acid is made from the widespread ore calcium phosphate by a reaction with sulphuric acid.

Calcium phosphate + Sulphuric acid → Phosphoric acid + Calcium sulphate

## 23.6 Making NPK fertilisers

**SUMMARY**

The most popular fertilisers are those which contain compounds of nitrogen, phosphorus and potassium: the NPK fertilisers.

Ammonium nitrate and ammonium phosphate are manufactured. Potassium chloride is mined. The three salts are mixed to make NPK fertilisers.

Figure 23.6A shows a flow diagram for the manufacture of NPK fertilisers. It shows how ammonium nitrate and ammonium phosphate are manufactured. There is no need to make potassium chloride because there are plentiful deposits of it in the UK.

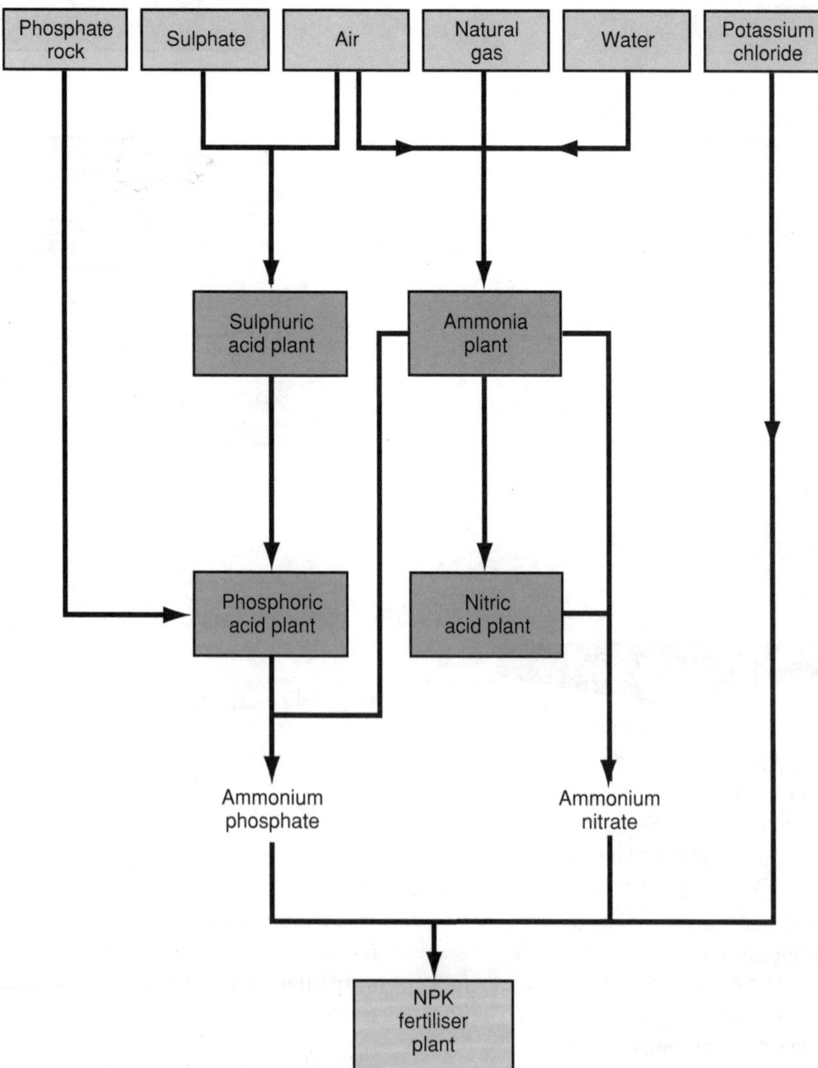

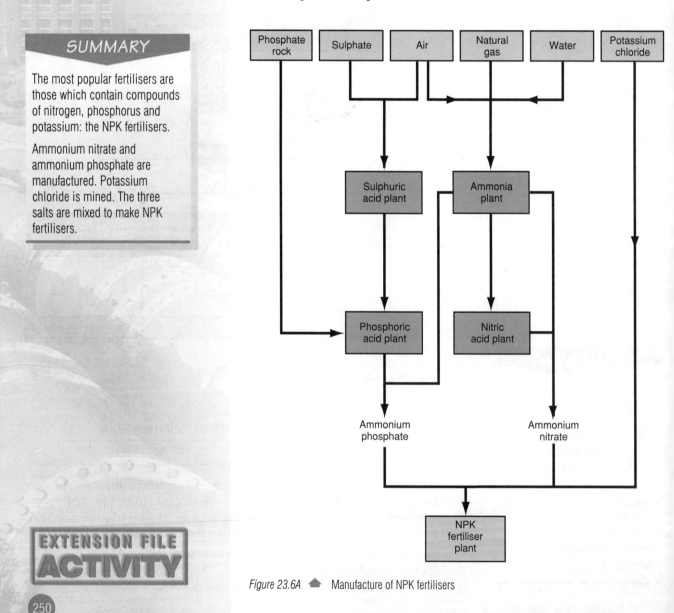

*Figure 23.6A* ▲ Manufacture of NPK fertilisers

## CHECKPOINT

1 State two advantages of the fertiliser ammonium phosphate over the insoluble salt calcium phosphate.

2 What methods do farmers use for applying (a) ammonia and (b) ammonium sulphate?

3 Why are farmers prepared to pay a lot for fertilisers? Why do all fertilisers contain nitrogen?

4 What is the advantage of using nitrogen as a starting material in the manufacture of fertilisers? What difficulty had to be overcome before manufacture started? Who solved the difficulty and how?

5 NPK fertilisers are popular. What do the letters NPK stand for? Explain why it is important that fertilisers contain (a) N, (b) P and (c) K.

6 State three routes by which nitrogen from the air finds its way into the soil. Why is the nitrogen content of the air not used up?

7 The figure below shows how the yields of winter wheat and grass increase as more nitrogenous fertiliser is used. After studying the graphs, say what mass of nitrogen you would apply to a 100 hectare field to avoid waste and to give a maximum yield of (a) winter wheat and (b) grass.

Grass response to nitrogen fertiliser

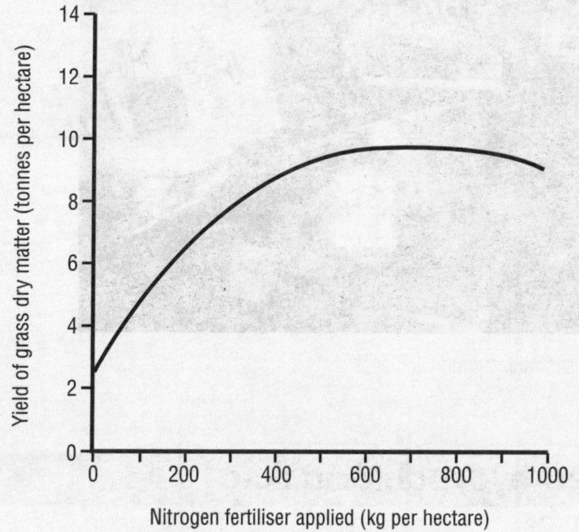


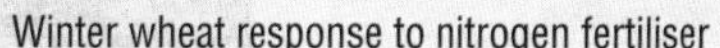
Winter wheat response to nitrogen fertiliser

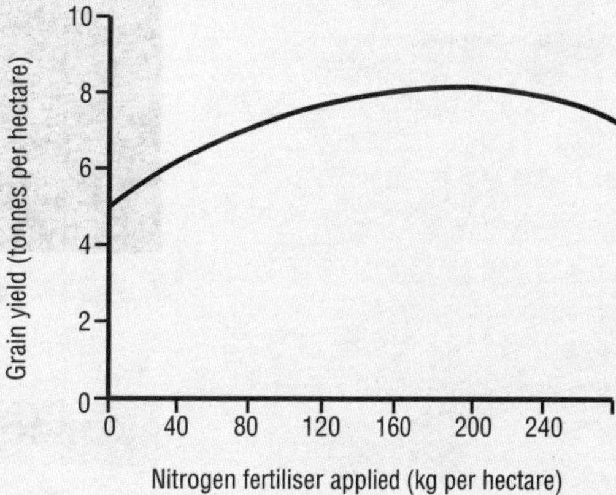


8 A dairy farmer spreads 240 tonnes of nitrochalk on his pasture. It costs £50 per tonne. If his milk cheque is £6000 a month, how long will it take him to recoup the cost of the fertiliser?

9 (a) The table opposite shows the mass of nitrogen, phosphorus and potassium (in kg/hectare) removed from the soil when different crops are grown. Name the crops which take out a large quantity of potassium. How do these crops differ from the rest?

| *Crop* | *Mass* (kg/hectare) *N* | *P* | *K* |
|---|---|---|---|
| Wheat grain | 115 | 22 | 26 |
| Oat grain | 72 | 13 | 18 |
| Sugar beet root | 86 | 14 | 302 |
| Potatoes | 109 | 14 | 133 |
| Pasture grass | 128 | 14 | 100 |

(b) The table opposite shows the composition of three NPK fertilisers that a farmer uses. Which fertiliser should she use (i) on her pasture (ii) on her wheat and (iii) on her sugar beet? Explain your choices.

| *Fertiliser* | *Composition* (percentage mass) *N* | *P* | *K* |
|---|---|---|---|
| Fertiliser A | 62 | 13 | 25 |
| Fertiliser B | 22 | 27 | 51 |
| Fertiliser C | 44 | 19 | 37 |

# 23.7 Concentrated acids

*Figure 23.7A* ▲ A Hazard sign for a corrosive substance

Check that you understand the difference between a concentrated acid, a strong acid (see 9.1) and a corrosive acid.
The photograph shows a spillage of corrosive acid being diluted and washed away.

A concentrated solution is one that contains a large amount of solute per litre of solution. Concentrated acids take part in the same chemical reactions as dilute acids but react more vigorously. In addition, concentrated sulphuric acid and concentrated nitric acid take part in some reactions which are not shared by the dilute acids (see below). Concentrated solutions of acids are extremely corrosive. They attack skin and clothing. Great care must be taken in handling concentrated acids and in transporting them. Occasionally, a lorry carrying a corrosive acid is involved in an accident (see Figure 23.7B). Two methods are employed to deal with a spill. The acid may be diluted with plenty of water and then washed away or it may be diluted and then neutralised with a base.

*Figure 23.7B* ▲ Tanker in simulated road accident

## Concentrated hydrochloric acid

Concentrated hydrochloric acid has the same reactions as dilute hydrochloric acid. The concentrated acid reacts more vigorously with metals, bases and carbonates.

## Concentrated sulphuric acid

Concentrated sulphuric acid takes part in the same reactions as dilute sulphuric acid and has two additional properties. It is an **oxidising agent** and a **dehydrating agent**.

### Oxidising agent

1 Copper and other metals low in the reactivity series do not react with dilute sulphuric acid. Concentrated sulphuric acid reacts with these metals. With copper it forms copper(II) sulphate, sulphur dioxide and water. Sulphuric acid has acted as an oxidising agent; it has taken electrons from copper atoms to form copper ions.

Copper + Conc. sulphuric acid → Copper(II) sulphate + Sulphur dioxide + Water

$$Cu(s) + 2H_2SO_4(aq) \rightarrow CuSO_4(aq) + SO_2(g) + 2H_2O(l)$$

2 Hot, dilute sulphuric acid reacts with iron to form iron(II) sulphate, $FeSO_4$. With concentrated sulphuric acid, the product is iron(III) sulphate, $Fe_2(SO_4)_3$, showing that concentrated sulphuric acid has acted as an oxidising agent, taking electrons from $Fe^{2+}$ ions to form $Fe^{3+}$ ions.

## Dehydrating agent

Ask your teacher to let a drop of concentrated sulphuric acid fall on to a piece of cloth. You will see a brown patch develop, which looks exactly as though the cloth has been burnt. **Don't try it** – but people who have accidentally spilt concentrated sulphuric acid on their skins will tell you that the same thing happens: it feels very much like being burnt by fire. The reason is that concentrated sulphuric acid removes water from the cloth, from skin and from other substances. A reagent which removes water from a compound is called a **dehydrating agent** (hudor = water in Greek).

Concentrated sulphuric acid has acidic properties. It is also an oxidising agent, as in its reactions with e.g. copper and iron(II) salts.

Concentrated sulphuric acid is a dehydrating agent removing the elements of water from e.g. copper(II) sulphate-5-water and sucrose.

1 Copper(II) sulphate crystallises with water of crystallisation as copper(II) sulphate-5-water, $CuSO_4.5H_2O$. Concentrated sulphuric acid can remove the water of crystallisation.

$$\text{Copper(II) sulphate-5-water (blue crystals)} \xrightarrow{\text{Conc. sulphuric acid}} \text{Copper(II) sulphate (white powder)} + \text{water;}$$

Heat is given out

The water of crystallisation combines with the sulphuric acid to leave anhydrous (without water) copper(II) sulphate, which is a white powder. Evidently, the water of crystallisation gives copper(II) sulphate-5-water its colour and its crystalline form.

2 Cane sugar, sucrose, is a carbohydrate (see Topic 30.4) of formula $C_{12}H_{22}O_{11}$. Concentrated sulphuric acid dehydrates sucrose (removes the elements of water from it) to leave a form of carbon known as 'sugar charcoal'.

$$\text{Sucrose} \xrightarrow{\text{Conc. sulphuric acid}} \text{Carbon} + \text{Water; Heat is given out}$$

$$C_{12}H_{22}O_{11}(s) \rightarrow 12C(s) + 11H_2O(l)$$

3 The reaction between concentrated sulphuric acid and water itself is very exothermic. This makes it difficult to prepare a dilute solution of sulphuric acid from concentrated sulphuric acid. Never attempt to make a dilute solution by adding water to concentrated sulphuric acid. Heat will be generated in a small volume of acid and will make the added water boil and splash out of the container, bringing with it a shower of concentrated sulphuric acid. Instead, add concentrated sulphuric acid slowly, with stirring, to a large volume of water. Then the heat generated will be spread through a large volume of water.

The figure shows the care that must be taken in diluting concentrated sulphuric acid.

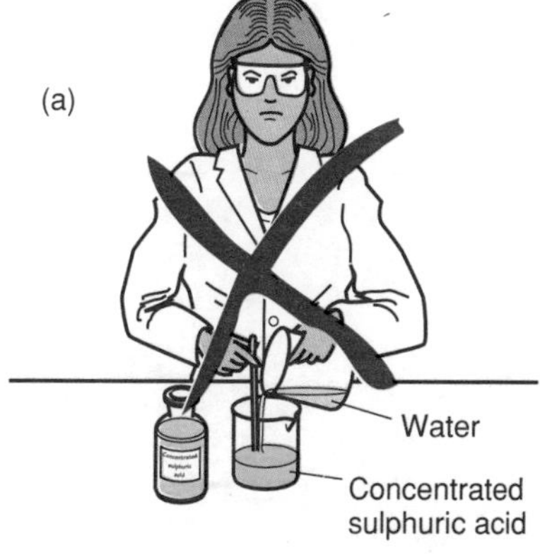


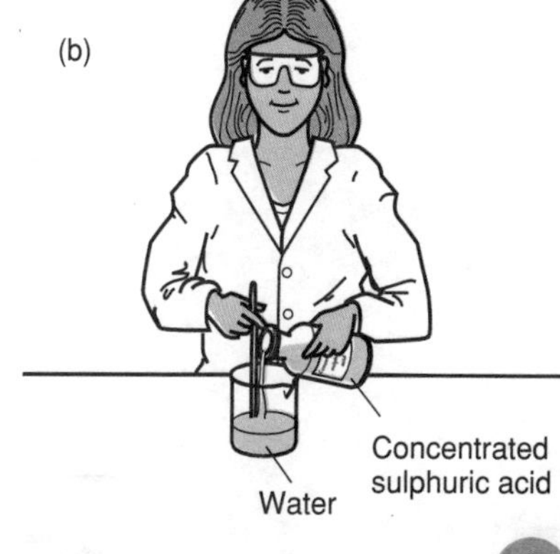


*Figure 23.7C* ▶
(a) Don't try it this way
(b) This is how to dilute a concentrated acid safely – but wear safety glasses

## Nitric acid – concentrated and dilute

Concentrated nitric acid takes part in the same reactions as dilute nitric acid and also acts as an oxidising agent. It is an even more powerful oxidising agent than concentrated sulphuric acid.

Dilute nitric acid is an oxidising agent. When heated, it reacts with copper to form copper(II) nitrate. Hydrogen is not formed. The colourless gas nitrogen oxide, NO, and the pungent brown gas nitrogen dioxide, $NO_2$, are formed. This gas is toxic, and the reaction must therefore be carried out in a fume cupboard.

When you need to work with a concentrated acid or alkali it is essential to wear safety glasses (see Figure 23.7C). Spilling a concentrated acid or alkali on your skin is painful and dangerous, and you should immediately wash off the spill with plenty of cold water. The danger to your eyes is very much greater. Firstly, the tissues of the eye are very delicate and easily damaged. Secondly, if you get something in your eye, you cannot see your way to the cold water tap to wash it out. Concentrated alkalis are dangerous too.

### SUMMARY

Concentrated sulphuric acid is an oxidising agent and a dehydrating agent. Nitric acid, both concentrated and dilute, is an oxidising agent. In addition, the concentrated acids take part in the same chemical reactions as the dilute acids.

### CHECKPOINT

1 Name two types of reaction which concentrated sulphuric acid takes part in but dilute sulphuric acid cannot bring about. Give one example of each.

2 'Here lies Susan still and placid.
She added water to the acid.'

Suggest which acid Susan could have been using to result in an accident. Explain why her method was dangerous. Describe how she should have diluted the acid.

3 'Dilute nitric acid is an oxidising agent.' Give two reactions which support this statement.

4 Explain why the following reactions are oxidation reactions (see Topic 12).

(a) Copper is converted into copper(II) sulphate
(b) Iron is converted into iron(II) sulphate
(c) Iron is converted into iron(III) sulphate
(d) Iron(II) sulphate is converted into iron(III) sulphate.

# Topic 24 Fuels

## 24.1 Methane

FIRST THOUGHTS

Developing countries need more fuel. Industrial countries have difficulty in disposing of all their waste. One solution to both problems is biogas. You can find out about biogas in this section.

Father Conlon and Father Williams are monks at Bethlehem Abbey in Northern Ireland. In 1987 they won a 'pollution abatement technology award'. They feed manure from their farm into a **biomass digester**. Biomass is material of plant or animal origin. Bacteria feed on biomass and make it decay. Under anaerobic conditions (in the absence of air), a gas called **biogas** forms. Biogas is a fuel, and the monks burn it to provide the monastery with heating. The solid remains of the biomass form an odourless compost which they bag and sell as a fertiliser for use on gardens, potato farms and golf courses.

Many farms in the UK now have biomass digesters. There are millions of such digesters in India and China. India's huge cattle population supplies plenty of dung for digestion. The biogas produced is used as a fuel for cooking and heating. It burns with a hotter, cleaner flame than cattle dung. The residual sludge which gathers in the digester is a better fertiliser than raw dung. Both the gas and the sludge are clean and odourless.

Science at work

Every year, 25 million tonnes of organic waste goes into landfill sites in the UK. Biogas forms as the rubbish decays. Most landfill operators burn biogas to get rid of it. Others sink pipes into the landfill and pump out biogas for sale. In Bedfordshire, biogas from a landfill is used to heat the kilns in a brickworks. On Merseyside, biogas is used to heat the ovens in a Cadburys' biscuit factory.

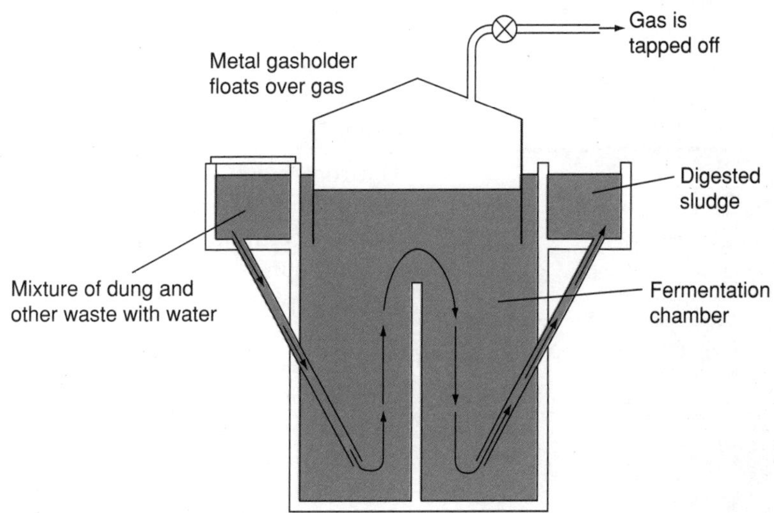

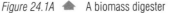

*Figure 24.1A* ▲ A biomass digester

SUMMARY

Biogas is formed when biomass (plant and animal matter) decays in the absence of air. The chief component of biogas is methane. Biogas can be collected from landfill rubbish sites, from sewage works and on farms. It can be burned as a fuel. Biogas generators are common in India and China. Methane accumulates in coal mines, where it can cause explosions.

*Figure 24.1B* ▲ A gas rig in the North Sea

The chief component of biogas is **methane**. Methane is the gas we burn in Bunsen burners and gas cookers. It is the valuable fuel we call **natural gas** or **North Sea gas**. Reserves of North Sea gas are usually found together with North Sea oil. Methane is also the gas called **marsh gas** which bubbles up through stagnant water. It collects in coalmines, and methane explosions have been the cause of many pit disasters.

## CHECKPOINT

1 (a) What are the advantages of biogas generators for developing countries?
(b) Why does India have plenty of cattle dung?
(c) In what ways is biogas a more convenient fuel than dung?
(d) What advantages do biogas generators have for industrial countries?

2 (a) What is biomass? Briefly describe how biomass can be fermented to produce biogas. What other product is formed?
(b) What are the problems in adapting the process to produce fuel gas for domestic use?
(c) Can you point out any situations in which a biogas generator might both save money and also benefit the environment?

# 24.2 Alkanes

**FIRST THOUGHTS**

Methane is one member of a family of compounds, the alkanes. Composed of hydrogen and carbon only, the alkanes are hydrocarbons. You will meet another family of hydrocarbons, the alkenes, in Topic 30.1.

Methane is a **hydrocarbon**, a compound of hydrogen and carbon. With formula $CH_4$, it is the simplest of the hydrocarbons. Figure 24.2A shows how, in a molecule of methane, four covalent bonds join hydrogen atoms to a carbon atom.

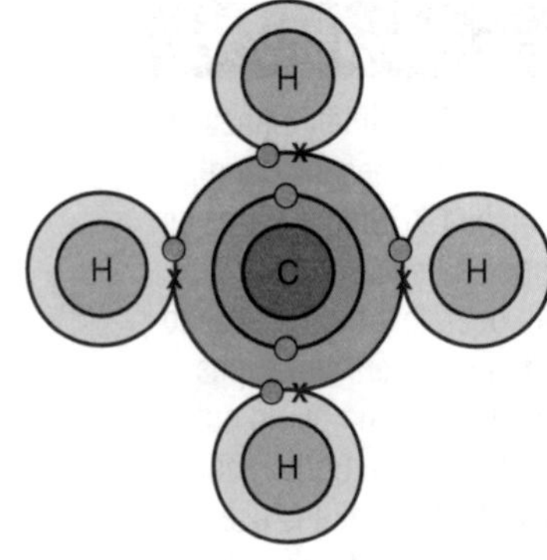


*Figure 24.2A* Bonding in methane

Hydrocarbons are **organic compounds**. Originally, the term 'organic compound' was applied to compounds which were found in plant and animal material, for example sugars, fats and proteins. All these compounds contain carbon, and the term 'organic compound' is now used for all carbon compounds, whether they have been obtained from plants and animals or made in a laboratory. However, simple compounds like carbon dioxide and carbonates are not usually described as organic compounds. Most organic compounds are covalent. The salts of organic acids contain ionic bonds.

You often meet the term 'organic compound'. It means a carbon compound, with the exception of simple compounds such as carbonates.

Methane is one of a **series** of hydrocarbons called the **alkanes**. The next members of the series are ethane, $C_2H_6$, propane, $C_3H_8$, and butane, $C_4H_{10}$ (see Figure 24.2B). Many hydrocarbons have much larger molecules.

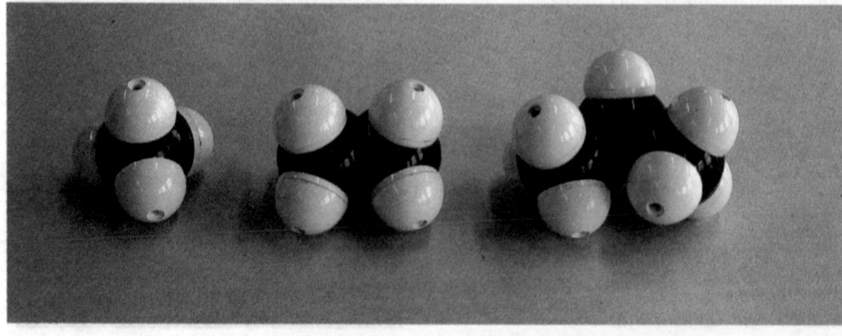

*Figure 24.2B* Models of methane, ethane and propane

As well as molecular formulas, $CH_4$, $C_2H_6$ and $C_3H_8$, the compounds have **structural formulas**. A structural formula shows the bonds between atoms.

The srtructural formula of a compound shows the bonds between atoms.

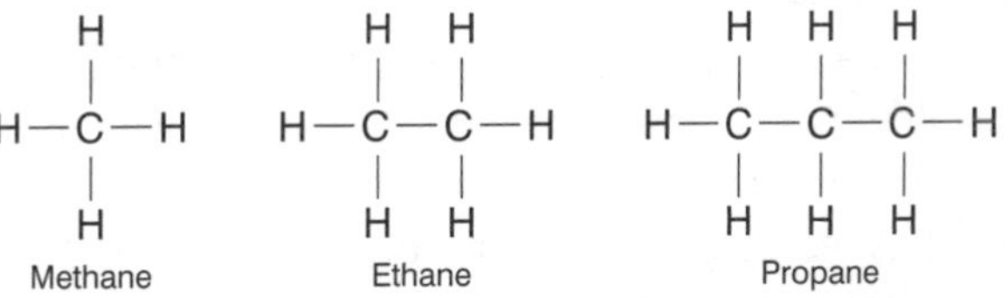


Each compound differs from the next in the series by the group

```
  H
  |
—C—
  |
  H
```

*Table 24.1* ▼ The alkanes

| Alkanes | Formula $C_nH_{2n+2}$ |
|---|---|
| Methane | $CH_4$ |
| Ethane | $C_2H_6$ |
| Propane | $C_3H_8$ |
| Butane | $C_4H_{10}$ |
| Pentane | $C_5H_{12}$ |
| Hexane | $C_6H_{14}$ |
| Heptane | $C_7H_{16}$ |
| Octane | $C_8H_{18}$ |

**SUMMARY**

Carbon compounds are called organic compounds.
Hydrocarbons are compounds of carbon and hydrogen. The alkanes are hydrocarbons with the general formula $C_nH_{2n+2}$. They are a homologous series: each member differs from the next by a $CH_2$ group.

A set of chemically similar compounds in which each member differs from the next in the series by a $CH_2$ group is called a **homologous series**. Table 24.1 lists the first members of the **alkane series**. Their formulas all fit the general formula $C_nH_{2n+2}$, for example for pentane $n = 5$, giving the formula $C_5H_{12}$. The alkanes are described as **saturated** hydrocarbons. This means that they contain only single bonds between carbon atoms. This is in contrast to the **alkenes**. The alkanes are unreactive towards acids, bases, metals and many other chemicals. Their important reaction is combustion.

## Isomerism

Sometimes it is possible to write more than one structural formula for a molecular formula. For the molecular formula $C_4H_{10}$, there are two possible structures

```
    H  H  H  H              H  H  H
    |  |  |  |              |  |  |
 H—C—C—C—C—H          H—C—C—C—H
    |  |  |  |              |  |  |
    H  H  H  H              H  |  H
                                |
      Butane                 H—C—H
                                |
(a)                         (b) H      Methylpropane
```

**SUMMARY**

Isomers have the same molecular formula and different structural formulas.

The difference is that (a) has an unbranched chain of carbon atoms and (b) has a branched chain. The formulas belong to different compounds, which differ in boiling point and other physical properties. The compound with formula (a) is called butane; the compound with formula (b) is called methylpropane. These compounds are **isomers**. Isomers are compounds with the same molecular formula and different structural formulas.

## CHECKPOINT

1 Explain what is meant by a 'homologous series'.

2 (a) Explain what is meant by the term 'alkane'.
(b) Why are alkanes very important in our way of life?
(c) Where do we obtain alkanes?
(d) The molecular formula for propane is $C_3H_8$. Write the structural formula for propane.
(e) What information does the structural formula of a compound give that the molecular formula does not give?
(f) Explain what is meant by the term 'isomerism'. Illustrate your answer by referring to pentane, $C_5H_{12}$.

## Substitution reactions

Alkanes react with chlorine in sunlight. The products are a **chloroalkane** and hydrogen chloride. For example,

| Methane | + | Chlorine | → | Chloromethane | + | Hydrogen chloride |
|---|---|---|---|---|---|---|
| $CH_4(g)$ | + | $Cl_2(g)$ | → | $CH_3Cl(g)$ | + | $HCl(g)$ |

```
     H                      H
     |                      |
H—C—H + Cl—Cl → H—C—Cl + H—Cl
     |                      |
     H                      H
```

The reaction is called a **substitution reaction** because one atom in the molecule of alkane is replaced by another atom. It is called **chlorination** because it is a chlorine atom that is substituted into the alkane molecule. It is described as a **photochemical reaction** because it takes place only in sunlight. The reaction can continue further.

| Chloromethane | + | Chlorine | → | Dichloromethane | + | Hydrogen chloride |
|---|---|---|---|---|---|---|
| $CH_3Cl(g)$ | + | $Cl_2(g)$ | → | $CH_2Cl_2(g)$ | + | $HCl(g)$ |

A mixture of products, including trichloromethane, $CHCl_3$ (chloroform) and tetrachloromethane, $CCl_4$ (also called carbon tetrachloride), is formed. Trichloromethane, $CHCl_3$, is chloroform, which was used as an anaesthetic. The mixture of products formed in the chlorination of an alkane may make a useful solvent without the need to isolate the individual compounds. A much used solvent is 1,1,1-trichloroethane. The 1,1,1- in the name tells you that the three chlorine atoms are all on the same carbon atom, and the structure is

```
     Cl  H
     |   |
Cl—C—C—H
     |   |
     Cl  H
```

Bromine also reacts with alkanes in sunlight. The substitution product which is formed is called a **bromoalkane**. Hydrogen bromide is the other product, e.g.

| Hexane | + | Bromine | → | Bromohexane | + | Hydrogen bromide |
|---|---|---|---|---|---|---|
| $C_6H_{14}(l)$ | + | $Br_2(l)$ | → | $C_6H_{13}Br(l)$ | + | $HBr(g)$ |

Iodination is very slow and is a reversible reaction. Fluorination is a dangerously violent reaction. The substitution of a halogen in a compound is called **halogenation**. Chlorination and bromination are

In a substitution reaction, an atom or atoms in a molecule are replaced by another atom or atoms. An example is the halogenation of alkanes. This includes chlorination to form chloroalkanes and bromination to form bromoalkanes.

SUMMARY

Alkanes take part in substitution reactions, e.g. halogenation.

both halogenation reactions. Compounds in which halogen atoms have been substituted for one or more of the hydrogen atoms in an alkane are called **halogenoalkanes**. Chloromethane, tetrachloromethane and bromohexane are all halogenoalkanes.

CHECKPOINT

1 (a) Write the equation for the reaction between ethane and chlorine to form chloroethane.
(b) Write the structural formulas for all the substances.
(c) Construct molecular models of all the substances.

2 (a) Write the structural formulas for iodomethane, dibromomethane, fluoroethane, tetrachloromethane.
(b) Make molecular models of the molecules.

3 Is the chlorination of methane exothermic or endothermic? Explain your answer.

## 24.3 Combustion

SUMMARY

The combustion of hydrocarbons is exothermic. The products of complete combustion are carbon dioxide and water. Incomplete combustion produces carbon and poisonous carbon monoxide.

Hydrocarbons burn in a plentiful supply of air to form carbon dioxide and water. The reaction is **exothermic**: energy is released.

Methane + Oxygen → Carbon dioxide + Water; Energy is released
$CH_4(g) + 2O_2(g) \rightarrow CO_2(g) + 2H_2O(l)$

Butane + Oxygen (in Camping Gaz) → Carbon dioxide + Water; Energy is released

Octane + Oxygen (in petrol) → Carbon dioxide + Water; Energy is released

CHECKPOINT

Revise what you learned about combustion in Topic 14.6.

1 (a) What harmless combustion products are formed when hydrocarbons burn in plenty of air?
(b) When does incomplete combustion take place?
(c) What are the products of incomplete combustion? What is dangerous about incomplete combustion? How can it be avoided?

## 24.4 Petroleum oil and natural gas

FIRST THOUGHTS

Prospecting for oil is an even more important business than prospecting for gold. *How do prospectors find oil? And how was oil formed in the first place?* You can find out in this section.

In some parts of the world, a black, treacle-like liquid seeps out of the ground. At one time, farmers in Texas, USA, used to burn this substance to get rid of it when it formed troublesome pools on their land. The black liquid was called **petroleum oil** or **crude oil**. It is now regarded as one of the most valuable resources a country can have. The petrochemicals industry obtains valuable fuels such as petrol, and thousands of useful materials such as plastics, from crude oil. Natural gas is found in the same deposits as crude oil.

*We call oil and gas fossil fuels because they were formed by slow decay of the remains of dead sea creatures.*

Oil and gas are **fossil fuels**. They were formed millions of years ago when a large area of the Earth was covered by sea. Tiny sea creatures called plankton died and sank to the sea bed, where they became mixed with mud. Bacteria in the mud began to bring about the decay of the creatures' bodies. Decay took place slowly because there is little oxygen dissolved in the depths of the sea. As the covering layer of mud and silt grew thicker over the years, the pressure on the decaying matter increased. Bacterial decay at high pressure with little oxygen turned the organic matter into crude oil and natural gas.

The sediment on top of the decaying matter became compressed to form rock. Rocks formed in this way are called **sedimentary rocks**. Some of these rocks are porous: they contain tiny passages through which liquid and gas can pass. Others are impermeable: they do not let any substances through. Ground water carries crude oil and natural gas upward through porous rocks. They may reach the surface, but more often they become trapped by a **cap rock** of impermeable rock (see Figure 24.4A). There the crude oil and natural gas remain unless an oil prospector drills down to them.

## It's a fact!

*Have you heard of the petrol tree?* The gopher tree grows well in desert areas. Its sap contains about 30% hydrocarbons. Petrol could be made from the sap of this tree.

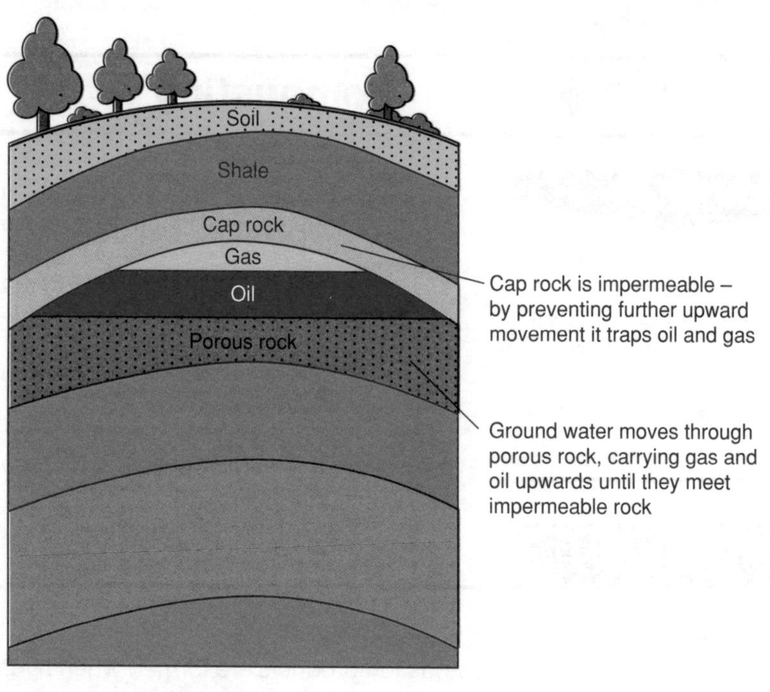
Soil
Shale
Cap rock
Gas
Oil
Porous rock
Cap rock is impermeable – by preventing further upward movement it traps oil and gas
Ground water moves through porous rock, carrying gas and oil upwards until they meet impermeable rock

*Figure 24.4A* ▲ An oil trap

Many countries have deposits of oil and gas, for example the USA, the USSR, Iran, Nigeria and the countries in the Arabian Gulf. The UK and Norway have oil and gas beneath the North Sea. The UK has piped ashore oil and gas since 1972. There are two ways of getting oil ashore from an oil well in the sea. One way is to lay a pipeline along the sea bed and pump the oil through it. The other method is to use 'shuttle tankers' which pick up the oil from the oilfield and transport it to a terminal on land. Natural gas is almost always brought ashore by pipeline.

Crude oil is transported from oil wells to refineries. Pipelines carry oil overland. Tankers carry oil overseas. When a tanker has an accident at sea, oil can escape to form a huge oil slick. The damage which the oil may cause to wildlife and coastlines is described in Topic 17.6.

Crude oil does not burn very easily. **Fractional distillation** is used to separate crude oil into a number of important fuels. The fractions are separated on the basis of their boiling point range. They are not pure compounds: they are mixtures of alkanes with similar boiling points.

## SUMMARY

Crude oil and natural gas were formed by the slow bacterial decay of animal and plant remains in the absence of oxygen. These fuels are found in many parts of the world.

*Figure 24.4B* An oil rig being towed out to sea

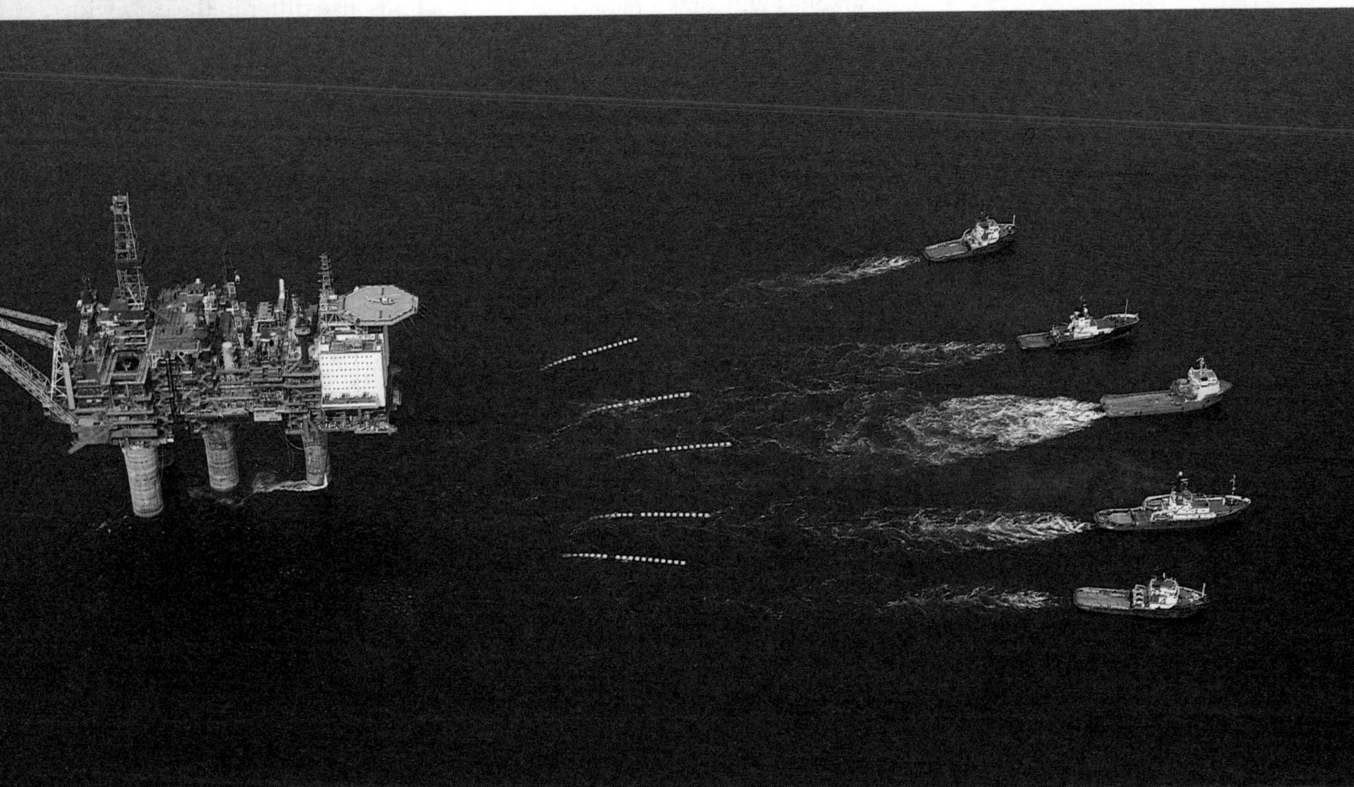

*Table 24.2* Petroleum fractions and their uses

| *Fraction* | *Approximate boiling point range (°C)* | *Approximate number of carbon atoms per molecule* | *Use* |
|---|---|---|---|
| Petroleum gases | below 0 | 1–4 | Petroleum gases are liquefied and sold in cylinders as 'bottled gas' for use in gas cookers and camping stoves. They burn easily at low temperatures. Smelly sulphur compounds must be removed to make bottled gas pleasant to use and non-polluting. |
| Gasoline (petrol) | 0–65 | 5–6 | Petrol is liquid at room temperature, but vaporises easily at the temperature of vehicle engines. |
| Naphtha | 65–170 | 6–10 | Naphtha is used by the petrochemicals industry as the source of a huge number of useful chemicals, e.g. plastics, drugs, medicines, fabrics (see Figure 24.4C overleaf). |
| Kerosene | 170–250 | 10–14 | Kerosene is a liquid fuel which needs a higher temperature for combustion than petrol. The major use of kerosene is as aviation fuel. It is also used in 'paraffin' stoves. |
| Diesel oil | 250–340 | 14–19 | Diesel oil is more difficult to vaporise than petrol and kerosene. The diesel engine has a special fuel injection system which makes the fuel burn. It is used in buses, lorries and trains. |
| Lubricating oil | 340–500 | 19–35 | Lubricating oil is a viscous liquid. With its high boiling point range, it does not vaporise enough to be used as a fuel. Instead, it is used as a lubricant to reduce engine wear. |
| Fuel oil | 340–500 | Above 20 | Fuel oil has a high ignition temperature. To help it to ignite, fuel oil is sprayed into the combustion chambers as a fine mist of small droplets. It is used in ships, heating plants, industrial machinery and power stations. |
| Bitumen | Above 500 | Above 35 | Bitumen has too high an ignition temperature to be used as a fuel. It is used to tar roads and to waterproof roofs and pipes. |

Physical properties of alkanes which depend on the size of the molecules include

- viscosity
- flash point
- ignition temperature
- boiling point.

They decide the uses that are made of petroleum fractions

Alkanes with small molecules boil at lower temperatures than those with large molecules. Alkanes with large molecules are more viscous than alkanes with small molecules. The fractions also differ in the ease with which they burn: in their **flash points** and **ignition temperatures**.

- When a fuel is heated, some of it vaporises. When the fuel reaches a temperature called the **flash point**, there is enough vapour to be set alight by a flame. Once the vapour has burned, the flame goes out.
- The **ignition temperature** of a fuel is higher than its flash point. It is the temperature at which a mixture of the fuel with air will ignite and continue to burn steadily.

The use that is made of each fraction depends on its boiling point range, flash point, ignition temperature and viscosity; see Table 24.2.

## It's a fact!

A petroleum fraction called white spirit is widely used as a solvent. In November, 1988, a fire at King's Cross underground station in London killed 30 people. Under the escalators, investigators found empty drums that had held 150 litres of white spirit. The staff use white spirit for cleaning the treads of the escalators. The presence of flammable white spirit could help to explain the speed with which the fire spread.

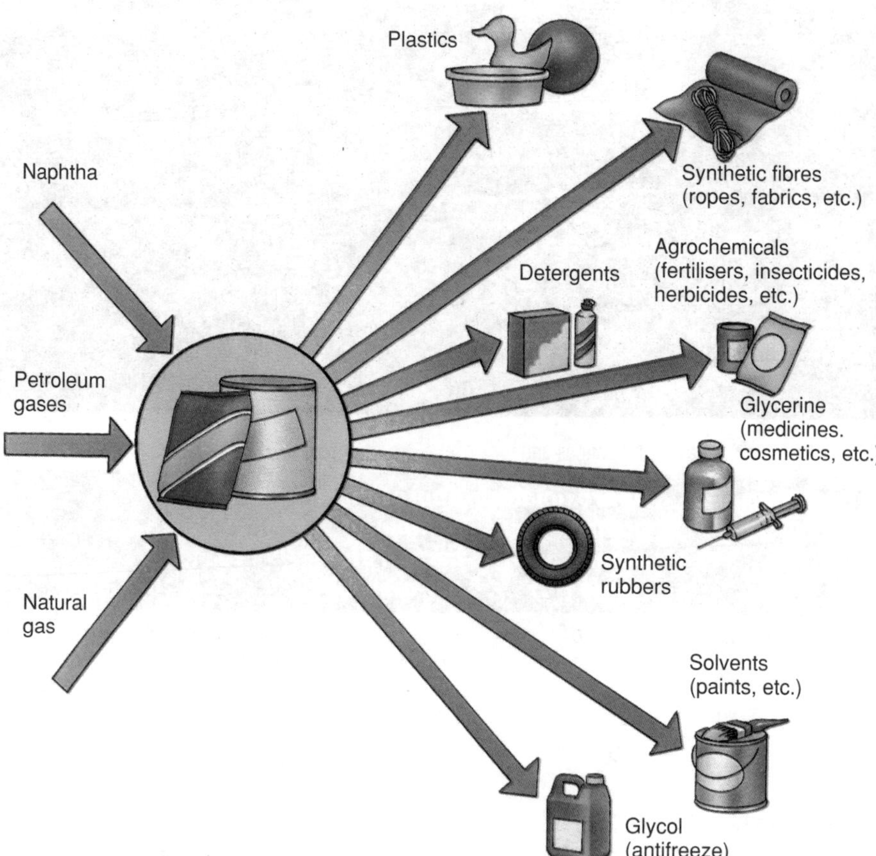


*Figure 24.4C* ▲ Petrochemicals from naphtha

## Cracking

We use more petrol, naphtha and kerosene than heavy fuel oils. Fortunately, chemists have found a way of making petrol and kerosene from the higher-boiling fractions, of which we have more than enough. The technique used is called **cracking**. Large hydrocarbon molecules are cracked (split) into smaller hydrocarbon molecules. A heated catalyst (aluminium oxide or silicon(IV) oxide) is used.

It's nice to meet a term that means what it sounds like. 'Cracking' means cracking large molecules into smaller molecules.

Vapour of hydrocarbon with large molecules and high b.p.

Mixture of hydrogen and hydrocarbons with smaller molecules and low b.p.

## SUMMARY

Crude oil is separated into useful components by fractional distillation. The use that is made of each fraction depends on its boiling point range, ignition temperature and other properties. The fuels obtained from crude oil are listed in Table 24.2. Cracking is used to make petrol and kerosene from heavy fuel oils. The petrochemicals industry makes many useful chemicals from petroleum.

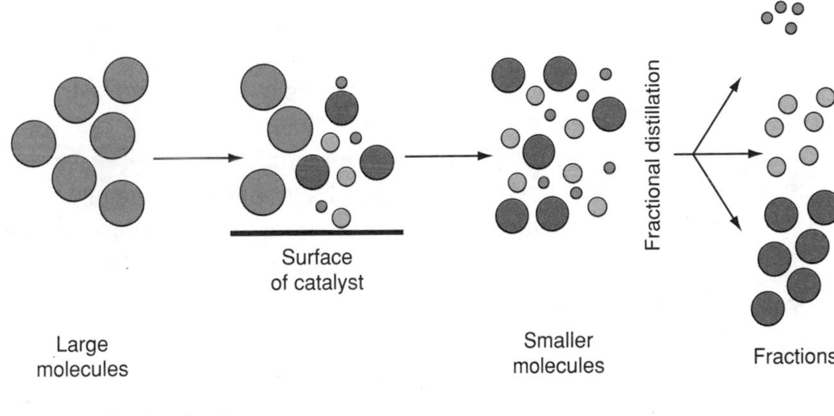


*Figure 24.4D* ▲ Cracking

## CHECKPOINT

▶ **1** The use of crude oil fractions in the UK is as follows:

Road transport 37%
Heating and power for industry 21%
Heating and power for other consumers 19%
Making chemicals 11%
Power stations 8%
Heating houses 4%

Show this information in the form of a pie-chart or a bar graph.

▶ **2** The figure below shows the percentage of crude oil that distils at various temperatures.

(a) From the graph, read the percentage of crude oil that distils:

(i) below 70 °C,
(ii) between 120 and 170 °C,
(iii) between 170 and 220 °C,
(iv) between 220 and 270 °C,
(v) between 270 and 320 °C,
(vi) above 320 °C.

(b) Draw a pie-chart to show these figures.

(c) The percentages vary from one sample of oil to another. North Sea oil contains a larger percentage of lower boiling point compounds than Middle East oil does. Which oil should sell for a higher price? Explain your answer.

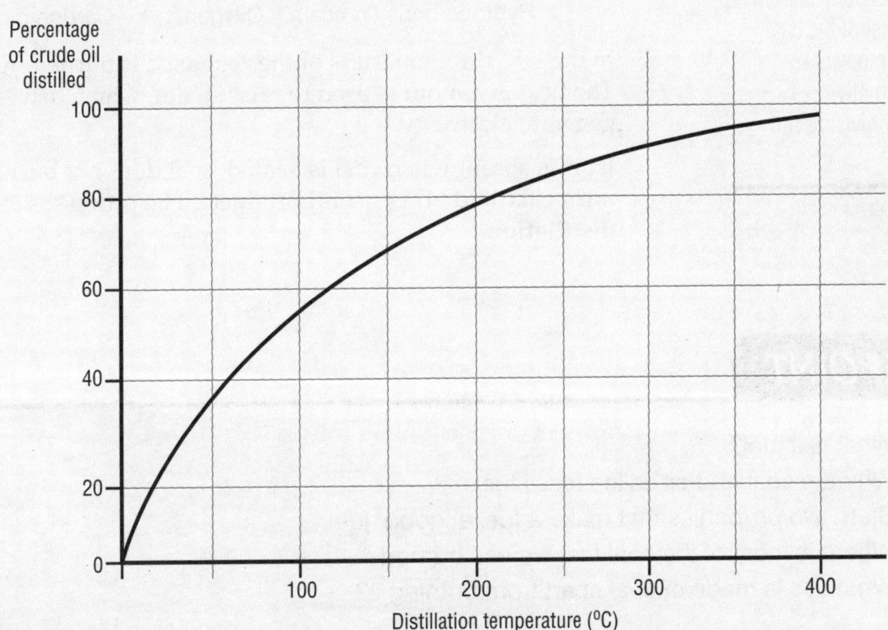

3 A barrel of Middle East oil contains 160 litres. From it are obtained:
natural gas (7 litres) kerosene (7 litres)
petrol (9 litres) gas oil (41 litres)
naphtha (23 litres) fuel oil (73 litres)
Show these figures as a pie-chart.

4 (a) How are petrol and kerosene obtained from crude oil?
(b) What is the name of the process for converting heavy fuel oil into petrol? What are the economic reasons for carrying out this reaction?

5 (a) What is crude oil?
(b) Why is it described as a fossil fuel?
(c) What useful substances are obtained from crude oil?
(d) Why is an increase in the price of crude oil such a serious matter?

## 24.5 Coal

### FIRST THOUGHTS

Coal is an important source of energy. It fuelled the Industrial Revolution and is still a major source of energy in the UK today.

Between 200 and 300 million years ago, the Earth was covered with dense forests. When plants and trees died, they started to decay. In the swampy conditions of that era, they formed peat. Gradually, the peat became covered by layers of mud and sand. The pressure of these layers and high temperatures turned the peat into coal. The mud became shale and the sand became sandstone. Coal is a fossil fuel: it was formed from the remains of living things.

Many countries have coal deposits. The countries with the largest coal-mining industries are the countries of the former USSR, the USA, China, Poland and the UK. Coal is a mixture of carbon, hydrocarbons and other compounds. When it burns, the main products are carbon dioxide and water.

Carbon (in coal) + Oxygen → Carbon dioxide
Hydrocarbons (in coal) + Oxygen → Carbon dioxide + Water

In the UK, three-quarters of the coal used is burned in power stations. The heat given out is used to raise steam, which drives the turbines that generate electricity.

If air is absent when coal is heated, coal does not burn. It decomposes to form coke and other useful products. The process is called **destructive distillation**.

### SUMMARY

Coal is a fossil fuel derived from plant remains. Most of the coal mined is burned in power stations. The destructive distillation of coal gives useful products.

### CHECKPOINT

1 (a) What is a fuel?
(b) Why are coal and oil called fossil fuels?
(c) State two properties that make a fuel a 'good' fuel.
(d) Where is most of the coal that we use burned?
(e) What use is made of coal apart from burning it?

# Theme Questions

**1** Iron is extracted from its ores in a blast furnace. Most iron is converted into steel in a basic oxygen furnace.

(a) Name the two substances, in addition to iron ore, which are fed into the top of the blast furnace.

(b) Explain how the ore haematite, $Fe_2O_3$, is reduced to iron.

(c) Molten iron and slag form at the base of the blast furnace. Explain how slag is formed.

(d) What impurities are removed from iron in the basic oxygen furnace?

(e) Explain the chemical reaction that removes them.

**2** Aluminium is mined as its oxide. It is extracted from its ore in a plant called an aluminium smelter.

(a) Name this aluminium oxide ore.

(b) State what process is used to extract aluminium from its oxide.

(c) Why can a blast furnace not be used for the extraction of aluminium?

(d) An aluminium ore called cryolite is used in the extraction of aluminium from aluminium oxide. What part does it play in the process?

(e) What environmental damage can be caused by an aluminium smelter?

(f) What two economies are made when aluminium is recycled?

**3** The table shows part of the reactivity series of metals.

| |
|---|
| Aluminium |
| Zinc |
| Iron |
| Tin |
| Lead |
| Copper |

Use the table to explain the following.

(a) Iron food cans are coated with tin.

(b) Zinc bars are attached to the hulls of ships below the waterline.

(c) When zinc powder is dropped into a solution of copper(II) sulphate, the colour of the solution fades.

(d) Galvanised (zinc-coated) steel does not rust.

(e) A mixture of aluminium and iron oxide is used to mend gaps in railway lines.

**4** Explain why:

(a) electrical wiring is made of copper

(b) saucepans are made from aluminium

(c) aeroplanes are made from aluminium alloys

(d) bells are made from bronze

(e) trumpets are made from brass

(f) bridges are made from steel

(g) baking foil is made from aluminium

(h) solder is made from brass and tin

(i) lead was for many years used for water pipes

(j) lead is no longer used for water pipes

(k) dental amalgams are made from mercury

(l) teeth can be fitted with gold caps.

**5** A geologist finds a green compound in a sample of rock. He does a number of experiments on the sample. The results are shown in the diagram.

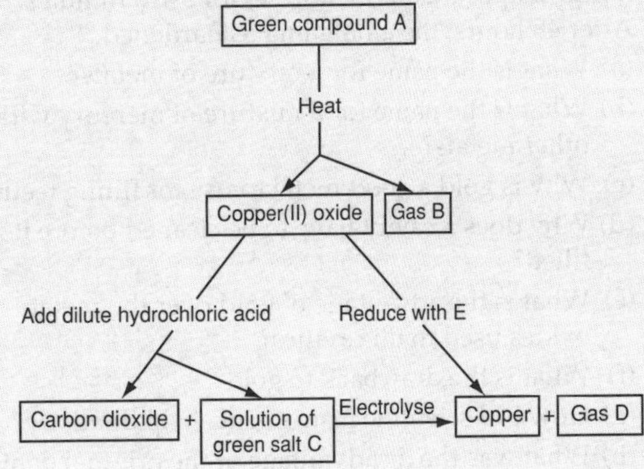


(a) Name the substances A, B, C and D.

(b) Name one substance which could be E. Describe (with a diagram if you wish) how this reaction could be carried out. Say, with an explanation, whether this reaction is an oxidation or a reduction.

(c) Is the formation of copper from C by electrolysis oxidation or reduction? Explain your answer.

**6** (a) The table shows the prices and the lifetimes of two kinds of AA-size dry cell.

| | *Alkaline manganese dry cell* | *Zinc-carbon dry cell* |
|---|---|---|
| *Price of cell* | £0.45 | £0.20 |
| *Life when used in* | | |
| *(i) electronic flash* | 180 flashes | 20 flashes |
| *(ii) walkman* | 14 hours | 7 hours |

Which of the two dry cells is the better buy when used in (i) an electronic flash (ii) a walkman? Explain your choices.

**7** Read the following passage and answer the questions on it.

**Mercury**

In ancient times, the material used for filling cavities in teeth was a cement made of metal oxides. In medieval times, the method of clearing the cavity of decayed matter and then filling it with gold leaf was employed. Early in the nineteenth century the first dental amalgam was used. (An amalgam is an alloy of mercury with other metals.) The original dental amalgam of bismuth, lead, tin and mercury melted at about 100 °C and was poured into the cavity at this temperature. Later the composition was altered to make an amalgam which melted at 66 °C.

In the late nineteenth century, dentistry was revolutionised by G.V. Black. He measured the force required to chew different foods and the pressure which fillings could stand without cracking. Black's formula, 65% silver, 27% tin, 6% copper and 2% zinc is still used today, although many manufacturers use more copper and less tin. Dentists mix this powder with mercury, put it into the cavity, 'carve' it into shape, and allow it to set for 5–10 minutes. After 48 hours, the amalgam has hardened.

(a) What is the name for a mixture of metals?
(b) What is the name for a mixture of mercury with other metals?
(c) Why is gold a good metal to use for filling teeth?
(d) Why does a cavity have to be cleaned before it is filled?
(e) What is the advantage of gold over the metal oxides used to fill cavities?
(f) What is the drawback to gold?
(g) How did scientific instruments help dentistry?
(h) What was the disadvantage of the original dental amalgam?
(i) Give three reasons why mercury is a good basis for a dental amalgam.
(j) Your little brother finds out that his fillings contain mercury. He becomes very alarmed because he has heard of mercury poisoning. How would you explain to him why the mercury in his teeth does not poison him?

8 The following metals are listed in order of decreasing reactivity. X and Y are two unknown metals.

K X Ca Mg Al Zn Y Fe Cu

(a) Will X react with cold water?
(b) Will Y react with cold water?
(c) Will Y react with dilute hydrochloric acid?

Explain how you arrive at your answers.

(d) What reaction would you expect between zinc sulphate solution and (i) X (ii) Y?
(e) Which is more easily decomposed, $XCO_3$ or $YCO_3$?

9 Look at the following pairs of chemicals. If a reaction happens, copy the word equation, and complete the right hand side.

(a) Copper + Oxygen ➔
(b) Calcium + Hydrochloric acid ➔
(c) Copper + Sulphuric acid ➔
(d) Carbon + Lead(II) oxide ➔
(e) Hydrogen + Calcium oxide ➔
(f) Aluminium + Tin(II) oxide ➔
(g) Gold + Oxygen ➔
(h) Zinc + Copper(II) sulphate solution ➔
(i) Magnesium + Sulphuric acid ➔
(j) Hydrogen + Silver oxide ➔
(k) Carbon + Magnesium oxide ➔
(l) Lead + Copper(II) sulphate solution ➔
(m) Hydrogen + Potassium oxide ➔

10 Explain these statements about aluminium.

(a) Although aluminium is a reactive metal, it is used to make doorframes and windowframes.
(b) Although aluminium conducts heat, it is used to make blankets which are good thermal insulators.
(c) Although aluminium oxide is a common mineral, people did not succeed in extracting aluminium from it until seven thousand years after the discovery of copper.
(d) Recycling aluminium is easier than recycling scrap iron.

11 (a) Explain how steel is made from cast iron.
(b) Explain what advantages steel has over cast iron.
(c) Explain how the following methods protect iron against rusting: painting, galvanising, tin-plating, sacrificial protection.

12 Imagine that you live in a beautiful part of Northern Ireland. A firm called Alumco wants to build a new aluminium plant in your area so that they can use a river as a source of hydroelectric power.

(a) Write a letter from a local farmer to the Secretary of State for Northern Ireland. Say what you fear may happen as a result of pollution from the plant.
(b) Write a letter from a group of environmentalists to the Secretary of State, opposing the plan and giving your reasons.
(c) Write a letter from an unemployed couple to the Secretary of State saying that you welcome the coming of new industry to the area.
(d) Write a letter from the local Council to the Secretary of State. Tell him or her that there is very little unemployment in the area. Say that the new plant would have to bring in workers from outside the region. Explain that there is not enough housing in the area for newcomers.
(e) Write a letter from Alumco to the Secretary of State for Northern Ireland. Tell the Secretary of State of the importance of aluminium. Point out the many uses of aluminium. Explain that to keep up with increasing demand you have to build another plant to supply aluminium.

(If five letters are too many for you, divide the work among the class. Then get together to read out the letters. Have a discussion to decide what the Secretary of State ought to do.)

**13** Concentrated sulphuric acid has some properties which are not shared by dilute sulphuric acid. It is a dehydrating agent and an oxidising agent.

(a) Give two examples of the action of concentrated sulphuric acid as a dehydrating agent.

(b) Give one example of concentrated sulphuric acid acting as an oxidising agent.

(c) Give two examples of the reactions of dilute sulphuric acid.

(d) Name three substances which are attacked by the sulphuric acid in acid rain.

**14** Calcium carbonate is quarried as limestone. It is heated in lime kilns to form calcium oxide (quicklime) and a second product.

(a) What is the second product?

(b) State two uses of this product.

(c) Explain why quicklime is spread by some farmers on their fields.

(d) Explain the advantage of using powdered limestone in place of quicklime.

(e) Most of the limestone which is quarried is used in the manufacture of concrete. Briefly explain the process involved.

(f) What are the properties of concrete that make it a useful building material?

(g) When all the limestone has been extracted from a quarry, what should be done to the quarry?

**15** (a) What type of plants can use atmospheric nitrogen as a nutrient?

(b) Why can't most plants use atmospheric nitrogen in this way?

(c) Name the nitrogen-containing compounds that are built by plants.

(d) Name one natural process that produces nitrogen compounds that enter the soil and are used by plants.

(e) Explain why there is not enough nitrogen from natural sources to support the growth of crop after crop on the same land.

**16** (a) Explain why ammonium salts are used as fertilisers even though plants cannot absorb them (see Topic 13.6).

(b) Describe how you could make ammonium sulphate crystals (see Topic 10.5).

**17** (a) Explain what is meant by the **nitrogen cycle**.

(b) Are we likely to run out of nitrogen?

(c) Say how the human race alters the natural nitrogen cycle (i) by taking nitrogen out of the cycle and (ii) by adding nitrogen.

(d) Explain how fertilisers have created a problem concerning nitrogen compounds.

(e) Suggest two ways in which this problem could be attacked.

**18** Farmer Short had a field in which his crops did not grow well. One year, he added fertiliser to the soil and his crops grew better. However, a pond near the field became stagnant and full of algae, and the fish in the pond died. When the pond was tested, it was found to contain fertiliser.

Farmer Long did not use fertiliser. She rotated her crops between fields so that every few years each field grew peas, beans or clover.

(a) Explain how fertiliser got into the pond.

(b) Why did the algae in the pond increase?

(c) Why did the fish die?

(d) How do crops of peas, beans and clover make up for the lack of fertiliser?

(e) Give one advantage and one disadvantage of Farmer Long's system compared with Farmer Short's.

**19** Explain each of the following statements:

(a) Prolonged use of artificial fertilisers is bad for the soil.

(b) Extensive use of fertilisers on arable land can lead to the pollution of waterways.

(c) Growing clover improves the fertility of the soil.

**20** The diagram shows the nitrogen cycle. The labelled arrows represent different processes.

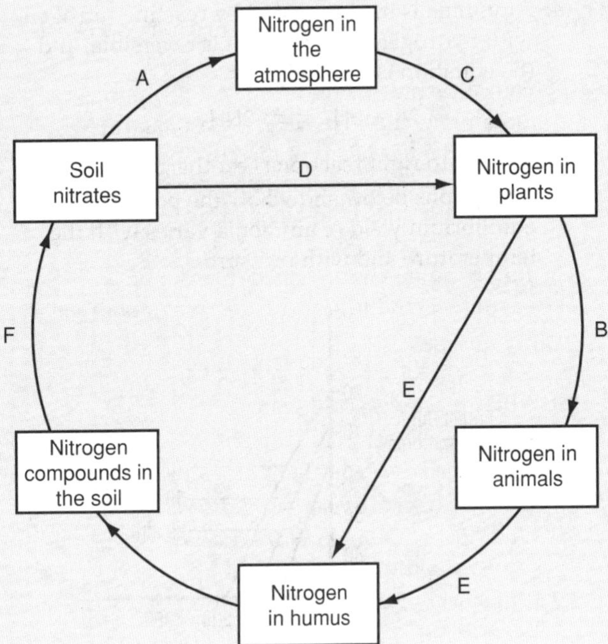


(a) Name the processes A–F, e.g. A = denitrification.

(b) (i) Why is process C useful to plants?

(ii) Process C takes place in clover but not in grass. Why is this so?

(c) What organisms are responsible for process E?

(d) In what form is nitrogen passed in process B?

**21** Match up the type of glass with the use.

| *Type of glass* | *Use* |
|---|---|
| A Heat-resistant glass, which can be heated or cooled quickly without cracking | 1 Windowpanes |
| B Light-sensitive glass, 'Reactolite', which darkens in bright light | 2 Oven dishes and laboratory glassware |
| C Slow-dissolving glass which slowly dissolves in water | 3 Glass-reinforced plastic, GRP |
| D Float glass (sheets of glass) | 4 Spectacle lenses |
| E Glass fibres (thin strands) | 5 Pellets of glass containing a substance which kills the snails which spread the disease Bilharzia and which live in canals |

**22** (a) What is meant by 'recycling' glass?

(b) The value of used glass from a Bottle Bank just pays for the cost of its collection. Is it worth the trouble of collecting glass?

(c) Why do Bottle Banks collect brown, green and colourless glass separately?

(d) Why are people asked to remove metal bottle tops before placing bottles in the Bottle Banks?

(e) In which section of the Periodic Table would you be likely to find metal oxides to colour glass blue, green, brown and other colours?

**23** (a) Ammonia is manufactured by reacting nitrogen with hydrogen. This reaction is reversible, and the equation is shown below.

$$N_2 + 3H_2 \rightleftharpoons 2NH_3$$

The left to right reaction is exothermic.

The graphs below show how the percentage equilibrium yield of ammonia varies with the temperature and with pressure.

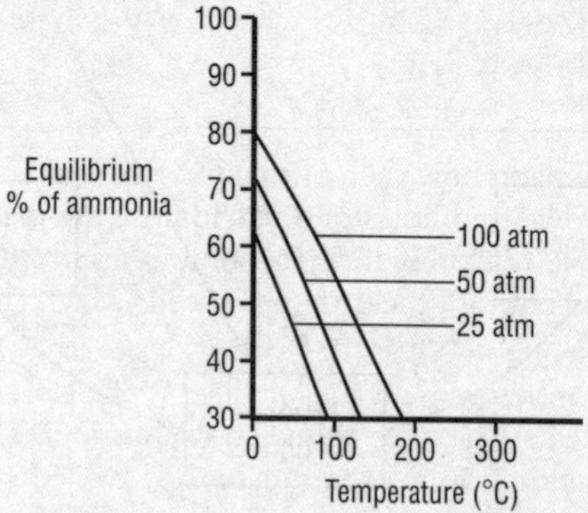


From the graphs, state the effect of increasing (i) temperature and (ii) pressure on the percentage equilibrium yield of ammonia and justify each answer in terms of Le Chatelier's Principle.

(b) The graph below shows the amount of ammonia produced world-wide since 1900. Most of this is converted into ammonium salts, which are used as fertilisers.

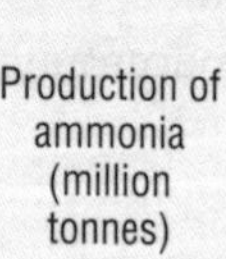


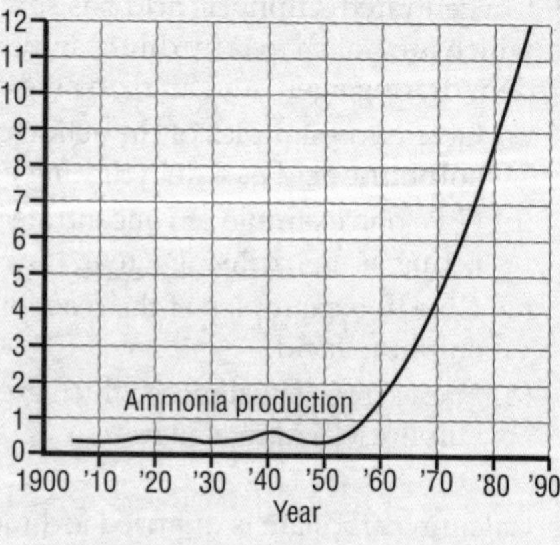


Suggest **three** reasons why the demand for fertilisers produced from ammonia has risen so sharply since about 1950.

(c) Up to 15% of the nitrogen applied to a field can be 'lost' and not taken up by the plants. This figure will be greatly increased if an ammonium salt fertiliser is applied shortly after the field has been treated with an alkali, such as lime (calcium hydroxide).

Suggest:

(i) why less nitrogen will be available for the plants to use if the fertiliser is applied after the field has been treated with lime;

(ii) one other reason why some of the nitrogen available in the fertiliser may not be available for the plants and the problems which can be caused by this. (ULEAC)

**24** Four metals **A, B, C and D** were each heated separately with the oxides of the other metals. The results are given below.

| *metal oxide* | *metal* A | B | C | D |
|---|---|---|---|---|
| **A** oxide | – | No reaction | No reaction | No reaction |
| **B** oxide | **B** displaced | – | No reaction | **B** displaced |
| **C** oxide | **C** displaced | Box 1 | – | **C** displaced |
| **D** oxide | Box 2 | No reaction | Box 3 | – |

To help you understand the table, in the first line none of the metals **B, C** or **D** displaced **A** from its oxide.

(a) (i) Write an equation for the reaction between metal **A** and **C** oxide.

(ii) From the results of the experiments, predict the order of reactivity of metals **A, B, C** and **D** starting with the most reactive.

(iii) Copy and complete the table, by writing what happens in each of Box 1, Box 2, and Box 3.

(b) One way of extracting iron from iron(III) oxide is to heat the oxide with aluminium powder. The equation is given below.

$$2Al + Fe_2O_3 \rightarrow Al_2O_3 + 2Fe$$

(i) Explain this reaction in terms of oxidation and reduction.

(ii) Why is this reaction unsuitable for the large scale manufacturing of iron? Give two reasons.

(c) Leaflets like the one below have been pushed through letter boxes.

COLLECT CANS
FOR CASH
EVERY ALUMINIUM DRINKS CAN IS WORTH MONEY!
Cash -a- can
THE ALUMINIUM RECYCLING CAN–PAIGN


(i) Other than to make money, suggest a reason why people should collect aluminium cans.

(ii) Explain how you can tell if a can used to contain a drink is made out of aluminium or steel.

(iii) How is the steel, which is used to make cans, protected from corrosion?

(d) People are paid 40p for collecting 50 used aluminium cans, but not nearly as much for steel cans. 50 used aluminium cans weigh 1 kg.

(i) Calculate the scrap value of a tonne of aluminium (1000 kg). Show your working.

(ii) Suggest a reason why scrap aluminium cans are more valuable than scrap steel cans.

(MEG)

**25** (a) Concentrated sulphuric acid reacts with sugar. The equation for the reaction is:

$$C_{12}H_{22}O_{11} \rightarrow 12\,C + 11\,H_2O$$

(i) What type of reaction is this?

(ii) What observations would be expected in this reaction?

(iii) What would happen if **dilute** sulphuric acid was used in this reaction? Explain your answer.

(iv) Write a balanced ionic equation to represent the neutralisation reaction between dilute sulphuric acid and aqueous sodium hydroxide.

(b) A blue solution is formed when copper carbonate is added to dilute sulphuric acid.

(i) Give the name of the blue solution.

(ii) What type of reaction is this?

(c) State **one** major use of sulphuric acid. (SEG)

**26** (a) Sulphuric acid is manufactured by a process in which the most important change may be regarded as the exothermic oxidation of sulphur dioxide

$$2SO_2(g) + O_2(g) \rightleftharpoons 2SO_3(g)$$

State and explain the conditions of temperature and pressure under which the reaction is carried out and explain the part played by the catalyst.

(b) Much sulphuric acid is used in the manufacture of synthetic soapless detergents. The detergent is made by treating a hydrocarbon with sulphuric acid, followed by neutralisation of the resulting sulphonic acid with sodium hydroxide.

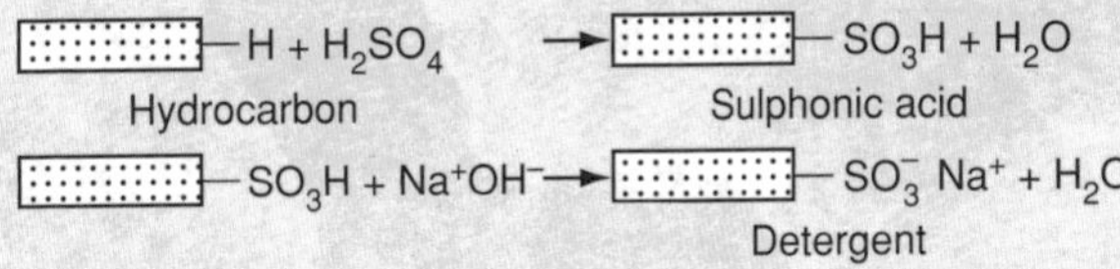


The detergent has a similar molecular shape to a soap made from fats or oils.

$—CO_2^-\ Na^+$

Soap

(i) Suggest how the structures of the synthetic detergent and the soap might enable them to lift grease and dirt from clothing.

(ii) What environmental factor might influence the manufacturer in the choice of hydrocarbon used to make the detergent?

(ULEAC)

**27** (a) Starting from limestone, describe how calcium oxide and calcium hydroxide are manufactured. Give word equations as part of your answer.

(b) The pH of soil in a field was 5.2. The field was divided into two halves. One half was treated with powdered calcium hydroxide at a rate of 500 g/m². The other half was treated with powdered limestone at the same rate. The pH of the soil in each half at different times is shown in the table below.

| Number of months after addition | 1 | 2 | 3 | 4 | 5 | 6 |
|---|---|---|---|---|---|---|
| pH of soil after adding limestone | 5.3 | 5.4 | 5.5 | 5.6 | 5.7 | 5.8 |
| pH of soil after adding calcium hydroxide | 5.4 | 5.6 | 5.8 | 6.0 | 6.2 | 6.2 |

Explain how each of calcium hydroxide and powdered limestone acts to raise the pH of this soil. Use their different solubilities in water to account for the different speeds at which the pH changes occur.

(c) Ammonium nitrate fertiliser is also spread onto fields. The table below shows the annual mean concentration of nitrate in the River Rhine in Germany in various years since 1970 (mg/dm³).

| Year | 1970 | 1975 | 1980 | 1985 |
|---|---|---|---|---|
| Nitrate concentration (mg/dm³) | 1.82 | 3.02 | 3.59 | 4.20 |

*(WWF UK Data Support for Education Service)*

Suggest and explain one possible reason for the rising concentration of nitrates in the river. Describe **two** disadvantages of increasing nitrate concentration in the river. (ULEAC)

# Theme F

# Measurement

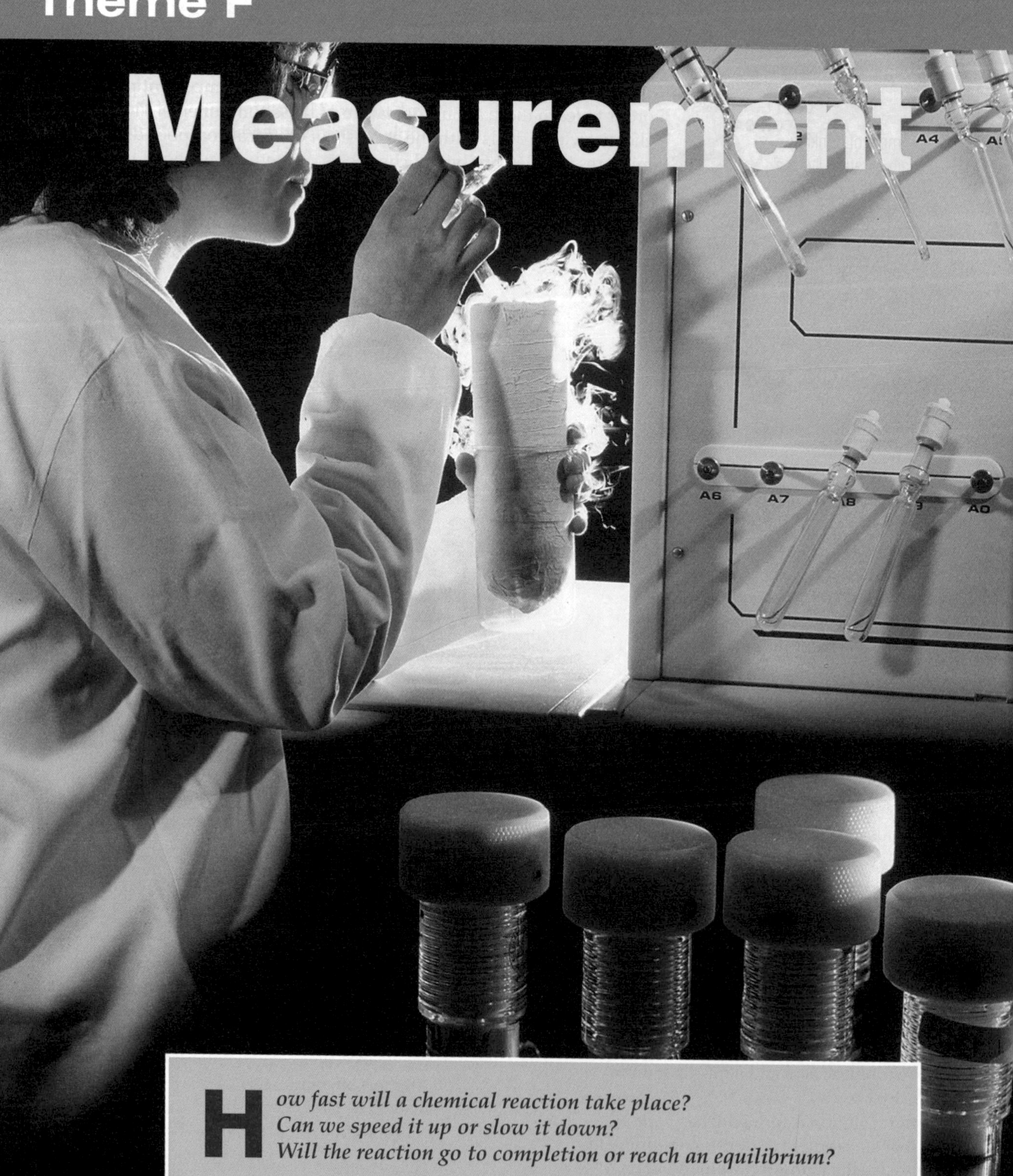


*How fast will a chemical reaction take place?*
*Can we speed it up or slow it down?*
*Will the reaction go to completion or reach an equilibrium?*

*How much product will be formed?*

*Can we increase the yield of product?*

*Which fuel shall we use to heat the plant?*

**These are questions to which the manager of an industrial plant must find answers. In this theme, we show how the answers can be found.**

# Topic 25 Analytical chemistry

**FIRST THOUGHTS**

Does a sample of drinking water contain nitrates? Does a brand of paint contain lead compounds? Which metals are present in an alloy steel? To find out the answers to such questions, we use analytical chemistry.

To **analyse** means to separate something into its component parts in order to learn more about the nature of these components. The branch of chemistry that deals with the analysis of chemical compounds and mixtures is called **analytical chemistry**. Sometimes we need to know only which substances are present in a sample; not the quantities of those substances. Which elements are present in this sample of soil? Which pigments are present in this food dye? Does this sample of zinc contain a trace of zinc sulphide? To find answers to such questions, we use **qualitative analysis**. If we need to know the precise quantity of one or more components of a sample, we use **quantitative analysis**. This might entail finding the percentage of nickel in a nickel ore or the number of parts per million of mercury in a fish. Quantitative analysis can be done by means of **volumetric analysis**, in which reactions take place in solution. An alternative is **gravimetric analysis**, in which the masses of reacting substances are found by the use of an accurate balance. In this topic, we shall be dealing with qualitative analysis.

## 25.1 Flame colour

EXTENSION FILE ACTIVITY

Some metals and their compounds give characteristic colours when heated in a Bunsen flame. The colour can be used to identify the metal ion present in a compound.

*Table 25.1* Some flame colours

| Metal | Colour of flame |
|---|---|
| Barium | Apple-green |
| Calcium | Brick-red |
| Copper | Green with blue streaks |
| Lithium | Crimson |
| Potassium | Lilac |
| Sodium | Yellow |

## 25.2 Tests for ions in solution

The following tests can be done on a solution. A soluble solid can be analysed by dissolving it in distilled water and then applying the tests.

*Table 25.2* Tests for anions in solution

| Anion | Test and observation |
|---|---|
| Chloride, $Cl^-(aq)$ | Add a few drops of dilute nitric acid followed by a few drops of silver nitrate solution. A white precipitate of silver chloride is formed. The precipitate is soluble in ammonia solution. |
| Bromide, $Br^-(aq)$ | Add a few drops of dilute nitric acid followed by a few drops of silver nitrate solution. A pale yellow precipitate of silver bromide is formed. The precipitate is slightly soluble in ammonia solution. |

(continued on following page)

Table 25.2 ▼ Tests for anions in solution (continued)

| Anion | Test and observation |
|---|---|
| Iodide, $I^-(aq)$ | Add a few drops of dilute nitric acid followed by a few drops of silver nitrate solution. A yellow precipitate of silver iodide is formed. It is insoluble in ammonia solution. |
| Sulphate, $SO_4^{2-}(aq)$ | Add a few drops of barium chloride solution followed by a few drops of dilute hydrochloric acid. A white precipitate of barium sulphate is formed. |
| Sulphite, $SO_3^{2-}(aq)$ | Test as for sulphate. A white precipitate of barium sulphite appears and then reacts with acid to give sulphur dioxide. |
| Carbonate, $CO_3^{2-}(aq)$ | Add dilute hydrochloric acid to the solution (or add it to the solid). Bubbles of carbon dioxide are given off. |
| Nitrate, $NO_3^-(aq)$ | Make the solution strongly alkaline. Add Devarda's alloy and warm. Ammonia is given off. (If the solution contains ammonium ions, warm with alkali to drive off ammonia before testing for nitrate.) |

Analytical chemists have traditionally made good use of these tests. Now they also have instrumental methods of analysis to help them.

Table 25.3 ▼ Tests for cations in solution

| Cation | Add sodium hydroxide solution | Add ammonia solution |
|---|---|---|
| Ammonium, $NH_4^+(aq)$ | Warm. Ammonia is given off | – |
| Copper, $Cu^{2+}(aq)$ | Blue jelly-like precipitate, $Cu(OH)_2(s)$ | Blue jelly-like ppt., dissolves in excess to form a deep blue solution |
| Iron(II), $Fe^{2+}(aq)$ | Green gelatinous ppt., $Fe(OH)_2(s)$ | Green gelatinous ppt., $Fe(OH)_2(s)$ |
| Iron(III), $Fe^{3+}(aq)$ | Rust-brown gelatinous ppt., $Fe(OH)_3(s)$ | Rust-brown gelatinous ppt., $Fe(OH)_3(s)$ |
| Lead(II), $Pb^{2+}(aq)$ | White ppt., $Pb(OH)_2(s)$ dissolves in excess $NaOH(aq)$ | White ppt., $Pb(OH)_2$ |
| Zinc, $Zn^{2+}(aq)$ | White ppt., $Zn(OH)_2(s)$ dissolves in excess $NaOH(aq)$ | White ppt., $Zn(OH)_2(s)$ dissolves in excess $NH_3(aq)$ |
| Aluminium, $Al^{3+}(aq)$ | Colourless ppt., $Al(OH)_3(s)$ dissolves in excess $NaOH(aq)$ | Colourless ppt., $Al(OH)_3(s)$ |

Note: ppt. = precipitate

EXTENSION FILE
ACTIVITY

## 25.3 Identifying some common gases

If you are testing the smell of a gas, you should do so cautiously (see Figure 25.3A). Tests for common gases are given in Table 25.4.

Figure 25.3A How to smell a gas

*Table 25.4* ▼ Testing gases

| Gas | Colour and smell | Test and observation |
|---|---|---|
| Ammonia† | Colourless<br>Pungent smell | Hold damp red litmus paper or universal indicator paper in the gas. The indicator turns blue. |
| Bromine | Reddish-brown<br>Choking smell | Hold damp blue litmus paper in the gas. Litmus turns red and is then bleached. |
| Carbon dioxide | Colourless<br>Odourless | Bubble the gas through limewater (calcium hydroxide solution). A white solid precipitate appears, making the solution appear cloudy. |
| Chlorine† | Poisonous<br>Green gas<br>Choking smell | Test a very small quantity in a fume cupboard. Hold damp blue litmus paper in the gas. Litmus turns red and is quickly bleached. Chlorine turns damp starch-iodide paper blue-black. |
| Hydrogen | Colourless<br>Odourless when pure | Introduce a lighted splint. Hydrogen burns with a squeaky 'pop'. |
| Hydrogen chloride | Colourless<br>Pungent smell | Hold damp blue litmus paper in the gas. Litmus turns red. With ammonia, a white smoke of $NH_4Cl$ forms. |
| Iodine | Black solid or purple vapour | Dissolve in trichloroethane (or other organic solvent). Gives a violet solution. |
| Nitrogen dioxide† | Reddish-brown<br>Pungent smell | Hold damp blue litmus paper in the gas. Litmus turns red but is not bleached. |
| Oxygen | Colourless<br>Odourless | Hold a glowing wooden splint in the gas. The splint bursts into flame. |
| Sulphur dioxide† | Colourless<br>Choking smell | Dip a filter paper in potassium dichromate solution, and hold it in the gas. The solution turns from orange through green to blue. Potassium manganate(VII) solution turns very pale pink. |

†These gases are poisonous. Test with care.

## 25.4 ▶ Solubility of some compounds

The solubilities of some compounds are tabulated in Table 25.5.

*Table 25.5* ▼ Soluble and insoluble compounds

| Soluble | Insoluble |
|---|---|
| All sodium, potassium and ammonium salts | |
| All nitrates | |
| Most chlorides, bromides and iodides | Chlorides, bromides and iodides of silver and lead |
| Most sulphates | Sulphates of lead, barium and calcium |
| Sodium, potassium and ammonium carbonates | Most other carbonates |
| Sodium, potassium and calcium oxides | Most other oxides |
| Sodium, potassium and calcium hydroxides | Most other hydroxides |

## CHECKPOINT

1. A body is found at the bottom of a clay pit. The dead man is known to have quarrelled violently with a neighbour, and the neighbour has clay on his shoes. The clay in the clay pit contains a high percentage of iron(II). How can you test the clay on the neighbour's shoes for $Fe^{2+}$?
2. After a visit to Somerset, some students bring home a sample of rock from an underground cavern. Luke says that the rock is calcium carbonate. Natalie believes that it may be magnesium carbonate. Rosalie suggests that it may be barium carbonate. How can the students investigate whether the rock is (a) a carbonate (b) a calcium compound (c) a barium compound (d) a magnesium compound?
3. A packet is labelled 'bicarb of soda'. How can you test to see whether it contains (a) a sodium compound (b) either a carbonate or a hydrogencarbonate?
4. A solution contains the metal ions, $Q^{2+}(aq)$ and $R^{2+}(aq)$. When a solution of sodium hydroxide is added, a precipitate is obtained. Addition of excess of sodium hydroxide to this precipitate gives a green precipitate containing Q and a colourless solution containing R. When dilute hydrochloric acid is added to this solution, a white precipitate is obtained. Deduce what Q and R may be, explaining your reasoning.
5. Old Sir Joshua Vellof often falls asleep after meals. He is a suspicious old gentleman, and he wonders whether his no-good nephew Jake is putting some of the tranquilliser potassium bromide into the salt cellar. Sir Joshua asks you to analyse the contents of the salt cellar to see whether it contains sodium chloride or potassium bromide. What do you do?
6. A factory orders calcium oxide, magnesium oxide and barium oxide. A lorry driver deposits three sacks at the factory and drives off without saying which is which. The works chemist has to sort it out. Suggest two tests which he could use to tell which sack is which.
7. A solution contains the sulphates of three metals, A, B and C. Explain the reactions shown in the flow chart and identify A, B and C.

Solution of sulphates of A, B and C

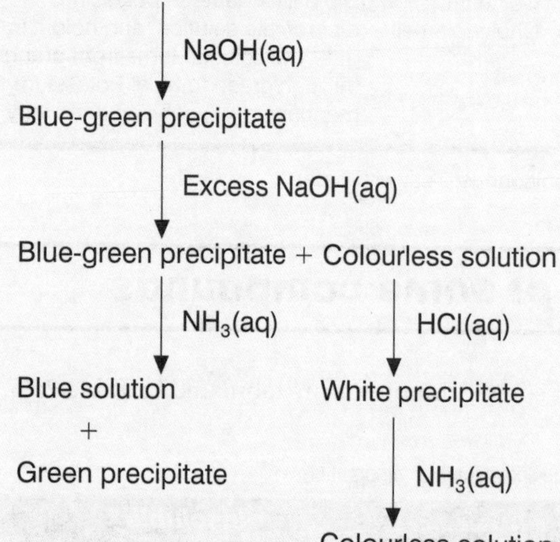


## 25.5 Methods of collecting gases

Gases can be bought in cylinders from industrial manufacturers or made in the laboratory by a chemical reaction. You may need to obtain gas jars full of gas. The following methods can be used to collect gases from either source.

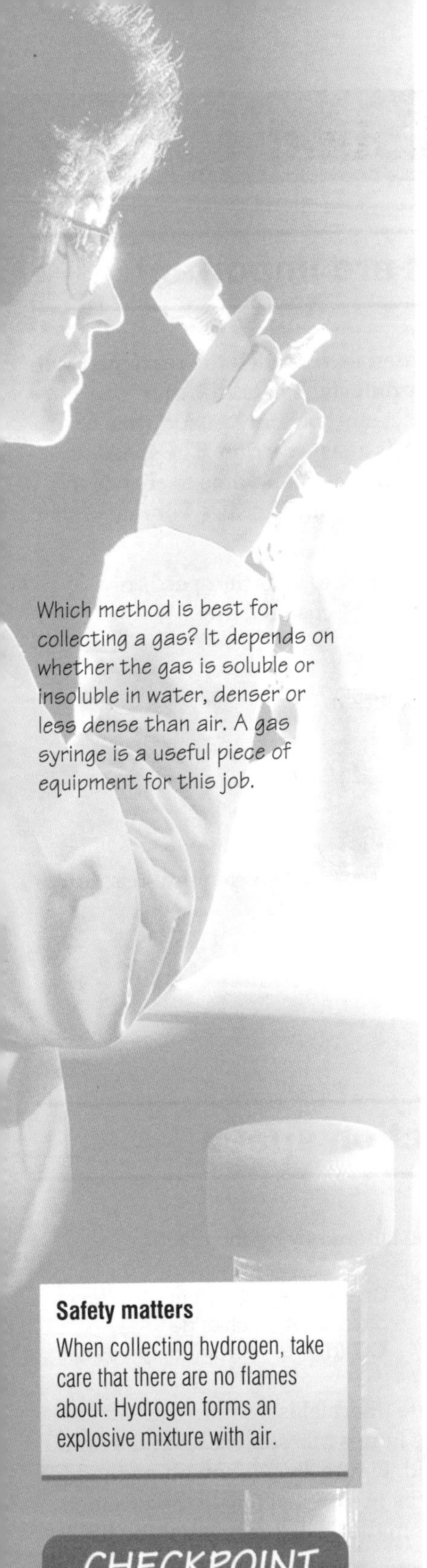

## Collecting over water

A gas which is insoluble in water can be collected over water (see Figure 25.5A). The gas displaces the water in the gas jar. When the gas jar is full, it must be replaced by another. This method can be used for hydrogen, oxygen, chlorine, carbon dioxide and other gases. A little of the gas will dissolve in the water.

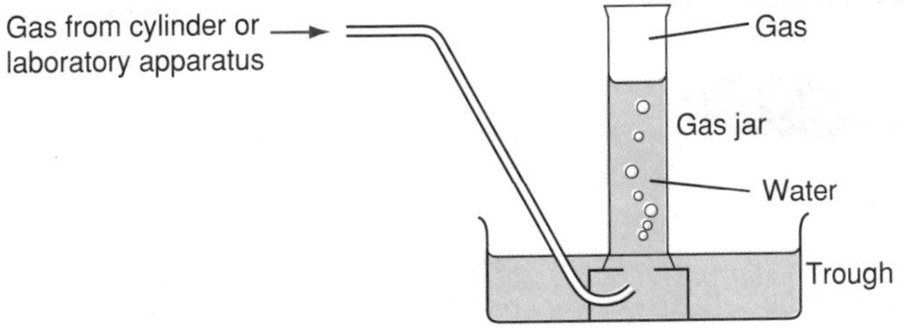


*Figure 25.5A* ▲ Collecting over water (for an insoluble gas)

Which method is best for collecting a gas? It depends on whether the gas is soluble or insoluble in water, denser or less dense than air. A gas syringe is a useful piece of equipment for this job.

## Collecting downwards by displacement

A gas which is denser than air can be collected downwards by displacement of air (see Figure 25.5B). The gas displaces the air in the gas jar. This method can be used for chlorine, carbon dioxide, sulphur dioxide and other dense gases.

## Collecting upwards by displacement

A gas which is less dense than air can be collected upwards by displacement of air (see Figure 25.5C). This method can be used for ammonia.

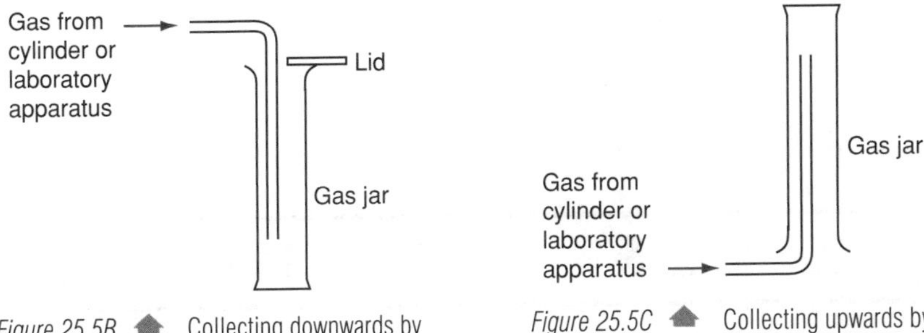


*Figure 25.5B* ▲ Collecting downwards by displacement (for a dense gas)

*Figure 25.5C* ▲ Collecting upwards by displacement (for a gas of low density)

## Collecting in a gas syringe

A gas supply can be connected to an empty gas syringe. As gas enters, it drives the plunger down the barrel (see Figure 25.5D). When the syringe is full, it must be replaced by another.

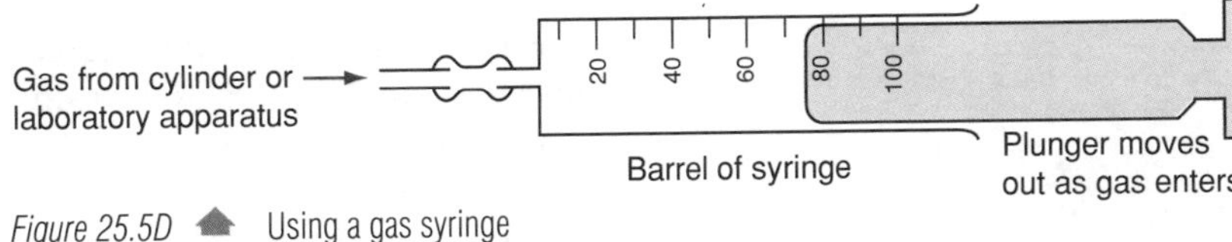


*Figure 25.5D* ▲ Using a gas syringe

**Safety matters**

When collecting hydrogen, take care that there are no flames about. Hydrogen forms an explosive mixture with air.

## CHECKPOINT

1. (a) Why is the method shown in Figure 25.5A not used for collecting ammonia?
   (b) Why is the method shown in Figure 25.5C not used for collecting sulphur dioxide?
2. Hydrogen can be collected by either of the methods shown in Figures 25.5A and C.
   (a) State an advantage of the method shown in Figure 25.5A over that in Figure 25.5C.
   (b) Sketch the apparatus which you would use to collect dry hydrogen.

# Topic 26 Chemical reaction speeds

## 26.1 Why reaction speeds are important

**FIRST THOUGHTS**

*Who is interested in the speeds of chemical reactions?* If you were a cheese manufacturer, you would be interested in speeding up the chemical reactions which produce cheese. The more tonnes of cheese you could produce in a month, the more profit you would make. If you were a butter manufacturer, you would want to slow down the chemical reactions which make your product turn rancid.

In a chemical reaction, the starting materials are called the **reactants**, and the finishing materials are called the **products**. It takes time for a chemical reaction to happen. If the reactants take only a short time to change into the products, that reaction is a **fast reaction**. The **speed** or **rate** of that reaction is high. If a reaction takes a long time to change the reactants into the products, it is a **slow reaction**. The speed or rate of that reaction is low.

Many people are interested in knowing how to alter the speeds of chemical reactions. The factors which can be changed are:

- the size of the particles of a solid reactant,
- the concentrations of reactants in solution,
- the temperature,
- the presence of light,
- the addition of a catalyst.

A **catalyst** is a substance which can alter the rate of a chemical reaction without being used up in the reaction.

## 26.2 Particle size and reaction speed

Carbon dioxide can be prepared in the laboratory by the reaction

| Calcium carbonate | + | Hydrochloric acid | → | Carbon dioxide | + | Calcium chloride | + | Water |
|---|---|---|---|---|---|---|---|---|
| $CaCO_3(s)$ | + | $2HCl(aq)$ | → | $CO_2(g)$ | + | $CaCl_2(aq)$ | + | $H_2O(l)$ |

One of the reactants, calcium carbonate (marble) is a solid. You can use this reaction to find out whether large lumps of a solid react at the same speed as small lumps of the same solid. Figure 26.2A shows a method for finding the rate of the reaction. It can be used when one of the products is a gas. As carbon dioxide escapes from the flask, the mass of the flask and contents decreases.

In some reactions one of the products is a gas. When the gas escapes, there is a loss in mass. By measuring the speed at which the mass decreases, we can measure the speed of the reaction.

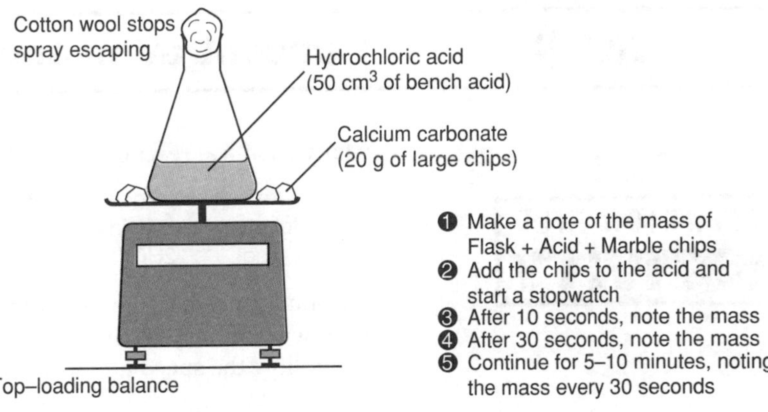


*Figure 26.2A* ▲ Apparatus for following the loss in mass when a gas is evolved

EXTENSION FILE ACTIVITY

The reaction starts when the marble chips are dropped into the acid. The mass of the flask and contents is noted at various times after the start of the reaction. The mass can be plotted against time. Figure 26.2B shows typical results.

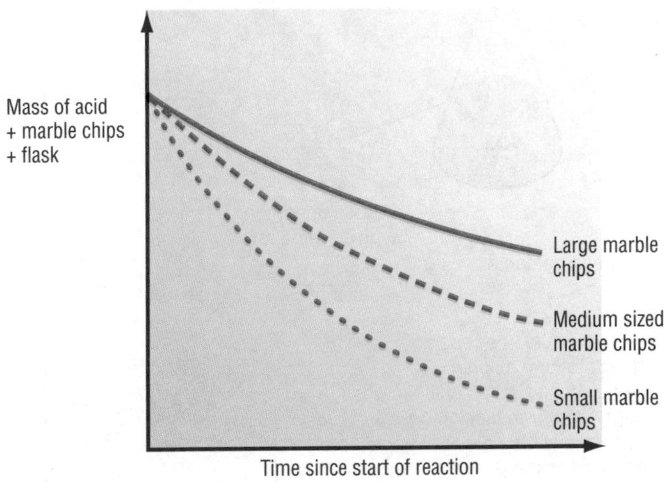


*Figure 26.2B* ▲ Results obtained with different sizes of marble chips

The results show that the smaller the size of the particles of calcium carbonate, the faster the reaction takes place. The difference is due to a difference in surface area. There is a larger surface area in 20 g of small chips than in 20 g of large chips. The acid attacks the surface of the marble. It can therefore react faster with small chips than with large chips.

## SUMMARY

Reactions in which one reactant is a solid take place faster when the solid is divided into small pieces. The reason is that a certain mass of small particles has a larger surface area than the same mass of large particles.

## CHECKPOINT

1. When potatoes are cooked, a chemical reaction occurs. What can you do to increase the speed at which potatoes cook?
2. There is a danger in coal mines that coal dust may catch fire and start an explosion. Explain why coal dust is more dangerous than coal.
3. 'Alko' indigestion tablets and 'Neutro' indigestion powder are both alkalis. Which do you think will act faster to cure acid indigestion? Describe how you could test the two remedies in the laboratory with a bench acid to see whether you are right.

# 26.3 Concentration and reaction speed

Many chemical reactions take place in solution. One such reaction is

| Sodium thiosulphate | + | Hydrochloric acid | → | Sulphur | + | Sodium chloride | + | Sulphur dioxide | + | Water |
|---|---|---|---|---|---|---|---|---|---|---|
| $Na_2S_2O_3(aq)$ | + | $2HCl(aq)$ | → | $S(s)$ | + | $2NaCl(aq)$ | + | $SO_2(g)$ | + | $H_2O$ |

Sulphur appears in the form of very small particles of solid. The particles do not settle: they remain in suspension. Figure 26.3A shows how you can follow the speed at which sulphur is formed.

Observer

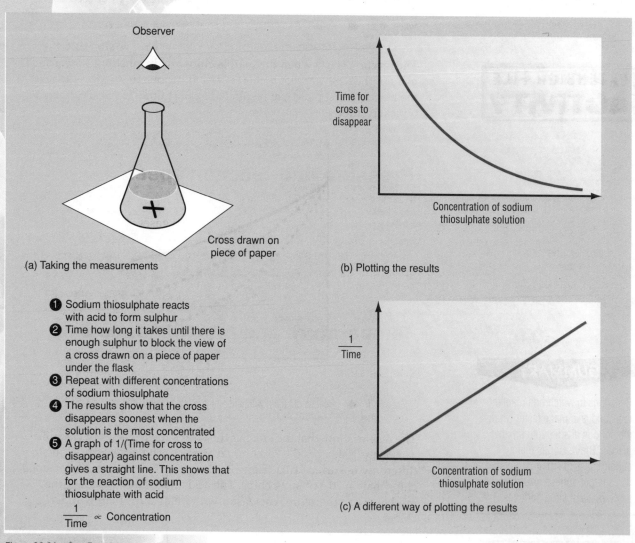

Cross drawn on piece of paper

(a) Taking the measurements

Time for cross to disappear

Concentration of sodium thiosulphate solution

(b) Plotting the results

1. Sodium thiosulphate reacts with acid to form sulphur
2. Time how long it takes until there is enough sulphur to block the view of a cross drawn on a piece of paper under the flask
3. Repeat with different concentrations of sodium thiosulphate
4. The results show that the cross disappears soonest when the solution is the most concentrated
5. A graph of 1/(Time for cross to disappear) against concentration gives a straight line. This shows that for the reaction of sodium thiosulphate with acid $\frac{1}{\text{Time}} \propto$ Concentration

$\frac{1}{\text{Time}}$

Concentration of sodium thiosulphate solution

(c) A different way of plotting the results

*Figure 26.3A* Experiment on reaction speed and concentration

*The speed at which the precipitate of a solid product appears can be used to measure the speed of a reaction.*

The faster a reaction takes place, the shorter is the time needed for the reaction to finish. To be more precise, the speed of the reaction is **inversely proportional** to the time taken for the reaction to finish

$$\text{Speed of reaction} \propto 1/\text{Time}$$

You can see from Figure 26.3A(c) above that

$$1/\text{Time} \propto \text{Concentration}$$

Therefore

$$\text{Speed of reaction} \propto \text{Concentration}$$

SUMMARY

For the reactions in solution mentioned here, the speed of the reaction is proportional to the concentration of the reactant (or reactants). That is, the speed doubles when the concentration is doubled.

In this experiment, only one concentration was altered. A variation is to keep the concentration of sodium thiosulphate constant and alter the concentration of acid. Then the speed of the reaction is found to be proportional to the concentration of the acid. If the acid concentration is doubled, the speed doubles. The reason for this is that the ions are closer together in a concentrated solution. The closer together they are, the more often the ions collide. The more often they collide, the more chance they have of reacting.

CHECKPOINT

1 Molly is asked to investigate the marble chips–acid reaction. She must find out what effect changing the concentration of the acid has on the speed of the reaction. Explain how Molly could adapt the experiment shown in Figure 26.2A to carry out her investigation.

## 26.4 Pressure and reaction speed

SUMMARY

The speed of a reaction between gases increases when the pressure is increased.

Pressure has an effect on reactions between gases. The speed of the reaction increases when the pressure is increased. The reason is that increasing the pressure pushes the gas molecules closer together. The molecules therefore collide more often, and the gases react more rapidly.

## 26.5 Temperature and reaction speed

EXTENSION FILE ACTIVITY

You met the reaction between sodium thiosulphate and acid in Topic 26.3. This reaction can also be used to study the effect of temperature on the speed of a chemical reaction. Warming the solutions makes sulphur form faster. There is a steep increase in the speed of the reaction as the temperature is increased.

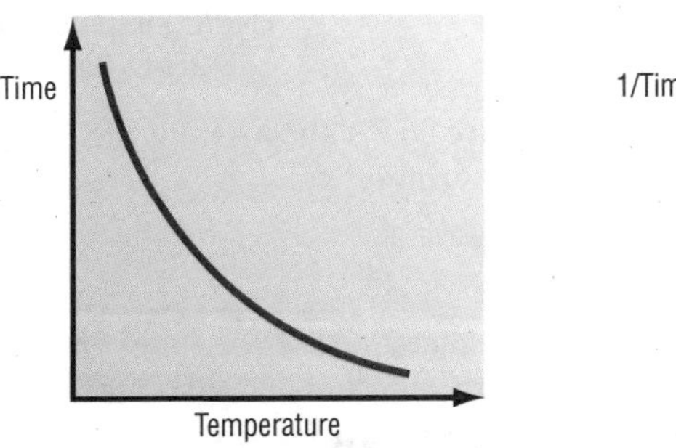

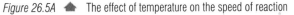

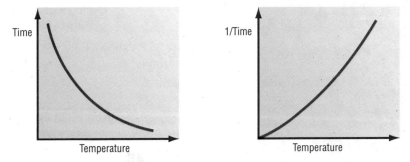

*Figure 26.5A* ▲ The effect of temperature on the speed of reaction

SUMMARY

The speed of a reaction increases when the temperature is raised.

This reaction goes approximately twice as fast at 30 °C as it does at 20 °C. It doubles in speed again between 30 °C and 40 °C and so on.

At the higher temperature, the ions have more kinetic energy. Moving through the solution more rapidly, they collide more often and more vigorously and so there is a greater chance that they will react.

## 26.6 Light and chemical reactions

SUMMARY

Some chemical reactions are speeded up by light.

Heat is not the only form of energy that can speed up reactions. Some chemical reactions take place faster when they absorb light. The formation of silver from silver salts takes place when a **photographic film** is exposed to light. In sunlight, green plants are able to carry on the process of **photosynthesis**.

### CHECKPOINT

1. Polly's project is to find out what effect temperature has on the speed at which milk goes sour. Suggest what measurements she should make.
2. Ismail's project is to find out whether iron rusts more quickly at higher temperatures. Suggest a set of experiments which he could do to find out.
3. You are asked to study the reaction

   Magnesium + Sulphuric acid → Hydrogen + Magnesium sulphate

   You are provided with magnesium ribbon, dilute sulphuric acid, a thermometer and any laboratory glassware you need. Describe how you would find out what effect a change in temperature has on the speed of the reaction. Say what apparatus you would use, what measurements you would make and what you would do with your results.
4. Magnesium reacts with cold water slowly

   Magnesium + Water → Magnesium hydroxide + Hydrogen

   If there is phenolphthalein in the water, it turns pink, showing that an alkali has been formed. Describe experiments which you could do to find the effect of increasing the temperature on the speed of this reaction. Say what you would measure and what you would do with your results. With your teacher's approval, try out your ideas.

## 26.7 Catalysis

A reaction used to prepare oxygen is

Hydrogen peroxide → Oxygen + Water

$$2H_2O_2(aq) \rightarrow O_2(g) + 2H_2O(l)$$

Figure 26.7A shows how the oxygen can be collected and measured in a gas syringe.

EXTENSION FILE ACTIVITY

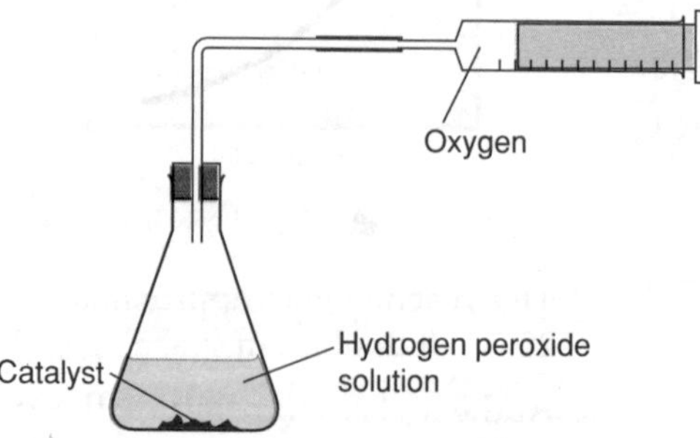


Oxygen collects in the syringe. The volume can be read and the time since the start of the reaction can be noted. Readings of volume and time can be tabulated.

*Figure 26.7A* ▲ Collecting and measuring a gas

The apparatus shown in Figure 26.7A can be used whenever you want to collect the gas that is formed in a reaction. The speed of the reaction can be measured by noting the volume of the gas formed at different time intervals.

A substance which speeds up a reaction without being used up in the reaction is called a catalyst.

The formation of oxygen is very slow at room temperature. The reaction can be speeded up by the addition of certain substances, for example manganese(IV) oxide. When manganese(IV) oxide is added to hydrogen peroxide, the evolution of oxygen takes place much more rapidly (see Figure 26.7B). Manganese(IV) oxide is not used up in the reaction. At the end of the reaction, the manganese(IV) oxide can be filtered out of the solution and used again. A substance which increases the speed of a chemical reaction without being used up in the reaction is called a **catalyst**. Different reactions need different catalysts.

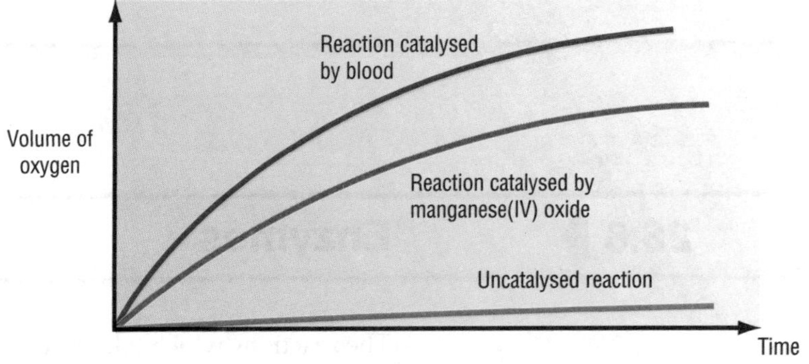


*Figure 26.7B* ▲ Catalysis of the decomposition of hydrogen peroxide

Any individual catalyst will only catalyse a certain reaction or group of reactions. Platinum catalyses a number of oxidation reactions. Nickel catalyses hydrogenation reactions. Industries make good use of catalysts. If manufacturers can produce their product more rapidly, they make bigger profits. If they can produce their product at a lower temperature with the aid of a catalyst, they save on fuel. The reactions which make plastics take place under high pressure. The construction of industrial plastics plants which are strong enough to withstand high pressure is expensive. A plastics manufacturer therefore tries to find a catalyst which will enable the reaction to give a good yield of plastics at a lower pressure. Then the plant will not have to withstand high pressures. Less costly materials can be used in its construction. Industrial chemists are constantly looking for new catalysts.

Some reactions only give good yields of product at high temperatures. Such a reaction costs a manufacturer high fuel bills. Industrial chemists look for a catalyst which will make the reaction take place more readily at a lower temperature. This will cut running costs and increase profits.

## SUMMARY

Some chemical reactions can be speeded up by adding a substance which is not one of the reactants, which is not used up in the reaction. Such a substance is called a catalyst.

A certain catalyst will catalyse a limited number of reactions. Catalysts are of great importance to industry. They may diminish the need to carry out a reaction at high temperature or high pressure and so cut production costs.

## CHECKPOINT

1 Catalysts A and B both catalyse the decomposition of hydrogen peroxide. The following figures were obtained at 20 °C for the volume of oxygen formed against the time since the start of the reaction.

| *Time* (minutes) | 0 | 5 | 10 | 15 | 20 | 25 | 30 | 35 |
|---|---|---|---|---|---|---|---|---|
| Volume of oxygen with catalyst A ($cm^3$) | 0 | 4 | 8 | 12 | 16 | 17 | 18 | 18 |
| Volume of oxygen with catalyst B ($cm^3$) | 0 | 5 | 10 | 15 | 16.5 | 18 | 18 | 18 |

(a) Plot a graph to show both sets of results.
(b) Say which is the better catalyst, A or B.
(c) Explain why both experiments were done at the same temperature.
(d) Explain why both sets of figures stop at 18 $cm^3$ of oxygen.
(e) Add a line to your graph to show the shape of the graph you would obtain for the uncatalysed reaction.

2 Someone tells you that nickel oxide will catalyse the decomposition of hydrogen peroxide to give oxygen. How could you find out whether this is true? Draw the apparatus you would use and state the measurements you would make.

## 26.8 Enzymes

The reactions which take place inside the cells of living organisms are catalysed by **enzymes**. Enzymes are proteins. An enzyme is **specific** for a certain reaction or type of reaction, that is, it will catalyse only that reaction or type of reaction. For example, the enzyme maltase catalyses the reaction of maltose and water to form glucose. The substance which the enzyme helps to react, e.g. maltose, is called the **substrate**.

Enzymes have large molecules of complicated shapes. The substrate has to fit into the contours of a certain region in the enzyme called the **active site**. Only a substance which will fit the active site can be enabled to react by the enzyme; this is the reason why enzymes are specific. The fit between substrate and enzyme has been described as 'like a lock and key'. Figure 26.8A shows how the digestive enzyme amylase is thought to catalyse the hydrolysis of starch into sugars. The hydrolysis of starch to sugars can be carried out in the laboratory by boiling a solution of starch with hydrochloric acid for an hour. Catalysed by an enzyme, the reaction takes place rapidly at body temperature.

### SUMMARY

Enzymes are proteins. They catalyse reactions which take place in living organisms. An enzyme is specific for a certain reaction or type of reaction. Enzymes are powerful catalysts.

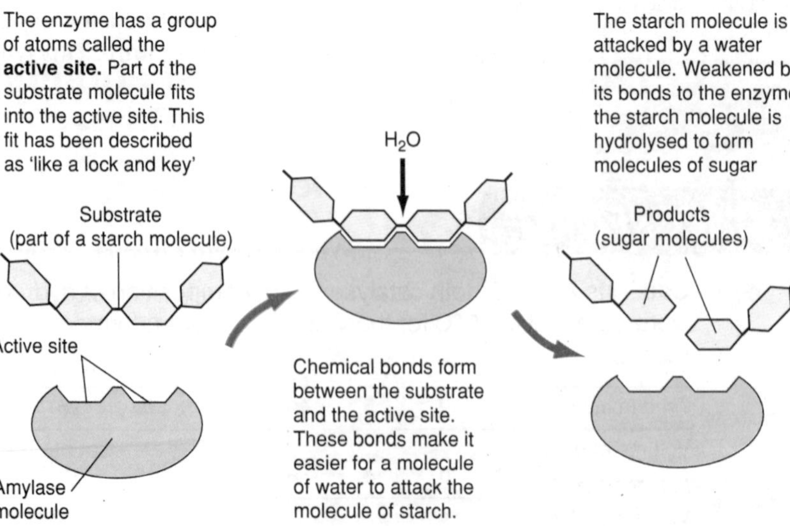

The enzyme has a group of atoms called the **active site.** Part of the substrate molecule fits into the active site. This fit has been described as 'like a lock and key'
Substrate (part of a starch molecule)
Active site
Amylase molecule
$H_2O$
Chemical bonds form between the substrate and the active site. These bonds make it easier for a molecule of water to attack the molecule of starch.
The starch molecule is attacked by a water molecule. Weakened by its bonds to the enzyme, the starch molecule is hydrolysed to form molecules of sugar
Products (sugar molecules)


*Figure 26.8A* How the enzyme amylase catalyses the breakdown of starch. Notice that the enzyme is unchanged at the end of the reaction and able to catalyse the breakdown of more starch

SUMMARY

Protein molecules have a three-dimensional structure. If this structure is lost, an enzyme is 'denatured' and loses its catalytic activity.

Like other proteins, enzymes have a complex three-dimensional structure. A protein can lose this structure at high pH, low pH and high temperature. These conditions are said to **denature** the protein. The active site of an enzyme changes if the enzyme is denatured, and the enzyme loses its catalytic power.

## Examples of the use of enzymes

1 *Baking:* Enzymes catalyse the hydrolysis of the starch in flour to sugars. Then enzymes in yeast catalyse the fermentation of sugars with the production of carbon dioxide. This makes bread and cakes 'rise'.

2 *Dairy produce:* The enzyme rennin is present in the extract called rennet which is obtained from the digestive juices in calves' stomachs (or from bacteria by genetic engineering). It is added to milk to speed up the clotting of milk to form cheese.

3 *Brewing:* Enzymes from barley are used in beer production; they convert starches and proteins into sugars and amino acids. Enzymes in yeast ferment sugars to ethanol.

4 *Sweets:* Have you ever wondered how they get a liquid centre into a chocolate? To start with, the centre is solid. Enzymes incorporated in the solid centre convert the solid into a liquid.

SUMMARY

Enzymes have many uses in industry.

CHECKPOINT

1 (a) What is meant by the 'substrate' of an enzyme?
(b) What is meant by the 'active site' of an enzyme?
(c) What is meant by the statement that an enzyme is 'specific' for a certain substrate?
(d) How does the possession of an active site explain why the enzyme is specific for a certain substrate?
(e) How does the active site enable the enzyme to do its job?

# 26.9 Chemical equilibrium

## Reversible reactions

When calcium carbonate is heated strongly, it decomposes:

$$\text{Calcium carbonate} \rightarrow \text{Calcium oxide} + \text{Carbon dioxide}$$
$$CaCO_3(s) \rightarrow CaO(s) + CO_2(g)$$

However, the base calcium oxide reacts with the acidic gas carbon dioxide to form calcium carbonate.

$$CaO(s) + CO_2(g) \rightarrow CaCO_3(s)$$

We say that this reaction is **reversible**; it can go from right to left as well as from left to right. When a substance decomposes to form products which can recombine, we say that the substance **dissociates**. When the dissociation is brought about by heat, it is called **thermal dissociation**.

Sometimes the products of a reaction can combine to form the reactants. Such a reaction is a reversible reaction. An example is the thermal decomposition of calcium carbonate. In a reversible reaction a state of equilibrium is reached when

rate of forward reaction = rate of backward reaction

Think about what happens when calcium carbonate is heated inside a closed container. When heating starts, there is no carbon dioxide or calcium oxide, and the dissociation of calcium carbonate is the only reaction that takes place. Soon, the amounts of calcium oxide and carbon dioxide build up, and the combination of calcium oxide and carbon dioxide begins. As the concentration of carbon dioxide builds up inside the closed container, the rate of combination increases. Eventually a state is reached in which the rate of recombination is equal to the rate of dissociation of calcium carbonate. Then the amounts of calcium carbonate, calcium oxide and carbon dioxide remain constant. The system is described as being at **equilibrium**. We use reverse arrows to denote an equilibrium:

$$CaCO_3(s) \rightleftharpoons CaO(s) + CO_2(g)$$

The reaction from left to right (as written) is called the forward reaction. The reaction from right to left is called the backward reaction.

In a system at equilibrium, both the forward reaction and the backward reaction are taking place. The rate of the forward reaction is equal to the rate of the backward reaction. Therefore, once equilibrium is established, the amounts of the three substances remain constant.

A number of factors can disturb the equilibrium. If the container is opened, carbon dioxide escapes, and more calcium carbonate dissociates. When limestone is heated in a kiln (see Figure 22.1F on p. 241) to make calcium oxide (quicklime) and carbon dioxide, the kiln is open, and a draft of air is blown through the kiln to remove carbon dioxide and prevent equilibrium from becoming established. Then more and more calcium carbonate dissociates in an attempt to reach equilibrium, and more quicklime is produced.

## The Haber process

Nitrogen and hydrogen combine to form ammonia (See Topic 23.2):

$$N_2(g) + 3H_2(g) \rightleftharpoons 2NH_3(g)$$

Equilibrium is reached as ammonia dissociates into nitrogen and hydrogen. Nitrogen is an unreactive gas, and the position of equilibrium is very far towards the left-hand side. Fritz Haber developed this reaction to manufacture ammonia. He had to try a number of ways to get the equilibrium to move towards the right-hand side and give a bigger yield of ammonia. He took note of **Le Chatelier's Principle**. This states that: **When conditions are changed, a system in a state of equilibrium adjusts itself in such a way as to minimise the effects of the change.** You can see that when the reaction goes from left to right there is a decrease in the number of molecules of gas. As a result (see Topic 27.8) if the reaction went to completion, the volume of ammonia would be only half that of the mixture of nitrogen and hydrogen used. So if the pressure on the mixture is increased, the system can absorb the increase in pressure by reducing its volume, that is, by reacting to form ammonia. The reaction is exothermic; that is, heat is given out in going from left to right. The reverse reaction, the dissociation of ammonia, is endothermic. If the temperature is raised, the system adjusts so as to absorb heat. It does this through the dissociation of ammonia because this reaction is endothermic. Running the process at high temperature therefore reduces the percentage conversion of the elements into ammonia. If the reactants are at a very low temperature, the system takes a long time to come to equilibrium. In practice, a high pressure (about 200 atm) and a moderate temperature (about 450 °C) are employed. A catalyst (iron or iron(III) oxide) is used to increase the speed of the reaction.

The Haber process for making ammonia uses a reversible reaction which comes to equilibrium. The right conditions move the position of equilibrium towards the product side of the equation. High pressure and low temperature favour the forward reaction.

The conditions employed are high pressure, moderate temperature and catalyst

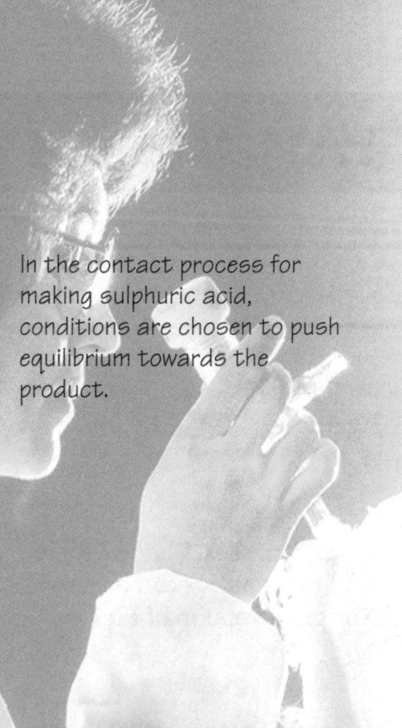

In the contact process for making sulphuric acid, conditions are chosen to push equilibrium towards the product.

## The contact process

In the manufacture of sulphuric acid (See Topic 23.4), an important reaction is:

$$\text{Sulphur dioxide} + \text{Oxygen} \rightleftharpoons \text{Sulphur trioxide}$$
$$2SO_2(g) + O_2(g) \rightleftharpoons 2SO_3(g)$$

The formation of sulphur trioxide involves a decrease in volume as three moles of gas form two moles of gas. High pressure should favour the forward reaction. However, sulphur dioxide is easily liquefied, and industry uses a pressure of only 2 atm. The reaction from left to right is exothermic. If the temperature is raised, the reverse reaction, the dissociation of sulphur trioxide is favoured. Therefore the industry uses a moderate temperature (about 450 °C). The use of a catalyst, vanadium(V) oxide, helps to compensate for not using a very high temperature.

## CHECKPOINT

1. Reactions take place more rapidly as the temperature rises. In view of this, why is the manufacture of ammonia carried out at 450 °C, rather than at a very high temperature?
2. A gas, A, reacts to form a mixture of two gases, B and C. The reaction is reversible, and the system reaches equilibrium.

   $$A(g) \rightleftharpoons B(g) + 2C(g)$$

   (a) Will an increase in pressure increase or decrease the yield of B and C?

   (b) What information would you need before you could predict the effect of raising the temperature on the rate of equilibrium?
3. An aqueous solution of bromine is called 'bromine water'. Some bromine molecules react with water molecules:

   $$\text{Bromine} + \text{Water} \rightleftharpoons \text{Hydrobromic acid} + \text{Bromic(I) acid}$$
   $$Br_2(aq) + H_2O(l) \rightleftharpoons HBr(aq) + HBrO(aq)$$

   The products are both strong acids. The reaction is reversible, and a solution of bromine in water reaches an equilibrium state in which the concentrations of all the species are constant.

   Predict what change will happen if you add a small amount of sodium hydroxide. In which direction will the equilibrium be displaced, from left to right or from right to left? Predict what colour change you will see. How could you reverse the colour change?

   Do an experiment to check your predictions.

# Topic 27 Chemical calculations

## 27.1 Relative atomic mass

FIRST THOUGHTS

Every dot of ink on this page is big enough to have a million hydrogen atoms fitted across it from side to side.

The masses of atoms are very small. Some examples are:

- Mass of hydrogen atom, H = $1.4 \times 10^{-24}$ g.
- Mass of mercury atom, Hg = $2.8 \times 10^{-22}$ g.
- Mass of carbon atom, C = $1.7 \times 10^{-23}$ g.

Chemists find it convenient to use **relative atomic masses**. The hydrogen atom is the lightest of atoms, and the masses of other atoms can be stated *relative to* that of the hydrogen atom. On the original version of the relative atomic mass scale:

- Relative atomic mass of hydrogen = 1.
- Relative atomic mass of mercury = 200 (a mercury atom is 200 times heavier than a hydrogen atom).
- Relative atomic mass of carbon = 12 (a carbon atom is 12 times heavier than a hydrogen atom).

Chemists now take the mass of one atom of carbon-12 as the reference point for the relative atomic mass scale. On the present scale

$$\text{Relative atomic mass of element} = \frac{\text{Mass of one atom of the element}}{(^1/_{12})\text{ Mass of one atom of carbon-12}}$$

Since relative atomic mass (symbol $A_r$) is a ratio of two masses, the mass units cancel, and relative atomic mass is a number without a unit.

The relative atomic masses of some common elements are listed in Table 27.1.

SUMMARY

The relative atomic mass, $A_r$, of an element is the mass of one atom of the element compared with $^1/_{12}$ the mass of one atom of carbon-12.

*Table 27.1* Some relative atomic masses

| Element | $A_r$ | Element | $A_r$ |
|---|---|---|---|
| Aluminium | 27 | Magnesium | 24 |
| Barium | 137 | Mercury | 200 |
| Bromine | 80 | Nitrogen | 14 |
| Calcium | 40 | Oxygen | 16 |
| Chlorine | 35.5 | Phosphorus | 31 |
| Copper | 63.5 | Potassium | 39 |
| Hydrogen | 1 | Sodium | 23 |
| Iron | 56 | Sulphur | 32 |
| Lead | 207 | Zinc | 65 |

CHECKPOINT

1 Refer to Table 27.1.
(a) How many times heavier is one atom of nitrogen than one atom of hydrogen?
(b) What is the ratio
Mass of one atom of mercury/Mass of one atom of bromine?
(c) How many atoms of oxygen are needed to equal the mass of one atom of bromine?
(d) How many atoms of sodium are needed to equal the mass of one atom of lead?

## 27.2 Relative molecular mass

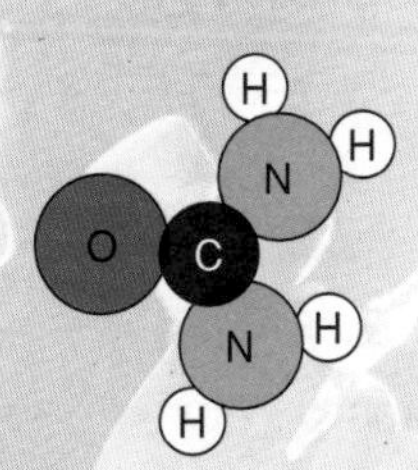


*Figure 27.2A* ▲ Atoms in a molecule of urea

**SUMMARY**

The relative molecular mass of a compound is equal to the sum of the relative atomic masses of all the atoms in one molecule of the compound or in one formula unit of the compound.

$M_r$ = Sum of $A_r$ values

The mass of a molecule is the sum of the masses of all the atoms in it. The relative molecular mass (symbol $M_r$) of a compound is the sum of the relative atomic masses of all the atoms in a molecule of the compound (see Figure 27.2A).

**Worked example** Find the relative molecular mass of urea.

**Solution** Formula of compound is $CON_2H_4$

1 atom of C ($A_r$ 12) = 12
1 atom of O ($A_r$ 16) = 16
2 atoms of N ($A_r$ 14) = 28
4 atoms of H ($A_r$ 1) = 4
Total = 60
Relative molecular mass, $M_r$, of urea = 60

Many compounds consist of ions, not molecules. The formula of an ionic compound represents a **formula unit** of the compound; for example, $CaSO_4$ represents a formula unit of calcium sulphate, not a molecule of calcium sulphate. The term relative molecular mass can be used for ionic compounds as well as molecular compounds.

**CHECKPOINT**

1 Work out the relative molecular masses of these compounds:
$CO$ $CO_2$ $SO_2$ $SO_3$ $NaOH$ $NaCl$ $CaO$ $Mg(OH)_2$ $Na_2CO_3$ $CuSO_4$ $CuSO_4.5H_2O$ $Ca(HCO_3)_2$

## 27.3 Percentage composition

How are you at percentages? 'Per cent' means 'per hundred'. What is the percentage of boys in your class? Say the class is 21 boys and 9 girls. This means that 21 out of 30 students are boys. If the class were 100 students, there would be (100/30) × no. of boys = (100/30) × 21 = 70% or

% of boys = (no. of boys/total no of students) × 100%

From the formula of a compound you can find the percentage by mass of the elements in the compound.

**Worked example 1** Find the percentage by mass of (a) calcium (b) chlorine in calcium chloride.

**Solution** First find the relative molecular mass of calcium chloride, formula $CaCl_2$.

$$M_r = A_r(\text{Ca}) + 2A_r(\text{Cl}) = 40 + (2 \times 35.5) = 111$$

$$\text{Percentage of calcium} = \frac{40}{111} \times 100 = 36\%$$

$$\text{Percentage of chlorine} = \frac{71}{111} \times 100 = 64\%$$

You can see that the two percentages add up to 100%.

**Worked example 2** Find the percentage of water in crystals of magnesium sulphate-7-water.

**Solution** Find the relative molecular mass of $MgSO_4.7H_2O$.

| | |
|---|---|
| 1 atom of magnesium ($A_r$ 24) | = 24 |
| 1 atom of sulphur ($A_r$ 32) | = 32 |
| 4 atoms of oxygen ($A_r$ 16) | = 64 |
| 7 molecules of water = 7 × [(2 × 1) + 16] | = 126 |
| Total = $M_r$ | = 246 |

$$\text{Percentage of water} = \frac{\text{Mass of water in formula}}{\text{Relative molecular mass}} \times 100$$

$$= \frac{126}{246} \times 100 = 51.2\%$$

The percentage of water in magnesium sulphate crystals is 51%.

SUMMARY

You can calculate the percentage by mass composition of a compound from its formula.

CHECKPOINT

You do not need calculators for these problems.

1 Find the percentage by mass of;
   (a) calcium in calcium bromide, $CaBr_2$,
   (b) iron in iron(III) oxide, $Fe_2O_3$,
   (c) carbon and hydrogen in ethane, $C_2H_6$,
   (d) sulphur and oxygen in sulphur trioxide, $SO_3$,
   (e) hydrogen and fluorine in hydrogen fluoride, HF,
   (f) magnesium, sulphur and oxygen in magnesium sulphate, $MgSO_4$.

2 Calculate the percentage by mass of water in:
   (a) copper(II) sulphate-5-water, $CuSO_4.5H_2O$ (take $A_r$ (Cu) = 64),
   (b) sodium sulphide-9-water, $Na_2S.9H_2O$.

## 27.4 The mole

FIRST THOUGHTS

Chemical equations tell us which products are formed when substances react. Equations can also be used to tell us what mass of each product is formed. The key to success is the mole concept.

A reaction of industrial importance is

$$\text{Calcium carbonate} \rightarrow \text{Calcium oxide} + \text{Carbon dioxide}$$
$$CaCO_3(s) \rightarrow CaO(s) + CO_2(g)$$

Cement manufacturers use this reaction to make calcium oxide (quicklime) from calcium carbonate (limestone). The mole concept makes it possible to calculate what mass of calcium oxide will be formed when a certain mass of calcium carbonate dissociates.

The mole concept dates back to the nineteenth century Italian chemist called Avogadro. This is how he argued:

You are familiar with **mass** of substance and **volume** of substance. These are physical quantities. Mass is stated in the unit g or kg. Volume may be stated in the unit $cm^3$, $dm^3$ or l (litre) or $m^3$.

The relative atomic masses of magnesium and carbon are: $A_r$ (Mg) = 24, $A_r$ (C) = 12.

Therefore we can say:

Since one atom of magnesium is twice as heavy as one atom of carbon, then one hundred Mg atoms are twice as heavy as one hundred C atoms, and five million Mg atoms are twice as heavy as five million C atoms.

Imagine a piece of magnesium that has twice the mass of a piece of carbon. It follows that the two masses must contain equal numbers of atoms:

two grams of magnesium and one gram of carbon contain the same number of atoms; ten tonnes of magnesium and five tonnes of carbon contain the same number of atoms.

The same argument applies to the other elements. Take the relative atomic mass in grams of any element:

Another physical quantity is **amount** of substance. It depends on the number of particles present. The unit in which it is stated is the mole. Equal amounts of elements contain equal numbers of atoms.

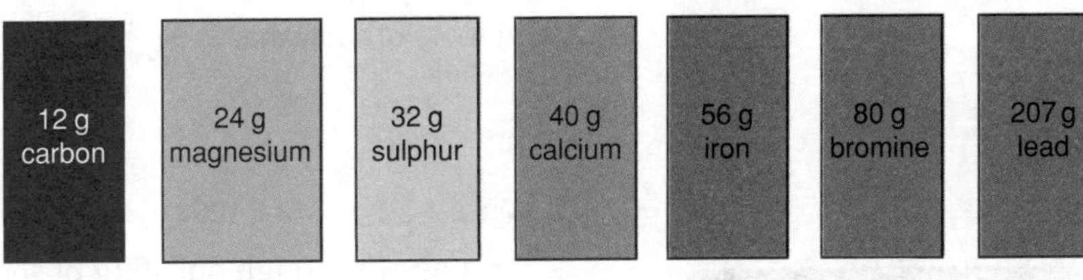


All these masses contain the same number of atoms. The number is $6.022 \times 10^{23}$.

> The amount of an element that contains $6.022 \times 10^{23}$ atoms (the same number of atoms as 12 g of carbon-12) is called one **mole** of that element.

The symbol for mole is **mol**. The ratio $6.022 \times 10^{23}$/mol is called the Avogadro constant. When you weigh out 12 g of carbon, you are counting out $6 \times 10^{23}$ atoms of carbon. This amount of carbon is one mole (1 mol) of carbon atoms. Similarly, 46 g of sodium is two moles (2 mol) of sodium atoms. You can say that the **amount** of sodium is two moles (2 mol). One mole of the compound ethanol, $C_2H_6O$, contains $6 \times 10^{23}$ molecules of $C_2H_6O$, that is, 46 g of $C_2H_6O$ (the molar mass in grams). To write 'one mole of oxygen' will not do. You must state whether you mean one mole of oxygen atoms, O (with a mass of 16 grams) or one mole of oxygen molecules, $O_2$ (with a mass of 32 grams).

## Molar mass

The mass of one mole of a substance is called the **molar mass**, symbol $M$. The molar mass of carbon is 12 g/mol. The molar mass of sodium is 23 g/mol. The molar mass of a compound is the relative molecular mass expressed in grams per mole. Urea, $CON_2H_4$, has a relative molecular mass of 60; its molar mass is 60 g/mol. Notice the units: relative molecular mass has no unit; molar mass has the unit g/mol.

$$\text{Amount (in moles) of substance} = \frac{\text{Mass of substance}}{\text{Molar mass of substance}}$$

Molar mass of element = Relative atomic mass in grams per mole
Molar mass of compound = Relative molecular mass in grams per mole

**Worked example 1** What is the amount (in moles) of sodium present in 4.6 g of sodium?

**Solution** $A_r$ of sodium = 23. Molar mass of sodium = 23 g/mol

$$\text{Amount of sodium} = \frac{\text{Mass of sodium}}{\text{Molar mass of sodium}} = \frac{4.6\text{ g}}{23\text{ g/mol}} = 0.2\text{ mol}$$

The amount (moles) of sodium is 0.2 mol.

### SUMMARY

The number of atoms in 12.000 g of carbon-12 is $6.022 \times 10^{23}$. The same number of atoms is present in a mass of any element equal to its relative atomic mass expressed in grams. This amount of any element is called **one mole** (1 mol) of the element. The ratio $6.022 \times 10^{23}$/mol is called the Avogadro constant. The number of moles of a substance is called the **amount** of that substance. The mass of one mole of an element or compound is the **molar mass**, $M$, of that substance.

$M$ of an element = $A_r$ expressed in g/mol
$M$ of a compound = $M_r$ expressed in g/mol

**Worked example 2** If you need 2.5 mol of sodium hydroxide, what mass of sodium hydroxide do you have to weigh out?

**Solution** Relative molecular mass of NaOH = 23 + 16 + 1 = 40

Molar mass of NaOH = 40 g/mol

$$\text{Amount of substance} = \frac{\text{Mass of substance}}{\text{Molar mass of substance}}$$

$$2.5 = \frac{\text{Mass}}{40}$$

Mass = 40 × 2.5 = 100 g

You need to weigh out 100 g of sodium hydroxide.

## CHECKPOINT

1 State the mass of:
 (a) 1.0 mol of aluminium atoms,
 (b) 3.0 mol of oxygen molecules, $O_2$,
 (c) 0.25 mol of mercury atoms,
 (d) 0.50 mol of nitrogen molecules, $N_2$,
 (e) 0.25 mol of sulphur atoms, S,
 (f) 0.25 mol of sulphur molecules, $S_8$.

2 Find the amount (moles) of each element present in:
 (a) 100 g of calcium,
 (b) 9.0 g of aluminium,
 (c) 32 g of oxygen, $O_2$,
 (d) 14 g of iron.

3 State the mass of:
 (a) 1.0 mol of sulphuric acid, $H_2SO_4$,
 (b) 0.5 mol of nitrogen dioxide molecules, $NO_2$,
 (c) 2.5 mol of magnesium oxide, MgO,
 (d) 0.10 mol of calcium carbonate, $CaCO_3$.

## 27.5 The masses of reactant and product

### FIRST THOUGHTS

*Have you grasped the mole concept?* In this section you will find out how it is used to obtain information about chemical reactions.

As well as knowing what products are formed in a chemical reaction, chemists want to know **what mass** of product is formed from a given mass of starting material. For example, calcium oxide (quicklime) is made by heating calcium carbonate (limestone). Manufacturers need to know what mass of limestone to heat to yield the mass of quicklime they want.

**Worked example 1** What mass of limestone (calcium carbonate) must be decomposed to yield 10 tonnes of calcium oxide (quicklime)?

**Solution** First write the equation for the reaction

| Calcium carbonate | → | Calcium oxide | + | Carbon dioxide |
|---|---|---|---|---|
| $CaCO_3(s)$ | → | $CaO(s)$ | + | $CO_2(g)$ |

The equation tells us that

1 mol of calcium carbonate forms 1 mol of calcium oxide.

You have met calculations of this type before.

For example,
To make 12 buns a cook needs 3 eggs. How many eggs does she need to make 20 buns?

The method you would use is:
To make 1 bun needs 3/12 eggs, therefore to make 20 buns needs 20 × 3/12 eggs = 5 eggs.

This chemical calculation uses the same method.

Using the molar masses $M(CaCO_3) = 100\,g/mol$, $M(CaO) = 56\,g/mol$,

100 g of calcium carbonate forms 56 g of calcium oxide.

The mass of calcium carbonate needed to make 10 tonnes of calcium oxide is therefore given by

$$\text{Mass of } CaCO_3 = \frac{100}{56} \times (\text{Mass of CaO}) = \frac{100}{56} \times 10 = 17.8\,\text{tonnes}$$

**Worked example 2** What mass of aluminium can be obtained by the electrolysis of 60 tonnes of pure aluminium oxide, $Al_2O_3$?

**Solution** The equation comes first.

$$\text{Aluminium oxide} \rightarrow \text{Aluminium} + \text{Oxygen}$$
$$2Al_2O_3(s) \rightarrow 4Al(s) + 3O_2(g)$$

From the equation you can see that

1 mole of aluminium oxide forms 2 mol of aluminium.

Using the molar masses $M(Al) = 27\,g/mol$, $M(Al_2O_3) = 102\,g/mol$,

102 g of aluminium oxide form 54 g of aluminium.

The mass of aluminium obtained from 60 tonnes of aluminium oxide is therefore given by

$$\text{Mass of aluminium} = \frac{54}{102} \times \text{Mass of aluminium oxide}$$
$$= \frac{54}{102} \times 60 = 31.8\,\text{tonnes}$$

## SUMMARY

The equation for a chemical reaction shows how many moles of product are formed from one mole of reactant. Using the equation and the molar masses of the chemicals, you can find out what mass of product is formed from a certain mass of reactant. In chemical calculations, the balanced equation for the reaction is the key to success.

## CHECKPOINT

You do not need calculators for these problems.

1. What mass of magnesium oxide, MgO, is formed when 4.8 g of magnesium are completely oxidised?
2. Hydrogen will reduce hot copper(II) oxide, CuO, to copper:

   Hydrogen + Copper(II) oxide → Copper + Water

   (a) Write the balanced chemical equation for the reaction.
   (b) Calculate the mass of copper that can be obtained from 4.0 g of copper(II) oxide. Use $A_r\,(Cu) = 64$
3. What mass of sodium bromide, NaBr, must be electrolysed to give 8 g of bromine, $Br_2$?
4. Ammonium chloride can be made by neutralising hydrochloric acid with ammonia:

   $$HCl(aq) + NH_3(aq) \rightarrow NH_4Cl(aq)$$

   What mass of ammonium chloride is formed when 73 g of hydrochloric acid are completely neutralised by ammonia?

## 27.6 Percentage yield

If you managed the work on percentage composition you will have no difficulty here.

Calculations based on chemical equations give the **theoretical yield** of product to be expected from a reaction. Often the actual yield is less than the calculated yield of product. The reason may be that some product has remained in solution or on a filter paper and has not been weighed with the final yield. The percentage yield of a product is given by:

$$\text{Percentage yield} = \frac{\text{Actual mass of product}}{\text{Calculated mass of product}} \times 100$$

**Worked example** A student calculates that a certain reaction will yield 7.0 g of a salt. Her product weighs 6.3 g. What percentage yield has she obtained?

**Solution**

$$\text{Percentage yield} = \frac{\text{Actual mass of product}}{\text{Calculated mass of product}} \times 100$$

$$= \frac{6.3}{7.0} \times 100 = 90\%$$

### CHECKPOINT

You will need a calculator to solve some of these problems.

1 When 6.4 g of copper were heated in air, 7.6 g of copper(II) oxide, CuO, were obtained.

$$2Cu(s) + O_2(g) \rightarrow 2CuO(s)$$

(a) Calculate the mass of copper(II) oxide that would be formed if the copper reacted completely. (Use $A_r(Cu) = 64$)

(b) Calculate the percentage yield that was actually obtained.

2 When 28 g of nitrogen and 6 g of hydrogen were mixed and allowed to react, 3.4 g of ammonia formed.

$$N_2(g) + 3H_2(g) \rightarrow 2NH_3(g)$$

(a) What is the maximum mass of ammonia that could be formed?

(b) What percentage of this yield was obtained?

3 A student passed chlorine over heated iron until all the iron had reacted. He collected 16.0 g of iron(III) chloride, $FeCl_3$. What percentage yield had he obtained?

4 A student neutralised 98 g of sulphuric acid, $H_2SO_4$, with ammonia, $NH_3$. On evaporating the solution until the salt crystallised, she obtained 120 g of ammonium sulphate, $(NH_4)_2SO_4$.

(a) Write a balanced chemical equation for the reaction.

(b) Calculate the theoretical yield of ammonium sulphate.

(c) Calculate the actual percentage yield.

(d) What do you think happened to the rest of the ammonium sulphate?

5 Copper(II) sulphate can be made by neutralising sulphuric acid with copper(II) oxide:

$$CuO(aq) + H_2SO_4(aq) \rightarrow CuSO_4(aq) + H_2O(l)$$

The salt crystallises as copper(II) sulphate-5-water, $CuSO_4.5H_2O$.

(a) Calculate the mass of crystals that can be made from 8.0 g of copper(II) oxide and an excess (more than enough) of sulphuric acid. Use $A_r$ (Cu) = 64.

(b) A student obtained 22 g of crystals from this preparation. What percentage yield was this?

# 27.7 Finding formulas

**FIRST THOUGHTS**

All those fascinating formulas in chemistry books – where do they come from? In this section you can find out.

The formula of a compound is worked out from the percentage composition by mass of the compound.

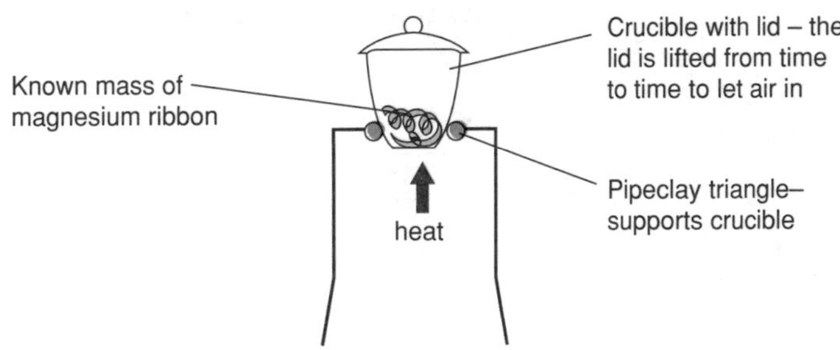


*Figure 27.7A* ▲ Heating magnesium

The steps in finding the formula are

- Find the mass of each element in a sample of the compound.
- Calculate the amount of each element present.
- Deduce the number of atoms of each element present, which gives the formula.

**Worked example** Finding the formula of magnesium oxide.
First, an experiment must be done to find the mass of oxygen that combines with a weighed amount of magnesium. Figure 27.7A shows a weighed quantity of magnesium being heated until it has been converted completely into magnesium oxide. Then the magnesium oxide must be weighed. The mass of oxygen that has combined with the magnesium is found by subtraction. A typical set of results is given below.

[1] Mass of crucible = 19.24 g
[2] Mass of crucible + magnesium = 20.68 g
[3] **Mass of magnesium** = [2] − [1] = 1.44 g
[4] Mass of crucible + magnesium oxide = 21.64 g
[5] **Mass of oxygen combined** = [4] − [2] = 0.96 g

**Solution** The results are used in this way:

| *Element* | *Magnesium* | *Oxygen* |
|---|---|---|
| *Mass* | 1.44 g | 0.96 g |
| $A_r$ | 24 | 16 |
| *Amount in moles* | 1.44/24 | 0.96/16 |
| | = 0.060 | = 0.060 |
| *Divide through by 0.060* | 1 mole Mg | 1 mole O |
| *Formula is* | MgO | |

The formula MgO is the simplest formula which fits the results. Other formulas, such as $Mg_2O_2$, $Mg_3O_3$, etc. also fit the results. MgO is the **empirical formula** for magnesium oxide.

**SUMMARY**

The empirical formula of a compound shows the symbols of the elements present and the ratio of the number of atoms of each element present in the compound.

> The empirical formula of a compound is the simplest formula which represents the composition by mass of the compound.

## Empirical formula of hydrate

The formula of a hydrate is worked out in the same way.

**Worked example** Copper(II) sulphate crystallises as the well-known blue hydrate.

The percentage of water in crystals is 36%. Find the formula of the hydrate $CuSO_4.nH_2O$.

**Method**

| | *Molar mass* | *Mass* | *Amount* | |
|---|---|---|---|---|
| $CuSO_4$ (taking Cu = 64) | 160 | 64% | 64/160 | = 0.4 mol |
| $H_2O$ | 18 | 36% | 36/18 | = 2.0 mol |
| Ratio = 0.4 $CuSO_4$ : 2.0 $H_2O$ = 1 $CuSO_4$ : $5H_2O$ | | | | |

**Answer** Formula = $CuSO_4.5H_2O$

## CHECKPOINT

You do not need calculators for these problems.

**1** Find the empirical formulas of the following compounds:
(a) a compound of 3.5 g of nitrogen and 4.0 g of oxygen,
(b) a compound of 14.4 g of magnesium and 5.6 g of nitrogen,
(c) a compound of 5.4 g of aluminium and 9.6 g of sulphur.

**2** Calculate the percentage by mass of water in
(a) sodium sulphide-9-water, $Na_2S.9H_2O$
(b) chromium(III) nitrate-9-water, $Cr(NO_3)_3.9H_2O$.

**3** Calculate the empirical formulas of the compounds which have the following percentage compositions by mass:
(a) 40% sulphur, 60% oxygen,
(b) 50% sulphur, 50% oxygen,
(c) 20% calcium, 80% bromine,
(d) 39% potassium, 1% hydrogen, 12% carbon, 48% oxygen.

**4** Find the empirical formulas of the compounds formed when
(a) 18 g of beryllium form 50 g of beryllium oxide,
(b) 11.2 g of iron form 16.0 g of an oxide of iron,
(c) 2.800 g of iron form 8.125 g of an iron chloride.

**5** The element titanium ($A_r$ = 48.0) combines with oxygen to form two different oxides. The mass of oxygen combining with 1.00 g of titanium in oxide A is 0.50 g and in oxide B is 0.67 g. Deduce the empirical formulas of the two oxides.

The molecular formula is a multiple of the empirical formula:

Molecular formula = $n \times$ Empirical formula

The molecular formula gives the correct relative molecular mass.

## Finding the molecular formula from the empirical formula

The empirical formula is the simplest formula that represents the composition of a compound. It may not be the **molecular formula**, that is the formula that shows the number of atoms present in a molecule of the compound. The molecular formula of ethanoic acid is $CH_3CO_2H$, which can be written as $C_2H_4O_2$. The empirical formula is $CH_2O$. To work out the relative molecular mass for ethanoic acid, you have to use the molecular formula $C_2H_4O_2$, and then you obtain a value of 60.

**Worked example 1** What is the molecular formula of a compound which has an empirical formula of $CH_2$ and a relative molecular mass of 70?

**Method** Relative molecular mass = 70
Empirical formula mass = 12 + 2 = 14
The relative molecular mass is 5 × the relative empirical formula mass
The molecular formula is 5 × the empirical formula

**Answer** The molecular formula is $C_5H_{10}$.

**Worked example 2** What is the molecular formula of a compound which has an empirical formula of $C_2H_6O$ and a relative molecular mass of 46?

**Method** Relative molecular mass = 46
Relative empirical formula mass = 24 + 6 + 16 = 46
The empirical formula gives the correct relative molecular mass. Therefore the molecular formula is the same as the empirical formula.

**Answer** The molecular formula is $C_2H_6O$.

### CHECKPOINT

1 Find the molecular formula of each of the following compounds from the empirical formula and the relative molecular mass.

| *Compound* | *Empirical formula* | *$M_r$* | *Compound* | *Empirical formula* | *$M_r$* |
|---|---|---|---|---|---|
| A | $CH_4O$ | 32 | B | $CH_2$ | 42 |
| C | $C_2H_4O$ | 88 | D | $CH_3O$ | 62 |
| E | $CH_3$ | 30 | F | $CH_2Cl$ | 99 |
| G | CH | 78 | H | $C_2HNO_2$ | 213 |

2 Calculate (a) the empirical formula and (b) the molecular formula of:
(i) a hydrocarbon which contains 80% by mass of carbon and has a relative molecular mass of 30
(ii) a hydrocarbon which contains 85.7% of carbon and has a relative molecular mass of 28.

3 What is (a) the empirical formula and (b) the molecular formula of a compound which contains 4.04% H, 24.24% C, 71.72% Cl and has a relative molecular mass of 99?

# 27.8 Reacting volumes of gases

FIRST THOUGHTS

Calculations on reacting volumes of gas are simple if you understand one fact.

Whatever the gas, the volume occupied by one mole of the gas is the same: 24.0 $dm^3$ at r.t.p. (the gas molar volume).

The volume of a certain mass of gas depends on its temperature and on its pressure. We therefore state the temperature and the pressure at which a volume was measured. It is usual to state gas volumes either at standard temperature and pressure (s.t.p., 0 °C and 1 atm) or at room temperature and pressure (r.t.p., 20 °C and 1 atm).

One mole of a gas is the amount of the gas that contains $6 \times 10^{23}$ molecules of that gas. To measure one mole of gas, you take the molar mass expressed in grams, e.g. 2 g of hydrogen, $H_2$. Measurements show that one of mole of gas occupies 24.0 $dm^3$ at r.t.p. For all gases, the volume is the same.

- 2 g of hydrogen
- 28 g of nitrogen
- 64 g of sulphur dioxide

} All these are one mole of gas (the molar mass expressed in grams) and all occupy 24 $dm^3$ at r.t.p.

The volume of one mole of gas, 24.0 $dm^3$ at r.t.p., is called the **gas molar volume**.

Calculations on the reacting volumes of gases start with the equation for the reaction. After that, it's as easy as one, two, three. Take the reaction:

$$A(g) + 3B(g) \rightarrow 2C(g)$$

The equation tells you that

1 mole of A reacts with 3 moles of B to form 2 moles of C

therefore, at r.t.p.,

24 $dm^3$ of A react with $3 \times 24$ $dm^3$ of B to form $2 \times 24$ $dm^3$ of C

and in general,

1 volume of A reacts with 3 volumes of B to form 2 volumes of C.

*Write the equation.*

*Then the calculation becomes a simple ratio-type calculation.*

**Worked example 1** Nitrogen and hydrogen combine to form ammonia:

$$N_2(g) + 3H_2(g) \rightleftharpoons 2NH_3(g)$$

If hydrogen is fed into the plant at 12 $m^3$/second, at what rate should nitrogen be fed in?

**Method** From the equation, 1 mole of nitrogen combines with 3 moles of hydrogen, therefore 1 volume of nitrogen combines with 3 volumes of hydrogen, therefore, the rate of flow of nitrogen should be one third that of hydrogen, that is $\frac{1}{3} \times 12\,m^3/s = 4\,m^3/s$.

**Answer** Nitrogen should be fed in at a rate of 4 $m^3$/s.

**Worked example 2** What volume of carbon dioxide (at r.t.p.) is formed by the complete combustion of 3 g of carbon?

**Method** From the equation,

$$C(s) + O_2(g) \rightarrow CO_2(g)$$

we can see that 1 mole of carbon forms 1 mole of carbon dioxide
that is 12 g carbon form 24 $dm^3$ of carbon dioxide
therefore 3 g carbon form $\frac{3}{12} \times 24\,dm^3 = 6\,dm^3$ of carbon dioxide.

**Answer** The combustion of 3 g of carbon produces 6 $dm^3$ of carbon dioxide at r.t.p.

SUMMARY

Calculations on reacting volumes of gases always start with the equation for the chemical reaction. One mole of any gas occupies 24.0 $dm^3$ at r.t.p. This volume is called the gas molar volume.

## CHECKPOINT

1 Calculate the volume at r.t.p. of oxygen needed for the complete combustion of 8 g of sulphur.

2 Calculate the volume at r.t.p. of oxygen needed for the complete combustion of 250 $cm^3$ of methane, $CH_4$. What volume of carbon dioxide is formed?

3 A power station burns coal which contains sulphur. The sulphur burns to form sulphur dioxide, $SO_2$. If the power station burns 28 tonnes of sulphur in its coal every day, what volume of sulphur dioxide does it send into the air? (1 tonne = 1000 kg)

4 What volume of hydrogen at r.t.p. is formed when 7.0 g of iron react with an excess of sulphuric acid? The equation is

$$Fe(s) + H_2SO_4(aq) \rightarrow H_2(g) + FeSO_4(aq)$$

5 A cook puts 3 g of sodium hydrogencarbonate into a cake mixture. Calculate the volume at r.t.p. of carbon dioxide that will be produced in the reaction:

$$2NaHCO_3(s) \rightarrow Na_2CO_3(s) + CO_2(g) + H_2O(g)$$

6 The equation for the reaction between marble and hydrochloric acid is

$$CaCO_3(s) + 2HCl(aq) \rightarrow CaCl_2(aq) + CO_2(g) + H_2O(l)$$

What mass of marble is needed to give 6.0 $dm^3$ of carbon dioxide?

## 27.9 Calculations on electrolysis

If you need to refresh your grasp of electrolysis, refer to Topic 7.

When a current passes through a solution of a salt of a metal which is low in the electrochemical series, metal ions are discharged and metal atoms are deposited on the cathode (see Figure 27.9A).

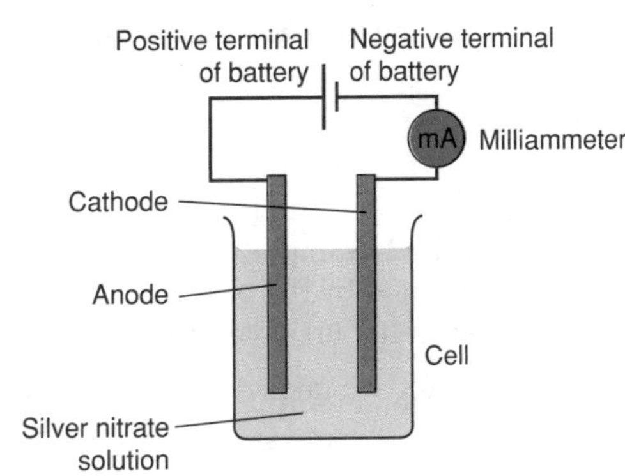


*Figure 27.9A* ▲ The electrolysis of silver nitrate solution

The cathode process is

$$Ag^+(aq) + e^- \rightarrow Ag(s)$$

This equation tells us that 1 silver ion accepts 1 electron to form 1 atom of silver. Therefore 1 mole of silver ions accept 1 mole of electrons to form 1 mole of silver atoms.

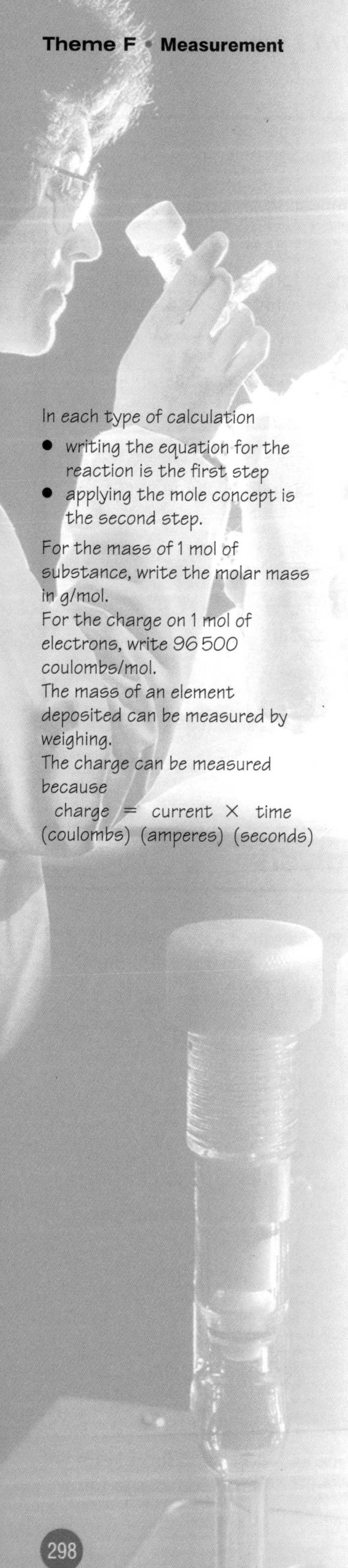

In each type of calculation
- writing the equation for the reaction is the first step
- applying the mole concept is the second step.

For the mass of 1 mol of substance, write the molar mass in g/mol.
For the charge on 1 mol of electrons, write 96 500 coulombs/mol.
The mass of an element deposited can be measured by weighing.
The charge can be measured because
charge = current × time
(coulombs) (amperes) (seconds)

When 1 mole of silver has been deposited, the cathode has increased in mass by 108 g (the mass of 1 mole of silver). This process must have needed the passage of 1 mole of electrons through the cell. We can measure the quantity of electric charge that passes through the cell when 108 g of silver are deposited. This quantity is the charge on 1 mole of electrons. Electric charge is measured in coulombs (C). One coulomb is the electric charge that passes when 1 ampere flows for 1 second.

$$\text{Charge in coulombs} = \text{Current in amperes} \times \text{Time in seconds}$$
$$Q = I \times t$$

Using a cell such as that in Figure 27.9A, one can pass a current through a milliammeter in the circuit for a known time. By weighing the cathode before and after the passage of a current, the mass of silver deposited can be found. Experiments of this kind show that to deposit 108 g of silver requires the passage of 96 500 coulombs. This quantity of electricity must be the charge on one mole of electrons ($6 \times 10^{23}$ electrons). The value 96 500 C/mol is called the Faraday constant after Michael Faraday who did much of the early work on electrolysis.

Now in the deposition of copper in electrolysis, the cathode process is

$$Cu^{2+}(aq) + 2e^- \rightarrow Cu(s)$$

1 copper ion needs 2 electrons to become 1 atom of copper;
1 mole of copper ions need 2 moles of electrons for discharge;
1 mole of copper ions need 2 × 96 500 coulombs for discharge.

When gold is deposited during the electrolysis of gold(III) salts, the cathode process is:

$$Au^{3+}(aq) + 3e^- \rightarrow Au(s)$$

1 gold ion needs 3 electrons to form 1 mole of gold atoms;
1 mole of gold ions need 3 moles of electrons for discharge;
3 × 96 500 coulombs are needed to deposit 1 mole of gold.

It's as easy as one-two-three. First work out whether

1 mole of electrons discharge 1 mole of the element, e.g. silver, or

$\frac{1}{2}$ mole of the element, e.g. copper, or

$\frac{1}{3}$ mole of the element, e.g. gold.

$$\text{No. of moles of element discharged} = \frac{\text{No. of moles of electrons}}{\text{No. of charges on one ion of the element}}$$
$$= \frac{\text{No. of coulombs}/96\,500}{\text{No. of charges on one ion of the element}}$$

**Worked example 1** What masses of the following elements are deposited by the passage of one mole of electrons through solutions of their salts? (a) silver (b) copper (c) gold

**Method** Start with the equations.

$$Ag^{+}(aq) + e^- \rightarrow Ag(s)$$
$$Cu^{2+}(aq) + 2e^- \rightarrow Cu(s)$$
$$Au^{3+}(aq) + 3e^- \rightarrow Au(s)$$

As already argued,

1 mol electrons deposit 1 mol silver = 108 g silver

1 mole electrons deposit $\frac{1}{2}$ mol copper = $\frac{1}{2} \times 63.5$ g = 31.8 g copper

1 mole electrons deposit $\frac{1}{3}$ mol gold = $\frac{1}{3} \times 197$ g = 65.7 g gold

**Answer** (a) 108 g silver, (b) 31.8 g copper, (c) 65.7 g gold

If you work patiently through the worked examples, you will overcome any difficulties. Guaranteed!

**Worked example 2** A current of 10.0 milliamps (mA) passes for 4.00 hours through a solution of silver nitrate, a solution of copper(II) sulphate and a solution of gold(III) nitrate connected in series. What mass of metal is deposited in each?

**Method** Charge in coulombs = Current in amperes × Time in seconds

Charge = $0.010 \times 4.00 \times 60 \times 60 = 144\,C$

Charge = $\frac{144}{96\,500}$ moles of electrons

Equations are:

$$Ag^{+}(aq) + e^{-} \rightarrow Ag(s)$$
$$Cu^{2+}(aq) + 2e^{-} \rightarrow Cu(s)$$
$$Au^{3+}(aq) + 3e^{-} \rightarrow Au(s)$$

1 mol electrons deposit 1 mol silver; therefore $\frac{144}{96\,500}$ mol electrons deposit $\frac{144}{96\,500}$ mol silver = $144 \times \frac{108}{96\,500}$ g silver = 0.161 g silver

1 mol electrons deposit $\frac{1}{2}$ mol copper; therefore $\frac{144}{96\,500}$ mol electrons deposit $\frac{1}{2} \times \frac{144}{96\,500}$ mol copper = $\frac{1}{2} \times 144 \times \frac{63.5}{96\,500}$ g copper = 0.0474 g copper

1 mol electrons deposit $\frac{1}{3}$ mol gold; therefore $\frac{144}{96\,500}$ mol electrons deposit $\frac{1}{3} \times \frac{144}{96\,500}$ mol gold = $\frac{1}{3} \times 144 \times \frac{197}{96\,500}$ g gold = 0.0980 g gold

**Answer** 0.161 g silver, 0.0474 g copper, 0.0980 g gold

**Worked example 3** Aluminium is extracted by the electrolysis of molten aluminium oxide. How many coulombs of electricity are needed to produce 1 tonne of aluminium (1 tonne = 1000 kg)?

**Method** The equation

$$Al^{3+}(l) + 3e^{-} \rightarrow Al(s)$$

shows that 3 mol electrons are needed to give 1 mol aluminium

$1.00 \times 10^{6} \times \frac{3}{27}$ mol electrons are needed to give 1 tonne of aluminium

$1.00 \times 10^{6} \times \frac{3}{27}$ mol of electrons $= 1.00 \times 10^{6} \times \frac{3}{27} \times 96\,500\,C$

$= 1.07 \times 10^{10}\,C$

**Answer** $1.07 \times 10^{10}$ C will deposit 1 tonne of aluminium.

**Worked example 4** A metal of relative atomic mass 27 is deposited by electrolysis. If 0.201 g of the metal is deposited when 0.200 A flow for 3.00 hours, what is the charge on the ions of this element?

**Method**
Charge = $0.200 \times 3.00 \times 60 \times 60\,C = 2160\,C$
If 2160 C deposit 0.201 g of the metal,
then 96 500 C deposit $96\,500 \times \frac{0.201}{2160}\,g = 8.98\,g$

Since 8.98 g metal are deposited by 1 mol electrons, 27.0 g of metal are deposited by $\frac{27}{8.98}$ mol electrons = 3 mol electrons

**Answer** The charge on the metal ions is +3.

## Calculate the volume of gas evolved during electrolysis

### Hydrogen

When a current is passed through a solution of a salt of a metal which is high in the electrochemical series, hydrogen ions are discharged at the cathode.

$$H^+(aq) + e^- \rightarrow H(g)$$
$$\text{followed by } 2H(g) \rightarrow H_2(g)$$

Each hydrogen molecule needs 2 electrons for its evolution, and 1 mole of hydrogen needs 2 moles of electrons for its evolution. Thus 2 moles of electrons ($2 \times 96\,500\,C$) will result in the evolution of $24\,dm^3$ at r.t.p. (the gas molar volume) of hydrogen.

When the product of electrolysis is a gas, for 1 mol of gas write $24.0\,dm^3$ at r.t.p.
The only way you can go wrong is to forget that, e.g. 1 mol of hydrogen is 1 mol of $H_2$ (not H).
1 mol H needs 1 mol of electrons
1 mol of $H_2$ needs 2 mol of electrons
1 mol of $OH^-$ needs 1 mol of electrons
1 mol of $O_2$ needs 4 mol of electrons.

### Chlorine

When chlorine is evolved at the anode,

$$Cl^-(aq) \rightarrow Cl(g) + e^-$$
$$\text{followed by } 2Cl(g) \rightarrow Cl_2(g)$$

Each chlorine molecule gives 2 electrons when it is evolved, and 1 mole of chlorine gives 2 moles of electrons when it is evolved. Thus 2 moles of electrons ($2 \times 96\,500\,C$) accompany the evolution of $24\,dm^3$ at r.t.p. (the gas molar volume) of chlorine.

### Oxygen

When oxygen is evolved at the anode,

$$OH^-(aq) \rightarrow OH(g) + e^-$$
$$\text{followed by } 4OH(g) \rightarrow O_2(g) + 2H_2O(l)$$

Thus 4 moles of electrons must pass with the evolution of 1 mole of oxygen ($24\,dm^3$ at r.t.p.).

**Worked example 5** Name the gases formed at each electrode when 15.0 mA of current passes for 6.00 hours through a solution of sulphuric acid and calculate their volumes (at r.t.p.).

**Method**
At the cathode hydrogen is evolved

$$H^+(aq) + e^- \rightarrow H(g)$$
$$\text{followed by } 2H(g) \rightarrow H_2(g)$$

so that 2 mol electrons discharge 1 mol hydrogen gas.

At the anode oxygen is evolved

$$OH^-(aq) \rightarrow OH(g) + e^-$$
$$\text{followed by } 4OH(g) \rightarrow O_2(g) + 2H_2O(l)$$

so that 4 mol electrons discharge 1 mol oxygen gas.

$$\text{Charge (C)} = \text{Current (A)} \times \text{Time (s)}$$
$$= 15.0 \times 10^{-3} \times 6.00 \times 60 \times 60 = 324\,C$$

$$\text{No. of mol electrons} = \frac{324\,C}{96\,500\,C/mol} = 3.36 \times 10^{-3}\,mol$$

Amount of hydrogen discharged = $\frac{1}{2} \times 3.36 \times 10^{-3}$ mol
Volume of hydrogen = $1.68 \times 10^{-3} \times 24.0\,dm^3 = 40.4\,cm^3$ at r.t.p.
Amount of oxygen = $\frac{1}{4} \times 3.36 \times 10^{-3}$ mol = $0.84 \times 10^{-3}$ mol
Volume of oxygen = $0.84 \times 10^{-3} \times 24.0\,dm^3 = 20.2\,cm^3$ at r.t.p.

**Answer** At the cathode 40.4 $cm^3$ (at r.t.p.) of hydrogen are evolved. At the anode 20.2 $cm^3$ (at r.t.p.) of oxygen are evolved.

## CHECKPOINT

(See relative atomic masses given opposite.)

| Some relative atomic masses | |
|---|---|
| Ag | 108 |
| Al | 27 |
| Br | 80 |
| Cd | 112 |
| Cl | 35.5 |
| Cu | 63.5 |
| Fe | 56 |
| H | 1 |
| Ni | 59 |
| O | 16 |
| Pb | 207 |
| Sn | 119 |

1 Calculate the mass of each element discharged when 0.250 mol of electrons passes through each of the solutions listed.
(a) copper from copper(II) sulphate solution
(b) nickel from nickel chloride solution
(c) lead from lead(II) nitrate solution
(d) bromine from potassium bromide solution
(e) tin from tin(II) nitrate solution

2 Calculate the volume (at r.t.p.) of each gas evolved when 48 250 C pass through a solution of (a) dilute hydrochloric acid (b) dilute nitric acid.

3 A current passes through two cells in series. The cells contain solutions of silver nitrate and lead(II) nitrate. In the first, 0.540 g of silver is deposited. What mass of lead is deposited in the second?

4 When a current passes through a solution of copper(II) sulphate, 0.635 g of copper is deposited on the cathode. What volume of oxygen is evolved at the anode?

5 A current passes through two cells in series. In the first, 0.2160 g of silver is deposited. In the second, 0.1125 g of cadmium is deposited. Use this information to calculate the charge on the cadmium ion.

6 A current of 2.00 A passes for 96.5 hours through molten aluminium oxide. What mass of aluminium is deposited?

7 A current of 2.01 A passed for 8.00 minutes through aqueous nickel sulphate. The mass of the cathode increased by 0.295 g.
(a) How many coulombs of electricity passed during the experiment?
(b) How many moles of nickel were deposited?
(c) How many moles of electrons are needed to discharge 1 mole of nickel ions?
(d) Write an equation for the cathode reaction.

**8** Choose the correct answers.

A student electrolysed a solution of silver nitrate in series with a solution of sulphuric acid. A steady current passed for 30 minutes, and 24 $cm^3$ of hydrogen collected at the cathode of the sulphuric acid cell.

(a) The volume of oxygen evolved at the anode is
(i) 6 $cm^3$ (ii) 12 $cm^3$ (iii) 24 $cm^3$ (iv) 32 $cm^3$ (v) 48 $cm^3$

(b) If the student doubled the current and passed it for 15 minutes, the volume of hydrogen evolved would be
(i) 6 $cm^3$ (ii) 12 $cm^3$ (iii) 24 $cm^3$ (iv) 48 $cm^3$ (v) 96 $cm^3$

(c) The mass of silver deposited on the cathode of the silver nitrate cell is
(i) 0.108 g (ii) 0.216 g (iii) 0.0270 g (iv) 0.0540 g (v) 0.0135 g

**9** Which one of the following requires the largest quantity of electricity for discharge at an electrode?
(a) 1 mol $Ni^{2+}$ ions (b) 2 mol $Fe^{3+}$ ions (c) 3 mol $Ag^+$ ions
(d) 4 mol $Cl^-$ ions (e) 5 mol $OH^-$ ions

**10** A current of 0.010 A passed for 5.00 hours through a solution of a salt of the metal M, which has a relative atomic mass of 52. The mass of M deposited on the cathode was 0.0323 g. What is the charge on the ions of M?

## 27.10 Concentration of solution

FIRST THOUGHTS

Remember:
One litre (1 l) is the same as one cubic decimetre (1 $dm^3$).
$1\,l = 1000\,cm^3 = 1\,dm^3$

One way of stating the concentration of a solution is to state the mass of solute in one litre of solution, for example in grams per litre, g/l. Chemists find it more useful to state the amount in moles of a solute present in one litre of solution (see Figure 27.10A).

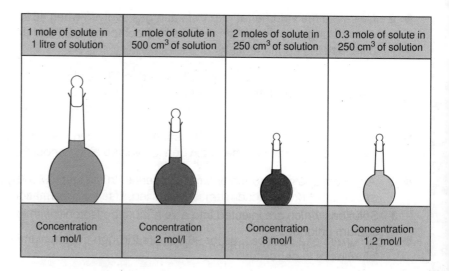

| 1 mole of solute in 1 litre of solution | 1 mole of solute in 500 $cm^3$ of solution | 2 moles of solute in 250 $cm^3$ of solution | 0.3 mole of solute in 250 $cm^3$ of solution |
|---|---|---|---|
| Concentration 1 mol/l | Concentration 2 mol/l | Concentration 8 mol/l | Concentration 1.2 mol/l |

*Figure 27.10A* Concentrations of solutions in moles per litre (mol/l)

A solution of concentration 1 mol/l (1 mol $l^{-1}$) or 1 mol/$dm^3$ (1 mol $dm^{-3}$) is called a 1 M solution.

$$\text{Concentration in moles per litre} = \frac{\text{Amount of solute in moles}}{\text{Volume of solution in litres}}$$

Rearranging,

Amount of solute (mol) = Volume of solution (l) × Concentration (mol/l)

Be sure you know that for a solution

$$\text{concentration} = \frac{\text{amount of solute}}{\text{volume of solution}}$$

Usually, with amount in mol and volume in l or $dm^3$, concentration is stated in mol/l or mol/$dm^3$. The worked examples will make everything clear. Have patience; do work through them.

**SUMMARY**

The concentration of a solute in a solution can be stated as:
(a) grams of solute per litre of solution, g/l,
(b) moles of solute per litre of solution, mol/l.

$$\text{Concentration (mol/l)} = \frac{\text{Amount of solute (mol)}}{\text{Volume of solution (l)}}$$

**Worked example 1** Calculate the concentration of a solution that was made by dissolving 100 g of sodium hydroxide and making the solution up to 2.0 litres.

**Solution** Molar mass of sodium hydroxide, NaOH = 23 + 16 + 1 = 40 g/mol

$$\text{Amount in moles} = \frac{\text{Mass}}{\text{Molar mass}} = \frac{100\,\text{g}}{40\,\text{g/mol}} = 2.5\,\text{mol}$$

Volume of solution = 2.0 litres

$$\text{Concentration} = \frac{\text{Amount in moles}}{\text{Volume in litres}} = \frac{2.5\,\text{mol}}{2.0\,\text{l}} = 1.25\,\text{mol/l}$$

**Worked example 2** Calculate the amount in moles of solute present in 75 $cm^3$ of a solution of hydrochloric acid which has a concentration of 2.0 mol/l.

**Solution** Amount (mol) = Volume (l) × Concentration (mol/l)

Amount of solute, HCl = $(75 \times 10^{-3}) \times 2.0 = 0.15$ mol

Note that the volume in $cm^3$ has been changed into litres to make the units correct:

Amount (**moles**) = Volume (**litres**) × Concentration (**moles per litre**)

**Worked example 3** What mass of sodium carbonate must be dissolved in 1 l of solution to give a solution of concentration 1.5 mol/l (a 1.5 M solution)?

**Solution** Amount (mol) = Volume (l) × Concentration (mol/l)
= 1.00 l × 1.5 mol/l
= 1.5 mol

Molar mass of sodium carbonate, $Na_2CO_3$ = (2 × 24) + 12 + (3 × 16)
= 106 g/mol

Mass of sodium carbonate = 1.5 mol × 106 g/mol = 159 g

## CHECKPOINT

1 Calculate the concentrations of the following solutions.
(a) 30 g of ethanoic acid, $C_2H_4O_2$, in 500 $cm^3$ of solution.
(b) 8.0 g of sodium hydroxide in 2.0 l of solution.
(c) 8.5 g of ammonia in 250 $cm^3$ of solution.
(d) 12.0 g of magnesium sulphate in 250 $cm^3$ of solution.

2 Find the amount of solute in moles present in the following solutions.
(a) 1.00 l of a hydrochloric acid solution of concentration 0.020 mol/l.
(b) 500 $cm^3$ of a solution of potassium hydroxide of concentration 2.0 mol/l.
(c) 500 $cm^3$ of sulphuric acid of concentration 0.12 mol/l.
(d) 100 $cm^3$ of a 0.25 mol/l solution of sodium hydroxide.

3 Solutions which are injected into a vein must be isotonic with the blood. They contain 8.43 g/l of sodium chloride.
(a) What is the concentration of sodium chloride in mol/l?
(b) What does 'isotonic' mean? Why must the solutions be isotonic with blood?

4 A woman mixes a drink containing 9.2 g of ethanol, $C_2H_6O$, in 100 $cm^3$ of solution. What is the concentration of ethanol in the solution (in mol/l)?

5 Calculate the concentrations of the following solutions.
(a) 4.0 g of sodium hydroxide in 500 $cm^3$ of solution
(b) 7.4 g of calcium hydroxide in 5.0 l of solution
(c) 49.0 g of sulphuric acid in 2.5 l of solution
(d) 73 g of hydrogen chloride in 250 $cm^3$ of solution

➧ **6** Find the amount of solute in moles present in the following solutions.
(a) 1.00 l of a solution of sodium hydroxide of concentration 0.25 mol/l
(b) 500 $cm^3$ of hydrochloric acid of concentration 0.020 mol/l
(c) 250 $cm^3$ of 0.20 mol/l sulphuric acid
(d) 10 $cm^3$ of a 0.25 mol/l solution of potassium hydroxide

## 27.11 Volumetric analysis

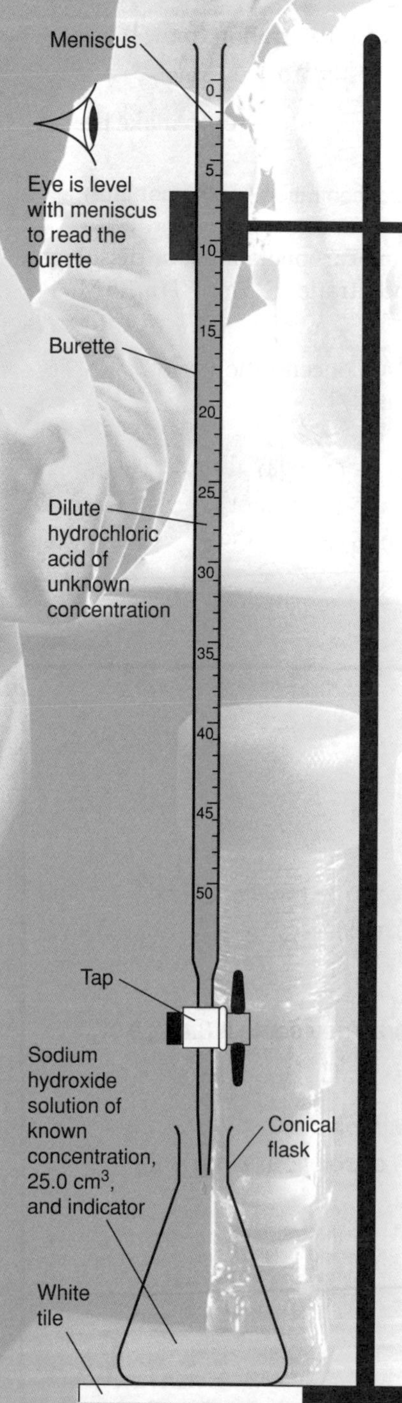


**FIRST THOUGHTS**

In your laboratory periods, you will learn how to do careful practical work in order to make precise measurements of concentration. This topic summarises the technique you will learn.

**Volumetric analysis** is a means of finding out the concentration of a solution. The method is to add a solution of, say an acid, to a solution of, say a base in a careful way until there is just enough of the acid to react with the base. This method is called **titration**. The concentration of one of the two solutions must be known, and the volumes of both must be measured. You can use a standard solution of a base to find out the concentration of a solution of an acid. You have to find out what volume of the acid solution of unknown concentration is needed to neutralise a known volume, usually 25 $cm^3$, of the standard solution of a base. The steps in volumetric analysis are as follows.

1 Pipette 25.0 $cm^3$ of the alkali solution into a clean conical flask. Add a few drops of indicator.
2 Read the burette, $V_1$ $cm^3$. Run the acid solution from the burette drop by drop into the alkali until the indicator just changes colour. This is the 'end-point' of the titration.
3 Read the burette again, $V_2$ $cm^3$. The volume of acid used, $(V_2 - V_1)$ $cm^3$, is the volume of acid needed to neutralise 25.0 $cm^3$ of the alkali.

**Worked example 1** A solution of hydrochloric acid is titrated against a standard sodium hydroxide solution. 15.0 $cm^3$ of hydrochloric acid neutralise 25.0 $cm^3$ of a 0.100 mol/l solution of sodium hydroxide. What is the concentration of hydrochloric acid?

**Solution**

(1) First write the equation for the reaction:

Hydrochloric acid + Sodium hydroxide → Sodium chloride + Water

$$HCl(aq) + NaOH(aq) \rightarrow NaCl(aq) + H_2O(l)$$

The equation tells you that 1 mole of HCl neutralises 1 mole of NaOH.

(2) Now work out the amount in moles of base. You must start with the base because you know the concentration of base, and you do not know the concentration of acid.

$$\text{Amount (mol)} = \text{Volume (l)} \times \text{Concentration (mol/l)}$$

$$\text{Amount (mol) of NaOH} = \text{Volume } (25.0\ cm^3) \times \text{Concentration } (0.100\ \text{mol/l})$$

$$= 25.0 \times 10^{-3}\ \text{l} \times 0.100\ \text{mol/l} = 2.50 \times 10^{-3}\ \text{mol}$$

*Fig. 27.11A* Titration. Hydrochloric acid is added to the conical flask until the indicator shows that the contents of the flask are neutral. The volume of acid added can be read off the burette

Here we go again with the worked examples. They will make everything clear if you will work through them.

## SUMMARY

A standard solution is a solution of known concentration. The concentration of a solution can be found by volumetric analysis. The concentration of a solution of an acid can be found by titrating it against a standard solution of a base. Similarly, the concentration of a solution of a base can be found by titrating it against a standard solution of an acid.

(3) Now work out the concentration of acid

Amount (mol) of HCl = Amount (mol) of NaOH = $2.50 \times 10^{-3}$ mol

also Amount (mol) of HCl = Volume of HCl(aq) × Concentration of HCl(aq)

Therefore, if $c$ is the concentration of HCl,

$$2.50 \times 10^{-3}\,\text{mol} = 15.0 \times 10^{-3}\,\text{l} \times c$$

$$c = \frac{2.50 \times 10^{-3}\,\text{mol}}{15.0 \times 10^{-3}}$$

$$= 0.167\,\text{mol/l}$$

The concentration of hydrochloric acid is 0.167 mol/l.

**Worked example 2** In a titration, 25.0 cm$^3$ of sulphuric acid of concentration 0.150 mol/l neutralised 30.0 cm$^3$ of potassium hydroxide solution. Find the concentration of the potassium hydroxide solution.

**Solution**

(1) First write the equation:

Sulphuric acid + Potassium hydroxide → Potassium sulphate + Water

$H_2SO_4(aq) + 2KOH(aq) \rightarrow K_2SO_4(aq) + 2H_2O(l)$

The equation tells you that 1 mole of $H_2SO_4$ neutralises 2 moles of KOH.

(2) Now work out the amount (moles) of acid. You start with the acid because you know the concentration of the acid.

Amount (mol) of acid = Volume (25.0 cm$^3$) × Concentration (0.150 mol/l)

$= 25.0 \times 10^{-3}\,\text{l} \times 0.150\,\text{mol/l} = 3.75 \times 10^{-3}\,\text{mol}$

(3) Now work out the concentration of base.

Amount (mol) of KOH = 2 × amount (mol) of $H_2SO_4$

$= 7.50 \times 10^{-3}$ mol

also Amount (mol) of KOH = Volume of KOH(aq) × Concentration of KOH(aq)

Therefore, if $c$ mol/l is the concentration of KOH

$$c\,\text{mol/l} = \frac{7.50 \times 10^{-3}\,\text{mol}}{30.0 \times 10^{-3}\,\text{l}}$$

$$= 0.250\,\text{mol/l}$$

The concentration of potassium hydroxide is 0.250 mol/l.

## CHECKPOINT

1 25.0 cm$^3$ of sodium hydroxide solution are neutralised by 15.0 cm$^3$ of a solution of hydrochloric acid of concentration 0.25 mol/l. Find the concentration of the sodium hydroxide solution.

2 A solution of sodium hydroxide contains 10 g/l.
(a) What is the concentration of the solution in mol/l?
(b) What volume of this solution would be needed to neutralise 25.0 cm$^3$ of 0.10 mol/l hydrochloric acid?

3 25.0 cm$^3$ of hydrochloric acid are neutralised by 20.0 cm$^3$ of a solution of 0.15 mol/l sodium carbonate solution.
(a) How many moles of sodium carbonate are neutralised by 1 mol HCl?
(b) What is the concentration of the hydrochloric acid?

4 The class decide to test some antacid indigestion tablets. They dissolve tablets and titrate the alkali in them against a standard acid. The table shows their results.

| *Brand* | *Price of 100 tablets* (£) | *Volume of 0.01 mol/l acid required to neutralise 1 tablet* (cm$^3$) |
|---|---|---|
| Paingo | 0.60 | 2.8 |
| Relievo | 0.70 | 3.0 |
| Alko | 0.80 | 3.3 |
| Clearo | 0.90 | 3.6 |

(a) Which antacid tablets offer the best value for money?
(b) What other factors would you consider before choosing a brand?

# Topic 28 Chemical energy changes

## 28.1 Exothermic reactions

FIRST THOUGHTS

In many places in this book, we study the importance of energy. One means by which energy is converted from one form into another is by chemical reactions.

Check that you know the meanings of these terms:

- exothermic reaction
- oxidation
- combustion
- burning
- fuel
- respiration
- neutralisation
- hydration.

Our bodies obtain energy from the combustion of foods, and our vehicles obtain energy from the combustion of hydrocarbons. Reactions of this kind, which give out energy, are **exothermic reactions** (ex = out; therm = heat). You will meet many such reactions. Some are described below.

### Combustion

Topic 24.3 described the combustion of the fuels:

- coal (carbon + hydrocarbons),
- natural gas (largely methane, $CH_4$),
- petrol (a mixture of hydrocarbons, e.g. octane, $C_8H_{18}$)

We use the heat from these exothermic reactions to warm our homes and buildings and to drive our vehicles.

*Figure 28.1A* ▲ They all use exothermic reactions

- An oxidation reaction in which heat is given out is **combustion**. The oxidation of sugars in our bodies is combustion.
- Combustion accompanied by a flame is **burning**. When coal is burned in a fireplace, you see the flames.
- A substance which is oxidised with the release of energy is a fuel. Both sugars and coal are fuels.

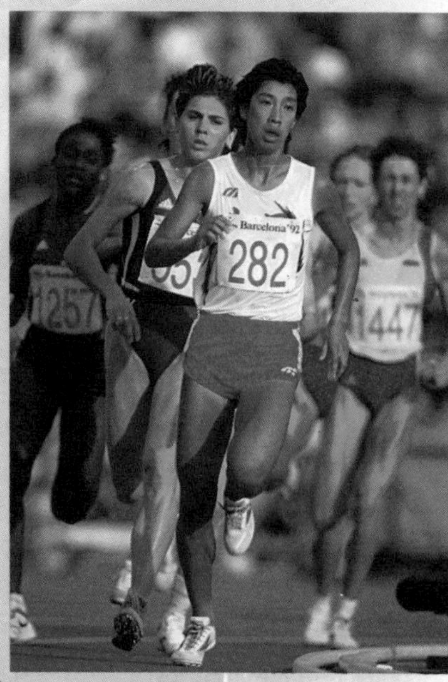


*Figure 28.1B* ▲ Energy from respiration

### Respiration

We use the combustion of foods to supply our bodies with energy. Important among 'energy foods' are **carbohydrates**. Glucose, a sugar, is oxidised to carbon dioxide and water with the release of energy. This exothermic reaction, which takes place inside plant cells and animal cells, is called **aerobic respiration**.

### Neutralisation

When an acid neutralises an alkali, heat is given out.

| Hydrogen ion + | Hydroxide ion | → | Water | Heat is given out |
|---|---|---|---|---|
| $H^+(aq)$ + | $OH^-(aq)$ | → | $H_2O(l)$ | |

EXTENSION FILE ACTIVITY

## SUMMARY

In exothermic reactions, energy is released. Examples are:

- the combustion of hydrocarbons,
- respiration,
- neutralisation,
- hydration of anhydrous salts.

## Hydration

Many anhydrous salts react with water to form hydrates. During hydration, heat is given out.

| Anhydrous copper (II) sulphate | + | Water | → | Copper(II) sulphate-5-water | Heat is given out |
|---|---|---|---|---|---|
| $CuSO_4(s)$ | + | $5H_2O(l)$ | → | $CuSO_4.5H_2O(s)$ | |

# 28.2 Endothermic reactions

*Check that you know the meanings of these terms:*

- *endothermic reaction*
- *photosynthesis*
- *thermal decomposition.*

## Photosynthesis

Plants manufacture sugars in the process of **photosynthesis**. They convert the energy of sunlight into the energy of the chemical bonds in sugar molecules. Photosynthesis is an **endothermic reaction**: it takes in energy.

Carbon dioxide + Water —(catalysed by chlorophyll, in the leaves of green plants)→ Glucose + Oxygen   Energy is taken in

$$6CO_2(g) + 6H_2O(l) \rightarrow C_6H_{12}O_6(aq) + 6O_2(g)$$

## Thermal decomposition

Many substances decompose when they are heated. An example is calcium carbonate (limestone).

| Calcium carbonate | → | Calcium oxide | + | Carbon dioxide | Heat is taken in |
|---|---|---|---|---|---|
| $CaCO_3(s)$ | → | $CaO(s)$ | + | $CO_2(g)$ | |

This is an important reaction because it yields **calcium oxide** (quicklime).

## SUMMARY

In endothermic reactions, energy is taken in from the surroundings. Examples are:

- photosynthesis,
- thermal decomposition,
- the reaction between steam and coke.

## The reaction between steam and coke

The endothermic reaction between steam and hot coke produces carbon monoxide and hydrogen. This mixture of flammable gases is used as fuel.

| Carbon (coke) | + | Steam | → | Carbon monoxide | + | Hydrogen | Heat is taken in |
|---|---|---|---|---|---|---|---|
| $C(s)$ | + | $H_2O(g)$ | → | $CO(g)$ | + | $H_2(g)$ | |

## CHECKPOINT

1. Give an example of a reaction which is of vital importance in everyday life and which is (a) exothermic and (b) endothermic. Explain why the reactions you mention are so important.
2. The reaction shown below is exothermic.

   Anhydrous copper(II) sulphate + Water → Copper(II) sulphate-5-water

   Describe or illustrate an experiment by which you could find out whether this statement is true.

# 28.3 Heat of reaction

**FIRST THOUGHTS**

*Why is energy (heat and other forms of energy) given out or taken in during a chemical reaction?* That is the question for this section.

**Chemical bonds** are forces of attraction between the atoms or ions or molecules in a substance. To break these bonds, energy must be supplied. When bonds are created, energy is given out. In a chemical reaction, bonds are broken, and new bonds are made.

Figure 28.3A shows the breaking and making of bonds when methane burns.

$$\text{Methane} + \text{Oxygen} \rightarrow \text{Carbon dioxide} + \text{Water}$$
$$CH_4(g) + 2O_2(g) \rightarrow CO_2(g) + 2H_2O(l)$$

The energy given out when the new bonds are made is greater than the energy taken in to break the old bonds. This reaction is therefore **exothermic**.

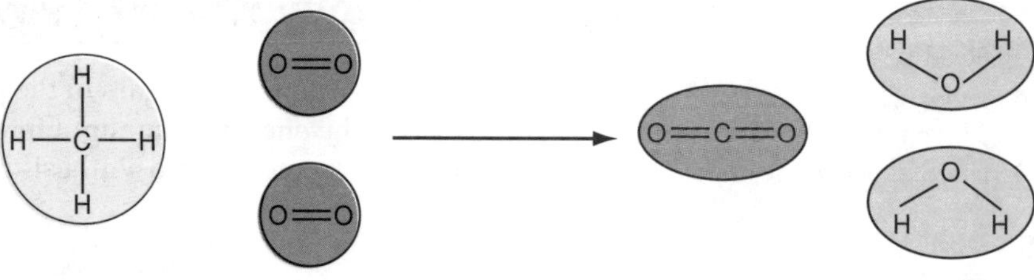
H
H—C—H
H
O═O
O═O
O═C═O
H O H
These bonds are broken.
Energy must be taken in
These new bonds are made.
Energy is given out
In this reaction, the energy given out is greater than the energy taken in –
this reaction is exothermic

*Figure 28.3A* Bonds broken and made when methane burns

To break chemical bonds in a compound, energy must be supplied. Heat or light or electricity may be supplied. The other side of the coin is that when bonds are made energy is given out.

The difference:
energy supplied – energy given out is the heat of reaction.

Figure 28.3B shows the breaking and making of bonds in the reaction

$$\text{Carbon (coke)} + \text{Water (steam)} \rightarrow \text{Carbon monoxide} + \text{Hydrogen}$$
$$C(s) + H_2O(g) \rightarrow CO(g) + H_2(g)$$

In this reaction, the energy taken in to break the old bonds is greater than the energy given out when the new bonds form. The reaction is therefore **endothermic**.

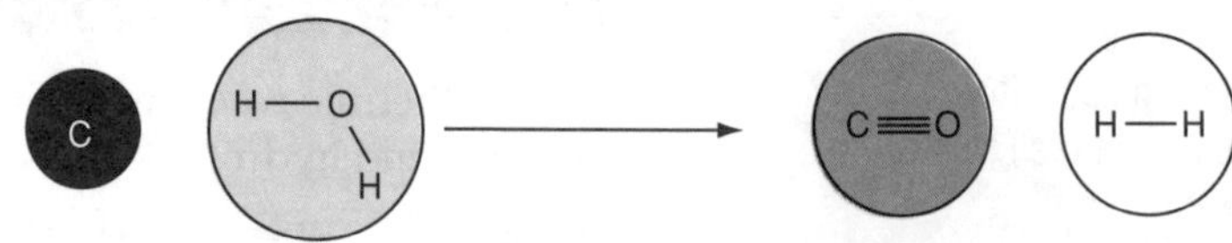
C
H—O
H
C≡O
H—H
The bonds in $H_2O$ must be broken.
Energy must be taken in
As the bonds in CO and $H_2$ are made,
energy is given out
In this reaction, the energy taken in is greater than the energy given out.
This reaction is endothermic

*Figure 28.3B* Bonds broken and made when steam reacts with hot coke

## Energy diagrams

In a chemical reaction, the reactants and the products possess different chemical bonds. They therefore possess different amounts of energy. A diagram which shows the energy content of the reactants and the products is called an **energy diagram**.

Figure 28.3C is an energy diagram for an exothermic reaction. It shows the energy content of the reactants and the products.

An energy diagram aids our understanding by showing what is happening in pictorial form.

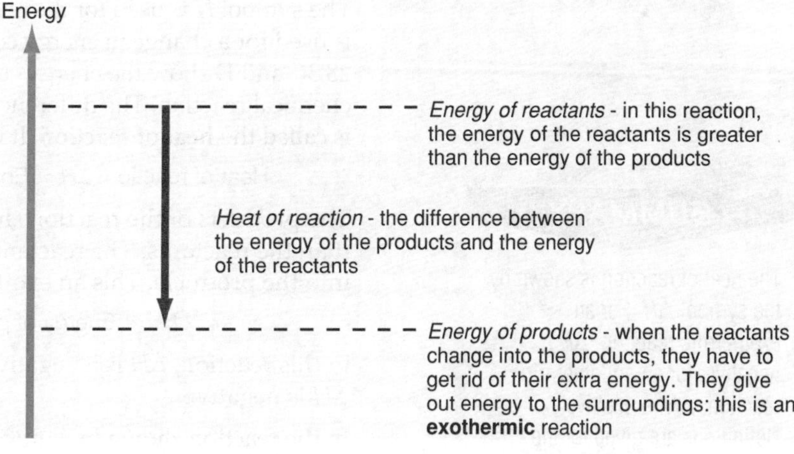


*Figure 28.3C* ▲ An energy diagram for an exothermic reaction (e.g. the combustion of methane)

Figure 28.3D shows an energy diagram for an endothermic reaction. Marked on both energy diagrams is the heat of reaction

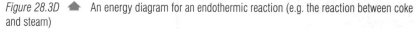

Heat of reaction = Energy of products − Energy of reactants

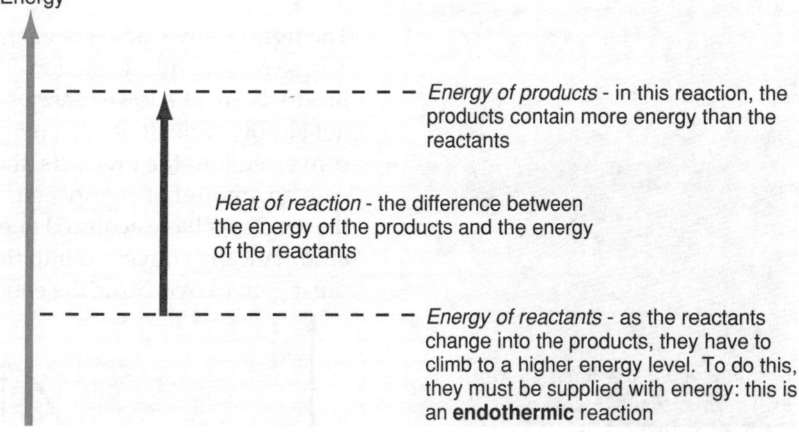


*Figure 28.3D* ▲ An energy diagram for an endothermic reaction (e.g. the reaction between coke and steam)

## SUMMARY

In a chemical reaction, energy must be supplied to break chemical bonds in the reactants. Energy is given out when new chemical bonds are made during the formation of the products. The difference between the energy content of the products and the energy content of the reactants is the heat of reaction.

## CHECKPOINT

1. Draw an energy diagram for the neutralisation of hydrochloric acid by sodium hydroxide solution. If you have forgotten whether it is exothermic or endothermic, see Topic 28.1. Mark the heat of reaction on your diagram.
2. Draw an energy diagram for the combustion of petrol to form carbon dioxide and water. Mark the heat of reaction on your diagram.
3. Is photosynthesis exothermic or endothermic? What are the reactants and what are the products? Illustrate your answer by an energy diagram.
4. Describe the reaction that takes place when magnesium ribbon is heated in air. What do you see that shows the reaction is exothermic? What forms of energy are released? Draw an energy diagram for the reaction.

## 28.4 The value of the heat of reaction

SUMMARY

The heat of reaction is shown by the symbol $\Delta H$. For an exothermic reaction, $\Delta H$ is negative; for an endothermic reaction, $\Delta H$ is positive. Definitions are given for the heats of neutralisation, combustion, formation and reaction.

The symbol $H$ is used for the energy content of a substance. The symbol $\Delta H$ is used for a change in energy content. The energy diagrams in Figures 28.3C and D show the energy content of the reactants and the products in a chemical reaction. The difference (energy of products – energy of reactants) is called the **heat of reaction**. It is given the symbol $\Delta H$. By definition,

$$\text{Heat of reaction, } \Delta H = \text{Energy of products} - \text{Energy of reactants}$$

The products of the reaction shown in Figure 26.3C contain less energy than the reactants. The reactants must give out energy as they change into the products. This an exothermic reaction. Since by definition,

$$\Delta H = \text{Energy of products} - \text{Energy of reactants}$$

in this reaction, $\Delta H$ is a negative quantity. For an exothermic reaction, $\Delta H$ is negative.

In the reaction shown in Figure 26.3D, the products contain more energy than the reactants. In order to change into the products, the reactants must take energy from the surroundings: they cool the surroundings. Since

$$\Delta H = \text{Energy of products} - \text{Energy of reactants}$$

in this reaction $\Delta H$ is positive. For an endothermic reaction, $\Delta H$ is positive.

## 28.5 Activation energy

Perhaps you are wondering why exothermic reactions do not happen spontaneously – without any help from heat or electricity.

The reason is that there is a barrier to be overcome: a barrier called the activation energy.

The figure shows how the energy of the reactants changes during the course of a chemical reaction. This particular reaction is exothermic; the products are at a lower energy level than the reactants. The reactants do not simply slide down an energy hill, however. Before the reactants can be converted into the products, they must overcome an energy barrier, and gain an amount of energy equal to that shown at P on the diagram. Once the reactants have acquired energy equal to that at the peak P they are automatically converted into the products. The energy which the reactants must gain to overcome the energy barrier is called the **activation energy**.

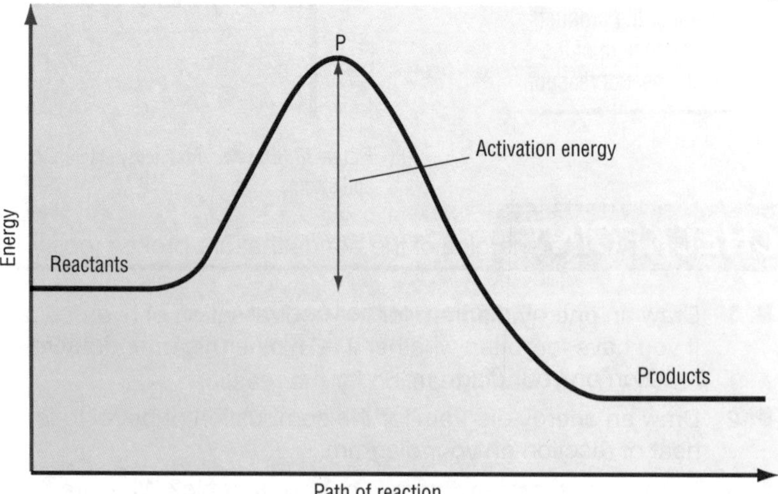


*Figure 28.5A* ▲ Activation energy

Molecules of the reactants must collide before they can react. If they are moving fast enough, part of their kinetic energy is converted into the activation energy.

The way in which a catalyst increases the speed of a reaction is by lowering the activation energy. Colliding molecules then have a lower energy barrier to climb, and more of the collisions between molecules result in reaction.

If there were no energy barrier, all exothermic reactions would happen spontaneously. A piece of iron would turn into iron oxide before your very eyes. A can of petrol would burst into flames as soon as you unscrewed the cap. The world would be a very unstable place.

## 28.6 Using bond energy values

*Table 28.1* Bond energy values

| Bond | Bond energy (kJ/mol) |
|---|---|
| H—H | 436 |
| O—O | 496 |
| C—C | 348 |
| C=C | 612 |
| C—H | 412 |
| C—O | 360 |
| C=O | 743 |
| H—O | 463 |
| Br—Br | 193 |
| C—Br | 276 |

Chemists have drawn up tables which tell exactly how much energy it takes to break different chemical bonds. To break 1 mole of C—C bonds requires 348 kJ. We can say that the bond energy of the C—C bond is 348 kJ/mol. Table 28.1 lists the bond energies of some bonds in kJ/mol. To break a chemical bond, energy must be supplied so all the bond energies are positive. There are no bonds that fly apart by magic and release energy.

We can use bond energy values to calculate a value for the heat of reaction.

**Worked example** Use bond energies to calculate $\Delta H$ for the reaction,

$$CH_4(g) + 2O_2(g) \rightarrow CO_2(g) + 2H_2O(l)$$

**Solution** First show the bonds that are broken and the bonds that are made.

```
     H
     |
H—C—H + 2O=O  →  O=C=O + 2H—O—H
     |
     H
```

Next list the bonds that are broken and the bonds that are made and their bond energies.

Bonds broken are 4(C—H) bonds; energy = 4 × 412 = 1648 kJ/mol
2(O=O) bonds; energy = 2 × 496 = 992 kJ/mol
Total energy required = +2640 kJ/mol

Bonds made are 2(C=O) bonds; energy = −2 × 743 = −1486 kJ/mol
2(H—O) bonds; energy = −4 × 463 = −1852 kJ/mol
Total energy given out = −3338 kJ/mol

Heat of reaction = Energy required to break old bonds + Energy given out when new bonds are made
= +2640 − 3338 = −698 kJ/mol

### CHECKPOINT

1 Find the heat of reaction for the reaction

```
  H  H              H  H
  |  |              |  |
H—C=C—H + H2  →  H—C—C—H
                    |  |
                    H  H
```

(Hint: add the energies of the bonds that are broken (positive) and the energies of the bonds created (negative).)

2 Find the heat of reaction for the reaction

$$2H_2(g) + O_2(g) \rightarrow 2H_2O(l)$$

3 Calculate the heat of reaction for the reaction

```
       O                        O—H
       ||                       |
CH3—C—CH3(g) + H2(g)  →  CH3—C—CH3(g)
                                |
                                H
```

4 Calculate the heat of reaction for the reaction

$$CH_2{=}CHCH_3(g) + Br_2(g) \rightarrow CH_2BrCHBrCH_3(g)$$

(Hint: draw structural formulas to show all the bonds.)

# Theme Questions

1 The energy sources used in the UK are as follows:
Coal 31%
Oil 30%
Natural gas 21%
Nuclear power 17%
Hydroelectric power 1%

(a) Show these figures in the form of a pie-chart or a bar graph.
(b) Say which of these fuels are fossil fuels.
(c) Explain the meaning of the term 'fossil fuel', and say why such fuels are non-renewable.
(d) Name a product which is formed when all fossil fuels burn.
(e) Why is the contribution of hydroelectric power small in the UK?

2 You are provided with a solution of dilute hydrochloric acid of concentration 0.100 mol/l and a dilute solution of sodium hydroxide of unknown concentration.

(a) Briefly describe how you would find out what volume of the acid solution is needed to neutralise 25.0 $cm^3$ of the sodium hydroxide solution.
(b) A student following your method finds that 35.0 $cm^3$ of 0.100 mol/l hydrochloric acid neutralise 25.0 $cm^3$ of sodium hydroxide solution. Calculate the concentration of the sodium hydroxide solution.

3 (i) What amount (in moles) of hydroxide ion is present in 25.0 $cm^3$ of 0.050 mol/l sodium hydroxide solution?
(ii) Which of the following will neutralise exactly 25.0 $cm^3$ of 0.050 mol/l sodium hydroxide solution? (More than one answer is correct.)
(a) 25.0 $cm^3$ of 0.050 mol/l hydrochloric acid
(b) 25.0 $cm^3$ of 0.050 mol/l nitric acid
(c) 25.0 $cm^3$ of 0.050 mol/l sulphuric acid
(d) 50.0 $cm^3$ of 0.025 mol/l hydrochloric acid
(e) 25.0 $cm^3$ of 0.025 mol/l sulphuric acid

4 Aluminium is obtained by the electrolysis of molten aluminium oxide.

(a) Write the equation for the electrode process which yields aluminium.
(b) A current of 20 000 A is passed through a cell containing molten aluminium oxide for 24 hours. What mass of aluminium is produced? (Take the Faraday constant as 96 000 C/mol and $A_r$ (Al) = 27)

5 You are given a sample of copper(II) oxide mixed with carbon. You are asked to find the percentage purity of the copper oxide. You decide to add an excess of dilute sulphuric acid to 2.00 g of the mixture to react with all the copper(II) oxide. You filter, wash and dry the carbon that remains. The mass is 0.090 g. What is the percentage purity of the sample of copper(II) oxide?

6 The manufacture of ammonia uses the reaction:

$$N_2(g) + 3H_2(g) \rightleftharpoons 2NH_3(g)$$

The reaction is reversible, and the forward reaction is exothermic.

(a) What can be done to increase the rate of the reaction?
(b) What steps can be taken to increase the yield of ammonia?
(c) Is there any conflict between the requirements of (a) and (b)?
(d) What conditions does the industrial plant employ?

7 Zinc reacts with sulphuric acid to give hydrogen. Mention three ways in which you could speed up the reaction.

8 In a set of experiments, zinc was allowed to react with sulphuric acid. Each time, 0.65 g of zinc was used. The volume of acid was different each time. The volume of hydrogen formed each time is shown in the table.

| *Volume of sulphuric acid* ($cm^3$) | *Volume of hydrogen* ($cm^3$) |
|---|---|
| 5 | 45 |
| 15 | 135 |
| 20 | 180 |
| 25 | 215 |
| 30 | 235 |
| 35 | 240 |
| 40 | 240 |

Plot, on graph paper, the volume of hydrogen produced against the volume of acid used. From the graph, find out:

(a) where the reaction is most rapid (and explain why),
(b) what volume of sulphuric acid will produce 100 $cm^3$ of gas,
(c) what volume of gas is produced if 10 $cm^3$ of sulphuric acid are used,
(d) what volume of sulphuric acid is just sufficient to react with 0.65 g of zinc.

9 A student collected samples of two minerals, **A** and **B**, whilst on a field course. Both minerals were colourless and crystalline, though **B** was obviously mixed with small amounts of a red-brown solid.

(a) Fragments of each sample were moistened with concentrated hydrochloric acid and held in a Bunsen flame. The results were as shown.

| *Sample* | *Flame test colour* | *Ion present* |
|---|---|---|
| **A** | Dull red | |
| **B** | Yellow | |

Copy and complete the table above.
(b) **A** was insoluble in water. It dissolved in dilute hydrochloric acid with the evolution of a gas which turned limewater cloudy (milky).
(i) What was the gas?
(ii) Write the name or formula of **A**.

(iii) Write a balanced equation for the action of acid on **A**.

(c) **B** was found to be almost completely soluble in water. When dilute nitric acid was added to the solution followed by aqueous silver nitrate a white precipitate was obtained.
(i) Write the name or formula for **B**.
(ii) Write a balanced equation to show what happened in this test.

(d) The only part of **B** which did not dissolve in water was the red-brown solid. This was filtered off and washed well. It dissolved in hot dilute hydrochloric acid but did not effervesce. The resulting pale yellow solution gave a brown gelatinous precipitate with aqueous sodium hydroxide.
(i) Give the formula of the **cation** in the red-brown solid and the pale yellow solution.
(ii) Suggest a possible identity for the red-brown solid impurity in **B**. (ULEAC)

**10** (a) The bonding in the molecules of nitrogen, hydrogen and ammonia is as follows.

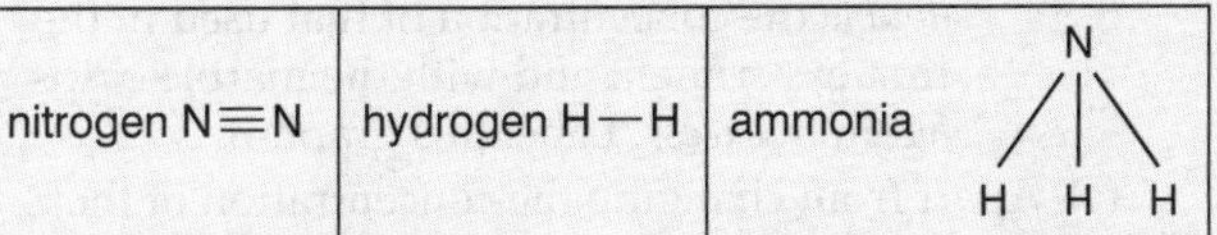


Bond energies in kJ/mol: N≡N 945, H—H 436, N—H 391.

The equation showing the bonds for the reversible reaction in which ammonia is made could be written as below.

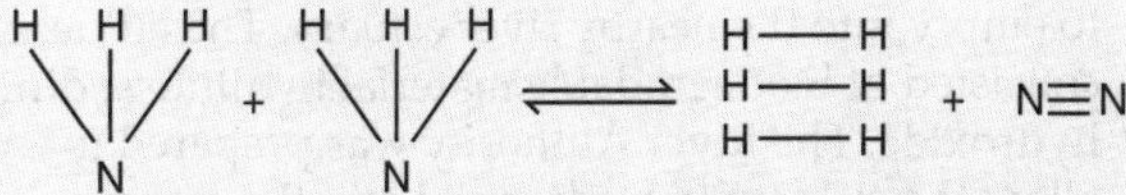

(i) Calculate the total bond energy needed to break the bonds in one mole of nitrogen and three moles of hydrogen.
(ii) Calculate the energy given out when the bonds in two moles of ammonia are formed.
(iii) Use the results from (i) and (ii) to explain why the reaction to produce ammonia is exothermic.

(b) The graphs show the percentage of ammonia present at equilibrium in the Haber process at particular temperatures and pressures. When deciding the conditions to be used, a manufacturer has to consider the costs of the process.

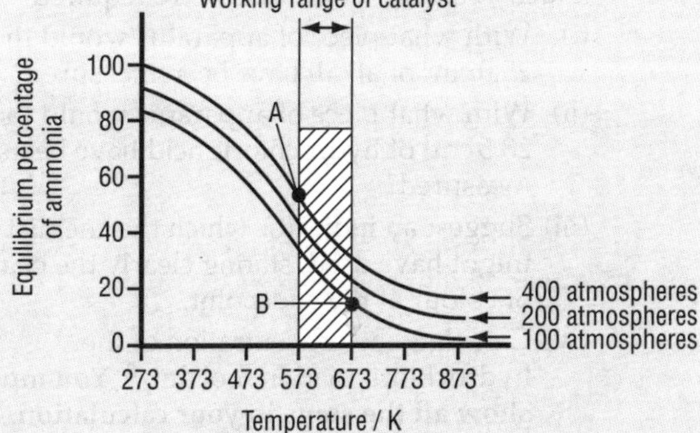


Assume that
1 the hourly energy cost of maintaining pressure is £1 per atmosphere,
2 the hourly energy cost of maintaining temperature is £5 per kelvin,
3 all other costs remain constant irrespective of the pressure and temperature chosen.

(i) Compare the energy costs of operating the process at the conditions represented by points **A** and **B** on the graph.
Suggest **two** other factors which the manufacturer has to consider in deciding which set of conditions to adopt.
(ii) A particular manufacturer chose 660 K and 200 atmospheres with an hourly energy cost of £3500 as the working conditions.
Assuming that all other costs remained constant, what can be deduced about the factors you have mentioned in **(b) (i)** which make these conditions economically viable?
(iii) The graphs show that a temperature of 373 K and a pressure of 100 atmospheres would give an equilibrium of 80% ammonia. Explain why it is unlikely that these conditions would be chosen.
(iv) Starting from a given mass of nitrogen and hydrogen at 660 K and 200 atmospheres, and without changing the catalytic conditions, what should a manufacturer do to obtain as complete a conversion into ammonia as is possible?

(c) The annual production of ammonia in the UK is about $2.5 \times 10^6$ tonnes. Explain why ammonia is in such high demand. (MEG)

**11** (a) Objects made from pure aluminium can be protected by anodising them. Describe how a piece of aluminium may be anodised and what happens during the process.

In the extraction of aluminium approximately 4 tonnes of bauxite give 2 tonnes of aluminium oxide and from this 1 tonne of aluminium is obtained. Recycling aluminium uses only 5% of the energy required to extract aluminium from bauxite.

(b) Part of the energy cost of recycling aluminium is the heat energy needed to melt the scrap materials such as cans. Give another major energy cost involved when aluminium cans are recycled.

(c) Describe **three** of the major energy costs involved when aluminium is produced from bauxite.

(d) In the extraction of aluminium, the electrolyte is aluminium oxide dissolved in molten cryolite. Cryolite is not decomposed during the electrolytic process.
(i) Write equations for the electrode reactions which occur in the cell at the anode and at the cathode.
(ii) How many moles of electrons are needed to produce 1 mol of aluminium?
(iii) How many moles of electrons are needed to produce 1 kg of aluminium? (Relative atomic mass: Al = 27.)

(iv) A current of 30 A is used for the electrolysis. How long does it take to produce 1 kg of aluminium? (Charge on 1 mol of electrons = 96 500 coulombs.)

(v) Calculate the maximum volume of oxygen, measured at room temperature and atmospheric pressure, that could be evolved when 1 kg of aluminium is formed. (1 mol of gas occupies 24 $dm^3$ at room temperature and atmospheric pressure.)

(vi) The oxygen evolved reacts with the carbon electrode to produce a mixture of carbon monoxide and carbon dioxide in the ratio 1 : 3. Write a balanced equation for this reaction, and calculate the **total** volume of gases produced when 1 kg of aluminium is produced. (ULEAC)

**12** (a) A student was given three reagents.

**dilute nitric acid**
**aqueous sodium hydroxide**
**aqueous silver nitrate**

Using only these three reagents, the student was able to show that a solution **X** was iron(III) chloride. Describe the tests and observations the student did to prove the presence of

(i) iron(III) ions,

(ii) chloride ions.

(b) Tests on another substance **Q** produced the following table of results.

| *test* | *observation* |
|---|---|
| 1 To solid **Q** add dilute hydrochloric acid. | greeny-blue solution, and a gas **R** which turns lime-water cloudy |
| 2 To a portion of the solution from test 1 add an excess of aqueous sodium hydroxide. | permanent blue precipitate **W** |

(i) What is the name of gas **R** in test 1?

(ii) Suggest the name of the metal ion present in precipitate **W** in test 2.

(iii) Suggest the name of substance **Q**. (MEG)

**13** Marble chips (calcium carbonate) react with dilute hydrochloric acid to produce carbon dioxide. The volume of carbon dioxide produced is measured using a gas syringe.

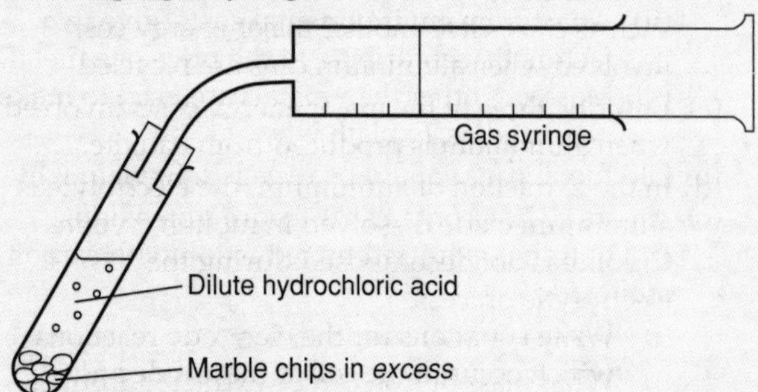


(a) On a copy of the grid below plot a graph of the following results:

| *Time, in seconds* | 0 | 10 | 20 | 30 | 40 | 50 | 60 | 70 | 80 |
|---|---|---|---|---|---|---|---|---|---|
| *Volume of gas, in* $cm^3$ | 0 | 19 | 32 | 43 | 50 | 56 | 59 | 60 | 60 |

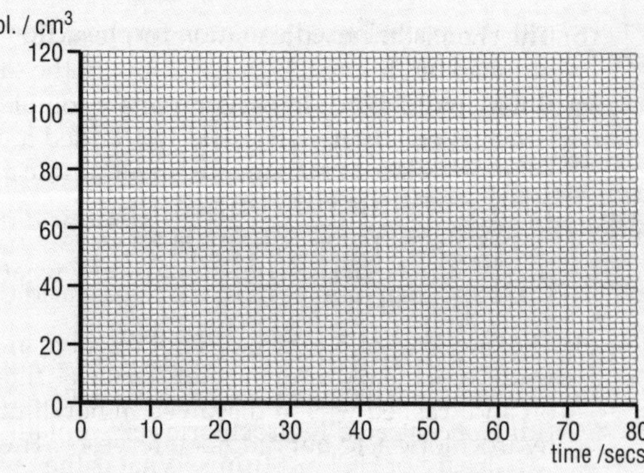


(b) On the same grid sketch carefully the graph that would be obtained if the acid had been replaced by:

(i) An equal volume of hydrochloric acid of *half* the concentration of that used in the first experiment and with the marble chips again in *excess*. Label this graph **A**.

(ii) An equal volume of hydrochloric acid of *double* the concentration of that used in the first experiment and with the marble chips again in *excess*. Label this graph **B**.

(c) Apart from changing the concentration of the acid, give **two** ways in which the first experiment could have been made to take place faster. (WJEC)

**14** A manufacturing company had 25 000 $dm^3$ of industrial effluent in a storage tank which it wished to empty into the nearby river estuary. The effluent consisted of biodegradable material in dilute sodium hydroxide. The River Authority was prepared to allow the effluent to be dumped into the estuary provided that it was neutralised with hydrochloric acid first. The works chemist was asked to determine the amount of concentrated hydrochloric acid which must be added to the effluent to neutralise it. In the laboratory were some potassium hydroxide solution of concentration 0.100 mol $dm^{-3}$ and some very dilute hydrochloric acid.

(a) The chemist first determined the exact concentration of the dilute hydrochloric acid by measuring out 25.0 $cm^3$ of the potassium hydroxide solution and titrating it with the hydrochloric acid.

$$KOH + HCl \rightarrow KCl + H_2O$$

Exactly 28.5 $cm^3$ of the acid were required.

(i) With what piece of apparatus would the 25.0 $cm^3$ of alkali have been measured?

(ii) With what piece of apparatus would the 28.5 $cm^3$ of hydrochloric acid have been measured?

(iii) Suggest an indicator which the chemist might have used, stating clearly the change of colour at the end-point.

(iv) Calculate the concentration of the hydrochloric acid in mol $dm^{-3}$. **You must show all the steps in your calculation.**

(b) The chemist then withdrew a sample of the effluent and titrated 25.0 $cm^3$ of it with the same dilute hydrochloric acid.

$$NaOH + HCl \rightarrow NaCl + H_2O$$

12.8 $cm^3$ of the acid were required.

(i) How many moles of hydrochloric acid (HCl) were used in this titration?

(ii) How many moles of hydrochloric acid (HCl) would be required to neutralise all 25 000 **$dm^3$** in the tank? (Ignore the small sample which had been withdrawn).

(iii) Concentrated hydrochloric acid contains 365.0 g of HCl in 1.0 $dm^3$. What is the concentration of concentrated hydrochloric acid in mol $dm^{-3}$? (Relative atomic masses: H = 1.0, Cl = 35.5)

(iv) What volume of concentrated hydrochloric acid should be added to the effluent to neutralise it?

(v) What should be done to the effluent during the addition of this acid? (ULEAC)

**15** Limestone is impure calcium carbonate. The purity can be found by adding a measured amount of limestone to a measured excess of dilute hydrochloric acid. The following reaction takes place.

$$CaCO_3(s) + 2HCl(aq) \rightarrow CaCl_2(aq) + H_2O(l) + CO_2(g)$$

(a) (i) State **two** methods by which you could tell that pure calcium carbonate had completely reacted with dilute hydrochloric acid.

(ii) One of these methods is not reliable with impure samples of calcium carbonate such as limestone. State which one, and give a reason for its unreliability.

(b) After the limestone has reacted, sodium hydroxide can then be added to neutralise the remaining excess acid. The following reaction takes place.

$$NaOH(aq) + HCl(aq) \rightarrow NaCl(aq) + H_2O(l)$$

Write the ionic equation for this reaction.

(c) In a determination of the percentage purity of limestone, the following results were obtained.

Mass of hydrochloric acid in solution = 20.00 g
Mass of limestone = 25.00 g
Mass of hydrochloric acid left unreacted = 2.48 g

(i) What mass of hydrochloric acid reacted with the limestone?

(ii) Using the equation

$$CaCO_3(s) + 2HCl(aq) \rightarrow CaCl_2(aq) + H_2O(l) + CO_2(g),$$

calculate the mass of calcium carbonate with which this mass of hydrochloric acid would react.

(iii) Calculate the percentage purity of the limestone (the percentage of pure calcium carbonate in the limestone). (ULEAC)

**16** (a) In terms of *ions*, what do you understand by the expression *"an acid"*?

(b) In terms of *ions*, what do acids react with when neutralisation occurs?

(c) Ethanoic acid is a **weak** acid and is present in vinegar.

(i) What is meant by a weak acid?

(ii) Give a **named** example of another weak acid.

(d) Describe how you could distinguish between a weak acid and a strong acid using magnesium ribbon.

(e) Vinegar contains ethanoic acid. Different concentrations of ethanoic acid are found in different brands of vinegar. A 10 $cm^3$ sample of vinegar was placed in a flask and titrated with sodium hydroxide solution containing 20 g per litre ($dm^3$) of sodium hydroxide.

(i) Describe how you would carry out a titration to find accurately the concentration of ethanoic acid in the vinegar.

(ii) Sodium hydroxide, NaOH, has a relative formula mass of 40. Calculate the concentration of the sodium hydroxide in moles per litre ($dm^3$).

(iii) 20 $cm^3$ of the sodium hydroxide solution is required to neutralise the 10 $cm^3$ sample of vinegar. Calculate the number of moles of sodium hydroxide reacting.

(iv) How many moles of ethanoic acid are present in 10 $cm^3$ of vinegar if the acid and alkali react in a 1 : 1 ratio?

(v) Hence calculate the moles per litre ($dm^3$) of ethanoic acid in the vinegar. (WJEC)

**17** The extraction and use of a metal, such as iron, involves these stages.

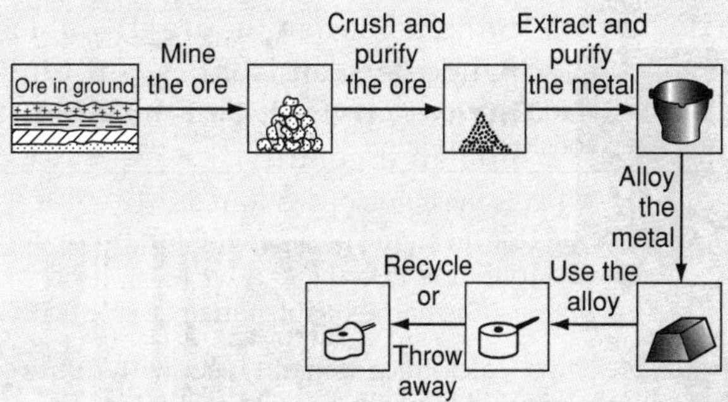


(a) (i) Give **two** ways in which the extraction of metals can affect the environment.

(ii) Apart from environmental costs, give **two** factors that contribute to the price of a metal.

(iii) Materials other than metals are used to make pan handles. Explain why.

(b) Old metal pans should be recycled. Explain why.

(c) An ore of iron contains 80% pure iron oxide, $Fe_2O_3$. Iron is extracted by reduction with carbon monoxide.

$$Fe_2O_3 + 3CO \rightarrow 2Fe + 3CO_2$$

(i) What mass of iron can be extracted from 1000 kilograms (kg) of this iron ore? The relative atomic masses are C 12; Fe 56; O 16.

(ii) What mass of carbon dioxide will be released into the atmosphere by the extraction of iron from 1000 kg of this ore? (SEG)

Theme G

# Organic Chemistry

The chemistry of carbon compounds is known as organic chemistry. Many of the substances which it includes are obtained from living organisms. This explains how this branch of the subject got its name.

Organic chemistry includes plastics and fibres. Many of our leisure activities need equipment such as footballs, tennis racquets, golf balls, climbing ropes, canoes and sailing dinghies, all made from synthetic organic chemicals. Much of the clothing we wear is composed of synthetic fibres such as nylon, polyester and acrylics which are made by the chemical industry.

Organic chemistry includes all the foodstuffs: carbohydrates, fats and proteins. When we are ill we need the help of organic chemicals such as painkillers, antiseptics, antibiotics and anaesthetics. The chemical industry provides all these substances by using the Earth's resources and human ingenuity.

# Topic 29 Alkenes and plastics

**FIRST THOUGHTS**

The toys you see in Figure 29A are made of plastics. The clothes you see in Figure 29B are made of synthetic fibres. Both types of material are polymers. In this topic, you will find out what a polymer is.

*Figure 29A* ▲ Plastics

*Figure 29B* ▲ Fibres

## 29.1 ▶ Alkenes

You studied alkanes in §24.2. They are saturated hydrocarbons with single carbon-carbon bonds in the molecules. They take part in substitution reactions. Some hydrocarbons have carbon–carbon double bonds in the molecule. They are unsaturated hydrocarbons. They take part in addition reactions.

Ethene is a hydrocarbon of formula $C_2H_4$ (see Figure 29.1A). There is a double bond between the carbon atoms.

```
H       H
 \     /
  C=C
 /     \
H       H
```

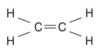

*Figure 29.1A* ▲ (a) A model of ethene (b) The formula of ethene

Ethene and other hydrocarbons which contain double bonds between carbon atoms are described as **unsaturated** hydrocarbons. The double bond will open to allow another molecule to add on. Unsaturated hydrocarbons will add hydrogen to form **saturated** hydrocarbons. For example, ethene adds hydrogen to form ethane, which is an **alkane**. Reactions of this kind are called **addition reactions**; see Figure 29.1B Saturated hydrocarbons, such as alkanes, contain only single bonds between carbon atoms.

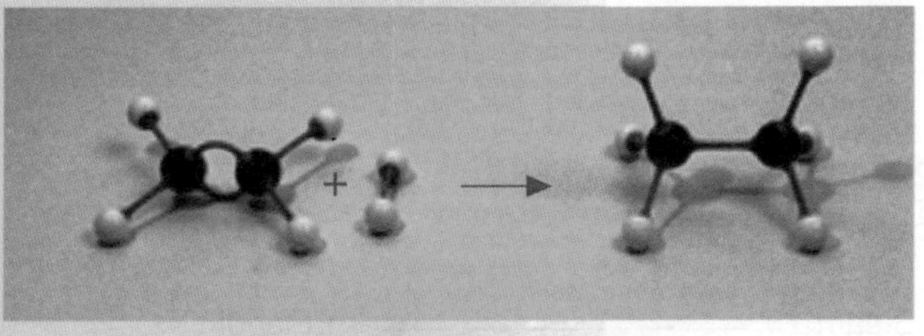

*Figure 29.1B* ▲ The addition reaction between ethene and hydrogen

Ethene + Hydrogen → Ethane

```
    H                     H
    |                     |
 H—C      H           H—C—H
   ||  +  |    ——▶        |
 H—C      H           H—C—H
    |                     |
    H                     H
```

Propene is an unsaturated hydrocarbon which resembles ethene (see Figure 29.1C). Ethene and propene are members of a **homologous series**, that is, a set of similar compounds whose formulas differ by $CH_2$. They are called **alkenes** (see Table 29.1). The general formula is $C_nH_{2n}$.

SUMMARY

Alkenes are unsaturated hydrocarbons. They possess a double bond between carbon atoms. They are a homologous series of general formula $C_nH_{2n}$.

```
H   H   H
|   |   |
C = C - C - H
|       |
H       H
```

*Figure 29.1C* ▲ The formula of propene

*Table 29.1* ▼ The first members of the alkene series

| Alkene | Formula, $C_nH_{2n}$ |
|---|---|
| Ethene | $C_2H_4$ |
| Propene | $C_3H_6$ |
| Butene | $C_4H_8$ |
| Pentene | $C_5H_{10}$ |
| Hexene | $C_6H_{12}$ |
| Heptene | $C_7H_{14}$ |
| Octene | $C_8H_{16}$ |

## Reactions of alkenes

### Hydrogenation

Animal fats, such as butter, are solid. Vegetable oils, such as sunflower seed oil, are liquid. More vegetable oil is produced than we need for cooking, and insufficient butter is produced to satisfy the demand for solid fat. It is therefore profitable to convert vegetable oils into solid fats. Manufacturers make use of the fact that fats are saturated, while oils are unsaturated. Hydrogenation (the addition of hydrogen) is used to convert an unsaturated oil into a saturated fat. The vapour of an oil is passed with hydrogen over a nickel catalyst.

$$\text{Vegetable oil (unsaturated)} + \text{Hydrogen} \xrightarrow[\text{nickel catalyst}]{\text{pass over heated}} \text{Solid fat (saturated)}$$

The product is margarine. The process can be modified to leave some of the double bonds intact and yield soft margarine.

### Hydration

Water will add to alkenes. A molecule of water can add across the double bond. Combination with water is called **hydration**. The product formed by the hydration of ethene is ethanol, $C_2H_5OH$.

$$\text{Ethene} + \text{Water} \rightarrow \text{Ethanol}$$

```
    H                     H
    |                     |
H - C    H            H - C - H
    ||  + |    ---->       |
H - C    O - H        H - C - O - H
    |                     |
    H                     H
```

Ethanol is the compound we commonly call **alcohol**. It is an important industrial solvent. The industrial manufacture is

$$\text{Ethene} + \text{Steam} \xrightarrow[\text{(phosphoric acid), under pressure}]{\text{pass over a heated catalyst}} \text{Ethanol}$$

These unsaturated hydrocarbons are called alkenes. Addition reactions include

- hydrogenation of vegetable oils to form solid fats, e.g. margarine
- hydration to form alcohols, e.g. ethene → ethanol

and …

Only about 10% of ethene is converted to ethanol. The unreacted gases are recycled over the catalyst.

...
- addition polymerisation in which molecules of monomer, e.g. ethene, join to form a molecule of polymer, a poly(alkene), e.g. poly(ethene) and ...

## Addition polymerisation

The double bond in ethene enables many molecules of ethene to join together to form a large molecule.

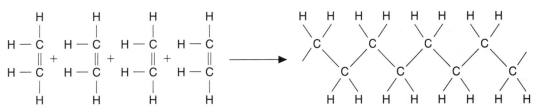

$$4\ CH_2{=}CH_2 \longrightarrow -CH_2-CH_2-CH_2-CH_2-CH_2-CH_2-CH_2-CH_2-$$

This reaction is called **addition polymerisation**. Many molecules (30 000–40 000) of the **monomer**, ethene, **polymerise** (i.e. join together) to form one molecule of the **polymer**, poly(ethene). The conditions needed for polymerisation are high pressure, a temperature of room temperature or higher and a catalyst.

## SUMMARY

The double bonds makes alkenes reactive. They take part in addition reactions, e.g. with hydrogen to form alkanes and with water to form alcohols. Hydrogenation is used to turn vegetable oils into saturated fats.

$$\text{Ethene} \xrightarrow[\text{over a heated catalyst}]{\text{pass at high pressure}} \text{Poly(ethene)}$$

$$nCH_2{=}CH_2 \xrightarrow[\text{catalyst}]{\text{heat, pressure,}} -\!(CH_2-CH_2)\!-_n$$

Poly(ethene) is better known by its trade name of **polythene.** It is used for making plastic bags, kitchenware (buckets, bowls, etc.), laboratory tubing and toys. It is flexible and difficult to break.

...
- addition of bromine. The decolourisation of a bromine solution is used as a test for a carbon–carbon double bond.
- addition of chlorine

## Addition of bromine

A solution of bromine in an organic solvent or in water is brown. If an alkene is bubbled through a solution of bromine, the solution loses its colour. Bromine has added to the alkene to form a colourless compound. The reaction can be shown as

Ethene + Bromine → 1,2-Dibromoethane

$$CH_2{=}CH_2 + Br-Br \longrightarrow CH_2Br-CH_2Br$$

The product has single bonds: it is a saturated compound. With two carbon atoms in the molecule, it is named after ethane. With two bromine atoms in the molecule, it is a dibromo-compound, 1,2-dibromoethane. The numbers 1,2- tell you that one bromine atom is bonded to one carbon atom and the second bromine atom is bonded to the second carbon atom. The decolourisation of a bromine solution is used to distinguish between an alkene and an alkane. Chlorine adds to alkenes in a similar way.

# 29.2 Plastics

### Science at work

When someone breaks a bone, the broken ends must be held in place by a plaster cast until new bone grows. Sometimes, metal pins are used to join the broken ends together. Now there is an easier way. Poly(methylmethacrylate) can be used to 'glue' the ends of a broken bone together without the need for pins or plaster.

*Thermosoftening plastics and thermosetting plastics. Again the uses to which they are put depend on their properties. And the properties depend on the structures.*

Polymers such as poly(ethene) and other poly(alkenes) are **plastics**. Plastics are materials which soften on heating and harden on cooling. They are therefore useful materials from which to mould objects. There are two kinds of plastics: **thermosoftening plastics** and **thermosetting plastics**.

Thermosoftening plastics can be softened by heating, cooled and resoftened many times. Thermosetting plastics are plastic during manufacture, but once moulded they set and cannot be resoftened.

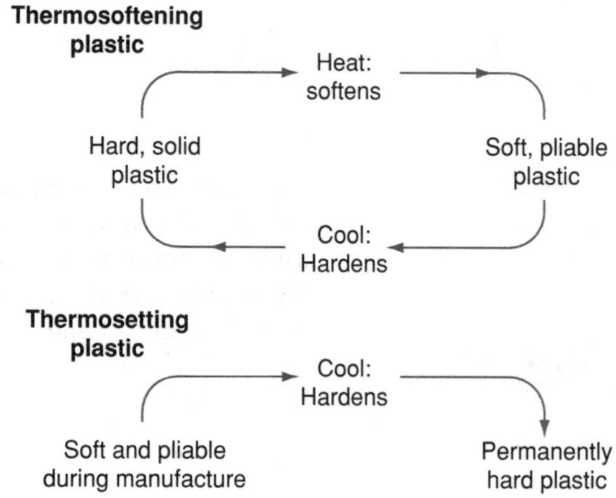


The reason for the difference in behaviour is a difference in structure (see Figure 29.2A).

Thermosoftening plastics consist of long polymer chains. The forces of attraction between chains are weak

(a) Part of the structure of a thermosoftening plastic

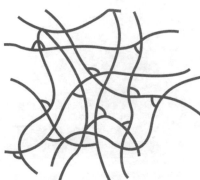

When a thermosetting plastic is softened and moulded, the chains react with one another. Cross-links are formed and a huge three-dimensional structure is built up. This is why thermosetting plastics can be formed only once

(b) Part of the structure of a thermosetting plastic

*Figure 29.2A*

### SUMMARY

Addition polymerisation is the addition of many molecules of monomer to form one molecule of polymer. Thermosoftening and thermosetting plastics have advantages for different uses.

## Condensation polymerisation

The chief thermosetting polymers are not poly(alkenes). Many of them are made by another type of polymerisation called **condensation polymerisation**. For this to occur, the monomer must possess two groups of atoms which can take part in chemical reactions. When the reactive groups in one molecule of monomer react with the groups in other molecules of monomer, large polymer molecules are formed (see Figure 29.2B). In the reaction, small molecules are eliminated, e.g. $H_2O$, HCl, $NH_3$. The elimination of water gave the name **condensation** to this type of polymerisation.

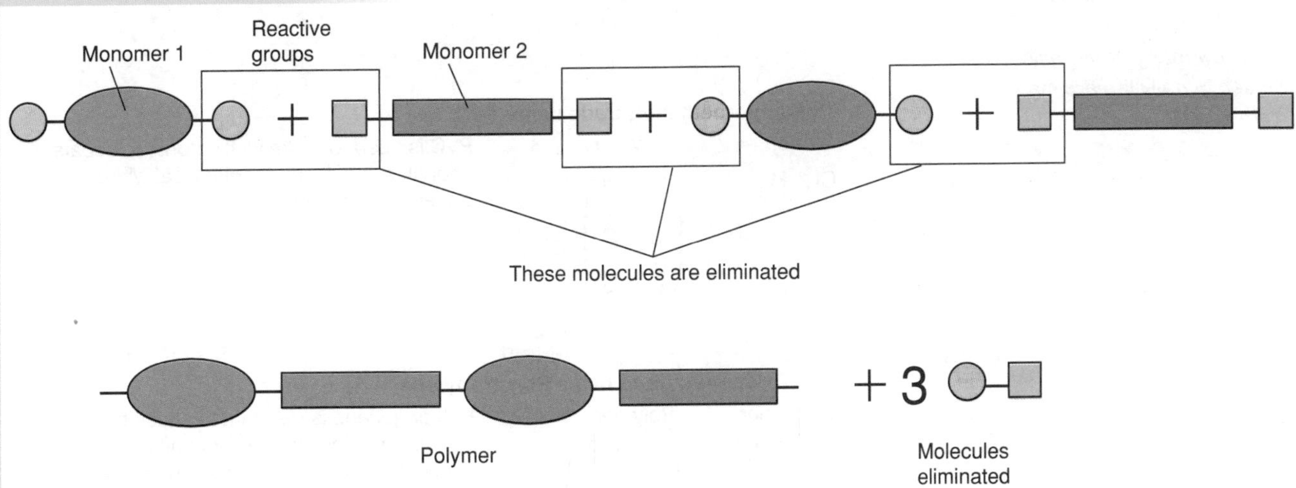


*Figure 29.2B* ▲ Condensation polymerisation

> **SUMMARY**
>
> Condensation polymerisation is the reaction between many molecules of monomer to form one molecule of polymer with the elimination of a small molecule such as a molecule of water.

Examples of condensation polymers are:

- epoxy resins, which are used in glues,
- polyester resins, which are used in glass-reinforced plastics,
- polyurethanes, which are used in varnishes.
- See Table 29.3 for nylon, terylene and rayon.

## Methods of moulding plastics

Different methods are used for moulding thermosoftening and thermosetting plastics.

Manufacturers find thermosoftening plastics very convenient to use. They can buy thermosoftening plastic in bulk in the form of granules, then melt the material and press it into the shape of the object they want to make. Thermosoftening plastics are easy to colour. When a pigment is added to the molten plastic and thoroughly mixed, the moulded objects are coloured all through. This is much better than a coat of paint which can become chipped. Thermosetting plastics have important uses too. Materials used for bench tops must be able to withstand high temperatures without softening. 'Thermosets' are ideal for this use.

Some manufacturers mix gases with plastics to make low density plastic foams. These foams are used in packaging, for thermal insulation of buildings, for insulation against sound and in the interior of car seats. Sometimes it is necessary to strengthen plastics by the addition of other materials. Boat hulls, plastic panels in cars, instrument panels and wall-mounted hand-driers are just a few of the articles which are made of GRP (glass-fibre-reinforced plastic).

> **SUMMARY**
>
> Different methods are used for moulding thermosoftening plastics (which can be softened by heat many times) and thermosetting plastics (which are softened by heat during manufacture and then set permanently).

## Polyalkenes

You have met all these plastics many times. You will see even from the small selection listed here how many of the objects we use in daily life are made from plastics.

*Table 29.2* ▼ Some poly(alkenes) and their uses

**Poly(ethene); trade name Polythene**

```
Monomer        Polymer
H   H          / H   H \
|   |          | |   | |
C=C           -+-C--C-+-
|   |          | |   | |
H   H          \ H   H / n
```

Polythene is used to make plastic bags. High density polythene is used to make kitchenware, laboratory tubing and toys.

**Poly(chloroethene); trade name PVC**

```
Monomer        Polymer
Cl  H          / Cl  H \
|   |          | |   | |
C=C           -+-C--C-+-
|   |          | |   | |
H   H          \ H   H / n
```

PVC is used to make wellingtons, raincoats, floor tiles, insulation for electrical wiring, gutters and drainpipes.

**Poly(propene); trade name Polypropylene**

```
Monomer        Polymer
H   CH3        / H   CH3 \
|   |          | |   |   |
C=C           -+-C--C--  |
|   |          | |   |   |
H   H          \ H   H   / n
```

Polypropylene is resistant to attack by chemicals. Since it does not soften in boiling water, it can be used to make hospital equipment which must be sterilised. Polypropylene is drawn into fibres which are used to make ropes and fishing nets.

**Poly(tetrafluoroethene); trade names PTFE and Teflon**

```
Monomer        Polymer
F   F          / F   F \
|   |          | |   | |
C=C           -+-C--C-+-
|   |          | |   | |
F   F          \ F   F / n
```

PTFE is a hard plastic which is not attacked by most chemicals. Few substances can stick to its surface. It is used to coat non-stick pans and skis.

**Perspex**

```
Monomer
H   CO2CH3
|   |
C=C
|   |
H   CH3
```

Perspex finds many applications because it is transparent and can be used instead of glass. It is more easily moulded than glass and less easily shattered.

**Polystyrene**

```
Monomer
H   H
|   |
C=C
|   |
H   C6H5
```

Polystyrene is a hard, brittle plastic used for making construction kits. Polystyrene foam is made by blowing air into the softened plastic. It is used for making ceiling tiles, insulating containers and packaging materials.

**EXTENSION FILE ACTIVITY**

### SUMMARY

Some important poly(alkenes) are:

- poly(ethene),
- poly(chloroethene),
- poly(propene),
- poly(tetrafluoroethene),
- perspex,
- polystyrene.

*Table 29.3* ▼ Some condensation polymers and their uses

| Polymer | Properties | Uses |
|---|---|---|
| Nylon | A polyamide<br>Thermosoftening<br>m.p. 200 °C<br>High tensile strength | Textile fibre used in clothes; also in ropes, fishing lines and nets |
| Terylene | A polyester<br>Thermosoftening | Important in clothing manufacture |
| Bakelite | Thermosetting<br>An electrical insulator<br>Insoluble in water and organic solvents<br>Not attacked by chemicals | Electrical appliances, e.g. switches, sockets, plugs. Casings for radios and telephones |
| Melamine | Similar to bakelite | Kitchen surfaces, bench tops |
| Rayon | Made from cellulose by reshaping, that is breaking the long chains of sugar molecules in cellulose into shorter chains and then rejoining them | Important in clothing manufacture |

SUMMARY

Some important condensation polymers are nylon, terylene, bakelite, melamine and rayon.

## Some drawbacks

Nylon and terylene do not absorb water well. Clothes made of these fibres do not absorb perspiration and allow it to evaporate. Nylon, terylene and other polyesters are usually mixed with natural fibres such as wool and cotton. The mixtures absorb water and are therefore more comfortable to wear.

Many refreshment stands serve coffee and soft drinks in disposable cups. These are often made of polystyrene. All the plastic cups, plates and food containers which people use and throw away have to be disposed of. This is difficult to do because plastics are **non-biodegradable**. They are not decomposed by natural biological processes. Some plastic waste is burned in incinerators and the heat generated is used. Other plastics cannot be disposed of in this way because they burn to form poisonous gases, for example, hydrogen chloride, carbon monoxide and hydrogen cyanide. Much plastic waste is buried in landfill sites. One third of the plastics manufactured is used for packaging: it is made to be used once and thrown away. As the mass of non-biodegradable plastic waste increases, more land is used up to bury the waste.

Chemists are working on the problem. They have invented some **biodegradable** plastics, that is plastics which can be broken down by micro-organisms. One type has starch incorporated into the plastic. When bacteria feed on the starch, the plastic partially breaks down. Other new types of plastic are completely biodegradable.

There are some dangers in the use of plastics. Plastic foams are used as insulation in many buildings. Furniture is often stuffed with plastic foam. If there is a fire, burning plastics spread the fire rapidly. This is because plastics have lower ignition temperatures than materials like wood, metal, brick and glass. There is another danger too: some burning plastics give off poisonous gases.

SUMMARY

Some disadvantages in the use of plastics are:

- Clothes made of synthetic fibres do not absorb perspiration.
- Most plastics are non-biodegradable.
- They ignite easily, and some burn to form toxic products.

SUMMARY

Plastics are petrochemicals obtained from oil. Earth's resources of oil are limited. Should we use oil as fuel or save it for the petrochemicals industry?

## Oil: a fuel and a source of petrochemicals

Industry is constantly finding new uses for plastics. The raw materials used in their manufacture come from oil. At first the price of oil was low and plastics were cheap materials. As the price of oil has risen, this is no longer true. The Earth's resources of oil will not last for ever. We should be thinking about whether we ought to be burning oil as fuel when we need it to make plastics and other petrochemicals.

## CHECKPOINT

**1** Say what materials the following articles were made out of before plastics came into use. Say what advantage plastic has over the previous material, and state any disadvantage.

(a) gutters and drainpipes
(b) toy soldiers
(c) dolls
(d) motorbike windscreens
(e) lemonade bottles
(f) buckets
(g) electrical plugs and sockets
(h) wellingtons
(i) furniture stuffing
(j) electrical cable insulation
(k) dustbins

**2** (a) What does the word 'plastic' mean?
(b) Plastics can be divided into two types, which behave differently when heated. Name the two types. Describe how each behaves when heated. Say how the difference in behaviour is related to (i) the use made of the plastics and (ii) the molecular nature of the plastics.

**3** Study the following list of substances.
nylon sucrose ethane styrene olive oil starch
margarine glass silk rubber melamine
(a) List the naturally occurring polymers.
(b) List the synthetic polymers.
(c) Name the substance which is not a polymer but can easily be converted into one.
(d) List the substances which are not polymers and which cannot easily polymerise.

**4** PVC is used in the manufacture of drainpipes and plastic bags.
(a) Calculate the relative molecular mass of the monomer, $CH_2{=}CHCl$.
(b) The $M_r$ of the polymer is 40 000. Calculate the number of molecules of monomer which combine to form one molecule of polymer.
(c) What method could be used to mould PVC pipes?
(d) If you needed to mould a straight length of PVC pipe to make it fit round a curve in a drainage system, how would you do this?
(e) Used PVC bags have to be disposed of. Burning PVC bags in an incinerator would cause pollution. Name two pollutants that would be released into the atmosphere.

# 29.3 Review of materials

**FIRST THOUGHTS**

You have now studied materials of many types. In this section, we compare the properties of different solid materials. We look at how these properties make different materials useful for different purposes. We look at the types of structure which give rise to the different properties.

## Metals

Metals are crystalline materials. Metals:

- are generally hard, tough and shiny
- change shape without breaking (i.e. are ductile and malleable)
- are generally strong in tension and compression
- are good thermal and electrical conductors
- are in many cases corroded by water and acids.

These properties result from the nature of the metallic bond (see Topic 20.2). Metals are giant structures in which some electrons are free to move. These electrons hold the atoms together and also allow atoms to slide over one another when the metal is under stress and to conduct heat and electricity. The bonds are non-directed in space. Dislocations (see Topic 20.2) allow atoms in one plane of a crystal to move relative to an adjacent plane. This is why metals can be bent so easily.

## Ceramics

Ceramics are crystalline compounds of metallic and non-metallic elements. Ceramics:

- are very hard and brittle
- are strong in compression but weak in tension
- are electrical insulators
- have very high melting points
- are chemically unreactive.

The bonds are mainly covalent, but some are ionic. The bonds are directed in space, and the structure of a ceramic is therefore more rigid and less flexible than a metallic structure. As a result of this structure, ceramics are both harder than metals and also more brittle. Ceramics have lower densities and higher melting points than metals.

**It's a fact!**

The ceramic of formula $YBa_2Cu_3O_7$ is a superconductor. (A superconductor has practically zero resistance to the flow of an electric current.) The ceramic is metallic in appearance and becomes a superconductor at 90 K. It has a potential for use in zero-loss power transmission lines. Schemes for cooling underground superconducting cables in liquid air have been put forward.

## Glasses

Glasses are similar to ceramics. Glasses:

- are generally transparent
- have lower melting points than ceramics.

Plate glass is transparent because it is non-crystalline so there are no reflecting surfaces, such as grain boundaries, to make the material opaque. It is brittle because of the rigid bonds.

You will remember that what makes metals strong and ductile is the metallic bond; Topic 20.2.

Ceramics find important uses because of their very high melting points and very low chemical reactivity.

Glasses are used when a transparent material is needed.

Plastics are used when the demand is for a material that can be easily moulded. There is a variety of plastics with a range of properties.

## Plastics

Plastics consist of a tangled mass of a very long polymer molecules. Plastics:

- are usually strong in relation to their mass
- are usually soft, flexible and not very elastic
- soften easily when heated and melt or burn
- are thermal and electrical insulators.

The structure is described as *amorphous* (shapeless) because the polymer chains take up a random arrangement. The bonds between chains are

## It's a fact!

Polymers can be made to behave like metals. Polymers which conduct electricity are made by including salts in certain polymers. They have up to $10^{11}$ times the conductivity of typical polymers.

weak in thermosoftening plastics and strong in thermosetting plastics. Most polymers are poor conductors due to the lack of free electrons.

In some polymers there are regions where the chains are packed together in a regular way (see Figure 29.3A). Many polymers are a mixture of ordered regions and amorphous regions (see Figure 29.3B). In high-density poly(ethene), the chains pack closely together to give a material which is stronger than low-density poly(ethene) and is not as easily deformed by heat. High-density poly(ethene) is used to make water tanks and kitchen equipment, e.g. buckets and food containers. Articles made from high-density poly(ethene) can be heated to sterilise them so they are used for hospital equipment.

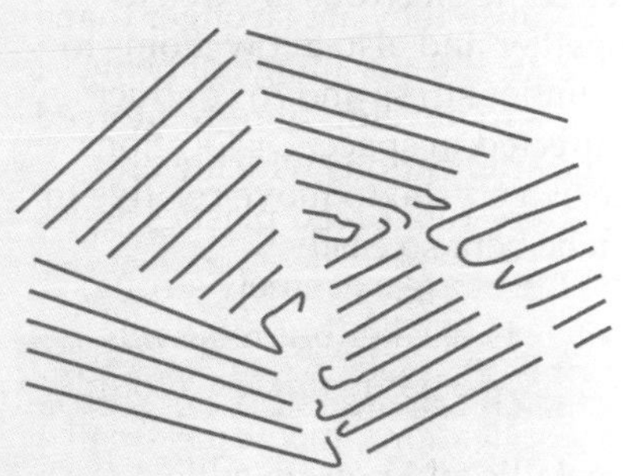

*Figure 29.3A* ▲ The arrangement of chains in high-density poly(ethene)

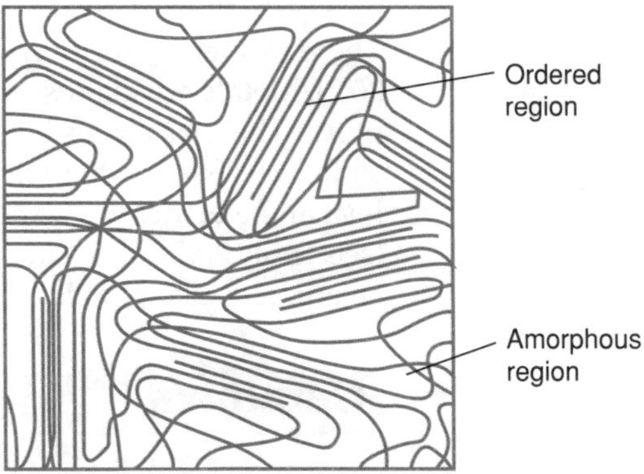


*Figure 29.3B* ▲ A polymer with ordered and amorphous regions

## Synthetic fibres

Fibres are polymers which have been stretched to increase their tensile (stretching) strength.

Fibres are made by drawing (stretching) plastics. Molten plastic is forced through a fine hole. As the fibre cools and solidifies, it is stretched to align its molecules along the length of the fibre. Fibres therefore have greater tensile strength along the length of the fibre than plastics. Some polymers can be deformed by straightening out the polymer chains. Examples are rubber, polyesters, polyamides and polycarbonates.

## Concrete and most types of rock

Concrete and most types of rock are:

- strong in compression (loading)
- weak in tension (stretching).

Reinforced concrete is described below.

## Composite materials

Composite materials combine the properties of their components.

Concrete reinforced with steel girders is strengthened.

Fibre reinforced plastics are stronger than plastics but can still be moulded, e.g. glass-fibre-reinforced-polyester e.g. carbon-fibre-reinforced epoxy resin.

Some materials are not homogeneous. They include composite materials, e.g. reinforced concrete, and cellular materials, e.g. coral. Composite materials combine the properties of more than one material. The resulting material is more useful for particular purposes than the individual components.

### ■ Reinforced concrete

Concrete is weak in tension, strong in compression and brittle. It can be reinforced with steel wires or bars. Steel is strong in tension, and the combination is a relatively cheap, tough material. It is essential for the construction of large structures, e.g. high-rise buildings, bridges and oil platforms.

The reason why unreinforced concrete has low tensile strength is that it contains microscopic pores. To reduce porosity and increase tensile strength, one solution is to drive air out of the powder by vibrating it before mixing with water. Another method is to add to the cement-water mixture materials, e.g. sulphur or resin, which will fill the pores. Alternatively, a water-soluble polymer can be added to the cement-water mixture to fill the spaces left between cement particles. Other filler materials, e.g. glass fibre, silicon carbide, aluminium oxide particles or fibres have been used in the polymer-cement mixture.

The strengths of the different materials depend on the types of chemical bond in them: the metallic bond, covalent bond, ionic bond, intermolecular forces of attraction.

## Fibre-reinforced plastics

Fibre composites such as epoxy resin and polyester resin contain thermosetting plastics. These are chosen because they are stronger than thermoplastics. However, thermoplastics have the advantage of being easy to mould, and this encouraged chemists to develop fibre-reinforced thermoplastics. These can be used for some applications which were traditionally filled by metals. Examples of fibre-reinforced plastics are:

1. *Glass-fibre-reinforced-polyester*: The glass fibres are extremely strong, though relatively brittle. The polymer matrix is weaker but relatively flexible. The combination of materials and properties results in a tough, strong material. The composite has many applications, including small boats, skis and motor vehicle bodies.

### Science at work

Kevlar fibres are five times as strong as steel. They are flexible, fire-resistant and of low density. Long, rigid, linear molecules of the polymer

```
 /    CH=CH         O   \
|    /     \        ‖    |
-+-C          C—C—N-+-
|    \\    //        |   |
 \    CH—CH          H  / n
```

are packed tightly together by forces of attraction between >C=O groups in one chain and >N—H groups in a parallel chain. Kevlar fibres are used to strengthen car tyres.

Figure 29.3C ▲ Glass-fibre-reinforced polyester (GRP)

2. *Carbon-fibre-reinforced epoxy resin*: This is used in tennis racquets and in aircraft. It is stronger and lower in density than conventional materials, e.g. aluminium alloys.

Figure 29.3D ▲ Carbon-fibre-reinforced resin in use

## Fibres

The fibres may be glass, carbon, silicon carbide, poly(ethene), kevlar and other substances. The tensile strength of carbon fibres is not high but they are stiff and therefore included in materials used for the frames of aircraft.

## Economics

Composites are expensive. However, a reduction in fuel consumption by lightweight transport vehicles could increase the use of composites.

## Why do metals bend, polymers bend, stretch or break and ceramics and glasses break?

**Metals** are strong in tension and in compression because the bonds between atoms are strong. Metals can change shape without breaking because the metallic bond allows layers of atoms to slide over one another when the metal is under stress. Metal crystals contain dislocations which will move under stress and allow the metal to change shape. Metals can withstand smaller loads without changing shape than can glass.

**Glass** shatters easily. Glass should be a strong material because the bonds between atoms are strong, and it does not contain dislocations. It is able to withstand mechanical loading without breaking. The brittleness of glass is due to surface defects or scratches. Glass breaks easily if a scratch is first made on its surface. Cracks find it harder to grow in metals because of the movement of dislocations.

**Glass fibres** resist cracking when they are bent because they are free of surface scratches.

**Polymers**. If you break a plastic ruler, made of poly(methyl methacrylate), only 10% of the fracturing is due to the breaking of covalent bonds; the rest is due to individual chains being pulled out of the material. In other polymers chain pull-out occurs less because the molecular chains are more densely packed and more entangled. Polymers such as rubber are elastic because the polymer chains can adjust in position and shape relative to one another.

### SUMMARY

The properties of metals, ceramics, glasses, plastics, fibres and concrete are revised. Composite materials, e.g. reinforced concrete, combine the properties of their components. The physical properties of the different materials are related to their structures.

### CHECKPOINT

1 Explain why it is possible to bend a piece of poly(ethene) tubing more easily than a piece of glass tubing.

2 Explain why a piece of metal is dented by a hammer blow but a piece of pottery shatters.

3 Explain why a rubber band stretches more than a length of metal wire.

4 For each of the materials below list the uses to which it can be put. Each material may have more than one use.

| *Material* | *Use* |
|---|---|
| 1. Metal | **A** Windows |
| 2. Ceramic | **B** Ovenware |
| 3. Glass | **C** Electrical plugs |
| 4. Thermosoftening plastic | **D** Carrier bags |
| 5. Thermosetting plastic | **E** Ropes |
| 6. Synthetic fibre | **F** Bridge |
| 7. Concrete | **G** Rifle |

# Topic 30 The variety of organic compounds

## 30.1 Alcohols

FIRST THOUGHTS

Timothy's idea of a good time is to go down to the pub and have a few beers with his friends. Sometimes, however, he wakes up the next day with a throbbing headache and a feeling of nausea. He has a 'hangover'. The substance which has produced these effects is **ethanol**, a liquid which we usually call **alcohol**.

### Ethanol

Ethanol is the best-known member of a series of compounds called alcohols. Ethanol is a drug. It is classified as a depressant. This means that it depresses (suppresses) feelings of fear and tension and therefore makes people feel relaxed. The body can absorb ethanol quickly because it is completely soluble in water.

*Figure 30.1A* ▲ Ethanol and relaxation

*Figure 30.1B* ▲ Ethanol

### The alcohols

The compounds of carbon are called organic compounds, with the exception of some simple compounds such as carbonates. Organic compounds were originally obtained only from natural sources.

Alcohols are a homologous series with the formula $C_nH_{2n+1}OH$.

Ethanol is a member of a homologous series of compounds called **alcohols**. Alcohols possess the group

$$-\overset{|}{\underset{|}{C}}-O-H$$

and have the general formula $C_nH_{2n+1}OH$ (see Table 30.1). The formulas are written as $CH_3OH$, etc. rather than $CH_4O$ to show the —OH group and show that they are alcohols. The members of the series have similar physical properties and chemical reactions. Ethanol is the only alcohol that is not poisonous. Methanol is very toxic: drinking only small amounts of methanol can lead to blindness and death.

*Table 30.1* ▼ The alcohols

| Alcohol | Formula, $C_nH_{2n+1}OH$ |
|---|---|
| Methanol | $CH_3OH$ |
| Ethanol | $C_2H_5OH$ |
| Propanol | $C_3H_7OH$ |
| Butanol | $C_4H_9OH$ |
| Pentanol | $C_5H_{11}OH$ |

SUMMARY

Alcohols are a homologous series of formula $C_nH_{2n+1}OH$. Ethanol is the only alcohol which is safe to drink in moderate quantities. Regular abuse of alcohol ruins your health. The alcohols have important industrial uses.

## Drinking ethanol

Ethanol is the only alcohol that can be safely drunk (in moderation). Methanol, $CH_3OH$, is very toxic. and drinking even small amounts can cause permanent harm. Ethanol dissolves completely in water and is therefore rapidly absorbed through the stomach and intestines. It can take up to 6 hours for the ethanol in a single drink to be absorbed when the stomach is full, but only about 1 hour when the stomach is empty. This is why people feel the effects of a drink faster on an empty stomach than on a full one. After ethanol enters the bloodstream, it diffuses rapidly into the tissues until the concentration of ethanol in the tissues equals that in the blood. This is why measuring the concentration of ethanol in the breath or in urine will indicate the level of ethanol in the blood. As the concentration of ethanol in the blood increases, speech becomes slurred, vision becomes blurred and reaction times increase. This is why it is so dangerous to drive 'under the influence' of alcohol. A driver needs short reaction times. Drinking large amounts of ethanol regularly causes damage to the liver, kidneys, arteries and brain. Many people abuse alcohol; they do not use it properly, that is, in moderation. Such people become addicted to alcohol, and their health suffers.

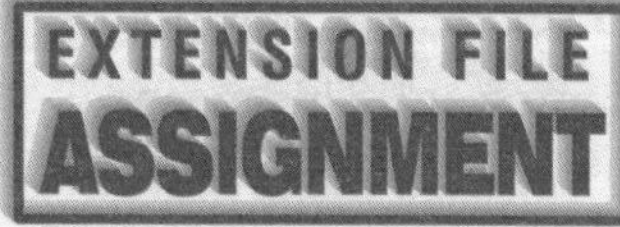


SUMMARY

Alcohols have the general formula $C_nH_{2n+1}OH$. The group $C_nH_{2n+1}$ is called an alkyl group, e.g. $—CH_3$ the methyl group and $—C_2H_5$, the ethyl group. The group —OH is called the hydroxyl group. This is the group which gives alcohols their reactions: it is the **functional group** of the alcohols.

## Uses of alcohols

The use of ethanol is not restricted to drinking. Ethanol is an important solvent. It is used in cosmetics and toiletries, in thinners for lacquers and printing inks. Being volatile (with b.p. 78 °C), the solvent evaporates and leaves the solute behind. Other alcohols also are used as solvents for paints, lacquers, shellacs and industrial detergents. The big advantages of alcohols as solvents is that they are miscible with water and many organic liquids.

CHECKPOINT

**1** (a) Write the structural formulas of (i) methane and methanol (ii) ethane and ethanol (iii) propane and propanol.
(b) What is the difference in structure between the members of each pair of compounds in (a)?
(c) What general formula can be written for the members of (i) the alkane series and (ii) the alcohol series?

## Manufacture of ethanol

### Fermentation

People have known for centuries how to obtain ethanol from sugars and starches. These substances are carbohydrates. They are compounds of carbon, hydrogen and oxygen, e.g., the sugar glucose, $C_6H_{12}O_6$, and starch, $(C_6H_{10}O_5)_n$. Glucose can be converted into ethanol by an enzyme called zymase, which is found in yeast. The conversion of glucose into ethanol is called **fermentation**.

$$\text{Glucose} \xrightarrow{\text{Enzyme in yeast}} \text{Ethanol} + \text{Carbon dioxide}$$
$$C_6H_{12}O_6(aq) \longrightarrow 2C_2H_5OH(aq) + 2CO_2(g)$$

Wine is made by adding yeast to fruit juices, which contain sugars. When the content of ethanol produced by fermentation reaches 14%, it kills the yeast. More concentrated solutions of ethanol (up to 96% ethanol) can be obtained by fractional distillation.

Beer is made from a number of starchy foods, such as potatoes, rice, malt, barley, hops and others. Germinated barley (malt) contains an enzyme which hydrolyses starch to a mixture of sugars. The addition of malt is followed by yeast which ferments the sugars.

$$\text{Starch} + \text{Water} \xrightarrow{\text{Enzyme in malt}} \text{Sugar}$$
$$(C_6H_{10}O_5)_n(aq) + nH_2O(l) \longrightarrow nC_6H_{12}O_6(aq)$$

Ethanol is sold in four main forms.

- Absolute alcohol: 96% ethanol, 4% water.
- Industrial alcohol or methylated spirit: 85% ethanol, 10% water, 5% methanol. The methanol is added to make the liquid unfit to drink.
- Spirits: gin, rum, whisky, brandy, etc. which contain about 35% ethanol.
- Fermented liquors: wines (12–14% ethanol), beers and ciders (3–7% ethanol). These contain flavourings and colouring matter.

Baking, like wine-making and brewing, involves fermentation. In baking, the important product of fermentation is carbon dioxide. In bread-making, flour, water, yeast, salt, sugar and fat are mixed to form a dough. The yeast starts to act on the sugar in the dough. The dough is divided into portions and allowed to stand in a warm container while the carbon dioxide produced makes it 'rise'. Then it is passed on a conveyor belt through a hot oven, from which emerge baked loaves.

*Figure 30.1C* ▲ Ethanol in beer, wine, brandy, after-shave lotion

## Hydration

Ethanol which is to be used as an industrial solvent is made from ethene by **catalytic hydration**.

$$\text{Ethene} + \text{Steam} \xrightarrow{\text{Heated catalyst (phosphoric acid)}} \text{Ethanol}$$
$$CH_2{=}CH_2(g) + H_2O(g) \longrightarrow C_2H_5OH(g)$$

About 10% of the ethene is converted; the unreacted ethene and steam are recycled.

EXTENSION FILE ACTIVITY

### It's a fact!

One of the effects of drinking alcohol is to increase reaction times. This is why it is dangerous to drive 'under the influence' of alcohol. A survey of 17- and 18-year-old drivers involved in road accidents showed that half of them had drunk too much.

### SUMMARY

Alcoholic drinks are made from sugars by fermentation. The reaction is catalysed by an enzyme in yeast. Starches can be hydrolysed to sugars and then fermented to give ethanol. Fermentation also takes place in baking when carbon dioxide makes dough rise. Industrial ethanol is made by the hydration of ethene.

The two chief methods of manufacture of ethanol are the fermentation of sugars and the hydration of ethene.

You should weigh up whether it is better to use a renewable resource (sugars) or a finite resource (oil).

Another question is whether it is more convenient to use a batch process or a continuous process.

Table 30.2 shows a comparison of the two methods.

*Table 30.2* A comparison of two methods of manufacturing ethanol

| Method | Raw materials | Rate | Quality of product | Type of process |
|---|---|---|---|---|
| Fermentation | Sugars (a renewable resource) | Slow | Impure ethanol | Batch process |
| Catalytic hydration | Ethene (from oil, a finite resource) | Faster, but needs heat and pressure | Pure ethanol | Continuous process |

## Reactions of alcohols

### Oxidation

Ethanol is oxidised by air if the right micro-organisms are present. The reason why wine goes sour if it is open to the air is that the ethanol in it is oxidised to ethanoic acid. Vinegar is 3% ethanoic acid.

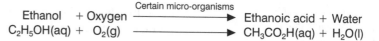

$$\text{Ethanol} + \text{Oxygen} \xrightarrow{\text{Certain micro-organisms}} \text{Ethanoic acid} + \text{Water}$$

$$C_2H_5OH(aq) + O_2(g) \longrightarrow CH_3CO_2H(aq) + H_2O(l)$$

This reaction is used as an industrial method of making ethanoic acid.

A faster method of oxidising ethanol is to use the powerful oxidising agent, acidified potassium dichromate(VI), $K_2Cr_2O_7$. Dichromate(VI) ions, $Cr_2O_7^{2-}$, which are orange, are reduced by ethanol and other reducing agents to chromium(III) ions, which are blue. As the reaction proceeds, the colour changes from orange through green to blue. This colour change was the basis of the first 'breathalyser' test. A motorist suspected of having too much ethanol in his or her blood had to breathe out through a tube containing some orange potassium dichromate(VI) crystals. If they turned green, or even blue, he or she was 'over the limit' (Figure 30.1D).

Now you can explain why wine goes sour. Ethanol is oxidised by air to ethanoic acid.

Powerful oxidising agents, e.g. acidified potassium dichromate(VI), also oxidise alcohols.

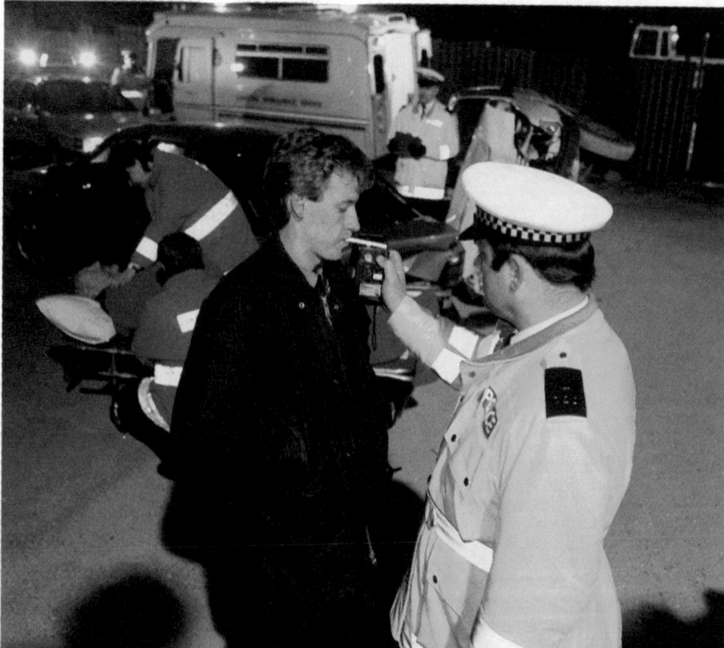

*Figure 30.1D* A modern breath analysis

Ethanol can be dehydrated to ethene.

Alcohols and organic acids react to form esters.

Figure 30.1F gives a summary of the reactions of ethanol.

## Dehydration

When ethanol is dehydrated, it gives ethene (see Figure 30.1E). Catalysts for the reaction include aluminium oxide, unglazed porcelain, porous pot and pumice stone.

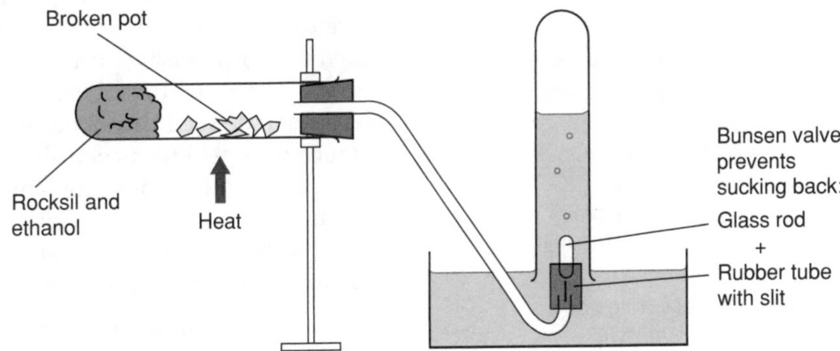


*Figure 30.1E* ▲ Making ethene by the dehydration of ethanol

## Esterification

Alcohols react with organic acids to form sweet-smelling liquids called **esters** (see Topic 31.3). For example,

$$\text{Ethanol} + \text{Ethanoic acid} \xrightarrow{\text{Conc. sulphuric acid}} \text{Ethyl ethanoate} + \text{Water}$$

$$C_2H_5OH(l) + CH_3CO_2H(l) \longrightarrow CH_3CO_2C_2H_5(l) + H_2O(l)$$

Concentrated sulphuric acid is added to catalyse the **esterification** reaction and to combine with the water that is formed. The structural formula for ethyl ethanoate is:

```
    H     O
    |    //
H — C — C      H   H
    |    \     |   |
    H     O — C — C — H
               |   |
               H   H
```

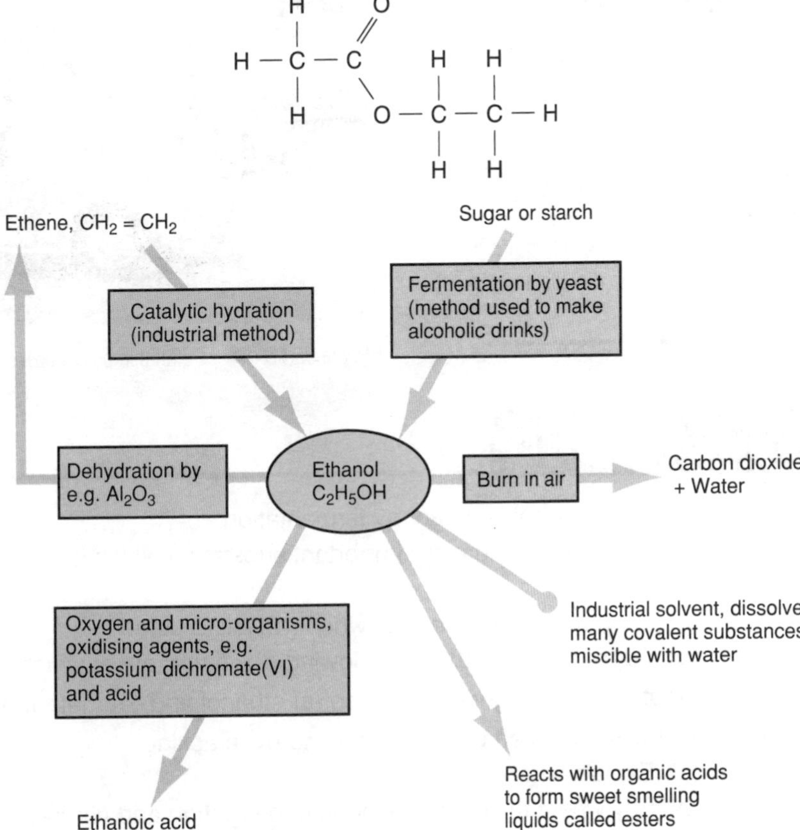


*Figure 30.1F* ▲ Some reactions of ethanol

## SUMMARY

Ethanol is an important industrial solvent. For this use, ethanol is made from ethene by catalytic hydration. Ethanol is oxidised to ethanoic acid. Alcohols react with organic acids to form esters, e.g. ethyl ethanoate, $CH_3CO_2C_2H_5$.

## Gasohol

The photograph shows 'alcool', a mixture of ethanol and petrol, for sale in Brazil. Sugar cane is grown as a source of sugar for fermentation to produce ethanol. To obtain more agricultural land for growing sugar cane, Brazil is clearing rain forest. Topic 13.8 discusses how cutting down forests will increase the greenhouse effect.

The cost of petroleum oil rose shaply after the oil crisis of 1973. Countries which have to import oil want to find alternatives. The major use of petroleum oil fractions is in vehicle engines. Petrol engines are designed to operate over a temperature range at which petrol will vaporise. Any fuel added to petrol must vaporise at the engine temperature and it must dissolve in petrol. Ethanol has the same boiling point as heptane, and it dissolves in petrol. Ethanol burns well in vehicle engines, producing 70% as much heat per litre as petrol does. A petrol engine will take 10% ethanol in the petrol without any adjustments to the carburettor (which controls the ratio of air to fuel in the cylinders). A mixture of petrol and ethanol is described as **gasohol**. The combustion of ethanol produces carbon dioxide and water: there is little atmospheric pollution.

To invest in the production of ethanol by fermentation, a country needs land available for growing suitable crops. It needs plenty of sunshine to ripen the crops quickly and supply sugar or starch for fermentation. Brazil has already started using ethanol as a vehicle fuel. Brazil has very little oil, but plenty of land and sunshine. Most of the petrol sold there contains 10% ethanol. This reduces Brazil's expenditure on oil imports. By the year 2000, Brazil hopes to provide for 75% of its motor fuel needs by using 2% of its land for growing crops for fermentation (Figure 30.1G).

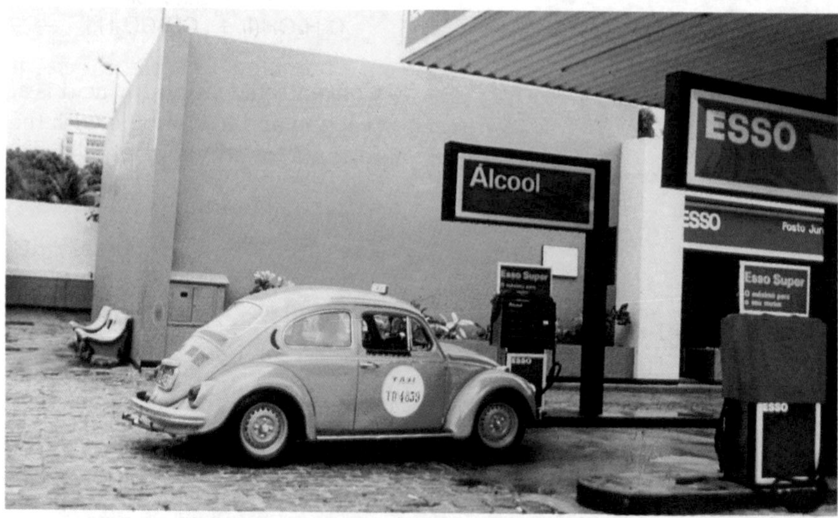


*Figure 30.1G* ▲ Petrol pumps in Brazil selling 'Alcool' gasohol

### SUMMARY

Ethanol can be added to the petrol in vehicle engines. Brazil is a country with no oil but with plenty of arable land and a sunny climate. Brazil is growing crops which can be fermented to give ethanol. Ethanol is a clean fuel: it burns to form carbon dioxide and water.

### CHECKPOINT

1 (a) Explain what is meant by fermentation.
 (b) Name a commercially important substance which is made by this method. Say what it is used for.

2 (a) Why does wine turn sour when it is left to stand?
 (b) Suggest two methods of slowing down the rate at which wine turns sour.

3 What are the dangers of drinking (a) ethanol and (b) methanol?

4 Petrol is produced by distillation and by 'cracking'.
 (a) Explain what cracking is.
 (b) Say where the energy used in (i) distillation and (ii) cracking comes from.
 (c) Say where the energy used in fermentation comes from.

▶ **5** Europe has a surplus of grain. Someone proposes building plants to obtain ethanol from the surplus cereals. Say what advantages this would bring to
(a) the environment (b) the farmer (c) the motorist (d) industry (e) the tax payer.
Can you see anything wrong with the idea? Explain your answers.

▶ **6** Ethanol is made by fermentation and by catalytic hydration.
(a) Why is it an advantage to make ethanol for solvent use by a continuous process, rather than a batch process?
(b) Why are alcoholic drinks not made by catalytic hydration?
(c) Which method is more economical with energy?
(d) Which method is more economical with the Earth's resources?

# 30.2 ▶ Carboxylic acids

The reactions of carboxylic acids are due to the carboxyl group, $-CO_2H$. In most organic acids this group is only slightly ionised and they are therefore weak acids. Their reactions are summarised in Figure 30.2A.

Ethanoic acid has the structural formula:

```
    H     O
    |    //
H — C — C
    |    \
    H     O — H
```

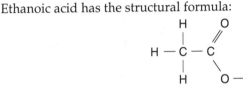

The formula is usually written as $CH_3CO_2H$. Ethanoic acid is a member of a homologous series called **carboxylic acids** (or alkanoic acids). Other members of the series are

methanoic acid, $HCO_2H$
propanoic acid, $C_2H_5CO_2H$
general formula, $C_nH_{2n+1}CO_2H$.

The carboxylic acids possess the group $-CO_2H$, which is called the carboxyl group. This is the functional group of the carboxylic acids. The carboxyl group ionises in solution to give hydrogen ions:

$$\text{Carboxylic acid} + \text{Water} \rightleftharpoons \text{Hydrogen ions} + \text{Carboxylate ions}$$
$$RCO_2H(aq) \rightleftharpoons H^+(aq) + RCO_2^-(aq)$$

R is an alkyl group, e.g. $CH_3$, $C_2H_5$. Since they are only partially ionised, carboxylic acids are weak acids. Some of the reactions of ethanoic acid are shown in Figure 30.2A.

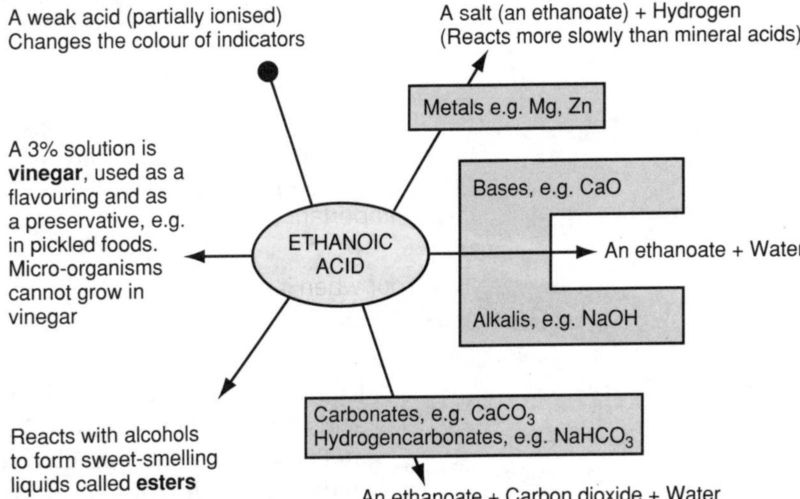


*Figure 30.2A* ▲ Reactions of ethanoic acid

## SUMMARY

Ethanoic acid is a carboxylic acid. It is a weak acid. It reacts in the same way as mineral acids but more slowly. Carboxylic acids react with alcohols to form esters.

## CHECKPOINT

1 (a) Name three substances which will react with ethanoic acid. Name the products of the reactions.
(b) Explain why a solution of ethanoic acid is less reactive than a solution of hydrochloric acid of the same concentration.
(c) Describe an experiment which you could do to demonstrate that ethanoic acid is weaker than hydrochloric acid.

## 30.3 Esters

Many esters are liquids with fruity smells. They occur naturally in fruits. They are used as food additives to improve the flavour and smell of processed foods. Other esters are used as solvents, for example in glues. People can get 'high' by inhaling esters. Some people enjoy the sensation so much that they become 'glue sniffers'. It is really the solvent, which may be a hydrocarbon or an ester, that they want to inhale. This dangerous habit is called **solvent abuse**. It produces the same symptoms as ethanol abuse. In addition, sniffers who are 'high' on solvents become disoriented. They may believe that they can jump out of windows or off bridges or walk through traffic. Many deaths occur from solvent abuse both through disoriented behaviour and from sniffers passing out and suffocating on their own vomit.

### SUMMARY

Esterification is the reaction between carboxylic acids and alcohols to form esters. These compounds are used as food additives and as solvents. Solvent abuse is a dangerous habit.

*Figure 30.3A* ▲ Esters in thinner, nail varnish remover, UHU glue

Animal fats and vegetable oils are esters. They are liquids or solids, depending on the size and shape of their molecules. They are esters of glycerol, an alcohol which has three hydroxyl groups:

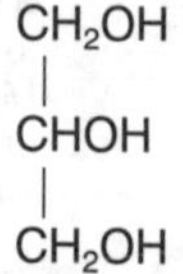


Each —OH group can esterify with a molecule of a carboxylic acid. The acids hexadecanoic acid, $C_{15}H_{31}CO_2H$, octadecanoic acid, $C_{17}H_{35}CO_2H$ and others combine with glycerol to form fats. These acids are known as 'fatty acids'. (Hexadecane means 16, and octadecane means 18. Count the number of carbon atoms, and you will see how they get their systematic names.)

### SUMMARY

Fats and oils are esters of glycerol and fatty acids. Saponification converts fats and oils into soaps, the sodium salts of fatty acids.

In soap-making, fats and oils are boiled with a concentrated solution of sodium hydroxide. The reaction can be represented by:

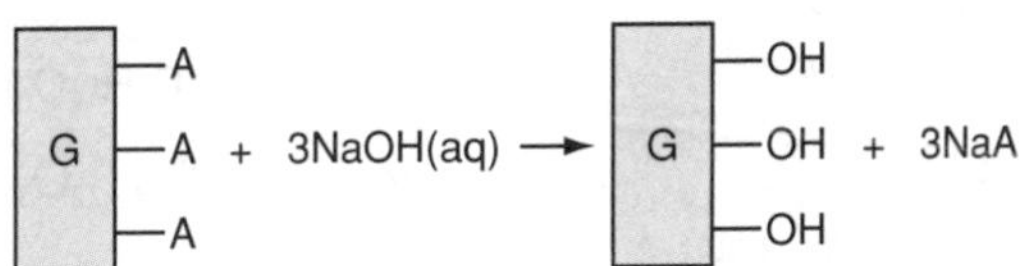


where $G(OH)_3$ represents glycerol and HA represents the fatty acid. The sodium salt of the fatty acid, e.g. sodium hexadecanoate or sodium octadecanoate, is a soap. This important reaction is called **saponification**.

## CHECKPOINT

1. Explain the terms 'esterification' and 'solvent abuse'.
2. (a) How are soaps manufactured?
   (b) How do soaps emulsify oil and water?

# 30.4 Carbohydrates

**FIRST THOUGHTS**

What do sugar and starch, the cell walls of plants and the exoskeletons of insects have in common? This topic will tell you!

## Sugars

The sugar in the sugar bowl is a compound called **sucrose**. There are many other sugars. One of the simplest is **glucose**, $C_6H_{12}O_6$. It is a **carbohydrate**, that is, a compound of carbon, hydrogen and oxygen only, in which the ratio (number of hydrogen atoms/number of oxygen atoms) = 2. A molecule of glucose contains a ring of six atoms, five carbon atoms and one oxygen atom; see Figure 30.4A(a). A shorthand way of writing the formula is shown in Figure 30.4A(b).

You will understand the formula better if you construct a model.

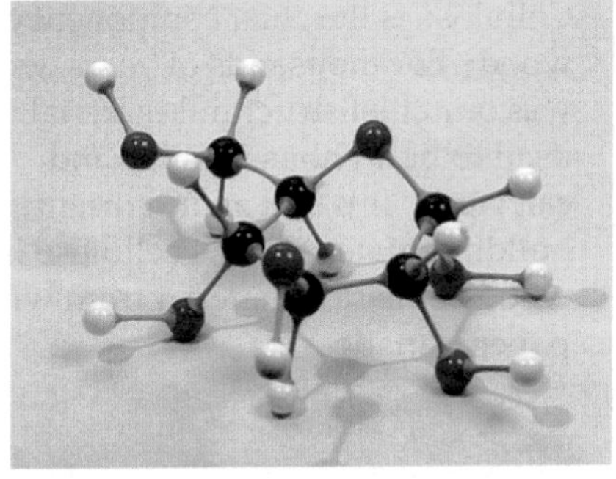

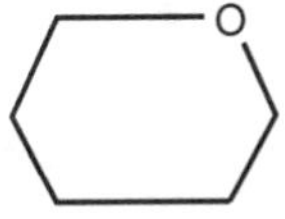


*Figure 30.4A* ▲ (a) Model of a glucose molecule (b) The formula in shorthand form

**Fructose**, $C_6H_{12}O_6$, is another sugar. It is an isomer of glucose and, as with glucose, its molecules contain a ring of six atoms. Sugars whose molecules contain one ring structure are called **monosaccharides** (mono = one).

A monosaccharide has one sugar ring per molecule, e.g. glucose and sucrose. A disaccharide has two sugar rings per molecule, e.g. sucrose and maltose.

The cellular respiration of monosaccharides and disaccharides provides energy in all plants and animals.

Other biologically important carbohydrates are starches, glycogen and cellulose. All these are polysaccharides with many sugar rings per molecule.

Sucrose, $C_{12}H_{22}O_{11}$, is a **disaccharide**: its molecules contain two ring structures. It is formed from glucose and fructose:

| Glucose | + Fructose | → | Sucrose | + Water |
|---|---|---|---|---|
| $C_6H_{12}O_6(aq)$ | $+ C_6H_{12}O_6(aq)$ | → | $C_{12}H_{22}O_{11}(aq)$ | $+ H_2O(l)$ |

Maltose, $C_{12}H_{22}O_{11}$, is a disaccharide formed from glucose:

| Glucose | → | Maltose | + Water |
|---|---|---|---|
| $2C_6H_{12}O_6(aq)$ | → | $C_{12}H_{22}O_{11}(aq)$ | $+ H_2O(l)$ |

The shorthand form of the formula of maltose is shown in Figure 30.4B.

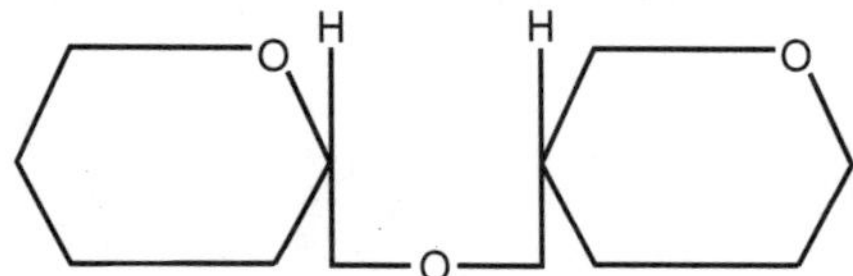


*Figure 30.4B* ▲ The formula for maltose in shorthand form

Carbohydrates have a vital function in all living organisms: they provide energy. When carbohydrates are oxidised, carbon dioxide and water are formed, and energy is released.

| Glucose | + Oxygen | → | Carbon dioxide | + Water |
|---|---|---|---|---|
| $C_6H_{12}O_6(aq)$ | $+ 6O_2(aq)$ | → | $6CO_2(g)$ | $+ 6H_2O(l)$; Energy is released |

If you burn glucose in the laboratory, you will observe a very rapid release of energy. Inside the cells of a plant or animal, however, the oxidation takes place in a controlled manner so that energy is made available to the organism as needed. The process is called cellular respiration. Living organisms also respire fats, oils and proteins.

## Starch, glycogen and cellulose

Starch is a food substance which is stored in plant cells. Glycogen is a food substance which is stored in animal cells. Cellulose is the substance of which plant cell walls are composed. All these substances are carbohydrates. They are **polysaccharides**, that is, their molecules contain a large number (several hundred) of sugar rings. These are glucose rings (see Figure 30.4C). Polysaccharides differ in the length and structure of their chains (see Figure 30.4D).

Cellulose is the chief component of wood. For thousands of years, wood was our chief structural material, used to build houses, ships and carriages. It is still an important building material. The cellulose in wood is the raw material from which paper is made.

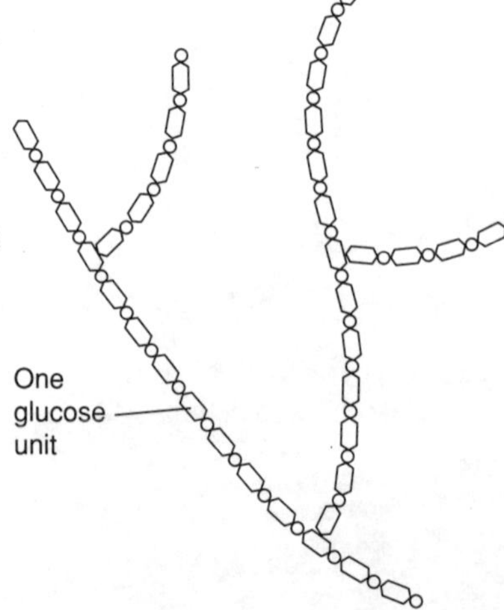

*Figure 30.4C* ▲ Part of a starch molecule

*Figure 30.4D* ▲ Structure of a starch molecule

**SUMMARY**

Carbohydrates contain carbon, hydrogen and oxygen only. Monosaccharides are sugars, including hexoses, with six-atom rings in the molecule, e.g. glucose and fructose, and pentoses, with five-atom rings, e.g. ribose. Disaccharides have two sugar rings in the molecule, e.g. sucrose and maltose. Polysaccharides have molecules with a large number of sugar rings. They include starch and glycogen (food stores), cellulose (in plant cell walls) and chitin (in insect exoskeletons).

Starch and glycogen are only slightly soluble in water. They can remain in the cells of an organism without dissolving and therefore make good food stores. When energy is needed, cells convert starch and glycogen into glucose, which is soluble, and then oxidise the glucose.

Cellulose is insoluble. The walls of plant cells consist of a tough framework of cellulose fibres (see Figure 30.4E). Another polysaccharide, chitin, forms the exoskeleton that encloses the body of some invertebrates (see Figure 30.4F).

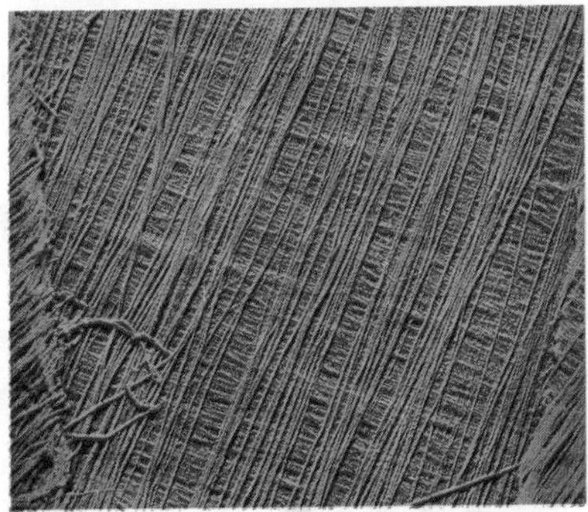

*Figure 30.4E* ▲ Plant cell wall (×10 000) showing the framework of cellulose fibres

*Figure 30.4F* ▲ An invertebrate exoskeleton of chitin

**CHECKPOINT**

1. Runners in the London marathon ate pasta the evening before. Why did they think this would give them energy?
2. Give examples of carbohydrates which are used as structural materials in plants and animals.
3. Glucose, maltose and starch are carbohydrates. Answer the following questions about them.
   (a) Which one tastes sweet?
   (b) Which two are very soluble in water?
   (c) List the elements that make up all three.
   (d) The formula of glucose is $C_6H_{12}O_6$. What is the ratio number of atoms of hydrogen : number of atoms of oxygen?
   (e) The formula of maltose is $C_{12}H_{22}O_{11}$. What is the ratio number of atoms of hydrogen : number of atoms of oxygen?
   (f) Using your answers to (d) and (e), try to explain where the name 'carbohydrate' comes from.
   (g) The 'shorthand' formula for glucose is O
   Draw the shorthand formulas for maltose and starch.

## 30.5 Fats and oils

**FIRST THOUGHTS**

Have you seen brands of margarine and cooking oil described as 'high in polyunsaturates' and wondered what it means? This section will tell you.

Fats and oils are together called **lipids**. They contain carbon, hydrogen and oxygen only. At room temperature, most fats are solid and most oils are liquid. Fats and oils are insoluble in water. Their functions are as follows:

*Source of energy*: Lipids are important stores of energy in living organisms. The oxidation of fats and oils gives carbon dioxide and water with the release of energy. Lipids provide about twice as much energy per gram as do carbohydrates.

**SUMMARY**

Fats and oils are together called lipids. As foods, they provide energy and dissolved vitamins. They also provide thermal insulation and protection for delicate organs.

*Thermal insulation*: Mammals have a layer of fat under the skin. This acts as a thermal insulator.
*Protection*: Delicate organs, e.g. kidneys, are protected by a layer of fat.
*Food*: Some vitamins, e.g. vitamins A, D and E, are insoluble in water and soluble in lipids. Foods containing lipids provide these vitamins.
*Cell membranes*: Lipids are incorporated in cell membranes.

## Saturated and unsaturated fats and oils

*Do you watch your diet?* Many people are concerned about having too much fat in their diet, particularly **saturated fats**. Let us look at what the term 'saturated' means here.

(a) Model of a molecule of glycerol

(b) Model of a molecule of a fatty acid (hexadecanoic acid)

```
     H
     |
 H — C — O — H
     |
 H — C — O — H
     |
 H — C — O — H
     |
     H
```

(c) The formula of glycerol

```
     H   H   H   H   H   H   H   H   H   H   H   H   H   H   H    O
     |   |   |   |   |   |   |   |   |   |   |   |   |   |   |   //
 H — C — C — C — C — C — C — C — C — C — C — C — C — C — C — C — C
     |   |   |   |   |   |   |   |   |   |   |   |   |   |   |    \
     H   H   H   H   H   H   H   H   H   H   H   H   H   H   H     O — H
```

(d) The formula of hexadecanoic acid

*Figure 30.5A* ▲

Fats and oils contain esters of glycerol and fatty acids (carboxylic acids with long alkyl groups) (see Topic 30.3). The fatty acid may be saturated or unsaturated. If there is one carbon-carbon double bond in the molecule, the compound is described as **unsaturated**. If there is more than one carbon-carbon double bond, the compound is described as **polyunsaturated**. Below is part of the carbon chain in (a) a saturated fatty acid, (b) an unsaturated fatty acid and (c) a polyunsaturated fatty acid.

```
       H   H   H   H   H   H   H   H   H
       |   |   |   |   |   |   |   |   |
(a)  — C — C — C — C — C — C — C — C — C —
       |   |   |   |   |   |   |   |   |
       H   H   H   H   H   H   H   H   H

       H   H   H   H   H   H   H   H   H
       |   |   |   |   |   |   |   |   |
(b)  — C — C — C — C — C — C = C — C — C —
       |   |   |   |   |           |   |
       H   H   H   H   H           H   H

       H   H   H   H   H   H   H   H   H
       |   |   |   |   |   |   |   |   |
(c)  — C — C = C — C — C — C = C — C — C —
       |           |   |           |   |
       H           H   H           H   H
```

You read a lot of discussion about the value of polyunsaturated fats in the diet. Now you know what they are!

## SUMMARY

Lipids (fats and oils) are mixtures of esters of glycerol with carboxylic acids (called fatty acids). The carboxylic acid may be saturated or unsaturated or polyunsaturated. Unsaturated fats and oils can be converted into saturated fats and oils by catalytic hydrogenation.

Fats and oils are mixtures of esters. Fats which contain esters of glycerol and saturated fatty acids are called saturated fats. Fats which contain esters of glycerol and unsaturated fatty acids are called unsaturated fats or polyunsaturated fats, depending on the number of double bonds in a molecule of the acid. Animal fats contain a large proportion of saturated esters and are solid. Plant oils contain a large proportion of unsaturated esters and have lower melting points. Many scientists believe that eating a lot of saturated fats increases the risk of heart disease.

Unsaturated oils can be converted into saturated fats by catalytic hydrogenation. The vapour of the oil is passed with hydrogen over a nickel catalyst (see Topic 29.1).

## CHECKPOINT

1. (a) Say which are the fats and which are the oils.
   soft margarine, butter, lard, hard margarine, cooking oil
   (b) What is the main physical difference between fats and oils?
2. The manufacturers of *Flower* margarine claim that their product encourages healthy eating because it is 'low in saturated fats'.
   (a) What do they mean by 'low in saturated fats'?
   (b) Name a product which contains much saturated fat.
   (c) What particular benefit to health is claimed for products such as *Flower*?
3. The following equation represents the formation of a fat.

   A + 3 (B) = C (with three circles attached) + 3 (D)

   (a) Name the substances A and D.
   (b) Name one substance which could be B.
   (c) Name one fat and one oil which could be C.

# 30.6 Proteins

## FIRST THOUGHTS

Muscle, skin, hair and haemoglobin; what do they have in common? You can find out in this section.

Proteins are compounds of carbon, hydrogen, oxygen, nitrogen and sometimes sulphur. They are vitally important in living things for the following reasons:

1. Proteins are the compounds from which new tissues are made. When organisms need to grow or to repair damaged tissues, they need proteins.
2. Enzymes are proteins which catalyse reactions in living things.
3. Hormones are proteins which control the activities of plants and animals.
4. Proteins can be used as a source of energy in respiration.

Amino acids have two functional groups: the amino group $-NH_2$ and the carboxyl group $-CO_2H$. These groups enable them to polymerise to form peptides, polypeptides and proteins.

*Figure 30.6A* ▲ Model of a glycine molecule

Proteins have large molecules. A protein molecule consists of a large number of amino acid groups. There are about 20 **amino acids**. The simplest is glycine, of formula

$$H_2N-CH_2-C(=O)-O-H$$

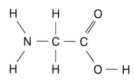

The general formula of an amino acid is

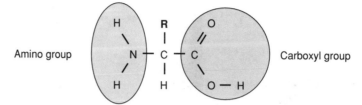

Amino group $H_2N-$ $CHR$ $-C(=O)-O-H$ Carboxyl group

**R** is a different group in each amino acid; it can be —H, $—CH_3$, $—CH_2OH$, —SH and many other groups. The carboxyl group ($—CO_2H$) is acidic, and the amino group ($—NH_2$) is basic (like ammonia, $NH_3$). The amino group of one amino acid can react with the carboxyl group of another amino acid. When this happens, the two amino acids combine to form a **peptide** and water.

$$H_2NCH_2CO_2H + H_2NCH_2CO_2H \rightarrow H_2NCH_2CONHCH_2CO_2H + H_2O$$

The bond which has formed between the two amino acids, —CONH—, is called the **peptide link**. Its structure is

$$-C(=O)-N(H)-$$

The peptide which has been formed has an amino group at one end and a carboxyl group at the other end so it can form more peptide links. We can show this by an equation, using a different shape, e.g. ○ and □, for each different amino acid.

□ + ○ → □—○ + $H_2O$
□—○ + ▲ + ○ → ▲—□—○—○ + $2H_2O$
▲—□—○—○ + ▲ + □ → ▲—▲—□—○—○—□ + $2H_2O$

In this way, many amino acids can combine to form a long chain. Each link in the chain is a peptide bond.

**Peptides** have molecules with up to 15 amino acid groups.
**Polypeptides** have molecules with 15–100 amino acid groups.
**Proteins** have still larger molecules.

EXTENSION FILE ACTIVITY

## SUMMARY

Many tissues in living organisms are made of proteins. The molecules of proteins are long chains of amino acid groups. Peptides and polypeptides have smaller molecules.

## It's a fact!

Fifteen million children a year die of starvation and disease. You will have seen pictures of children with swollen abdomens. They are suffering from a disease called kwashiorkor. They are starved of protein. In parts of the world where rice is the staple diet, kwashiorkor is common.

## Diet

An animal can break down the proteins in its food into amino acids and then use the amino acids to build the proteins which it needs. Animals are able to make some amino acids in their bodies. The other amino acids which they need must be supplied in the diet. These are called **essential amino acids**. A protein which contains all the essential amino acids is called a **first class protein**. Meat, fish, cheese and soya beans supply first class protein. A protein which lacks some essential amino acids is called a **second class protein**. Such proteins are in flour, rice and oatmeal.

## Structure

A protein molecule is twisted into a three-dimensional shape (see Figures 30.6B and C). It is kept in this shape by bonding between groups in different parts of the chain.

Enzymes are proteins; see Topic 26.8.

*Figure 30.6B* ▲ Insulin. In 1953, Dr Frederick Sanger of the UK reported the complete sequence of amino acids in the protein insulin. This was the first time that a protein structure had been worked out. Insulin controls blood sugar levels. Its molecule is one of the smallest protein molecules.

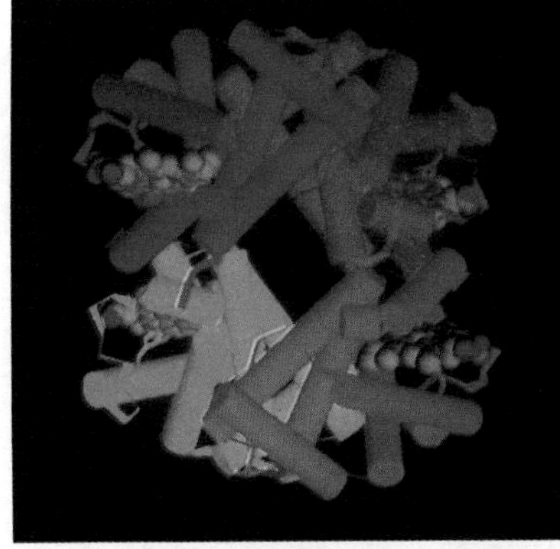

*Figure 30.6C* ▲ Haemoglobin. In 1959, Dr Max Perutz worked out the shape of the haemoglobin molecule. Haemoglobin is the red pigment in blood which transports oxygen round the body. The haemoglobin molecule is larger than the insulin molecule.

### CHECKPOINT

1. What is the difference between (a) a peptide and a polypeptide (b) a polypeptide and a protein?
2. What compounds are formed when molecules of the following compounds combine? (a) monosaccharides (b) amino acids (c) glycerol and fatty acids?
3. (a) What type of compound is this?

$$\begin{array}{c} H \\ | \\ H_2N-C-CO_2H \\ | \\ CH_2-SH \end{array}$$

(b) Draw the structure of the compound (other than water) that is formed when two molecules of the above substance react.
(c) What type of bond is formed in the reaction?
(d) What type of compound is the substance in (b)?
(e) What further reaction or reactions can this substance undergo?

# 30.7 Food additives

### FIRST THOUGHTS

We eat some foods exactly as they are harvested, e.g. apples and cucumbers. Most foods, however, go to the food industry to be processed, that is changed in some way. Three quarters of the food eaten in industrialised countries is processed.

## Why use additives?

In food processing, many substances are added. They are called **food additives**. A food additive is defined as a substance which is not normally eaten or drunk as a food either by itself or as a typical ingredient of food. Salt and sugar are added to foods but are not called additives. No substance may be used as an additive unless the food manufacturer can give a good reason for its use. The reasons for using additives are:

- to flavour food
- to colour food
- to alter the texture of food
- to preserve food.

Additives are chemicals, but so are proteins, vitamins and all the other substances which occur naturally in foodstuffs. In 1950, 50 additives were in general use; now 3500 additives are used in our food.

## Types of food additives

### Additives which alter the taste of food

*Flavourings* Flavourings are the largest group of food additives, with about 3000 in use. The large number is not so surprising when it takes a mixture of up to 50 substances to produce a natural flavour such as apple or peach.

*Sweeteners* The commonest sweetener is sucrose (sugar); this is a food, not an additive. Some people want to cut down on sucrose either because it causes tooth decay or because they are overweight. Diabetics cannot cope with sucrose. Substitutes are saccharin, sorbitol and mannitol.

*Flavour enhancers* Flavour enhancers are not flavourings; they are substances which make existing flavours seem stronger. The best known is monosodium glutamate, MSG. It stimulates the taste buds.

*Figure 30.7A* ▲ They all contain MSG

### Additives which alter the colour of food

When food is processed, it may lose some of its colour; then the manufacturer will want to restore the original colour of the food. Different countries vary widely in the number of colourings allowed. Colourings are not added to baby foods.

### Additives which alter the texture of food

*Emulsifiers and stabilisers* When oil and water are added, they form two layers. Some substances, called **emulsifiers**, can make oil and water mix. The mixture is called an emulsion. Any substance which helps to prevent the emulsion from separating out again is called a **stabiliser**. Margarine, ice cream and salad dressings all use emulsifiers and stabilisers.

**SUMMARY**

Food additives are used to alter
- the taste of food
- the colour of food
- the texture of food.

*Figure 30.7B* ▲ All these contain emulsifiers and stabilisers

*Thickeners* You will see that 'modified starch' appears on many labels. It is used to thicken foods. It can be a main ingredient of instant soups and puddings.

*Figure 30.7C* ▲ These contain modified starch

*Anti-caking agents and humectants* Anti-caking agents are substances which can absorb water without becoming wet. They are added to powdery or crystalline foods, such as cake mixes and table salt, to prevent lumps from forming. Humectants keep products moist. They are added to products such as bread and cakes.

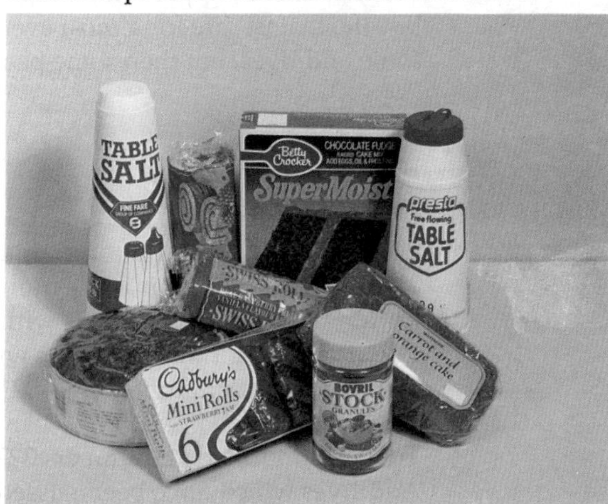


*Figure 30.7D* ▲ Which products contain an anti-caking agent? Which contain a humectant?

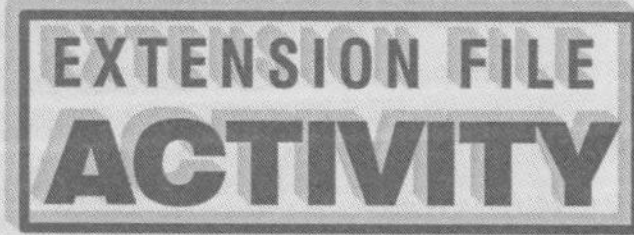


*Gelling agents* To make jams, desserts etc. set, a gelling agent is added. Pectin is the commonest.

### Anti-oxidants

Foods which contain fats and oils can turn rancid on exposure to the air. The fats and oils are oxidised to unpleasant-smelling acids. Anti-oxidants are added to prevent oxidation. Two common ones are BHA and BHT (butylated hydroxyanisole and butylated hydroxytoluene). Sulphur dioxide and sulphites are widely used as anti-oxidants. They have two effects: as well as preventing oxidation of fats and oils, they also deprive micro-organisms of the oxygen they need and delay their growth.

**SUMMARY**

Food additives used to preserve food include antioxidants and preservatives, e.g. BHA, BHT, sulphur dioxide and sulphites.

*Figure 30.7E* ▲ Preserved by sulphur dioxide and sulphites

### Preservatives

Food is stored in the warehouse, in the shop and in the home before it is eaten. Chemical preservatives are added to stop the growth of micro-organisms. Preservatives increase the food's shelf-life, i.e. the length of time the food will keep before it deteriorates. Longer shelf-lives mean less wastage on the shelves. This allows the shopkeeper to charge lower prices and to stock a wider range of foods. Some people are doubtful about whether the customer receives all the benefit of the lower prices. Some people believe that the savings from the widespread use of additives go to the food manufacturers rather than the customers.

## Controls on additives

It is illegal to put anything harmful into food. Before an additive may be used, it must be approved by the Government. By law, all additives must be safe, that is, safe for almost everyone. Some people are made ill by some additives, but then some people are made ill by some natural foods.

## E Numbers

On the labels of some foods you will see E numbers. E numbers are a device of the European Community (EC). The EC has drawn up a list of 314 safe additives. This is the numbering system:
Colourings: E number begins with 1, e.g. E150, caramel
Preservatives: E number begins with 2, e.g. E221 sodium sulphite
Anti-oxidants: E numbers 310–321, e.g. E320, butylated hydroxyanisole, BHA
Texture controllers: E numbers 322–494, e.g. E461, methylcellulose.
Additives which have been passed in the country of origin but not yet in the EC have a number only, e.g. 107 yellow 2G; 524 sodium hydroxide; 925 chlorine.

Food additives are classified by E numbers.

## It's a fact!

We eat between 3 and 7 kg of additives a year. Some people think that over a long period some additives could be a threat to health. Medical doctors believe that additives make little, if any, contribution to serious illness.

## Are additives good for you?

Some foods make some people ill. When a person reacts to a food by becoming ill, the reaction is called an **intolerance reaction** or an **allergic reaction**. Allergic reactions can take the form of asthma (breathing difficulty), eczema (a skin complaint), digestive troubles, rhinitis (like hay fever), headaches, migraines and hyperactivity. Putting hyperactive children on a diet free from additives often produces a dramatic improvement. Tartrazine (E102), a yellow dye, is the one that is most under suspicion. It is used in sweets, fizzy drinks and packet desserts. If you know that you are allergic to tartrazine, you can read the labels on the foods you fancy and reject any which list E102. Many people prefer to buy additive-free foods. Many of the large supermarket chains are reducing the number of additives in their products, and some firms are offering additive-free items.

## CHECKPOINT

▶ **1** Which types of additives are present in this Apricot Pie?

**APRICOT PIE**

Ingredients: Flour, Sugar, Apricot, Animal and Vegetable fat, Dextrose, Modified starch, Sorbitol syrup, Glucose syrup, Salt, Citric acid, Preservative E202, Flavouring, Emulsifiers E465, E471, E475, Whey powder, Colours E102, E110

▶ **2** A ham manufacturer wants to increase the weight of his hams by injecting water into them. He needs an additive to keep the water and other ingredients well mixed. What type of additive does he use? Who benefits from the use of this additive, the manufacturer or the consumer?

▶ **3** The labels on two soft drink bottles are shown below.

**Kooler-cola**

*Soft drink with vegetable extracts*

Ingredients: Carbonated water, Sugar, Colour (caramel), Phosphoric acid, Flavourings, Caffeine

***BEST SERVED ICE COLD***

**diet cola**

*Low calorie soft drink with vegetable extracts*

Ingredients: Carbonated water, Colour (caramel), Artificial sweetener (Aspartame), Phosphoric acid, Flavourings, Citric acid, Preservative (E211)
Contains phenylamine

***TASTES GREAT ICE COLD!***

(a) Which sweetener is present (i) in A (ii) in B?
(b) Which preservative is present in both A and B?
(c) Why is a preservative needed?
(d) Why do the labels recommend serving the drinks ice-cold?
(e) What is the difference between A and B that enables one to describe 'diet cola' as 'low calorie'?
(f) What pH would you expect the two drinks to have?

# 30.8 Dyes

FIRST THOUGHTS

Visible light, which consists of light of many different wavelengths, is called 'white light'. An object may absorb light of a particular wavelength and transmit the rest of the spectrum. The colour of the object which the eye sees is the result of the transmitted light. The colour of the transmitted light and the colour of the absorbed light together add up to white light.

## What is a dye?

Colouring materials may be **pigments** or **dyes**. Both are coloured substances which can bond to the material to be coloured. The difference between the two lies in their physical state. Dyestuffs are usually present as single molecules or as clusters of a few molecules. Pigments are completely insoluble and are applied as very small solid particles containing millions of molecules. The colourants in printing inks, paints, enamels, rubber and plastics are pigments.

Dyes have been used since ancient times. The dye Tyrian purple was extracted from molluscs in the days of ancient Rome and used to dye the robes of wealthy people. The ancient Egyptians extracted the blue dye indigo from plants 5000 years ago. The red dye alizarin was extracted from the roots of madder. The yellow dye saffron was extracted from a crocus. Cochineal, a scarlet dye, was extracted from a species of insect. These dyes are now synthesised by the chemical industry.

It's a fact!

A century ago, indigo plantations in India occupied over one million acres of land. In 1897 a commercial method of manufacturing the dye was invented. A good result of this was that vast areas of land in India were freed for growing food. A bad result was that many people lost their jobs.

## Synthetic dyes

These are the properties which a commercial dye must have:

1 *Colouration:* A wide range of bright colours and dark colours, such as black and navy blue is required. A strong dye has the advantage that little of it is needed.

2 *Safety:* No dye may present a hazard to health, safety or the environment, either during manufacture or in use.

3 Reliability: Dyes must be easy to use and give reliable and reproducible results.

4 *Fastness:* The dye must show:
- light-fastness: it must not fade in daylight
- wet-fastness: it must not be removed by washing
- perspiration-fastness: it must not stain the skin
- heat-fastness: it must not be removed by ironing.

5 *Economy:* Factors affecting cost are:
- the cost of the dyestuff
- the fraction of the dye applied which is retained by the fabric
- the temperature and time needed for the dyeing process.

## Mordants

The dye molecules must be retained by the fabric. Some dyes form strong covalent bonds to fibres. Others are bonded to fibres by weaker intermolecular forces of attraction. Natural dyes are not fast; they do not bind well to fabrics. The fastness is improved by treating the fabric with a mordant before it is dyed. A mordant (from a French word meaning 'to bite') is a substance which helps the dye to 'bite' the fabric. The most commonly used mordant is alum; either aluminium ammonium sulphate, $NH_4Al(SO_4)_2.12H_2O$, or aluminium potassium sulphate, $KAl(SO_4)_2.12H_2O$. The fabric is soaked in a dilute solution of ammonia and then in a solution of alum. The reaction between the alkaline

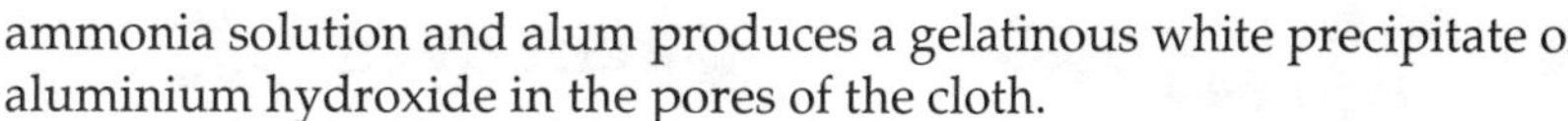

ammonia solution and alum produces a gelatinous white precipitate of aluminium hydroxide in the pores of the cloth.

$$\underset{\text{(from alum solution)}}{Al^{3+}(aq)} + \underset{\text{(from ammonia solution)}}{3OH^-(aq)} \rightarrow \underset{\text{(precipitated in the fibres of the cloth)}}{Al(OH)_3(s)}$$

When the mordanted fabric is immersed in a solution of a dye, the dye can bond to particles of aluminium hydroxide dispersed between the fibres of the cloth. Different mordants can give different colours with the same dye. Modern synthetic dyes do not need mordants.

**SUMMARY**

Since ancient times, dyes have been extracted from plant and animal sources. Now, dyes are manufactured by the chemical industry.

A mordant is a substance which is used to treat a fabric so that dyes will bond to it more strongly. Modern synthetic dyes do not need mordants.

## Vat dyes

An example is indigo, which is used for blue denim. Vat dyes are used especially for cotton and are colour-fast. Indigo is insoluble, and must be converted into a soluble substance before it can be used to dye cloth. This is done by reducing indigo to a soluble yellow compound. The reduction was traditionally carried out in vats, and gave the name **vat dyes** to indigo and similar dyes. Oxidation restores the original blue colour.

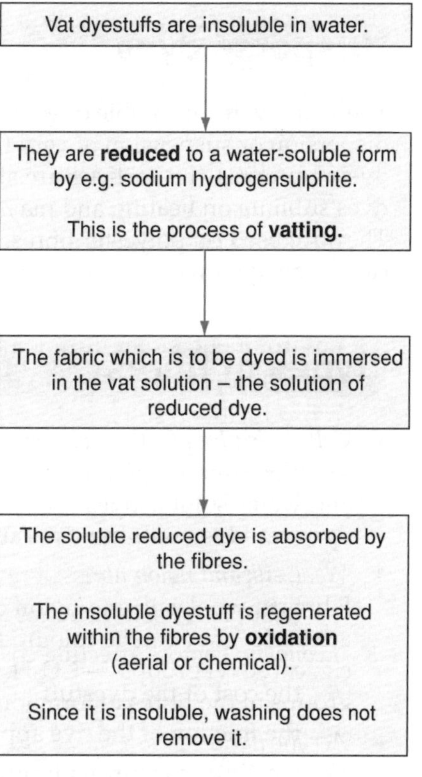

Vat dyestuffs are insoluble in water.
They are reduced to a water-soluble form by e.g. sodium hydrogensulphite.
This is the process of vatting.
The fabric which is to be dyed is immersed in the vat solution – the solution of reduced dye.
The soluble reduced dye is absorbed by the fibres.
The insoluble dyestuff is regenerated within the fibres by oxidation (aerial or chemical).
Since it is insoluble, washing does not remove it.

*Figure 30.8A* ▲ Using a vat dye, e.g. indigo

## Sulphur dyes

The sulphur dyes are organic compounds containing sulphur. They are applied in the same way as vat dyes, in a reduced state. Oxidation produces an insoluble dye within the fibres. Sulphur dyes are colour-fast. They are limited to blue and black dyes.

## Azo dyes

An insoluble coloured compound is formed by allowing two chemicals to react within the fabric. Azo dyes are very water-fast. Some are used as food colourings.

**SUMMARY**

Types of dyes include:

1 vat dyes, which are applied in reduced form, then oxidised on the fabric
2 sulphur dyes, applied in the same way as vat dyes, colour-fast
3 azo dyes, which are formed in the fabric
4 direct dyes, applied from an aqueous solution
5 fibre-reactive dyes, which possess reactive groups which can bond to groups in the fibres and are therefore very fast
6 disperse dyes, which are applied as a suspension.

## Direct dyes

Direct dyes are water-soluble. They are so named because they can dye cellulose fibres straight from an aqueous dye bath. No mordant is used. The forces which bind the dye to the fibre are fairly weak, and direct dyes do not have great water-fastness.

## Fibre-reactive dyes

The first commercial fibre-reactive dyes were invented in 1956. They are now known as Procion dyes. These dyes were developed in a search for new dyes which would form covalent bonds to the new synthetic fibres. This bonding makes them much faster than dyes which bond to fabrics by weak intermolecular forces. They are therefore not removed by washing. They are water-soluble and therefore easy to use. They offer a very wide range of shades, many of which are much brighter than those obtained from earlier dyes.

## Disperse dyes

These are water-insoluble dyes. They are added to the dye bath as a dispersion or suspension of very fine particles of solid dyestuff. In the dyeing process, the particles are absorbed into the fibres. Some disperse dyes sublime on heating and may come off the fabric during ironing. They are used on polyester fibres, which are impermeable to water and cannot be dyed with water-soluble dyes.

## Types of fibres

- *Cellulosic fibres (cotton, viscose, rayon)*
  Cellulose fibres contain many hydroxyl groups which absorb water and water-soluble dyestuffs. The dyes which can be used are direct dyes, vat dyes, sulphur dyes and fibre-reactive dyes.
- *Wool, silk and nylon fibres*
  Fibres of wool, silk and nylon contain amino groups, $—NH_2$, which are basic. They can be coloured by dyes which have acidic groups, e.g. carboxyl groups, $—CO_2H$. These groups combine with the amine groups in the fibres to form ionic bonds.
- *Acrylic fibres*
  Acrylic fibres contain acidic groups, e.g. carboxyl groups. They are often dyed with basic water-soluble dyes.

**SUMMARY**

Different types of fabrics are dyed with different kinds of dyes. Fibres which contain basic groups need an acidic dye. Fibres which contain acidic groups need a basic dye.

**CHECKPOINT**

1 How does a dye differ from a pigment?
2 Comment on the relative merits of natural and synthetic dyes with respect to colour-fastness.
3 In addition to being colour-fast what other properties should a dye possess?
4 Why does a fibre-reactive dye work better on some fabrics than on others?
5 Iron(II) sulphate is used as a mordant. Suggest how it may work.
6 Why have fibre-reactive dyes proved so popular?

# 30.9 ▶ Chemistry helps medicine

FIRST THOUGHTS

Chemotherapy is the treatment of illness by chemical means. Chemicals can cure diseases and relieve pain. Advances in chemotherapy have made life safer, longer and more free from pain than it was a century ago. Some of the chemicals which are used in medicine are mentioned in this section.

## Painkillers

**Aspirin** is the most popular pain-killer. It was first sold in 1899. Each person in the UK swallows an average of 200 aspirins a year. Many other medicines, e.g. APC, contain aspirin. When you swallow aspirin, there is slight bleeding of the stomach wall. To reduce irritation of the stomach wall, you should always swallow plenty of water with aspirin. **Codeine** is a stronger pain-killer, used in headache tablets and in cough medicines.

**Morphine** is a substance with an amazing ability to relieve intense pain. The problem with morphine is that it is **addictive**. It is only used in cases of dire necessity, for example when soldiers are wounded in battle. Morphine can only be obtained by doctors. **Heroin** is a pain-reliever which is even more potent and more addictive than morphine. Heroin is not used in medicine because it is so addictive. Drug-dealers like to get their customers to try heroin because they will soon become 'hooked' and come back to buy again and again.

It's a fact!

Many of the victims of AIDS are drug addicts who used infected needles to inject themselves with heroin. Never be tempted to give drugs a try. The dangers of addiction and AIDS are very real. It's not worth the risk to satisfy your curiosity.

## Tranquillisers and sedatives

One person in twenty takes **tranquillisers** every day. Tranquillisers are substances which relieve tension and anxiety. With so many people taking tranquillisers over long periods of time, some doctors are worried about long-term effects. Research workers are now investigating whether there is any danger in taking tranquillisers for a long time.

Many people take **sedatives** (sleeping tablets). Drugs called **barbiturates** are used for this purpose. They are habit-forming. People who rely on barbiturates sometimes kill themselves accidentally by taking an overdose. These drugs are also used in the treatment of high blood pressure and mental illness.

## Stimulants

**Adrenalin** is a substance which the body produces when it needs to prepare for action: for 'fight or flight'. It is a **stimulant**: it makes the heart beat faster and makes a person ready for strenuous action.

**Amphetamines** (pep pills) have similar effects on the body. Some people take amphetamines to keep themselves awake; other people want to pep themselves up and make themselves more entertaining. People can become addicted to amphetamines. A person taking amphetamines becomes excitable and talkative, with trembling hands and enlarged pupils.

KEY SCIENTIST

Laughing gas was discovered by Sir Humphry Davy. At first it was used only as a curiosity at parties. One of these parties was attended by a dentist, Horace Morton. He decided to try laughing gas as an anaesthetic on his patients. Morton obtained better results by using ether. The use of chloroform spread rapidly after Queen Victoria took it during the birth of one of her children in 1853.

## Anaesthetics

Anaesthetics made modern surgery possible. Before the days of anaesthetics, a surgeon asked his patient to drink some brandy or smoke some opium to deaden the pain. Then the surgeon did the operation as quickly as he could. Sometimes patients died from pain and shock. After anaesthetics came into use, surgeons were able to take time to do the best operation they could, rather than the fastest. They were able to explore new techniques.

The first anaesthetics to be used were chloroform (1846), ether (1847) and dinitrogen oxide (laughing gas). There were some drawbacks to using these gases. In large doses, chloroform is harmful. Ether is very

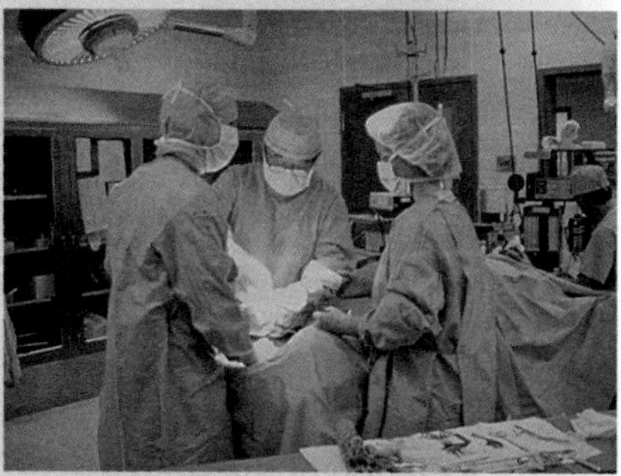

*Figure 30.9A* ▲ A modern operating theatre

flammable, and it has been the cause of fires in hospitals in the past. Dinitrogen oxide (laughing gas) does not produce a very deep anaesthesia; however it is suitable for use in dentistry. Research workers have found a better anaesthetic, **fluothane**, which was first used in 1956. It has been so successful and free from side-effects that most operations are now done under fluothane.

## Antiseptics

Another advance in surgery came with antiseptics. A century ago, many patients survived operations but died later in the wards. A surgeon called James Lister realised that the patients' wounds were becoming infected. He sprayed the operating theatre and the wards with a mist of phenol and water. His experiment produced immediate results: the death rate fell. Phenol had killed micro-organisms that would otherwise have infected surgical wounds. Phenol is an **antiseptic**. It is unpleasant to use. Solid phenol will burn the skin, and its vapour is toxic. Research chemists made other compounds which work as well as phenol and are safer to use. TCP® and Dettol® are antiseptics which contain trichlorophenol.

## Antibiotics

Infectious diseases, such as tuberculosis and pneumonia, used to kill thousands of people every year. The grim picture changed in 1935 with the discovery of the **sulphonamides**. They are antibiotics: substances which fight disease carried by bacteria. Infectious diseases are no longer a serious threat. In surgery too, powerful antibiotics have made having an operation safer than it was at the beginning of this century.

Alexander Fleming was a research bacteriologist in a London hospital. When the First World War began, he joined the Medical Corps. He was distressed by what he saw in his field hospital. Soldiers who did not seem to be mortally wounded when they arrived in the hospital died later from infections. Bacteria in mud and dirty clothing had infected their wounds, and gangrene set in. Watching men die a slow, painful death, Fleming wished that his work as a bacteriologist would enable him to discover a substance that would kill bacteria: a **bactericide**. After the war, Fleming went back to his research. One day, he found a mould called *Penicillium* on a dish of bacteria which he was culturing. To his amazement, he saw that the mould had killed bacteria. From the mould, he prepared an extract which he called **penicillin**. Would this be the powerful bactericide he had been hoping for? Sadly, although penicillin worked in the laboratory, it did not work in patients. Substances in the blood made the bactericide inactive.

In 1940, work on penicillin recommenced. The Second World War had started, and a powerful bactericide was needed urgently. Two chemists, Howard Florey and Ernst Chain, succeeded in making a stable extract of penicillin. They tested it, first on mice and then on human patients. The tests were successful, and the USA built a plant for the mass-production of penicillin. In 1942, penicillin was used in hospitals on the battlefield. The results were spectacular. No longer did soldiers die from minor wounds. In 1944, Fleming was knighted. Later, Fleming, Florey and Chain shared the Nobel Prize for medicine.

Penicillin has been widely used to treat a variety of infections. One disadvantage is that penicillin cannot be taken by mouth. It is broken down by acids, in this case the hydrochloric acid in the stomach. A more recent antibiotic, **tetracycline**, does not share this drawback. Tetracycline is a 'broad spectrum' antibiotic which can be used against many kinds of bacteria.

### KEY SCIENTIST

The power of sulphonamide drugs was discovered by Gerhardt Domagh. The first human being on whom he tested the drug was his own daughter. She was dying of 'child-bed fever', a bacterial infection which can attack women who have just had a baby. Happily, the drug worked. Domagh received the Nobel prize in 1939.

### KEY SCIENTIST

The old treatment for syphilis was to apply a paste of heavy metal salts (salts of arsenic, lead, mercury and antimony) to the skin. The metal salts killed the micro-organism that caused the disease, but many patients experienced metal poisoning. Paul Ehrlich reasoned that, if he could attach one of these metals to a dye which was able to stain the micro-organism, the metal–dye compound might target the micro-organism 'like a magic bullet' without attacking body tissues. Ehrlich and his team of research workers set to work. At the 606th attempt they were successful. They called their new compound salvarsan. Ehrlich received the Nobel prize in 1908.

## CHECKPOINT

1. Your Uncle Bert is always swallowing aspirins. Explain to him (a) why he should not take too many aspirins and (b) why he should take water with aspirins.
2. (a) A singer you know feels so tired that she is thinking of taking pep pills before giving a performance. Explain to her why this is not a good idea in the long run.
   (b) What is the body's natural stimulant? How does it work?
3. Briefly explain how surgery has changed as a result of (a) the discovery of anaesthetics and (b) the discovery of antiseptics.
4. Morphine is a very powerful pain reliever. Why do doctors prescribe it for so few patients? For what types of patient is morphine prescribed?
5. Why do doctors never prescribe heroin as a pain-killer?
6. Someone you know takes barbiturates to help her to get to sleep. Suggest to her what else she could do, instead of taking pills, to get to sleep.
7. (a) Before the time of James Lister, surgeons did not change their operating gowns between patients. What was wrong with this practice?
   (b) What did James Lister do to improve surgery?
8. (a) What did Sir Alexander Fleming do to merit a Nobel prize for medicine?
   (b) He had a piece of luck in his research, but his success was not due to luck. What else went into his discovery?

## 30.10 The chemical industry

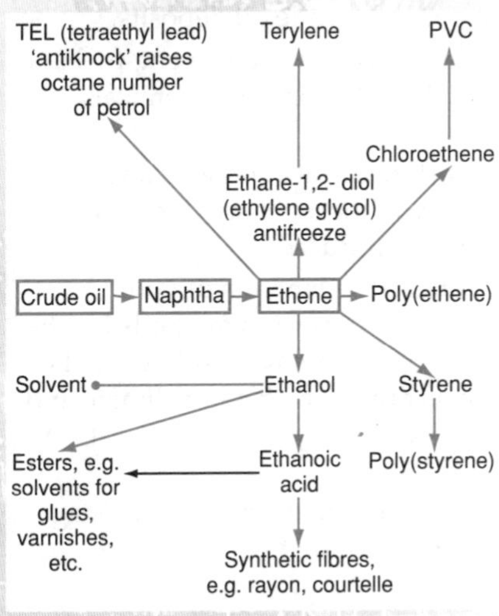


*Figure 30.10A* ▲ Some petrochemicals

The chemical industry can be divided roughly into ten sections.

1. **Heavy chemical industry:** oils, fuels, etc. (see Topic 24).
2. **Agriculture:** fertilisers, pesticides, etc. (see Topic 23).
3. **Plastics:** poly(ethene), poly(styrene), PVC, etc. (see Topic 29.2).
4. **Dyes** (see Topic 30.8).
5. **Fibres:** nylon, rayon, courtelle, etc. (see Topic 29.2).
6. **Paints, varnishes,** etc.
7. **Pharmaceuticals:** medicines, drugs, cosmetics, etc. (see Topic 30.9).
8. **Metals:** iron, aluminium, alloys, etc. (see Topic 20).
9. **Explosives:** dynamite, TNT, etc. (see Topic 23.3).
10. **Chemicals from salt:** sodium hydroxide, chlorine, hydrogen, hydrochloric acid, etc. (see Topic 7.6 and Topic 10.1).

Figure 30.10A shows some of the petrochemicals which are obtained by the route

Petroleum oil → Naphtha → Ethene → Petrochemical

# Theme Questions

1 (a) Describe the manufacture of aqueous ethanol from a carbohydrate. Name the starting material, state the conditions required and give a chemical equation for the reaction.

(b) Name a process which can be used to concentrate the ethanol formed. Explain the principle on which the process depends.

(c) Suggest one disadvantage of drinking ethanol.

(d) What is 'methylated spirit'? Why must it never be drunk?

2 (a) Ethene is an unsaturated hydrocarbon. Describe how you could test it for unsaturation.

(b) Ethene polymerises to form poly(ethene). Write an equation for this reaction.

(c) Give two examples of articles commonly made from poly(ethene). In each case name an alternative material and suggest why poly(ethene) has been chosen in preference to the other material.

3 (a) State the conditions under which ethene is formed from ethanol.

(b) State the conditions under which ethanol is manufactured from ethene.

(c) Give a brief description of the fermentation process for the manufacture of ethanol.

(d) Compare the ethene process of (b) with the fermentation process of (c) with regard to

(i) the purity of the product

(ii) whether the process can be run continuously

(iii) the availability of the starting materials

(iv) the rate of formation of ethanol.

4 Tennis racquets were traditionally made of wood. Starting in the 1970s, aluminium alloys have been used for tennis racquets. Many popular tennis racquets are now made from composite materials, consisting of plastic reinforced with graphite fibres and mounted on a metal core. Suggest how such a material compares with (a) wood and (b) aluminium alloy with respect to (i) weight of racquet, (ii) strength of racquet and (iii) retaining its shape in all weathers.

5 The structures of plastics A and B are shown below.

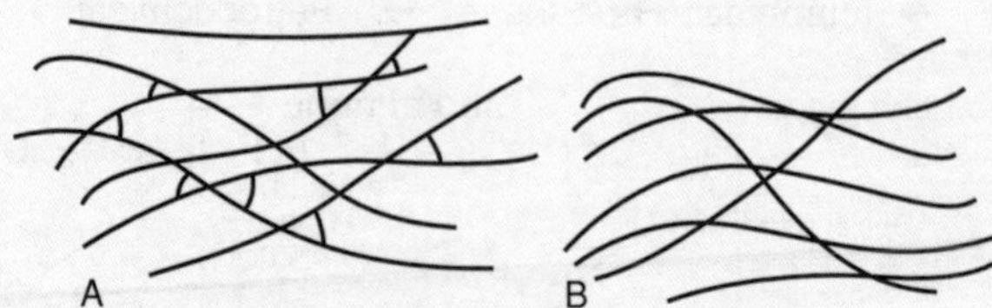


(a) Say what will happen when each of the plastics is heated strongly. Explain the reason for the difference in behaviour.

(b) Suggest a use for which plastic A would be more suitable than plastic B.

(c) Suggest a use for which plastic B would be more suitable than plastic A.

6 (a) Explain what is meant by

(i) a food additive

(ii) a food preservative

(iii) an emulsifier

(iv) an anti-oxidant

(v) a permitted colouring

(vi) an E number

(b) Suggest reasons why the number of food additives in use has grown rapidly during the last 20 years.

7 (a) When current reserves of oil have been used up, it will be possible to use less conventional sources such as the Athabascan Tar Sands in America. These consist of sand which is saturated with a very thick, tarry liquid composed of the heavier fractions of crude oil.

(i) Explain how these tar sands could be treated to recover the tar and obtain a full range of hydrocarbons, including petrol, from it.

(ii) Suggest **three** reasons why it is more expensive to obtain a full range of hydrocarbons from these tar sands than from conventional crude oil deposits.

(b) Ethanol is used extensively as a solvent, fuel and general feed-stock for the organic chemicals industry.

(i) Describe, with relevant equations and conditions, **two** different ways by which ethanol can be produced.

(ii) Suggest, with an explanation of your answer, which of these two methods for the manufacture of ethanol is most likely to be adopted in a country with a hot climate and few natural resources. (ULEAC)

8 A student found the following recipes for making elderberry wine in two different wine-making books.

| *Recipe A* | *Recipe B* |
|---|---|
| 1.5 kg elderberries | 2 kg elderberries |
| 0.75 kg sugar | 1 kg sugar |
| 20 g wine yeast | 25 g wine yeast |
| 10 g ginger | 5 $dm^3$ water |
| 5 $dm^3$ water | |

The student decided to make trial quantities of each, using 1 $dm^3$ of water, in the apparatus shown in the diagram on the next page.

(a) What is the purpose of the air lock?

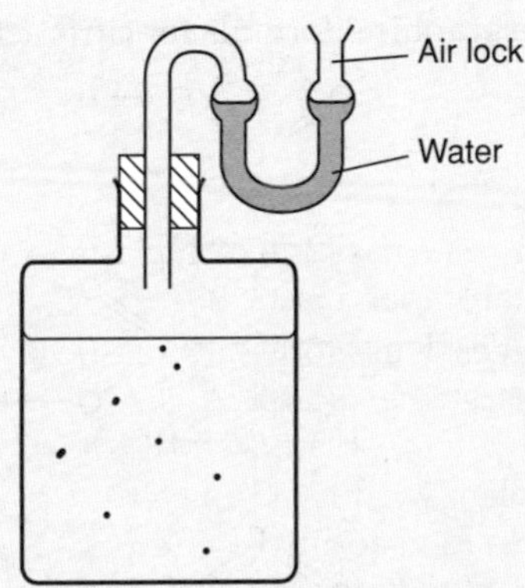


(b) Sucrose (sugar) has the formula $C_{12}H_{22}O_{11}$. The conversion of an aqueous solution of sucrose to glucose, $C_6H_{12}O_6$, occurs in the presence of an enzyme present in yeast.
   (i) Suggest a balanced equation to show the conversion of sucrose to glucose in the presence of water.
   (ii) What is an enzyme?

(c) When the student set up the experiments, the elderberries, ginger and sugar were added to the water and the mixture boiled for five minutes. The mixtures were then filtered into a flask and the temperature of the liquid taken. The yeast was not added until the liquid had cooled to room temperature.
Explain why the yeast was not added when the mixture was at a higher temperature.

(d) After two days, the reaction had started in both flasks, but it was going much more quickly in one of the flasks.
   (i) Suggest how the rate at which the reaction was taking place could be measured.
   (ii) Suggest, with a reason, which of the two recipes was fermenting faster.

(e) Suggest one other method by which the rate of fermentation of both experiments could be increased.

(f) Which process, other than wine-making or brewing, also uses the fermentation of sugar in the presence of yeast? (ULEAC)

**9** Some brands of face cleaning pads contain ethanol and salicylic acid.

(a) The formula for salicylic acid is:

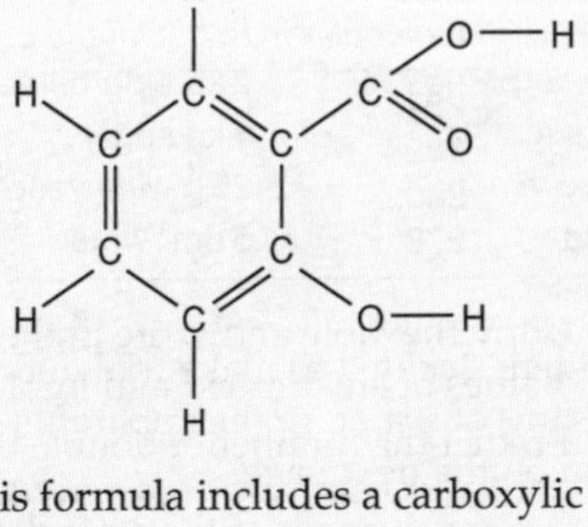

   (i) This formula includes a carboxylic acid group.
Copy the formula and put a ring around the carboxylic acid group.
   (ii) Copy and complete the formula below to show the substance formed when salicylic acid reacts with sodium hydroxide.

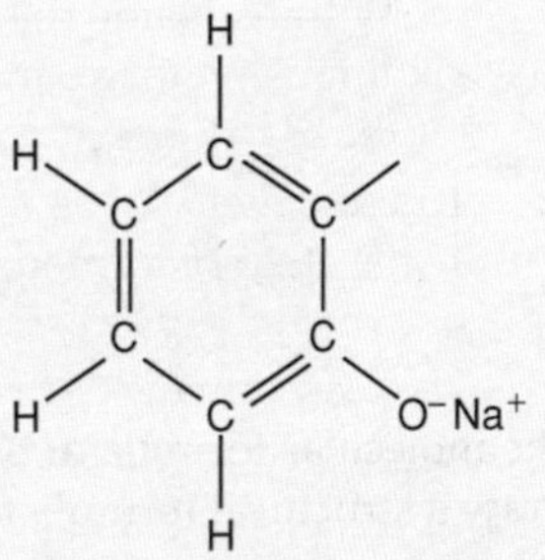

(b) When ethanol, $CH_3CH_2OH$, is heated with acidified potassium dichromate, ethanoic acid, $CH_3CO_2H$, is formed.
   (i) Name the type of reaction which occurs when ethanoic acid is formed from ethanol.
   (ii) Draw the graphical formula for ethanol.
   (iii) Name the group of atoms which gives ethanol the characteristic chemical properties of an alcohol.

(c) Ethanol reacts with ethanoic acid to form ethyl ethanoate and water.
22 g of ethyl ethanoate was analysed and was found to contain 12 g of carbon, 2 g of hydrogen and 8 g of oxygen.
   (i) Calculate the empirical formula of ethyl ethanoate.
Show all the stages in your calculations.
   (ii) The molar mass of ethyl ethanoate is 88 g/mol. Show how your calculations in part (c)(i) lead to the molecular formula for ethyl ethanoate. (MEG)

**10** Starch and protein are natural polymers.

(a) Starch can be *hydrolysed* to form sugars. Hydrolysis of starch in water is very slow.
   (i) Explain what happens when starch is hydrolysed. You will be awarded more marks if you write your ideas clearly.
   (ii) How could you increase the rate of the hydrolysis of starch? Give **two** ways.

(b) Proteins are formed from amino acids joining together with the elimination of water. Two amino acids are

glycine, and alanine

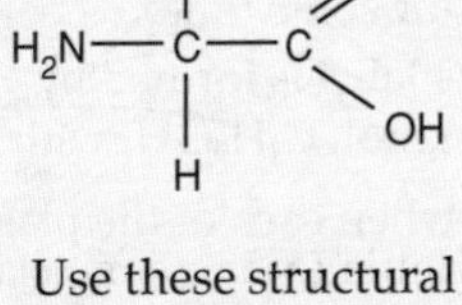

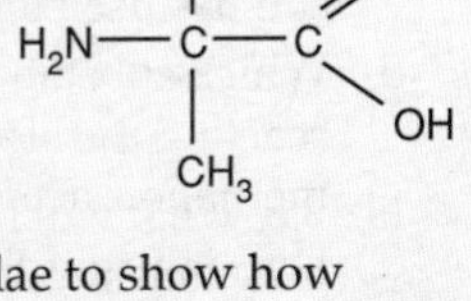

Use these structural formulae to show how proteins are formed. (SEG)

**11** The flow chart outlines some useful products which can be obtained from crude oil. Crude oil is a mixture of hydrocarbons.

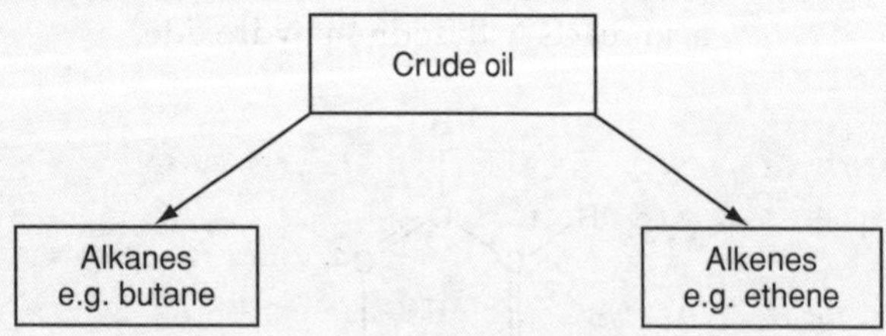

(a) (i) The molecular formula of butane is $C_4H_{10}$. Draw a structural formula for butane.

(ii) The molecular formula of ethene is $C_2H_4$. Draw a structural formula for ethene.

(b) Ethene and butane react with bromine:

$$C_2H_4 + Br_2 \rightarrow C_2H_4Br_2$$

$$C_4H_{10} + Br_2 \rightarrow C_4H_9Br + HBr$$

(i) State the conditions needed for each reaction.

(ii) Name the type of reaction when bromine reacts with: (1) ethene and (2) butane.

(iii) Draw the full structural formula of each organic product, showing all bonds.

(c) (i) Write a balanced chemical equation for the combustion of butane in oxygen.

(ii) Use your equation to calculate the volume of oxygen needed to burn 1 $dm^3$ of butane at room conditions. (1 mole of gas occupies 24 $dm^3$ at room conditions). (SEG)

**12** A fizzy raspberry-flavoured drink has the following contents panel on its label:

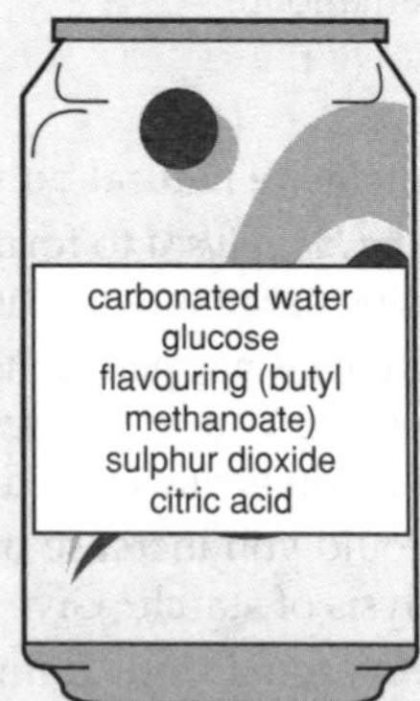

(a) Carbonated water contains carbon dioxide gas dissolved in water under pressure. Explain why the raspberry drink becomes less fizzy after the can has been left open for some time.

(b) Which substance on the label is formed by reacting the alcohol butanol, $C_4H_9OH$, with methanoic acid, $HCO_2H$?

(c) Draw the graphical formula for methanoic acid.

(d) Some drinks contain citric acid.

The graphical formula for citric acid is:

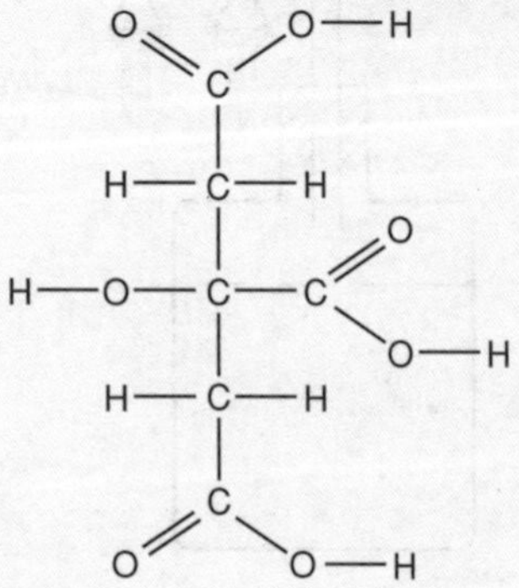

Copy the formula and put a ring around a group of atoms which cause citric acid to be acidic.

(e) A student was given 20 $cm^3$ of a solution of citric acid.
The citric acid was completely neutralized by 0.003 moles of sodium hydroxide.

(i) How many moles of citric acid react with the 0.003 moles of sodium hydroxide?

(ii) Calculate the concentration of the solution of citric acid.

amount of substance in solution (mol) = volume of solution (litres) × concentration (mol/litre) (MEG)

**13** Fats or oils may break down to produce fatty acids and glycerol. The equation below represents the break down of glyceryl tristearate, a fat.

| | | | |
|---|---|---|---|
| $C_{17}H_{35}COOCH_2$ | | | $CH_2OH$ |
| $C_{17}H_{35}COOCH$ + 3**X** | → | $3C_{17}H_{35}COOH$ + | $CHOH$ |
| $C_{17}H_{35}COOCH_2$ | | | $CH_2OH$ |
| glyceryl tristearate | | stearic acid | glycerol |

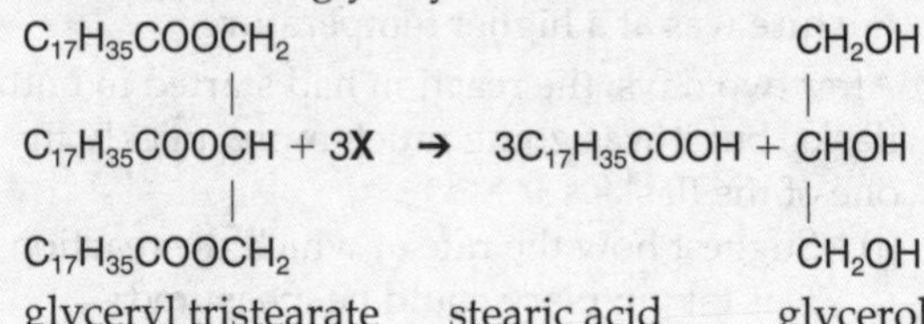

(a) (i) Name the compound, **X**.

(ii) What is the main difference between fats and oils?

(b) The hydrocarbon chain in the stearic acid contains no double bonds. What word is used to describe hydrocarbon chains with no double bonds?

(c) The number of double bonds in fatty acids is expressed in a quantity known as the Iodine Value. This is defined as the number of grammes of iodine, $I_2$, which will react with 100 g of the fatty acid. The table gives information about a number of fatty acids.

| *Fatty acid* | *Relative molecular mass* | *Number of double bonds* | *Iodine Value* |
|---|---|---|---|
| stearic acid | 284 | 0 | |
| oleic acid | 282 | 1 | 90 |
| linoleic acid | 280 | 2 | |
| linolenic acid | 278 | | 270 |

(i) Copy the table and write in the Iodine Values of stearic acid and linoleic acid.

(ii) Predict the number of double bonds in linolenic acid. Write your prediction in your copy of the table.

(iii) 328 g of erucic acid react with 254 g of iodine. Calculate the iodine value of erucic acid.

(d) Soaps can be made in the laboratory from fats such as glyceryl tristearate. Describe how this is done and how the soap can be separated from the other products. (MEG)

**14** Beer is a complex mixture of substances.

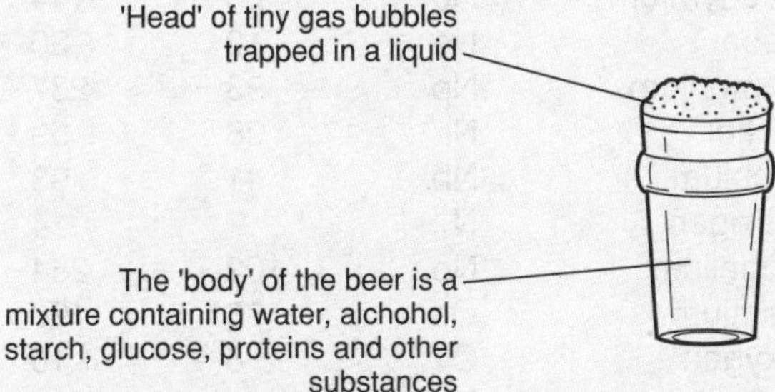


(a) During the brewing process, the 'body' of the beer often becomes cloudy due to the presence of starch particles.
Enzymes are used to clarify the beer. Explain how these enzymes make the beer less cloudy.

(b) The 'head' on the beer is a foam.
What is the continuous phase in the 'head' on the beer?

(c) The diagram shows an enzyme molecule and its substrate.

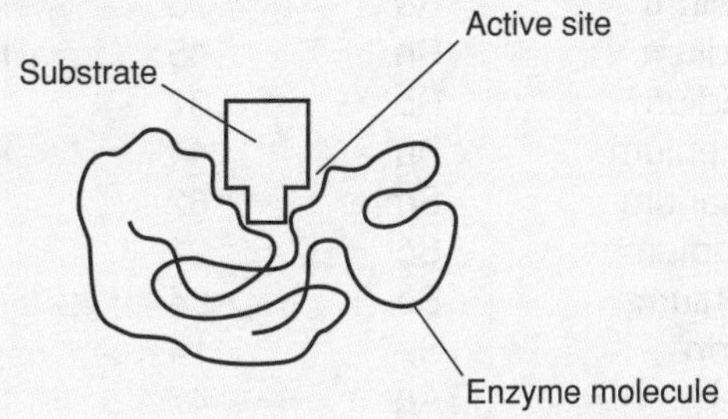


Use the diagram to help explain the following statements.

(i) Enzymes act as catalysts.

(ii) Enzymes are inactivated at temperatures above 50 °C.

(d) Proteins are present in the liquid which surrounds the gas bubbles in the 'head' of the beer. The protein chains are unfolded.
Suggest how protein molecules stabilise the 'head' of the beer. (MEG)

**15** The following list shows some substances added to fruit to make jam.

sugar
pectin (gelling agent)
citric acid
sodium citrate (acidity regulator)
water

(a) Which substance in the list thickens the jam?

(b) (i) Which substance in the list is a preservative?

(ii) State the job of a preservative.

(c) Part of a pectin molecule is shown below.

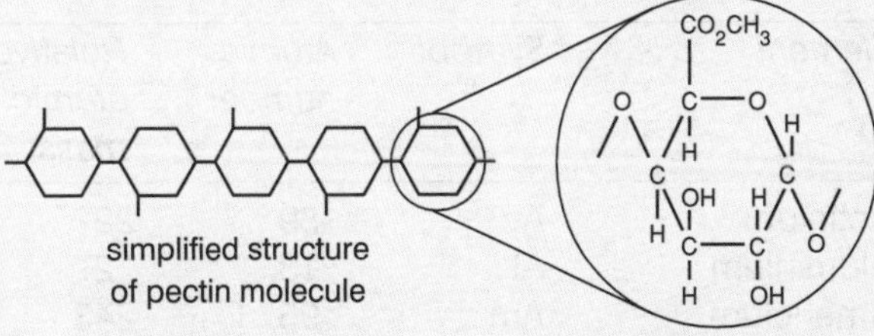


(i) What type of compound is pectin?
Choose the best answer from the following list:
**amino acid carbohydrate mineral salt protein**

(ii) Which word describes the structure of pectin?
Choose the correct answer from the list:
**atom ion monomer polymer**

(d) Pectin links together the cells in plants.
When vegetables are cooked in boiling water, pectin becomes more soluble.

(i) Suggest what effect cooking has on the texture of vegetables.

(ii) Give a reason for your answer.

(e) Warm jam containing pectin is a sol. When the jam is cooled, it forms a gel.
Copy the following sentences and use the words below to fill the spaces. You may use each word once, more than once or not at all.
**citric acid emulsion gel pectin sol**
The _______ contains _______ dispersed through water.
The _______ has water droplets trapped in a network of pectin. (MEG)

**16** This question is about alcohols.

(a) The alcohols are members of a homologous series.

(i) Write down the general formula for this homologous series.

(ii) Write down the names and draw the structures of the two monohydric alcohols containing **three** carbon atoms per molecule.

(b) On analysis, a 10.00 g sample of an alcohol was found to contain 3.75 g of carbon, 1.25 g of hydrogen, and 5.00 g of oxygen.
Use this data to find the empirical formula of this alcohol.
Suggest its molecular formula.

(c) The table shows some bond energies, E.

| *bond* | *E/kJ per mole* |
|---|---|
| C—H | 413 |
| C—C | 347 |
| O—H | 464 |

The dissociation energy of ethanol is shown below.

$C_2H_5OH(g) \rightarrow 2C(g) + 6H(g) + O(g)$ $\Delta H = 3234$ kJ per mole of ethanol

Calculate the bond energy of C – O per mole. (MEG)

## Table of symbols

**Table of symbols, atomic numbers and relative atomic masses**

| Element | Symbol | Atomic number | Relative atomic mass |
|---|---|---|---|
| Actinium | Ac | 89 | 227 |
| Aluminium | Al | 13 | 27 |
| Americium | Am | 95 | 243 |
| Antimony | Sb | 51 | 122 |
| Argon | Ar | 18 | 40 |
| Arsenic | As | 38 | 75 |
| Astatine | At | 85 | 210 |
| Barium | Ba | 56 | 137 |
| Berkelium | Bk | 97 | 247 |
| Beryllium | Be | 4 | 9 |
| Bismuth | Bi | 83 | 209 |
| Boron | B | 5 | 11 |
| Bromine | Br | 35 | 80 |
| Cadmium | Cd | 48 | 112 |
| Caesium | Cs | 55 | 133 |
| Calcium | Ca | 20 | 40 |
| Californium | Cf | 98 | 251 |
| Carbon | C | 6 | 12 |
| Cerium | Ce | 58 | 140 |
| Chlorine | Cl | 17 | 35.5 |
| Chromium | Cr | 24 | 52 |
| Cobalt | Co | 27 | 59 |
| Copper | Cu | 29 | 63.5 |
| Curium | Cm | 96 | 247 |
| Dysprosium | Dy | 66 | 162.5 |
| Einsteinium | Es | 99 | 254 |
| Erbium | Er | 68 | 167 |
| Europium | Eu | 63 | 152 |
| Fermium | Fm | 100 | 253 |
| Fluorine | F | 9 | 19 |
| Francium | Fr | 87 | 223 |
| Gadolinium | Gd | 64 | 157 |
| Gallium | Ga | 31 | 70 |
| Germanium | Ge | 32 | 72.5 |
| Gold | Au | 79 | 197 |
| Hafnium | Hf | 72 | 178 |
| Helium | He | 2 | 4 |
| Holmium | Ho | 67 | 164 |
| Hydrogen | H | 1 | 1 |
| Indium | In | 49 | 115 |
| Iodine | I | 53 | 127 |
| Iridium | Ir | 77 | 192 |
| Iron | Fe | 26 | 56 |
| Krypton | Kr | 36 | 84 |
| Lanthanum | La | 57 | 139 |
| Lawrencium | Lw | 103 | 257 |
| Lead | Pb | 82 | 207 |
| Lithium | Li | 3 | 7 |
| Lutecium | Lu | 71 | 175 |
| Magnesium | Mg | 12 | 24 |
| Manganese | Mn | 25 | 55 |
| Mendelevium | Md | 101 | 256 |
| Mercury | Hg | 80 | 201 |
| Molybdenum | Mo | 42 | 96 |
| Neodymium | Nd | 60 | 144 |
| Neon | Ne | 10 | 20 |
| Neptunium | Np | 93 | 237 |
| Nickel | Ni | 28 | 59 |
| Niobium | Nb | 41 | 93 |
| Nitrogen | N | 7 | 14 |
| Nobelium | No | 102 | 254 |
| Osmium | Os | 76 | 190 |
| Oxygen | O | 8 | 16 |
| Palladium | Pd | 46 | 106 |
| Phosphorus | P | 15 | 31 |
| Platinum | Pt | 78 | 195 |
| Plutonium | Pu | 94 | 242 |
| Polonium | Po | 84 | 210 |
| Potassium | K | 19 | 39 |
| Praesodymium | Pr | 59 | 141 |
| Promethium | Pm | 61 | 147 |
| Protactinium | Pa | 91 | 231 |
| Radium | Ra | 88 | 226 |
| Radon | Rn | 86 | 222 |
| Rhenium | Re | 75 | 186 |
| Rhodium | Rh | 45 | 103 |
| Rubidium | Rb | 37 | 85 |
| Ruthenium | Ru | 44 | 101 |
| Samarium | Sm | 62 | 150 |
| Scandium | Sc | 21 | 45 |
| Selenium | Se | 34 | 79 |
| Silicon | Si | 14 | 28 |
| Silver | Ag | 47 | 108 |
| Sodium | Na | 11 | 23 |
| Strontium | Sr | 38 | 87 |
| Sulphur | S | 16 | 32 |
| Tantalum | Ta | 73 | 181 |
| Technetium | Tc | 43 | 99 |
| Tellurium | Te | 52 | 127 |
| Terbium | Tb | 65 | 159 |
| Thallium | Tl | 81 | 204 |
| Thorium | Th | 90 | 232 |
| Thulium | Tm | 69 | 169 |
| Tin | Sn | 50 | 119 |
| Titanium | Ti | 22 | 48 |
| Tungsten | W | 74 | 184 |
| Uranium | U | 92 | 238 |
| Vanadium | V | 23 | 51 |
| Xenon | Xe | 54 | 131 |
| Ytterbium | Yb | 70 | 173 |
| Yttrium | Y | 39 | 89 |
| Zinc | Zn | 30 | 65 |
| Zirconium | Zr | 40 | 91 |

# The Periodic Table of the Elements

| Group | | | | | | | | | | | | | | | | | |
|---|---|---|---|---|---|---|---|---|---|---|---|---|---|---|---|---|---|
| 1 | 2 | | | | | | | | | | | 3 | 4 | 5 | 6 | 7 | 0 |
| | | | | | 1 **H** Hydrogen 1 | | | | | | | | | | | | 4 **He** Helium 2 |
| 7 **Li** Lithium 3 | 9 **Be** Beryllium 4 | | | | | | | | | | | 11 **B** Boron 5 | 12 **C** Carbon 6 | 14 **N** Nitrogen 7 | 16 **O** Oxygen 8 | 19 **F** Fluorine 9 | 20 **Ne** Neon 10 |
| 23 **Na** Sodium 11 | 24 **Mg** Magnesium 12 | | | | | | | | | | | 27 **Al** Aluminium 13 | 28 **Si** Silicon 14 | 31 **P** Phosphorus 15 | 32 **S** Sulphur 16 | 35·5 **Cl** Chlorine 17 | 40 **Ar** Argon 18 |
| 39 **K** Potassium 19 | 40 **Ca** Calcium 20 | 45 **Sc** Scandium 21 | 48 **Ti** Titanium 22 | 51 **V** Vanadium 23 | 52 **Cr** Chromium 24 | 55 **Mn** Manganese 25 | 56 **Fe** Iron 26 | 59 **Co** Cobalt 27 | 59 **Ni** Nickel 28 | 64 **Cu** Copper 29 | 65 **Zn** Zinc 30 | 70 **Ga** Callium 31 | 73 **Ge** Germanium 32 | 75 **As** Arsenic 33 | 79 **Se** Selenium 34 | 80 **Br** Bromine 35 | 84 **Kr** Krypton 36 |
| 85 **Rb** Rubidium 37 | 88 **Sr** Strontium 38 | 89 **Y** Yttrium 39 | 91 **Zr** Zirconium 40 | 93 **Nb** Niobium 41 | 96 **Mo** Molybdenum 42 | 99 **Tc** Technetium 43 | 101 **Ru** Ruthenium 44 | 103 **Rh** Rhodium 45 | 106 **Pd** Palladium 46 | 108 **Ag** Silver 47 | 112 **Cd** Cadmium 48 | 115 **In** Indium 49 | 119 **Sn** Tin 50 | 122 **Sb** Antimony 51 | 128 **Te** Tellurium 52 | 127 **I** Iodine 53 | 131 **Xe** Xenon 54 |
| 133 **Cs** Caesium 55 | 137 **Ba** Barium 56 | 139 **La** Lanthanum 57 * | 178 **Hf** Hafnium 72 | 181 **Ta** Tantalum 73 | 184 **W** Tungsten 74 | 186 **Re** Rhenium 75 | 190 **Os** Osmium 76 | 192 **Ir** Iridium 77 | 195 **Pt** Platinum 78 | 197 **Au** Gold 79 | 201 **Hg** Mercuy 80 | 204 **Tl** Thallium 81 | 207 **Pb** Lead 82 | 209 **Bi** Bismuth 83 | **Po** Polonium 84 | **At** Astatine 85 | **Rn** Radon 86 |
| **Fr** Francium 87 | 226 **Ra** Radium 88 | 227 **Ac** Actinium 89 † | | | | | | | | | | | | | | | |

*58–71 Lanthanum series
†90–103 Actinium series

| 140 **Ce** Cerium 58 | 141 **Pr** Praseodymium 59 | 144 **Nd** Neodymium 60 | 147 **Pm** Promethium 61 | 150 **Sm** Samarium 62 | 152 **Eu** Europium 63 | 157 **Gd** Gadolinium 64 | 159 **Tb** Terbium 65 | 162 **Dy** Dysprosium 66 | 165 **Ho** Holmium 67 | 167 **Er** Erbium 68 | 169 **Tm** Thulium 69 | 173 **Yb** Ytterbium 70 | 175 **Lu** Lutetium 71 |
|---|---|---|---|---|---|---|---|---|---|---|---|---|---|
| 232 **Th** Thorium 90 | **Pa** Protactinium 91 | 238 **U** Uranium 92 | **Np** Neptunium 93 | **Pu** Plutonium 94 | **Am** Americium 95 | **Cm** Curium 96 | **Bk** Berkelium 97 | **Cf** Californium 98 | **Es** Einsteinium 99 | **Fm** Fermium 100 | **Md** Mendelevium 101 | **No** Nobelium 102 | **Lr** Lawrencium 103 |

Key: a **X** b

a = relative atomic mass
**X** = atomic symbol
b = atomic number

# Numerical answers

**CHECKPOINT 1.3**
**1** 2.7 g/cm$^3$
**2** 7500 g (7.5 kg)
**3** 2720 g (2.72 kg)
**4** A 0.86 g/cm$^3$: floats B 4.3 g/cm$^3$: sinks

**CHECKPOINT 1.6**
**8** (a) (i) 55 g (ii) 100 g (b) 45 g crystallise (c) 200 g (d) 300 g (e) (i) 3 g (ii) 25 g

**CHECKPOINT 2.4**
**3** (a) E (b) C (c) D (d) A, B

**THEME A QUESTIONS**
**6** (b) 48 g/100 g (c) 50 C (d) 55 g $KNO_3$ crystallise

**CHECKPOINT 6.2**
**2** 9p, 9e, 10n
**3** (a) 17, 35 (b) 27, 59 (c) 50, 119

**CHECKPOINT 6.4**
**3** N 7, 7, 7; Na 11, 11, 12; K 19, 20, 19; U 92, 143, 92

**THEME B QUESTIONS**
**12** 11p, 12n
**13** 82p, 124n, No

**CHECKPOINT 10.3**
**3** (a) 4 (b) 9 (c) 3 (d) 11 (e) 18
**4** (a) C (b) A (c) B

**THEME C QUESTIONS**
**2** (b) 45%
**8** (a) blue (b) orange (c) orange (d) 4–6 (e) 7.5 (f) purple (g) orangè

**CHECKPOINT 15.7**
**1** 20 g KCl crystallises
**2** 20 g NaCl crystallises
**3** 20 g KCl crystallises
**4** 10 g $K_2SO_4$ + 10 g KBr crystallise

**CHECKPOINT 16.8**
**6** (a) (i) 80–100 km/h (ii) 30 km/h (iii) 100 km/h (b) (i) 80–100 km/h (ii) 50–60 mph

**CHECKPOINT 18.2**
**2** 2 mm

**CHECKPOINT 18.4**
**1** (a) Philippine or Caribbean (b) Andes (c) Basaltic (d) Sliding
**2** (a) A = Continental plate, B = Oceanic trench, C = Subduction zone D = Sedimentary (c) (i) W to E (ii) E to W (iii) W to E (d) Mantle, sufficiently fluid to transport plates (e) Sea-floor spreading
**3** (a) 4 m (b) 1.5 m–1.9 m (c) a > 2b
**6** (a) (i) 44 million years (ii) 7 million years (iii) 1 million years

**CHECKPOINT 19.5**
**1** (a) D (b) C (c) F

**THEME D QUESTIONS**
**13** (a) $3.75 \times 10^5$ tonnes (b) 0.075
**19** 2400 tonne
**20** 1600 tonne

**CHECKPOINT 23.6**
**7** (a) 65 tonnes (b) 18 tonnes
**8** 2 months

**THEME E QUESTIONS**
**6** (a) (i) Alkaline manganese dry cell gives 400 flashes/£; zinc–carbon cell gives approximately 100 flashes/£. (ii) Zinc–carbon cell gives 35 hours/£; alkaline manganese cell gives 31 hours/£
**24** (d) £400

**CHECKPOINT 27.2**
**1** 28, 44, 64, 80, 40, 58.5, 56, 58, 106, 159.5, 249.5, 162

**CHECKPOINT 27.3**
**1** (a) 20% (b) 70% (c) 80% C, 20% H (d) 40% S, 60% O (e) 5.0% H, 95.0% F (f) 20% Mg, 27% S, 53% O
**2** (a) 36% (b) 67.5%

**CHECKPOINT 27.4**
**1** (a) 27 g (b) 96 g (c) 50 g (d) 14 g (e) 8 g (f) 64 g
**2** (a) 2.5 mol (b) 0.33 mol (c) 1.0 mol (d) 0.25 mol
**3** (a) 98 g (b) 23 g (c) 100 g (d) 10 g

**CHECKPOINT 27.5**
**1** 8.0 g
**2** (b) 3.2 g
**3** 10.3 g
**4** 107 g

**CHECKPOINT 27.6**
**1** (a) 8.0 g (b) 95%
**2** (a) 34 g (b) 10%
**3** 98%
**4** (b) 132 g (c) 91%
**5** (a) 25 g (b) 88%

**CHECKPOINT 27.7A**
**1** (a) NO (b) $Mg_3N_2$ (c) $Al_2O_3$
**2** (a) 67.5% (b) 40.5%
**3** (a) $SO_3$ (b) $SO_2$ (c) $CaBr_2$ (d) $KHCO_3$
**4** (a) BeO (b) $Fe_2O_3$ (c) $FeCl_3$
**5** $Ti_2O_3$ and $TiO_2$

**CHECKPOINT 27.7B**
**1** A $CH_4O$, B $C_3H_6$, C $C_4H_8O_2$, D $C_2H_6O_2$, E $C_2H_6$, F $C_2H_4Cl_2$, G $C_6H_6$, H $C_6H_3N_3O_6$
**2** (i) (a) $CH_3$ (b) $C_2H_6$ (ii) (a) $CH_2$ (ii) $C_2H_4$
**3** (a) $CH_2CI$ (b) $C_2H_4Cl_2$

**CHECKPOINT 27.8**
**1** 6 dm$^3$
**2** 500 cm$^3$ $O_2$, 250 cm$^3$ $CO_2$
**3** 21 million dm$^3$
**4** 3 dm$^3$
**5** 0.43 dm$^3$
**6** 25 g

**CHECKPOINT 27.9**
**1** (a) 15.9 g (b) 14.8 g (c) 52.0 g (d) 20.0 g (e) 29.8 g
**2** (a) 6.00 dm$^3$ $H_2$, 6.00 dm$^3$ $Cl_2$ (b) and (c) 6.00 dm$^3$ $H_2$, 3.00 dm$^3$ $O_2$
**3** 0.518 g
**4** 120 cm$^3$
**5** +2
**6** 64.8 g
**7** (a) 965 C (b) $5 \times 10^{-3}$ mol (c) 2 mol (d) $Ni^{2+}(aq) + 2e \rightarrow Ni(s)$
**8** (a) 12 cm$^3$ (b) 24 cm$^3$ (c) 0.216 g
**9** (b)
**10** +3

**CHECKPOINT 27.10**
**1** (a) 1.00 mol/l (b) 0.10 mol/l (c) 2.0 mol/l (d) 0.40 mol/l
**2** (a) 0.02 mol (b) 1.00 mol (c) 0.06 mol (d) 0.025 mol
**3** (a) 0.144 mol/l
**4** 2 mol/l
**5** (a) 0.2 M (b) 0.02 M (c) 0.2 M (d) 8 M
**6** (a) 0.25 mol (b) 0.010 mol (c) 0.05 mol (d) 0.0025 mol

**CHECKPOINT 27.11**
**1** 0.15 M
**2** (a) 0.25 M (b) 10.0 cm$^3$
**3** (a) 0.5 mol (b) 0.24 M
**4** (a) Paingo (b) speed of action, taste

**CHECKPOINT 28.6**
**1** –124 kJ/mol
**2** –484 kJ/mol
**3** –56 kJ/mol
**4** –95 kJ/mol

**THEME F QUESTIONS**
**2** (b) 0.14 mol/l
**3** (i) $1.25 \times 10^{-3}$ mol (ii) a,b,d,e
**4** (a) $Al^{3+} + 3e \rightarrow Al(I)$ (b) 162 kg
**5** 95.5%
**8** (b) 11 cm$^3$ (c) 90 cm$^3$ (d) 35 cm$^3$
**10** (a) (i) 2253 kJ (ii) 2346 kJ (iii)Energy given out in reaction > Energy required to break bonds (b) (i) A £3265, B £3465 (ii) the higher yield of ammonia (iii) the higher cost of plant that would withstand the pressure (iv) use a catalyst
**11** (d) (ii) 3 (iii) 111 mol electrons (iv) 99.2 hours (v) 667 dm$^3$ (vi) 762 dm$^3$
**14** (a) (iv) 0.0877 mol dm$^3$ (b) (i) $1.123 \times 10^{-3}$ mol (ii) 1123 mol (iii) 10.0 mol dm$^{-3}$ (iv) 112.3 dm$^3$ (v) stir
**15** (c) (i) 17.5 g (ii) 24 g (iii) 96%
**16** (e) (ii) 0.50 mol dm$^{-3}$ (iii) $1.0 \times 10^{-2}$ mol (iv) $1.0 \times 10^{-2}$ mol (v) 1.0 mol dm$^{-3}$.
**17** (c) (i) 560 kg (ii) 660 kg

**CHECKPOINT 29.2**
**4** (a) 62.5 (b) 640

**THEME G QUESTIONS**
**9** (b) (i) $C_2H_4O$ (ii) $C_4H_8O_2$
**11** (c) (i) $2C_4H_{10}(g) + 13O_2(g) \rightarrow 8CO_2(g) + 10H_2O(g)$ (ii) 6.5 dm$^3$
**12** (e) (i) 0.001 mol (ii) 0.050 mol dm$^{-3}$
**13** (c) (i) stearic acid 0, linoleic acid 180 (ii) 3 (iii) 77
**16** (b) $CH_4O$ (c) 358 kJ/mol

# Index

Page numbers in bold indicate the main entry for that topic.

## A

## B



## C

## D

## E



## F

## I



## J



## K



## L

## M



## N

## O



## P

## Q



## R



## S

## T

surgery – but there is a deeper examination of death, and an angrier exposition of the shameful betrayal of the NHS by successive generations of politicians . . . Honesty is abundantly apparent here – a quality as rare and commendable in elite surgeons as one suspects it is in memoirists'

Gavin Francis, *Guardian*

'[Marsh] is wise and insightful about the balance and confidence, truth and uncertainty faced by doctors . . . His insights about life, death and professional purpose are irresistible'
Hannah Beckerman, *Sunday Express*

'I particularly relished his descriptions of the anatomy of the brain itself, as well as his can-do accounts of freeing cancerous masses from their baroque architecture – but I enjoyed (if this is the correct word) still more his willingness to delve as fearlessly into his own, troubled being . . . Perhaps most disarming of all is Marsh's frankness about his own fears of growing older and dying . . . should be distributed to every care home in Britain' Will Self, *New Statesman*

'Extraordinary . . . both exhilarating and alarming . . . harrowing but fascinating . . . It is a privilege to dance with [Marsh] through these engrossing, revealing pages'
Libby Purves, *Daily Mail*

'Wonderful . . . eloquent . . . a testament to the tenacity of the human spirit' Adrian Woolfson, *Financial Times*

# ADMISSIONS

*A Life in Brain Surgery*

Henry Marsh

*To William, Sarah, Katharine and Iris*

First published in Great Britain in 2017
This paperback edition first published in 2018
by Weidenfeld & Nicolson
an imprint of The Orion Publishing Group Ltd
Carmelite House, 50 Victoria Embankment
London EC4Y 0DZ

An Hachette UK Company

1 3 5 7 9 10 8 6 4 2



A CIP catalogue record for this book is available from the British Library.

ISBN 978 1 4746 0387 4

Typeset by Input Data Services Ltd, Somerset

Printed and bound by CPI Group (UK) Ltd, Croydon, CR0 4YY

www.orionbooks.co.uk

*'Neither the sun nor death can be looked at steadily'*

La Rochefoucauld

*'We should always, as near as we can, be booted and spurred, and ready to go . . .'*

Michel de Montaigne

*'Medicine is a science of uncertainty, and an art of probability . . .'*

Sir William Osler

# CONTENTS

# PREFACE

I like to joke that my most precious possession, which I prize above all my tools and books, and the pictures and antiques that I inherited from my family, is my suicide kit, which I keep hidden at home. It consists of a few drugs that I have managed to acquire over the years. But I don't know whether the drugs would still work – they came with neither a 'Use By' nor a 'Best Before' date. It would be embarrassing to wake up in Intensive Care after a failed suicide attempt, or to find myself having my stomach pumped out in Accident and Emergency. Attempted suicides are often viewed by hospital staff with scorn and condescension – as failures in both living and dying, and as the agents of their own misfortune.

There was a young woman, when I was a junior doctor and before I started training to be a brain surgeon, who was saved from a barbiturate overdose. She had been determined to die in the wake of an unhappy love affair, but had been found unconscious by a friend and taken to hospital, where she was admitted to the ITU – the Intensive Therapy Unit – and ventilated for twenty-four hours. She was then transferred to the ward where I was a houseman – the most junior grade of hospital doctor – when she started to wake up. I watched her regain consciousness, coming back to life, surprised and puzzled at first still to be alive, and then not quite sure whether she wanted to return to the land of the living or not. I remember sitting on the edge of her bed and talking with her. She was very thin, and was obviously anorexic. She had short, dark-red hair, which was matted and dishevelled after a day in a coma on a ventilator. She sat with her chin resting on the hospital blanket over her drawn-up knees. She was quite calm; perhaps this was still the effect of the overdose, or perhaps it was because she felt that here, in hospital, she was in limbo, between heaven and hell – that she had been given a brief reprieve from her unhappiness. We became friends of a kind for the two days that she was on the ward and before she was transferred to the care of the psychiatrists. It turned out that we had acquaintances in common from Oxford in the past, but I do not know what happened to her.

I have to admit that I'm not at all sure that I would ever dare to use the drugs in my suicide kit when – and it may well happen quite soon – I am faced with the early signs of dementia, or if I develop some incurable illness such as one of the malignant brain tumours with which I am so familiar from my work as a brain surgeon. When you are feeling fit and well, it is relatively easy to entertain the fantasy of dying with dignity by taking your own life, as death is still remote. If I don't die suddenly, from a stroke or a heart attack, or from being knocked off my bicycle, I cannot predict what I will feel when I know that my life is coming to an end – an end which might well be distressing and degrading. As a doctor, I cannot have any illusions. But it wouldn't entirely surprise me if I started to cling desperately to what little life I had left. Apparently, in countries where so-called doctor-assisted suicide is legal many people, if they have a terminal illness, having initially expressed an interest in being able to die quickly, do not take up the option as the end approaches. Perhaps all that they wanted was the reassurance that if the end was to become particularly unpleasant, it could be brought to a quick conclusion and, in the event, their final days passed peacefully. But perhaps it was because, as death approached, they started to hope that they might yet still have a future. We develop what psychologists call 'cognitive dissonance', where we entertain entirely contradictory thoughts. Part of us knows, and accepts, that we are dying but another part of us feels and thinks that we still have a future. It is as though our brains are hardwired for hope, or at least that part of them is.

As death approaches, our sense of self can start to disintegrate. Some psychologists and philosophers maintain that this sense of self, of being coherent individuals free to make choices, is little more than a title page to the great musical score of our subconscious, a score with many obscure, often dissonant voices. Much of what we think of as real is a form of illusion, a consoling fairy story created by our brains to make sense of the myriad stimuli from inside and outside us, and of the unconscious mechanics and impulses of our brains.

Some even claim that consciousness itself is an illusion – that it is not 'real', that it is a trick played on us by our brains – but I do not understand what they mean by this. A good doctor will speak to both the dissonant selves of a dying patient – the part that knows that it is dying, and the part that hopes that it will yet live. A good doctor will neither lie nor deprive the patient of hope, even if the hope is only of life for a few more days. But it is not easy, and it takes time, with many long silences. Busy hospital wards – where most of us are still doomed to die – are not good places in which to have such conversations. As we lie dying, many of us will keep a little fragment of hope alive in a corner of our minds, and only near the very end do we finally turn our face to the wall and give up the ghost.

# I

# THE LOCK-KEEPER'S COTTAGE

The cottage stands on its own by the canal, derelict and empty, the window frames rotten and hanging off their hinges and the garden a wilderness. The weeds were as high as my chest and hid, I was to discover, fifty years of accumulated rubbish. It faces the canal and the lock, and behind it is a lake, and beyond that a railway line. The property company that owned it must have paid somebody to clear out the inside of the cottage, and whoever had done the work had simply thrown everything over the old fence between the garden and the lake, so the lake side was littered with rubbish – a mattress, a disembowelled vacuum cleaner, a cooker, legless chairs and rusty tins and broken bottles. Beyond the junk, however, lay the lake, lined by reeds, with two white swans in the distance.

I first saw the cottage on a Saturday morning. A friend had told me about it. She had seen that it was for sale and knew that I was looking for a place where I could establish a woodworking workshop in Oxford to help me cope with retirement. I parked my car beside the bypass and walked along the flyover, deafening cars and trucks rushing past me, to find a small opening in the hedge, almost invisible, at the side of the road. There was a long line of steps covered in leaves and beechmast, under a dark archway formed by the low, bending branches of beech trees, leading down to the canal. It was as though I was suddenly dropping out of the present and returning to the past. The roar of the traffic became abruptly muted as I descended to the quiet and still canal. The cottage was a few hundred yards away along the towpath, over an old, brick-built humpback canal bridge.

There were several plum trees in the garden, one of them growing up through an obsolete and rusty old machine with reciprocating blades like a hedge-trimmer, for cutting heavy undergrowth. It had two big wheels with *Allens* and *Oxford* stamped on the rims in large letters. My father had had exactly the same model of machine, which he used in the two-acre garden and orchard where I had grown up less than one mile away in the 1950s. He once accidentally ran over a little shrew in the grass of the orchard as I stood watching him, and I remember my distress at seeing its bleeding body and hearing its piercing screams as it died.

The cottage looks out over the still and silent canal and the heavy black gates of the narrow lock. There is no road access – it can only be reached along the towpath on foot or by barge. There is a brick wall with drinking troughs for horses along one side of the garden, facing the canal – I found later the metal rings to which the horses which towed the barges along the canal would have been tethered. A long time ago the lock-keeper would have been responsible for the gates, but the lock-keepers' cottages along the canal have all been sold off and the gates are now left to be operated by whoever is on the passing barges. I am told that a kingfisher lives here and can be seen flashing across the water, and that there are otters as well, even though only a few hundred yards away there is the roar of the bypass traffic crossing the canal on the high flyover on its concrete stilts. But if I turn away from the road, all I can see are fields and trees, and the reed-lined lake behind the house. I can imagine that I am in ancient, deep countryside, as it was when I was growing up nearby, before the bypass was built sixty years ago.

The young woman from the estate agents was sitting on the grass bank in the sunshine beside the entrance to the cottage, waiting for me. She opened the bolted and padlocked front door. I stepped over a few letters on the floor inside, covered in muddy footprints. The estate agent saw me looking down at them and told me that an old man had lived here by himself for almost fifty years – the deeds for the property described him as a canal labourer. When he died the property developers, who had bought the house some years ago, put it up for sale. She did not know whether he had died here or in hospital or in a nursing home.

The place smelt damp and neglected. The cracked and broken windows were covered by torn, dirty lace curtains and the window sills were black with dead flies. The rooms had been stripped out and had the sad and despondent air of all abandoned homes. Although there was water and electricity, the facilities were primitive, and there was only an outside toilet, smashed into pieces, with the door off its hinges. The dustbin by the front door contained plastic bags full of faeces.

The ancient farmouse nearby where I had spent my childhood was said to have been haunted – at least, according to the Whites, the elderly couple who lived across the road and whom I liked to visit. An improbable tale of a sinister coach and horses in the yard at night and also of a 'grey lady' in the house itself. It was easy to imagine the old man's ghost haunting the cottage.

'I'll take it,' I said.

The girl from the estate agents looked at me sceptically. 'But don't you want to get a survey?'

'No, I do all my own building work and it looks OK to me,' I replied confidently, but wondering whether I was still capable of the physical work that would be required and how I would manage without any road access. Perhaps I should stop being so ambitious and abandon my obsessive conviction that I must do everything myself. Perhaps it no longer mattered. I ought to employ a builder. Besides, although I wanted a workshop, I wasn't sure that I wanted to live in this small and lonely cottage, with a possible ghost.

'Well, you'd better make an offer to Peter, the manager in our local office,' she replied.

I drove back to London the next day – with the uneasy thought that perhaps this little cottage would be where I myself would eventually end my days and die, and where my story would end. Now that I am retiring, I am starting all over again, I thought, but now I am running out of time.

I was back in the operating theatre on Monday – I was in my blue theatre scrubs, but expected to be only an observer. In three weeks' time I was to retire – after almost forty years of medicine and neurosurgery. My successor, Tim, who had started off as a trainee in our department, had already been appointed. He is an exceptionally able and nice man, but not without that slightly fanatical determination and attention to detail that neurosurgery requires. I was more than happy to be replaced by him and it seemed appropriate to leave most of the operating to him, in preparation for the time when – and it would probably be something of a shock for him – he suddenly carried sole responsibility for what happened to the patients under his care.

The first case was an eighteen-year-old woman who had

been admitted for surgery the previous evening. She was five months pregnant but had started to suffer from severe headaches, and a scan showed a very large tumour – almost certainly benign – at the base of her brain. I had seen her as an emergency in my outpatient clinic a few days earlier; she came from Romania and her English was limited, but she smiled bravely as I tried to explain things to her via her husband, who spoke a little English. He told me that they came from Maramures, the area of northern Romania on the border with Ukraine. I had been there myself two years ago on a journey from Kiev to Bucharest with my Ukrainian colleague Igor. The landscape was exceptionally beautiful, with ancient wooden farms and monasteries – it seemed that the modern world had scarcely caught up with the place at all. There were haystacks in the fields and hay wagons drawn by horses on the roads, with the drivers wearing traditional peasant costumes. Igor was outraged that Romania had been allowed to join the European Union whereas Ukraine had been kept out. My Romanian colleague, who had come to collect us from the border with Ukraine, wore a tweed cloth cap and leather driving gloves, and drove us at high speed on the terrible roads in his son's souped-up BMW all the way to Bucharest, almost without stopping. We did, however, spend a night on the way at Sighisoara, where the house still stood where Vlad the Impaler – the prototype for Dracula – had been born. It was now a fast-food joint.

The operation on the woman was not an emergency in the sense that it did not need to be done at once, but it certainly had to be done within a matter of days. Such cases do not fit easily into the culture of targets which now defines how the National Health Service in England is supposed to function. She was not a routine case but nor was she an emergency.

My own wife Kate, a few years ago, had fallen into the

same trap when awaiting major surgery after many weeks of intensive care at a famous hospital. She had been admitted as an emergency and underwent emergency surgery without any difficulty, but then needed further surgery after several weeks of intravenous feeding. I became accustomed to the sight of a large foil-wrapped bag of glutinous fluid hanging above her bed, dripping into her central line – a catheter inserted into the great veins leading to her heart. Kate was now no longer an emergency but nor was she a routine admission, so there was no provision for her to undergo surgery. For five days in a row she was prepared for surgery – very major surgery, with all manner of frightening potential complications – and each day by midday the operation was cancelled. Eventually, in despair, I rang her surgeon's secretary. 'Well, it's not really up to Prof as to who goes on the routine operating lists,' she explained apologetically. 'It's a manager – the *List Broker*. Here's the number to ring . . .'

So I rang the number only to receive a message that the voice mailbox was full and I could not leave a message. At the end of the week the decision was made to make Kate into a routine case by sending her home with a large bottle of morphine. She was readmitted a week later, presumably now with the List Broker's permission. The operation was a great success, but I mentioned the problem we had encountered to one of my neurosurgical colleagues at the same hospital when we met at a meeting shortly afterwards.

'I find it very difficult being a medical relative,' I said. 'I don't want people to think my wife should get better treatment just because I'm a surgeon myself, but it really was getting pretty unbearable. Having your operation cancelled is bad enough – but five days in a row!'

My colleague nodded. 'And if we can't look after our own, what about Joe Bloggs?'

So I had gone to work on Monday morning worried that there would be the usual shambles of trying to find a bed for the young girl into which she could go after surgery. If her condition was life-threatening I would be able to start the operation without having to seek the permission of the many hospital staff involved in trying to allocate an insufficient number of beds to too many patients, but her condition was not life-threatening – at least not yet – and I knew that I was going to have a difficult start to the day.

At the theatre reception area there was an animated group of doctors and nurses and managers looking at the day's operating lists sellotaped to the top of the desk, discussing the impossibility of getting all the work done. I saw that several of the cases were routine spinal operations.

'There are no ITU beds,' the anaesthetist said with a grimace.

'Well why not just send for the patient anyway?' I asked. 'A bed always turns up later.' I always say this, and always get the same reply.

'No,' she said. 'If there's no ITU bed I will end up having to recover the patient in theatre after the op and it could take hours.'

'I'll try to go and sort it out after the morning meeting,' I replied.

There was the usual collection of disasters and tragedies at the morning meeting.

'We admitted this eighty-two-year-old man with known prostate cancer yesterday. He had gone first to his local hospital because he was going off his legs and was in retention of urine. They wouldn't admit him and sent him home,' Fay, the on-call registrar, told us as she put up a scan. This was met with sardonic laughter in the darkened room.

'No, no, it's true,' Fay said. 'They catheterized him and

wrote in the notes that he was now much better. I have seen the notes.'

'But he couldn't fucking walk!' somebody shouted.

'Well, that didn't seem to trouble them. At least they must have achieved their four-hour target by sending him home. He spent forty-eight hours at home and the family got the GP in, who sent him here.'

'Must have been a very uncomplaining and long-suffering patient,' I observed to my colleague sitting next to me.

'Samih,' I said to one of the other registrars, 'what do you see on the scan?' I had first met Samih some years earlier on one of my medical visits to Khartoum. I had been very impressed by him and did what I could to help him to come to England to continue his training. In the past it had been relatively easy to bring trainees over to my department from other countries, but the combination of European Union restrictions on doctors from outside Europe and increasing bureaucratic regulations in recent years has made it very difficult, even though the UK has fewer doctors per capita than any country in Europe other than Poland and Romania. Samih passed all the required examinations and hurdles with flying colours. He was a joy to work with, a large and very gentle man, utterly dedicated to our craft, who was loved by the patients and nurses. He was now to be my last registrar.

'The scan shows metastatic posterior compression of the cord at T3. The rest of the scan looks OK.'

'What's to be done?' I asked.

'Well, it depends on how he is.'

'Fay?'

'He was sawn off when I saw him at ten o'clock last night.'

This is the brutal but accurate phrase to describe a patient who has a spinal cord so badly damaged that they have no feeling or movement of any kind below the level of the damage and when there is no possibility of recovery. T3 means the third thoracic vertebra, so the poor old man would have no movement of his legs or trunk muscles. He would even have difficulties just trying to sit upright.

'If he's sawn off he's unlikely to get better,' Samih said. 'It's too late to operate now. It would have been a simple operation,' he added.

'What's this man's future?' I asked the room at large. Nobody replied so I answered the question myself.

'It's very unlikely he'll be able to get home as he'll need full twenty-four-hour nursing, with being turned every few hours to prevent bed sores. It takes several nurses to turn a patient, doesn't it? So he will be stuck in some geriatric ward somewhere until he dies. If he's lucky the cancer elsewhere in his body will carry him off soon, and he may make it into a hospice first, nicer than a geriatric ward, but the hospices won't take people if their prognosis is that they might live for more than a few weeks. If he's unlucky, he may hang on for months.'

I wondered if that was how the old man in the cottage had died, alone in some impersonal hospital ward. Would he have missed his home, the little cottage by the canal, even though it was in such a sorry state? My trainees are all much younger than I am; they still have the health and self-confidence of youth, which I too had at their age. As a junior doctor you are pretty detached from the reality that faces so many of the older patients. But now I am losing my detachment from patients as I prepare to retire. I will become a member of the underclass of patients – as I was before I became a doctor, no longer one of the elect.

The room remained silent for a while.

'So what happened?' I asked Fay.

'He came in at ten in the evening and Mr C. planned to

operate but the anaesthetists refused – they said there was no prospect of his getting better and they weren't willing to do it at night.'

'Well, there's not much to be lost by operating – we can't make him any worse,' somebody said from the back of the room.

'But is there any realistic prospect of making him better?' I asked, but I went on to say: 'Although, to be honest, if it was me I'd probably say go and operate . . . just in case . . . The thought of ending my days paraplegic on a geriatric ward is so awful . . . indeed, if the operation killed me, I wouldn't complain.'

'We decided to do nothing,' Fay said. 'We're sending him back to his local hospital today – if there's a bed there, that is.'

'Well, I hope they take him back – we don't want another Rosie Dent.' Rosie had been an eighty-year-old woman earlier in the year with a cerebral haemorrhage whom I had been forced to admit by a physician at my own hospital – at least, so many complaints and threats were made if I didn't admit her to an acute neurosurgical bed that I gave in – even though she did not need neurosurgical treatment. It proved impossible to get her home and she sat on the ward for seven months, before we eventually managed to persuade a nursing home to accept her. She was a charming, uncomplaining old lady and we all became quite fond of her, even though she was 'blocking' one of our precious acute neurosurgical beds.

'I think it will be OK,' Fay said. 'It's only our own hospital which refuses to take patients back from the neurosurgical wards.'

'Any other admissions?' I asked.

'There's Mr Williams,' Tim said. 'I was hoping to do him at the end of your list after the girl with a meningioma.'

'What's the story?' I asked.

'He's had some epileptic fits. Been behaving a bit oddly of late. Used to be pretty high-functioning – engineer or something like that. Fay, could you put the scan up please?'

The scan flashed up on the wall in front of us. 'What's it show, Tiernan?' I asked one of the most junior doctors, known as SHOs, short for senior house officer.

'Something in the left frontal lobe.'

'Can you be a bit more precise? Fay, put up the Flair sequence.'

Fay showed us some different scan images, sequences that are good for indicating tumours which are invading the brain rather than just displacing it.

'It looks as though it's infiltrating all of the left frontal lobe and most of the left hemisphere,' Tiernan said.

'Yes,' I replied. 'We can't remove the tumour, it's too extensive. Tiernan, what are the functions of the frontal lobes?'

Tiernan hesitated, finding it hard to reply.

'Well, what happens if the frontal lobes are damaged?' I asked.

'You get personality change,' he replied immediately.

'What does that mean?'

'They become disinhibited – get a bit knocked off . . . ', but he found it difficult to describe the effects in any more detail.

'Well,' I said, 'the example of disinhibition loved by doctors is the man who pisses in the middle of the golfing green. But the frontal lobes are where all our social and moral behaviour is organized. You get a whole variety of altered social behaviours if the frontal lobes are damaged – almost invariably for the worse. Sudden outbursts of violence and irrational behaviour are among the commonest. People who were previously kind and considerate become coarse and selfish, even though their intellect can be perfectly well preserved. The person with frontal-lobe damage rarely has any insight into it – how can the "I" know that it is changed? It has nothing to compare itself with. How can I know if I am the same person today as I was yesterday? I can only assume that I am. Our selves are unique and can only know ourselves as we are now, in the immediate present. But it's terrible for the families. They are the real victims. Tim, what do you hope to achieve?'

'If we take some of it out, create some space, we'll buy him a bit more time,' Tim replied.

'But will surgery get his personality change any better?'

'Well, it might,' Tim said. I was silent for a while.

'I rather doubt it,' I eventually commented. 'But it's your case. And I haven't seen him. Did you discuss all this with him and his family?'

'Yes.'

'It's nine o'clock,' I said. 'Let's see what's happening about beds and find out if we are allowed to start operating.'

An hour later, Tim and Samih started the operation on the Romanian woman. I spent most of the time sitting on a stool, my back propped up against the wall behind me, while Tim and Samih slowly removed the tumour. The lights in the theatre were dimmed as they were using the microscope, and I dozed, listening to the familiar sounds and muted drama of the theatre – the bleeping of the anaesthetic monitors, the sighing of the ventilator, Tim's instructions to Samih and the scrub nurse Agnes and the hiss of the sucker which Tim was using to suck the tumour out of the woman's head. 'Toothed forceps . . . Adson's . . . diathermy . . . Agnes, pattie please . . . Samih, can you suck here? . . . there's a bit of a bleeder . . . ah! got it . . .'

I could also hear the quiet conversation between the two

anaesthetists at the far end of the table, where they sat on stools next to the anaesthetic machine with its computer screen showing the girl's vital functions, as they are called – the functioning of her heart and lungs. These appear as a series of pretty, bright-coloured lines and numerals in red and green and yellow. In the distance, from the prep area between the theatres, there would be occasional bursts of laughter and chatter from the nurses – all good friends of mine, with whom I had been working for many years – as they prepared the instruments for the next cases.

Will I miss this? I asked myself. This strange, unnatural place that has been my home for so many years, a place dedicated to cutting into living bodies and, in my case, the human brain – windowless, painfully clean, air-conditioned and brilliantly lit, with the operating table in the centre, beneath the two great discs of the operating lights, surrounded by machines? Or when the time comes in a few weeks, will I just walk away without any regrets at all?

A long time ago, I thought brain surgery was exquisite – that it represented the highest possible way of using both hand and brain, of combining art and science. I thought that brain surgeons – because they handle the brain, the miraculous basis of everything we think and feel – must be tremendously wise and understand the meaning of life. When I was younger I had simply accepted the fact that the physical matter of brains produces conscious thought and feeling. I thought the brain was something that could be explained and understood. As I have got older, I have instead come to realize that we have no idea whatsoever as to how physical matter gives rise to consciousness, thought and feeling. This simple fact has filled me with an increasing sense of wonder, but I have also become troubled by the knowledge that my brain is an ageing organ, just like the organs of the rest of my body. That my 'I' is ageing and that I have no way of knowing how it might have changed. I look at the liver spots on the wrinkled skin of my hands, the hands whose use has been the dominant theme of my life, and wonder what my brain would look like on a brain scan. I worry about developing the dementia from which my father died. On the brain scan that was done some years before his eventual death, his brain had looked like a Swiss cheese – with huge holes and empty spaces. I know that my excellent memory is no longer what it was. I often struggle to remember names.

My understanding of neuroscience means that I am deprived of the consolation of belief in any kind of life after death and of the restoration of what I have lost as my brain shrinks with age. I know that some neurosurgeons believe in a soul and afterlife, but this seems to me to be the same cognitive dissonance as the hope the dying have that they will yet live. Nevertheless, I have come to find a certain solace in the thought that my own nature, my I – this fragile, conscious self writing these words that seems to sail so uncertainly on the surface of an unfathomable, electrochemical sea into which it sinks every night when I sleep, the product of countless millions of years of evolution – is as great a mystery as the universe itself.

I have learnt that handling the brain tells you nothing about life – other than to be dismayed by its fragility. I will finish my career not exactly disillusioned but, in a way, disappointed. I have learnt much more about my own fallibility and the crudity of surgery (even though it is so often necessary), than about how the brain really works. But as I sat there, the back of my head resting against the cold, clean wall of the operating theatre, I wondered if these were just the tired thoughts of an old surgeon about to retire.

The woman's tumour was growing off the meninges – the thin, leathery membrane that encases the brain and spinal cord – in the lower part of the skull known as the posterior cranial fossa. It was immediately next to one of the major venous sinuses. These are drainpipe-like structures that continuously drain huge volumes of deep-purple, deoxygenated blood – blood which would have been brilliant red when it first reached the brain, pumped up from the heart. Blood flashes through the brain in a matter of seconds, one quarter of all the blood from the heart, darkening as the brain takes the oxygen out of it. Thinking, perceiving and feeling, and the control of our bodies, most of it unconscious, are energy-intensive processes fuelled by oxygen. There was some risk that removing the tumour might tear the transverse venous sinus and cause catastrophic haemorrhage, so I scrubbed up and helped Tim with the last twenty minutes of the operation, carefully burning and peeling the tumour off the side of the sinus without puncturing it.

'I think we can call that a complete removal,' I said.

'I don't think I'm going to have time to do Mr Williams – the man with the frontal tumour,' Tim said. 'I've got a clinic starting at one. I'm terribly sorry. Could you possibly do him? And take out as much tumour as you can? Get him some extra time?'

'I suppose I'll have to,' I replied, disliking having to operate on patients I had not spoken to in detail myself, and not at all sure as to whether surgery was really in the patient's best interests.

So Tim went off to do his outpatient clinic and Samih finished the operation, filling the hole in the girl's skull with quick-setting plastic cement and stitching together the layers of her scalp. An hour later, Mr Williams was wheeled into the anaesthetic room next to the operating theatre. He was in his forties, I think, with a thin moustache and a pale, rather vague expression. He must have been quite tall as his feet, clad in regulation white anti-embolism stockings with the bare toes coming out at the ends, stuck out over the edge of the trolley.

'I'm Henry Marsh, the senior surgeon,' I said, looking down at him.

'Ah,' he said.

'I think Tim Jones has explained everything to you?' I asked.

It was a long time before he replied. It looked as though he had to think very deeply before replying.

'Yes.'

'Is there anything you would like to ask me?' I said.

He giggled and there was another long delay.

'No,' he eventually replied.

'Well, let's get on with it,' I said to the anaesthetist and left the room.

Samih was waiting for me in the operating theatre, beside the wall-mounted computer screens where we can look at our patients' brain scans. He already had Mr Williams's scan on the screens.

'What should we do?' I asked him.

'Well, Mr Marsh, it's too extensive to remove. All we can do is a biopsy, just take a small part of the tumour for diagnosis.'

'I agree, but what's the risk with a biopsy?'

'It can cause a haemorrhage, or infection.'

'Anything else?'

Samih hesitated, but I did not wait for him to reply.

I told him how if the brain is swollen and you only take a little bit of tumour out, you can make the swelling worse. The patient can die after the operation from 'coning': the swollen brain squeezes itself out of the confined space of the skull, part of it becoming cone-shaped where it is forced out of the skull through the hole at its base called the foramen magnum ('the big hole' in Latin), where the brain is joined to the spinal cord. This process is invariably fatal if it is not caught in time.

'We have to take enough tumour out to allow for any post-op swelling,' I said to Samih. 'Otherwise it's like kicking a hornet's nest. Anyway, Tim said he was going to remove as much of the tumour as possible as this might prolong his life a bit. What sort of incision do you want to make?'

We discussed the technicalities of how to open Mr Williams's head while waiting for the anaesthetists to finish anaesthetizing him, and to attach the necessary lines and tubes and monitors to his unconscious body.

'Get his head open,' I told Samih, 'and give me a shout when you've reached the brain. I'll be in the red leather sofa room.'

The scan had shown that the left frontal lobe of Mr Williams's brain was largely infiltrated by tumour, which appeared on the scan as a spreading white cloud in the grey of his brain. Tumours like this grow into the brain instead of displacing it, the tumour cells pushing into the brain's soft substance, weaving their way between the nerve fibres of the white matter and the brain cells of the grey matter. The brain can often go on working for a while even though the tumour cells are boring into it like deathwatch beetles in a timber building, but eventually, just as the building must collapse, so must the brain.

I lay on the red leather sofa in the neurosurgeons' sitting room, slightly anxious, as I always am when waiting to operate, longing to retire, to escape all the human misery that I have had to witness for so many years, and yet dreading my departure as well. I am starting all over again, I said to myself once more, but am running out of time. The phone rang and I was summoned back to the theatre.

Samih had made a neat left frontal craniotomy. Mr Williams's forehead had been scalped off his skull and was reflected forward with clips and sterile rubber bands. His brain, looking normal but a little 'full', as neurosurgeons describe a swollen brain, bulged gently out of the opening Samih had sawn in his skull.

'We can't miss it, can we?' I said to Samih. 'The tumour's so extensive. But the brain's a bit full – we'll have to take quite a lot out to tide him over the post-operative period. Where do you want to start?'

Samih pointed with his sucker to the centre of the exposed surface of brain.

'Middle frontal gyrus?' I asked. 'Well, maybe, but let's go and look at the scan.' We walked the ten feet across the room to the computer screens.

'Look, there's the sphenoid wing,' I said to Samih. 'We should go in just a little above it, but you'll have to go deeper into the brain than you think from the scan as his brain is bulging out a bit.'

We returned to the table and Samih burned a little line across Mr Williams's brain with the diathermy forceps – a pair of forceps with electrical tips that we use for cauterizing bleeding tissue.

'Let's bring in the scope,' I said, and once the nurses had positioned the microscope, Samih gently pushed downwards with sucker and diathermy.

'It looks normal, Mr Marsh,' Samih said, a little anxiously. Even though there are all manner of checks and cross-checks to make sure we have opened the correct side of the patient's head, I always experience a moment of complete panic at times like this, and have to quickly reassure myself that we are indeed operating on the correct side – in this case the left side – of Mr Williams's brain.

'Well, the trouble with low-grade tumours is that they can look and feel like normal brain. Let me take over.'

So I started to cautiously prod and poke the poor man's brain.

'Yes, it looks and feels entirely normal,' I said, feeling a little sick as I looked through the microscope at the smooth, unblemished white matter. 'But we've *got* to be in tumour – there's so much of it on the scan.'

'Of course we are, Mr Marsh,' Samih said respectfully. 'Would Stealth or a frozen section have helped?'

These are techniques that would have reassured me that I was in the right place. Rationally I knew that I had to be in tumour – at least in brain infiltrated by tumour – but the man's brain looked and felt so normal that I could not suppress the fear that some bizarre mistake had occurred. Perhaps the wrong name was on the brain scan, or it hadn't been a tumour in the first place and the problem had got better on its own since the brain scan had been done. The thought of removing normal brain – however unlikely – was terrifying.

'Well, you're probably right, but it's too late now and, having started, I can't stop,' I said to Samih. 'I'll have to remove a lot of normal-looking brain to stop him swelling and dying post-op.'

The brain becomes swollen with the least provocation, and Mr Williams's brain was already ominously enlarging and starting to bulge out of his opened skull. At the end of a craniotomy – the medical name for opening a person's head – the skull is closed with little metal screws and plates and the scalp stitched back together over it. The skull becomes once again a sealed box. If there is very severe post-operative swelling as a reaction to the surgery, the pressure inside the skull will become critically raised and the brain will, in effect, suffocate and the patient can die. Surgery, especially for tumours within the actual substance of the brain like Mr Williams's, where you cannot remove all of the tumour, will inevitably cause swelling, and it is always important to remove enough tumour – to create space within the skull to allow for the swelling. The pressure in the patient's head after the operation will then not become dangerously high. But you always worry that you might have removed too much tumour and that the patient will wake up damaged and worse than before the operation.

I can remember two cases – both young women – from the early years of my career where my inexperience made me too timid and I failed to remove enough tumour. They both died from post-operative brain swelling within twenty-four hours after surgery. I learnt to be braver with similar cases in future – in effect, to take greater risks when operating on such tumours, because the deaths of the two women had taught me that the risks of not removing sufficient tumour were even greater. And yet both the tumours were malignant and the patients had a grim future ahead of them, even if the operations were to have been successful. Looking back now after thirty years, having seen so many people die from malignant brain tumours since then, these two tragic cases do not seem quite as disastrous as they did at the time.

This is about as bad as it gets, I thought with disgust as I started to remove several cubic centimetres of Mr Williams's brain, the sucker slurping obscenely. What's the glory in this? This coarse and crude surgery. This evil tumour, changing this man's very nature, destroying both himself and his family. It's time to go.

As I watched my sucker down the microscope, controlled by my invisible hands, working on the poor man's brain, teasing and pulling out the tumour, I told myself that I wouldn't have panicked in the past. I would just have shrugged and got on with it. But now that my surgical career was coming to an end, I could feel the defensive psychological armour that I had worn for so many years starting to fall away, leaving me as naked as my patients. Bitter experience of similar cases to Mr Williams's told me that the best outcome for this man would be if the operation killed him – but I felt unable to let that happen. I knew of surgeons in the distant past who would have done just that, but we live in a different world now. At moments like this I hate my work. The physical nature of our thought, the incomprehensible unity of mind and brain, is no longer an awe-inspiring miracle but instead a cruel and obscene joke. I think of my father slowly dying from dementia and his brain scan, and I look at the age-wrinkled skin of my hands, which I can see even through the rubber of my surgical gloves.

As I worked the sucker, Mr Williams's brain started slowly to sink back into his skull.

'That's enough space now, Samih,' I said. 'Close please. I'll go and find his wife.'

Later in the day I went up to the ITU to see the post-operative patients. The young Romanian woman was well, though she looked pale and a little shaken. The nurse at the end of her bed glanced up from the mobile computer where she was inputting data and told me that everything was as it should be. Mr Williams was three beds further down the row of ITU patients. He was sitting upright, awake, looking straight ahead.

I sat by his bedside and asked him how he felt. He turned to look at me and said nothing for a while. It was hard to know if his mind was blank or whether he was struggling to organize the thoughts in his disrupted, infiltrated brain. It was hard even to know what 'he' had now become. Once I would have waited only a short time for an answer. Many of my patients have lost – sometimes permanently, sometimes transiently – language or the ability to think and there is a limit to how long you can put up with waiting. But on this occasion, perhaps because I knew that this would probably never happen again and perhaps also as a silent apology to all the patients I must have hurried by in the past, I sat quietly for what felt like a long time.

'Am I going to die?' he suddenly asked.

'No,' I said, alarmed at the way he seemed to know what was going on after all. 'And if you were I promise I would tell you. I always tell my patients the truth.'

He must have understood that because he laughed – an odd, inappropriate sort of laugh. No, you are not going to die just yet, I said to myself, it is going to be much worse than that. I sat beside him for a while longer but it seemed he had nothing further to say.

Samih was waiting for me as usual at 7.30 the next morning at the nurses' desk. He was a junior doctor in the traditional mould and could not bear to think that he might not be in the hospital when I was there. When I was a junior it was inconceivable that I might leave the building before my consultant, but in the new world of shift-working doctors the master-and-apprentice form of medical training has largely disappeared.

'She's in the interview room,' he said. We walked down the corridor and I sat down opposite Mrs Williams. I introduced myself.

'I'm sorry we haven't met before. Tim was going to do

the operation but I ended up doing it. I'm afraid this is not going to be good news. What did Tim tell you?'

As a doctor you get used to patients and their families looking so very intently at you as you talk that sometimes it feels as though nails are being driven into you, but Mrs Williams smiled sadly.

'That it was a tumour. That it couldn't all be removed. My husband was pretty bright, you know,' she added. 'You're not seeing him at his best.'

'In retrospect, looking back, when do you think things started to go wrong?' I asked gently.

'Two years ago,' she said immediately. 'It's a second marriage for both of us – we married seven years go. He was a lovely man, but two years ago he changed. He was no longer the man that I had married. He started playing strange, cruel tricks on me . . .'

I did not ask what these might have been.

'It became so bad,' she went on, 'that we had more or less decided to go our separate ways. And then the fits started . . .'

'Do you have children?' I asked.

'He has a daughter from his first marriage but we have no children from our marriage.'

'I'm afraid I have to tell you that treatment won't get him better,' I said, very slowly. 'We can't undo the personality change. All we can do is possibly prolong his life and he may yet live for years anyway, but he will slowly get worse.'

She looked at me with an expression of utter despair – she could not have helped but hope that the operation would undo the horrors of the past, that her nightmare would come to an end.

'I thought it was the marriage that had gone wrong,' she said. 'His family all blamed me.'

'It was the tumour,' I said.

'I realize that now,' she replied. 'I don't know what to think . . .'

We talked for a while longer. I explained that we would have to wait for the pathology report on what I had removed. I said it was just possible I might have to operate again if the analysis showed that I had missed the tumour. The only potential further treatment would be radiation and, as far as I could tell, this had no prospect of making him any better.

I left her in the little interview room with one of the nurses – most of my patients' families prefer, I think, to cry after I have left the room, but perhaps that is wishful thinking on my part – perhaps they would prefer me to stay.

Samih and I walked back down the corridor.

'Well,' I said, 'at least the marriage was coming to an end, so I suppose it's a bit easier for her, but how can anybody know how to deal with something like this?'

I thought of the end of my first marriage fifteen years earlier and how cruel and stupid my wife and I had been to each other. Neither of us had had frontal brain tumours, though I wonder what deep and unconscious processes might have been driving our behaviour. I look back with horror at how little attention I paid to my three children during that time. The psychiatrist I was seeing at the time told me to become more of an observer, but I simply could not detach myself from the raging intensity of my feelings at being forced to leave my own home, so much of which I had built with my own hands. I feel that I have learnt a certain amount of wisdom and self-control as a result of that terrible time, but also wonder whether it might in part be simply because the emotional circuits in my brain are slowing down with age.

I went to see Mr Williams. The nurses had told me, when I had come onto the ward, that he had tried to abscond during the night, and they had had to keep the ward door locked. It was a fine morning and low sunlight streamed into the ward through the east-facing windows, over the slate roofs of south London. I found him standing in front of the windows in his pyjamas. I noticed that they were decorated with teddy bears. His arms were stretched out on either side as though to welcome the morning sun.

'How are you?' I said, looking at his slightly swollen forehead and the neatly curved incision behind it across his shaven head.

He said nothing in reply and gave me a vague, cryptic smile, slowly lowered his arms and shook my hand politely without saying a word.

The pathology report came back two days later and confirmed that all the specimen I had sent was infiltrated by a slow-growing tumour. It was going to take a long time to find any kind of long-term placement for Mr Williams and it seemed unlikely he could be managed at home, so I told my juniors to send him back to the local hospital to which he had first gone after the epileptic fits had started. The doctors and nurses there would have to find a solution to the problem. The tumour was certainly going to prove fatal, but it was impossible to know whether this would be a matter of months or longer. When I went round the ward early next morning I saw that there was a different patient in his bed and Mr Williams had gone.

# 2

# LONDON

I had decided to resign from my hospital in London in a fit of anger in June 2014, four months before I came across the lock-keeper's cottage. Three days after handing in my letter of resignation I was in Oxford, where I live with my wife Kate at weekends, running along the Thames towpath for my daily exercise. I was panic-stricken about what I would do with myself once I no longer had my work as a neurosurgeon to keep me busy and my mind off the future. It was in exactly the same place, on the same towpath, but walking, not running, many years earlier, in a much greater state of distress, that I had decided to abandon my degree in politics, philosophy and economics at Oxford University – much to my parents' distress and dismay when they got to hear of it.

While I ran beside the river, I suddenly remembered a young Nepali woman with a cyst in her spine that had been slowly paralysing her legs. I had operated on her two months previously. The cyst turned out to be cysticercosis, a worm infection common in impoverished countries like Nepal but almost unheard of in England. She had returned to the outpatient clinic a few days earlier to thank me for her recovery; like so many Nepalis, she had the most perfect, gentle manners. As I ran – it was late summer, the river level was low and the dark-green water of the Thames seemed to be almost motionless – I thought of her and then thought of Dev, Nepal's first and foremost neurosurgeon, more formally known as Professor Upendra Devkota. We had been friends and surgical trainees together in London thirty years ago.

'Ah!' I thought. 'Perhaps I can go to Nepal and work with Dev. And I will see the Himalayas.'

Both decisions, separated by forty-three years – to abandon my first degree and to resign from my hospital – had been provoked by women. The first was a much older woman, a family friend, with whom I was passionately and wholly inappropriately in love. Although twenty-one years old, I was immature and sexually entirely inexperienced, and had had a repressed and prudish upbringing. I can see now that she seduced me, although only with one passionate kiss – it never went beyond that. She burst into tears immediately afterwards. I think she had been attracted by my combination of intellectual precocity and awkwardness. Perhaps she thought that she could help me overcome the latter. She probably later felt ashamed, and perhaps embarrassed, by my passionate, poetic response – the poems now long forgotten and destroyed. She died many years ago, but my intense embarrassment about this episode is still with me now, even though the kiss resulted in my finding a sense of meaning and purpose to my life. I became a brain surgeon.

I was confused and ashamed by the pangs of my frustrated and absurd love, and overwhelmed by feelings of both love and rejection. I felt there were two armies fighting within my head and I wanted to kill myself to escape them. I tried to compromise by pushing my hand through a window in the flat where I had student digs beside the Thames in Oxford, but the glass would not break or, rather, a deeper part of my self showed a sensible caution.

Unable to translate my unhappiness into physical injury, I decided to run away. I made the decision while walking along the Thames towpath in the early hours of the morning of 18 September 1971, having fortunately failed to hurt myself. The towpath is narrow, in summer dry and grassy, in winter muddy and with many puddles. It passes through Oxford and past Port Meadow, the wide flood meadow to the north of the city. My childhood family home was a few hundred yards away. I might even have seen it as I walked miserably along the river – the area was deeply familiar. If I had gone a little further and followed a narrow cut, linking the river to the Oxford canal, I would have come across the lock-keeper's cottage, but I think I had already turned back by then, having made my decision. The old man, though young at that time, would already have been living there.

I abandoned my university degree for unrequited love, but it was also a rebellion against my well-meaning father, whose belief in the virtue of attending Oxford or Cambridge university was an article of faith. He had been an Oxford don before moving to London. He deserved better from me, but such rebellious behaviour is buried deep within the psyche of many young people; and my father, the kindest of men, but who had himself once rebelled against his own father, resigned himself to my decision. I left my predictable professional career path to work as a hospital theatre porter in a mining town north of Newcastle. I hoped that by seeing other people suffering with 'real', physical illness I would somehow cure myself. My subsequent life as a neurosurgeon was to teach me that the distinction between physical and psychological illness is false – at least, that illnesses of the mind are no less real than those of the body, and no less deserving of our help. A friend's father, John Maud, was the general surgeon in the hospital, and although he had never met me, at his daughter's request he got me a job in his operating theatre. I find it quite extraordinary that he did this, just as I find it remarkable that my Oxford college agreed that I could return after a year's truancy. It is impossible to know how my life would have developed without so much help and kindness from others.

It was my experience as a theatre porter, watching surgeons operate, that led me to become a surgeon. It was a decision that came quite suddenly to me, while talking to my sister Elisabeth – a nurse by training – as she did her family's ironing, when I returned to London for a weekend. I had gone to visit her to hold forth at great length about my unhappiness. It somehow became clear to me – I can't remember how – that the solution to my unhappiness was to study medicine and become a surgeon. Perhaps Elisabeth suggested it to me. I took the train back to Newcastle on the Sunday evening. As I sat in the carriage, seeing myself reflected in the dark glass of the window, I knew that I had now found a sense of purpose and meaning. It would be another nine years, however, when I was already a qualified doctor, before I discovered the all-consuming love of my life – the practice of neurosurgery. I have never regretted that decision, and have always felt deeply privileged be a doctor.

I am not sure, however, if I would take up medicine or neurosurgery now, if I could start my career all over again. So many things have changed. Many of the most challenging neurosurgical operations – such as operating on cerebral aneurysms – have become redundant. Doctors are now subject to a regulatory bureaucracy that simply did not exist forty years ago and which shows little understanding of the realities of medical practice. The National Health Service in

England – an institution I passionately believe in – is chronically starved of funds, since the government dares not admit to the electorate that they will need to pay more if they want first-class health care. Besides, there are other, more pressing problems now facing humanity than illness.

As I returned to Newcastle with my new-found sense of having a future, I read the first issue of a magazine called *The Ecologist*. It was full of gloomy predictions about what was going to happen to the planet as the human population continued to grow exponentially, and as I read it I wondered whether becoming a doctor, healing myself by healing others, might not be a little self-indulgent. There might be more important ways of trying to make the world a better place – admittedly less glamorous ones – than by being a surgeon. I have never entirely escaped the view that being a doctor is something of a moral luxury, by which doctors are easily corrupted. We can so easily end up complacent and self-important, feeling ourselves to be more important than our patients.

A few weeks later, back at work as a theatre technician, I watched a man undergoing surgery to his arm. He had deliberately pushed his hand through a window in a drunken rage and his hand had been left permanently paralysed by the broken glass.

The other woman who quite unintentionally played a pivotal role in my life – at the end of my neurosurgical career – was the medical director of my hospital. She was sent one day by the hospital's chief executive to talk to the consultant neurosurgeons. I believe that we had the reputation of being arrogant and uncooperative. We were too aloof and not playing our part. I was probably considered to be one of the worst offenders. She came into our surgeons' sitting room – the one with the red leather sofas that I had bought some years previously – accompanied by a colleague who was called, I think, the Service Delivery Unit Leader (or some similarly absurd title) for the neurosurgery and neurology departments. He was a good colleague and on several occasions had saved me from the consequences of some of my noisier outbursts. He was suitably solemn on this occasion, and the medical director was looking perhaps a little anxious at the prospect of disciplining eight consultant neurosurgeons. She sat down and carefully placed her large pink handbag beside her on the floor. Our Service Delivery Unit Leader made a little introductory speech and handed over to the medical director.

'You have not been following the Trust dress code,' she declared. Apparently this meant that the consultant neurosurgeons had been seen wearing suits and ties. I had always thought that dressing smartly was a sign of courtesy to my patients, but apparently it now posed a deadly risk of infection to them. A more probable, albeit unconscious, explanation for the ban – which came from high up the NHS hierarchy – was that the senior doctors should not look any different from the rest of the hospital staff. It's called teamwork.

'You have not been showing leadership to the juniors,' the medical director continued. This meant, she told us, that we had not been making sure that the junior doctors had been completing the Trust computer work on time when patients were discharged. In the past we had had our own neurosurgical discharge summaries, which had been exemplary, and I had always taken some pride in them, but they had now been replaced by a Trust-wide, computerized version of such appallingly poor quality that I, for one, had lost all interest in making sure that the juniors completed them.

'If you do not follow Trust policies, disciplinary action will be taken against you,' she concluded. There was no discussion, no attempt to persuade us. The problem, I knew, was that the hospital was about to be inspected by the Care Quality Commission, an organization that puts great store by dress policy and the completion of paperwork. She could have said that she knew this was all rather silly, but could we please help the hospital, and I am sure we would all have agreed – but no, it was to be disciplinary action. She picked up her pink handbag and left, followed by the Service Delivery Unit Leader, who looked a little embarrassed. So I sent off my letter of resignation the next day, unwilling to work any longer in an organization where senior managers could demonstrate such a lack of awareness of how to manage well, although I prudently postponed the date of my departure until my sixty-fifth birthday so that my pension would not suffer.

It is often said that it is better to leave too early rather than too late, whether it is your professional career, a party, or life itself. But the problem is to know when that might be. I knew that I was not yet ready to give up neurosurgery, even though I was so anxious to stop working in my hospital in London. I hoped to go on working part-time, mainly abroad. This would mean that I would need to be revalidated by the General Medical Council if I were to remain a licensed doctor.

Aircraft pilots need to have their competence reassessed every few years and, it is argued, it should be no different with doctors, because both pilots and doctors have other people's lives in their care. There is a new industry called Patient Safety, which tries to reduce the many errors that occur in hospitals and which are often responsible for patients coming to harm. Patient Safety is full of analogies with the aviation industry. Modern hospitals are highly complex places, and many things can go wrong. I accept the need for checklists and trying to instil a blame-free culture, so that mistakes and errors are identified and, hopefully, avoided. But surgery has little in common with flying an aircraft. Pilots do not need to decide what route to fly or whether the risks of the journey are worth taking, and then discuss these risks with their passengers. Passengers are not patients: they have chosen to fly, patients do not choose to be ill. Passengers will almost certainly survive the flight, whereas patients will often fail to leave the hospital alive. Passengers do not need constant reassurance and support (apart from the little charade where the stewardesses and stewards mime the putting-on of life jackets and point confusingly to the emergency exits). Nor are there anxious, demanding relatives to deal with. If the plane crashes, the pilot is usually killed. If an operation goes wrong, the surgeon survives, and must bear an often overwhelming feeling of guilt. The surgeon must shoulder the blame, despite all the talk about blame-free culture.

To revalidate doctors is important but not easy, and it took the General Medical Council in Britain many years to decide how to do it. As well as being 'appraised' by another doctor, I had to complete a '360-degree' assessment by several colleagues, and one by fifteen consecutive patients. I was tempted, when instructed to provide the names of colleagues, to name ten people who disliked me (alas, not very difficult), but I chickened out, and instead listed various people who would be unlikely to find great fault with me. They ticked the online boxes, saying how good I was, and how I achieved a satisfactory 'work–life balance', and I returned the favour when they sent me their 360-degree forms.

I was provided with fifteen questionnaires to hand out to patients. The exercise was managed by a private company – one of the many profitable businesses to which much NHS work is now outsourced. These companies prey off the NHS like hyenas off an elderly and disabled elephant – disabled by the lack of political will to keep it alive.

I was told to ask the patients to complete the lengthy, two-sided form after I had seen them in my outpatient clinic and to have them return the forms to me. Not surprisingly, I was on my best behaviour. Besides, the patients would probably have been reluctant to criticize me to my face. My patients obediently filled in the forms. It seemed to me that whoever would be examining them might well suspect that I had fraudulently completed them myself, as all the completed forms were both eulogistic and anonymous. I was tempted to do this but to accuse myself of being impatient and unsympathetic – in short, of being a typical surgeon – and see if this made any difference to the absurd charade.

My first neurosurgical post had been as a senior house officer in the hospital where I had trained as a medical student. There were two consultant neurosurgeons, the younger one very much my mentor and patron. The senior surgeon retired shortly after I started working in the department. He rang me once at night when I was on call, seeking advice about a friend of his who had passed out at home, asking whether this might be due to his blood-pressure drugs. It was fairly obvious that the friend was himself. I remember once standing with him in front of an X-ray screen looking at an angiogram – an X-ray that shows blood vessels – of a patient with a difficult aneurysm, and him telling me to ask his younger colleague to take over the case.

'By my age, aneurysm surgery is not good for the coronaries,' he said. I knew that recently one of the senior neurosurgeons in Glasgow had clipped an aneurysm and then immediately collapsed with a major heart attack.

My senior consultant's career ended gloriously with a

successful operation on a large benign brain tumour in a young girl. She recovered perfectly and a few days later, still in her hospital gown and with a shaven head, came to his retirement party to present him with a bouquet of flowers. I believe that he died a few months afterwards. My own surgical career, thirty-four years later, was to end ignominiously.

I had two weeks left before retiring and I was looking at a brain scan with my registrar, Samih.

'Fantastic case, Mr Marsh!' he said happily, but I did not reply. Until recently, I would have said exactly the same myself. The difficult and dangerous operations were always the most attractive and exciting ones, but as my career approached its end I was finding that my enthusiasm for such cases, and for the risk of disaster, was rapidly diminishing. The thought of the operation going badly, and of my leaving a wrecked patient behind me after my retirement, filled me with dismay. Besides, I thought, as I am soon to give all this up, why must I go on inflicting it on myself? But the patient had been referred to me personally by one of the senior neurologists. Suggesting that one of my colleagues do the operation instead was out of the question: it was just not compatible with my self-esteem as a surgeon.

'It should separate away from all the vital bits,' I said to Samih, pointing to the tumour on the scan. The tumour was growing at the edge of the foramen magnum. Damage to the brainstem or the nerves branching off it can be catastrophic for the patient, including paralysis of swallowing and coughing. This can lead to fluid in the mouth getting into the lungs and causing a very severe form of pneumonia that can easily be fatal. At least the tumour appeared benign. It did not look as though it would be stuck to the brainstem and spinal nerves so, at least in theory, it should be possible to remove the tumour without causing severe damage. But you can never be certain.

It was Sunday evening and Samih and I were sitting in front of the computer at the nurses' station on the men's ward. We both regretted the fact that our work together was soon to end. The close relationship you can have with your trainees is one of the great pleasures of a surgeon's life.

It was early March, and it was dark outside but the sky was clear; there was a very bright full moon, low over south London, which I could see through the ward's long line of windows. There had been a scent of spring in the air as I had bicycled in to work, along the back streets, the moon cheerfully racing along beside me over the slate roofs of the terraced houses.

'I haven't met him yet,' I said. 'So we had better go and talk to him.'

We found the patient in one of the six-bed bays, the curtains drawn around the bed.

'Knock, knock,' I said, drawing the curtain aside.

Peter was sitting up. There was a young woman in the chair beside the bed. I introduced myself.

'I'm so pleased to see you at last,' he said, looking much happier than most of my patients when I first meet them. 'The headaches have really been getting awful.'

'Have you seen the scan?' I asked.

'Yes, Dr Isaacs showed it to me. The tumour looked huge.'

'It's not that big,' I replied. 'I have seen many bigger, but then one's own tumour always looks enormous.'

Samih had pulled along one of the new mobile computer stations from the corridor and placed it at the end of Peter's bed. He summoned up the brain scan while we talked.

'That's a centrimetric scale there,' I explained, pointing to the edge of the scan. 'Your tumour is four centimetres in diameter. It's causing hydrocephalus – water on the brain – it's acting like a cork in a bottle and trapping the spinal fluid in your head where it is supposed to drain out at the bottom of your skull. Without treatment – although I apologize for terrorizing you – you only have a few weeks to live.'

'I can believe that,' he said. 'I've been feeling really lousy, though the steroids Dr Isaacs started me on helped a bit.'

We talked for a while about the risks of the operation – death or a major stroke were possible but unlikely, I said, and he might have difficulty swallowing. He nodded and told me that in recent weeks he had sometimes choked when eating. We talked also about his work, and about his children. I asked his wife what they knew about their father's illness.

'They're only six and eight,' she said. 'They know their Daddy is coming to hospital and that you are going to make his headaches better.'

While we talked, Samih filled up the long consent form and Peter signed it quickly.

'I'm not at all frightened,' he said, 'and I'm really glad I've got you to do it just before you retire.' I let this pass – patients want to think their surgeon is the best and don't particularly like it when I tell them that I am not and that I am dispensable. Samih noted his wife's phone number down on the edge of the consent form.

'I'll ring you after the op,' I said to Peter's wife. 'See you tomorrow.' I waved to Peter and slipped out between the curtains. There were five other men in the room who looked up at me as I left – no doubt they had all listened to the conversation with great interest.

As I cycled to work next morning, I reflected on the strange fact that almost forty years of working as a surgeon were coming to an end. I would no longer have to feel constantly anxious, with my patients so often on the edge of disaster, yet for almost forty years I had never had to worry about what to do each day. I had always loved my work, even though it was often so painful. Every day was interesting; I loved looking after patients, I loved the fact that I was – at least in my own little hospital pond – quite important, indeed my work had frequently felt more like a glorious opportunity for adventure and self-expression than mere work. It had always felt profoundly meaningful. But in recent years this love had started to fade. I attributed this to the way in which working as a doctor felt increasingly like being an unimportant employee in a huge corporation. The feeling that there was something special about being a doctor had disappeared – it was just another job, I was just a member of a team, many of whose members I did not even know. I had less and less authority. I felt less and less trusted. I had to spend more and more time at meetings stipulated by the latest government edicts that I felt were often of little benefit to patients. We spent more time talking about work rather than actually working. We would often look at brain scans and decide whether the patient should be treated or not without any of us having ever seen the patient. Like almost all the doctors I knew, I was becoming deeply frustrated and alienated.

And yet despite this, I was still burdened with an overwhelming sense of personal responsibility for my poor patients. But perhaps my discontent was because I had less and less operating to do – although I was lucky compared to many other surgeons in that I still had two days of operating a week. Many of my colleagues are now reduced to a single day each week; you may well wonder what they are supposed to do for the rest of the week. Recent increases in the number of surgeons have not been matched by any increase in the facilities we need in order to operate. Or then again, perhaps it was simply because I was getting old and tired and it was time to go. Part of me longed to leave, to be free from anxiety, to be master of my own time, but another part of me saw retirement as a frightening void, little different from the death, preceded by the disability of old age and possibly dementia, with which it would conclude.

There had been fewer emergency admissions than usual over the course of the weekend and there were empty beds on the ITU, so I was told that my list could start on time. The anaesthetist, Heidi, had been away on prolonged leave to look after her young son and was now back at work part-time. We were old friends and I was relieved to see her. The relationship between anaesthetist and surgeon is critical, especially if there is going to be trouble, and having colleagues who are friends is all-important. I walked into the anaesthetic room where Heidi and her assistants had Peter already asleep. The ODA – the operating department assistant, whose job is to help the anaesthetist – was stretching a wide band of Elastoplast across his face to keep the endotracheal tube – the tube which Heidi had inserted through his mouth, down his throat and into his lungs – in place. His face now disappeared beneath the Elastoplast, and the process of depersonalization that starts as the intravenous anaesthetic takes effect and the patient becomes unconscious was now complete.

I have watched that process thousands of times – it is, of course, one of the miracles of modern medicine. One moment the patient is talking, wide awake and anxious – although a good anaesthetist like Heidi will be soothing and reassuring – and the next instant, as the intravenously injected drug travels up the veins of the arm via the heart to the brain, the patient sighs, the head falls back a little, and he or she is suddenly and deeply unconscious. As I watch, it still looks to me as though the patient's soul is leaving the body to go I know not where and all I now see is an empty body.

'It might bleed a bit,' I said to Heidi, 'and the brainstem might be a problem.' Sudden and alarming changes in the patient's heart rate and blood pressure, even cardiac arrest, can occur if you get into trouble with the lower part of the brainstem, known as the medulla oblongata.

'Not to worry,' said Heidi. 'We're prepared. Big IV and plenty of blood cross-matched, ready in the fridge.'

Peter was wheeled into the operating theatre and, having assembled the theatre staff, we rolled him off the trolley face-down onto the operating table with Samih holding his head.

'Prone, neutral position, head well flexed,' I told him. 'Get him in the pins. Midline incision with the craniectomy more to the left and take out the back of C1. Give me a shout when you've done that and you're down to the dura and I'll come and join you.'

I left the operating theatre and went round to the surgeons' sitting room for the regular Monday morning meeting with my consultant colleagues. The meeting had already started, with our two line managers in attendance – both of whom, I might add, I liked and got on well with. The meetings were to discuss the day-to-day business of the neurosurgical department and the managers would sometimes tell us about the department's 'financial position'. Much of the meeting was spent letting off steam about all the petty frustrations and inefficiencies of working in a large hospital. There was a sky-blue cushion in the shape of a brain that had been given to me by the sister of one of my American trainees and sometimes we would throw it around the room as we talked, rather like holding the conch shell in Golding's

*Lord of the Flies*. Sean, the senior of the two managers, was talking. He declined to hold the cushion when I threw it at him.

'I'm afraid that this last year we made only one million pounds' profit for the Trust whereas the year before we made four million, even though we did not do any more work. We used to be one of the most profitable departments in the Trust but that is no longer the case.'

'But where on earth did the three million go?' somebody asked.

'It's not very clear,' Sean replied. 'We spent a lot on agency nurses. And you're spending a lot more on putting metalwork into people's spines and you're doing too many emergencies – we get only thirty per cent payment if you exceed the target for emergency work.'

'It's so bloody ridiculous,' I snorted. 'What would the public say if they knew we got penalized for saving too many lives?'

'You know the reason,' Sean said. 'It's to stop hospitals making cases into emergencies when they're not emergencies and over-claiming.'

'Well, we never did that,' I replied.

I should explain that 'profit' in an NHS department is not profit in the usual sense – instead it is whether we have exceeded our 'financial target' or not, which is based on previous performance and is an arcane process that I find entirely incomprehensible. Any 'profit' that we make goes to prop up less profitable parts of the Trust, so, despite the introduction into the NHS of the incentives and penalties so loved by economists, there is little real motive at a clinical level, on the shop floor, to work more efficiently. Besides, whenever there does seem to be any extra money, it all appears to be spent on employing more and more members of staff, as though to encourage the existing members of staff to do less work.

The conversation meandered on for a while, discussing the problem of spinal implants. There is no easy answer to this question. As intracranial neurosurgery has declined, replaced by non-surgical methods such as the radiological treatment of aneurysms and highly focused radiation for tumours, neurosurgeons (and there are ever-increasing numbers of them, all keen to operate) have moved into spinal surgery. This is largely about inserting all manner of very expensive titanium nuts and bolts and bars into people's backs, for cancer or for backache, although the evidence base and justification for such surgery, at least for back pain, are very weak. Even with the cancer patients – metastatic cancer often spreads to the spine – it can be a moot point as to whether to operate or not as the poor patient is going to die anyway, sooner or later, from the underlying cancer. Spinal implant surgery is major surgery and is a six-billion-dollar-a-year business in the US. It is a prime example of the 'over-treatment' that is a growing problem in modern health care, and especially in commercial, marketized health-care systems such as in America.

I stopped doing such surgery myself some years ago in order to concentrate on brain surgery, so I was happy to abandon the conversation when I was summoned back to the operating theatre, where Samih had started the operation.

'Let's have a look,' I said, and I leant forward, taking care not to touch the sterile drapes, to peer into the large hole in the back of Peter's head. 'Very good,' I commented. 'Open the dura and I'll go and put some gloves on. Jinja,' I said to the circulating nurse (the nurse who is not scrubbed up and does the fetching and carrying while the operation proceeds), 'can you get the scope in please?'

While Jinja shoved the heavy scope up to the operating table I scrubbed up at the large sink in the corner of the room – a soothing and deeply familiar act, although always accompanied by a feeling of tension in the pit of my stomach. I must have done this many thousands of times over the years and yet now I knew that it was soon to end – at least in my home country.

Jinja came and tied up the laces at the back of my blue gown and I marched up to the table where Peter lay hidden under the sterile blue drapes, with only the gaping and bloody hole in the back of his head to be seen, brilliantly lit by the operating lights. Samih opened the dura – the leathery, outer layer of the meninges – with a small pair of scissors while I watched. I then took over. I sat down on the operating chair with its arm rests. The first rule of microscopic surgery, I tell my trainees, is to be comfortable, and I usually sit when operating, although in some departments this is not considered to be very manly, and the surgeons stand throughout the procedure, often for very many hours on end.

It was easy enough to find the tumour – a bright-red ball shining in the microscope's light – a few millimetres beneath the back part of the brain, the cerebellum. To the left would be the all-important brainstem, and to the right and deep down the lower cranial nerves, scarcely thicker than thread; but all this was hidden by the tumour. I would not be able to see them until the very end of the operation, when I had removed most of the tumour. As soon as I touched the tumour with the sucker, blood spurted up out of it.

'Heidi,' I said, 'it's going to bleed.'

'No problem,' came the encouraging reply, and I settled down to attack the tumour.

'If the blood loss gets too much,' I said to Samih, 'your anaesthetist might ask you to stop and pack the wound, but then you worry you might damage the brain with the packing. If it looks as though the patient is going to bleed to death – to exsanguinate – sometimes you just have to operate as quickly as possible, get the tumour out before the patient dies and just hope you haven't damaged anything. The bleeding usually stops once the tumour is all out.'

'I saw you do a case like that when you came to Khartoum,' Samih commented.

'Ah yes. I'd forgotten that. He did OK though . . .'

It took four hours of intense concentration to get the tumour out. Down the three-centimetre-wide hole in Peter's head, all I could see was bright-red arterial blood, welling endlessly upwards. There was no way I could see the brain and no way I could delicately dissect the tumour off it. To my disappointment I did not enjoy the operation, which I think I would have done in the past. I should have arranged to do the operation jointly, I told myself, with a colleague. This greatly reduces the stress of operating, but I had not expected the tumour to bleed quite so much, and it is always difficult as a surgeon to ask for help, as bravery and self-reliance are seen as such an important part of the job. I would hate my colleagues to think that I was getting old and losing my nerve.

'Look, Samih,' I said, 'the damn thing did separate away.' With the tumour finally removed and the bleeding stopped, we could see the brainstem, and the lower cranial nerves and the vertebral artery all perfectly preserved. It made me think of the moon, appearing from behind clouds and transforming the night. It was a good sight.

'We were lucky,' I said.

'No, no,' said Samih, obeying the first rule for all surgical trainees, flattering me. 'That was fantastic.'

'Well, it didn't feel it,' I replied, and then shouted across to the far end of the table, 'Heidi, what was the blood loss?'

'Only a litre,' she said happily 'No need to transfuse him. His haemoglobin is still one hundred and twenty.'

'Really? It felt like a lot more,' I said, thinking that maybe I had been unnecessarily nervous during the operation. I consoled myself with the thought that perhaps all the years of experience counted for something after all. But Peter was going to be all right and that was all that mattered, and his young children would be happy that I had cured their Daddy's headaches.

'Come on, Samih,' I said, 'let's close.'

Peter awoke well from the operation. His voice was hoarse but I checked that he could cough, so I was not worried that he was at risk of aspiration.

I went back to the hospital late in the evening to see the post-op cases. I went in most evenings: I live nearby so it was easy for me and I knew that my patients liked seeing me on the evenings both before and after their surgery. It was also a private protest against the way in which doctors now are expected to work shifts with fixed hours, and medicine is no longer perceived as a vocation, a true profession. Many doctors now seem to have the same expectation.

I walked onto the ITU and found Peter among the two long lines of beds on either side of the warehouse-like room, each with its own nurse at the foot end, and a little forest of high-tech monitoring equipment at the head end.

'How is he?' I asked the nurse.

'He's OK,' came the reply. There are so many ITU nurses that I know only a few of them and I did not recognize this one. 'We had to put a nasogastric down in case he aspirated . . .'

To my surprise, when I looked at Peter, who was sitting upright in his bed, wide awake, I saw that somebody had indeed put a nasogastric tube up his nose and then taped it to his face. I was angry that he had been subjected to the unpleasant procedure of having the tube inserted; it should not have been done, as he did not need it. The tube is pushed up the nose and then down the back of the throat into the stomach – a very unpleasant experience, I am reliably informed by my wife Kate, who has personal experience of it. Nor is it an entirely harmless procedure, and cases have been recorded of the end of the tube getting into the lungs and causing aspiration pneumonia and death, or even getting into the brain. These are, admittedly, rare complications, but after such a difficult but successful operation I was furious that it had been done. The decision to insert it had been made by one of the ITU doctors, clearly less experienced than I was, and the doctor on duty on the ITU for the night denied all knowledge of it. There seemed little point in blaming the nurse. I asked Peter how he felt.

'Better than I expected,' he said in a slightly hoarse voice, and then proceeded to thank me again and again for the operation. I bid him goodnight and told him that we'd remove the wretched nasogastric tube in the morning.

I went into work next morning, and immediately went with Samih to the ITU. There was a different nurse at the end of Peter's bed whom, once again, I did not recognize. Peter was awake and told me that he'd managed to sleep a little – quite an achievement in all the inhuman noise and bright lights of the ITU. I turned to the nurse.

'I know you didn't insert the nasogastric tube, but please take it out,' I said.

'I'm sorry, Mr Marsh, but he will have to be checked by SALT.'

SALT are the speech and language therapists who some years ago started to assume responsibility for patients with swallowing problems as well as speech difficulties. I had had several disagreements with speech therapists in the past when they had refused to sanction removal of nasogastric tubes which in my opinion the patients did not need. As a result several patients had been kept in hospital being unnecessarily tube-fed, despite my protests. I was not the speech therapists' favourite neurosurgeon.

'Take the tube out,' I said, between gritted teeth. 'It should never have been inserted in the first place.'

'I'm sorry Mr Marsh,' the nurse replied politely, 'but I won't.'

I was seized by a furious wave of anger.

'He doesn't need the tube!' I shouted. 'I will take responsibility. It is perfectly safe. I did the operation – the brainstem and cranial nerves were perfectly intact at the end, he's got a good cough . . . take the bloody tube out.'

'I'm sorry Mr Marsh,' the hapless nurse began again. Overcome with rage and almost completely out of control, I pushed my face in front of his, took his nose between my thumb and index finger and tweaked it angrily.

'I hate your guts,' I shouted, turning away, impotent, furious and defeated, to wash my hands at the nearest sink. We are supposed to clean our hands after touching patients, so I suppose the same applies to assaulting members of staff. Years of frustration and dismay at my steady loss of authority, at the erosion of trust and the sad decline of the medical profession, had suddenly exploded – I suppose because I knew I was to retire in two weeks' time and suddenly could no longer restrain my rage and feeling of intense humiliation. I stormed off the ward followed by Samih, leaving a little group of amazed nurses standing at the end of Peter's bed. I do not often lose my temper at work and have certainly never laid a hand on a colleague before.

I slowly calmed down and returned later in the day to the ITU to apologize to the nurse.

'I'm very sorry,' I said. 'I shouldn't have done that.'

'Well, what's done is done,' he replied, though I did not know what he meant and wondered whether he would be making an official complaint – to which I felt he was fully entitled. Towards the end of the day I received an email from the matron for the ITU saying that she had learnt that there had been an 'incident' on the ITU and asking me to come and talk to her the next day.

I went home in a state of craven and abject panic, the like of which I had scarcely ever known before. It took me a long time to calm down – I was so pathetically frightened by the prospect of some kind of official disciplinary action being launched against me. Where's the brave surgeon now? I asked myself as I lay on my bed, shaking with fear and anger. It's time to go, it really is.

Next morning I duly reported to the ITU matron – a colleague I knew well and had been working with for many years. It brought back memories of being summoned to the headmaster's office at school for some misdemeanour in the past, and of my intense anxiety as I waited outside the door. Sarah, the ITU matron, and I had been together at the old hospital which had been closed twelve years earlier. It had become something of an anomaly: a single-specialty hospital, with a staff of about 180, dealing only with neurosurgery and neurology in a garden suburb surrounded by trees and gardens. There were some good clinical reasons for integrating us into the major hospital where we now work, with a staff of many thousands; and the site of the old hospital, Atkinson Morley's in Wimbledon (AMH), was of course far too beautiful to be a mere hospital. It was sold for commercial development and the hospital turned into apartments that now cost millions of pounds.

But we lost a lot as well – above all the friendly working relationships that can come when you work in a small organization where everybody knows each other on a personal level and work together on the basis of personal obligation and friendship. The efficiency of the hospital was a perfect illustration of Dunbar's number – that magic number of 150. The size of our brain, Robin Dunbar, an eminent evolutionary anthropologist at Oxford University, has argued (and the brain size of other primates), is determined by the size of our 'natural' social group, when humans and their brains evolved in small hunting and gathering groups. We have the largest brains among primates, and the largest social group. We can relate to about 150 people on an informal, personal basis, but beyond that leadership, impersonal rules and job descriptions become necessary.

So Sarah knew me quite well. Some of the comradely atmosphere of the old hospital had been preserved, despite the best efforts of the management to merge our department into the anonymous collective of the huge hospital where we now worked. I think anybody else in the nursing hierarchy of the hospital would have initiated some kind of formal disciplinary procedure against me.

'I'm ashamed of myself,' I told her. 'I suppose it happened partly because I know I'm leaving . . .'

'Well, he wasn't to know that SALT for you is like a red rag to a bull. He doesn't want to make a formal complaint but he said you were very frightening and it brought back memories of an assault he suffered some months ago.'

I hung my head in shame and remembered how my first wife had told me how terrifying I could look, as our marriage fell apart with furious arguments.

'He handled me very well and kept admirably calm,' I

said. 'Please thank him when you next see him. It won't happen again,' I added with a slight smile. Sarah knew well enough that I was about to retire. I left her office and went round to the men's ward where Peter had been sent the previous evening. At least the senior nurse there had been happy to remove the wretched nasogastric tube at my request and it was nice to find Peter drinking a cup of tea without any problems, although he certainly had a very hoarse voice.

'I'm not supposed to attack the nurses in front of the patients,' I said. 'I'm really sorry.'

'No, no, not at all,' he replied with a croaking laugh. 'I told them I didn't need the tube and could swallow perfectly well but they wouldn't listen to me and just shoved it in. I was on your side.'

My last operation here, I thought, as I cycled home in the evening.

I finally left my hospital two weeks later, having cleared my office. I disposed of the accumulated clutter that a consultant surgeon acquires over the course of his career. There were letters and photographs from grateful patients, presents and plaques, and outdated textbooks, some of which had belonged to the surgeon whom I had replaced almost thirty years earlier. There were even some books, and an ophthalmoscope, that had belonged to his predecessor, the famous knighted surgeon who seventy years ago had created the neurosurgical department in which we worked. I spent days emptying eight filing cabinets, occasionally stopping to read with amusement some of the pronouncements and plans and protocols, reports and reviews, generated by a labyrinth of government offices and organizations, mostly now defunct, renamed, reorganized or restructured. And there were files dealing with cases where I had been sued, or bitter letters of complaint, from which I quickly averted my eyes – the memory was so painful. Having done all this I left my office, empty, for my successor. I had no regrets whatsoever.

# 3

# NEPAL

There was a minor earthquake in the evening, small enough to be exciting rather than frightening. We were sitting in the garden, in the dusk, the crescent moon in the west blood-red with the city's polluted air, when there was a sudden low sound, almost like a breath of wind or a subterranean thought – a fleeting presence of something of immense size and distance. The bench I was sitting on in the garden briefly shook as though somebody had nudged it, and thousands of voices rose up all around us in the night from the dark valley below, wailing, crying out in fear like the damned on hearing that they are to go down to hell, and all the dogs of Kathmandu started barking furiously. And then, when it became clear there was not to be a major quake like the one which had killed thousands of people the year before, everything fell quiet and we could hear the cicadas again.

I slept very well that night and woke to the dawn chorus of the birds singing in the garden. A pair of syncopated cuckoos were calling, while the hooded crows croaked and quarrelled in the camphor tree and all the cocks in the valley crowed. At ten past eight I set off for the hospital – it is a walk of which I never tire and, for reasons I struggle to understand, I feel more deeply content as I go to work each morning than I have ever felt before. The rising sun casts long and peaceful shadows. The air is often hazy with pollution, but sometimes I am lucky and I can see the foothills that surround the city and, just peaking above them in the distance, the snow-covered summit of Mount Ganesh, the elephant god.

At first I walk in silence, apart from the birdsong, past houses with cascades of crimson and magenta bougainvillea at the entrance, and Buddhist prayer flags, like coloured handkerchiefs on a washing line, on the roof. The houses are all built of rendered brick and concrete, painted cheerful colours and look like stacked-up matchboxes with balconies and roof terraces and the occasional added gable or Corinthian column. Sometimes there is a peasant woman watching over a couple of cows, peacefully grazing on the thin and scruffy grass at the side of the cracked, uneven road. There is rubbish everywhere, and stinking open drains. Dogs lie sleeping on the road, probably worn out by a night's barking. Sometimes I walk past women carrying huge baskets of bricks on their backs, supported by straps across their foreheads, to a nearby building site. After the houses there are then many small shops, all open at the front; looking into them is like opening a storybook, or peeking into a doll's house.

Life here is lived on the street. There is the barber shaving a man with a cut-throat razor; another customer reads a newspaper while waiting, and the meat shop with ragged lumps of fresh meat and the severed head of a mournful, lop-eared goat looking blankly at me as I pass. There is the cobbler sitting cross-legged on the ground while he cuts soles from rubber sheeting, with cans of adhesive stacked against the wall. Cobblers are *dalits*, the untouchables of the Hindu caste system, and second only to the sweepers and cleaners, who are at the very bottom of society. He once repaired my brogue boots which have accompanied me all over the world, and which I polish assiduously every morning – the only practical activity I have when in Nepal, other than operating. He did a very good job of it and it was only when I later learned that he was a *dalit* that I understood why he at first looked awkward and embarrassed when I politely greeted him each morning as I passed his open workshop. There is the metalworker welding metal in a shower of blue sparks and a seamstress, with clothes hanging up at the front of her shop, while she sits at the back. I can hear the whirring of her sewing machine as I walk past. Motorbikes wind their way between the children in smart uniforms going to school. The children will look slightly askance at me – this is not a part of town to which expats normally come – and if I smile at them they give me a happy smile in return and wish me a good morning. I would not dare to smile at children back in England. There is a rawness, a directness to life here, with intense and brilliant colours, which was lost in wealthy countries a long time ago.

I walk past all these familiar sights to reach the main road, a melee of cars, trucks and pedestrians, with swarms of motorbikes weaving their way between them in a cloud of pollution, all blowing their horns. The broken gutters are full of rubbish, and next to them there are fruit vendors selling apples and oranges from mobile stalls that rest on bicycle wheels. There are long lines of colourful, ramshackle shops, and everywhere you look, hundreds of people going about their daily business, many of them wearing face masks which are, of course, useless against vehicle fumes. Electric cables droop like tangled black cobwebs from the pylons, which lean at drunken angles, and there are often broken ends with exposed wires, hanging down onto the pavement. I cannot even begin to imagine how any repair work is ever carried out. The women, with their fine faces, their jet-black hair swept back from their foreheads and their spectacularly colourful dresses and gold jewellery, transform what would otherwise often be depressing scenes of grinding poverty.

I have to cross the road to reach the hospital. I found this at first an unnerving experience. The traffic is chaotic and if you wait for a break in it, you will be there for a very long time. You must calmly step out onto the road, join the traffic, and walk slowly and predictably across, trusting the buses, vans and motorcycles to weave their way around you. Some of the motorcyclists have their helmets pushed back over their heads, so they look like the ancient Greek warriors to be seen on Attic vases. If you break into a run they are more likely to hit you by mistake. My guidebook to Nepal helpfully told me that 40 per cent of victims of road traffic accidents – RTAs, as they are called in the trade – are pedestrians. We admitted such cases every day to the hospital. I was to witness several fatal accidents. On one such occasion I passed a dead pedestrian on the Kathmandu ring road. He was sprawled on his face across the gutter, his legs bent out akimbo at an improbable angle like a frog's, with a group of curious onlookers watching silently as the police made notes. I have come to enjoy crossing the road – there is a feeling of achievement each time I get across it safely.

When I was a student almost fifty years ago, Kathmandu had been the fabled, near-mystical destination for many of my contemporaries. This was partly because cannabis grows wild in Nepal – and still does on building plots and derelict land in the city – but also because it was a place of pristine beauty and still living a life of medieval simplicity. They would trek overland. The world was a different place: you could safely travel through Syria, Iran and Afghanistan. Since then Kathmandu has also changed, almost beyond recognition. The population of Kathmandu has gone from a few hundred thousand twenty years ago to two and half million and it is the fastest-growing city in South East Asia. The new suburbs are entirely unplanned, without any proper infrastructure, occasionally with a few pathetic scraps of rice paddy or wheat field left as an afterthought between the cheap concrete buildings. There are open drains and dirt tracks, with rubbish and building materials strewn about everywhere. The roads are chaotic and the air is dark with pollution. You can rarely, if ever, see the high Himalayas to the north.

Nepal is one of the poorest countries in the world, shattered by a recent earthquake and with the ever-present threat of another catastrophic one to come. There are minor tremors every week. I am working with patients with whom I have only the most minimal human contact. The work is neurosurgical, so there are constant failures and disasters, and the patients' illnesses are usually more advanced and severe than in the West. The suffering of the patients and their families is often terrible, and you have to fight not to become inured and indifferent to so much tragedy. I can rarely, if ever, feel pleased with myself. The work, if I care to think about it, is often deeply upsetting and, compared to Public Health, of dubious value in a country as poor as Nepal. The young doctors I am trying to train are so painfully polite that I am never sure what they really think. I do not know whether they understand the burden of responsibility that awaits them if they ever become independent neurosurgeons. Nor do I know what they feel about their patients, or how much they care for them, as their English is limited and I cannot speak Nepali. What I do know is that most of them want to leave Nepal if they possibly can. Their pay and professional prospects here are poor compared to what they can find in wealthier countries. It is a tragedy affecting many low-income countries such as Nepal and Ukraine – the educated younger generation, the countries' future, all want to leave. I am working in a very alien, deeply superstitious culture with a cult of animal sacrifice, centred on blood.

Few, if any, of the patients and their families understand the unique and overarching importance of the brain, of the physical nature of thought and feeling, or of the finality of death. Few of the patients or their families speak English, and I feel very remote from them. They have wholly unrealistic expectations of what medicine can achieve, and take it very ill if things go badly, although they think we are gods if we succeed. I lead a life of embarrassing luxury compared to most people here – in my colleague Dev's guest house, with its little paradise of a garden – but I live out of a suitcase, with none of the property and possessions that dominate my life back in England. I am in bed by nine in the evening and up by five, and spend ten hours a day in the hospital, six days a week. I miss my home and family and friends intensely. Yet when I am here I feel that I have been granted a reprieve, that I am in remission, with the future postponed.

The day before my flight to Nepal had not been uneventful. I had reported to the private hospital where, for many years, I had worked in my own time, in addition to my work for the NHS, although I had stopped all private practice two years earlier. Over the preceding weeks I had noticed a slightly scaly lump growing on my forehead. One of the privileges of being a doctor is that you know to whom to go if you have a problem, and a plastic surgeon I knew well, and greatly liked, had told me the lump should be removed.

'You must have got the supra-orbital nerve. I can't feel a thing. The top of my head feels like wood,' I said to David once he had started, although I could feel the pressure of the scalpel cutting into my forehead. I had often subjected my own patients to this – although usually with much longer incisions and more local anaesthetic. This had been in order to saw into their skulls and expose their brains for an awake craniotomy, an operation I had pioneered for brain tumours, where you operate on the patient's exposed brain while they are awake. This was the first time that I could understand a little of what they would have experienced. I could feel David mopping up my blood as it ran down into my ear.

'Hmm,' he said. 'There are two points to it. It looks a bit invasive. You may need wider resection and skin-grafting.'

I felt a sudden surge of anxiety: although he was avoiding the word, he was obviously talking about cancer. I had thought removal of the little lump growing on my forehead was all going to be very simple. I now imagined myself with a large and ugly skin graft on my forehead. Perhaps I would need radiotherapy as well. I couldn't help but remember some of the patients I had treated with malignant scalp tumours that had eventually eaten their way through their skulls and bored into their brains.

'But it is curable, isn't it? And they don't normally metastasize do they?'

'Henry, it will all be fine,' David said reassuringly, probably amused by my anxiety.

'And can it wait two months?' I asked.

'Yes, I'm sure it can, but we'll have to see what the microscopy shows. How invasive it is. I'll email you.'

Doctors traditionally pay their colleagues for their services in wine, and before I left I arranged for some to be sent to David. Many years ago I operated on a local GP's wife with a difficult cerebral aneurysm, and she died immediately afterwards; I felt I was to blame. I was deeply ashamed when he sent me a case of wine some weeks after her death but it was, I now understand, an act of great professional kindness.

So I was on the plane to New Delhi next day, en route to Kathmandu, sporting a large, sticky plaster on the right side of my forehead, which I inspected gloomily in the mirror whenever I went to the cramped little toilet on the eight-hour flight, cursing my prostatism and skin cancer.

Having braved the traffic, I walk down the steep drive to Neuro Hospital, as it is called, set in a small valley off the main road. When the hospital was built ten years ago this was a rural area of paddy fields, but now it is almost entirely built up, although there is still one small paddy field left stranded, with a banana tree, next to the hospital.

The full name of the hospital Dev built is the National Institute for Neurology and Allied Sciences. It is large and spacious and spotlessly clean, with good natural light almost everywhere. The hospital is surrounded by gardens, just like AMH, the old hospital in Wimbledon, where Dev and I had trained together many years ago. Many of the patients – the women in brilliantly coloured dresses, deep reds, blues and greens, often with gold decorations – wait on the benches in front of the entrance. Dev planted a magnolia tree there, in memory of the magnolia tree that grew in front of AMH (that particular tree has now been felled as part of the conversion of the old and famous hospital into luxury flats). At night there will be many families sleeping on mats outside the side entrance. It is strange to come to a country as poor as Nepal and find such a sympathetic hospital, with so many windows and so much space, and so clean and well cared for. It incorporates all the lessons Dev learnt from working in small, specialist hospitals in Britain. It is a perfect embodiment of the architectural adage – so neglected in the hospital construction in Britain of recent years – that the secret of a successful building is an informed client. Dev knew exactly what would make his hospital work efficiently.

There are uniformed guards in military caps at the entrance, who snap to attention as I enter.

'Good morning sir!' they say, whipping off smart salutes. The receptionists, in elegant blue saris, smile at me while pressing their hands together in respectful greeting.

'*Namaste*, Mr Marsh!'

This is rather different from entering my hospital in London in the morning.

Nepal has a very strong caste system. Ritual burning of widows and slavery were abolished only in 1924. Although discrimination on the grounds of caste or ethnicity is illegal, caste is still very important. Nepal was entirely closed to outsiders until the 1950s, and ruled by an absolute, feudal monarchy where the king was believed to be the incarnation of the god Vishnu. The end of the monarchy was precipitated in 2001 by the crown prince taking a submachine gun to his own parents, killing them and several other family members. He was then shot in the head – there are conflicting accounts as to whether he did this himself or not. Dev operated on him, carrying out a decompressive craniectomy, but – I suspect to everybody's relief – he died. There are over a hundred ethnic groups, often with their own languages and castes. It is a nation of immigrants – Mongols from the north and Indians from the south, often living in isolated mountain valleys. It remains a deeply divided and hierarchical society, although most people still look up to foreigners, who are treated with respect, verging on servility. Landlocked, stuck between China and India – described by one of its most famous kings as a 'yam between two rocks' – ethnically so diverse and hierarchical, desperately poor and damaged by the recent earthquake, over-dependent on foreign aid and NGOs, Nepal is a tragic mess. The politics of the country is largely the politics of patronage and corruption, with little sense of the public good and public service which we take for granted in the West. The towns are festooned with advertisements for foreign language courses, promising work abroad. Most Nepalis, if they possibly can, want to leave Nepal. And yet, as an outsider, it is almost impossible not to fall in love with the land and its people.

Can you really fall in love with a country, with a people? I thought that you could only fall in love with a person, but in my first weeks there I started to feel for Nepal as I felt for the women with whom I have fallen in love – seven in total – over the course of my life. Yet I knew that the intensity of my feelings for Nepal would be just as ephemeral as my feelings for the women with whom I had been in love (and much of the love was unrequited anyway). Furthermore, I was leading an utterly spoilt and luxurious life, waited on hand and foot, and in one of the poorest countries on the planet. Some people would probably view my feelings with disdain. But at least I am trying to be helpful and of service, I told myself – not so much with the operating but with trying to help the young doctors become better doctors.

When I was told one morning that the MOs (medical officers) wanted me to stay for ever, I felt very happy and proud. But of course disillusion – or at least a more realistic understanding of Nepal and its sad and intractable problems – was to come quite quickly. There were periods of intense frustration and long periods of inactivity. At times I became deeply despondent. I felt that I was living in self-imposed exile. I often longed to return home, to my family and friends, and wondered why I had abandoned them. I thought of how I had always put work first, ahead of my wife and children, when I was younger, and now I was doing it all over again. But the deep contentment I experienced each day as I walked to the hospital in the low morning sunlight never faded.

I climb up the stairs to the third floor, past the locked suite with the letters VVIP over the door – built in case the president or prime minister falls ill – and go to the library. There are wide windows and on clear mornings you can just see the glittering snow-covered peak of Mount Ganesh, like a broken white tooth above the green hills of the Shivapuri National Park to the north of the city. There is an army base in the park, which was once a TB sanatorium. Some claim that during the recent civil war people were taken there to be tortured, and that many of them disappeared, but others deny it. Nepal has yet to come to terms with its civil war, and the atrocities carried out by both sides. I sit down and wait for the junior doctors to arrive.

The juniors drift in one by one – although the more enthusiastic of the registrars will already be waiting for me. Nepalis are not good time-keepers. About half of the ten medical officers have turned up.

'Good morning everybody,' Salima, the duty MO, says. She is wearing a short white coat and standing in front of a white board on an easel on which there is a handwritten list of the hospital's admissions and discharges. Salima is rather nervous as she knows I am going to quiz her about the cases. She looks a little Chinese, but with enormous black eyes behind a large pair of spectacles. I was to see her at a hospital get-together a few days later, dancing exquisitely to Nepali music. The Nepalese, both men and women, are almost all very good-looking, with a complex mix of Indian, Mongol and Chinese faces. There has been a population explosion in the last thirty years as a result of declining infant mortality, so the streets are full of young people. So many of the men work abroad – 30 per cent of Nepal's national income comes from remittances – that you see far more women than young men on the streets.

'Eighty inpatients, seven admissions, mortality one and no morbidity,' Salima rattles off quickly.

'Well, what's the first case?' I ask.

'Fifty-year-old lady present with loss of consciousness two days ago. Bowels open every day. Known hypertensive and alcoholic. On examination . . .'

'No, no, no! What's she do for a living?' I ask. I have noticed they never describe the patient's occupation, which is supposed to be a normal part of presenting a patient's history, although in Nepal it seems that everybody is either a farmer, a driver, a shopkeeper or a housewife. Mentioning the patient's occupation is important: not so much for the traditional reason, which is to alert us to possible occupational diseases, but more to remind us that the patient is a person, an individual, and has a life and a story beyond being a mere anonymous patient with a disease.

Salima looks embarrassed and fumbles with the sheet of paper in her hand. She probably hadn't seen the patient herself and relied on what had been written by one of the other junior doctors, so I was being unfair.

'Shopkeeper,' she says after a while.

'You're guessing!' I say and everybody laughs, Salima included.

'Now tell us about this loss of consciousness.'

'She comes from other hospital . . .'

'So we have no real history? Whether she had a headache first, whether she fitted?'

Salima looks awkward and says nothing.

Protyush, the registrar who had been on call, takes pity on her.

'Her husband found her on the floor at home. She was intubated at the other hospital and the family wanted her brought here.'

Dev's hospital is a private hospital; patients only come here by choice, or by their family's choice, and only if they can afford it. On the other hand they also have to pay if they go to a government hospital, where the treatment is only free in theory, and possibly worse.

'OK,' I said. 'Salima, what did you find on examination?'

'She localize to pain, not eye opening. Make sounds. Pupils equal and reacting. Cranial nerves intact. Power one on right, plantars up-going,' she continued in high-speed Nepali English, 'CT scan show . . .'

'No, no,' I interrupt again. 'What's your one-line summary?'

'Fifty-year-old lady present with loss of consciousness with known hypertension. Bowels open regularly. On examination pupils equal and reacting, and . . .'

'Salima – one line, not three!'

After a while we agree on a one-line summary. Presenting cases is a hugely important part of medical practice – about both communication and analysis. A short summary after presenting the details of a case forces the doctor to think about the diagnosis. I quickly learnt that most of the doctors were so shy in front of me that they found it very difficult to think analytically. It took them a long time to overcome this in my presence. I also suspected that much of their teaching had been entirely by rote.

'Right, now we can look at the CT scan.'

The scan showed that almost all of the left side of the woman's brain was dark grey, almost black. The woman had clearly suffered a massive and irreversible stroke – an 'infarct' caused by a blood clot forming in the left carotid artery. The left cerebral hemisphere, along with all her language and much of her intellect and personality as well as her ability to move the right side of her body, was dead, with no chance of recovery. Such damage cannot be undone. Some surgeons favour opening up the patient's skull to allow the dead, infarcted brain to swell outwards and stop the patient from dying from the build-up of pressure in the skull, as infarcted brain swells and severe brain swelling kills you.

Helping a patient survive a stroke with this operation of 'decompressive craniectomy' is perhaps justifiable if the stroke is on the right side of the brain (so that they do not lose the ability to communicate, speech usually being on the left side) and if the patient is young, but it seems a strange thing to do in patients who are going to be left dreadfully disabled if they survive. And yet it is recommended in articles in various learned journals claiming that such patients are happy to be alive, and is widely practised. You might wonder how the victims' happiness can be established if they have lost much of their intellect and personality, the part of their brain responsible for self-respect, or the ability to speak. You might also wonder whether their families are of the same opinion as the patients. Patients with severe brain damage, as far as you can tell, will often have little insight or understanding of their plight, whereas those that do are often deeply depressed. In a way, the true victims are the families. They must either devote themselves to caring, twenty-four hours a day, for somebody who is no longer the person that they once were, or suffer the guilt of consigning them to institutional care. Many marriages fail when faced with problems of this sort. It is worst for parents, who are tragically bound to their brain-damaged children, whatever their age, by unconditional love.

'So the patient's going to die?' I ask the room at large.

'We operated,' Protyush says. I express surprise.

'I spent half an hour trying to persuade the family that we shouldn't operate but they wouldn't accept it,' he adds.

After the morning meeting I go downstairs, take my shoes off outside the operating theatre and ITU area, get the uniformed guard to open the locked door for me and choose a pair of ill-fitting pink rubber clogs from a rack in the theatre corridor. Nepali feet are mostly small so I hobble uncomfortably to Dev's office, which is conveniently located between the ITU and the theatres.

Dev and I had always got on well together as colleagues when we were training together thirty years earlier, but it had been little more than that. I regret to say that I was far too ambitious and concerned for my own career at that time to take much interest in my colleagues, although I suppose that working a 120-hour week and having three young children at home left me with little spare time. And yet as soon as I came to Kathmandu, Dev and his wife Madhu were so welcoming that it felt as though we had always been the oldest of friends, even though we had only seen each other briefly at a few conferences over the intervening years. Dev is also charismatic, a man of great integrity and very determined. Like most Nepalis he is quite short and slight, although now a little rounder (which he blames on my presence and the beer we drink in the evening). He has a prominent, stubborn chin but slightly hunched shoulders, so that he looks like a cross between a bulldog and a bird. His intensely black, wavy hair has now turned grey. He has a chronic cough which he attributes to breathing the polluted air of the city centre when he worked for many years in the government hospital known as the Bir. He speaks very fast, with great animation, as though in a permanent state of excitement, about his past achievements and the great difficulties he had to overcome in trying to bring neurosurgery to Nepal. He also talks of how difficult it is to run a major neurosurgical practice more or less single-handedly.

He told me that it had been much easier when he was the only neurosurgeon in Nepal – if he gave bad news to patients they had little choice but to accept it. But now there are other neurosurgeons, most of whom have worked with him, from whom they get second opinions, and it would seem that there is little love lost between the professor and some of his former trainees. So he sometimes now has major problems with patients' families when things have gone badly, which they so often do with neurosurgery. When he told me this, I pointed out that in England there was more and more litigation against doctors, and this always involved doctors giving evidence against each other as expert witnesses.

'Yes, but here the families threaten us with violence, demand money and have even said that they'll burn the hospital down,' he retorted. 'Of course, we don't really have malpractice litigation here – it's almost unheard of to sue doctors.'

Doctors, especially surgeons, are often intensely competitive, and we all worry that other doctors might be better than we are, although I can think of a few famous international surgeons who are so supremely arrogant that they seem to have suppressed this problem by completely forgetting their bad results. We need, of course, self-confidence to cope with the fact that surgery is dangerous and we sometimes fail. We also need to radiate confidence to our frightened patients, but deep down most of us know that we might not be as good as we make out. So we feel easily threatened by our colleagues and often disparage them, accusing them of having the faults that we fear we have ourselves. It is made all the worse if we surround ourselves with junior colleagues whose careers depend on us, and only tell us what they think we want to hear. But it is also because, as the French surgeon René Leriche observed, we all carry cemeteries within ourselves. They are filled with the headstones of all the patients who have come to harm at our hands. We all have guilty secrets, and silence them with self-deception and exaggerated self-belief.

Dev remembers all sorts of details of our time spent working together in London, which I have long forgotten. His determination and energy are remarkable and I quickly came to understand why he has had such an extraordinary and brilliant career and is famous throughout Nepal. This has not been without its disadvantages. Driven and ambitious people can achieve great things, but often make many enemies in the process. Patients come to his outpatient clinic with all manner of non-neurosurgical problems, hoping that he can cure everything. A few years ago one of his daughters was abducted from the family home at gunpoint and Dev had to pay a large ransom. Since then he goes everywhere with a bodyguard.

The ITU is a large room with good natural light, as there are windows all the way along two of the walls. There are ten beds; they are rarely empty. The hospital admits strokes as well as head injuries and many of these patients have undergone decompressive craniectomies. Most of the patients are on ventilators, with pink bandages around their heads and the usual array of monitors and drip-stands and flashing lights and noisy alarms beside them. I had forgotten how grim neurosurgical ITUs can be – in London I had only been responsible for a small proportion of the patients since I was only one of many consultants.

Many of the patients on the ITU here would not survive, few would make a good recovery, especially in Nepal.

'You do far more decompressive craniectomies here than I would do,' I say to Dev. 'Only in America have I seen so much treatment devoted to so many people with such little chance of making a useful recovery. And yet Nepal is one of the poorest countries in the world.'

'I have to compete with many other neurosurgeons – trained in India or China – and they'll operate on anything, and it's always for the money. Like in America. If I tell the family now that no treatment is possible, they'll go and see somebody else who'll tell them the opposite and then they'll kick up a big fuss. So I am forced to operate now when in the past I wouldn't have. I often wish I still worked for the NHS,' he adds.

My colleague Igor in Ukraine often faced similar problems. I have been in countries where the surgeons sometimes have to operate with the patients' families outside the operating theatre wielding guns, threatening to kill the surgeon if the operation is unsuccessful. As a visiting doctor from the West it is hard, at first, to understand the difficulties our colleagues face working in countries with very different cultures and without the rule of law. It is easy to feel superior, to pass condescending judgement. I hope that over the years I have learnt to observe, and no longer to judge. I want to be useful, not to criticize. Besides, so often I find that I have misunderstood or misinterpreted what I have seen or been told – I have learnt not to trust myself. All knowledge is provisional.

'Many of these patients are going to die anyway, aren't they?' I say as we look at the next comatose patient with a bandaged head, labelled 'No Bone Flap'. After a decompressive craniectomy the patients are left, for a few weeks or months, with a large hole in their skull, like a giant version of the fontanelle with which we are all born. The 'No Bone Flap' label is to remind the medical and nursing staff that part of the brain is no longer protected by overlying bone. This particular patient – like so many in Nepal – has been involved in a motorbike accident.

'Cultural case,' Dev says. 'The family ties here are so strong. The family just can't accept that there is no treatment. If I hadn't got the boys to operate last night the family would say: "Oh Neuro Hospital doesn't want to operate!" Can you imagine the situation? Next thing they take the patient out of my hospital and somebody else will operate. The patient will be a vegetable but the family are happy and my reputation will be rubbished . . .'

Dev turns to look at me.

'When I was the Minister of Health under the last king – before the Maoists abolished the monarchy – I saved more lives by making crash helmets for motorcyclists compulsory than I will ever save as a neurosurgeon. Most of the families are uneducated,' he goes on. 'They have no conception of brain damage. They are hopelessly unrealistic. They think that if the patient is alive they might recover, even if the patient is just about brain-dead. And even if they are brain-dead they still won't accept it.' I was to learn more about this later.

So much for the value of commercial competition in health care, I thought, in a poor country like Nepal. And all this on one man's shoulders, day in, day out, with never a day off, for thirty years.

Neurosurgery is something of a luxury for poor countries. Illnesses requiring neurosurgical treatment are relatively rare compared to problems affecting other parts of the body. It requires very expensive equipment, and for problems such as cancer and severe head injuries treatment often fails or achieves little. We operate in the hope that patients will make a good recovery, and many will. There can be wonderful triumphs, but the triumphs wouldn't be triumphant if there weren't disasters. If the operations never went wrong, there would be nothing very special about them. Some patients will be left more disabled than they were before surgery and others, who would have died if we had not operated, will survive, but terribly disabled. At times, in my more despondent moments, it is not always clear to me whether we are reducing the sum total of human suffering or adding to it. So for countries like Nepal and Ukraine, with impoverished and weak governments and poor primary health care, it makes little sense to spend large sums of money on neurosurgery. Dev in Nepal and Igor in Ukraine have had little choice other than to move into private practice, albeit reluctantly, and yet both feel a little tainted by it, even though they often treat poor patients for free. But there is a limit to how often you can do that if your hospital is to survive.

There has always been a tension at the heart of medicine, between caring for patients and making money. It involves, of course, a bit of both, but it's a delicate balance and very easily upset. High pay and high professional standards are essential if this balance is to be maintained. The rule of law, after all, in part depends on paying judges so well that they will not be tempted to accept bribes.

Many medical decisions – whether to treat, how much to investigate – are not clear-cut. We deal in probabilities, not certainties. Patients are not consumers who, by definition, always know what is best for themselves, and instead must usually accept their doctors' advice. Clinical decision-making is easily distorted by the possibility of financial gain for the doctor or hospital, without necessarily being venal (although it certainly can be). Increasing litigation against doctors also drives over-investigation and over-treatment – so-called 'defensive medicine'. It is always easier to do every possible test and treat 'just in case' rather than run the risk of missing some very obscure and unlikely problem and being sued. This combination of paying doctors on a 'fee for service' basis – the more we do, the more we get paid – and increasing litigation against doctors in many countries is one of the reasons why health-care costs are running out of control.

On the other hand, a fixed salary can breed complacency and an irritating moral righteousness, to be found in some doctors who disdain private work. It is indeed a delicate balance, and Dev and Igor, both doctors of great integrity, have mixed feelings about running private hospitals.

'I am the country's highest taxpayer,' Dev tells me with a laugh, pointing to a photograph on his office wall of the Finance Minister recently handing him a certificate to this effect. Yet it seems highly unlikely to me that Dev is the highest earner in Nepal.

In Nepal and Ukraine – and many other countries – government is widely seen as corrupt and, understandably, people are reluctant to pay taxes, doing everything they can to evade them. There's another parallel here between Dev and Igor: both are scrupulous in paying their taxes. But it is difficult to be honest in a dishonest society, and many people will hate you for it.

Low tax revenues mean that governments in poor countries like Nepal and Ukraine have little money to spend on health care and infrastructure projects that would benefit the country. Besides, Ukraine is involved in a war and Nepal is still recovering from a vicious civil war. The lack of government spending on welfare and infrastructure only serves to reinforce the public's reluctance to pay taxes. It is a vicious circle from which it is very hard to escape. Driving in

Kathmandu can be a vision of hell and Hobbesian anarchy, especially at night in the suburbs. There is no street lighting. Trucks, cars and motorbikes are crammed together in narrow, rough lanes, driving in a cloud of dust and diesel fumes, eerily lit by undipped, dazzling headlights. Nobody gives way, each driver tries to go first – if you give way you will never move. There is no argument or shouting, nobody loses their temper, there is only the occasional blowing of horns. Everybody is resigned to the grotesque struggle which they have no power to end. Pedestrians join the crush to cross the road like ghosts in the dust. The unfortunate traffic police must inhale the poisoned air all day when they stand at the crossroads, trying to direct the chaotic vehicles. The city is asphyxiating, but the government appears to be utterly helpless and apparently has no plans to do anything about it at all.

The only certainties in life, as Benjamin Franklin once observed, are death and taxation. We all try to avoid both. But health care is getting more and more expensive – in most countries the population is ageing and needs more medical attention, and high-tech modern medicine is ever more extravagant. We all want to see cancer cured, but this will only drive costs up and not down. Not just because the complex genetic and drug treatments involved are so costly but because more of us will then live longer, to die later from some other disease, or slowly from dementia, requiring constant and expensive care. And rather than discover new antibiotics – the human race, especially in poor countries, faces decimation within a few decades from bacterial antibiotic resistance – the pharmaceutical companies concentrate on drugs for cancer and the diseases, such as diabetes and obesity, of affluence.

So health care is becoming ever more expensive, but most

governments fear that putting up taxes or insurance premiums will lose them the next election. So instead, in the West, a small fortune is spent on management consultants who subscribe to the ideology that marketization, computers and the profit motive will somehow solve the problem. The talk is all of greater efficiency, reconfiguring, downsizing, outsourcing and better management. It is a game of musical chairs where, in England at least, the music is constantly being changed but not the number of chairs, and yet there are more and more of us running around the chairs. The politicians seem unable to admit to the public that the health-care system is running out of money. I fear that the National Health Service in England, a triumph of decency and social justice, will be destroyed by this dishonesty. The wealthy will grab the chairs, and the poor will have to doss out on the floor.

As the weeks went by I took to absenting myself from the ITU rounds, unless there was a patient with whose operation I had been involved. I found the rounds too depressing.

After the ITU round Dev spends up to an hour on 'counselling'. The patients' families will stay in or near the hospital throughout the time their family member is there. There is a small hall in the centre of the hospital's first floor, well lit by a glass roof and decorated with palms in large planters. A prayer room with colourful Hindu and Buddhist icons is on one side. The families of the patients on the ITU wait here to be seen, one by one, by Dev and his colleagues in the counselling room next to the prayer room. They are updated on their relative's condition, questions are answered, and then they sign the medical notes, confirming what they have been told.

'I had problems to begin with,' Dev said. 'Some of the

families denied that they had had things explained to them, so I now do it formally every day.'

Although it was all in Nepali, it was fascinating to see Dev at work. As all good doctors do, he adjusted his style to the people he was talking to – sometimes joking, sometimes grave, sometimes consoling, sometimes dictatorial. On one occasion the patient's daughter was a nurse who had been working in England and spoke good English. Her elderly mother had suffered a huge stroke and the whole of the right side of her brain had died. She had undergone decompressive surgery and had therefore not died within the first few days but was now lying in the ITU, half paralysed and unconscious.

'You talk to her,' Dev muttered to me, 'and you'll see the problem.'

So I spoke to the daughter as I would speak to the families of my patients in England. I told her that if her mother survived she would be utterly dependent and disabled, with grave damage to her personality and intellect.

'Would she want to survive like that?' I asked. 'That's the question you and the rest of the family should be asking yourselves. I would not want to live like that,' I added.

'I hear what you are saying,' she replied, 'but we want you to do everything possible.'

'You see?' Dev said to me later. 'They're all like that. I've even had it with families of doctors. They just can't face reality.'

The child's head was completely shaven and had already been fixed to the operating table with the pin headrest. The juniors had had problems inserting a central intravenous line into one of the major veins in the neck and had ended up hitting the carotid artery. They then decided to rely on two large peripheral lines in the smaller veins of her arms for blood transfusion, in case there was heavy bleeding from the tumour. So there had been long delays before I came into the operating theatre. Much of her face was hidden by the plaster strapping holding the endotracheal tube in place, but despite all this and the disfiguring shave, she looked painfully sweet and vulnerable, with a broad Tibetan face, light-brown skin and slightly red-tinted cheeks.

Dev was standing by the patient's head. 'You and I trained together,' he said. 'We think along the same lines.' He has six trainees whom he has trained to do the simple emergency work and the 'opening and closing' of the routine surgery. Dev, however, does almost all the major operating himself. Occasionally he has been joined by foreign surgeons, but only for short periods of time. There are major cases to be done every day, six days a week, and the pressure is relentless. In six weeks working in Kathmandu I saw more major operations than I would have done in six months in London.

This was the first time I had seen the child, although I had looked carefully at her brain scan with Dev earlier that morning.

'She was operated on by one of the other neurosurgeons here in Kathmandu,' Dev told me, 'but I don't think he removed much of it. Just did a biopsy. It's said to be a Grade Two astrocytoma.'

'It's not a good tumour,' I said, looking unhappily at the scan. 'It may be benign but it's involving all the structures around the third ventricle and God knows where the fornices are.'

'I know,' said Dev.

The fornices are two narrow bands of white matter, a few millimetres in size, that are crucial for memory. White matter consists of the billions of insulated fibres – essentially electrical cables – that connect the eighty or so billion nerve cells of the human brain together. If the fornices are damaged, people lose a large part of their ability to take in new information – a catastrophic disability.

Average income in Britain is forty times greater than in Nepal. Primary health care in Nepal is poor (although better than in many other low-income countries) and diagnosis of rare problems such as brain tumours is invariably delayed. The tumours, therefore, by the time they are diagnosed, are much larger than in the West and treatment is more difficult, more dangerous and less likely to achieve a useful result. Brain tumours in children are very rare but very emotive, and although the rational part of myself considered that operating on this child was a waste of time and money, it is almost always impossible, wherever you are in the world, to say this to the desperate parents. And I myself had once been the parent of a child with a brain tumour. But the decision was Dev's responsibility and not mine.

Once I had checked that they had positioned the child correctly, I left them to start the operation, returning when a nurse came to Dev's office and silently beckoned me to come to the operating theatre and join Dev.

I am becoming little better than a vet, I told myself as I scrubbed up at the long zinc sink with its row of taps and iodine dispensers. I am operating on patients without knowing anything about them, without even seeing them other than as unconscious, impersonal heads in a pin headrest.

# 4

# AMERICA

One year before I went to Nepal, and before I had retired, I attended a cerebrovascular workshop in Houston, intended to help trainee surgeons learn how to operate on the brain's blood vessels. I was to be one of the instructors. I arrived from London after a ten-hour flight. The workshop started next morning at eight, after I had delivered a lecture at seven to my colleagues in the neurosurgical department which I was visiting. American hospitals start early – the interns, the most junior doctors, often begin their ward rounds before five in the morning. I once asked a group of them about the physiological effects of sleep deprivation on their patients and they seemed quite startled by the suggestion that their immensely hard work might actually be harming the patients.

My lecture was about how to avoid mistakes in neurosurgery, but only a handful of people had turned up to listen to me, presumably because they felt that they had little to learn from the mistakes made by an obscure English surgeon such as myself. The large breakfast laid out in the room outside the lecture theatre remained uneaten. There was a short briefing at the beginning of the workshop. We sat on tiered seats in a small room with three enormous LED screens in front of us. Everything looked new and immaculately clean. A businesslike woman dressed in scrubs told us that under no circumstances was photography permitted and that everything we would be doing was regulated by federal law. She cited various specific statutes which were flashed up on the screens, each one with a long reference number. She also told us that we must respect the subjects of the workshop. Different-coloured hats were then handed out – mine was blue as I was a member of the faculty. The medical students' were yellow and the neurosurgical residents' were green. We were then ushered through a pair of large security doors into the research facility.

This looked like a cross between an operating theatre and an open-plan office, with several bays. Floor-to-ceiling windows looked out onto the many glittering skyscrapers that form the Texas Medical Center, the largest concentration of hospitals anywhere on the planet. There are 8,000 hospital beds here – fifty-one clinical institutions in total, I was told – practising some of the most advanced medical care anywhere in the world. There were half a dozen shapes lying on operating tables; I suppose each one was about the size of a ten-year-old child. They lay entirely hidden under blue surgical drapes, with anaesthetic tubing and cables coming out of one end, connected to the same ventilators and monitors with colourful digital displays that I see every day at work. I walked up to one of them and put a hesitant hand out – it was strange to feel the hoofed trotters under the drapes at the end of the operating table.

'Isn't this just fantastic!' said my colleague, a trainee of mine from many years ago, who had recently become the chairman of the neurosurgical department which was staging the workshop, which he had organized. 'Nobody anywhere else is doing anything like this! Come on, guys!' he said to the residents in their green hats. 'Enjoy!'

One of the faculty pulled back the blue drapes off the head of one of the pigs and started to operate. The pig was lying on its back with its broad, pink neck stretched out. It had probably been shaved, and although it was clearly not a human neck – it was far too flat and wide – the skin looked disconcertingly similar. He used cutting diathermy to dissect down to the carotid artery, one of the main arteries for the brain – a smaller vessel than in a human. The plan was to dissect out a vein and graft it to the artery, creating an aneurysm, a model for the life-threatening aneurysms that occur in people and cause fatal haemorrhages. The artificial aneurysm can then be treated – with an 'endovascular' or 'coiling' technique where a microscopic wire is inserted into the aneurysm via the artery, involving only a simple puncture in the skin, and the aneurysm is blocked off from the inside. Alternatively the aneurysm can be treated with the more old-fashioned method of open surgery, where it is clipped off from the outside. Most aneurysms in people in the modern era are treated with coils, but a few still need clipping. The purpose of the workshop was to give trainee surgeons some practice in the techniques without putting a human life at risk. I am sentimental about animals, and felt sorry for the pigs, but reminded myself that they were doing more for humanity by being used for surgical practice than by being turned into bacon – and there were all those federal statutes protecting them, after all.

My fellow instructor started to stitch the vein graft to the artery. It was a rather slow business and I wandered off towards a group of doctors gathered in a corner of the room. A blue-capped faculty member was talking with great enthusiasm.

'This is awesome! This is so much better than specimens preserved in formaldehyde!'

I looked over his shoulder. Two trainees were operating on a severed human head. It was held in the steel head clamp most neurosurgeons use when operating and the skin of the neck had been formed into two flaps; these had been stitched together with a few broad sutures to form a stump, although some slightly obscure fluid was dripping out between the sutures. If I had not done my year of cadaveric dissection as a medical student forty years ago I think the sight would have given me nightmares for many days afterwards. It was bizarre and disturbing to see a head in a standard head clamp – something I must have experienced thousands of times with living patients when I operate – and yet with no body attached to it.

So I joined the small group standing around the two trainees who were carrying out a craniotomy under the guidance of a fellow instructor – sawing open the severed head with surgical tools, looking down an expensive operating microscope. I was staggered by all the equipment which surrounded the various stations, six with anaesthetized pigs and now one with a dead person's head. All of it had been provided by the manufacturers – hundreds of thousands of dollars' worth, all to be used for practice. As I watched the two trainees uncertainly drilling into the severed head, a young man behind me – dressed, to my surprise, entirely in black scrubs like a ninja – accosted me.

'Professor!' he said, with the passionate conviction of an equipment rep. 'Have a look at this.' He pointed to the beautiful array of miniature titanium plates and screws and tools, each in its own perfect moulded cavity on a black plastic tray in front of him. These plates are screwed in place to reassemble the skull after sawing it open – although in this case only for practice.

'Have you tried our latest electric screwdrivers?' he asked, handing me a neat little battery-powered screwdriver which I suppose would save about five seconds, and needed only marginally less effort than the manual screwdriver I normally use when putting patients' skulls back together again with titanium plates. I switched the electric screwdriver off and on, marvelling at the extravagance of the American medical system.

'How d'ya like it?' asked the rep.

'Outstanding,' I replied, thinking of how, on my flight the previous day, the pilot had told us over the intercom when the plane was about to begin its descent that now would be an outstanding time to visit the restrooms.

'Guys! We have a master here!' the instructor called out when he saw me. 'Professor, can you give us some surgical pearls?' I thought a little apologetically of the swine in the nearby bay undergoing surgery.

Happy to have something useful to do, I pulled on a pair of gloves and went up to the microscope to reposition it and look down into the dead brain.

'Have you got any brain retractors?' I asked. '"Ribbons" you call them here in the US.' It seemed they did not, so I used a small chisel to gently lift up the frontal lobe. There was, of course, no bleeding, but the consistency of the dead tissue was not unlike that of the living thing.

'Formaldehyde makes it all stiff and solid, and it smells awful,' I said. 'But where do they get these freshly dead heads?' I asked of nobody in particular.

'Maybe a John Doe scraped off the sidewalk,' somebody offered.

Using the small chisel I dissected out the anterior cerebral arteries, explaining how you approach an anterior communicating artery aneurysm by resecting – that is, removing – part of the brain called the gyrus rectus to find the aneurysm.

'The gyrus rectus serves olfaction,' I told my small audience. 'The patients are better off with perhaps some impairment of smell than dying from another haemorrhage if they don't have the aneurysm treated.'

I handed over the operation to the two residents and walked round to look at the dead face: head shaved, eyes closed, stubble on his cheeks, blackened stumps of a few remaining teeth. He clearly had never seen a dentist. As far as I could tell he was not – or rather had not been – that old before he died. It was impossible not to wonder for a moment who he had been and what sort of life he had led, and to think that once he had been a child, with all his future in front of him.

Workshops like this are not unusual, but I had never been at one before and I found it rather distressing. I would consider this to be a weakness on my part – it is clearly much better that trainee surgeons should practise in workshops like this than on living patients. When I was back in England two weeks later I mentioned this to a colleague who had recently organized a similar workshop in the UK.

'Ah!' he said with a laugh. 'Only one? I had fifteen heads, freeze-dried, flown in from the US for my skull base workshop last year. I needed to put them all through the MRI scanner before the meeting and drove to the hospital with the heads in the boot of my car. I wasn't quite sure what I would say if I was stopped by the police. The other problem was that they were starting to thaw. I don't know where they get them from,' he added.

I left the room with its severed head and anaesthetized pigs and found another huge breakfast laid out next to the lecture theatre where we had started. After breakfast I was taken on a whirlwind tour of the hospital.

The hospital consisted of a series of multi-storey towers, and we went through what seemed to be an almost endless series of huge lobbies and halls. The hospital had its own twelve-floor hotel; patients came from all over the world for treatment, not just from America. There were twenty – twenty! – other hospitals next to my colleague's, as well as many other medical and clinical research institutions. The Medical Center occupies more than a square mile, and when I looked out of my twelfth-floor hotel window all I could see was hospital after hospital, all built of glittering glass, receding into the distance like a mountain range. Medicine in the USA is notoriously extravagant. I saw one hospital in Chicago which had a luxurious restaurant, bar and garden on its roof. The hospitals are locked in fierce competition for business and many are designed to look as little like hospitals as possible. They resemble instead luxury hotels or shopping malls or first-class airport lounges. They are the peacocks' tails of health care.

That evening my colleague took me to his country club. We drove there through the city's suburbs, past large mansions with pillared porticoes and extensive lawns. The club too was built on a grand scale and the clubhouse – icy-cold with air-conditioning – had a massive baronial fireplace in the Scottish style decked out with mounted stags' heads on either side, and a large reproduction of the famous Victorian painting by Landseer of a stag, known as *The Monarch of the Glen*, hung above the grand staircase. We had an excellent dinner there. The waiters were elderly Mexican men with solemn and expressionless Aztec faces. They were dressed in black suits with white aprons and they moved with slow dignity as they served the clientele, nearly all of whom were dressed in baggy shorts and long T-shirts. Over dinner there was the usual surgical gossip – mainly about a colleague who had been sacked for having an affair with a rep, and whether the rep was enhanced with silicone or not. Opinions differed as to this latter question. After his dismissal she had apparently sued him successfully for sexual harassment but now, my colleague told me, they were back together again. I also learnt that the operation on the pig to create an aneurysm had not been a success: one of the technicians had forgotten to give the animal an anti-coagulant injection and the pig had suffered a major stroke as a result of the surgery to its carotid artery. It would, however, have presumably been sacrificed – as it is called – in any case, even if the mistake had not occurred.

After dinner we went out into the sweltering, humid heat to inspect a car show outside the clubhouse. Thirty or so classic cars were drawn up in the car park, shiny and polished, many with their bonnets up so that you could see the spotless, chromed engines inside. A red Ferrari inched its way past us to find a parking place.

My colleague nudged me and said with awe: 'That's a seven-million-dollar car. And the guy driving it is a billionaire.'

It transpired later that the car was only a reproduction, but was still worth a million dollars. The billionaire apparently was a real billionaire but looked a fairly ordinary sort of guy. A group of people gathered admiringly around the car once the billionaire had parked it, and they took photographs of each other in front of it.

I went out for a run next morning as the sun was rising. I was streaming with sweat within a matter of minutes as I ran along the street beneath the tall hospital towers, past neatly tended flowerbeds. At the edge of the great block of hospitals there was a large park, with a miniature railway line running round it. Several dozen homeless people were dossing out on the benches and sidewalks in one corner of the park. I was told later that there was a church nearby which gave out free meals. As I ran back to the hotel the sun rose behind me, over the dozens of buildings of the Medical Center, and I was almost blinded by its dazzling reflection in the thousands of hospital windows facing me.

# 5

# AWAKE CRANIOTOMY

For a surgeon to help operate on patients he did not know, whom he would scarcely ever see again, for whom he carried no practical responsibility – if there were problems Dev would have to deal with them – had always been anathema to me. And yet I had already discovered, to my surprise, that my lack of human contact with the Nepali patients both before and after surgery had not reduced my anxiety when I was operating. It did not seem to matter after all. Operating in Kathmandu I was in the same state of tense concentration as I was in London and it seemed that I cared just as much for the patients, even though my concern for them had now become entirely abstract and impersonal. I used to feel critical of surgeons if they were remote and detached from their patients but now, very late in my career, I was forced to recognize that some of this had perhaps been vanity on my part, and simply yet another attempt to feel superior to other surgeons.

Surgeons describe operating on patients with whom they have no personal or emotional contact as being veterinary. There was a veterinary surgery near the old hospital in Wimbledon and one of the vets there – Clare Rusbridge

– specialized in veterinary neurological disorders. Devoted pet owners can take out insurance for their pets which includes the cost of MRI brain and spinal scans. Clare would bring to our weekly X-ray meetings fascinating scans of cats and dogs with neurological disorders. We would look at the scans at the end of the meeting and called it Pets' Corner. They provided a bizarre contrast in anatomy to the images of human brains and spines with which we were so familiar. Cavalier King Charles spaniels, we learnt, often suffer from the brain abnormality known as a Chiari malformation, which humans also get. Labradors can develop malignant meningiomas. The spaniels' problem is the result of selective breeding aimed to produce the small round head which wins points at dog shows. The malformation leads to spinal cord damage, and the poor creatures suffer from intractable pain and scratch themselves incessantly.

I operated with Clare on a couple of occasions, though she was unable to find an owner of a King Charles spaniel who was willing to let us operate on their pet. We did, however, once operate on a badger, which had been found confused and wandering on Epsom Downs and had been rescued and brain-scanned by an animal charity. The brain scan suggested that she might have hydrocephalus, although, to be honest, not much is known about badger brains. She was a beautiful creature and once she had been anaesthetized, I held her on my lap for a few minutes, stroking her grey and white fur, before Clare removed most of it with a pair of clippers in preparation for the surgery. I tried to carry out an operation for the possible hydrocephalus. I already had an article published entitled 'Brain Surgery in Ukraine' and I hoped I would be able to add to my CV 'Brain Surgery in Badgers', but the operation was not a success and the poor creature died. Or rather, she was 'euthanased'.

'At least our patients don't have to suffer,' one of Clare's colleagues, who had watched the operation, commented afterwards. 'Unlike yours.'

The first case I had done with Dev – two days before the operation on the child – had been an awake craniotomy for a tumour. This was the first time that such surgery had been carried out in Nepal. I had brought the equipment for cortical brain stimulation from London in my suitcase. Many years ago I had been the first surgeon in Britain to use the technique of awake craniotomy for treating a particular type of brain tumour known as a low-grade glioma. It was unorthodox at the time, but is now standard practice in most neurosurgical departments. It is, in fact, a very simple way of operating which allows you to remove safely more of a tumour in the brain than with the patient asleep under a general anaesthetic. The problem is that the 'tumour' is in fact part of the brain which has tumour growing in it – brain and tumour are muddled up together. The abnormal area, especially at its edges, looks almost identical to normal brain and only by having the patient awake, so that you can see what is happening to them as you remove the tumour, can you tell if you are straying into normal brain and running the risk of causing serious damage. Patients tolerate the procedure much better than you might expect, once they understand why it has been recommended.

The brain cannot feel pain: pain is a sensation created within the brain in response to electrochemical signals sent to it from the nerve endings in the body. When I see patients with chronic pain, I try to explain to them that all pain 'is in the mind' – that if I pinch my little finger, it is an illusion that the pain is in my finger. It is not 'in' my finger but really in my brain – an electrochemical pattern in my brain, in a map that my brain has made of my body. I try to explain this in the hope that the patient will understand that a psychological approach to pain might be just as effective as a 'physical' treatment. Thought and feeling, and pain, are all physical processes going on within our brains. There is no reason why pain caused by injury to the body to which the brain is connected should be any more painful, or any more 'real', than pain generated by the brain itself without an external stimulus from the body. The phantom limb pain of an amputated arm or leg can be excruciating. But most patients with chronic pain problems or conditions like chronic fatigue syndrome find this hard to accept. They feel that their symptoms are being dismissed – as they often are – if it is suggested that there is a psychological component to their problem and that a psychological approach might help. The dualism of seeing mind and matter as separate entities is deeply ingrained in us, as is the belief in an immaterial soul which will somehow outlive our bodies and brains. My 'I', my conscious self, writing these words, does not feel like electrochemistry, but that is what it is.

So, for an awake craniotomy, only the scalp needs to be anaesthetized and the rest of the operation is painless, although patients find having their skull drilled into very noisy – the skull acts like a sounding board. I therefore usually do this part of the operation under a brief general anaesthetic. The patient is then woken up, but unlike normal operations, where you wake up in a bed back on a hospital ward, with an awake craniotomy you wake up in the operating theatre, in the middle of the operation. There are various ways of conducting the 'awake' part of the operation. All involve using an electrode to stimulate the patient's brain, which tells you where, in functional terms, you are on its surface. You will be able to produce limb movement or interfere with the ability to talk as the electrode momentarily stimulates or stuns the relevant part of the brain. It is a little like pulling the strings of a puppet. You also need to ask the patient to perform simple tasks or name and identify pictures if the tumour is near the speech centres of the brain. Some surgeons rely on speech therapists or physiotherapists to talk to the patient and assess them as the operation proceeds. I always relied on my anaesthetists, in particular Judith Dinsmore, whose highly skilled and reassuring manner never failed to keep the patients calm and cooperative.

I operate with a transparent screen between myself and the patient. Judith would sit facing the patient, talking to them and assessing the relevant functions – their ability to talk fluently, or to read, or to move the limbs on the opposite side of their body to the tumour (for obscure evolutionary reasons, each half of the brain controls the opposite side of the body). I would be standing behind and above the patient's opened head and exposed brain, and watch and listen to Judith through the transparent screen as she put the patient through their paces. When she started to look anxious, I knew it was time to stop. If the patient had been under a general anaesthetic for all of the operation, I would have had to stop much earlier and would have removed less of the tumour. There would have been no way of knowing whether I was still removing tumour or normal, functioning brain. Obviously, more subtle social or intellectual functions cannot be tested, but this is not usually a problem. It would seem that low-grade gliomas have to be very extensive indeed – and effectively inoperable – before the patient's personality is at risk.

I operate with a microscope which has a camera connected to a video monitor. The operation is mainly done with a simple sucker or an ultrasonic aspirator (which is a sucker with an ultrasonic tip that emulsifies what you are operating on). All you can see, as you look into the patient's brain with the microscope, is the brain's white matter, which is like a smooth, thick jelly. It is usually – but not always – slightly darker than normal because of the presence of tumour within it. It took me some time to learn to operate like this, with the patient awake. I am always a little anxious when operating and, at first, having the patient awake made this worse, especially as I had to affect a complete calm and confidence for their sake that I did not inwardly feel.

'Do you want to see your brain?' I will usually ask the patient. Some say yes and some say no. If they say yes, I go on to say: 'You are now one of the few people in the history of the human race who have seen their own brain!' And the patients gaze in awe at their brains on the monitor. I have even had the left visual cortex – the part of the brain responsible for seeing things on the right-hand side – looking at itself. You feel there should be some philosophical equivalent of acoustic feedback when this happens, a metaphysical explosion, but there is nothing, although one patient, having looked at his speech cortex, as I brushed it with a sucker and told him that was what was talking to me, commented: 'It's crazy.'

Towards the end of the first ever awake craniotomy in Nepal, the patient's leg had suddenly become paralysed.

'It's probably temporary,' I assured Dev. 'It can happen when operating in the supplementary motor area, which was where the tumour was.'

I nevertheless awoke next morning feeling miserable. But Dev came to find me as I sat with a cup of coffee in the garden of his home – I had only stayed two nights in the hotel to which I had first been taken, and was now living in the guest house at the end of Dev's garden. He told me that the juniors had rung to say that the patient had started to move his leg.

'I knew you were upset, though you said nothing,' he said. The morning was instantly transformed.

'Were there any admissions overnight?' I asked.

'Couple of head injuries,' Dev replied.

I would often be rung at night when I was on call in London, although unlike Dev I was not on call every night. The telephone would ring and I would be dragged out of sleep, often with the strange illusion that I had chosen to wake up before the phone started ringing. These emergency cases were usually cerebral haemorrhages – bleeding into the brain caused by head injuries or a weakened blood vessel. I had to decide whether the patient should be operated on or not. Sometimes it was obvious that they would die if they did not undergo surgery and that they would make a good recovery after surgery. Sometimes it was obvious that they did not need surgery and would survive without it, and sometimes it was obvious that they would die whatever we did. But often it was not clear whether to operate and, if you did, whether they would make a good recovery. If the haemorrhage had been a big one, the patient was going to be left disabled, however well the operation went, as the brain – being so intricate and delicate – has much less capacity for repair and recovery than other parts of the body. The question then was whether the disability might be so severe – the patient left a 'vegetable', as the saying goes – that it might be kinder to let them die.

You can rarely predict with absolute certainty from a brain scan what sort of recovery the patient might make, but if we operate on everybody (as some surgeons do), without any regard to the probable outcome, we will create terrible suffering for some of the patients, and even more for their families. It is estimated that there are 7,000 people in the UK in a 'persistent vegetative or minimally conscious state'. They are hidden from view in long-term institutions or cared for at home, twenty-four hours a day, by their families. There is a great underworld of suffering away from which most of us turn our faces. It is so much easier to operate on every patient and not think about the possible consequences. Does one good result justify all the suffering caused by many bad results? And who am I to decide the difference between a good result and a bad result? We are told that we must not act like gods, but sometimes we must, if we believe that the doctor's role is to reduce suffering and not just to save life at any cost.

'Twenty-six-year-old. Collapsed last night while in the shower. Looks like a spontaneous ICH. Probably an underlying AVM – there's some calcification. Blown a big hole in the left basal ganglia, a bit into the midbrain too. GCS four according to the paramedics. Blown left pupil but came down with mannitol and ventilation. Lots of shift on the CT. Basal cisterns just visible. Now tubed and ventilated.'

'Hang on a mo, I'll have a look at the scan,' I said. I pulled my laptop off the shelf by my bed and, balancing it on my knees, I spent a few minutes connecting to the hospital X-ray system over the internet. I looked at the scan.

'He's not going to do well, is he?'

'No,' my registrar replied.

'Have you spoken to the family?'

'Not yet. He isn't married. There's a brother, who's coming in. Should be here soon.'

'What's the time?'

'Six.'

'Well, we can probably wait until the brother's here.'

Translated, the story – or 'history', as doctors call it – was of a young man who had suffered an intracerebral haemorrhage (ICH) from an AVM, an arterio-venous malformation. This is a kind of rare birthmark, a tangle of weak, abnormal blood vessels that often burst and bleed into the brain. The haemorrhage was into the left side of his brain, and also into part of the brain called the midbrain, which is important in keeping us conscious. It looked unlikely to me, from the scan, that if we operated he would get back to any kind of independent life. It is rarely possible to be certain, but I doubted if he would ever regain consciousness, let alone walk or talk again. His GCS was four, which meant he was in a deep coma. The scan showed a critical build-up of pressure in his head ('lots of shift on the CT', as my registrar put it). The fact that the pupil of his left eye was 'blown' – enlarged and no longer reacting to light – was a warning sign that without surgery he would probably die within the next few hours. The pupil had become smaller with a drug called mannitol, which temporarily reduces intracranial pressure, so we had a little time to decide what to do.

I couldn't get back to sleep and went into the hospital an hour later. The sun was rising over south London, a long line of bright orange seen through the hospital windows. The corridors were quiet and empty as it was so early in the morning, but the ITU was very busy and full of noise. The twelve beds were all occupied and the nursing shift was about to change over, so there were many staff milling around the nurses' station. There was a forest of drip-stands for intravenous fluids and syringe pumps, and flashing monitors standing guard beside each bed, the constant sound of the monitors bleeping and the softer, sighing sound of the ventilators doing the patients' breathing for them. The nurses were all talking, handing over their patients to each other.

The unconscious patients lay immobile, covered by white sheets, with tubes in their mouths connected to ventilators, IV lines in the veins of their arms, nasogastric tubes in their noses and catheters in their bladders. Some had drainage tubes and pressure-monitoring cables coming out of their heads.

My patient was in the far corner and there was a young man sitting at the bedside. I went up to him.

'Are you his brother?'

'Yes.'

'I'm Henry Marsh, the consultant responsible for Rob. Can we go and talk?'

We shook hands, and left Rob's bed to go to a small room used for interviews, for breaking bad news. I signalled to one of the nurses to join us. My registrar appeared, slightly out of breath.

'I didn't know you were going to come in so early,' he said.

I gestured to the patient's brother to sit down and sat opposite him.

'We need to have a very difficult conversation,' I said.

'Is it bad?' the brother asked, but he would have known already from my tone of voice that it was.

'He's suffered a major bleed into his brain.'

'The doctor here,' he said, pointing to the registrar, 'said you had to operate.'

'Well,' I replied, 'I'm afraid it's a bit more complicated than that.'

I went on to explain that if we operated and he survived, there was very little chance of his getting back to an independent life.

'You know him better than I do,' I said. 'Would he want to be disabled, in a wheelchair?'

'He loved the outdoor life, went sailing . . . had his own boat.'

'Are you close to him?'

'Yes. Our parents died when we were kids. We were best mates.'

'Girlfriend?'

'Not at the moment. Broke up recently.' He sat with his hands between his knees, looking at the floor.

We sat in silence for several minutes. It is very important not to try to fill these sad silences with talking too much. I find it very difficult, but have got a little better at it over the years.

'No chance?' his brother asked me after a while, looking up at me, into my eyes.

'I doubt it,' I replied. 'But to be honest you can never be entirely certain.'

There was another long silence.

'He'd hate to be disabled. He told me that once. He'd rather be dead.'

I said nothing.

'Rob was my best friend.'

'I think it's the right decision,' I said slowly, even though neither of us had explicitly stated what we had decided. 'If he was a member of my family, that's what I would want. I've seen so many people with terrible brain damage. It's not a good life.'

So the decision was made and we did not operate. Rob died later that day – at least, he became brain-dead, the ventilator was switched off and his organs were used for transplantation. I suppose it was just possible that I might have been wrong and he might have got back to some semblance of an independent life, or perhaps his brother was wrong, and Rob would have come to terms with being disabled, or simply have had no insight into it and led a happy, minimally conscious existence, no longer the person that he once was. Perhaps, perhaps – but doctors deal with probabilities, not certainties. Sometimes, if you are to make the right decision, you have to accept that you might be wrong. You may lose one patient with a good outcome but save a far greater number – and their families – from great suffering. It's a difficult truth that even now I find hard to accept. When I received phone calls at night about cases like this, if I told the surgeon on call in the hospital to operate, I would roll over and get back to sleep. If I told him not to operate, and that it was better to let the patient die, I would lie awake until it was time to go to work.

The operation on the six-year-old child in Nepal was only two days after the awake craniotomy and not especially difficult. I had done many similar cases before, but rarely with tumours of this size. It involved separating the two halves of the brain – the cerebral hemispheres – using what is known as a transcallosal approach. It was an operation I had always taken a particular interest in because it was the one that had saved my son's life many years earlier when, at the age of only three months, he had undergone surgery for a brain tumour. But his tumour had been only a fraction of the size of this one.

Dev and I had already decided that my principal role in his hospital would be to help him train his juniors so that they could learn how to do more than just the opening and closing and the emergency work at night. Within the first few days of coming to Nepal I knew, with the blind confidence of a lover, that as long as I could still usefully work, I would want to spend as much time as possible in the country.

'You can't go on running your hospital single-handed like this for ever,' I told him, 'and you need to think about the succession. You're not that much younger than me. What do you want to leave behind?'

'I know,' he said. 'It's been worrying me a lot of late.'

'I'd like to help, if I can be useful,' I went on. 'But you must promise to tell me as soon as I stop being useful.'

'I agree,' he replied.

You always get more nervous operating on children than on adults because of the terrible anxiety of the parents waiting outside the operating theatre. I had trained in paediatric neurosurgery and for many years had done most of the paediatric surgery in the hospital in Wimbledon. When that was closed and we were moved to the huge teaching hospital three miles away, I was unhappy about the way the paediatric neurosurgical ward was arranged. It was a very long distance from the neurosurgical theatres and my office, and it would no longer be possible for me to visit the children's ward several times a day as I had done in the past. It had been my way of coping with the parents' anxiety, which I understood all too well from the time when my own son had undergone surgery for his brain tumour and before I had become a neurosurgeon myself. To my shame and slight dismay, I found that I did not miss the paediatric work. In fact it was a relief to stop doing it.

'Can I have the microscope please?' I asked. The microscope – brand-new and as good as anything I have in London – was pushed into place.

'Has it been counterbalanced properly?' I asked. The second operation I had been involved with after arriving in Nepal had almost ended in disaster. The microscope, with an optical head weighing at least thirty kilos, had not been properly counterbalanced. The registrar had accidentally pressed the button that releases the perfectly floating optical head, and instead of floating it had crashed down onto my hands, almost forcing the instruments I was holding into the patient's brain.

With Dev's agreement we introduced a checklist to be completed before each operation, making sure (hopefully) that this would not happen again. I found it ironic that despite my well-known hatred of paperwork and checklists in my own hospital, I was now trying to introduce them in the hospital in Nepal.

'Yes, sir,' said Pankash, the registrar assistant, in answer to my question about counterbalancing the microscope. The registrars are very polite and respectful. If they do not know the answer to a question, they find it quite impossible to admit this. Rather than say no, they will stand speechless. The silence can last for many minutes and makes teaching very difficult. I had quickly resigned myself to never knowing what they really thought about what I told them or the questions I asked them.

I positioned the microscope and cautiously pressed the release button. The optical head remained steady and I settled down in the operating chair, which had also been pushed up to the table.

'What is the first rule of microsurgery?' I asked Pankash.

'To be comfortable, sir,' he replied. I had told him this the day before.

'Look,' I said. Pankash was peering down the microscope's side-arm so he could see what I was doing. Dev was watching on the monitor. I put the retractor against the inner side of the right hemisphere and gently pulled it a few millimetres to the right, away from the thick midline membrane called the falx cerebri that separates the two hemispheres. As I looked down the binocular microscope, it was as though

I was descending a ravine or negotiating a narrow crevice, with the shiny, silvery-grey surface of the falx to the left and the pale surface of the brain, etched with thousands of fine blood vessels, glittering in the microscope's brilliant light, to the right. Even after thirty years I still find using an operating microscope profoundly exhilarating – the feeling of beauty and mystery and exploration has never left me. After years of practice, the perfectly balanced instrument is like an extension of your own body and you feel – until things go wrong – equipped with superhuman powers.

'If we are lucky we will quickly drop right down onto the corpus callosum – there it is!'

The white corpus callosum came into view at the floor of the chasm, like a white beach between two cliffs. Running along it, like two rivers, were the anterior cerebral arteries, one on either side, bright red, pulsing gently with the heartbeat, which you must not damage under any circumstances. The corpus callosum contains countless millions of nerve fibres joining the two halves of the brain. If all of the corpus callosum is divided – as is occasionally done for severe epilepsy – patients develop what is called the 'split-brain phenomenon'. Outwardly they appear normal enough, but if placed in an experimental situation where the two cerebral hemispheres each see a different image, the two halves of the brain can end up disagreeing about what they are seeing – in particular, what something is called and what it is used for: knowledge of names is in the left hemisphere, and knowledge of how to use things is in the right. The self has been split. Actual conflicts between the two halves of the brain are rare, but it is said that one patient, on losing his temper with his wife, attacked her with his right hand while his left hand tried to restrain it.

Who has not felt contradictory impulses within themselves?

The more you learn about the brain, about our true selves, the stranger it becomes. It is almost as if we have many competing and cooperating selves within our brains, and yet somehow they all resonate together to produce a coherent individual capable of thought and action. There was a famous experiment many years ago by the American neuroscientist Benjamin Libet – confirmed many times since – that showed that the conscious decision to move the hand is *preceded* by electrical activity in the hand area of the brain. Nobody has yet provided a satisfactory account of what this really means. It is as though the deciding self is no different from the sailor in the storm who is forced to steer his boat in the storm's direction, but then claims to have chosen the direction himself. And yet to claim, as some do, that the conscious self is an illusion, that it is no more than running before the wind, or a consoling fairy story, somehow seems as implausible as maintaining that pain is an illusion, and not 'really' painful.

I was only going to make a small hole in the child's corpus callosum. This would bring me straight to the centre of her brain where the tumour was growing. A small callosotomy, as it is called, does not seem to produce any obvious problems for the patient after the operation. Besides, there were other, much greater risks to the child from what I was going to do, just as there were from not removing the tumour in the first place. And yet I was not at all certain that it was worth it: the child was almost certainly doomed whatever we did. Although the tumour was benign it was simply too large to remove completely without causing awful damage, both to the fornices and to an area near them called the hypothalamus. The hypothalamus controls vital functions such as thirst, appetite and growth. Children with hypothalamic damage typically become morbidly obese dwarves.

It was easy enough to find the tumour. It was at least four or five centimetres in diameter and of a soft, grey consistency which 'sucked easily', as neurosurgeons say. It was difficult to see if the all-important fornices still existed or whether they had been obliterated by the tumour. I managed to preserve a thin thread of white matter on one side that might, or might not, have been all that was left of them.

'Have a look,' I said to Dev, gesturing to the side-arm on the microscope. 'Do we know what her memory was like before the operation? It may already have been very poor, so there is less to be lost.'

'No,' he replied, 'but it's not going to make much difference to what we do, is it?'

'No, I suppose not,' I said, a little reluctant to accept his brutal realism. He scrubbed up and, once he had taken over from me, I went to have a cup of coffee.

Dev's office has a broad window looking out onto the space in front of the hospital, one floor down. In the distance you can see the green foothills to the north, which, whenever I look at them, invariably make me think of the celestial, snow-covered high Himalayas that lie hidden beyond them and which I long to see.

I drank my coffee and went back to the operating theatre. I scrubbed up and joined Dev and, after removing more of the tumour, we found that we had passed all the way through the child's brain onto the base of the skull on the right side. The tumour was so large that it had, effectively, split the lower part of the front of her brain in half.

'I think it's time to stop,' I said to Dev. 'We've taken out at least one fornix and one half of the hypothalamus – any more damage and she'll be completely wrecked.'

'I agree,' said Dev. 'She can have radiotherapy for what's left.'

'It's hard to know how much of the tumour is still there,' I said 'Maybe twenty per cent.'

There is no pleasure or glory in this kind of operating, I thought to myself as I tidied up the bleeding in the large cavity we had left in the girl's brain by removing the tumour. There's tumour left behind, she's almost certainly going to be left very damaged, and all we have done is slow down her dying.

# 6

# THE MIND–BRAIN PROBLEM

'Patient is thirty-five-year-old man. He thinks there is insect in his head.'

'And you got an MRI scan?' I said.

'Yes, sir. No insect.' We looked at the scan.

'Well, you can tell him it's OK,' I said, though I had already seen so many cases of neurocysticercosis in the brain resulting in epilepsy or filariasis causing painful, swollen limbs and other problems that were entirely new to me that I had momentarily wondered if the man really did have some unusual skull-boring Nepali insect in his head.

'Shall we send him to see psychiatrist, sir?'

'Good idea,' I replied.

Once the day's operating is done the outpatient clinic is started. The patients will have been waiting all day, clerked by the juniors in the morning, and various investigations organized, and then seen by the more senior doctors, including the professor, once they have finished in the operating theatres.

I was ushered into the outpatient room on my first day to see a row of three patients and their families sitting next to the desk. In front of the desk stood five junior doctors. The patients looked startled and anxious. A receptionist brought some notes and one of the junior doctors, freshly out of a Chinese or Bangladeshi medical school, read out the history to me in stumbling but gabbled Nepali English, much of which I struggled to understand. The patient was an anxious-looking woman in a beautiful red dress.

'Patient is thirty-five-year-old and has headache for five years. Bowels and bladder normal. On examination pupils equal and reacting. Cranial nerves intact, reflexes equal and plantars downgoing. Had MRI scan.'

'Well, let's look at the MRI scan,' I suggested, which we did and which was, predictably, normal. How much does that cost? I wondered to myself. The answer, I learnt later, was an entire month's income. I was completely nonplussed. Uncertain as to what I was supposed to recommend, I asked the MOs.

After some hesitant discussion with them, I discovered that a huge variety of drugs were widely used in Nepal, often in a largely random manner. As it is, the patients can buy virtually any drugs themselves from small pharmacies on the streets. There is one on my walk to work, always with a queue. Steroids, I discovered, were popular for all manner of complaints, as was diazepam – Valium. After a few weeks of outpatient clinics, I began to suspect that the entire population of Nepal was on the pain-killing antidepressant amitriptyline.

The first patient was hustled off to be given a prescription and the next, who had been sitting next to her, was moved sideways onto the chair she had left. The clinic was clearly run on ergonomic, assembly-line principles. There was a long line of patients with headaches and backache, sore joints and one with rectal bleeding. I realized that the outpatient clinic functioned more as a GP surgery than a specialist neurosurgical clinic and I had to reach back into my basic medical knowledge from more than thirty years ago. This was both interesting – I was surprised at how much came back to me – and worrying. I was anxious that I might have forgotten something obvious and important after so many years spent specializing in neurosurgery. At least there was internet access, and it was helpful to find answers to most of my uncertainties on my laptop.

The next patient is a young woman with complete paralysis of half of her face after surgery for a huge acoustic tumour. It's a common complication and often inevitable if the tumour is as large as they usually are in Nepal because of delayed diagnosis. The patient and her husband are delighted when Dev comes into the room. They chatter happily. Dev puts his arm on the husband's shoulder.

'I was congratulating him on being a devoted husband. She was very ill after surgery but he stuck by her. They come from a part of the country where if the buffalo is ill, worth 63,000 rupees, they will spend money to treat it but not if the wife is ill. He's a good man!' And he slapped the man on the back again.

'Twenty-two-year-old woman with headache for three months. On examination pupils equal . . .'

'No, no, hang on a moment. What does she do for a living?'

There was a brief discussion between the MO and the patient.

'She counsels victims of torture, sir.'

'What? From the time of the Maoist insurgency?'

'Yes, sir.'

'Does she enjoy the work?'

Apparently she rather liked it. Had she received training for this? I asked.

'Yes,' came the reply.

'For how long?' I asked.

'Five days,' she said.

A skull X-ray was produced.

'This is a waste of time for headache,' I said.

'No, sir,' came the very polite reply. 'It is of her sinuses and she has sinusitis.' And now that I thought of it she certainly sounded as though she had a blocked nose.

'Ah, yes. I missed that. Shall we send her to the ENT clinic?'

'They are on holiday for *Dasain*, sir.'

'Well, you'd better prescribe her a decongestant then.'

And every so often there might be a patient with a brain tumour about whom Dev wanted my opinion, or another serious and often rare problem, but most of the patients had chronic headache or dizziness or the peculiarly Nepali symptom of total body-burning pain, and were determined to have MRI scans, despite my assurance that the scan would not help. As they would have to pay for the scan, it was not worth arguing over.

I quickly learnt that many of the patients were very disappointed to see me as opposed to the famous professor, even for the simplest of problems. I might have spent thirty minutes explaining things via one of the MOs but I had to resign myself to politely disappointed patients insisting on seeing him, although a few declared themselves happy with my opinion.

Meanwhile, in the room next door, Dev would be conducting his own high-speed clinic. The patients all expected to see him and he tried to see all the new ones himself. His room was full of doctors, receptionists and relatives, all standing, with the patient sitting in the middle of the melee. It made you think of a king, surrounded by courtiers and petitioners.

The door between our rooms was open and I could hear him coaxing, cajoling, declaiming, reassuring in rapid Nepali, depending on the class and education of the patients. They ranged from impoverished peasants from the mountains to teachers and politicians.

'How many patients actually have a neurosurgical problem?' I asked him.

'One point six per cent,' came the answer.

'Do other doctors refer you patients?'

'No, they all have their own connections and hate me. They try to refer them elsewhere but the patients come and see me anyway.'

As I left my first outpatient clinic I was stopped by a man I did not recognize.

'I am the girl's father,' he said in passable English. 'Thank you, sir, thank you so much,' pressing his hands together and holding them against his chest in Nepali greeting. Dev must have told him that I was involved in the surgery. I smiled, I hope not too sadly.

'My son had a brain tumour,' I told him, 'I know what you are feeling.' He thanked me profusely again, and I nodded in acknowledgement and sympathy and went to the management office to wait for Dev and to be driven home.

I have never enjoyed swimming – I was taught to swim at school at the age of eight, in the muddy river at the edge of the school playing fields, with a canvas belt around my waist attached to a rope and wooden pole, which one of the schoolmasters held like a heavy fishing rod. I dreaded having to climb down the slimy wooden ladder attached to the landing stage, with the cold, wet belt around me, seeing the master's shoes above me through the stage's planks, into the dark water. I would hang onto the ladder, half submerged, before being tugged by the master controlling the rope. I floundered into the water like a hooked fish. You were just expected to keep afloat by dog-paddling. There was no attempt to teach you to swim and the rope and pole were used to stop you drowning. I remember one of my schoolmates being flung into the river by the master when he was too frightened to descend the ladder. I used to wet myself with fear when changing into my swimming trunks for this character-building experience.

At my next school I was taught to swim properly by the kindly headmaster, but after that there was a notoriously sadistic ex-commando PE master who once hit my face so hard that it was numb for hours afterwards. I was so frightened of the man that I would slam my classroom desk's hinged lid on my hand to bruise it and claim that I had fallen and couldn't swim. That only worked once, so I then took to sticking my finger in one of my ears for many hours, mimicking an ear infection. The school doctor was very puzzled by this, as it only happened once a week. I was marched off to an ENT clinic at St Thomas's Hospital accompanied by the school matron. A sceptical consultant, with a row of medical students, looked in my ear and expressed some doubts. I can't remember what was said, but I do recall trying to persuade myself that there really was a problem with my ear even though I knew that I was malingering. It was my first experience of cognitive dissonance – entertaining entirely contradictory ideas – and the importance of self-deception in trying to deceive others. I then discovered that music lessons for playing the trumpet were on the same day and at the same time as the swimming class with the vile ex-commando, so I took up the trumpet but did not get on with it. Eventually I would just hide in a cupboard and skip the swimming lessons – an act of some bravery, I thought – and I got away with it.

I was at my weekly brain-tumour meeting twenty-five years later when a brain scan with a familiar name appeared on the screens in front of us. It was the PE master from my past and it showed a malignant brain tumour.

'He's a most unpleasant person,' my oncology colleague said. 'We've had no end of trouble with him but it's a frontal tumour, so maybe he's suffered personality change.'

'No, he hasn't,' I said, and explained my connection with the unfortunate man.

'The tumour needs to be biopsied,' my colleague said.

'I think it might be better if you got somebody else to do it,' I replied.

I wake with the dawn, the crack between the curtains facing my bed going from dark to light, to the sound of cocks crowing, dogs barking and birds singing. I go for a run every morning, but it took me a few weeks to overcome my fear of the local dogs – the guidebooks warn of rabies but my Nepali friends assured me this is more of a problem with the temple monkeys than the street dogs. So at first I ran in slightly absurd small circles and figures of eight in Dev and Madhu's garden, and up and down the many steps, for half an hour. Later, a little braver, I took to running for longer along the local lanes, between the tightly packed houses that didn't exist even ten years ago, past the rubbish and open drains, past sagging power and phone lines and bougainvillea hanging over garden walls. The road is uneven earth and rock, but there are a few short stretches of rough concrete, prettily patterned with the street dogs' pawprints. There is a small shrine on my usual route, and passers-by ring the bell that hangs by its entrance. All around me there is the sound of people coughing and hawking as they start the day. Neither the dogs nor the local people take any interest in me – it seems that there is nothing unusual in the sight of an elderly and breathless Englishman in football shorts stumbling along the road, but Nepalis are very polite and so perhaps are the dogs.

In England I run for longer. I used to run close on fifty miles a week, but one of my knees started to complain and now I only run twenty-five miles a week. I rarely enjoy it – I find it a considerable effort and my body feels stiff and leaden – but I do it for fear of old age and because exercise is supposed to postpone dementia. But there were occasionally wonderful moments when I was still running long distances – up to seventeen miles at weekends in the countryside surrounding Oxford. One early spring morning I was in Wytham Woods, the low sunlight falling diagonally through the trees, when I came across a leveret – a young hare – eating grass beside the path. It appeared completely unafraid of me and I was able to stand only three feet away as it quietly grazed, looking at me with its bright eyes. It was a unique moment of innocent trust from a wild animal, and I felt deeply moved. There is a beautiful ink and sepia drawing by the mystical early-nineteenth-century artist Samuel Palmer in the Ashmolean Museum in Oxford which shows the very same scene – a young hare in a wood, early in the morning, with the sun rising.

On another occasion, as I ran along the Thames, I noticed a duck desperately flapping in the water at the end of a broken-down pier. It appeared to be caught on something, so I crawled out along a steel beam projecting over the river, all that was left of the pier, feeling heroic. I found that the duck had a fish hook in its beak, with the fishing line wrapped around the beam. I managed to free it without falling into the river. The duck promptly dived into the water without stopping to thank me. Nevertheless, I like to think that if one day I ever get into trouble when swimming, the grateful duck – as in the fairy stories – will come and rescue me.

After running round Dev and Madhu's garden, I do fifty press-ups and a few other exercises, all of which I also hate doing, but I feel much better for it afterwards. I finish with a short swim in the small swimming pool outside the guest house. There is a very brief moment of ecstasy as I push out into the cold, mirror-calm water, which reflects the early-morning sky, with a view of the nearby Himalayan foothills in front of me. I momentarily forget my deep dislike of swimming. I complete this morning ritual with a cold shower – something I started doing two years ago. At first, admittedly in England in the winter, I thought I had discovered the elixir of life. A feeling of exhilaration, of intense well-being, would last for up to two hours afterwards. To my great disappointment, this wonderful feeling – acquired so easily within a couple of minutes – became shorter and shorter within a matter of weeks. I continue to have a cold shower every day, but the feeling now lasts only a few seconds at best, although the cold water still makes me jump about and gasp. I suppose my physiology has adapted, although health fanatics claim that cold is good for 'vagal tone' – the activity in the vagus nerve, which controls many of our body's functions in ways that we scarcely understand. It is a long nerve, which bypasses the spinal cord and reaches from the brain to the heart and many other organs, carrying information and instructions in both directions. It is an extraordinary nerve. Stimulation of the nerve with an electric current can help epilepsy, though nobody knows why. It can allow the generation of orgasms in women who are paralysed and have suffered complete destruction of the spinal cord. Apparently, people who have had it divided (an obsolete operation for gastric ulcer) will not develop Parkinson's disease.

After all this I sit beside the swimming pool in the little paradise of Dev and Madhu's garden, with flowers and birds all around me, and drink a cup of coffee before setting off for the hospital. Sometimes a bird with brilliant turquoise plumage dives down onto the surface of the pool, its wings and the splashing water flashing in the sunlight.

After a few weeks I decided to rearrange the way my clinic was run. I had the junior doctors sitting down, I would politely greet each patient when they entered, as I would do in England, which seemed to be less expected here. We would only have one patient in the room rather than a whole queue. The patients would usually come into the room looking expressionless, but my saying '*Namaste*' and pressing my hands together would almost invariably produce an utterly charming, slightly shy smile in reply. I insisted that every consultation had to end with asking the patient if they had any questions. This made the consultations feel a little less like assembly-line work but greatly reduced the number of patients I could see with the MOs, as the patients had so many questions to ask. They rarely spoke English and often were poor historians, as doctors call patients who have difficulties describing their symptoms. Many of them were subsistence farmers who could not read or write, and the MOs' English was often very limited as well. Making any kind of diagnosis could sometimes be impossible as the patients seemed so uncertain about their symptoms and were so determined to be given some new drug treatment. On the other hand, some of the patients had diseases such as TB and filariasis, with which I was unfamiliar. I found conducting the clinic extremely difficult, and had to be careful not to miss a serious problem in the constant stream of patients with chronic low back pain, headaches and total body-burning pain.

'Do you know what somatization is?'

'No, sir.'

'Well, it's the idea that if people are unhappy or depressed – marriage problems, things like that – rather than admit it to themselves, they develop headaches or total body pain, or strange burning feelings. They attribute their unhappiness to these symptoms, rather than consciously admit that they are unhappy in their marriage or that there is some similar problem. Such symptoms are called psychosomatic. You can see it as a sort of self-deception. Is the diagnosis of depression recognized here?'

'Not really, sir.'

'All pain is in the brain,' I explained as I pinched the little finger of my left hand in front of the MOs on the other side of the desk. 'The pain is not in the finger – it's in my brain. It's an illusion that the pain is in the finger. With psychosomatic symptoms, the pain is created by the brain without a stimulus from the peripheral nervous system. So the pain is perfectly real, but the treatment is different. But patients don't like being told this. They think they're being criticized.'

'Many of the women are seeking attention,' Upama, the MO said. 'Their husbands are away working abroad and they are unhappy.'

Amidst the flood of patients with minor problems, there are terrible cases as well – a young woman with much of her scalp infiltrated by a malignant skin tumour, a man dying from a brain tumour. There was a child, a thirteen-year-old girl, with half her face paralysed. The scan showed a complex congenital malformation of the joint between the spine and skull, which was the likely, though a very unusual, cause of her paralysis. Neither Dev nor I are very expert in such problems, and we had agreed that surgery was probably too difficult and dangerous. Upama explained this to the girl and her father, and the girl started sobbing silently.

'She is a girl,' Upama explained. 'Her face . . .'

While I watched the child cry, I thought about my detachment from her suffering – detachment both as a doctor and also because of the great gap of culture and language between us. I have to be detached, I thought, and it is something I learnt as soon as I qualified as a doctor. I cannot help this child, and there is little point in being emotional about it. But I also thought of the research into bonobos (previously known as pygmy chimpanzees), our closest evolutionary relatives, which shows that they have compassion and kindness, a sense of fairness and console each other over pain – at least for their own group. They have not been told to do this by priests or philosophers or teachers, it is part of their genetic nature, and it is reasonable to conclude that the same applies to us.

For most of us, when we become doctors, we have to suppress our natural empathy if we are to function effectively. Empathy is not something we have to learn – it is something we have to unlearn. Patients become part of the 'out-group' as anthropologists call it, people with whom we need no longer identify. But the child went on crying and I started to feel uncomfortable. Besides, I told myself, the only way that doctors can lay claim to any kind of moral superiority over other professions is that we treat – at least in theory – all our patients in the same way, irrespective of class or race or nationality, or even of wealth. So my detachment wilted as the child cried and I thought I might just see if Dev and I could be wrong. I used my smartphone to photograph the girl's scans and emailed them to a colleague on the other side of the world, an expert in problems of this sort, for an opinion. He replied thirty minutes later, saying he felt that surgery was both possible and relatively straightforward.

I showed his opinion to Dev.

'Isn't the internet wonderful!' I said. 'We can get a world-class opinion so quickly.'

'We'd better get the child back and talk to the family,' he replied, but the girl and her family had disappeared.

While the patients come and go, the day outside grows dark. The high Himalayan foothills on the horizon disappear. The ragged leaves of the banana tree in the paddy field next to the hospital start to shake and flap in the wind. A flock of small birds is suddenly flung up into the sky like a handful of leaves, to be quickly swept from sight. The windows of the outpatient room are open – the room fills with the intoxicating smell of wet earth and the patients' notes in front of me blow off the table. There are frequent power cuts and every so often the room plunges into darkness for a few minutes. Thunder crashes directly overhead, to echo away into the distance.

'Patient is sixty-five-year-old man with numbness in his fingers.'

The MRI scan shows slight compression of the sixth cervical nerve roots in his neck.

'How much is he troubled by his symptoms?'

'He has difficulty climbing trees and milking buffalo, sir.'

We decide to continue with conservative treatment.

'It is proxy case, sir. Father has brought scan. His two-month-old daughter is in other hospital. They have diagnosed bacterial encephalomeningitis. Child is fitting, and they grew enterobacter in the blood. They have recommended three weeks of IV antibiotics. He wants to know if the treatment is right.'

The CT scan was of poor quality and I found it difficult to interpret, but it looked as though the child might have suffered extensive brain damage.

'He wants to know if it is good idea to spend money treating the child.'

'How many other children does he have?'

'Three, sir.' But then we worked out that two of them had already died.

I looked at the scan for a long time, not knowing what to recommend.

'I think I'd get an MRI scan,' I eventually said. 'If it shows severe brain damage, perhaps it is better to let the child die.'

Jaman, the excellent MO, spoke to the father.

'It is economic problem for the MRI scan,' he told me.

'Then it's very difficult,' I said.

I left Jaman and the other MOs to have a long conversation with the father. I don't know what was decided, but the father said '*Namaste*' to me very politely as he left.

'Patient is forty-year-old lady who has had headache for twenty years, sir.'

My heart sinks a little.

'Well, tell me more about the headache.'

We discuss this for a few minutes. The patient has been on a long list of drugs over the years.

'She suffers from panic attacks. She finds diazepam helps, sir.'

I deliver a little lecture on the evils of diazepam and the way that millions of housewives became addicted to it in the past in Europe and America. It is very difficult to know what to suggest.

'Do you know the word stigma?'

'Yes, sir.'

'Is there stigma in Nepal against seeing psychiatrists?'

'Yes, there is, sir.'

'I think you should suggest she sees a psychiatrist. I find it helps if I tell patients that I had psychiatric treatment myself once. It was invaluable.'

There was a rapid exchange in Nepali.

'She wants MRI scan, sir.'

'It's a waste of her money.'

'But she lives in Nepalgunj.'

'How far away is that?'

'Two days by bad road.'

'Oh all right, get an MRI scan then . . . it won't show anything but I suppose she hopes that somehow it will make her unhappiness real.'

Afterwards the MO tells me that the patient has already tried to kill herself twice.

'How do people kill themselves in Nepal?'

'Usually by hanging, sir.'

The patients come from all over Nepal, often from remote mountain villages accessible only on foot. They come to the clinic hoping for an instant cure, and with an exaggerated faith in medicines, perhaps connected to their belief in prayer and sacrifice. The idea that drugs can have side effects, that there is a balance to be struck between cost and benefit, seems very alien to them. It is impossible to treat effectively chronic problems such as headache, epilepsy, raised blood pressure or low back pain on the basis of a single visit. So the patients end up on a bewildering variety of different drugs that they either acquire themselves or from different doctors over the years. They come with plastic bags full of shiny foil blister packs of coloured tablets of many shapes and sizes, which they spread out on the table in front of me and the MOs.

'She is thirty-year-old lady with headache, sir.'

Oh dear, I thought, not another one. She sat diffidently in front of me with her husband beside her.

'And she cannot stop laughing, sir.'

'Really? Pathological laughter? That's interesting.'

I was handed the scan. It was indeed very interesting, but very sad.

'What do you see, Salima?'

After a while, Salima, with my help, worked out that we were looking at a huge brain tumour – technically a petroclival meningioma. I had once had a similar case in London who also had the very rare symptom of uncontrollable, pathological laughter. I had operated, and had left her in a persistent vegetative state. It was one of the larger headstones in my inner cemetery.

'Tell them to come back tomorrow when Prof is here,' I said.

Once she had left the room I told the MOs that without surgery the poor young woman would die within a matter of years – slowly, probably from aspiration pneumonia. She already had difficulties swallowing, from the pressure of the tumour on the cranial nerves that controlled her throat, a sure harbinger of death. But surgery, I told them, was almost impossibly difficult – at least, it was very difficult to operate without, at best, inflicting lifelong disability on the patient. So what was better? To die within the next few years, or face a longer life of awful disability?

'Prof needs to talk to them,' I said, but she never came back.

'All is well apart from the child . . . the baby where we tried to do an endoscopic ventriculostomy yesterday.' This had been another patient, a baby only a few months old, with a huge hydrocephalic head.

'In what way?'

'Not doing well . . .'

'Does the mother have other children?'

'Yes.'

'It's best if we let her die, isn't it?'

Dev said nothing but silently conveyed his agreement.

'In England we wouldn't be allowed to do that,' I said. 'We'd raise heaven and earth and spend a fortune to keep the child alive even though she will have a miserable future with severe brain damage and a head the size of a football. My old boss, at the children's hospital where I trained, sometimes said to me, after we had operated on a particularly hopeless case who was doomed to live a miserable and disabled life, that he wished he could tell the parents to let the child die and go and have another one. But you're not allowed to say that.'

'The child died during the night,' the registrar told me when he saw me next morning looking at the space where she had been. The child had gone, leaving only a sad little huddle of sheets on the bed, as the nurses had not yet had time to change the bedding.

I had some difficulties setting up a patient for an MVD, an operation for facial pain which involves microscopically manipulating a small artery off the trigeminal nerve, the nerve for sensation over the face. It is an operation I have done hundreds of times in London – but doing it here feels very different. Turning the patient was somewhat problematic.

'In London we say one, two, three and then turn the patient,' I said. 'Do you do that here?'

'Yes, sir,' the registrar assured me happily.

'One,' I said, and he grabbed the patient and started pushing him off the trolley.

'No! No!' I shouted. 'One, two, three . . . and *then* roll.'

It felt more like a rugger scrum than a coordinated manoeuvre, but we managed to get the patient safely face-down onto the table.

*

It was a twenty-minute drive from Neuro Hospital to the Bir, past a few small demonstrations with heavily armed police in attendance. Nepal is in a constant state of political chaos. The civil war only ended a few years ago. The monarchy collapsed four years after the royal massacre. The democratically elected Marxist government which replaced it is riven by continuous political infighting. The streets around the hospital were packed with pedestrians and motorbikes. An emaciated young woman was selling a few halved cucumbers, daubed with a red relish, from an empty oil drum that served as a stall, at the hospital entrance. There was a row of ramshackle pharmacies across the road from the entrance, with crowds of people standing in front of them.

'That was the first pharmacy in Nepal,' Dev said, pointing to an old brick building behind the pharmacy shacks with wide cracks in its walls from the recent earthquake.

The hospital itself was more like a dirty old warehouse. It reminded me of some of the worst hospitals I had seen in Africa and rural Ukraine. It had been built in the 1960s by the Americans, and although some of the wards had plenty of windows, it was a typical example of the style of architecture that treats hospitals as being little different from factories or prisons, with long, dark corridors and lots of gloomy spaces. The wards were very crowded and everything felt sad and neglected. Dev was greeted with many delighted smiles and '*Namastes*' by the staff, but he told me afterwards that he had been deeply upset by the visit.

'I created my own neurosurgical unit here,' he told me. 'The first in Nepal. We had to build everything from scratch with second-hand equipment. I used to do my own cerebral angiograms by direct carotid puncture in the neck. Jamie Ambrose at AMH had shown me how to do it. We painted the ward every year – I paid for the paint myself – we had a painting party. And look at it all now! All gone, filthy, neglected.'

'When I came back here from the UK,' he continued, 'nobody would work after two in the afternoon. So I sat in the office by myself, the only senior doctor in the building. Eventually everybody else stayed as well. We had no money then. I was working all the time.'

We left the hospital and stood outside, waiting for Dev's driver. Dev was recognized by many people – he is famous throughout Nepal, let alone at the hospital where he used to work – and while he chatted and joshed with them I stood and watched the endless flow of people coming and going. There was a large pool of dirty water from a leaking water main, and rubbish and broken bricks – probably left over from the earthquake – in the gutter opposite. And yet, as the women picked their way across the road in their brilliantly coloured and elegant clothes, I thought, with a slight feeling of shame, that the scene was rather beautiful.

As Ramesh, Dev's driver, manoeuvred the car past the long and chaotic queues outside the petrol stations, Dev returned to the subject of the Bir.

'I need a rest after what I have just been through. It was terrible, terrible . . . people would come to appreciate just how good a ward could be . . . all gone. That floor was something different. Nice working environment. It was recognized by the Royal College in England for training. All gone, all gone.'

A few months later I met an English neurosurgeon in New Zealand who, when a medical student, had visited Dev's department at the Bir. He fully confirmed just how different the department had been from the rest of the hospital.

'It was a beacon of light in the darkness,' he said.

'We came back here with such high hopes,' Madhu told me over supper that evening, 'and everything has got so much worse.'

# 7

## AN ELEPHANT RIDE

We set off for the Terai – the flat lowlands in the south of Nepal, bordering India – early in the morning, just after dawn, the air very still and humid and hot. I was in a cheerful mood: the day before, I had received the pathology report on the skin tumour I had had removed before setting out for Nepal. The tumour was indeed cancerous but the 'excision margins were clear' – in other words, I had been cured and would not need further treatment.

There is an entire tourist village centred on elephant rides adjacent to the Chitwan National Park. Tourism in Nepal had suffered badly because of last year's earthquake and the nearby tourist town of Sauraha, with many bars and small hotels, looked almost empty when we drove through it.

'What will they live off?' I asked Dev.

'Hope,' he said, with a shrug.

There were only a few Western people to be seen, easily identified by their baggy shorts and T-shirts. I always wear a long-sleeved shirt and trousers, not just to be different but because the Catholic missionaries I lived with in Africa fifty years ago, when I was working as a volunteer teacher, taught me that this shows respect for the local people.

We were taken to the government elephant station at the edge of the jungle. We walked beneath tall, widely spaced trees, through patches of sunlight. It was remarkably quiet. There was a group of elephant-high, ramshackle shelters – roofs of battered and rusty corrugated-iron sheets on four posts, surrounded by tattered electric fencing. In the centre of each shelter there was a massive wooden pillar, with heavy chains and shackles hanging down. There were no elephants to be seen.

'They used to keep the elephants chained at night but an Englishman showed them they could use electric fencing,' Dev told me.

Beyond the sheds were a few low buildings, and in one with an open front two European teenage girls were sitting cross-legged in very short shorts on the ground next to a dark-skinned elderly Nepali man. They were rolling up handfuls of rice mixed with sweets – the plastic wrappers were carefully removed – into a ball, wound around with long grass, a packed lunch for elephants. They held the ball with one foot and used their hands to bind the long grass around the rice. The girls were very absorbed and everybody was silent. When I asked them where they came from, they smiled and said they came from Germany. I wasn't quite sure what to think about seeing these children of the affluent West playing at being peasants.

And then, slowly, coming out of the surrounding jungle, a huge elephant appeared with a mahout perched high on her neck, his feet behind her ears. The creature was enormous, solemn and stately, and strangely graceful for such a massive beast. The last of the megafauna on land to survive mankind.

'That's the one we'll be going on,' Dev told me.

The mahout brought the great creature to where we were waiting, and the elephant bent her huge knees and settled awkwardly down on the ground, back legs pointing backwards and front legs forward. The mahout and his helpers then spent some time fitting a wooden frame over a mattress onto the elephant's back, kept in place with a wide girth, which they heaved on with ropes to get it tight. While they did this I walked up to the elephant and looked into her small, thoughtful eyes and she looked back at me. She had elegantly curled the end of her trunk over her left foreleg. The day before I had been reading about elephants – of the 40,000 muscles in their trunks, and of their great brains, the largest brain of all land mammals. They are intensely social, with a complex social life. They can console each other, mourn the dead and have a language of sorts. They can also recognize themselves in mirrors (which is generally considered to mean that they have a sense of self).

Nobody knows how many brain cells are needed for consciousness. Recent work on insects suggests that even they might be capable of it; their brains show similarities to the midbrain of reptiles and mammals, where some authorities think conscious experience arises. To ask if a creature is conscious is equivalent to asking if it can feel pain, and nobody knows at what point pain arises in nervous systems. If you deliver a painful stimulus to one of a lobster's claws, it will rub the painful area with the other claw. Is this a mere reflex? It seems more likely that it feels pain. We boil lobsters alive, of course, before eating them.

When patients are unconscious, for instance after a head injury, we assess the depth of their coma by hurting them. You either squeeze the nail bed of one of their fingers with a pencil, or press very hard with your thumb over the supraorbital nerve just above one of the eyes. If they respond purposefully to the pain – trying to push you away or, just like the lobster, trying to get one of their hands to the painful area – we assume there is some kind of conscious perception of pain going on, even if the patient has no memory of it afterwards. On the other hand, if the patient is in deep coma, they do not move in response to the pain at all, or move their limbs in a reflex, purposeless way. We assume, then, that there is no conscious element to the response and that the patient is deeply unconscious.

And then there is the wonderful mystery, at the other end of the scale from insects, as to why whales have brains which are so much larger than ours. It is true that there are structural differences (whales lack our cortical layer IV and most of them have a much higher ratio of supportive glial cells to neuronal cells than we do), but nobody knows why they have evolved such massive brains, and to what use they are put. In recent years the floodgates have opened with new research into animal intelligence: cows have friends among other cows, pilot whales (a species of dolphin) have more neuronal cells in their brains than any other creature, manta rays can recognize themselves in mirrors, fish can communicate and work together when hunting. We are moving further and further away from Descartes's separation of mind from matter, and his dreadful claim that animals are mere automata.

Self-consciousness, however, to be aware of one's own awareness, to think about thinking, is probably a more complex phenomenon. I first discovered it at the age of fourteen on a school expedition to the ruins of Battle Abbey on the South Coast. With the other boys, I ended up larking about on the nearby shingle beach. I ran fully clothed into the sea and stood with the waves lapping about my knees, soaking my school uniform. As I stood there, I was suddenly struck by an overwhelming awareness of myself and of my own consciousness. It was like looking into a bottomless well, or seeing myself between a pair of parallel mirrors, and I was terrified. We returned to London in a coach and I came home in a state of considerable distress. I tried to explain what I felt to my father as he sat in his book-lined study. I started to shout about suicide, but I think he was rather confused by my hysterical outburst, as was I.

Clearly this sudden self-awareness was a philosophical version of the awkward self-consciousness which comes to boys with adolescence, when testosterone levels rise. I remember the shock I experienced on seeing my first, solitary pubic hair. Over the next two years I had a series of what are best described as mystical experiences – feelings of profound illumination and unity associated with intense visual effects, where shadows and colours acquired an extraordinary depth and beauty. My hands, and the veins on them, would look especially profound. I would gaze at them with wonder.

When I was a medical student many years later, studying anatomy, I was particularly fascinated by the anatomy of the human hand. There was a large polythene bag in the Long Room – the room with corpses for dissection – full of amputated hands in various degrees of dissection. The hand is a wonderfully complex mechanism, with a series of tendons and joints and muscles, a device of articulated levers and pulleys. I drew and painted careful and elaborate watercolour studies of these hands, but to my regret lost my anatomy notebooks many years ago. I subsequently discovered in Aldous Huxley's writings that my mystical experiences were identical to those he described while taking mescaline. There is a form of epilepsy, known as limbic epilepsy – Dostoevsky is thought to have had it – where people have an intense feeling of unity and transcendence, and often interpret it as being in the presence of God. The limbic system is part of the human brain involved in emotions, and in 'lower' mammals is mainly involved in olfaction – the sense of smell. When I was at Oxford University most of my friends were experimenting with LSD, but I never dared. I smoked cannabis occasionally but disliked the complacency it produced.

The mystical experiences faded as I grew older, replaced, perhaps, by sexual desire and sexual anxiety. While my contemporaries at school were going to parties and learning to kiss girls, I sat in my room at the top of the large house in Clapham, reading voraciously. I kept a diary which I destroyed in a fit of embarrassment and shame a few years later. I rather regret that: I suspect that many of the questions and problems that trouble me now, as I face retirement and old age, were already present then, when I was also trying to find a sense of purpose in my life, but had much more of it ahead of me. It would also amuse me to see what a prune I was when young, and how seriously I took myself.

My father recommended many books, ranging from Raymond Chandler to Karl Popper's *Open Society and Its Enemies* – this latter book, I think, had a great influence on my later life. Popper taught me to distrust unquestioned authority, and that our moral duty in life is *to reduce suffering*, by 'piecemeal social engineering' and not with grand schemes driven by ideology. This, of course, is very close to the Christian ethics and belief in social justice inculcated in me by my parents, and the understanding of the importance of evidence and honesty that I learnt as a doctor. Yet doctors get paid – usually very well – for their work, and we cannot but help people (unless singularly incompetent). So our work need not call for any particular moral effort on our part. It is easy for us to become complacent, the worst of all medical sins. The moral challenge is to treat patients as we would wish to be treated ourselves, to counterbalance with professional care and kindness the emotional detachment we require to get the work done. The problem is to find the correct balance between compassion and detachment. It is not easy. When faced with an unending queue of patients, so often with problems that we cannot help, it is remarkably difficult.

My experience as a hospital operating theatre porter had resulted in my deciding to become a surgeon. I had returned to Oxford to complete my degree before trying to get into medical school. My arrival back at Oxford was shortly followed by my first, and unsuccessful, attempt at sexual intercourse (with a sweet girl in Leicester who took pity on me). This precipitated a further crisis. I started to suffer from manic pressure of thought, seeing all sorts of wonderful connections between disparate things – at first rather exciting, but then very frightening. My ideas started to spin out of control, and the feeling of brilliant omniscience was replaced by a fear that there was some kind of evil presence beside me. I can see now that part of me was trying to force another part of me, through fear, to seek help. There is another form of limbic epilepsy, it is interesting to note, where people experience the presence of evil rather than the presence of God. At a friend's suggestion – another person to whom I am deeply indebted – I got in touch with the psychiatrist my father had unsuccessfully tried to persuade me to see a year earlier, when I had abandoned my degree. I was admitted for a short time to a psychiatric hospital.

I had a room to myself and lay there on the first night feeling miserable and tense. A friendly West Indian nurse came in and asked me if I wanted a sleeping pill.

'No, I don't need one,' I said defensively.

'Well, I'm called Charlie and I'm at the end of the corridor if you change your mind,' he said with smile.

I could not sleep. I had sunk so low that I had no future whatsoever. I had reached the bottom of a bottomless well, and there was no way up again. I had become a mental patient. I was utterly and completely alone. I cried and cried, but even as I cried I felt something frozen in my heart thawing, just like the fragments of the evil magician's mirror in the boy's heart in Hans Christian Andersen's story *The Snow Queen*. I had been fighting myself for so long, and for so long I had viewed other people only as mirrors in which I tried to see my own reflection (I am, alas, still prone to this). Was it that I had tried to freeze my heart, trying to suppress my hopeless and inappropriate love for the woman who had kissed me? I do not know, but I got out of bed in the early hours and walked down the dark hospital corridor to where Charlie was reading a newspaper, spread out on the table in front of him, in a little pool of light from a desk lamp. I asked him for a sleeping pill – it was Mogadon in those days, now banned. I fell asleep exhausted, and next morning I was rather pleased to see in the bathroom mirror that my inner suffering had finally become real, at least visible. Much better than trying to lacerate my hand and its veins with broken glass, I now had two enormous purple bruises under my eyes.

The next week I spent an hour a day with a deeply sympathetic elderly psychiatrist, unburdening myself. The feeling of being reborn, of being in love with everybody and everything, was intense. When I was discharged I drove up into the Chiltern Hills. It was a perfect autumnal afternoon. My body felt so stiff that it was as though I had run a marathon. I remember the difficulty I had climbing over a padlocked field gate. It was the happiest day of my life.

Research has shown that the ecstasy of being in love rarely lasts more than six months. It fades, and comes to be replaced by the more mundane practicalities of maintaining a successful relationship, but at least it fades a lot more slowly than the ecstasy I felt when I started to take cold showers. The intense feeling of illumination and optimism, of being part of a coherent whole that I felt after leaving hospital, was identical to many accounts I have read of religious conversion and revelation, except that I did not for a moment believe in any kind of divine presence in my life, or in the world. These intense feelings obviously involve the same cerebral mechanisms as when you are in love with a person, with the feeling of joyful unity, beauty and coherence all focused on that one person.

Zebra finches and other birds can grow new brain cells when the mating season begins, when they need to start singing to attract a mate. I wonder whether similar processes go on in our brains when we are in love. I also wonder whether other animals experience ecstasy. It has been suggested that the huge brains of dolphins and whales, creatures who also show great playfulness, mean that they do. It is easy to believe this if you watch a pod of dolphins swimming and leaping through the sea. I did not find God with my ecstatic experience, but instead I learnt that my own mind was a profound mystery, and that the sacred and the profane are inextricably linked. There must be a neural correlate for this, whereby the deep and basic instinct to procreate, present in almost all living things, becomes interwoven with the complex feelings and abstract reasoning of which our larger brains have evolved to be capable. This feeling of mystery about my own consciousness, but without any ecstasy, has grown stronger and stronger in recent years as my life starts to unwind and descend to its close. It is, I suppose, a substitute for religious faith and, in part, a preparation for death.

On one of my trips to the Sudan I had been taken to a small zoo in a huge sugar plantation in the desert, on the banks of the White Nile, a few hundred miles to the south of

Khartoum. There was an enclosure with five Nile crocodiles, who eyed me thoughtfully – they prey on humans – half submerged in their concrete pond. Next to it was a cage with a single young elephant in it. Deprived of its mother and its highly developed social life, it had clearly gone mad, and showed the same disturbed behaviour of grotesque and repetitive movements as severely autistic children, or the chronic schizophrenics I once cared for as a nursing assistant. And next to the poor elephant's cage there was a small enclosure with a young chimpanzee who seemed to have gone as mad as the young elephant. My Sudanese colleague – whom I greatly liked – roared with laughter when he saw my dismay.

'You English! You are so tender-hearted!' he said.

The look the Nepali elephant gave me, with her small, red-rimmed eyes – or so it seemed to me – as the girth was tightened round her was of deep and infinitely sad resignation.

We were taken to a twelve-foot-high mounting platform, with a staircase of rotten treads, overgrown with moss and climbing plants. The elephant was positioned alongside it, and Dev and I and two guides clambered into the wooden frame on her back, each of us sitting in a corner, facing outwards, our legs straddling one of the four corner posts. There was a thin cushion and it was less uncomfortable than I expected.

It was quite disconcerting at first, the slow, rocking movement, twelve feet off the ground, with the elephant gently placing her huge feet on the uneven track as we headed back into the jungle. This is going to be a bit boring, I thought, once I had got used to the swaying motion – there is nothing to do. But after a while I started to enjoy it, although I still wondered what the elephant thought about it all.

The mahout carried a sickle and a stick, and he used the

sickle from time to time to clear the way, as did the elephant with her trunk, coiling it expertly around branches and snapping them off. The sickle, I had read, could also be used to cut the elephant's ears if it became difficult to control. I have also read that training young elephants can involve considerable cruelty, although people also write of the close relationship between the mahouts and their elephants and of the benefit to conservation from the tourist income generated by the elephant rides.

Apparently you cannot get an elephant to do something it doesn't want to do, and watching the mahout and the animal as they chose which way to go through the dense jungle, it was clear some kind of negotiation was continuously going on. The mahout used his feet to kick gently behind her ears to steer her, like the pilot of an airplane using a rudder bar, but it was obvious that the elephant did not always agree with his suggestions. We crossed a river, the elephant effortlessly climbing the steep bank opposite, and went deeper and deeper into the tangled trees of the jungle, along paths that were almost invisible. In a small clearing we saw a herd of spotted deer which took fright and quickly disappeared, bounding with great elegance back into the trees. Apparently there are tigers and leopards in the reservation as well, but they are rarely seen. After an hour of this slow, rhythmic plodding between trees, the leaves brushing our faces, we emerged into grassland, some of the grasses almost as tall as the elephant. The mahout pointed out an area of flattened grass and said something to Dev.

'Rhino's bedroom,' Dev translated, and shortly afterwards, near the riverbank, we came across a rhino with a young calf, which quickly hid behind its mother as the monstrous shape of the great elephant with five human beings on top of it approached. Its mother took little notice of us, continuing to graze, as we admired the studded armour-plating of her skin and her single horn which the Chinese and Vietnamese so stupidly prize, ground up as an aphrodisiac or as a cure for cancer, resulting in the near-extinction of the creature from poaching.

'Why can't they use Viagra?' I complained, as we left the rhino and her calf behind and crossed back over the river. 'I'm sure it's cheaper.'

With Dev as translator, I asked the mahout about the elephant as we plodded majestically through the tall grass. He told me that she was forty-five years old and would probably live to seventy, but recently they had lost several elephants to TB. This elephant had had four calves, but three had died before they were three years old.

'When are the calves taken from their mothers to be trained?' I asked.

'At three years old,' I was told.

I asked if all the elephants were kept alone and was told that they were. As we crossed back over the river on the return journey the elephant suddenly let out a great trumpeting cry.

'What was that about?' I asked.

'She smells another elephant,' Dev translated.

Back at the elephant station, we climbed off our elephant's back and had to wait for a while for Dev's driver and bodyguard to appear. We sat in the sunshine outside a group of huts which clearly had been financed by some well-meaning foreign charity – a lopsided and mildewed notice announced that this was the Children and Women Promotion Centre. The notice was so faded that it was difficult to read, but there was a long list of projects and among these I could just make out 'Computer Letchur' (*sic*), 'Sports Coachers (any)', 'Environment', 'Caring Wildlife (injured)' and 'orphan animal',

'HIV/Aids awareness programme' and other projects funded by foreign aid. 'Unskill volunteers' were accepted. There was another notice, also worn and partly illegible, announcing a Vulture Recovery Programme, with the icons of all manner of international bird charities at the foot of it. The buildings were all dilapidated, with rusty corrugated-iron roofs. The shop was almost empty apart from some cheap imports from China and a single woman in attendance who, most unusually for Nepal, did not smile when I entered. All the world wants to help Nepal and vast sums of aid have been lavished on the country, yet much of it seems to have disappeared without trace, leaving only faded signs and notice-boards behind.

I cheerfully volunteered to join Dev operating on a large brain tumour in an eight-year-old boy and was soon regretting it. The tumour bled like a stuck pig right from the start and there were extraordinarily large arterialized veins running in it that bled furiously and too heavily for the diathermy to work. I started sweating. The problem is that when you worry about the patient bleeding to death you rely on a close working relationship with the anaesthetist, and she didn't speak English and was very uncommunicative. As I struggled to stop the child bleeding to death from a blood vessel in the centre of the tumour, I despaired of ever managing to train Dev's juniors to do such operating. It becomes almost impossible if you are trying to train your junior and you have to watch passively while they fumble and stumble, putting the patient's life at risk. It is easy to see why so often trainees are left to operate on their own, learning the hard way, on the poor and the destitute, who are unlikely to complain if things go badly. In all the countries where I have worked over the years, people with money or influence will make sure that they are not trained upon.

In poor countries such as Sudan and Nepal, there has been an explosion of private clinics and hospitals. The professional associations, largely based on the old British model, have become sidelined, and there is less and less effective maintenance of professional standards. Money and medicine have always gone together: what could be more precious than health? But patients are infinitely vulnerable, from both ignorance and fear, and doctors and health-care providers are easily corrupted by profit-seeking. It is true that socialized health care, as the Americans call it, has many faults. It tends to be slow and bureaucratic, patients can become mere items on an impersonal assembly line, clinical staff have little incentive to behave well and can grow complacent. It is often starved of resources. But these faults can be overcome if high morale and professional standards are maintained, if the correct balance between clinical freedom and regulation is found, and if politicians are brave enough to raise taxes. The faults of socialized health care are ultimately less than the extravagance, inequality, excessive treatment and dishonesty that so often come with competitive private health care.

Dev took over and I was able to go and have a sandwich. In fact the worst of the bleeding had stopped by then, but it was rather wonderful to be able to pause and have a break. And I thought, imagine running a practice like this single-handed for thirty years, with nobody to help out or relieve you – day in, day out and on call every night.

When I saw the boy on the ITU round next day he was awake and crying, and at first I thought that all was well. But something troubled me: his eyes were open, roaming and unfocused. I had missed it at first, but when I came back to see him after looking at the other patients it was clear that he was completely blind.

'What was his eyesight like before surgery?' I asked Dev.

'Not good,' he replied.

'He must have had severe papilloedema from the severe hydrocephalus,' I said. 'We know that some of them wake up blind after surgery.'

When I saw Dev later in the day he told me that he had seen the boy's mother.

'She said his eyesight was very bad indeed before the operation.'

'I have seen that happen twice before,' I said, 'it can't be avoided.' It was better not to think of the boy's future.

One of the first operations I did on a child, when I was a junior registrar, was on a nine-year-old boy with an acute subdural haematoma – a severe traumatic head injury – from a car crash. A neighbour was taking him with his own children to the zoo when another car drove into them. The neighbour was killed, as was his five-year-old daughter. The brain of the child I operated on became so swollen during the operation that I could scarcely get it back into his skull and I even had difficulties stitching the scalp back together over it. This can happen occasionally with acute subdurals. It turned out he was an only child, conceived after years of fertility and IVF treatment. There was no question of the mother having any further children. I had to tell his mother that he was going to die. I watched her as I told her this and realized I was delivering a death sentence, on her as much as on her child. It is not a good feeling to destroy somebody like this. The hospital where I was training was a high-rise building in the north of London, and the ITU had large windows with panoramic views of the city below. I remember how the light from the windows was brightly reflected on the ITU's polished floor as I led the mother to her son's bed, where he lay on a ventilator, a large, lopsided bandage hiding my rough stitching. I thought of how difficult it is to believe in a benign deity intervening in human life when you have to witness suffering like this. Unless, of course, in the words of the famous Victorian hymn, there really is a friend for little children, above the bright-blue sky, who will right all the wrongs we suffer in this life on earth in an afterlife in heaven.

But I am a neurosurgeon. I frequently see people whose fundamental moral and social nature has been changed for the worse, often grotesquely so, by physical damage to the frontal lobes of their brains. It is hard to believe in an immortal soul, and any life after death, when you see such things.

The outpatient clinic is usually finished by six in the evening. Dev's bodyguard will always materialize, as if by magic, at exactly the right time, and we are driven the short distance home. We then sit in the garden, drink beer and talk.

The kidnappers who had kidnapped their daughter six years ago had come up from the valley, poisoning one of Madhu's dogs with a piece of meat they threw over the fence, before climbing over the spiked fencing that surrounds the garden.

'It wasn't just being held at gunpoint and having your daughter kidnapped. There were extortion attempts as well. I used to carry my own mobile phone until I was telephoned one day. "You have heard of the Black Spider group?" the voice said. "You remember how we killed Dr So and So?" They wanted money but I took no notice, and now my driver Ramesh always carries my phone. And during the Maoist insurgency the Maoists often came demanding money. I always refused but said I was happy to give them free medical care.'

'But the deputy leader of the insurgency was a schoolfriend of yours, wasn't he?' I asked.

'Not exactly a friend,' he said, 'but we were at school together. He was very popular with the Christian missionary teachers. I wasn't.'

'What we call a little swot?' I asked.

'Something like that.'

I asked him what had happened to the men who had kidnapped his daughter.

'She was so brave!' Dev said, almost with tears in his eyes. 'When the kidnappers said they were going to take one of us away, she immediately got up and volunteered. My sixteen-year-old daughter! I felt so helpless. It was really very difficult for me. Why should she be tortured because of my success?'

'What happened?' I asked.

'I had to pay a ransom. But Medha had noticed some details of the place to where she was taken in Patan, because her blindfold had slipped. There was a big police operation and they caught the whole gang. But there was no established sentence for kidnapping – maybe just a year or two in prison. But then the police found drugs on them and they all got fifteen years.'

Dev knew well enough that I was longing to see the high Himalayas, but for much of the time that I was there both the foothills and the mountains beyond them were obscured by mist. We eventually managed to see them in the distance – briefly in the morning before the clouds moved in – from a town called Dhulikhel, after an hour's drive from Kathmandu at the crack of dawn.

The snow-covered mountains seemed to float in the sky, above the mist-hidden foothills and valleys, serene and celestial, and entirely detached from the world below in which I lived. It needed no imagination to think that gods lived there. I wept silently with happiness that I had lived long enough to see them. And then the clouds rose up from the west and in a matter of minutes the mountains had disappeared from view.

On a later trip to Nepal, I took a few days off work to go trekking in the mountains with my son William, who came out to join me for two weeks. We walked for five days, the first day ascending from Nayapul – a typically scruffy, dusty and rubbish-strewn Nepali town – up towards the foothills around Annapurna South, one of the several peaks of the Annapurna range. After a few miles the dust track ends, and from then on you climb up a path paved with rough-hewn stone and what feels like an endless flight of stone steps. We climbed more than 1,000 metres on the first day like this, the temperature in the 80s. William and our guide Shiva – a delightful man, both solicitous and discreet – climbed imperturbably while I streamed with sweat and had to stop at regular intervals to catch my breath. I had assumed that my daily exercise regime meant I was very fit. I am getting old, I thought, and remembered how so many of my elderly patients in England would protest when I explained that their problems were due to old age. 'But Mr Marsh, I still feel so young!'

As we climbed, the villages at first were all formed around small subsistence farms. Shiva would point out the various crops being grown – rice at the lower levels, potatoes and corn as we slowly climbed higher. Annapurna is a conservation area, and entirely free from the rubbish of the towns. There are medieval scenes to be seen – a farmer with a pair of oxen ploughing a sloping field, with steep hills and mountains in the distance, old women carrying firewood in large baskets on their backs, mule trains going up and down the stone steps. As you climb higher the hillsides become too steep and cold for any farming. The entire area is now based on the trekking industry, a very important part of the Nepalese economy. It is a little strange to see the local Nepalis slowly walking up and down the stone stairs, carrying enormous baskets on their backs or tree trunks or building materials, alongside the wealthy Westerners in their shorts and T-shirts and backpacks. I saw a couple of German tourists at one of the guest houses, slowly walking barefoot on the sharp gravel of the footpath outside our guest house. Later I saw them setting off with a group of trekkers all carrying yoga mats, so I suppose they were seeking enlightenment in the high mountains as well as on gravel. A grey-haired English woman, travelling alone, told us she was heading for a remote village.

'It's said that there are old lamas there,' she said with a note of awe in her voice, and then added, 'But they might not talk to me.'

'Twenty per cent houses now empty,' Shiva told us, pointing out yet another empty dwelling. Rural depopulation continues apace, with more and more people leaving for the nearby city of Pokhara. The houses in the mountains are usually built of stone, with wooden balconies, and some still with stone roofs. They can be very beautiful. The roofs are increasingly being replaced with incongruous bright-blue corrugated metal. His own house, Shiva told us, had been badly damaged in the earthquake. He had young children and elderly parents to care for, so he had to spend much of the year guiding trekkers, trying to accumulate enough money to build his family a new home. His life, he said, was rather difficult at the moment, and I thought he looked old and careworn for his age of thirty-three.

We passed many mule trains going up and down the trail,

the patient creatures carrying gas cylinders, concrete blocks, sacks of cement, food and crates of beer. They carefully picked their way over the rough stone steps, their neck bells daintily ringing. We longed to see the high, snow-covered mountains, but they remained stubbornly hidden in cloud and the view was only of the tree-covered foothills – mountains themselves by European standards, and thousands of feet high.

The guest house on the second night was at the trekking village of Ghorepani, at just under 11,000 feet in altitude. It seemed that William and I were the only guests. Our bedroom was like a large tea chest with just enough room for two hard beds, with walls made of plywood and the original manufacturers' stencils in black ink still in place. We spent a friendly evening sitting round a large stove with Shiva and our hosts. The stove – it was quite cold outside by now – was made from an oil drum with a flue going up through the ceiling, with metal bars welded to it for drying clothes. There was a tremendous thunderstorm, the first rain of the season. William and I fell asleep in our tea chest of a room to the sound of the rain beating a stereophonic tin symphony on the roof above our heads. Shiva had expressed the hope that the rain would clear the clouds and we would see the Himalayas in all their glory from nearby Poon Hill at dawn, before the clouds rise from the valleys and hide the mountains. So we had to get up next morning shortly after four.

Although our guest house was empty, many other trekkers – dim and silent figures in the pitch-dark night – suddenly appeared in single file, heading for the hill. We joined their silent procession; it felt more like a dark stampede and was a strangely sinister experience. At the foot of the stone stairs leading to the summit two dogs, snarling and fighting furiously, locked together, came tumbling down the steps in the dark and almost knocked me over. I was gasping for breath in the thin mountain air within minutes and felt as though I was having a panic attack. But I felt compelled upwards by the silent figures in the night around me. They seemed to be climbing the 300 metres of stone stairs to the summit quite effortlessly. All I could hear was my own panting breath and I was soon soaked in sweat. Or perhaps it was my deeply competitive nature that forced me upwards – the thought of being overtaken by anybody being unbearable, even though I was probably the oldest person on the hill. So I hurried, gasping, upwards.

It felt like an ascent to hell, rather than the more conventional descent. We hoped to see the sun rise over the high mountains but they would have none of it and had promptly wrapped themselves in dense cloud. William and I quickly left the crowd on the top of the hill, most of them clutching their smartphones and cameras, hoping to see the mountains. At the height of the trekking season, Shiva told me, there can be many hundreds of people on Poon Hill at dawn. There was some consolation in passing the late arrivals as we descended, and now that we could see them in the daylight, toiling up the stairs, they looked as breathless and tortured as I had felt.

The day was spent walking along a high ridge through a rhododendron forest. The trees were as large as oaks, with mottled and flaking trunks, and must have finished flowering a few days earlier, so we walked over a path of pink and red petals. The next night was spent in a guest house which was supposed to have fine views of the high mountains, but when we arrived there we could only see cloud and foothills. Our bedroom had unglazed windows with elaborately carved, black wooden shutters. I woke in the middle of the night. I could see a few stars through a half-open shutter and so could hope to see the mountains in the morning. I listened to my son's quiet breathing as he slept in the bed next to mine. I thought of his birth thirty-seven years earlier. How he had been placed on his mother's stomach and how he then opened two large, thoughtful blue eyes as he saw the outside world for the first time. I thought of how within a few months he had almost died. Years later we had come close to becoming estranged. He had gone through a difficult time, and I felt paralysed, knowing that I was part of the problem and that the past could not be undone, for all my regrets. His sister Katharine proved to be of much greater help than me. But that terrible time was also now in the past and I soon fell asleep again.

In the morning I woke from my one recurring nightmare – that I am back at university about to take my Finals, after my year of truancy working as a hospital theatre porter, but I have done no work whatsoever. I am filled with dread and panic. I am told anxiety dreams about examinations are quite common, but I find it curious that it is so locked into my subconscious. When I was allowed back to the university after running away, and after my brief stay in the psychiatric hospital – I continued to see my psychiatrist once a week – I worked frenetically hard and got a good degree, so I do not know why this fear of failure so often haunts me when I sleep.

I got out of bed to find that the mountains of the Annapurna range had miraculously appeared, where before there had been only cloud. It's more as though they had suddenly arrived, in complete silence, from somewhere else, that they had descended from heaven. They towered above us, brilliant white with ice falls, snow fields and glaciers, against the blue sky. They looked so close that you felt you could get to them with a walk of only a few hours, when in reality Annapurna base camp, at the foot of Annapurna South, is four days' walk away.

There was then a long walk downhill back to Nayapul, at first along a little-used track in a steep and peaceful wood, with the great mountains still to be seen between the trees, before the clouds rose up from the valleys and the mountains disappeared. We had to stop at regular intervals to kick the leeches off our boots. Later we rejoined the stone path and steps, passing many mule trains climbing up in the opposite direction. As we slowly descended we were accompanied by the sound of the glacial grey and white River Modi Kholi, rushing over rocks far below us.

'Twenty-two-year-old fell thirty metres. Caesarean section. On examination no movement in lower limbs and weak upper limbs.' The MO rattled and stumbled through the presentation.

'Oh, come on!' I shouted. 'That's a hopeless presentation. How much movement does she have in her arms? What's her spinal level?'

We worked out that she had only partial movement in her biceps and none in her triceps, that she could weakly shrug her shoulders and bend her elbows, but that everything below this – her hands, her spinal and abdominal muscles and her legs, her bowels and bladder – was all completely paralysed.

'So her spinal level is C5/C6. Yes? And don't you have any curiosity as to what's happened to your patients? Thirty metres? How can one survive that? Was it suicide? And was this after the caesarean section?'

'She fell off cliff while cutting grass with sickle. Foetal death, so caesarean section. Then she came to Neuro Hospital.'

'How many months pregnant?'

'Seven months. Husband working in Korea.'

'Ah,' I said, appalled. 'Well, let's look at the scan.'

The MRI scan showed fracture and complete translocation of the spinal column between the fifth and sixth cervical vertebrae. The spinal cord looked damaged beyond repair.

'She'll never recover from that,' I said. 'What's the next case?'

Dev and one of the registrars operated the next day, screwing the girl's broken spine back together again, although this could not undo the paralysis. Surgery would at least mean that she did not need to be kept flat on her back in one of those horrible cervical collars, and it would make the nursing and physio easier.

I saw her on the ITU next morning as Dev and I went round.

'Presumably here in Nepal she'll get bedsores and renal infections if she ever gets out of hospital?' I said.

Dev grimaced.

'She's unlikely to survive long. Christopher Reeve was a millionaire and lived in America and he eventually died from complications, so what chance a poor peasant in Nepal?'

I looked at the girl as we talked – at least she couldn't understand what we were saying. She was very beautiful in the way that so many Nepali women are, with large, dark eyes and high cheekbones and a perfectly symmetrical and outwardly serene face. Her eyes moved slowly, she spoke a few words when spoken to. Her head was immobilized in a large and uncomfortable pink plastic surgical collar. Dev agreed with my suggestion that it could be taken off now that she had her broken neck screwed and plated back together again.

'I put a locking plate in,' Dev said. 'Very expensive. Thousands of rupees.' He then launched into a tirade once again about the way the foreign equipment companies charged First World prices in Third World countries and how most surgeons using implants would be paid a 20 per cent kickback by the suppliers, the extra cost being passed on to the patient. He said he had always refused to get involved in this widespread, but thoroughly corrupt, practice. You can find it in many European countries as well, despite being illegal, although there the inflated extra cost can often be passed on to the taxpayer and government rather than to the patient.

'Well, medical-equipment manufacturers are businessmen, not altruists,' was all I could say.

After a few days on the ITU the paralysed girl was discharged to one of the wards, but shortly afterwards her breathing deteriorated – which often happens in these cases – and she had to be readmitted and put on a ventilator.

'I spoke to her husband again yesterday,' Dev told me. 'He's flown back from Korea. I think he is coming to accept that she might die. But it's very difficult in Nepal. If you are too honest and realistic it causes terrible trouble. The family will be shouting and screaming all over the hospital and causing all sorts of problems. You just can't tell them the truth straight out. I told him he was young. I said that if she dies he could at least start again.'

'It's easier now that she's on a ventilator, isn't it?' I replied, because it would be kinder if she died anaesthetized on a ventilator than from bed sores and infection on a bed in the hospital or back in her home – not that she was likely ever to get home.

On the morning round next day I noticed a group of doctors and nurses round the girl's bed. She was groaning terribly as an anaesthetist pushed a flexible, fibre-optic bronchoscope down her trache tube. Her chest X-ray looked awful. We watched the intriguing view of the ringed and ridged inside of her lungs' bronchi on the small monitor attached to the bronchoscope, while she groaned piteously as the anaesthetist tried to clear the fluid from her lungs. We agreed she was better off dying, but Dev was in an impossible situation. Should he have refused to operate and left the woman with her dislocated, broken neck untreated, leaving her to die without any treatment? The family would almost certainly have refused to accept this. Should he have left them to take her to another hospital where she would have undergone surgery that probably would not have been done as well as it would have been in his hospital? I had never had to face problems like this in my own career.

We get so used to most of our patients having brain damage and being unconscious that we forget that some of the paralysed patients on ITUs are wide awake, suffering horribly but unable to show it. Or perhaps it is wilful blindness on our part. I was painfully aware that I had found some of these cases so distressing during my career that I tended to avoid them and walk past them on the ward round. What do you say to somebody who is completely paralysed from the neck down, but awake, on a ventilator, so that they cannot talk?

I remembered an identical case in Ukraine many years ago. My colleague Igor was still working in the government emergency hospital at the time. He was very proud of the fact that he had managed to keep the patient alive, but on a ventilator.

'First case of long-term ventilation in Ukraine,' he declared.

The young man was in a bleak little side room and lived there for three years. Many religious icons surrounded him on the otherwise bare walls. He was equipped with a speaking tracheostomy tube and each time I visited Igor's department I would go and see him. His brother looked after him and spoke some English, so I communicated with the patient through him. Each time I saw him he had wasted away a little more. At the time of the injury – breaking his neck diving into shallow water – he had been quite heavily built, but by the time he died he was skin and bones. At first

I was able to have quite rational conversations with him, but it became more difficult with each visit. At least, he started to ask me about religious miracles and salvation, which he spoke about with intense passion (to the extent that you can speak passionately with a speaking tracheostomy tube), to which I had no answer. I was relieved on a later visit to see that the little side room was empty.

The young Nepali woman had fallen and broken her neck during *Dasain*, the most important of the many Nepali festivals, when upwards of fifty thousand goats and hundreds of buffalo are sacrificed to the goddess Durga. Blood is smeared everywhere in honour of the goddess, including, I noticed, on Dev's gold-coloured Land Rover. Animal rights activists, I read in a local newspaper, have recently suggested that the goats be replaced with pumpkins.

The festival goes on for two weeks. Two days earlier Dev had told me to accompany him to the gates in front of his house. A police jeep was parked there with a uniformed policeman standing beside it. Another policeman appeared, leading a beautiful goat with long, floppy ears on a rope from behind the garage.

'I give the local police a goat every year for *Dasain*,' Dev told me. The goat was bundled into the back of the jeep but immediately jumped out. So it was put back in, but now with a police escort. They drove away with the goat looking mournfully out at me over the tailgate, the policeman beside it.

'That goat will feed a hundred policemen,' Dev said approvingly.

'Nobody is in the mood for *Dasain*, this year, what with the earthquake and now the blockade and fuel crisis,' Dev commented as we drove back to Kathmandu from a visit to a nearby town. Yet in several places we passed the beautiful high swings – known as *pings* – which are a traditional part of *Dasain* celebrations. They are made simply of four bamboo poles lashed together, more than twenty feet high and decorated with colourful flags. I saw Nepalis – both adults and children – laughing ecstatically as they swung happily to great heights, although I thought the *pings* looked a little precarious.

The next day I sat in the library teaching the juniors and discussing how we could improve the MOs' jobs.

'I am going back to London tomorrow,' I told the new cohort of MOs, freshly out of medical school and, it seemed to me, pretty well out to lunch.

'You are good doctors. We want to make you better. I hope the registrars' – I looked pointedly at them – 'will try to continue the morning meetings in this spirit. Teasing, yes, but no bullying.' Pleased with this little speech, I then went down to Dev's office and was about to go downstairs to start the clinic when there was a sudden flurry of activity in the corridor outside.

I found Dev, looking grim, surrounded by several of his juniors at the theatre reception desk, all looking equally serious.

'The girl with a broken neck has just died,' Protyush told me. 'The husband is very angry.'

'Is Dev waiting to talk to him?'

'Yes, but we need backup – here in Nepal the families can assault us. We're waiting for the security guards.'

Thirty minutes later, I stood in a corner of the theatre reception area where I had a view into the counselling room, and I could see Dev, but not the angry husband. Dev listened to a long outburst in silence and spoke quietly in reply. I crept away, not liking to eavesdrop on so much tragedy and unhappiness.

'I wish I still worked for the NHS,' Dev said to me that

evening, as we sat in the garden. 'Or at least that I was still the only neurosurgeon here, or that I didn't have to worry about keeping the hospital afloat financially. It's yet to make a profit, you know, even after ten years. Twenty years ago I could simply have said that there was nothing to be done and the family would have accepted it.'

'How did the meeting with the family go?' I asked.

'Oh – the usual stuff. It happens now every few months. Never happened in the past. The husband said I had killed his wife by doing a tracheostomy. Nonsense of course – and in fact, in six months' time, he'll probably have a new wife. If she had survived it would have been terrible for both of them. And I spent so long, every morning, trying to explain. And he was so polite, as though I was a god, but now I'm a devil. But I'm sure you'll find there's another neurosurgeon in town who's told them that if he had treated her she'd have been OK.'

'You can't expect people to be reasonable immediately after a death like this one,' I said, trying to be helpful.

'Nepal is different,' Dev replied. 'I worry for the boys, when they become seniors, having to work in a country like ours where the people are so uneducated – they won't have my authority. All the hospitals have a permanent plain-clothes policeman stationed twenty-four hours a day because of problems like this. They said they would get all the other patients' families to blockade the hospital. Said they would burn it down. They want money. I know a lot of other doctors here who have had money extorted from them. That's the problem with having to run a private hospital – "We paid you to treat her," they said, "and now she's dead." It was so much easier in the past when I worked at the Bir. But the government medical service here now is terrible, almost completely broke. And so when I first see a patient the initial question is not what treatment would be best for them but "What can you afford?" You're so lucky to work in the NHS.'

'Well, she's better off dead,' I said.

It was sad to see Dev – usually so cheerful and enthusiastic – suddenly silent, looking grim.

'You can't really share it with anyone. It would only upset and frighten my wife,' he added.

'Only neurosurgeons understand,' I said, 'how difficult it is to be so hated, especially when you haven't even done anything wrong, and only tried to do your best.'

I remembered one of my first catastrophes as a consultant. A child who died as a result of my postponing an operation that should have been done urgently. I had thought it was safe to wait until the morning, but I had been wrong. I had to attend an external investigation. I did not have to meet the parents face to face but passed them in the corridor. The look of silent hatred the mother gave me was not easy to forget.

'You start,' he said, pointing to the bottle of beer I had already got out. 'The woman's MP might come round to the hospital – I don't want to smell of alcohol.'

I was summoned to supper two hours later. To my surprise, all the managerial team of the hospital were present – six people including the driver, all there to support Dev. I was rather touched. I'd never had support like this for my disasters.

Over a large Nepali dinner there was much animated discussion, most of it lost on me as they spoke in Nepali. But I was told that the family were threatening a hunger strike and a press conference, and planned to get the other patients' families to join them.

'Seven point five,' I heard the manager, Pratap, suddenly

say – he had been looking at his smartphone. This, it turned out, was the strength of an earthquake that had just hit Afghanistan and Pakistan. The catastrophic earthquake that had hit Nepal six months before my visit had been 7.8. This was discussed for a while, and then they resumed the conversation about the dead girl's family and what might happen.

'It's all because we now work for money,' Madhu, who was sitting next to me, said. 'We didn't want to, but had no choice. We can't provide free treatment to everybody.'

Next morning, the morning of my departure from Nepal, I sat drinking coffee in the garden, in Dev and Madhu's little Shangri-La. The pigeons were cooing and gurgling, the cocks were crowing, the hooded crows were quarrelling again in the camphor tree, although in truth for all I knew they might have been discussing their marital problems or the presence of the brown mongoose which can sometimes be seen, sinuous and graceful, running swiftly across the garden. Or perhaps they were excited about the prospect of the first day of the festival of *Tihar* in two weeks' time, the day of *kaag tihar*, when crows are worshipped and little dishes of food are put out for them. I probably understand as much about the crows as I do about the impenetratable complexities of Nepali society. Two birds with feathery trousers I couldn't identify waddled busily about on the small lawn in front of the gazebo.

I set off for work as usual but as it was the tenth and most auspicious day of *Dasain,* there was little traffic on the road. I passed women wearing their finest clothes – brilliants reds and blues and greens, decorated with gold and silver and paste jewels which flashed in the sunlight. They picked their way cautiously over the puddles and around the rubbish and stinking, open drains. When I got to work I found that there were twelve uniformed policemen with long iron-shod sticks in front of the hospital, sitting in the sunshine on the grass mound by the magnolia tree. The dead woman's family and supporters stood nearby. Dev and I looked down at them from his office window.

'How much longer will this go on for?' I asked.

'Oh, until the weather gets colder,' he said with a laugh, his cheerful good humour having returned.

'I'm not even sure the story about cutting grass on a cliff was true. Her husband has money – it's unlikely she'd be out gathering grass off a cliff,' he said. 'I'm pretty suspicious that it was another *ping* accident.'

We had admitted a sixty-five-year-old man two days earlier, also completely paralysed, with a broken neck, who had fallen from a *ping*.

'Happens all the time during *Dasain*,' Dev said.

I noticed that behind the policemen, the waiting outpatients and the dead woman's angry family, in the rice paddy next to the hospital, people were harvesting the rice – a picturesque and medieval sight, although in the background there was a long queue of dirty old trucks waiting at the petrol station. In the distance, the high Himalayas, beyond the foothills, were hidden.

# 8

# LAWYERS

I had to return to London from Nepal earlier than I had originally planned because I was due to appear in court. A patient was suing me. The case had been dragging on for four years. I had operated for a complex spinal condition causing progressive paralysis, and the patient had been initially left worse than he had been before the operation. As far as I could tell he had eventually ended up better than before the operation, but apparently he was deeply aggrieved. A neurosurgeon – justly famous for the very high opinion he had of himself, although less famous for his medico-legal pronouncements – was of the opinion that I had acted negligently. Just for once, I was as certain as I could be that I had not, and I had reluctantly felt obliged to defend myself. It was just like Nepal, I thought. All these surgeons attacking each other. I had had to attend various meetings about the case and many thousands, probably hundreds of thousands, of pounds must have been spent in legal fees. At the last moment, after I had come all the way back from Nepal, the claimant and his lawyers abandoned the case two days before the trial was due to start. The solicitor handling my defence was most apologetic about the waste of my time.

'But it's better than needing twelve policemen,' I replied cheerfully, without explaining what I meant.

Many doctors do what is called medico-legal work, providing reports for lawyers in cases involving personal injury or medical negligence. It is a lucrative but time-consuming business. I did a few such reports myself when I became a consultant, but quickly gave it up. I preferred operating and dealing with patients to the many meetings and lengthy paperwork which medico-legal work requires. I only became involved with lawyers if I was being sued myself – always a very distressing experience, whether I felt guilty or not.

This occurred four times during my career, including the case which had forced me to return from Nepal and which had now collapsed. The other three cases had all been settled, as I blamed myself for what had happened and did not want to defend myself. One case had been for a retained swab after a spinal operation (in the days when swab counts were not being done in the old hospital) which had not caused any severe injury, and the other two were cases where I had been slow to diagnose serious, although almost uniquely rare, post-operative infections. One of those patients had come to serious harm, the other to catastrophic harm.

But a few years ago I was subpoenaed to give evidence in a personal injury compensation case, which I regarded as an absurd and complete waste of my time. So I attended reluctantly, a series of High Court orders having been served on me over the three days before the hearing. The men serving them had never been able to serve them on me in person – which, strictly speaking, I believe the law requires. The first attempt had been made while I was operating and the second when I was away from London the following day. I returned the following evening to find that a copy of the order had been pushed through the letterbox of the front door of my home. I was operating the day after that until the evening, and came out of the theatre to be told that earlier in the morning a man had walked up to the hospital reception desk and had thrown down yet another copy of the High Court order in front of the receptionist and then stalked off.

This barrage of court orders had been unleashed upon me by a solicitor in a huge City law firm which was acting on behalf of an American law firm, which in turn was acting for the defendants in the compensation case.

An English woman had been involved in a minor car accident in the USA while on holiday, and had subsequently seen me as a patient about her 'whiplash' symptoms. I had confirmed with an MRI scan that there were no significant injuries to her neck and reassured her that she would get better in time. In practice it is not at all clear whether these whiplash syndromes do get better. Patients develop an array of aches and pains and altered sensations in their necks and arms which do not correspond to any known pathological processes such as bone fractures or torn muscles or trapped nerves, and do not spontaneously improve in the time it takes most proven 'soft-tissue' injuries to heal and become pain-free. It is well known that these syndromes do not occur in countries which do not have any legal recognition of whiplash injury as a consequence of minor car crashes.

The particular type of accident which is alleged to produce 'whiplash injury' is a 'shunt', when a car is driven into from behind by another car. These are typically low-speed accidents, where the driver or passengers are subjected to relatively slight forces, never enough to cause any obvious injuries, but which seem to produce severe and lasting symptoms without any evidence of injury such as bruising or swelling or changes on an X-ray or MRI scan. It has been pointed out that driving dodgem cars on fairgrounds involves near-continuous shunting as the cars are deliberately driven into each other, and yet there are no reports of whiplash symptoms afterwards. This discrepancy between the severity of the symptoms and the apparent triviality of the injury has been attributed to a putative 'whiplash' effect. The victim's neck is supposed to be cracked like a whip – something that has never in fact been demonstrated and is probably fallacious.

I used to see many of these patients every year in my outpatient clinic and it was clear to me that most of them were not consciously malingering – instead they were the willing, perhaps hapless, victims of a 'nocebo' effect, the opposite of the placebo effect. With the placebo effect, which is well understood, people will feel better, or suffer less pain, simply as a result of suggestion and expectation. With 'whiplash injury', the possibility of financial compensation for the victims, combined with the powerful suggestion that they have suffered a significant injury, can result in real and severe disability, even though it is, in a sense, purely imaginary. They are more the victims of the medico-legal industry and of the dualism that sees mind and brain as separate entities than of any physical injury outside the brain. It is the modern equivalent of the well-attested phenomenon of a witch doctor in tribal society casting a spell on somebody, causing the victim to fall ill, merely through the power of suggestion and belief. There was a further significant irony in this case, which I had mentioned in my original letter about the patient: the victim's husband was a lawyer specializing in personal injury compensation.

I had been given only two weeks' notice about the hearing – strictly speaking, the 'deposition of evidence before a Court-appointed Examiner'. I was told that I was required to attend but there was no mention of legal compulsion. My secretary had told the woman solicitor who had sent the letter that I could not attend as I was already committed to operations and outpatient clinics. As I had heard nothing more after my secretary had told the solicitor this, I had assumed that it had been accepted that I would not be coming. It seems that the solicitor, however, decided that I needed to be taught a lesson and served me with the court order. I had some urgent cases to do, which could not be postponed. I therefore started operating at seven in the morning on the day of the deposition, at high speed, something I hate doing; nor had I slept well, as I was angry that I was being dragged away from my work in this way.

I was not going to be paid, but doubtless the lawyers would be paid hundreds of pounds, probably thousands, for trying to extract a medical opinion from me for free. I knew the business would be absurd – I had seen the patient only twice, four years ago, had no memory of her whatsoever, and the lawyers already had copies of my correspondence. I clearly would have nothing to add. So I was angry, and had already telephoned the solicitor the day before and told her so.

The law firm's offices were housed in a huge postmodern marble and glass office block just beyond the Tower of London. I marched into the building full of righteous indignation, past the men in suits smoking cigarettes on the piazza outside, and clutching my folding bike and attaché case. I collected a laminated visitor's pass from a receptionist in a smart uniform, pushed past the barricade of the revolving stainless-steel turnstile and ascended to the seventh floor in one of the many tall, swift lifts lined with dark mirrors. If only my hospital had such lifts – how much time it would save!

I emerged into a three-storey-high atrium, walled and

floored in marble, even though already on the seventh floor. High plate-glass windows showed a panoramic view over the City towards the Lloyd's Building and the various high and imposing office blocks around it. Having announced myself, I had to wait for a while, and looked with sour awe at the City under a clear blue sky. Babylon! I thought – the heart of an extravagant culture, consuming itself and the planet, sheathed in glittering glass. A slim and polished barrister in a light-charcoal pinstriped suit, the Court-appointed Examiner, descended the elaborate glass, steel and hardwood spiral staircase at one side of the atrium and introduced himself. He was, perhaps, just a little apologetic and thanked me for coming.

'I am not pleased to be here,' I growled.

'Yes, so I heard,' he answered politely.

He led me to an anonymous, luxurious and windowless meeting room, the furniture all in white ash and chrome, where the English QC for the plaintiff and the American lawyer for the defendants were waiting for me. The American lawyer was in his fifties and was fit and trim, with short grey hair and a designer sports jacket. The elderly English QC, however, did not look as though he worked out in a gym every day and was rather overweight, with a florid face, and wore a crumpled white linen suit and half-moon glasses.

'Good morning gentlemen,' I said as I entered, feeling a little superior, knowing that they were not going to get anything out of me. I sat down and after the introductions a man with a video camera read out, in a bored voice, the description of the proceedings. I was sworn in (I affirmed rather than swore on the tatty little Bible on offer) and briefly cross-examined. This could only mean that I could agree that the notes I made four years ago were indeed mine and that I had no memory of the case. The American lawyer, of course, wanted to extract my opinion about whiplash injury, but I refused to be drawn.

'It is a medico-legal question,' I said, 'and I therefore have no opinion. I never give medico-legal opinions over personal injuries.' Whether they heard the disdain in my voice or not, I do not know.

I had seen the patient and had advised against surgery. The English QC wanted me to agree that if her symptoms had not got better as I had said they probably would, it was reasonable for her to seek a further opinion. I agreed that it was.

'Did you know,' the American lawyer then asked, 'that she did eventually undergo surgery on her neck?'

'No,' I said.

How much I could have said! I had affirmed that I would tell the truth, the whole truth and nothing but the truth, but not that I would not be economical with it. I could have explained the psychosomatic nature of whiplash injury, the nonsense written about the alleged mechanism, the fact that all the neurosurgical textbooks state that one should never operate on the spine of somebody involved in compensation litigation. They never, ever get better. Some greedy surgeon must have operated on her neck and now, most probably, her symptoms were even worse and the lawyers would be arguing over whether her disability was the result of the original trivial injury or the operation. I could have told the lawyers that they themselves were more responsible for her problems than the original minor car crash. The principal consequence of that trivial accident, and the millions of other ones like it, was not just the plaintiff's pain and suffering, but also the Babylonian marble offices where we were now meeting. The humourless men seated round the table before me were part of the great industry of personal injury compensation, with its army of suave and accomplished lawyers and assured expert witnesses, rooting in a great trough of insurance premiums.

At the end of the meeting the American lawyer went through my CV, which he had in his hand. His face was impassive but he seemed a little puzzled by it. I am rather proud of my CV and academic record, and I thought that perhaps he too would be impressed by it and would be arguing that since an English surgeon with such a brilliant CV had advised against surgery, the operation carried out by somebody else could not have been a good idea.

'How did you get all those prizes at college?' he eventually asked.

'By working very hard,' I replied, feeling deflated. He remained expressionless – perhaps he was just bored and wanted a little distraction – but the English QC smiled.

And that was that. The video camera was switched off and the Examiner thanked me for coming.

'Well, I'll get on with my day,' I said. I descended the spiral staircase, collected my folding bike from the reception desk and left.

# 9

# MAKING THINGS

A long time ago I had promised my daughter Sarah that I would make her a table. I am rather good at saying I'll make things, and then finding I haven't got the time, let alone getting round to make the many things I want to make or mend myself.

A retired colleague, a patient of mine as well, whose back I had once operated upon, had come to see me a year before I retired with pain down his arm. Another colleague had frightened him by saying it might be angina from heart disease – the pain of angina can occasionally radiate down the left arm. I rediagnosed it as simply pain from a trapped nerve in his neck that didn't need treating. It turned out that in retirement he was running his own oak mill, near Godalming, and we quickly fell into an enthusiastic conversation about wood. He suggested I visit, which I did, once I had retired. To my amazement I found that he had a fully equipped industrial sawmill behind his home. There was a stack of dozens of great oak trunks, twenty foot high, beside the mill. Eighty thousand pounds' worth, he told me when I asked. The mill itself had a fifteen-foot-long sawbed on which to put the trunks, with hydraulic jacks to align them, and a great motorized bandsaw that travelled along the bed. The tree trunks – each weighing many tons – were jostled into place using a specialized tractor. All this he did by himself, although in his seventies, and with recurrent back trouble. I was impressed.

I spent a happy day with him, helping him to trim a massive oak trunk so that it ended up with a neatly square cross-section, and then rip-sawing it into a series of thick two-inch boards. The machinery was deafening (we wore ear defenders), but the smell of freshly cut oak was intoxicating. I drove home that evening like a hunter returning from the chase, with the planks lashed to the roof rack of my ancient Saab – a wonderful car, the marque now, alas, extinct – that has travelled over 200,000 miles and only broken down twice. The roof rack was sagging under the weight of the oak and I drove rather slowly up the A3 back to London.

The next morning I went to collect my bicycle from the bicycle shop in Wimbledon Village, as it likes to be called, at the top of Wimbledon Hill. Brian, the mechanic there, has been looking after my bicycles for almost thirty years.

'I'm afraid the business is closing down,' Brian told me, after I had paid him.

'I suppose you can't afford the rates?'

'Yes, it's just impossible.'

'How long have you been here?'

'Forty years.'

He asked me for a reference, which I said I would gladly give. He is by far the best and most knowledgeable bike mechanic I have ever met.

'Have you got another job?' I asked.

'Delivery van driver,' he replied with a grimace. 'I'm gutted, completely gutted.'

'I remember the village when it still had real shops. Yours

is the last one to go,' I said. 'Now it's all just wine bars and fashion boutiques. Have you seen the old hospital just down the road where I worked? Nothing but rich-trash apartments. Gardens all built over, the place was just too nice to be a hospital.'

We shook hands and I found myself hugging him, not something I am prone to do. Two old men consoling each other, I thought, as I bicycled down the hill to my home. Twenty years ago I lived with my family in a house halfway up the hill. I assume that the only people who can afford to live in the huge Victorian and Edwardian villas at the top of the hill are bankers and perhaps a few lawyers. After divorce, of course, surgeons move to the bottom of the hill, where I now live when not in Oxford or abroad.

The oak boards needed to be dried at room temperature for six months before I could start working on them, so I clamped them together with straps to stop them twisting and left them in the garage at the side of my house (yet another of my handmade constructions with a leaking roof), and later brought them into the house for further drying.

Now that I was retired and back from Nepal, the wood was sufficiently dry for me to start work.

When my first marriage had fallen horribly apart almost twenty years earlier and I left the family home, I took out a large mortgage and bought a small and typical nineteenth-century semidetached house, two up and two down, with a back extension, at the bottom of Wimbledon Hill.

The house had been owned by an Irish builder, and his widow sold the house to me after his death. I had got to hear that the house was for sale from the widow's neighbours, who were very good friends of mine. So the house came with the best neighbours you could wish for, a wide and unkempt garden and a large garage in the garden itself, approached by a passage at the side of the house. Over the next eighteen years I subjected the property to an intensive programme of home improvements, turning the garage into a guest house (of sorts) with a subterranean bathroom, and building a workshop at the end of the garden and a loft conversion. I did much, but not all, of the work myself. The subterranean bathroom seemed a good idea at the time, but it floods to a foot deep from an underground stream if the groundwater pump I had to have installed beneath it fails.

The loft conversion involved putting in two large steel beams to support the roof and replacing the existing braced purlins (I had taken some informal advice from a structural engineer as to the size of steel beam required). With my son William's help I dragged the heavy beams up through the house and, using car jacks and sash cramps, manoeuvred them into position between the brick gables at either end of the loft. There was then an exciting moment when, with a sledgehammer, I knocked out the diagonal braces that supported the original purlins. I could hear the whole roof shift a few millimetres as it settled onto the steel beams. I was rather pleased a few years later to see a loft conversion being done in a neighbouring house – a huge crane, parked in the street, was lowering the steel beams into the roof from above. I suppose it was a little crazy of me to do all this myself, and I am slightly amazed that I managed to do it, although I had carefully studied many books in advance. The attic room, I might add, is much admired and I have preserved the chimney and the sloping roof, so it feels like a proper attic room. Most loft conversions I have seen in the neighbourhood just take the form of an ugly, pillbox dormer.

I have always been impatient of rules and regulations and sought neither planning nor building regulation permission for the conversion, something I should have done. This was to cause problems for me when I fell in love with the lock-keeper's cottage. I could only afford to buy it if I raised a mortgage on my house in London (I had been able to pay off the initial mortgage a few years earlier). The London house was surveyed and the report deemed it fit for a mortgage, 'subject to the necessary permits' for the loft extension from the local council, which, of course, I did not have.

With deep reluctance I arranged for the local building inspectors to visit. I expected a couple of fascist bureaucrats in jackboots, but they couldn't have been nicer. They were most helpful. They advised me how to change the loft conversion so as to make it compliant with the building regulations. The only problem was that the property developers who were selling the lock-keeper's cottage were getting impatient. So, over the course of three weeks, working mainly at night as I had not yet retired, I removed a wall and built a new one with the required fire-proof door, and installed banisters and handrails on the oak stairs – the stairs on which I had once slipped and broken my leg. I also installed a wirelessly linked mains-wired fire alarm system throughout the house. This last job was especially difficult as over the years I had laid oak floorboards over most of the original ones. Running new cables above the ceilings for the smoke alarms involved cutting many holes in the ceilings and then replastering them. But after three weeks of furious activity it was all done, and I am now the proud possessor of a 'Regularisation Certificate' for the loft conversion of my London home, and I also own the lock-keeper's cottage.

As soon as I had moved into my new home in London seventeen years ago, after the end of my first marriage, I had set about building myself a workshop at the bottom of the garden, which backs onto a small park and is unusually quiet for a London home. I was over-ambitious and made the roof with slates and, despite many efforts on my part, I have never been able to stop the roof leaking. I cannot face rebuilding the whole roof, so two plastic trays collect the water when it rains, and serve as a reminder of my incompetence. Here I store all my many tools, and it was here that I started work on Sarah's table. In the garden, which I have allowed to become a little wild, I keep my three beehives. London honey is exceptionally fine – there are so many gardens and such a variety of flowers in them. In the countryside, industrial agriculture and the use of chemical fertilizers, pesticides and herbicides have decimated the population of bees, as well as the wild flowers on which they once flourished.

It took many weeks to finish the table, sanded a little obsessionally to 400-grit, not quite a mirror finish, using only tung oil and beeswax to seal it. The critical skill in making tabletops is that the edges of the boards should be planed so flat – I do it all by hand – and the grain of the wood so carefully matched that the joints are invisible. You rest the planed edges of the boards on top of each other with a bright light behind them so that a gap of even fractions of a millimetre will show up. This requires a well-sharpened plane. A well-sharpened and adjusted plane – 'fettled' is the woodworker's traditional word for this – will almost sing as it works and minimal effort is required to push it along the wood.

It took me a long time to learn how to sharpen a plane properly. It now seems obvious and easy and I cannot understand why I found it difficult in the past. It is the same when I watch the most junior doctors struggling to do the simplest operating, such as stitching a wound closed. I cannot understand why they seem to find it so difficult – I become impatient. I start to think they are incompetent. But it is very easy to underestimate the importance of endless practice with practical skills. You learn them by doing, much more than by knowing. It becomes what psychologists call *implicit memory*. When we learn a new skill the brain has to work hard – it is a consciously directed process requiring frequent repetition and the expenditure of energy. But once it is learnt, the skill – the motor and sensory coordination of muscles by the brain – becomes unconscious, fast and efficient. Only a small area of the brain is activated when the skill is exercised, although at the same time it has been shown, for instance, that professional pianists' brains develop larger hand areas than the brains of amateur pianists. To learn is to restructure your brain. It is a simple truth that has been lost sight of with the short working hours that trainee surgeons now put in, at least in Europe.

The boards are glued together using what is called a rubbed joint – the edges rubbed against each other to spread the glue – and then clamped together for twenty-four hours with sash cramps. The frame and legs are held fast with pegs, and being oak, the table is very solid and heavy. I had taken great care, when sawing the wood with my friend, that it was 'sawn on the quarter', so that the grain would show the beautiful white flecks typical of the best oak furniture. Sarah was very happy with the result after I delivered it, and subsequently sent me a photograph of her eighteen-month-old daughter Iris sitting up to it, smiling happily at the camera as she painted pictures with paintbrush and paper. But, just like surgery, there can be complications, and to my deep chagrin a crack has recently developed between two of the jointed planks of the tabletop. I cannot have dried the wood sufficiently, I was impatient yet again. I will, however, be able to repair this with an 'eke' – a strip of wood filling the crack. It should be possible to make it invisible, but I will probably have to refinish the whole surface.

*

I'm not sure how my love of and obsession with making things arose. I hated woodwork at school: you had no choice as to what you made and you would come home at the end of term with some poorly fashioned identikit present for your parents – a wobbly little bookcase, a ridiculous egg-rack or a pair of bookends. I found these embarrassing; my father was a great collector of pictures, antiques and books and there were many fine things in the family home, so I knew how pathetic were my school woodwork efforts. He was also an enthusiastic bodger who loved to repair things, usually involving large quantities of glue, messily applied. The family made ruthless fun of his attempts, but there was a certain nobility to his enthusiasm, to his frequent failures and occasional successes.

He was a pioneer of DIY before the DIY superstores came into existence. I once found him repairing the rusted body-work of his Ford Zephyr by filling the holes with Polyfilla, gluing kitchen foil over the filler, and then painting it with gloss paint from Woolworth's. It all fell off as soon he drove the car out of the garage. My first attempts at woodwork away from school were made using driftwood from the beach at Scheveningen in Holland, where we lived when I was between the ages of six and eight. I sawed the wood, bleached white by the sea, into the shapes of boats. I made railings from small nails bought at the local hardware store. The only Dutch words I ever learnt were '*kleine spijkes, alsjeblieft*' – small nails, please. I would take these boats sailing with me in the bath, but they invariably capsized.

When I married my first wife, we had no furniture and little money. I made a coffee table from an old packing case with a hammer and nails. It was a wooden one from Germany, with some rather attractive stencilled stamps on it, a little reminiscent of some of Kurt Schwitters' *Merz* work. It had been sitting in my parents' garage for years and had contained some of the last possessions of my uncle, the wartime Luftwaffe fighter pilot and wonderful uncle who eventually died from alcoholism many years after the war.

My brother admired the coffee table and asked me to make one for him, and I said I would, for the price of a plane, which I could then buy and use to smooth the wood. I have not looked back since. My workshop is now stacked with tools of every description – for woodwork, for metalwork, for stone-carving, for plumbing and building. There are three lathes, a radial arm saw, a bandsaw, a spindle moulder and several other machine tools in addition to all the hand tools and power tools. I have specialist German bow saws and immensely expensive Japanese chisels, which are diabolically difficult to sharpen properly. One of my disappointments in life is that I have now run out of tools to buy – I have acquired so many over the years. Reading tool catalogues, looking for new tools to buy – 'tool porn', as my anthropologist wife Kate calls it – has become one of the lost pleasures of youth. Now all I can do is polish and sharpen the tools I already have, but I would hate to be young again and have to suffer all the anxieties and awkwardness that came with it. I have rarely made anything with which I was afterwards satisfied – all I can see are the many faults – but this means, of course, that I can hope to do better in future.

I once made an oak chest with which I was quite pleased. I cut the through dovetail joints at the corners by hand, where they could be seen as proof of my craftsmanship. The best and most difficult dovetail joints, on the other hand – known as secret mitre dovetail joints – cannot be seen. True craftsmanship, like surgery, does not need to advertise itself. A good surgeon, a senior anaesthetist once told me, makes operating look easy.

*

When I see the tidy simplicity of the lives of the people living in the boats moored along the canal by the lock-keeper's cottage, or the sparse homes of the Nepalese peasants William and I walked past on our trek, I cannot help but think about the vast amount of clutter and possessions in my life. It is not just all the tools and books, rugs and pictures, but the computers, cameras, mobile phones, clothes, CDs and hi-fi equipment, and many other things for which I have little use.

I think of the schizophrenic men in the mental hospital where I worked many years ago. I was first sent to the so-called Rehabilitation Ward, where attempts were being made to prepare chronic schizophrenics who had been in the hospital for decades for life in community care outside the hospital. Some of them had become so institutionalized that they had to be taught how to use a knife and fork. My first sight of the ward was of a large room with about forty men, dressed in shabby old suits, restlessly walking in complete and eerie silence, in circles, without stopping, for hours on end. It was like a march of the dead. The only sound was of shuffling feet, although occasionally there might be a shout when somebody argued with the voices in his head. Many of them displayed the strange writhing movements called 'tardive dyskinesias' – a side effect of the antipsychotic drugs that almost all of them were on. Those who had been treated with high doses of a drug called haloperidol – there had once been a fashion for high-dose treatment until the side effects became clear – suffered from constant and grotesque movements of the face and tongue. Over the next few weeks, before I was sent to work on the psychogeriatric ward, I slowly got to know some of them as individuals. I noticed how they would collect and treasure pebbles and twigs from the bleak hospital garden and keep them in their pockets.

They had no other possessions. Psychologists talk of the 'endowment effect' – that we are more concerned about losing things than gaining them. Once we own something, we are averse to losing it, even if we are offered something of greater value in exchange. The pebbles in the madmen's pockets became more valuable than all the other pebbles in the hospital gardens simply by virtue of being owned.

It reminds me of the way that I have surrounded myself with books and pictures in my home, rarely look at them, but would certainly notice their absence. These poor madmen had lost everything – their families, their homes, their possessions, any kind of social life, perhaps their very sense of self. It often seems to me that happiness and possessions are like vitamins and health. Severe lack of vitamins makes us ill, but extra vitamins do not make us healthier. Most of us – I certainly am, as was my father – are driven to collect things, but more possessions do not make us happier. It is a human urge that is rapidly degrading the planet: as the forests are felled, the landfill sites grow bigger and bigger and the atmosphere is filled with greenhouse gases. Progress, the novelist Ivan Klima once gloomily observed, is simply more movement and more rubbish. I think of the streets of Kathmandu.

My father may have been absent-minded and disorganized in some aspects of his life but he was remarkably shrewd when it came to property, even though as an academic lawyer he was never especially wealthy. When my family left Oxford for London in 1960 we moved to a huge Queen Anne terrace house, built in 1713, in the then run-down and unfashionable suburb of Clapham in south London. It was a very fine house with perfectly proportioned rooms, all wood-panelled and painted a faded and gentle green, with cast-iron basket fire grates (each one now worth a small fortune) in every room, and tall, shuttered sash windows looking out over the trees of Clapham Common. There was a beautiful oak staircase, with barley twist banisters. He had an eye for collecting antiques before it became a national pastime and impossibly expensive. So the new family home, with six bedrooms and almost forty windows – I painted them all once and then had a furious row with my father about how much he should pay me for the work – was filled with books, pictures and various objets d'art. I was immensely proud of all this when I was young. My father was also proud of his house and many possessions and liked to show them to visitors, but in an innocent and almost childlike way, wanting to share his pleasure with others. The family used to tease him that he was a wegotist, as opposed to an egotist – the word does exist in the *Oxford English Dictionary*.

My pride was of a more competitive and aggressive kind, albeit vicarious. When he eventually died at the age of ninety-six, my two sisters, brother and I were faced by a mountain of possessions. I discovered to my surprise that few, if any, of the many thousands of his books were worth keeping. It made me think about what would happen to all my books when I die. We divided everything else up on an amicable basis, but looking back I fear that I took more than my fair share, with my siblings acquiescing to their demanding younger brother so as to avoid disharmony. As for the house, with its forty windows and panelled rooms, I heard that it was recently sold for an astronomical sum, having been renovated. The estate agent's website showed the interior. It has been transformed: painted all in white, even the oak staircase, it now resembles an ostentatious five-star hotel.

When I am working in Nepal I live out of a suitcase, and have no belongings other than my clothes and my laptop. I have discovered that I do not miss my many possessions back in England at all – indeed I see them as something of a burden to which I must return, even though they mean so much to me. Besides, when I witness the poverty in Nepal, and the wretched effects of rapid, unplanned urbanization, I view my possessions in a different light. I regret that I did not recognize the virtues of trying to travel with hand luggage only at an earlier stage of my life. There are no pockets in the shroud.

'The first case is Mr Sunil Shrethra,' said the MO presenting the admission at the morning meeting. 'He was admitted to Norvik Hospital and then came here. Right-handed gentleman, sixty-six years old. Loss of consciousness five days go. On examination . . .'

'Hang on,' I cried out. 'What happened after he collapsed? Has he been unconscious since then? Did he have any neurological signs?'

'He was on ventilator, sir.'

'So what were his pupils doing?'

'Four millimetres and not reacting, sir. No motor response.'

'So he was brain-dead?'

The MO was unable to answer and looked nervously at me. Brain death is not recognized in Nepali law.

'Yes, sir,' said Bivec, the ever-enthusiastic registrar, helping the MO.

'So why was he transferred here from the first hospital if he was brain-dead?'

'No, sir. He came from home.'

I paused for a moment, unable to understand what this was all about.

'He went home from the hospital on a ventilator?' I asked, incredulous.

'No, sir. Family hand-bagged him, sir.' In other words,

the family took their brain-dead relative home, squeezing a respiratory bag all the time, connected to the endotracheal tube in his lungs to keep him oxygenated (after a fashion).

'And then they brought him here?'

'Yes.'

'Well, let's look at the scan.'

The scan appeared, shakily and a little dim, on the wall in front of us. It showed a huge and undoubtedly fatal haemorrhage.

'So what happened next?'

'We said there was no treatment so they took him home, hand-bagging him again.'

'Let's have the next case,' I said.

I had noticed that the sickest patients on the ITU, the ones expected to die or become brain-dead, had often disappeared by the next morning. I was reluctant to ask what had happened, and it was some time before I learnt that usually the families would take the patients home, hand-bagging them if necessary, so that they could die with some dignity within the family home, with their loved ones around them, rather than in the cruel impersonality of the hospital. It struck me as a very humane solution to the problem, although sadly unimaginable back home.

# 10

## BROKEN WINDOWS

Back in Oxford, I went to inspect the lock-keeper's cottage. I walked with mixed feelings along the towpath, rain falling from a dull grey sky, past the line of silent narrowboats moored beside the still, green canal. The air smelt of fallen wet leaves. Several friends had told me that I was mad to try to renovate the place: after fifty years of neglect, with fifty years of rubbish piled up in the garden, without any road access, the work and expense involved would be enormous. The plumbing had all been ripped out by thieves for a few pounds' worth of copper, the plaster was falling off the walls, the window frames were all rotten. The ancient Bakelite electrical sockets and light switches were all broken. The roof was intact, but the staircase and many of the floorboards in the three small bedrooms were crumbling with woodworm. The old man who had lived there was dead, and the cottage itself was dead. The only life was the green wilderness of the garden, where the rampant weeds flourished after fifty years of freedom.

I had spent months making new windows in my workshop in London, with fanciful ogee arches. Glazing them with glass panes cut into ogee curves had been, therefore, difficult and time-consuming. With the help of a Ukrainian colleague and friend, I had ripped out the old windows and carefully installed the new ones before leaving for Nepal. While I was away in Kathmandu they had all been smashed by vandals. This was presumably out of spite for the metal bars I had fitted on the inside. As it was, the thieves had managed to prise apart the metal bars on the window at the back of the cottage and get in. At least I had put the more valuable power tools in two enormous steel chests with heavy locks that I had had the foresight to install. Wheeling them along the narrow towpath on a sack trolley had not been easy and at one point one of them, weighing almost 100 kilograms, had come close to toppling into the canal.

Apparently the thieves had mounted one of the chests on the sack trolley and then abandoned the effort as they couldn't open the front door – I had spent many hours fitting a heavy-duty deadlock to it. On the other hand, my elder sister, an eminent architectural historian, had remarked that the ogee arches were not very authentic for a lock-keeper's cottage; perhaps the vandals had shared my sister's rather stern views about architectural heritage.

I had therefore arranged for rolling metal shutters to be fitted on the outside walls over the windows, which completely defeated the original purpose of decorating the cottage with pretty arched windows. So the vandals then turned their attention, once I was away again, to the expensive roof windows – triple-glazed with laminated glass – that I had installed last year. They had climbed onto the roof, breaking many roof slates in the process, and then heaved a heavy land drain through one of the windows. As far as I could tell, this was done simply for the joy of destruction rather than for burglary – for the love of the sound of breaking glass. I consoled myself with the thought that the frontal lobes in the adolescent brain are not fully myelinated – myelin being the insulating material around nerve fibres. This is thought to be the explanation for why young men enjoy dangerous behaviour: their frontal lobes – the seat of human social behaviour and the calculation of future risks and benefits – have not yet matured, while the rising testosterone levels of puberty impel them to aggression (if only against handmade windows), in preparation for the fighting and competition that evolution has deemed necessary to find a mate.

Each time I walk towards the cottage I feel a sinking feeling at what further damage I will find. Will they have broken the little walnut tree or snapped off the branches of the apple trees? Will they have managed to break open the metal shutters? In the past I always felt anxious when my mobile phone went off for fear that one of my patients had come to harm. Now I fear that it will be one of my friendly neighbours from the longboats nearby on the canal or the police, informing me of another assault on the cottage. I tell myself that it is absurd to worry about mere property, especially as the cottage only contains building tools, all locked up in steel site chests. I remind myself of what I have learnt from my work as a doctor, and from working in poor countries like Nepal and Sudan, but despite this the project of renovating the cottage has started to feel like a millstone. It fills me with a sense of despair and helplessness, when I had hoped it would give me a sense of purpose.

In the weeks before I left for Nepal I had started to clear the mountains of rubbish from the garden. At one end of the garden there is a brick wall, and on the side facing the canal there was a mass of weeds and brambles. I had cleared these to reveal a series of picturesque arched horse troughs made of red brick. The bricks had been handmade – you could see the saw marks on them. They would have been for the horses that pulled the barges along the towpath in the distant past, and there were rusty iron rings set into the bricks for tethering the horses. In front of the troughs, and still on my property, was a fine cobbled floor, which slowly appeared as I scratched away years of muck and weeds. Emma, one of the friendly boat people, stopped by to chat as I worked.

'There is a rare plant here,' she said. 'The local foragers were very excited, though I'm not sure what it's called. Fred and John [two other local boat dwellers] got into trouble with them a few years back when they tried to clear the area.'

'I'm worried that I might have dug it up,' I replied, anxious not to fall out with the local foraging community.

'Oh it will probably grow back,' she said. 'It has deep roots.'

We talked about the old man. He had been frightened of thieves, Emma told me, although as far as I could tell from the rubbish, he had owned little and lived off tinned sardines, cheap lager and cigarettes. He had also told her that the cottage was haunted. According to the locals he had been 'a bit of a wild one' when he was younger, but all I got to hear were stories of how he would sometimes come back to the cottage drunk on his bicycle and fall into the canal. He had a son who had once lived in the cottage with him for a while, but it seems that they had become estranged. There had been a few pathetic and broken children's toys in the rubbish in the garden. I had found shiny foil blister packs of antidepressants – selective serotonin reuptake inhibitors – in the piles of rubbish in the garden. Emma told me that he had died in the cottage itself.

'We didn't see him for several days and eventually got the police to break the door down. He was very dead – in an armchair.'

I slowly built up a huge mound of several hundred black plastic builders' bags, filled with fifty years of my predecessor's rubbish and discarded possessions. It included a matted pile of copies of the *Daily Mail* that was almost three foot thick and had acquired the consistency of wood, having been exposed to the elements for a long time. Rusted motorcycle parts, mouldy old carpets, plastic bags, tin cans, bottles galore (some still containing dubious-looking fluid), useless and broken tools, the pathetic children's toys – the list was endless. None of the rubbish was remotely interesting; even an archaeologist excavating it five hundred years from now, I thought as I laboured away, would find it dull and depressing. The more I dug down, the more rubbish I found.

The community of boat dwellers along the Oxford Canal is supplied with coal and gas cylinders by a cargo barge called *Dusty*. When I took possession of the cottage I found a cheerful note put through the letterbox from Jock and Kati, the couple who own *Dusty*, welcoming me to the cottage and offering me their services. This proved invaluable because with their help I was able to load two bargeloads of rubbish onto *Dusty* and take it up the canal a short distance to where a local farmer had agreed I could put a couple of big skips beside a farm track. It was heavy work, and when it was done I took Jock and Kati out to lunch in a nearby pub. Jock told me that he had backpacked round the world and then become an HGV driver, but he had always wanted to live in a boat from an early age. Kati was a primary school teacher who had taken a year off work and was now reluctant to return. They spent the day travelling slowly up and down the canal, delivering sacks of coal and cylinders of gas to the boat dwellers, all of whom they knew. It was like living in a village. They were very happy, they told me, with their slow and peaceful life, uncluttered by possessions, living in a second barge moored further along the canal.

I had to fell several trees – mainly thorn trees, over thirty feet tall – which had taken over one corner of the garden. Much as I love trees, to the point of worship, I must confess that I also love felling them and I own several splendid chainsaws. After some years I have finally mastered the art of sharpening the chains myself. I suppose tree surgery has a certain amount in common with brain surgery – in particular, the risk and precision. If you don't make the two cuts on either side of the trunk in precisely the right place the tree might fall on you, or fall to become jammed in the surrounding trees, which makes further work extremely difficult; or the bar of the chainsaw can get completely stuck in the tree trunk. And the chainsaw must be handled with some care – I once saw a patient whose chainsaw had kicked back into his face. But there is also the smell of the cut wood – oak is especially fine – mixed with the chainsaw's petrol fumes and, depending on where you are working, the silence and mystery of being in a forest. One of the first books I read as a child – perhaps because my mother was German – was *Grimms' Fairy Tales*, with its many stories of devils, bloody death and punishment, set in dark woods. Felling trees is also a little cruel – like surgery. There is your joy in mastery over a living creature. To see a tall tree fall to its death, especially if you have felled it yourself, is a profoundly moving sight. But what makes brain surgery so exciting is your intense anxiety that the patient should wake up well, and you fell trees to provide wood for making things or for firewood, or to help the growth of other trees. And, of course, you should always plant new ones.

Twenty-five years ago I acquired twenty acres of land around the farmhouse in Devon where my parents-in-law from my first marriage lived. I planted a wood of 4,000 trees in eight acres – native species, oak and ash, Scots pine, willow and holly. For a few short years I could happily tend the trees when I visited Devon, carefully pruning the lower branches of the young oaks, so that after a hundred years they would provide long lengths of knot-free, good-quality timber. I made an owl box and put it up in the branches of an old oak tree growing in one of the hedges that lined my land. I once saw an owl sitting thoughtfully in the box's large opening, which was a very happy moment, but to my disappointment the owl did not take up residence in it. I hoped that I would be buried in the wood after my death, and that eventually the molecules and elements of which I am made would be rearranged as leaves and wood. I had no idea at all of the disaster that awaited my marriage. I lost the land and the trees with divorce, and they were soon sold off. You can still see the wood, now overgrown and neglected, on Google Earth. A third of the trees should have been felled to allow the remaining ones to grow stronger, but this has not been done.

I miss the place greatly – not only the fields and the wood, but the workshop I set up in one of the ancient cob-built barns opposite the farmhouse. The windows, which I had made myself, in front of the workbench, which I had also made, looked out over the low hills of north Devon towards Exmoor. Swallows nested in the rafters above my head, and the young ones would learn to fly by fluttering from beam to beam. Their parents would dart in through the open doors and, if they saw me, would at first turn a somersault directly in front of me – I could feel the air under their wings in my face – and then shoot out again, but after a while they became used to my presence. By late summer the young birds would be flying outside in the farmyard and gather on the cables that stretched from the farmhouse to the barns, little crotchets and minims, making a sheet of sky music. Before autumn came, they would leave for Africa. I returned to have a look at the farm twenty years later, explaining to the new owner my connection with the place. He proudly showed me all the improvements that he had made. I probably should not have gone: the barns and my workshop had been converted into hideous holiday chalets and the swallows evicted, never to return.

Once I had cleared the tons of rubbish from the garden of the lock-keeper's cottage, I planted five apple trees and one walnut tree. The apple trees were traditional varieties such as Cox's Orange Pippin and Blenheims – the same kind of trees as those in the orchard of my childhood home nearby, where I had grown up sixty years earlier.

It had been a working farm until only a few years before my father bought it in 1953, when I was three years old. It was a very fine Elizabethan stone building with a stone roof. There was a farmyard with thatched stables, a large pantiled barn and the orchard and garden, with sixty apple and other fruit trees and a small copse – a paradise, and an entire world for a child such as myself. It was on the outskirts of Oxford, where open fields met the city. There is now a bypass running over the fields. The neighbouring farm has been replaced by a petrol station and hotel. Most of the orchard has been felled, and a dull housing estate has replaced it. The barn and stables have been demolished. There is still a pine tree there, stranded among the maisonettes and parked cars, which had stood guard at the entrance to the copse. I was frightened of the copse, and thought it was full of the witches and devils I loved reading about in fairy stories. I remember how I would stand by the pine tree, sixty years ago – the tree must have been much smaller then, but looked enormous to me. I was too scared to enter the deep and dark forest it guarded, despite longing to be a brave knight errant. Sometimes, as I stood there, I could hear the sound of the wind in its dark branches above me, and it filled me with a sense of deep and abiding mystery, of many things felt, but unseen.

We had many pets, one a highly intelligent Labrador called Brandy. He belonged to my brother but I wanted to train him to sit and beg. I'm not sure why – I now hate to see animals trained to do tricks. But I did it with great cruelty, using a whip made from electric cable, combined with biscuits. He learnt quickly, and I enjoyed the feeling of power over him until my mother found me once with the poor creature. The dog would never stay alone with me in a room for the rest of his life, a constant reminder of what I had done, however much I now tried to persuade him of my love for him. I was filled with a deep feeling of shame that has never left me, and a painful understanding of how easy it is to be cruel. This was also an early lesson in the corrupting effect of power and I wonder, sometimes, if this has perhaps made me a kinder surgeon than might otherwise have been the case.

I had a slightly similar experience when I started work as an operating-theatre porter in the northern mining town. There was an elderly anaesthetist who I now realize was appallingly incompetent. On the first day that I was on duty to assist him, he seemed to be having difficulties intubating the patient, who started to turn a deep-blue colour (known as cyanosis, the consequence of oxygen starvation). In all innocence, I asked him if patients normally went blue when he anaesthetized them. I do not remember his response – but the other theatre porters fell about laughing when they heard the story. A few weeks later he was having difficulties intubating another patient, who started to struggle – the poor man clearly had not been anaesthetized properly. He told me to hold the patient down, which I did with enthusiasm. I had always liked a good fight when I was at school (although there are more shameful episodes there as well, when my strength and aggression got the better of me and my schoolmates started crying). At that point Sister Donnelly, the theatre matron, entered the anaesthetic room and saw how I was restraining the patient. 'Henry!' was all she said, looking genuinely shocked. I cannot forget it. Perhaps it was these experiences that make me cringe when I sometimes see how other doctors can handle patients.

When I worked as a psychogeriatric nursing assistant many years later, it was obvious that the atmosphere on each of the wards was largely determined by the example set by the senior nurses in charge, most of whom understood the duty of care, and how difficult it can sometimes be, as it was a real and daily obligation. As authority in hospitals has gradually passed from the clinical staff to non-clinical managers, whose main duty is to meet their political masters' need for targets and low taxes, and who have no contact with patients whatsoever, we should not be surprised if care suffers.

At my home in Oxford, with its ancient house and garden – my little paradise – I ran a bit wild, the spoilt youngest child of a family of four. When we moved to London when I was ten years old, it was as though I had been evicted from the Garden of Eden.

As I walked along the towpath towards the cottage I also thought about why I had bought it in the first place and why I felt the need to do it up myself. Most of my life was behind me and I found the physical work involved increasingly difficult and much of it positively depressing. Work seemed to be going backwards, not forwards, let alone the damage caused by the vandals. When I cut into the plaster to install a new power socket, huge pieces of it fell off the wall. The lath-and-plaster ceiling of the room downstairs collapsed in a cloud of dust when I tried to strip the polystyrene tiles that had been glued to it. The new windows that I had made myself had all been smashed and I would have to reglaze all of them, and now one of the roof windows as well. Besides, if I ever finished the work, what would I do then? I had to conclude that what I was doing was not just to prove that I was capable of such work despite growing old, but also an attempt to ward off the future. A form of magic, whereby if I suffered now, I would somehow escape future suffering. It was as though the work involved was a form of penance, a secular version of the self-mortification found in many religions, like the Tibetans who crawl on all fours around Mount Kailash in the Himalayas. But I felt embarrassed by the way in which I was doing all this in the cause of home improvements – it seemed a little fatuous when there was so much trouble and suffering in the world. Perhaps I am just a masochist who likes drawing attention to himself. I always was a tremendous show-off.

Thinking these depressive thoughts, I arrived at the cottage but, just as on my previous visits, as soon as I saw it I had no doubts. There was the wild garden and the old brick horse troughs with the quiet canal in front and the lake behind, lined on one side with tall willows. The two swans were there, perfectly white on the dark water and, beyond them, reeds faded brown with the winter, and then the railway line, along which I had once watched steam trains roaring past as a child. When I went inside – in darkness, now that the broken windows were all boarded up – the light from the open door fell on broken glass, shining and scattered all over the floor, which crunched underfoot as I entered. But it no longer troubled me.

I would restore this pretty and humble building, I would exorcize the old man's death and all the sad rubbish he had left behind. The six apple trees and one walnut tree would flourish. I would put up nesting boxes in the trees, and an owl box, like the one I had installed in the old oak tree in the hedge beside the wood in Devon.

I would leave the cottage behind, for somebody else to enjoy.

I decided to put motion-detecting floodlights high up on the cottage walls, and also CCTV cameras – a reluctant concession to the thieves and vandals. This involved working up a ladder, high under the eaves of the roof. I have lost count of the number of elderly men I have seen at work with broken necks or severe head injuries sustained by falling off ladders: a fall of only a few feet can be fatal. And there is a clear connection between head injuries and the later onset of dementia. I therefore drilled a series of ringbolts into the cottage wall, like a climber hammering pitons into a rock face, and tied the ladder to the ringbolts and, wearing a safety arrest harness, attached myself to the ladder with carabiners as I fixed the lights and wretched CCTV cameras, wielding a heavy-duty drill to bore through the cottage walls for the cables.

While I was doing this work I received a visitor. I climbed down the ladder. He was a man my age, walking with a golden retriever, which happily explored the wild garden while we talked.

'I lived here as a child sixty years ago,' he said, 'in the 1950s, before Dennis the canal labourer took it over. My brother and I lived here with our parents. It was the happiest time of their lives.'

We worked out that we were of the same age and had lived at the same time in our respective homes less than one mile apart. He produced an old black and white photograph showing the cottage looking tidy and well cared for, with a large flowering plum tree in the front garden. You could just see that the garden had many vegetables growing in neat rows. His mother was standing at the garden gate, wearing an apron.

'I scattered my parents' ashes over there,' he told me, pointing to the grassy canal bank on the other side of the little bridge across from the cottage. 'I come here to talk to them every so often. I told them today that their grandson had just got a university degree. They would have been so proud.'

I showed him around the inside of the cottage. He gazed at it in silent amazement – so many memories must have come back.

'My dad used to sit in the corner over there in the kitchen,' he said, pointing to the place where there had once been a stove. 'He had a handful of lead balls. He'd throw them at the rats when they came in through the front door, but I don't know if he ever got one.'

# 11

# MEMORY

By the time that he died at the age of ninety-six my father had become profoundly demented. He was an empty shell, although his gentle and optimistic good nature remained intact. The live-in carers my brother had organized to look after him in his flat often remarked on how easy it was to look after him. Many of us, as we dement, increasingly confused and fearful as our memory fades, become aggressive and suspicious. I had seen this myself when I worked briefly as a geriatric nursing assistant, although in the grim and hopeless environment of a long-term psychiatric hospital – an environment which must have made the poor old men's problems many times worse than they might otherwise have been. He was famously eccentric – the porters at the Oxford college where he had been a don after the Second World War regaled me – when I went there as an undergraduate myself many years after he had left – with stories of his many eccentricities. He once met one of his former pupils, who told him that he had always been terrified when having a tutorial with him. My father, the mildest of men, was painfully surprised, until his former pupil went on to explain that he had rarely had any matches with him when giving tutorials. He would light the gas fire in his room by turning on an electric fire – one of those old models with a red-hot bar – and pressing it against the gas fire. The tutorial would therefore start with an alarming explosion. There had been an electric fire like that in my bedroom in the old farmhouse. There was no central heating: in winter there would often be frost flowers on the bedroom windows in the mornings and I would lean out of bed, trying to stay under the blankets, heating my clothes in front of the fire before getting up to bicycle to school.

I used to read late at night with a torch under the blankets after my mother had kissed me goodnight and turned off the lights in the room. At the age of seven I borrowed a school friend's book about King Arthur and his knights. I became slightly obsessed by these stories and read everything I could find about knights and chivalry, including Malory's *Morte d'Arthur*. I considered Lancelot and Galahad to be hopeless goody-goodies but greatly admired Sir Bors, who was tough, loyal and reliable. He would have had no time for women or religion, I thought. My edition of Malory had many coloured illustrations by Sir William Russell Flint, the popular late-nineteenth-century artist, who was famous for his erotic paintings of women. His illustrations for Malory had heroic knights and beautiful maidens with long Pre-Raphaelite hair in tresses and wearing long and flowing robes, which I found highly attractive. This night-time reading probably contributed to my severe short-sightedness, which resulted in retinal detachments many years later.

I found my weekly attendances at Sunday School extremely boring, and the illustrations in the little books of Bible stories for children very dull compared to the pictures in the *Morte d'Arthur*. Both my parents were sincere – although relaxed – Christians. I received a traditional English middle-class Christian education at Westminster School, including morning service six days a week in Westminster Abbey when I was a teenager. Every so often the organist would play the last movement of Widor's Fifth Symphony – the only piece of French organ music I could abide – at the end of the service. I would stay behind in the now empty building as the music crashed and boomed under the great Gothic roof and round all the marble statues and monuments, until my anxiety about being late for the first class would overcome me and I would run back to the school through the empty cloisters, over the worn gravestones, with the music fading behind me.

I was bitterly unhappy in my first year at the school. I was a boarder for the first time in my life. I think my parents thought it would be good for me, and it was a fairly traditional part of a middle-class boy's education at the time. I missed having my own room and, being very innocent and prudish, was shocked by the other boys' endless talk about sex. I once went to the housemaster to complain about this – I squirm at the memory. After a year I finally dared to tell my parents how unhappy I was. I remember the overwhelming sense of relief when I realized that it was going to be possible for me to become a day boy.

In my last year at the school I spent Friday afternoons in the Abbey Muniments Room, filing nineteenth-century inquest reports from the Westminster Coroner's Court. The Cadet Corps had been abolished. Until then we had spent Friday afternoons in military uniforms marching round the school yard with ancient rifles, .303 Lee Enfields said to date back to the Boer War, but converted to .22 bore. We were offered various alternatives to the Corps and I chose the Muniments Room. The room was part of the south transept of the Abbey, above the aisle, and you looked directly down into the Abbey itself. I was tasked with filing a large number of

Coroners' Inquests from the Westminster Court in the 1860s. The reports, bound in crumbling green tape, were kept in a huge, semicircular medieval chasuble chest made of oak which was black with age. I found an old sword on top of the chest. Henry V's sword, I was told, which indeed it was. I liked to wave it around my head while quoting the appropriate lines from Shakespeare. The Keeper of the Muniments was a small, round and bird-like man who wore bright yellow socks and rolled rather than walked. He took very long – probably liquid – lunch breaks so I was left largely to my own devices. Although I found the inquests fascinating – stories of death in Dickensian London in perfect copperplate writing – I was more excited to discover a spiral stone staircase leading up from the Muniments Room to the triforium and roof of the Abbey. I therefore spent most of my Friday afternoons exploring all the empty spaces and the roof of Westminster Abbey, with wonderful views of central London.

As far as I can remember, I never believed in God, not even for a moment. At one morning service in the Abbey I remember seeing the school bursar – a retired air commodore – praying. He was kneeling opposite me on the other side of the gilded choir stalls. There was a look of the most terrible pain and pleading on his face. He disappeared from the school shortly afterwards and I heard later that he had died from cancer.

My father's dementia was probably avoidable. He had suffered two significant head injuries in his seventies – once when falling between the rafters of a friend's attic and knocking himself unconscious on a marble fireplace in the room below, and once falling off a ladder when trying to read the gas meter in his huge eighteenth-century house in London. He already had form for losing his footing between attic rafters: in the ancient house in Oxford where we lived in the 1950s, his leg, much to the surprise of the family's au pair, once appeared in a shower of plaster through the ceiling above her bed, but fortunately not the rest of him. At the time of the two head injuries he had seemed to make a reasonable recovery, but they probably contributed to his slow deterioration in old age.

I was not a good son. In his declining years, after our mother's death, although I lived quite nearby I rarely went to visit him. I was impatient with his forgetfulness and distressed by the fact that he was no longer the man that he had been. My siblings went to see him more often than I did. Both my parents expected very little from me – they took pleasure in any success that I had, and were always anxious to help me in any way that they could – yet rarely, if ever, seemed to ask for anything in return for themselves and rarely, if ever, complained. I exploited their love, although their love was certainly the principal source of my feeling of self-importance, something which has been both a strength and a weakness throughout my life.

My father had been an eminent lawyer, although his career is not easy to categorize. An Oxford don for fourteen years, he left Oxford to run various international legal organizations and finally worked for the British government on reforming and modernizing British law as one of the first Law Commissioners. When I was young I had no interest in the law or in my father's work – it seemed terribly boring. It was only in the closing years of my own career as a doctor, mainly from my work overseas, and having to witness so much corruption and the abuse of power, that I came to understand the fundamental importance of the rule of law to a free society – the principle that lay at the heart of my father's work and view of the world. Democratic elections, for instance, mean little without an independent judiciary. His obituary in the London *Times* filled an entire page and I was filled in turn with both filial pride and guilt. My own career, as a doctor, now seems to me rather slight in comparison to his.

The profoundly serious nature of his work – and his deeply moral and almost austere view of the world – were completely at odds with the way he did not seem to take himself at all seriously. The family followed his lead, and I fear that we treated him more as a figure of fun than of authority. There were only a few rare occasions when he would briefly lose his temper with us over the way that we treated him with such a singular lack of respect. He enjoyed telling stories against himself and of his – not entirely unselfconscious – eccentricities. He often talked of writing his memoirs but he never got beyond the first page, which described how he pulled the lanyard on an artillery piece in Victoria Park in Bath in 1917, at the age of four, as a reward for his parents buying war bonds to help finance the First World War. I often said that I would sit down with a tape recorder and record his many memories and stories – he had led a most unusual and very interesting life and was an excellent raconteur – but I never did, and this is something I deeply regret. His past, and many of the stories that he had heard in turn about his family's origins in the countryside of Somerset and Dorset, faded as his brain decayed and are now lost for ever. I know only a few fragments.

During the war he had been in Military Intelligence, interrogating high-ranking German prisoners of war, as he spoke German. His preferred technique, he once told us, was to treat the interrogation sessions like an Oxford tutorial, and encourage the prisoners to write essays for him on democracy and the rule of law.'The hardened Nazis were a lost cause,' he said. 'But it worked for some of the others.' One U-boat captain, he discovered, was something of an anti-Nazi, so my father dressed him in a British Army greatcoat and smuggled him out of the prison camp to take him on a tourist trip round London, although he said he was a little worried as to what he would say if they were stopped by the police. He was outraged when stories emerged of the British Army hooding – effectively torturing – IRA suspects in Northern Ireland at the start of the Troubles. Like many experienced interrogators, he was of the opinion that kindness and persuasion worked much better than torture. His particular interest, when interrogating prisoners, was in German morale. He wrote a report arguing that the carpet bombing of German cities was strengthening it rather than breaking it. Apparently, when 'Bomber' Harris – the head of RAF Bomber Command – saw the report, he was so enraged that he wanted to have my father court-martialled. This did not happen, fortunately, and history, of course, has entirely vindicated my father.

He liked to claim – I suspect with some exaggeration – that there were only three books in his parents' house in Bath, where his father ran a jewellery business and his mother had a dress-making business until she had children. She was a farmer's daughter – one of eight children – and used to walk eight miles to work every day as a seamstress, eventually owning her own shop, which, our father assured us, had been highly fashionable. He had a difficult relationship with his father, and once had even come close to blows.

'Did you feel bad about that?' I asked him, when he once told me this.

'No,' he replied calmly. 'Because I knew I was right.'

His certainty was not arrogance, but was based on a coherent and stubbornly moral outlook which was usually correct, and which was most frustrating for the rebellious, selfish teenager that I once was. His own father – whom I never knew, as he died shortly after my birth – could not understand how his son had become a left-leaning liberal intellectual, with a house packed with many thousands of books and a German refugee for a wife.

Before reading law at Oxford he had been educated at a minor public school near Bath which, he once told me, specialized in turning out doctors and evangelical missionaries. In later life he said he still had nightmares about the place – *Tom Brown's Schooldays* were nothing, he once told me, compared to what he had to put up with. So he hated the place, yet he became the Sergeant Major for the school's army cadets, played rugger in the First XV and rowed in the First Eight. He said that he owed everything that he became to one inspirational history teacher. His attitude to success and conventional authority remained deeply ambiguous throughout his life. One of the few times I saw him seriously distressed (other than the various occasions when I caused him great pain) was when the teacher who had so inspired him wrote to him asking for a contribution to a fundraising campaign for his old school. My father was a deeply generous man and involved in much charitable activity, but after a great deal of painful heart-searching, he wrote to his old teacher and mentor saying that he felt unable to send any money.

He had been secretary to the Oxford League of Nations Association – an organization both idealistic and doomed – and had met my mother when he went to Germany to learn German in 1936. He stayed in the town of Halle, where he found himself in the same lodgings as my mother, who was training to be a bookseller. Her strongly anti-Nazi political views had prevented her from going to university and so she chose bookselling as the career closest to her love of philology. My father was the first person she met to whom she could pour out her heart about her deep unhappiness as to what was happening in Germany. Her views eventually got her into trouble with the Gestapo. Her colleagues at work, with whom she had been overheard exchanging anti-Nazi views, were tried and imprisoned but my mother was let off on the grounds that she was – as one of the two Gestapo men who interrogated her put it – a 'stupid girl'. They told her, however, that she would be cross-examined as a witness in her colleagues' trial and she felt that she would not be able to cope with this. So, in brief, my father married her and brought her to England, a few weeks before the Second World War started.

My mother's sister was an enthusiastic supporter of Hitler and the Nazis, and her brother joined the Luftwaffe, although from a love of flying and not from any political conviction. I do not know how my mother knew that Hitler's regime was evil. It is only since her death, by reading about Germany at that time (in translation, given my shameful lack of German) that I have come to understand just how remarkable was her defection. Her decision to leave Germany – a country with the deepest respect for authority and on the brink of war – would have been seen by many as treason. It seems obvious and easy in retrospect, but how I wish she was still here so that I could talk to her about this.

Sixty years later, I asked my parents about their decision to get married. My first marriage was falling violently apart, and sometimes I went to speak to them about my unhappiness. I should not have burdened them with this, but it was a strange experience to converse with them both almost as equals, as fellow adults, about the difficulties of married life. I learnt that the decision to marry my mother had been a difficult one for my father, although he did not specify exactly why. My mother suggested he was on the verge of marrying somebody else in England, but my father did not confirm this. He told me that he was in such a state of despair that one day, working as a young lawyer in London, he was walking up Tottenham Court Road and saw a doorway with a sign advertising a counselling service. I think it was some kind of Christian mission – whatever it was, my father said that he went in and found a man there who helped him greatly.

In Nepal marriages are still arranged. My Nepalese friends tell me that it usually works. Sometimes it seems to me that my parents' highly successful marriage was, in a way, also arranged – arranged on the basis of the rule of law, morality and liberal democracy. They were closely involved from the 1960s onwards in the creation of Amnesty International, and my mother ran – in her quiet and wonderfully efficient way – the registry of political prisoners in dozens of countries. At the beginning the offices were in the chambers of the lawyer Peter Benenson, who had come up with the idea for the organization. I would go in sometimes to help – mainly to lick and stamp envelopes. I would like to say these were for letters sent to dictators all over the world, but they were mainly newsletters to the small groups of volunteers – organized in cells like revolutionaries – who adopted particular prisoners and would write in protest to the dictatorial regimes that had imprisoned them.

By the time that I was born in 1950 my mother had developed a strange, disabling condition which resulted in excruciatingly painful bruising over many of her joints. She consulted a wide number of specialists but none could come up with a diagnosis. One suggested it was allergic, so they had the family pets put down. The only treatment that seemed to help was arsenic (from which she developed a rare skin cancer called Bowen's Disease many years later).

Eventually in despair – I think I was only a few months

old at the time – they sought a psychiatric opinion and my mother was admitted for six weeks' inpatient psychoanalytic treatment in the Park Hospital in Oxford under the care of a fellow German émigré. He was, she once told me, very much like her father, who had died quite suddenly in 1936 from metastatic bowel cancer when she was nineteen. As a fellow émigré, he must also have understood her deep sense of loss – of her family, of her past and much of her identity. Her mother had died from breast cancer during the war and her sister perished in a British air raid on the city of Jena. Her sister had been an enthusiastic Nazi and she and my mother had parted on bitter terms when my mother left for England, although my mother was told after the war that her sister had changed her mind before her death. And perhaps, although I never asked her directly, she also felt that she had betrayed her principles by fleeing from Germany and by not standing up for them and her colleagues in court.

The treatment worked and the stigmata-like bruises disappeared. There is no way of knowing whether it was the psychoanalysis which helped, or whether it was the way in which my father, by his own admission, became a more considerate husband because of her illness, or whether it was simply the rest in hospital from bringing up four children with minimal help and from a husband who was completely dedicated to his work. My parents once told me – as a joke – that my own personality, which at times caused them great problems, was partly to be explained by the famous child psychologist Bowlby's work on maternal deprivation. This may or may not be true, but only as I reach old age myself have I come to understand just how completely I am my parents' creation, and whatever good is in me came from them.

Thirty years later, by which time I was a medical student, my mother's purpuric swellings, or 'bumps' as she called them, reappeared, much to my parents' alarm. 'You must be stressed and anxious about something!' I remember my father saying to her, close to despair, as she lay on her bed in severe pain. But on this occasion a specialist prescribed the drug dapsone – a drug normally used for leprosy – and the bruises immediately disappeared. I still do not know what the underlying diagnosis might have been. If dapsone, a mere chemical, had been available in 1950, perhaps I would not be the person I am now, and I might not be sitting in a remote Nepalese valley, at the foot of Mount Manaslu, writing about my parents as I listen to the water of the glacial River Budhi Gandaki, rushing noisily past over its many rapids on its journey from the Himalayas to join the River Trishuli and then the Ganges, to end in the Indian Ocean.

My father was a tremendous optimist – even as his memory deteriorated he would express the hope that things would get better. Before his dementia became profound, while he was still living in the grand eighteenth-century house overlooking Clapham Common, I once engraved a brass plate with the family name to go by the bell push for the front door to distinguish it from the bell for the basement flat, which was now rented out. My engraving was very clumsy and the letters became smaller as they went from left to right. 'Like my faculties,' he said sadly as we looked at it together after I had screwed it in place. He still had some insight then, into what was happening to him. 'But I think things will get better.'

On my second trip to Nepal I accompanied Dev to a Health Camp he had organized in a remote corner of the country. The metalled road ended at Gorkha, and we then took three hours to travel the thirty-six kilometres over the mountains to the small town of Arughat on a wildly uneven dirt track – though not uneven enough to dissuade large trucks and buses from crawling along it as well, throwing up huge clouds of ochre dust. In places there was barely room for the vehicles to squeeze past each other, often with a precipitous drop, only a few centimetres away, to the valley below. On a clear day we would have seen Mount Manaslu at the head of the valley, one of the most beautiful of the great Himalayan peaks and the eighth-highest mountain in the world, but the haze was intense and there was no view at all.

The Health Camp took place in what had been a brand-new primary care hospital, but which had been badly damaged in the earthquake a few days before it was due to open. It had been abandoned until Dev had come to inspect it and found that most of the building was serviceable, although in a terrible mess. An advance party had arrived a day before us and I was very surprised to find a clean and tidy building – although with great cracks in the walls and one wing partially collapsed – when we arrived. The Neuro Hospital team of over thirty doctors, nurses and technicians had come with enough equipment to run two operating theatres, a pharmacy and five outpatient clinics, with plain X-rays and ultrasound and a laboratory. It was an impressive piece of organization – but they had done similar camps elsewhere before, especially after the earthquake, and had learnt from experience.

Next morning there was a long queue of patients – several hundred, mainly women, all dressed in brilliant red – waiting outside the hospital. All the treatment would be free. They were sheltering under equally colourful umbrellas, as the temperature was soon in the 90s. They were held back at the hospital gates by armed policemen and allowed in, one by one, to be registered at the entrance and directed to the appropriate clinic – such as orthopaedic, plastic-surgical or gynaecological. Fifteen hundred patients were seen in three days and many, relatively minor, operations performed, some under general anaesthetic. Difficult cases were advised to go to larger hospitals long distances away. Patients came from far and wide – the Health Camp had been advertised for many days in advance.

'Some patients are coming from the Tibetan border,' I was told.

'How far away is that?'

'Four or five days' walk. No roads. Ten days if you or I tried to do it.'

Sick patients arrived on stretchers. Several old people arrived carried piggyback.

Although I was treated like visiting royalty, and presented with bougainvillea garlands and silk scarves at the lengthy opening and closing ceremonies by the local people, and a framed certificate entitled 'Token of Love', I was completely useless. My days of general surgery and general medicine are long behind me. I was disappointed to find, as I watched Dev happily operating on inguinal hernias, hydrocoeles and similar lesions, that I had completely forgotten how to do them, even though I had spent a year doing dozens of such cases when working as a general surgeon. Thirty-five years ago, a year of general surgery had been a necessary part of qualifying for the final FRCS examination, which I had to pass before I could train as a neurosurgeon. I sat in on some of the Health Camp clinics, and I found that the junior doctors knew much more than I did.

Dev is doing a clinic – a chair has been put next to him for me. A tidal wave of patients now flows in: an old woman with elaborate gold ornaments in her nose and a rectal prolapse, old men with inguinal hernias, old women with haemorrhoids, many patients with varicose veins – cases which remind me why I was pleased when my year of general surgery came to an end almost forty years ago, and I could devote myself to neurosurgery. But it also reminded me how modern medicine is not just about prolonging life – it has probably achieved as much good by finding treatments for all the chronic non-fatal conditions from which we would otherwise suffer and from which people in poor countries like Nepal still suffer.

I remembered a rectal clinic I had done on Friday afternoons when I did my mandatory year of general surgery. Neither I, nor my patients, enjoyed the experience. They knew, and I knew, that I was going 'to give them a ride on the silver rocket' – the medical procedure of sigmoidoscopy, where an illuminated long stainless steel tube is used to examine the inside of the rectum. But I was happy enough when I was in the operating theatre.

There is a young woman with unilateral proptosis – her right eye is bulging outwards; we will send her to Kathmandu for a scan. There is a girl with what are probably pseudo-seizures, hurried in by her anxious mother. If people have fits in front of the doctor – which is what is happening here – it usually means, although not always, that the problem is psychological rather than epileptic. Dev prescribes the antidepressant amitriptyline. There's no question of any follow-up in such a remote country area. It is impossible to know what will happen to the patients. Many of them – most of whom are illiterate – produce plastic bags full of the many medicines they have been taking.

It's all in Nepali, of course. I am half asleep, lulled by the sound of hundreds of voices outside and the whirring of the ceiling fan. The temperature outside must be in the high 90s. The patients at the front of the queue are pressed up against the metal gates, and the police guards push themselves into the crowd from time to time to stop fights breaking out, or to allow urgent cases to be brought into the hospital. But inside the building everything is highly organized.

There is a man with huge, wart-like growths on his hands and feet. Next we see a five-year-old boy and his ten-year-old sister, who both went blind at the age of two. They are led into the room and sit sightlessly while my colleagues thumb through the stained and dog-eared pieces of paper that comprise their medical notes. All we can do is confirm that there is nothing to be done. I ask whether there are schools for blind children in Nepal and am told that there are, but that it is unlikely that these children, from a remote mountain village, could go to one.

At lunch on the baking-hot roof, sitting in the shade of a bright-blue UNHCR tarpaulin left over from earthquake relief, I talk to the gynaecologist.

'How many PVs [vaginal examinations] have you done so far?'

'Over five hundred.'

'Do the women know any anatomy?'

'Most know none at all. It's a waste of time trying to explain anything to them. A few of them can understand. But usually I just say take the medicine and this is the name of your illness. The women in the queue outside the room,' she adds, 'were starting to fight each other, trying to get in . . .'

One room is reserved for people who are too weak to sit or stand. There is a young woman with diabetes who is now in severe ketoacidosis. She lies on a stretcher, with dulled eyes and a deeply resigned expression on her face, coughing and retching into a plastic bowl from time to time. The MO fails to get a drip up on her and I reconfirm my uselessness by also failing. One of the anaesthetists succeeds. We give her IV fluids and find some insulin from another patient.

'What's her outlook?' I ask.

'Not good. Poor peasant in a remote village. Can't afford insulin. Diabetes is still a fatal disease here for many people. But we've told her husband to take her to the nearest big hospital. They may be able to help.'

I find an empty room in the ruined part of the hospital. There are large cracks in the wall from the earthquake. The many windows look out onto some tall mango trees and the room is full of the sound of the wide River Budhi Gandaki rushing past, coming down from the glaciers of the invisible Mount Manaslu. I sit there quietly for a while, trying to write, until two playful Nepali boys find me and peer over my shoulder at what I am doing and will not leave me alone, so I return to the clinic to watch more patients come and go.

After three days, Health Camp comes to an end. By the early evening there are only a few patients left waiting outside the entrance. I sit outside on a white plastic chair looking at the dim, blue hills around me. It is still very hot, and there is a strong wind, so the giant mango trees wave and shake. A noisy wedding party passes on the nearby road, dust swirling about them, the women all brilliantly dressed, with two men at the head of the procession blowing on long, curved horns. The bride is carried in a palanquin, and is veiled and dressed in red and flashing gold. The groom walks behind, his face heavily made up, wearing an elaborate and decorated coat. Three young girls are playing in the hospital courtyard and come up to me. We exchange a few words, which we don't understand, and they happily laugh and dance around me for a while, before running away. I so wish I spoke their language and could talk to them. My inability to speak any language other than English is the deepest of all my regrets. So I sit by myself and watch dust devils spiral up off the dry ground, driven by the wind, as the light slowly fades.

# 12

# UKRAINE

Igor, as always, is waiting for me at the airport, his head in a tight woollen cap, bobbing up and down in the crowd outside the exit, trying to spot me. His characteristically stern expression breaks into a brief smile when he sees me, but in recent years the smile has become briefer and briefer. His serious enthusiasm, which had so impressed me when we first met, seems to have changed into something grim and rather different.

I awkwardly accept his kisses. We argue over who should carry my suitcase (often full of second-hand surgical equipment) and climb into his van. We drive into the city to the hospital, and Igor holds forth in his broken, staccato English. It was only when I heard Ukrainian spoken by the Ukrainian poet Marjana Savka in Lviv, in the west of the country, that I realized Ukrainian could be very beautiful, and not the rather harsh, declamatory language spoken by Igor.

'Financial crisis terrible. Everybody have money problems. Everybody unhappy. Before crisis my doctors make maybe two thousand dollars in month, now only four, maybe five hundred.' This monologue continues until we reach the hospital. I know that there will be a long queue of patients in the corridor outside the small, cramped office waiting to see me, almost all of them with large and terrible brain tumours and other, often hopeless, neurosurgical problems.

'Two acoustics to see,' he says, as we pass the tall and ugly apartment blocks on the city's outskirts, looking bleak and unwelcoming in the winter mist. There is a thin layer of snow on the ground. I think, not for the first time, how grim Ukraine can be, and how tough its inhabitants have to be to survive. 'Many interesting cases, Henry,' he says happily.

'Well, interesting for you,' I reply grumpily.

'You lose enthusiasm since you retire,' he replies in a disapproving tone of voice.

'Maybe I'm just getting old.'

'No, no, no!' he cries, and then, reverting to his favourite topic, goes on to tell me that twenty Ukrainian banks have collapsed in the preceding year.

We cross the great River Dnieper on one of the many massive bridges built during the Soviet era. The river is frozen, but only in places, and I can see below us the small figures of dozens of people on the shelves of ice, close to the oil-black water, fishing through holes they have cut.

'People drown every day,' Igor observes. 'Twenty this year. It is disaster. It is silly.'

We drive up the steep cobbled street leading from the banks of the River Dnieper to the centre of Kiev and turn onto Institutskaya Street, where a few months earlier dozens of protesters had been killed by snipers in the Maidan demonstration. The SBU hospital where Igor rents space for his private clinic is just round the corner on Lipska Street. The SBU, once known as the KGB, being an important organ of the State, naturally had its hospital in the centre of Kiev. I had been in Kiev on several occasions during Maidan. I spent as much time as I could mingling with the thousands of demonstrators, proud to feel a small part of it.

I had first gone to Ukraine in 1992, just after the collapse of the Soviet Union. Entirely by chance I had met Igor in one of the hospitals I had visited. We became friends and I had been travelling to Ukraine for a few days each year ever since to help him with his surgery. At that time medicine in Ukraine was decades behind the West. I found many second-hand instruments and microscopes for him and taught him everything I knew. At first this was all spinal surgery, and Igor was soon probably the most accomplished spinal surgeon in Ukraine. As his fame spread, more and more patients came to his outpatient clinic with problems in the brain. He badgered me constantly to help him develop brain surgery, insisting that the large and difficult acoustic tumours he had seen me operating on in London could not be treated properly in Ukraine. In Ukraine, because of delays in diagnosis, these tumours are usually very large and the operations correspondingly difficult and dangerous. These are tumours which grow off the hearing nerves within the skull and can become large enough to compress the brain and slowly kill the patient. On my very first visit to Kiev in 1992 I went with two other colleagues from my hospital, one an anaesthetist and the other a pathologist. We visited the major State neurosurgical hospital, one of the two major centres for brain surgery in all of the Soviet Union, where we delivered lectures. My pathology colleague was taken on a tour of the pathology department and came back afterwards to tell us, looking a little shaken, of a series of buckets he had been shown containing the brains of patients who had died after surgery for acoustic tumours.

When Igor and his wife Yelena came to London to Kate's

and my marriage in 2004, all he could talk about was the need to develop acoustic neuroma surgery. Eventually Kate told him that she could stand it no longer. 'Igor,' she said, 'I'm sorry, but we're going to have a moratorium on the word acoustic. We have to talk about something else. We cannot spend every mealtime, every day, with you trying to persuade Henry to show you how to operate on acoustics.' Just for once, Igor did what he was told for a few days. Even I found his intense enthusiasm for neurosurgery and utter commitment to it rather tiring at times. Eventually I agreed to help him, but not without misgivings, as the treatment of patients with brain tumours involves much more than just operating.

In the years before Maidan I would return to England and enthusiastically tell people: 'Ukraine is a really important country!'

This was usually met with a puzzled expression.

'It's part of Russia, isn't it?'

And I would then deliver a little lecture on how Ukraine was one of the great historical watersheds, where Europe met Asia, where democracy met despotism.

I think most of my colleagues and friends in England regarded my slight obsession with Ukraine as an eccentric hobby, but when Maidan started and all of Europe saw the images of the fighting between the demonstrators and the *Berkut* riot police, resembling medieval battle scenes with staves and shields and catapults, and blazing car tyres filling Independence Square with flames and black smoke, I think I could claim some prescience. Igor had had many problems during the twenty-four years we had been working together. He had been something of a medical revolutionary and dissident, using what he had learnt from me to try to improve neurosurgery in Ukraine. The medical system in Ukraine was as authoritarian as the political system and he made many enemies and had many difficulties. But his patients did very well and eventually his clinic became well established. The many attempts by senior colleagues and administrators to thwart him failed. There was something heroic about what he had achieved, and I felt that my work with him over the years was part of the same struggle against corrupt autocracy as the Maidan protests.

There is a turnstile in the small lobby at the entrance to the hospital. The tiled floor is wet with thawing snow brought in on people's boots. Patients and their families can come and go quite freely, but as I am a foreign doctor I am regarded with suspicion by the SBU. When I arrive I have to show my passport to the unsmiling young soldiers behind the glass window next to the turnstile.

'You might be terrorist!' Igor says as the turnstile is unlocked and it clanks in an authoritarian sort of way as I push through it.

'They are SBU soldiers, and hospital has no control of them,' he added.

'It must be an awfully boring job.'

'No, no, no. They are happy not to be at frontline.'

I feel imprisoned once I am inside – imprisoned by my lack of Ukrainian or Russian, and intimidated by the soldiers at the entrance. It is probably quite unnecessary. I once arranged to wait for a film-maker to meet me at the entrance to the hospital. One of the soldiers had to keep me company on the pavement outside until she arrived. When she came she translated what the soldier had started to say to me. I thought he was threatening me with arrest or something similar, but apparently it was a long speech of thanks for my helping Ukrainian patients.

I look back with some shame on the years of my training. It is complete torment to assist a less experienced surgeon than yourself to do a difficult and dangerous operation. Some of the senior surgeons I had worked for simply couldn't do it and left me to get on with it – the so-called 'see one, do one' method of surgical teaching which was one of the more egregious aspects of some English surgical training in the past. I look back with horror on some of the mistakes I made when I was a trainee and, even worse, on some of the mistakes made by my trainees – for which I must hold myself responsible – once I became a senior surgeon myself. But some of my trainers, I now realize, had shown great patience and kindness (and courage) in taking me through operations. I had not thought for a moment how difficult it might be for them, so self-important was I, and so engrossed in what I was doing. Igor, I now realize, is no different. I don't think he ever saw how difficult I found the long outpatient clinics, which could easily last ten or twelve hours, or the agonies I went through as he operated on major brain tumour cases. The more I let him do, the more he would learn, but the greater the risk to the patient and the more anxious I would become. If it seems that it is safe for him to carry on, I retire to the recovery room next to the operating theatre and stretch myself out on the trolley by the window, resting my head on a cardboard box. I would be simultaneously bored and tense, going into the theatre at regular intervals to see how he was getting on and whether I should take over.

'Do you want me to scrub up?' I ask.

'No, no, not yet,' is the usual reply, but sometimes he asks for my help, and sometimes I insist that I take over.

On one previous visit in winter, some years before Maidan, the view through the window as I lay there was uncommonly beautiful – of fine snowflakes drifting down from a grey sky, and of the tall pine and silver birch trees in the hospital courtyard bending with the weight of the snow on their branches. The courtyard itself was virginal white, with only a few footprint tracks on the paths. We were doing a very difficult tumour in a young woman. Both Igor and I thought that I would have to help him, but in the event Igor did almost all of it and she awoke perfectly.

I had passed the long hours by the window watching the snow fall. I thought it was the crowning moment of the many years I had spent teaching him. But on one of my regular visits two years later I heard, only by accident, that the woman had died some months after the operation, from a post-operative infection in her brain. His silence over this, and the way in which he had not sought my advice when she fell ill after the surgery, made me furious and I came close to telling him that I would never return to Ukraine, but eventually I thought better of it. I was told later that he had thought I would refuse to come back to Ukraine if he *did* tell me, which was, of course, the complete opposite of the truth and showed how little he understood me. It reminded me of the long delays before the Soviet government admitted to the catastrophe at Chernobyl.

I told Igor how angry I was, and it was a long time before he grudgingly apologized, but the words of apology seemed almost to choke him, so difficult did he find it to speak them. I went on to tell him how I had once made a similar mistake myself with a post-operative infection, with catastrophic consequences for the patient, and had also failed to ask for help. I still have a photograph of the young Ukrainian woman, which I had taken when I first met her in the cramped little clinic room. She is looking pleadingly at me. Something died within me when I heard of her death, although I still wanted to come out to Ukraine – it had become such an important part of my life. It was only some years later that I understood how wilfully blind I had become. I came to regret bitterly that I had not left him then. I should never have agreed to help him with brain tumour surgery.

# 13

# SORRY

I had gone into work on Sunday as usual. I had been rather depressed about my work in recent weeks and, as I cycled down Tooting High Street in the dark, I decided that it was time to think more positively. Some of my patients came to grief, I told myself, but most did not, and I should remember the successes, not dwell on the failures. I had recently read an article which suggested that stress and anxiety made you more prone to develop Alzheimer's, and also that positive thinking was good for your immune system. I therefore strode into the hospital that Sunday evening full of good intentions.

There were four patients waiting for me – all with brain tumours. I talked to the first three for quite a long time and it was late by the time I reached the fourth. This last patient was a diabetic Asian woman, a few years younger than myself. Her family had brought her to see me two weeks earlier. Her English was limited and the family told me that she had been behaving increasingly strangely over the preceding two years. She had now started to become very drowsy. I had been unable to get much of a history from her, and discussed her problem and its treatment with her family instead. The scan showed that she had a small and benign meningioma at the front of her brain which was producing a great deal of reactionary swelling in the brain, and it was this swelling – medically called oedema – that was causing her symptoms. Surgery would almost certainly cure her. She would return to being the woman she had been, before her brain had started to become oedematous and her personality to change. Brain-swelling can be a major problem with brain surgery for tumours, and it is standard practice to put patients on steroids before surgery to reduce the swelling. In severe cases, such as with this woman, I start the steroids a week before surgery, and I wrote to her GP asking him to do this and warning him that the steroids would make her diabetes worse.

It was ten o'clock at night by the time I came to her room and found her asleep. A little apologetically, I shook her gently. She woke quickly and looked confusedly at me.

'It's Mr Marsh,' I said. 'Is everything OK?'

'Very sleepy,' she said and rolled over away from me.

'Are you OK?' I asked again.

'Yes,' she said, and went back to sleep.

The on-call registrar was standing beside me and I turned to him.

'She was pretty knocked off by the brain-swelling when I saw her two weeks ago,' I said, shrugging. 'And it's late at night, so we'd better leave her alone. The family are in the picture.'

Anxiety is contagious – doctors dislike anxious patients because anxious patients make anxious doctors – but confidence is also contagious, and as I walked out of the hospital I felt buoyed up by the way the first three patients had all expressed such great confidence in me. It allowed me to dismiss the last patient's sleepiness. I felt like the captain of a ship: everything was in order, everything was shipshape and the decks were cleared for action, for the operating list tomorrow. Playing with these happy nautical metaphors, I went home.

The next day got off to a good start. I had slept well and awoke feeling more enthusiastic and less anxious than I usually do on a Monday morning. The morning meeting went well; there were some interesting cases to discuss and I made a few good jokes at the patients' expense, which had the junior doctors laughing. The list got off to a prompt start and the first three tumour operations all went perfectly.

I went into the operating theatre, where my registrar had just positioned the fourth patient on the operating table. As I came in the anaesthetist looked at me with an expression that managed to be both accusing and apologetic at the same time. She had the printout from the blood gas analysis done routinely once the patients are asleep, as doctors say, in her hand.

'Do you know that her blood sugar is forty?'

'Bloody hell!'

'And her potassium is seven. And her pH is seven point two. She must be horribly dehydrated as well. Her diabetes is completely out of control.'

'It must be the steroids – but what was her blood sugar when she came in last night?'

'It seems that the night staff didn't check it. Her blood sugar was only slightly up three days ago when she was seen in the pre-admission clinic.'

'But it should have been checked yesterday anyway, shouldn't it, since she was a known diabetic?'

'Yes. It should have been. I thought she seemed a bit slow this morning when I saw her,' the anaesthetist said, 'but I thought it was because of the tumour, when I realize now it was because she was going into diabetic coma . . .'

'I made the same mistake last night,' I said unhappily. 'But I've never, ever seen anything like this happen before. Presumably we'll have to cancel the case?'

'I'm afraid so.'

'Bloody hell . . .'

'We'll get her round to the ITU and sort out her diabetes – it will take a few days. She needs to be rehydrated. To carry on with the operation now would be hopelessly dangerous.'

'Well, she's had a wacky haircut under GA, even though we haven't removed the tumour,' I said to my registrar as we unpinned her head from the operating table and admired the way he had shaved a couple of inches of hair off the front of her forehead.

'It's very typical of things going wrong in medicine,' the anaesthetist commented from the other end of the operating table. 'It's lots of little things coinciding together . . . If she had spoken better English, if she hadn't already been a bit confused because of the tumour, we'd all have realized that something was going wrong. The failure to check her glucose on admission wouldn't have mattered so much then . . . And if she hadn't been seen in the pre-admission clinic a few days earlier, and instead had just been clerked in and her bloods checked when she was admitted, as we used to do in the past . . .'

'The management introduced the pre-admission clinic as an efficiency measure,' I said.

'But that was because of the lack of beds and increasing workload,' she replied. 'Patients were being admitted later and later on the evening before surgery and there was no time to clerk them in properly.'

'Should we be doing anything about this?'

'I think it's an AIR rather than an SUI,' she said.

'A what?'

'Adverse Incident Report as opposed to a Serious Untoward Incident.'

'What's the difference?'

'An Adverse Incident is anonymous and gets filed somewhere.'

'But where do you send it?'

'Oh, some central office somewhere.'

'But shouldn't I just go and talk to the ward nurses? Do we really want the bloody managers beating them up?'

'Well, you could do that. But it's possible that this is a HONK coma and she might die.'

'What the hell is HONK?'

'Hyperosmolar non-ketotic diabetic coma.'

'Oh,' was all I could say in reply, realizing that my medicine was getting out of date.

So I went to find the nurses. The ward sister was very upset – she is the most conscientious of nurses, and looks perpetually anxious.

'I'll talk to the night staff about it,' she said, looking desperately unhappy. I was worried that she might burst into tears. 'She should have had her glucose checked.'

'Don't get upset,' I said cheerfully. 'These things happen. And the patient hasn't actually come to any harm. Just talk to the night staff about it. Mistakes happen. We're only human. I myself have started an operation on the wrong side you know . . . the important thing is not to make the same mistake twice.'

It had been many years earlier, before we had checklists – an operation on a man's neck for a trapped nerve in his arm. The operation is done through a midline incision, and you dissect down onto one side of the spinal column to drill out the trapped nerve. As I walked down the operating theatre corridor a few hours later, something was nagging me. My heart lurched when I suddenly realized that I had operated on the wrong side. I could quite easily have covered up the mistake – the incision was midline and post-operative scans do not clearly show where you have operated. The pain is not always relieved by surgery, and I could have told the patient a few weeks later that I would have to operate again, without telling him why. I knew of many stories of surgeons lying to patients in similar circumstances. But instead I went to the ward to see the man. It was the old hospital, and he was in one of the few single rooms, which had a window looking out onto the hospital gardens. It was spring, and you could see the many daffodils I had planted some years earlier. I had planted them while in the throes of a passionate – but one-sided – adulterous love affair. It had quickly fizzled out but it laid the foundations for the end of my marriage three years later. I sat down beside his bed.

'I'm afraid I've got some bad news for you.' He looked questioningly at me.

'And what's that, Mr Marsh?'

'I'm terribly sorry but I've gone and operated on the wrong side,' I said.

He looked at me in silence for a while.

'I quite understand,' he then said. 'I put in fitted kitchens for a trade. I once put one in back to front. It's easily done. Just promise me you'll do the right side as soon as possible.'

As I say to my juniors, when you make a stupid mistake, pick your patient carefully.

I went back to the operating theatre and looked up the phone numbers of the various patients' relatives and rang them to report on the day's events. The first three patients were all well: an old man with a pituitary tumour whose eyesight was already better, a monosyllabic garage mechanic on whom I had carried out an awake craniotomy and who had become less monosyllabic now that the operation was over, and a young woman with a tumour at the base of the skull who now had a stiff and painful neck but was otherwise surprisingly well. The last patient was on a ventilator on the ITU having her diabetic coma treated.

So I went home. The day had perhaps been a little chaotic, but none of the patients had come to serious harm. I remained determined that my resolution to see the positive side of neurosurgery should not be broken.

By eleven o'clock I was just about to go to bed when my mobile phone rang. I was upstairs brushing my teeth, and as I had left my phone charging on the kitchen table, I had to stumble naked down the stairs in a hurry, cursing and swearing. Neurosurgeons do not enjoy phone calls in the evening after a day's operating – it usually means that something has gone seriously wrong. The phone had stopped ringing by the time I got to it. The landline phone then started ringing just as the mobile phone also started ringing again with a voicemail message, so I cursed even more as I answered the landline phone.

'The meningioma has blown both her pupils – scan shows severe swelling,' I heard the voice of Vlad, the on-call registrar, telling me.

For a moment I was too surprised to reply.

'But I didn't even operate! I was expecting the other patients to get into trouble . . .'

'Maybe the diabetes, or the rehydration, made the brain swelling worse,' Vlad said. 'What do you want to do?'

'I don't know,' I replied. I sat down on a kitchen chair, stark naked, completely lost for an answer.

'Her biochemistry is OK now. The anaesthetists have corrected it,' Vlad added.

'She's probably had it,' I said after a while. 'But I really don't know what to do . . . if we operate, do a decompressive craniectomy to allow room for the swelling, she might survive but then be left wrecked – and if we don't operate she might just pull through and we can operate at a later date. I don't know what to do,' I repeated.

Vlad didn't comment as I stared vacantly at the kitchen wall.

'It's a unique set of circumstances,' I said. 'It's a toss-up either way.'

I thought of the man with a hammer to whom everything tends to look like a nail.

'We're surgeons,' I went on. 'We tend to see surgical solutions to everything. It doesn't necessarily mean operating is the right thing to do.'

Again Vlad said nothing, waiting for a decision. I sat silently for a while, willing my unconscious to tell me what to do, as the problem was far too unusual for reason and science to give me an answer, although an immediate decision was called for. Vlad was very experienced and shortly to become a consultant himself. The operation was well within his abilities. I could go to bed, I told myself.

'Take her to theatre. Take the front of her skull off and remove the tumour. If the brain is hopelessly swollen leave the bone out.'

'OK,' Vlad said, pleased that a decision had been made and happy at the prospect of operating.

I remained on the kitchen chair for a long time, staring at the kitchen wall. There was no great need for me to go to the hospital, but I realized that I would not get to sleep with the thought that the operation was going on while I lay in bed, so I quickly dressed and drove the short distance, along dark and deserted streets, back to the hospital. I ran up the stairs to the operating theatres on the second floor, but to my annoyance found that the patient had not yet arrived. I went round to the ITU. The patient was lying unconscious on her bed, on a ventilator, surrounded by doctors and nurses.

'Don't bloody wait for the bloody porters!' I said furiously. 'I'll take her to theatre.' So there was the usual hustle and bustle as all the machinery connected to the patient – the syringe drivers, monitors, catheters, IV and arterial lines and ventilator – was disconnected or reconnected to portable equipment and then we set off, a clumsy procession of doctors and nurses, bent double, pulling the bed or pushing or carrying equipment down the long corridor to the operating theatre.

Once there, I had the poor woman's head open within minutes. I rested my gloved finger on her brain.

'Brain's very slack,' I muttered.

'It's probably the ventilation and the drugs she's been given,' Vlad said. 'Look – the brain's pulsatile.' He pointed to the way that the yellow and light-brown mass, covered in blood vessels, that was her brain was gently expanding and contracting in synchrony with the bleeping of the cardiac monitor on the anaesthetic machine. 'She'll be OK.'

'That's what they say,' I said. 'But it doesn't mean much in my experience. When she blew her pupils she might well have suffered catastrophic infarction and most of her brain died. That could be why it's not swollen now. By tomorrow it may well start swelling again as it finally dies off. But she might just pull through . . .' I added, hoping against hope. I picked up a dissector and sucker and it took only a matter of minutes to remove the tumour which had caused all the trouble. It was absurdly easy.

We finished the operation quickly. At the end of these emergency operations, the critical moment is when one anxiously holds open the unconscious patient's eyelids to see if the pupils have started to constrict in reaction to light again. If they constrict, the patient will live.

'I think the left pupil is reacting,' the anaesthetist said happily, peering at the blank black pupil of the patient's left eye. The right eye was hidden by the head bandage I had wound around the patient's head after stitching up the scalp.

I looked. I had to look very closely, my face almost touching the woman's, as I had forgotten to bring my reading spectacles with me in my hurry to get to the hospital.

'I don't think so,' I replied. 'Wishful thinking.'

I wrote a brief operating note, asked Vlad to ring the family once she was out of theatre, and drove home.

I slept badly, waking frequently, hoping against hope, like a rejected lover, that all would be well, that when the dawn came Vlad would ring me to say that she was showing signs of recovery, but the phone remained silent. I went into work next morning and up to the ITU. The ITU consultant was standing beside the patient's bed.

'She's no better,' he said, and launched into a technical account of how he was managing the patient's complicated metabolic problems. He had always struck me as being a bit of a heartless technician.

'I couldn't sleep at all last night,' he suddenly said.

'It's not your fault.'

'I know that,' he replied. 'But it just feels so awful.'

The family were waiting outside the ITU, and we went to talk to them, and prepare them for the worst. There was still some hope, I told them. I said that she might survive but that it was also possible that she might die.

'She might have suffered a catastrophic stroke before I

operated. It's too early to tell,' I told them. I went on to explain that if she had suffered a catastrophic stroke, it would only show up clearly on a scan done the next day. So I said we would get a scan later in the day.

I went in that evening. In the X-ray viewing room I looked at the woman's brain scan on the computer screens. It was mottled and dark – clear evidence of catastrophic damage. Her brain had obviously swollen so badly while her diabetic coma was treated that she had suffered a major stroke. The operation had been too late. I walked round to the ITU where all the family were gathered in the interview room, waiting for me. Their eyes were fixed on me as I told them that there was no longer any hope. I told them that her death had been avoidable, because her blood sugar hadn't been checked on admission. I promised that this would be investigated and that I would report back to them in due course.

As I said this I wanted to scream to high heaven that it was not *my* fault that her blood sugar had not been checked on admission, that none of the junior doctors had checked her over, and that the anaesthetists had not realized this. It was not *my* fault that we were bringing patients into the hospital in such a hurry that they were not being properly assessed. I thought of the army of managers who ran the hospital, and their political masters, who were no less responsible than I was, who would all be sleeping comfortably in their beds tonight, perhaps dreaming of government targets and away-days in country house hotels, and who rarely, if ever, had to talk to patients or their relatives. Why should I have to shoulder the responsibility for the whole damn hospital like this, when I had so little say in how the hospital was run? Why should I have to apologize? Was it my fault that the ship was sinking? But I kept these thoughts to myself, and told them how utterly sorry I was that she was going to die and that I had failed to save her. They listened to me in silence, fighting back their tears.

'Thank you, Doctor,' one of them said to me eventually, but I left the small waiting room feeling all the worse for it.

I left the ITU staff to turn off the ventilator the next day.

I had told the family to sue the hospital – what had happened was indefensible – but they did not. Probably because of my apology.

I wish the authorities responsible for regulating doctors in the UK understood just how difficult it is for a doctor to say sorry. They show little sign of it. The General Medical Council recently produced a document on the Duty of Candour, which is now a statutory obligation. It orders us to tell patients whenever a mistake has been made, both in person and in writing. It would, the document told us, usually be the duty of the senior clinician responsible for the patient to do this, and to apologize, irrespective of who had made the mistake. It went on to add helpfully: 'for an apology to be meaningful it must be genuine,' seemingly unaware of the contradiction between an apology being compulsory and yet at the same time genuine. There was no discussion of how this contradiction can be resolved. It is resolved, of course, if senior doctors like myself feel trusted and respected, and if they have authority, and if they are not compelled to do meaningless things like asking patients to fill in a questionnaire about their behaviour. And if they are given the resources with which to do their work effectively.

I agree with everything in the document about the importance of honesty and apology, but I view with sadness and anger the increasing alienation and demoralization of doctors in England. The government, driven as always by the latest tabloid headlines, has set up an increasingly complex system of bureaucratic regulation based on distrust of the medical profession and its professional organizations. Of course doctors need regulating, but they need to be trusted as well. It is a delicate balance and it is clear to me that in England the government has got it terribly wrong.

# 14

# THE RED SQUIRREL

The two patients with acoustics were waiting for us in Igor's small and cramped office. One was a woman in her fifties, the other in her thirties. Both tumours were very large and both women were starting to lose their balance as a result of the tumours pressing on the brainstem. Patients with tumours this size gradually deteriorate, becoming more and more disabled, and will eventually die – but it can take many years. Surgery, if done competently, has a low risk of killing the patient, but, with the very large tumours, a high risk of leaving them with half their face paralysed, which is very disfiguring and for most people is a life-changing experience.

By this time I had operated on several acoustic tumours with Igor and there had been no disasters, so I was not too troubled about agreeing to operate. The older woman was certain she was happy to proceed with surgery, the younger was very frightened and indecisive. We spent almost two hours talking to her – all of it in Ukrainian, of course, and I spoke little. There was no question that she needed surgery, but she could choose between surgery done by Igor and myself and going to the State Institute. I was in no position to judge how our results compared to theirs.

'What would it cost to have the operation in Germany?' she asked me, Germany being a popular place for wealthy Ukrainians to go for medical treatment.

'At least thirty thousand dollars, probably much more.'

Although she did not say as much, it was clear she could not raise the money to go abroad, but I couldn't deny that probably it would be safer if she did. After two hours of discussion, it was agreed that we would operate. We would do the older patient on Monday and the younger one on Tuesday. There were a few other patients to see, but as it was a Sunday the outpatient clinic was quicker than usual. One was a young woman from a village in the west of the country, with a huge suprasellar meningioma compressing her optic nerves. She had long hair and a very pale face. I was told that she was losing her eyesight, but the details were a little vague.

'She'll go blind without surgery,' I told Igor when I looked at her brain scan. 'But the risks of surgery making her blind are also very high.'

'Oh, I have done several suprasellar meningiomas with good results. You show me how,' he declared confidently.

'Igor, this tumour is enormous. It's the biggest I've ever seen. It's a completely different problem from the usual, smaller ones.'

Igor said nothing in reply, but he looked unconvinced and was obviously itching to operate.

The first acoustic operation next day went well. I cannot remember the details – my memory has been so overlaid by what was to happen later. But I remember that the operation took many hours and, as usually happens when I work with Igor, we were not home until after nine o'clock in the evening.

'It is wonderful when you come!' Igor said to me as we

drove back home, the van bumping over the cobbled street that leads down to the Dnieper. 'It is like holiday for me. I scrub my brain. Recharge my batteries!' I was tempted to reply that my batteries felt correspondingly discharged and flat, and my brain-scrubbing brush worn and bent, but I kept quiet.

I sleep on a sofa bed in the living room of Igor and Yelena's apartment. It's not exactly comfortable but I always sleep well. It is on the sixteenth floor of a typical Soviet apartment block. The sofa bed once opened up in the middle of the night and deposited me on the floor. The view from the window is grimly impressive: a huge circle of identical, shabby high-rise blocks, with a dilapidated school and health centre in the middle of the ring – a Soviet Stonehenge. There is a large flat-screen television in the room, a few religious icons on the wall and a glass-fronted bookcase containing mainly medical books. Like the rest of their apartment, it is very plain and tidy and almost puritanical. Yelena is also a doctor, and works as a cardiologist in the Kiev Emergency Hospital, where I first met Igor in 1992. The family's life is devoted to work.

I usually rise early, woken by the hollow rumbling of the apartment block's battered elevator going up and down as people set off early for work. We get up at six and drive to work forty-five minutes later. The traffic is already very heavy but moves quite quickly, over the high Moskovskyi suspension bridge across the Dnieper. In the west I can see the golden domes of the churches of the Lavra Monastery shining, and to the east the rising sun is reflected in the windows of the garish apartment blocks built in the years of the property boom before the crash of 2007.

We go to see the woman with the acoustic tumour we operated on the day before. She is remarkably well and her face is not paralysed. I am always amazed at how tough the Ukrainians are – she is already standing out of bed, albeit unsteadily. We all laugh and smile happily. In the bed next to her is the young woman with long hair.

'Suprasellar meningioma for surgery tomorrow!' Igor announced. 'The second acoustic patient has sore throat. There is law in this country, if sore throat you not allowed to operate.'

So the day was spent seeing outpatients. In the late afternoon Igor drove me to an empty field on the outskirts of Trojeschina, the bleak suburb of Soviet apartment blocks where he lives. He had bought the field some years ago with a view to building his own hospital there, but then came the financial crash of 2007 and the site remained undeveloped. He was now instead converting an apartment block in the west of the city into a hospital, but had also bought himself a large, unfinished house. I wondered whether his increasingly grim expression was because he had started to overreach himself. The grass was still brown from the winter, burnt black in places, but with a few green shoots to be seen. There was rubbish everywhere. In the distance were the drab buildings of Trojeschina and the tall chimney of a power station. There was a small and dirty stream, partly blocked with plastic bags and tin cans and lined by weary-looking, bare willow trees. Igor produced a breadknife from his jacket pocket and proceeded to cut off a willow branch and stick it in the ground.

'Will that really grow?' I asked sceptically.

'Yes, one hundred per cent,' he replied, waving at the dozen or so willows along the litter-covered bank which had grown over the last ten years.

Opposite us, on the other side of the filthy stream, were a concrete ruin and a restaurant, with piles of rubbish and a dog which barked furiously when it saw me.

'Local people tell me how to do. People cut them down,' he said, pointing to the many stumps and burnt trunks. 'You have to plant five trees if one to live.'

As he planted more willow branches he discussed his endless problems with corrupt bureaucracy.

'This is land of losted possibilities,' he stated. 'First problem is Russia and second is corrupt Ukrainian bureaucrats. Everybody leave,' meaning that young people with any ambitions or dynamism have emigrated. 'I love and I hate this country. It is why I plant trees.'

The next day we operated on the girl who was losing her eyesight. One of the most difficult aspects of working with Igor is that it seems impossible to start the major cases until well after midday. I often complained to him that this was a serious problem as it meant that we ended up doing some of the most difficult and dangerous parts of the operation in the evening, when I, at least, was starting to feel tired. Dangerous and delicate brain surgery can be very intense and intensely draining. But Igor said it was impossible to get the theatre started sooner.

'I must do everything. Check equipment,' he said. 'If nurses or my doctors prepare case, they will make mistakes.'

When I suggested he should delegate more, and that by not trusting his team he was actually creating the problem, he disagreed violently.

'This is Ukraine,' he said. 'Silly people, silly country,' in his characteristically declamatory and assertive style. I had noticed over the years that few doctors or nurses lasted in his department for long. I still don't know if he was right and I was wrong, but I hated these late starts to difficult and dangerous operations.

By one o'clock he was finally starting to open the young

woman's head. I viewed the prospect of the operation continuing until late in the evening with dismay. I braved the guards at the entrance to the hospital and escaped for a walk round a small nearby park, something I have only dared to do since Maidan. The weather had turned suddenly mild and I slithered and slid over melting ice under the dull, grey sky. A red squirrel with high, pointed ears ran in circles around me, occasionally taking fright when I came too close, and darted up the nearest tree. I told the squirrel that I was finding it increasingly difficult to help Igor with these very major operations. I went back to the hospital, showed the guards my passport, and returned to the operating theatre.

The patient's head was almost open and Igor was scrubbed up at the table. I took my usual place on the hard trolley in the corner of the anaesthetic room. The door to the operating theatre was open and I could hear Igor shouting at his staff from time to time. He used a garage compressor for the air-powered drill I gave him nine years ago, used for cutting the bone of the skull to open the patient's head, and the compressor was next to the trolley I was lying on. It went off with a deafening explosion every few minutes when more compressed air was needed. I drifted in and out of sleep, woken at regular intervals by the compressor. It was half past two and they had only just started the bone work! Not for the first time, I told myself that this was the last visit. I had done this often enough before, but I knew it was not a good way to operate . . . And was it really necessary? Did Igor really need to do the occasional difficult case like this? I couldn't go on like this, I told myself, being angry all the time. It was like working in my hospital back in London.

Eventually Igor asked me to take over. He had been slow and careful, but had been unable to find the left optic nerve and needed my help. Once I had settled down in the chair my anger quickly dissipated and the operation seemed to go rather well. I felt happy and concentrated, full of fierce anxiety and excitement – full of the intense joy of operating. It took me several hours of delicate dissection to find the left optic nerve, working in a space only a centimetre wide. When I did, I realized I had wasted my time – it had been so thinned by the tumour that the woman could not possibly have had any vision in her left eye.

'Was she blind in the left eye?' I asked Igor and his staff. All I had been told was that she had only 20 per cent of her vision. But nobody knew. I realized at once that I had made a serious mistake. I should have asked about the nature of her visual loss before the operation. If I had known she was already blind in the left eye I would not have spent hours tiring myself out, trying to find and preserve the left optic nerve

After three hours of intense operating there was a fairly simple lump of tumour left, but I was tired – it was late, and I thought it would be easy for Igor to remove this last part. He had done pretty well so far, although a few of his comments later made me realize that I had not explained the anatomy of the optic nerves as clearly as I should have done. I went off to have a sandwich. When I came back I found that Igor had removed the tumour but had damaged the vital optic chiasm, the area where the two optic nerves meet and cross over. It was entirely my mistake – I should have stayed when Igor was removing that last part of the tumour, or done it myself.

I looked miserably down the microscope.

'She's going to be blind,' I said.

'But right optic nerve OK,' Igor said in surprise.

'But the chiasm has been damaged,' I replied. 'Well, it might well have happened if I had done that last part myself,'

I added. This was certainly true, but at least I would then have felt I had done my best.

Igor said nothing. I don't think he believed me.

We had finished by nine o'clock, and her head had been stitched back together again. She was still anaesthetized when we left but the pupils of both her eyes were large and black and did not react to light – a certain sign of blindness that Igor found hard to accept.

'Maybe she be better tomorrow.'

'I doubt it,' I replied.

Leaving somebody completely blind after surgery – it has happened to me twice – is a peculiarly unpleasant experience. It feels worse than leaving them dead: you cannot escape what you have done. Granted that in all these cases the patients were going to go blind without surgery and had already lost most of their vision, but it is a deeply painful experience to stand next to the patient at the bedside and see their blank, blind eyes fruitlessly casting about. Some of them at first go into a hallucinatory state and think that they can still see, and almost manage to persuade you that they can. You hesitate to disabuse them by demonstrating that they can't – the simplest test is to push your fist suddenly up to their eyes. They do not blink.

We sat at breakfast next morning, as we have done for many years.

'Did you sleep well?' Igor asked me.

'No.'

'Why not?'

'Because I was upset.'

He said nothing.

We made our way to work as usual. I went upstairs with Igor to see the young woman – there was no question that she was completely blind. I found it very hard to look at her while Igor leant over her with a bright desk lamp for a torch, trying to convince himself that her pupils still reacted a little to light, so I left the room.

We returned downstairs to Igor's office to see some of the outpatients already queuing up in the corridor.

'The acoustic woman we cancel is . . . oh!' – Igor waved his arms in the air – 'in terrible way. She hope very much you help with the operation and now you go away.'

After the long conversation and all the negotiations it seemed very cruel that her hopes – exaggerated although they might be – should be dashed and that I would not be involved in her operation after all. Besides, after the previous day's disastrous operating, I felt all the more the need to supervise Igor.

I looked at the diary on my smartphone, resigning myself to the inevitable.

'I can fly back in ten days' time,' I said, 'to do the op. Just for one day, but then I must leave.'

Igor simply nodded his head, and I could not help but feel I was being taken a little for granted. On the way back from work that evening, my guilt and despair about the blind girl finally overcame me. I started shouting angrily at Igor, not blaming him for the operation but for the way I felt he showed no understanding whatsoever for how difficult I found it to help him, and how utterly insensitive he seemed to other people's feelings. I ranted and raved for a while – we were crossing the Moskovskyi Bridge in the dark, and the black waters of the Dnieper, no longer frozen, were below us.

'Oh, I'd better shut up,' I finally said, worried that my outburst would distract him. He had never seen me behave like this before, and I was close to tears. 'Or you'll crash the car.'

'No,' he replied, in his best emphatic and Soviet style. 'I concentrate on driving.' And at that moment I felt an enormous gulf, as wide as the black river below, open up between us. But it was difficult not to be impressed by his apparent calm and detachment.

Later that week I went to the town of Lviv in the west of Ukraine. I had agreed to give a lecture at the Medical School. I spoke of how difficult it is for doctors to be honest. We learn this as soon as we put on our white coats after qualifying. Once we are responsible for patients, even at the lowest level of the medical hierarchy, we must start to dissemble. There is nothing more frightening for a patient than a doctor, especially a young one, who is lacking in confidence. Furthermore, patients want hope, as well as treatment.

So we quickly learn to deceive, to pretend to a greater level of competence and knowledge than we know to be the case, and try to shield our patients a little from the frightening reality they often face. And the best way of deceiving others, of course, is to deceive yourself. You will not then give yourself away with all the subtle signs which we are so good at identifying when people lie to us. So self-deception, I told the Ukrainians, is an important and necessary clinical skill we must all acquire at an early stage in our careers. But as we get older, and become genuinely experienced and competent, it is something we must start to unlearn. Senior doctors, just like senior politicians, can easily become corrupted by the power they hold and by the lack of people around them who will speak truth to power. And yet we continue to make mistakes throughout our careers, and we always learn more from failure than from success. Success teaches us nothing, and easily makes us complacent. But we will only learn from our mistakes if we admit to them – at least to ourselves, if not to our colleagues and patients. And to admit to our mistakes we must fight against the self-deception that was so necessary and important at the beginning of our careers.

When a surgeon advises a patient that they should undergo surgery, he or she is implicitly saying that the risks of surgery are less than those of not having the operation. And yet nothing is certain in medicine and we have to balance one set of probabilities against another, and rarely, if ever, one certainty against another. This involves judgement as much as knowledge. When I talk to a patient about the risks of surgery what I should really be talking about is the risks of surgery *in my hands*, in identical cases, and not just what is stated in the textbooks. Yet most surgeons are singularly poor at remembering their bad results, hate to admit to inexperience and usually underestimate the risks of surgery when talking to their patients. And even if the patient 'does well' and there are no complications after the operation, it can still be a mistake – it may well have been that the patient did not really need the operation in the first place and the surgeon, keen to operate, overestimated the risks of *not* operating. Over-treatment – unnecessary investigations and operations – is a growing problem in modern medicine. It is wrong, even if the patient comes to no obvious harm.

Critical to this is to understand that other people are better at seeing our mistakes than we are. As the psychologists Daniel Kahneman and Amos Tversky have shown, our brains are hardwired to fail to judge probabilities consistently. We are subject to many 'cognitive biases', as psychologists call them, which distort our judgement. We are too biased in our own favour and, under pressure, as doctors often are, we make decisions too quickly. However hard we try to admit to our mistakes, we will often fail. Safe medicine, I told them, is largely about having good colleagues who feel able to criticize and question us. As I said this, I thought of how difficult it is for surgeons like Igor and Dev who work, more or less, on their own.

I was told afterwards that for some of the Ukrainian audience this was almost a life-changing event – to hear a senior doctor admitting to fallibility, and stressing the importance of teamwork, of listening to criticism and of being a good colleague. It was an ironic counterpoint to my increasing problems with Igor.

It was a cold morning and the cars parked in the road outside my house were shrouded in frost that glittered in the moonlight as I bicycled to Wimbledon station nine days later. I sat on the train, wrapped in the heavy overcoat I wear when travelling to Ukraine in the winter, watching the sunrise over the slate roofs of south London alongside the railway line. I lost count many years ago of how many times I have made this journey. In the past I had been excited to return, but now I only felt the intense sadness and regret that you feel at the end of an affair. I felt obliged to keep my promise to operate on the second woman with an acoustic tumour, but I had decided that I could not go on helping Igor with these major cases. He was not the only surgeon doing these difficult operations in Kiev and I was pretty sure that the State Institute – a very large hospital compared to Igor's small, independent clinic – was doing many more such cases, and that it had not stood still since I first visited it twenty-four years earlier. Complex brain surgery, for me at least, is a question of teamwork – having the patient 'on the table' early in the morning, with colleagues and assistants you trust, and with whom you can share some of the burden of post-operative care.

It is a painful truth in medicine that we must expose some patients to risk, for the sake of future patients. As an experienced surgeon I have an ethical duty to the patient in front of me, but also to the future patients of the next generation of surgeons whom it is my duty to train. I cannot train surgeons less experienced than myself without exposing some patients to a degree of risk. If I did all the operating myself, if I instructed my trainees in every move, they would learn nothing, and their future patients would suffer. I had been willing to help Igor do dangerous cases and inflict the torment on myself of supervising his operating, in the belief that he was creating a sustainable and viable future for his clinic and that Ukrainian patients and his own trainees would benefit from it. I also believed him when he gave me to understand that other surgeons in Ukraine could not do these operations. That had probably been true twenty years ago, but I had come to doubt if it was still the case. I had been naive, perhaps worse than that. My own vanity, my wish for what looked like heroic action by working in Ukraine, had distorted my judgement.

I arrived back in Kiev to find that Igor had cancelled the operation on the young woman with an acoustic for a second time. He did not make it entirely clear to me why he had done so. We had a very unsatisfactory meeting with some of his doctors, which I had asked for. I thought I might be able to improve the working relationships in his department by getting them to talk together, but I was wrong. Igor became very angry. He clearly felt that his doctors had no right to criticize him or to complain, and saw the meeting as a plot against him, though of his colleagues' doing, and not mine. And I was a well-meaning but stupid outsider, interfering in a foreign country's internal affairs, having completely failed to understand them.

I returned to London next morning. I subsequently wrote to Igor, trying to explain why I felt unable to go on working with him unless he changed the way he ran his department, so that I felt it had a future, but received no reply. I don't know what happened to the young woman with the acoustic. I had been working with Igor for twenty-four years, for almost as long as my first marriage. In both cases I had clung to the wreckage for far too long, reluctant to open my eyes and admit that my marriage had ended, that my work with Igor no longer had a future. In both cases it was like waking from a nightmare, but one of my own making, and I felt ashamed.

Six months later I returned to Lviv as I had been invited to give some more lectures to the medical students. I talked once again of the importance of honesty and of being a good colleague. But I also told them how essential it is to listen to patients and how difficult it is to learn how to talk to patients as they will rarely, if ever, tell us whether we have spoken well to them or not, for fear of offending us. We get none of the negative feedback and criticism which is such an important part of learning how to do things better. I spoke to them of the importance of telling patients the truth, something most of us doctors find very difficult, as it often means admitting to uncertainty. I told them how the woman with the suprasellar meningioma whom we had left blind had heard I was in Lviv and had asked to come and meet me. I rather dreaded this but when she came, led into the room by her husband, she did not appear especially angry or unhappy. She told me how she had seen many doctors after the operation – it seemed that they had told her that she would have to wait longer for her eyesight to recover, and she wanted to know from me how long this might take.

'What should I tell her?' I asked the students rhetorically. 'I know she is never going to see again. And should she have been informed of that right from the start?'

I told them that it had seemed cruel to deprive her of all hope immediately after the operation, though I had warned her and her husband – but Igor might have chosen not to translate it – that I thought the chances of recovery were very small indeed. But after six months it seemed wrong to continue to lie to her. Up till then in the conversation she had been putting a brave face on things and even making a few jokes about her blindness. But then I told her, slowly, that I wanted her to know how sad I was that the operation had been such a disaster. And now she started crying, and her husband started crying, and I had difficulties not crying myself. And I told her that she never would see again and that she must learn to use a white stick and to read Braille. I delivered a little lecture on neuroscience – about how the visual areas of her brain would quickly be converting to the analysis of sound rather than vision, that blind people could lead almost normal lives, although it was very difficult. And so we talked, and at the end she asked when I was returning to Lviv, as she said she would like to come and talk with me again.

After three months of complete neglect, the weeds in the cottage garden had grown to an extraordinary size – there were stately thistles as tall as young trees, with purple flowers reaching over my head. The cow parsley was ten foot high. There were nameless, numberless plants, some with leaves as large as umbrellas. I was slightly ashamed that I did not know their names. The two rusted corrugated-iron sheds near the lake had almost disappeared under the wild, green tide pushing up against them. The abandoned garden had become an impenetrable jungle. There was a glorious, green freedom to the place, and I felt very reluctant to beat it into submission. But I wanted to know what had happened to the apple trees and single walnut tree I had planted in the winter – they had vanished.

Using my petrol-powered hedgecutter on a five-foot driveshaft, I swept a path towards where I had planted the young walnut tree. At first, to my dismay, all I could see was a dead stalk surrounded by the overbearing, giant weeds – even though I had put black plastic sheeting down around the tree to suppress them. But once I had cleared the surrounding weeds I found, to my joy, that the little walnut tree was alive and well, with large, tender green leaves lower down the stem. I then cut a path to the five apple trees in the opposite corner of the garden, and these I also found to be flourishing – there were even some small apples on their branches.

I spent five hours starting to cut back the weeds and also the overgrown hedge in front of the cottage, which was starting to block the towpath. This was the first hard, physical work I had done for many months and I found, once again, that although exhausting, it was a wonderful panacea. I forgot all my anxieties and preoccupations, I stopped thinking about my future, and my shame, anger and despair over the referendum on Britain's membership of the European Union. The air was full of the green scent of cut grass, the acrid smell of the giant cow parsley and crushed leaves. All I could think about was the next, painful sweep of the hedgecutter, which was balanced on a sling around my neck. My neck clicked and creaked as I worked, and there were constant showers of pins and needles into my right shoulder, from what I assume is a trapped nerve between the third and fourth vertebrae of my cervical spine – the problem has been troubling me for some months. My neck is so stiff that when I try to look up at the stars at night I tend to fall over backwards.

As my body ages, I notice all sorts of new symptoms. My left hip aches a little when I run, my right knee hurts when I sit cramped in airplanes. My prostatism wakes me at night. I am a doctor, so I know what these symptoms mean, and that they will get worse as I get older. I also know that sooner or later I will develop the first signs of a serious illness, which may well be my final illness. I will probably dismiss them at first, hoping that they will go away, but at the back of my mind I will be frightened. I was staying in an expensive hotel recently, and the multiple mirrors in the extravagant marble-clad bathroom not only showed me my elderly, sagging buttocks – a most offensive reminder of my age – but also a mole just in front of my right ear that I had not noticed before. It could not be seen in a single mirror, face-on. I lay in bed, convinced that I had developed melanoma – the most deadly of the skin cancers – and eventually had to get up and search through photographs on my laptop until I found one of myself in profile, which showed that the mole had been present years ago. Only then could I get back to sleep.

I came home from the cottage garden stiff and exhausted, and slept that night for nine hours. I lay in bed in the morning, my neck and back aching, and began to doubt whether I was still capable of all the work required to restore the cottage. I went back later in the day and started to cut out the broken glass from the windows smashed by the vandals. I had spent many hours one year earlier inserting the ogee-shaped glass panes with small nails – known as glazing sprigs – and putty and mastic.

It started to rain very heavily, and the green water of the usually still canal was so flailed by the rain that it seemed to be boiling. The sight distracted me, my hands slipped and I cut my left index finger badly, over the second metacarpal joint, on one of the glass fragments in the window frame, raising a flap of skin over the extensor tendons. It bled profusely, leaving a brilliant red trail on the window frame. I am so used to the sight of blood in the operating theatre that I had forgotten the fairy-tale beauty of its colour. I looked at it in wonder, until the rain started to wash it away. I probably should have taken myself off to hospital to have my finger stitched, but I did not like the thought of queuing for hours, so I went home with a bloody handkerchief wrapped around it. I improvised a repair with a series of plasters cut into strips, and a splint made with outsize matchsticks from an ornamental matchbox a friend had given me.

# 15

# NEITHER THE SUN NOR DEATH

Thirty-five years ago, when I started training as a neurosurgeon, you still had to take what was known as the 'general FRCS'. There was no specialized examination in neurosurgery, and instead you became a Fellow of the Royal College on the basis of an examination that was centred on 'general' surgery, which was mainly abdominal surgery. To qualify for the examination I had to spend a year as a junior registrar in general surgery, which I did in a district hospital in the suburbs of outer London.

It was a busy job, working 'three in seven' – meaning that I was on call in the hospital three nights a week and every third weekend, in addition to working a normal week. You were paid in *umtis* for work done over and above forty hours, an *umti* being a 'unit of medical time', a euphemism whereby four hours' overtime were paid little more than one hour at the basic rate. I was operating most nights – carrying out fairly simple operations for appendicitis or draining abscesses – but usually got enough sleep on which to get by. There were two consultants, both helpful and supportive and good teachers, but – probably like most junior doctors at that time – I took considerable pride in trying not to ask for their help unless absolutely necessary. I therefore learnt quickly, but still look back with deep shame and embarrassment at some of the mistakes I made, when I should have asked for help. At least none of my mistakes, as far I know, were lethal.

I have forgotten most of the patients I looked after during that year, just as I discovered in the Health Camp in Nepal that I have forgotten how to do the operations I did then. One patient, however, I remember very clearly, and even his name. He was a man in his fifties who turned up one evening with his wife in Casualty (as the Accident and Emergency departments were then called). He was smartly dressed in one of those fawn overcoats with a black velvet collar. He and his wife were perfectly polite, but quickly made me aware of the fact that he had previously been a private patient of one of my consultants. He had now run out of insurance and was back on the NHS. They looked very tense, and in retrospect I think they probably had a premonition of what the future might hold for him. He had developed increasing abdominal pain over the preceding two days. I asked him the usual questions about the pain: did it come and go in waves ('colicky' is the medical term), was he still able to pass gas, had he 'opened his bowels' – that clumsy and absurd phrase doctors still use. Had he vomited?

'Yes. That started today,' he said with a grimace, 'and it smelt horrible.' I noted silently that this was almost certainly faeculent vomiting, a sure sign of intestinal obstruction. I asked him to undress and he lay down on the trolley in the curtained cubicle.

His abdomen was criss-crossed with pale surgical scars and distended.

'I had an op for colon cancer three years ago,' he said. 'There were some problems afterwards and I was in hospital for many weeks and needed several more ops.'

'But then he was fine until two days ago,' his wife added, trying to find some grounds for hope. When I palpated – as doctors call examining by touch – his abdomen, I found that it was as tight as a drum. When I percussed it – pressing my left middle finger onto his stomach and then briskly tapping on it with my right middle finger – there was a deeply hollow sound. When I listened to his abdomen with a stethoscope I could hear the 'tinkling, tympanitic' bowel sounds characteristic of intestinal obstruction. He must have waited a long time at home, hoping and praying that the problem would go away.

Something was blocking his gut and the consequences of this are no different from the consequences of a blocked sewer. It is very painful, as the muscular intestines struggle to overcome the blockage.

'We'll need to admit you,' I said, using the reassuring plural that reduces doctors' feelings of personal responsibility and vulnerability. 'We'll get some abdominal X-rays and probably pass a nasogastric tube and put up a drip.'

'Is it serious?' his wife asked.

'Well, hopefully it's just post-op scarring and it will sort itself out,' I replied.

So he was admitted to the surgical ward. The X-rays confirmed intestinal obstruction, showing loops of bowel full of trapped gas. The regime of 'drip and suck' was instituted whereby he was kept nil by mouth and given intravenous fluid, and any fluid in his stomach was aspirated up a nasogastric tube. The idea is to 'rest' the bowel, and sometimes episodes of intestinal obstruction like this can indeed cure themselves without surgery. But this failed to happen and his condition deteriorated. Two days after I had admitted him, my consultant took him to the operating theatre. I was the assistant.

When we opened his abdomen, cutting through the scarred and distended skin and muscle, we found that his intestines were matted together by scar tissue. Parts of the bowel had turned black, meaning that they were dying from strangulation. Untreated, this leads to death within a few days.

'Oh dear. Not good,' my boss said with a sigh. He slowly freed up some of the scarred gut with a pair of scissors so that he could get his hand into the man's abdominal cavity and explore its contents by feeling his way round the liver and kidneys and down into the pelvis.

'Have a feel, Henry,' he said to me. 'Here – up by the liver.'

I put my gloved left hand – I was standing on the patient's left side, my boss on the right – into the gaping, warm hole and felt for the liver. Its normally smooth and firm surface had a large and craggy mass in it.

'A big met,' I said, a met being a metastasis, a secondary cancer which has broken off and spread from the original tumour. The presence of a met usually signifies the beginning of the end.

'Indeed, and there are mets in the mesentery as well. There's not much to do but I suppose we had better resect the gangrenous bits as he might live for a few months yet.'

So we spent the next two hours cutting out the three short, blackened lengths of intestine and then joined – 'anastomosed' – the healthy cut ends together again with stitches.

In brain surgery, if an operation has gone badly you usually know immediately after the operation: the patient wakes up disabled or does not wake up at all. In general surgery, the complications usually occur a few days later, when infections set in or suture lines fail and things fall apart. With this poor man an especially unpleasant complication developed: the anastomoses of his bowel broke down and he developed multiple faecal fistulae through his abdominal wall. In other words, several holes appeared in the operative incision – and in some of the old scars as well – and through these faeces steadily oozed out. The smell was truly awful and there were so many fistulae that it was impossible for the nurses to keep them clean. He had been put in a side room and you had to take a deep breath before going in.

I saw him each day on my regular morning ward round. There was nothing we could do to help him. It was simply a question of waiting for him to die. And he was, of course, wide awake and fully aware of what was happening, slowly dying, surrounded by the terrible smell of his own faeces. He must have seen the involuntary expressions on our faces as we entered his room, steeling ourselves to brave the awful stench.

We started a morphine drip with a pump 'to keep him comfortable', and over a period of many days he slowly died.

I had been a doctor for three years by then but felt utterly unequal to the task of discussing his death with him. I remember him looking sadly into my eyes as I stood above him while the nurses were carrying out the hopeless task of trying to keep his abdomen clean. I don't doubt that we exchanged a few words – probably I asked some banal questions about whether he was in much pain or not – but I know the one thing we did not discuss was his approaching death.

I will never know what he might have said if I had taken the time to sit down beside him and talk to him properly. Was I frightened that he might ask me to 'ease the passing', as it is called in medico-legal language? To increase the morphine and perhaps give him other drugs to end his life? I think it is what I would want if I have the misfortune to end up dying in the way that he died. Or would he have been in a state of denial and still somehow hoping that he might yet live? Or perhaps he would merely have wanted the comfort of talking about his past and memories, or even just about the weather. I remember how my eyes would drift away from him to the window as we talked, to where I could see the autumnal trees at the edge of the hospital car park. I had to make a conscious decision to force them back to look at him.

It is difficult to talk of death to a dying patient, it takes time, and it is difficult if the room stinks of shit. And I know that I let this man down and was a coward.

A few months after qualifying as a doctor, I was working as a medical houseman in a hospital in south London that in the nineteenth century had been a workhouse for the poor. It had not entirely managed to lose the atmosphere of its earlier incarnation. There were long, dark corridors and much of it was very dilapidated, as were so many English hospitals at that time.

It was my duty, as the most junior doctor, to go to the Casualty department and clerk in and admit the patients. If they needed to be admitted to the Intensive Unit, which was immediately above the Casualty department on the next floor, you had to push the patients in their beds a quarter of a mile down the main hospital corridor to the only lift and push them all the way back again along the corridor above to the ITU. If the patient was very ill you tried to run, pushing the bed as fast as you could with a porter and a nurse, but it still took a very long time.

One evening I admitted a man in his fifties with bleeding oesophageal varices. This is a condition when the veins that line the oesophagus – the tube that connects the mouth to the stomach – become enlarged as a result of cirrhosis of the liver. As far as I can remember, his cirrhosis resulted from previous infective hepatitis rather than alcoholism, the commonest cause. The enlarged veins are fragile and can bleed, often torrentially, and the patients come into hospital vomiting large volumes of blood. He was already 'in shock' from loss of blood when I went down to see him. To be in shock is a medical term which means that the blood pressure is falling and the patient will die if the bleeding continues and the blood is not replaced. The patient's pulse will be 'thin and thready', his face pale and his periphery – his hands and feet – cold. He will have an anxious, drawn expression and his breathing will be shallow and fast.

I quickly 'stuck in a large drip'– putting a wide-bored intravenous cannula into a vein in one of his arms – and telephoned my registrar, in some excitement, to say that we had a medical emergency. We transferred him the long distance to the ITU and started to give him blood transfusions. But just as quickly as we poured the blood in, he vomited it all up again. This was almost forty years ago and the only treatment available then, other than blood transfusion, was a device called a Sengstaken tube. This was a very large red rubber tube lined with balloons which you pushed down the patient's throat and then inflated, trying to squeeze the bleeding veins shut. We used to keep them in the ITU fridge, to make the rubber harder – pushing them down the patient's throat and into the oesophagus was very difficult because they were so large, and the patients would cough and gag, simultaneously throwing up blood. We wore aprons and boots. But despite all our efforts he continued to vomit his life away. I remember him silently looking into my eyes as I spent the night with my registrar and the ITU nurses trying to keep him alive. Every few minutes he would feebly try to roll to one side to vomit fresh blood into the grey cardboard vomit bowl beside him.

Despite giving him fresh frozen plasma and clotting factors, we were losing the battle – his blood was becoming visibly thinner and no longer clotting in the vomit bowls. It was quite obvious that he was going to die. My registrar drew up a syringe full of morphine to 'ease the passing'. I cannot remember which one of us administered it. By now it was dawn, and the harsh fluorescent light of the ITU at night was starting to be replaced by the kinder light of the day. Once I had certified him dead shortly afterwards, exhausted, I walked back to my scruffy on-call room, which always smelt of boiled cabbage from the hospital kitchen downstairs. It was summer, and a very beautiful, sunlit morning. That I remember very clearly.

I had been a consultant for only a year. I was responsible for a man with a malignant secondary brain tumour. It was a rare tumour that at first had developed as a solid mass, which I was able to remove. But it was malignant and a few months later recurred. It was now no longer a solid mass and instead individual tumour cells were growing in the spinal fluid, as though the tumour was dissolved in the fluid, making it thick and sticky, causing acute hydrocephalus and terrible headaches. So I operated again, inserting a drainage tube called a shunt to relieve the build-up of pressure in his head. In retrospect this was a mistake, and in later years I would very rarely operate on patients with carcinomatous meningitis, as the condition is called. It is kinder to let the patient die.

The operation was not a great success. His headaches were perhaps better, but he was profoundly confused and agitated. I had informed his family that he was going to die. There was no question of further treatment and I told the family that death was inevitable, which they accepted. But he stubbornly refused to die – his life prolonged, but pointlessly, by the operation. His family became increasingly distressed by the protracted death agony, although the agony was really more theirs than his. One morning the ward sister rang my secretary Gail and asked me to come up to the ward. It was in the old hospital in Wimbledon and my office was in the basement.

'The family are kicking up a terrible fuss,' I was told. 'They are demanding to see you.' So, feeling a little sick with anxiety, and deeply regretting the shunt operation, I walked quickly up the stairs.

It was one of the old Nightingale wards – a large, long room with tall windows and thirty beds arranged in two long rows on either side. The man was in the first bed on the left, and the curtains were drawn. I cautiously put my head between the curtains. The patient's wife was sitting beside the bed, sobbing silently, his middle-aged son standing beside her, his face red with anger. He did not give me a chance to say anything.

'You wouldn't treat your dog like this!' he said. 'You'd put him out of his misery, wouldn't you!'

For a moment I was quite lost for words.

'I don't think he's suffering much,' was all I managed to say, looking at the patient, who was staring silently at the ceiling, mute, awake but seemingly unaware of what was going on around him. I explained that we were giving him heroin (diamorphine is the pharmacological name) through a pump, pointing to the syringe driver bolted to the drip-stand at the bed's head.

'How much longer will this go on for?' asked his wife.

'I don't really know,' I replied. 'Not more than a few days . . .'

'You told us that a few days ago,' his son said.

'All we can do is wait,' I replied.

*

The word 'euthanasia' is used to describe the various ways in which doctors can deliberately bring about the death of patients. It ranges from the criminal mass-murder that went on in mental hospitals in Nazi Germany, to giving painkilling morphine to people in their final agony, to the 'doctor-assisted suicide' that is available in several countries for people faced with certain death from diseases such as terminal cancer or motor neurone disease, or, in a smaller number of countries such as Belgium, for people facing a life of intractable suffering from conditions such as paralysis or incurable depression. In Britain, and in most other countries, doctor-assisted suicide is illegal, although opinion polls in Britain have shown on many occasions a great majority in support of a change in the law. Doctors and members of parliament seem to have more of a problem with it than the public. Doctors rarely admit, even to each other, that they have ever helped patients die – but there can be no question that in the past it happened, and part of me hopes that it still does. You can never know, however, because no doctor wants to talk himself or herself into prison.

A doctor's duty is to relieve suffering as well as to prolong life, although I suspect this truth is often forgotten in modern medicine. Doctors are frequently accused of playing God but, in my experience, the opposite is more often the case. Many doctors shy away from decisions that might reduce suffering but which will hasten a patient's death. It is clear that some people can look death in the face and decide, quite rationally, that their life is no longer worth living. There was, for instance, a young Englishman who had been left quadriplegic by a neck injury while playing rugger, who went to Switzerland to die in the Dignitas clinic. There was an elderly woman who had spent her life looking after psychogeriatric patients and, as she entered old age herself, decided to end her life in the same way rather than run the risk of dementing. I consider these people to have been heroic, and can only hope that I might emulate them should I one day have to face a similar problem. There is no evidence that the moral fabric of these societies that permit euthanasia is being damaged by the availability of this form of euthanasia, or that elderly parents are being bullied into suicide by greedy children. But even if that occasionally happens, might it not be a price worth paying to allow a far greater number of other people a choice in how they die?

One senior politician told me that he was opposed to euthanasia because it would lead to 'targets' – presumably he fears that quotas of elderly people will be encouraged to kill themselves by doctors and nurses. This is an unnecessary anxiety: there are plenty of safeguards in the countries where euthanasia is permitted to prevent this happening. Besides, it is a question of people freely making a choice, *if they have mental capacity*, not of licensing doctors to kill patients who lack capacity. It will not solve the problem of the ever-increasing number of people with dementia in the modern world. They no longer have mental capacity. Besides, doctors do not want to kill patients – indeed most of us recoil from it and all too often go to the other extreme, not allowing our patients to die with dignity.

Clearly the suicidal young – as I once came close to being – need to be helped, as they have all of their lives ahead of them, and suicide is often impulsive, but in old age we no longer have much of a future to which to look forward. It is perfectly rational, on the balance of probabilities, as the lawyers call it, to decide to finish your life quickly and painlessly rather than run the risk of a slow and miserable decline. But neither the sun nor death can be looked at steadily, and I do not know what I will feel as I enter old age, if I start to become dependent. What will I decide if I begin to lose my eyesight, or the use of my hands?

Scientific medicine has achieved wonderful things, but has also presented us with a dilemma which our ancestors never had to face. Most of us in the modern world live into old age, when cancer and dementia become increasingly common. These are now usually diagnosed when we are still relatively well and of sound mind – we can predict what will happen to us, although not the exact timing. The problem is that we are condemned by our evolutionary history to fear death. In the remote past our ancestors – perhaps even the simplest life forms with some kind of brain – did not survive into frail old age, and extra years of healthy life were precious if the species was to survive.

Life, by its very nature, is reluctant to end. It is as though we are hardwired for hope, to always feel that we have a future. The most convincing explanation for the rise of brains in evolution is that brains permit movement. To move, we must predict what lies ahead of us. Our brains are devices – for want of a better word – for predicting the future. They make a model of the world and of our body, and this enables us to navigate the world outside. Perception is expectation. When we see, or feel, or taste or hear, our brains, it is thought, only use the information from our eyes, mouth, skin and ears for comparison with the model it has already made of the world outside when we were young. If, when walking down a staircase, there is one more or one less step than we expect, we are momentarily thrown off balance. The famous sea squirt, beloved of popular neuroscience lectures, in its larval stage is motile and has a primitive nervous system (called a notochord) so it can navigate the sea – at least, its own very small corner of it. In its adult stage it fastens limpet-like to a rock and feeds passively, simply depending on the influx of seawater through its tubes. It then reabsorbs its nervous system – it is no longer needed since the creature no longer needs to move. My wife Kate put this into verse.

*I wish I were a sea squirt,*
*If life became a strain,*
*I'd veg out on the nearest rock*
*And reabsorb my brain.*

The slow and relentless decline into the vegetative existence that comes with dementia cannot be stopped, although it can sometimes be slowed down. Some cancers in old age can be cured and most can be *treated*, but only a few of us will be long-term survivors, who then live on to die from something else. And if we are already old, the long term is short.

We have to choose between probabilities, not certainties, and that is difficult. How *probable* is it that we will gain how many extra years of life, and what might the *quality* of those years be, if we submit ourselves to the pain and unpleasantness of treatment? And what is the probability that the treatment will cause severe side effects that outweigh any possible benefits? When we are young it is usually easy to decide – but when we are old, and reaching the end of our likely lifespan? We can choose, at least in theory, but our inbuilt optimism and love of life, our fear of death and the difficulty we have in looking at it steadily, make this very difficult. We inevitably hope that we will be one of the lucky ones, one of the long-term survivors, at the good and not the bad tail-end of the statisticians' normal distribution. And yet it has been estimated that in the developed world, 75 per cent of our lifetime medical costs are incurred in the last six months of our lives. This is the price of hope, hope which, by the laws of probability, is so often unrealistic. And thus we often end up inflicting both great suffering on ourselves and unsustainable expense on society.

In every country, health-care costs are spiralling out of control. Unlike our ancestors, who had no choice in these matters, we can – at least in principle – decide when our lives should end. We do not *have* to undergo treatment to postpone fatal diseases in old age. But if we decide to let nature take its course, and refuse treatment for a fatal disease such as cancer, most of us are still faced with the prospect of dying miserably, as in only a few countries is euthanasia – *a good death* – allowed. So, if euthanasia is not permitted, we are faced with the choice of dying miserably now, or postponing it for a few months or longer, to die miserably at a later date. Not surprisingly, most of us choose the latter option and undergo treatment, however unpleasant it might be.

Our fear of death is deeply ingrained. It has been said that our knowledge of our mortality is what distinguishes us from other animals, and is the motive force behind almost all human action and achievement. It is true that elephants can mourn their dead and console each other, but there is no way of knowing whether this means that, in some way, they know that they themselves will die.

Our ancestors feared death, not just because dying in the past without modern medicine must have been so terrible but also for fear of what might come after death.

But I do not believe in an afterlife. I am a neurosurgeon. I know that everything I am, everything I think and feel, consciously or unconsciously, is the electrochemical activity of my billions of brain cells, joined together with a near-infinite number of synapses (or however many of them are left as I get older). When my brain dies, 'I' will die. 'I' am a transient electrochemical dance, made of myriad bits of information; and information, as the physicists tell us, is physical. What those myriad pieces of information, disassembled, will recombine to form after my death, there is no way of knowing. I had once hoped it would be oak leaves and wood. Perhaps now it will be walnut and apple in the cottage garden, if my children choose to scatter my ashes there. So there is no rational reason to fear death. How can you be afraid of nothing? But of course I am still frightened by the prospect. I also greatly resent the fact that *I will never know what happened* – to my family, my friends, to the human race. But my instinctive fear of death now takes the form of fear of dying, of the indignity of being a helpless patient at the mercy of impersonal doctors and nurses, working shifts in a factory-like hospital, who scarcely know me. Or, even worse, of dying incontinent and demented in a nursing home.

My mother was a deeply fastidious person. In the last few days of her life, as she lay dying in her bed in her room in the house at Clapham, with its wood-panelled walls and tall, shuttered sash windows that look out on the Common's trees, she became doubly incontinent. 'The final indignity,' she said, not without some rancour, as my sister and I cleaned her. 'It really is time to go.'

I doubt if she would have wanted to bring her life to a quick end with a suitable pill if she had been given the choice. She strongly disapproved of suicide. But for myself, I see little merit or virtue in the physical indignity which so often accompanies our last few days or weeks of life, however good the hospice care which a minority of us might be lucky enough to receive. Perhaps I am unrealistic and romantic to hope that in future the law in England will change – that I might be able to die in my own bed, with my family beside me, as my mother did, but quickly and peacefully, truly falling asleep, as the tombstone euphemisms put it, rather than incontinent and gasping with the death rattle – at first demonstrating the O-sign, as doctors call it, of the mouth open but with the tongue not visible, to be followed by the Q-sign, which heralds death, with the furred and dried tongue hanging out.

For those who believe in an afterlife, must we suffer as we lie dying, if we are to earn our place in heaven? Must the soul undergo a painful birth if it is to survive the body's death, and then ascend to heaven? Is it yet more magic and bargaining – if we suffer now, we will not suffer in the future? We will not go to hell or linger as unhappy ghosts? Is it cheating, to have a quick and easy death? But I do not believe in an afterlife – my concern is simply to achieve a good death. When the time comes, I want to get it over with. I do not want it to be some prolonged and unpleasant experience, presided over by terminal-care professionals, who derive their own sense of meaning and purpose from my suffering. The only meaning of death is how I live my life now and what I will have to look back upon as I lie dying. If euthanasia is legalized, this question of how we can have a good death, for those of us who want it, with pointless suffering avoided, can be openly discussed, and we can make our own choice, rather than have it imposed upon us. But too often we prefer to avoid these questions, as I did with the poor man at the beginning of my surgical career. It's as though it is better to die miserably than to admit to the inevitability of death and look it in the face.

Once again there is an aneurysm to do. I feel anxious, but also proud that I still have such difficult work to perform, that I might fail, that I am still held to account, that I can be of service. Each time I scrub up, I am frightened. Why am I continuing to inflict this on myself, when I know I can abandon neurosurgery at any time? Part of me wants to run away, but I scrub up nevertheless, pull on my surgical gown and gloves and walk up to the operating table. The registrars are opening the patient's head but I will not be needed yet, so I sit on a stool and lean the back of my head against the wall. I keep my gloved hands in front of my chest with the palms pressed together, as though I were praying – the pose of the surgeon, waiting to operate. Next to me the operating microscope also waits, its long neck folded back on itself, ready to help me. I don't know for how much longer I will feel able to be of use here, or whether I will return, but it seems I am still wanted.

It is hot and dusty outside in the city; the rains are late, the air is yellow with dust and pollution. The haze is so bad that even the nearby foothills are hidden. It gets worse every year. As for the celestial, snow-covered Himalayas, it's almost as though they had never existed. The glaciers are said to be retreating more quickly than in even the gloomiest predictions. It will not be so very long before the rivers run dry.

As I doze, leaning against the theatre wall, I long to return home. I think of the cottage. I walk around the wild, green garden in my imagination. The buds on the young apple trees, a few millimetres in size, are just starting to open, and I can see the miniature petals, tightly furled in layers of pink and white, starting to appear, full of enthusiasm to enter the outside world. It is raining, and the air is wet and smells of spring. The rain forms a million brief circles on the lake, and the two swans are there, looking a little disdainful as always, cruising regally past the reed beds. Perhaps they will make a nest this year among the reeds. I will have made an owl box and put it in the tall willow tree beside the lake. At night

I will practise on the owl whistle I recently bought, hoping to persuade an owl to make its home there. The weeds are starting to reclaim the garden again, but I will have used my book of wild flowers and I will have studied them with great care and learnt all their names. Grass is starting to appear between the red bricks that form the floor of the old pigsty which I spent many days clearing, and weeds are growing back between the cobblestones in front of the drinking troughs. Perhaps the rare vine loved by the foragers will be there as well. I will have brought up the beehives from the garden in London, and I can see the bees coming out of the hives to explore their new home and return with bright-yellow pollen on their legs, now that the winter is past. Perhaps I will buy a small boat, and when my granddaughter Iris is a little older, I will take her rowing on the lake. Even better, perhaps I will make the boat myself, if I have been given the time to build my workshop by then, and all my sharpened tools will be neatly hung and stored, and the place will smell of sawn oak and cedar wood. The windows of the workshop will look out over the lake. In summer there are yellow flags and lilies growing at the side of the lake, just beyond the window. And then there is the derelict, ramshackle cottage, to which I will have given a new life. Perhaps the vandals will have finally left it in peace.

There will be much that needs doing when I return. There will be many things to make or repair, and many things to give or throw away, as I try to establish what I will leave behind. But it seems to me now that it no longer matters if I never finish. I will try not to wait for the end, but I hope to be ready to leave, booted and spurred, when it comes. It is enough that I am well for a little longer, that I have been lucky to be part of a family – past, present and future – that I can still be useful, that there is still work to be done.

# ACKNOWLEDGEMENTS

It is only when you write a book yourself that you understand just how important and heartfelt are the acknowledgements. Whatever the quality of this book might be, it would have been many times worse without the comments and encouragement of many friends – in particular Robert McCrum, Erica Wagner, Geoffrey Smith, and my brother Laurence Marsh. My excellent agent Julian Alexander was always on hand to provide wise advice and my wonderful editor Bea Hemming transformed a rather chaotic manuscript into what is, I hope, a proper book. Alan Samson, Jenny Lord and Holly Harley at Weidenfeld & Nicolson gave me further help and advice with the manuscript. I am deeply in debt to my patients and colleagues in London, Kiev and Kathmandu and especially to Upendra and Madhu Devkota, whose exceptional hospitality and kindness make my trips to Nepal so rewarding. I am indebted to Catriona Bass who found the lock-keeper's cottage for me. Most important of all, however, has been the help of my wife Kate, who once again came up with the title and who has been involved in every aspect of the book, as both subject, critic, muse and wife.